AF508612

Robust Procedures for Estimating and Testing in the Framework of Divergence Measures

Robust Procedures for Estimating and Testing in the Framework of Divergence Measures

Editors

Leandro Pardo
Nirian Martin

MDPI • Basel • Beijing • Wuhan • Barcelona • Belgrade • Manchester • Tokyo • Cluj • Tianjin

Editors

Leandro Pardo
Department of Statistics and
Operations Research
Complutense University
of Madrid
Madrid
Spain

Nirian Martin
Department of Financial and
Actuarial Economics & Statistics
Complutense University
of Madrid
Madrid
Spain

Editorial Office
MDPI
St. Alban-Anlage 66
4052 Basel, Switzerland

This is a reprint of articles from the Special Issue published online in the open access journal *Entropy* (ISSN 1099-4300) (available at: https://www.mdpi.com/journal/entropy/special_issues/ Divergence_Measure).

For citation purposes, cite each article independently as indicated on the article page online and as indicated below:

LastName, A.A.; LastName, B.B.; LastName, C.C. Article Title. *Journal Name* **Year**, *Volume Number*, Page Range.

ISBN 978-3-0365-1460-4 (Hbk)
ISBN 978-3-0365-1459-8 (PDF)

Contents

About the Editors

Leandro Pardo

Leandro Pardo received the B.Sc. and Ph.D. degrees in mathematics from the Complutense University of Madrid, Madrid, Spain, in 1976 and 1980, respectively. He has been a Full Professor of Statistics and Operations Research with the Department of Statistics and Operational Research, Faculty of Mathematics, Complutense University of Madrid, Madrid, Spain, since 1993. He was elected "Distinguished Eugene Lukacs Professor" in Bowling Green State (Bowling Green, OH, USA) in 2004.

Nirian Martin

Nirian Martin received the B.Sc. and Ph.D. degrees in Statistics from the Complutense University of Madrid, Madrid, Spain, in 2003 and 2007, respectively. She is currently an Associate Professor of Statistics and Operations Research with the Department of Financial and Actuarial Economics and Statistics, Complutense University of Madrid, Madrid, Spain.

Preface to "Robust Procedures for Estimating and Testing in the Framework of Divergence Measures"

The approach for estimating and testing based on suitable divergence measures has become, in the last 30 years, a very popular technique not only in the field of statistics but also in other areas, such as machine learning, pattern recognition, etc. In relation to the estimation problem, it is necessary to minimize a suitable divergence measure between the data and the model under consideration. Some interesting examples of those estimators are the minimum phi-divergence estimators (MPHIE), in particular, the minimum Hellinger distance (MHD) and the minimum density power divergence estimators (MDPDE). The MPHIE are characterized by asymptotic efficiency (BAN estimators), the MHE by asymptotic efficiency and robustness inside the family of the MPHIE, and the MDPD by their robustness without a significant loss of efficiency as well as by the simplicity of getting them, because it is not necessary to use a non-parametric estimator of the true density function.

Based on these estimators of minimum divergence or distance, many people have studied the possibility to use them in order to obtain statistics for testing hypotheses. There are some possibilities to use them with that objective: (i) Plugging them in a divergence measure in order to obtain the estimated distance (divergence) between the model, whose parameters have been estimated under the null hypothesis and the model evaluated in all the parameter space; and (ii) extending the concept of the Wald test in the sense of considering MDPDE instead of maximum likelihood estimators. These test statistics have been considered in many different statistical problems: Censoring, equality of means in normal and lognormal models, logistic regression model, multinomial logistic regression, and GLM models in general, etc.

The scope of the contributions to this book will be to present new and original research papers based on MPHIE, MHD, and MDPDE, as well as test statistics based on these estimators from a theoretical and applied point of view in different statistical problems with special emphasis on robustness. Manuscripts given solutions to different statistical problems as model selection criteria based on divergence measures or in statistics for high-dimensional data with divergence measures as loss function are considered. Reviews making emphasis in the most recent state-of-the art in relation to the solution of statistical problems base on divergence measures are also presented.

Leandro Pardo, Nirian Martin
Editors

Editorial

Robust Procedures for Estimating and Testing in the Framework of Divergence Measures

Leandro Pardo [1,2,*] **and Nirian Martín** [2,3]

[1] Department of Statistics and O. R., Faculty of Mathematics, Universidad Complutense de Madrid, 28040 Madrid, Spain
[2] Interdisciplinary Mathematics Institute, Complutense University of Madrid, 28040 Madrid, Spain; nirian@estad.ucm.es
[3] Department of Financial and Actuarial Economics & Statistics, Faculty of Commerce and Tourism, Complutense University of Madrid, 28003 Madrid, Spain
* Correspondence: lpardo@mat.ucm.es

Citation: Pardo, L.; Martín, N. Robust Procedures for Estimating and Testing in the Framework of Divergence Measures. *Entropy* **2021**, *23*, 430. https://doi.org/10.3390/e23040430

Received: 25 March 2021
Accepted: 1 April 2021
Published: 6 April 2021

Publisher's Note: MDPI stays neutral with regard to jurisdictional claims in published maps and institutional affiliations.

The approach for estimating and testing based on divergence measures has become, in the last 30 years, a very popular technique not only in the field of statistics, but also in other areas, such as machine learning, pattern recognition, etc. In relation to the estimation problem, it is necessary to minimize a suitable divergence measure between the data and the model under consideration. Some interesting examples of those estimators are the minimum phi-divergence estimators (MPHIE), in particular, these minimum Hellinger distance (MHD) and the minimum density power divergence estimators (MDPDE). The MPHIE (Pardo [1], Morales et al. [2]) are characterized by asymptotic efficiency (BAN estimators), the MHE (Beran [3]) by asymptotic efficiency and robustness inside the family of the MPHIE, and the MDPDE (Basu et al. [4]) by their robustness without a significant loss of efficiency as well as by the simplicity of getting them, because it is not necessary to use a nonparametric estimator of the true density function.

Based on these estimators of minimum divergence or distance, many people have studied the possibility to use them to obtain statistics for testing hypotheses. There are some possibilities to use them with that objective: (i) plugging them in a divergence measure in order to obtain the estimated distance (divergence) between the model, whose parameters have been estimated under the null hypothesis and the model evaluated in all of the parameter space, see, for instance, Martín and Pardo [5], Menéndez et al. [6], Salicrú et al. [7], Morales et al. [8,9]; (ii) extending the concept of the Wald test in the sense of considering MDPDE instead of maximum likelihood estimators (MLE). These test statistics have been considered in many different statistical problems: censoring, equality of means in normal and lognormal models, logistic regression model, multinomial regression in particular, and GLM models in general, etc., see, for instance, Basu et al. [10–14], Ghosh et al. [15], Castilla et al. [16], Ghosh et al. [17], and references therein; and, (iii) extending the concept of the Rao's test in the sense of considering MDPDE instead of MLE, see Basu et al. [18] and Martín [19].

This Special Issue present new and original research papers that are based on MPHIE, MHD, and MDPDE, as well as test statistics that are based on these estimators from a theoretical and applied point of view in different statistical problems with special emphasis on robustness. Manuscripts give solutions to different statistical problems as model selection criteria based on divergence measures or in statistics for high-dimensional data with divergence measures as loss function are presented. It comprises nine selected papers that address novel issues, as well as specific topics illustrating the importance of the divergence measures or pseudodistances in statistics. In the following, the manuscripts are presented:

An important class of time-dynamic models is given by discrete-time integer-valued branching processes, in particular (Bienaymé-) Galton-Watson processes without immigration (GW), respectively, with immigration (GWI), which have numerous applications in

Entropy **2021**, *23*, 430. https://doi.org/10.3390/e23040430   https://www.mdpi.com/journal/entropy

biotechnology, population genetics, internet traffic research, clinical trials, asset price modelling, derivative pricing, and many others. As far as important terminology is concerned, they shall subsume both models as GW(I) and, simply as GWI in the case that GW appears as a parameter-special-case of GWI; recall that a GW(I) is called subcritical, respectively, critical, respectively, supercritical if its offspring mean is less than 1, respectively, equal to 1, respectively, larger than 1.

In "Some dissimilarity Measures of Branching Processes and optimal Decision Making in the Presence of Potential Pandemics", Kammerer and Stummer, [20], compute exact values respectively bounds of dissimilarity/distinguishability measures—in the sense of the Kullback-Leibler information distance (relative entropy) and some transforms of more general power divergences and Rényi divergences—between two competing discrete-time Galton-Watson branching processes with immigration for which the offspring and the immigration (importation) are arbitrarily Poisson-distributed; especially, they allow for an arbitrary type of extinction-concerning criticality and, thus, for non-stationarity. They apply this to optimal decision making in the context of the spread of potentially pandemic infectious diseases (such as, e.g., the current COVID-19 pandemic), e.g., covering different levels of dangerousness and different kinds of intervention/mitigation strategies. Asymptotic distinguishability behavior and diffusion limits are also investigated by them. In a more concrete way, this paper pursues the following main goals:

(A) for any time horizon and any criticality scenario (allowing for non-stationarities), to compute lower and upper bounds—and sometimes even exact values—of the Hellinger integrals $H_\lambda(P_\mathcal{A}||P_\mathcal{H})$, density power divergences $I_\lambda(P_\mathcal{A}||P_\mathcal{H})$, and Rényi divergences $R_\lambda(P_\mathcal{A}||P_\mathcal{H})$ of two alternative Galton-Watson branching processes $P_\mathcal{A}$ and $P_\mathcal{H}$ (on path/scenario space), where (i) $P_\mathcal{A}$ has Poisson $(\beta_\mathcal{A})$ distributed offspring as well as Poisson $(\alpha_\mathcal{A})$ distributed immigration, and (ii) $P_\mathcal{H}$ has Poisson $(\beta_\mathcal{H})$ distributed offspring as well as Poisson $(\alpha_\mathcal{H})$ distributed immigration; the non-immigration cases are covered as $\alpha_\mathcal{A} = \alpha_\mathcal{H} = 0$; as a side effect, they also aim for corresponding asymptotic distinguishability results;

(B) to compute the corresponding limit quantities for the context in which (a proper rescaling of) the two alternative Galton-Watson processes with immigration converge to Feller-type branching diffusion processes, as the time-lags between the generation-size observations tend to zero; and,

(C) as an exemplary field of application, to indicate how to use the results that are pointed out in A) for Bayesian decision making in the epidemiological context of an infectious-disease pandemic (e.g., the current COVID-19), where e.g., potential state-budgetary losses can be controlled by alternative public policies (such as e.g., different degrees of lookdown) for mitigations of the time-evolution of the number of infectious persons (being quantified by a GW(I)). Corresponding Neyman-Pearson testing will also be treated.

Because of the involved Poisson distributions, these goals can be tackled with a high degree of tractability, which is worked out in detail with the following structure they first introduce (i) the basic ingredients of Galton-Watson processes, together with their interpretations in the above-mentioned pandemic setup, where it is essential to study all types of criticality (being connected with levels of reproduction numbers), (ii) the employed fundamental information measures, such as Hellinger integrals, power divergences, and Rényi divergences, (iii) the underlying decision-making framework, as well as (iv) connections to time series of counts and asymptotical distinguishability. Thereafter, they start other detailed technical analyses by giving recursive exact values respectively recursive bounds-as well as their applications-of Hellinger integrals $H_\lambda(P_\mathcal{A}||P_\mathcal{H})$, density power divergences $I_\lambda(P_\mathcal{A}||P_\mathcal{H})$, and Rényi divergences $R_\lambda(P_\mathcal{A}||P_\mathcal{H})$. Explicit closed-form bounds of Hellinger integrals $H_\lambda(P_\mathcal{A}||P_\mathcal{H})$ that will be worked are obtained as well as Hellinger integrals and power divergences of the above-mentioned Galton-Watson type diffusion approximations.

The change point problem is a core issue in time series analysis because changes can occur in underlying model parameters, owing to critical events or policy changes, and ignoring such changes can result in false conclusions. Numerous studies exist on change point analysis in time series models; refer to Kang and Lee, see [21] and Lee and Lee, see [22], and the articles cited therein, for the background and history of change points in integer-valued time series models. Lee and Lee [22], conducted a comparison study of the performance of various cumulative sum (CUSUM) tests while using score vectors and residuals through the Monte Carlo simulations. In their work, the conditional maximum likelihood estimator (CMLE) is used for the parameter estimation and the construction of the CUSUM tests. However, the CMLE is often damaged by outliers, and so is the performance of the CMLE-based CUSUM test. In general, outliers easily mislead the CUSUM test, since they can be mistakenly taken for abrupt changes; in the opposite, they can misidentify change points in their presence on time series.

In the work "Monitoring Parameter Change for Time Series. Models of Counts Based on Minimum Density Power Divergence estimator", Lee and Kim [23] consider the CUSUM monitoring procedure to detect a parameter change for integer-valued generalized autoregressive heteroscedastic models (core area in time series analysis that includes diverse disciplines in social, physical, engineering, medical sciences, etc. Integer-valued autoregressive time series models and the integer-valued generalized autoregressive conditional heteroscedastic models have been widely studied in the literature and applied to various practical problems), whose conditional density of present observations over past information follows one parameter exponential family distributions. For this purpose, they use CUSUM of score functions that were deduced from the objective functions, constructed for the MDPDE that includes the MLE, to diminish the influence of outliers. It is well-known that, as compared to the MLE, the MDPDE is robust against outliers with little loss of efficiency. This robustness property is properly inherited by the proposed monitoring procedure. The CUSUM test has been a conventional tool to detect a structural change in underlying models, and it has been applied not only to retrospective change point tests, but also to on-line monitoring and statistical process control (SPC) problems, which were designed to monitor abnormal phenomena in manufacturing processes and health care surveillance. The CUSUM control chart has been popular due to its considerable competency in early detection of anomalies. A simulation study is conducted to affirm the validity of their method. Focus is placed on comparing the MDPDE-based CUSUM test with the MLE-based CUSUM test for Poisson INGARCH models to demonstrate the superiority of the former over the latter in the presence of outliers. A real data analysis of the return times of extreme events of Goldman Sachs Group (GS) stock prices is also provided to illustrate the validity of the proposed test. These authors, see [24], considered the CUSUM tests based on score vectors for the MLE and MDPDE in exponential family distribution INGARCH models.

In "Robust Change Point Test for General Integer-Valued Time Series Models Based on Density Power Divergence" by Kim and Lee [24], the problem of testing for a parameter change in general integer-valued time series models whose conditional distribution belongs to the one-parameter exponential family when the data are contaminated by outliers is considered. In particular, they use a robust change point test that is based on density power divergence (DPD) as the objective function of the MDPDE. The results show that, under regularity conditions, the limiting null distribution of the DPD-based test is a function of a Brownian bridge. Monte Carlo simulations are conducted to evaluate the performance of the proposed test and show that the test inherits the robust properties of the MDPDE and DPD. They compare the DPD-based test and the score-based CUSUM test to demonstrate the superiority of the proposed test in the presence of outliers. They provide a real data analysis of the return times of extreme events that are related to Goldman Sachs Group (GS) stock to illustrate the proposed tests.

MDPDE provides a general framework for robust statistics, depending on a parameter α, which determines the robustness properties of the method. The usual estimation method is numerical minimization of the power divergence. In "Robust Regression with Density Power Divergence: Theory, Comparisons, and Data Analysis", by Riani et al. [25], is considered to be the special case of linear regression developing an alternative estimation procedure using the methods of S-estimation. The so obtained rho function is proportional to one minus a suitably scaled normal density raised to the power α. We used the theory of S-estimation to determine the asymptotic efficiency and breakdown point for this new form of S-estimation. Two sets of comparisons were made. In one, S power divergence is compared with other S-estimators using four distinct rho functions. The plots of efficiency against breakdown point show that the properties of S power divergence are close to those of Tukey's biweight. The second set of comparisons is between S power divergence estimation and numerical minimization. Monitoring these two procedures in terms of breakdown point shows that the numerical minimization yields a procedure with larger robust residuals and a lower empirical breakdown point, thus providing an estimate of α, leading to more efficient parameter estimates.

Model selection is fundamental to the practical applications of statistics, and there is substantial literature on this issue. Classical model selection criteria include, among others, the Cp-criterion, the Akaike Information Criterion (AIC), based on the Kullback-Leibler divergence, and the Bayesian Information Criterion (BIC), as well as a General Information Criterion (GIC), which corresponds to a general class of criteria which also estimates the Kullback-Leibler divergence. These criteria have been proposed, respectively, in [26–28], and they represent powerful tools for choosing the best model among different candidate models that can be used to fit a given data set. On the other hand, many classical procedures for model selection are extremely sensitive to outliers and other departures from the distributional assumptions of the model. Robust versions of classical model selection criteria, which are not strongly affected by outliers, have been proposed, for example, in [29] and [30]. Some recent proposals for robust model selection are criteria based on divergences and minimum divergence estimators. Here, we recall the Divergence Information Criteria (DIC) based on the density power divergences that were introduced in [31], the Modified Divergence Information Criteria (MDIC) introduced in [32], and the criteria based on minimum dual divergence estimators introduced in [33]. In [34,35] some model selection criteria are presented. In "Robust Model Selection Criteria Based on Pseudodistances" by Toma et al. see [34], a new class of robust model selection criteria are introduced. These criteria are defined by estimators of the expected overall discrepancy using pseudodistances and the minimum pseudodistance principle. The theoretical properties of these criteria are proved, namely asymptotic unbiasedness, robustness, consistency, as well as the limit laws. The case of the linear regression models is studied and a specific pseudodistance based criterion is proposed. Monte Carlo simulations and applications for real data are presented to exemplify the performance of the new methodology. These examples show that the new selection criterion for regression models is a good competitor of some well known criteria and may have superior performance, especially in the case of small and contaminated samples.

Classical likelihood function requires the exact specification of the probability density function, but, in most applications, the true distribution is unknown. In some cases, where the data distribution is available in an analytic form, the likelihood function is still mathematically intractable due to the complexity of the probability density function. There are many alternatives to the classical likelihood function; one of them is the composite likelihood. Composite likelihood is an inference function that is derived by multiplying a collection of component likelihoods; the particular collection used is a conditional determined by the context. Therefore, the composite likelihood reduces the computational complexity, so that it is possible to deal with large datasets and very complex models, even when the use of standard likelihood methods is not feasible. Composite likelihood methods have been successfully used in many applications concerning, for example, genetics, generalized linear mixed models, spatial statistics, frailty models, multivariate survival analysis, etc. Asymptotic normality of the composite maximum likelihood estimator (CMLE) still holds with the Godambe information matrix to replace the expected information in the expression of the asymptotic variance-covariance matrix. This allows for the construction of composite likelihood ratio test statistics, Wald-type test statistics, as well as score-type statistics. Varin [36] provides a review of composite likelihood methods. They mentioned, at this point, that CMLE, as well as the respective test statistics are seriously affected by the presence of outliers in the set of available data. In this sense, [37–39] derived some new distance-based estimators and tests with good robustness behavior without an important loss of efficiency. In the context of the composite likelihood there are some criteria based on Kullback-Leibler divergence, see, for instance [40–42] and references therein. To the best of our knowledge, only Kullback-Leibler divergence was used to develop model selection criteria in a composite likelihood framework. To fill this gap, our interest is now focused on DPD. In "Model Selection in a Composite Likelihood Framework Based on Density Power Divergence", Castilla et al. see [35], consider the composite minimum density power divergence estimator (CMDPDE), as introduced in [37], in order to present a model selection criterion in a composite likelihood framework. The criterion introduced in [37] will be called composite likelihood DIC criterion (CLDIC). The motivation, as pointed out by the authors, of considering a criterion based on DPD instead of Kullback-Leibler divergence is due to the robustness of the procedures based on DPD in statistical inference, not only in the context of full likelihood, but also in the context of composite likelihood [37,38]. After introducing the new model selection criterion, CLDIC, based on CMDPDE, some of its asymptotic properties are studied. A simulation study is carried out and some numerical examples are also presented.

Bounding the best achievable error probability for binary classification problems is relevant to many applications, including machine learning, signal processing, and information theory. The Bayes error rate is the expected risk for the Bayes classifier, which assigns a given feature vector x to the class with the highest posterior probability. The Bayes error rate is the lowest possible error rate of any classifier for a particular joint distribution. The Bayes error rate provides a measure of classification difficulty. Thus, when known, the Bayes error rate can be used to guide the user in the choice of classifier and tuning parameter selection. In practice, the Bayes error is rarely known and it must be estimated from data. The estimation of the Bayes error rate is difficult due to the non-smooth in function within an integral. Thus, research has focused on deriving tight bounds on the Bayes error rate based on smooth relaxations of the min function. Many of these bounds can be expressed in terms of divergence measures between the pair of class distributions, such as the Bhattacharyya distance or Jensen-Shannon divergence measure. Many techniques have been developed for estimating divergence measures. These methods can be broadly classified into two categories: (i) plug-in estimators in which we estimate the probability densities and then plug them in the divergence function and (ii) entropic graph approaches, in which the relationship between the divergence function and a graph functional in Euclidean space is derived. Examples of plug-in methods include k-nearest neighbor (K-NN) and Kernel density estimator (KDE) divergence estimators. Examples of entropic graph approaches include methods that are based on minimal spanning trees (MST), K-nearest neighbors graphs (K-NNG), minimal matching graphs (MMG), traveling salesman problem (TSP), and their power-weighted variants. Recently, the Henze-Penrose (HP) divergence has been proposed for bounding classification error probability. In "Convergence Rates for Empirical Estimation of Binary Classification Bounds", by Sekeh et al. see [43], the problem of empirically estimating the HP-divergence from random samples is considered. The first contribution of this paper is that they obtain a bound on the convergence rates for the Friedman and Rafsky (FR) estimator of the HP-divergence, which is based on a multivariate extension of the non-parametric run length test of equality of distributions. This estimator is constructed using a multicolored MST on the labeled training set, where MST edges connecting samples with dichotomous labels are colored differently from edges connecting identically labeled samples. While previous works have investigated the FR test statistic in the context of estimating the HP-divergence, to the best of the author's knowledge, its minimax MSE convergence rate has not been previously derived. The bound on convergence rate is established by using the umbrella theorem, for which they define a dual version of the multicolor MST. The proposed dual MST in this work is different than the standard dual MST that was introduced by Yukich in [44]. They show that the bias rate of the FR estimator is bounded by a function of N, η and d, as $O\left(N^{-\eta^2/(d(\eta+1))}\right)$, where N is the total sample size, d is the dimension of the data samples $d > 2$, and η is the Hölder smoothness parameter $0 \leq \eta \leq 1$. They also obtain the variance rate bound as $O(N^{-1})$. The second contribution of this paper is a new concentration bound for the FR test statistic. The bound is obtained by establishing a growth bound and a smoothness condition for the multicolored MST. Because the FR test statistic is not a Euclidean functional, we cannot use the standard subadditivity and superadditivity approaches. Their concentration inequality is derived using a different Hamming distance approach and a dual graph to the multicolored MST. They experimentally validate their theoretic results comparing the MSE theory and simulation in three experiments with various dimensions d = 2, 4, 8. They observe that, in all three experiments, as sample size increases, the MSE rate decreases and, for higher dimensions, the rate is slower. Our theory matches the experimental results in all sets of experiments.

In "Distance-Based Estimation Methods for Models for Discrete and Mixed-Scale Data" by Sofikitou et al. [45], robust methods for mixed-scale data are developed. Mixed-scale measurements scenario have both discrete (categorical or nominal) and continuous type random variables. Initially, they reviews basic concepts in minimum disparity estimation (MDE), which has been extensively studied in models where the scale of the data is either interval or ratio ([3,12]). It has also been studied in the discrete outcomes case. Specifically, when the response variable is discrete and the explanatory variables are continuous, Pardo et al. [46] introduced a general class of distance estimators based on ϕ-divergence measures, the MPHIE, and they studied their asymptotic properties. The estimators can be viewed as an extension/generalization of the MLE. In Pardo et al. [47], the MPHIE is used in statistic to perform goodness-of-fit tests in logistic regression models, while Pardo and Pardo [48] extended the previous works to address solving problems for testing in generalized linear models with binary scale data. The case where data are measured on discrete scale (either on ordinal or generally categorical scale) has also attracted the interest of other researchers. For instance, Simpson [49] demonstrated that minimum Hellinger distance estimators fulfill desirable robustness properties and, for this reason, can be effective in the analysis of count data that are prone to outliers. Simpson [50] also suggested tests based on the minimum Hellinger distance for parametric inference that are robust as the density of the (parametric) model can be nonparametrically estimated. In contrast, Markatou et al. [51] used weighted likelihood equations to obtain efficient and robust estimators in discrete probability models and applied their methods to logistic regression, whereas Basu and Basu [52] considered robust penalized minimum disparity estimators for multinomial models with good small sample efficiency. Moreover, Gupta et al. [53], Martín and Pardo [54], and Castilla et al. [55] used the MPHIE to provide a solution to testing problems in polytomous regression models. Working in a similar fashion, Martín and Pardo [56] studied the properties of the family of MPHIE for log-linear models with linear constraints under multinomial sampling to identify the potential associations between various variables in multi-way contingency tables. Pardo and Martín [57] presented an overview of works that are associated with contingency tables of symmetric structure on the basis of MPHIE and ϕ-divergence test statistics. Additional works include Pardo and Pardo [58] and Pardo et al. [59]. Basu et al. [60] introduced alternative power divergence measures. Afterwards, define various Pearson residuals appropriate for the measurement scale of the data and study their properties. They further concentrate on the case of mixed-scale data, which is, data measured in both categorical and interval scale. We study the asymptotic properties and the robustness of MDE obtained in the case of mixed-scale data and exemplify the performance of the methods via simulation. The results show that, depending on the level of contamination and the type of contaminating probability model, the performance of the methods is satisfactory.

The asymptotic distributions of minimum Hellinger distance estimators has been well investigated; nevertheless, the probabilities of rare events that are induced by them are largely unknown. In "Event Analysis for Minimum Hellinger Distance Estimators via Large Deviation Theory" by Vidayashankar and Collamore [61], rare event probabilities, for the minimum Hellinger distance estimators of a family of continuous distributions satisfying an equicontinuous condition, using large deviation theory under a potential model misspecification, in both one and higher dimensions are analyzed. They show that these probabilities decay exponentially, characterizing their decay via a "rate function", which is expressed as a convex conjugate of a limiting cumulant generating function. In the analysis of the lower bound, in particular, certain geometric considerations arise, which facilitate an explicit representation, also in the case when the limiting generating function is non-differentiable. The analysis also involves the modulus of continuity properties of the affinity, which may be of independent interest. The results that are presented in this paper extend large deviation asymptotics for M-estimators that were given previously. In contrast to the case for M-estimators, our setting is complicated due to its inherent nonlinearity, leading to complications in the proofs of both the upper and lower bounds, and an unexpected subtlety in the form of the rate function for the lower bound. The results of Vidayashankar and Collamore (2021) suggest that one can, under additional hypotheses, establish saddlepoint approximations to the density of minimum Hellinger distance estimators, which would enable one to sharpen inference for small samples.

Similar results are expected to hold for discrete distributions. However, the equicontinuity condition is not required in that case, since ℓ_1, unlike $L_1(S)$ (the space of integrable functions on S), possesses the Schur property. Hence, the large deviation principle in the weak topology of ℓ_1 can be derived (more easily) using a standard Gartner-Ellis argument and, utilizing this, one can, in principle, repeat all of the arguments above to derive results that are analogous to Theorems 2.2 and 2.3. Large deviations for other divergences under weak family regularity (such as non-compactness of the parameter space) and their connections to estimation and test efficiency are interesting open problems that require new techniques beyond those that are described in this article.

Funding: This research is supported by Grant PGC2018-095194-B-I00 from Ministerio de Ciencia, Innovación y Universidades (Spain).

Conflicts of Interest: The authors declare no conflict of interest.

References

1. Pardo, L. *Statistical Inference Based on Divergence Measures*; Chapman & Hall: Boca Raton, FL, USA, 2006.
2. Morales, D.; Pardo, L.; Vajda, I. Asymptotic divergence of estimates of discrete distributions. *J. Stat. Plan. Inference* **1995**, *48*, 347–369. [CrossRef]
3. Beran, R. Minimum Hellinger Distance Estimates for Parametric Models. *Ann. Stat.* **1977**, *5*, 445–463. [CrossRef]
4. Basu, A.; Shioya, H.; Park, C. *Statistical Inference: The Minimum Distance Approach*; Chapman and Hall/CRC: Boca Raton, FL, USA, 2011.
5. Martín, N.; Pardo, L. New families of estimators and test statistics in log-linear models. *J. Multivar. Anal.* **2008**, *99*, 1590–1609. [CrossRef]
6. Menéndez, M.L.; Morales, D.; Pardo, L.; Vajda, I. Asymptotic distributions of φ-divergences of hypothetical and observed frequencies on refined partitions. *Stat. Neerl.* **1998**, *52*, 71–89. [CrossRef]
7. Salicrú, M.; Morales, D.; Menéndez, M.L.; Pardo, L. On the applications of divergence type measures in testing statistical hypotheses. *J. Multivar. Anal.* **1994**, *51*, 372–391. [CrossRef]
8. Morales, D.; Pardo, L.; Vajda, I. Rényi statistics in directed families of exponential experiments. *Stat. A J. Theor. Appl. Stat.* **2000**, *34*, 151–174. [CrossRef]
9. Morales, D.; Pardo, L.; Pardo, M.C.; Vajda, I. Rényi statistics for testing composite hypotheses in general exponential models. *Stat. A J. Theor. Appl. Stat.* **2004**, *38*, 133–147. [CrossRef]
10. Basu, A.; Mandal, A.; Martín, N.; Pardo, L. A Robust Wald-Type Test for Testing the Equality of Two Means from Log-Normal Samples. *Methodol. Comput. Appl Probab.* **2019**, *21*, 85–107. [CrossRef]

11. Basu, A.; Ghosh, A.; Mandal, A.; Martín, N.; Pardo, L. A Wald-type test statistic for testing linear hypothesis in logistic regression models based on minimum density power divergence estimator. *Electron. J. Stat.* **2017**, *11*, 2741–2772. [CrossRef]
12. Basu, A.; Lindsay, B.G. Minimum Disparity Estimation for Continuous Models: Efficiency, Distributions and Robustness. *Ann. Inst. Stat. Math.* **1994**, *46*, 683–705. [CrossRef]
13. Basu, A.; Ghosh, A.; Martín, N.; Pardo, L. Robust Wald-type tests for non-homogeneous observations based on minimum density power divergence estimator. *Metrika* **2018**, *81*, 493–522. [CrossRef]
14. Basu, A.; Mandal, A.; Martín, N.; Pardo, L. Robust tests for the equality of two normal means based on the density power divergence. *Metrika* **2015**, *78*, 611–634. [CrossRef]
15. Ghosh, A.; Basu, A.; Pardo, L. Robust Wald-type tests under random censoring. *Stat. Med.* **2020**. [CrossRef]
16. Castilla, E.; Martín, N.; Pardo, L. Pseudo minimum phi-divergence estimator for the multinomial logistic regression model with complex sample design. *AStA Adv. Stat. Anal.* **2018**, *102*, 381–411. [CrossRef]
17. Ghosh, A.; Mandal, A.; Martín, N.; Pardo, L. Influence analysis of robust Wald-type tests. *J. Multivar. Anal.* **2016**, *147*, 102–126. [CrossRef]
18. Basu, A.; Mandal, A.; Martín, N.; Pardo, L. A Robust Generalization of the Rao Test. *J. Bus. Econ. Stat.* **2021**, 1–12. [CrossRef]
19. Martín, N. Rao's Score Tests on Correlation Matrices. *arXiv* **2020**, arXiv:2012.14238.
20. Kammerer, N.B.; Stummer, W. Some dissimilarity Measures of Branching Processes and optimal Decision Making in the Presence of Potencial Pandemics. *Entropy* **2020**, *22*, 874. [CrossRef] [PubMed]
21. Kang, J.; Lee, S. Parameter change test for Poisson autoregressive models. *Scand. J. Stat.* **2014**, *41*, 1136–1152. [CrossRef]
22. Lee, Y.; Lee, S. CUSUM test for general nonlinear integer-valued GARCH models: Comparison study. *Ann. Inst. Stat. Math.* **2019**, *71*, 1033–1057. [CrossRef]
23. Lee, S.; Kim, D. Monitoring Parameter Change for Time Series. Models of Counts Based on MInimum Density Power Divergence estimator. *Entropy* **2020**, *22*, 1304. [CrossRef]
24. Kim, B.; Lee, S. Robust change point test for general integer-valued time series models based on density power divergence. *Entropy* **2020**, *22*, 493. [CrossRef]
25. Riani, M.; Atkinson, A.C.; Corbellini, A.; Perrotta, D. Robust Regression with Density Power Divergence: Theory, Comparisons, and Data Analysis. *Entropy* **2020**, *22*, 399. [CrossRef]
26. Akaike, H. Information theory and an extension of the maximum likelihood principle. In *Springer Series in Statistics, Proceedings of the Second International Symposium on Information Theory Petrov*; Springer: Berlin/Heidelberger, Germany, 1973; pp. 267–281.
27. Konishi, S.; Kitagawa, G. Generalised information criteria in model selection. *Biometrika* **1996**, *83*, 875–890. [CrossRef]
28. Schwarz, G. Estimating the dimension of a model. *Ann. Stat.* **1978**, *6*, 461–464. [CrossRef]
29. Agostinelli, C. Robust model selection in regression via weighted likelihood estimating equations. *Stat. Probab. Lett.* **2002**, *76*, 1930–1934. [CrossRef]
30. Ronchetti, E.; Staudte, R.G. A robust version of Mallows' Cp. *J. Am. Stat. Assoc.* **1994**, *89*, 550–559.
31. Mattheou, K.; Lee, S.; Karagrigoriou, A. A model selection criterion based on the BHHJ measure of divergence. *J. Stat. Plann. Inf.* **2009**, *139*, 228–235. [CrossRef]
32. Mantalos, P.; Mattheou, K.; Karagrigoriou, A. An improved divergence information criterion for the determination of the order of an AR process. *Commun. Stat.* **2010**, *39*, 865–879. [CrossRef]
33. Toma, A. Model selection criteria using divergences. *Entropy* **2014**, *16*, 2686–2698. [CrossRef]
34. Toma, A.; Karagrigoriou, A.; Trentou, P. Robust Model Selection Criteria Based on Pseudodistances. *Entropy* **2020**, *22*, 304. [CrossRef] [PubMed]
35. Castilla, E.; Martín, N.; Pardo, L.; Zografos, K. Model Selection in a Composite Likelihood Framework Based on Density Power Divergence. *Entropy* **2020**, *22*, 270. [CrossRef] [PubMed]
36. Varin, C.; Reid, N.; Firth, D. An overview of composite likelihood methods. *Stat. Sin.* **2011**, *21*, 4–42.
37. Castilla, E.; Martín, N.; Pardo, L.; Zografos, K. Composite Likelihood Methods Based on Minimum Density Power Divergence Estimator. *Entropy* **2018**, *20*, 18. [CrossRef]
38. Castilla, E.; Martín, N.; Pardo, L.; Zografos, K. Composite likelihood methods: Rao-type tests based on composite minimum density power divergence estimator. *Stat. Pap.* **2019**. [CrossRef]
39. Martín, N.; Pardo, L.; Zografos, K. On divergence tests for composite hypotheses under composite likelihood. *Stat. Pap.* **2019**, *60*, 1883–1919. [CrossRef]
40. Varin, C.; Vidoni, P. A note on composite likelihood inference and model selection. *Biometrika* **2005**, *92*, 519–528 [CrossRef]
41. Gao, X.; Song, P.X.K. Composite likelihood Bayesian information criteria for model selection in high-dimensional data. *J. Am. Stat. Assoc.* **2010**, *105*, 1531–1540. [CrossRef]
42. Ng, C.T.; Joe, H. Model comparison with composite likelihood information criteria. *Bernoulli* **2014**, *20*, 1738–1764. [CrossRef]
43. Sekeh, S.Y.; Noshad, M.; Moon, K.R.; Hero, A.O. Convergence Rates for Empirical Estimation of Binary Classification Bounds. *Entropy* **2019**, *21*, 1144. [CrossRef]
44. Yukich, J. Probability Theory of Classical Euclidean Optimization. In *Lecture Notes in Mathematics*; Springer: Berlin, Germany, 1998; Volume 1675.
45. Sofikitou, E.M.; Ray, L.; Wang, H.; Markatou, M. Distance-Based Estimation Methods for Models for Discrete and Mixed-Scale Data. *Entropy* **2020**, *23*, 107. [CrossRef]

46. Pardo, J.A.; Pardo, L.; Pardo, M.C. Minimum φ-Divergence Estimator in Logistic Regression Models. *Stat. Pap.* **2005** , *47*, 91–108. [CrossRef]
47. Pardo, J.A.; Pardo, L.; Pardo, M.C. Testing In Logistic Regression Models on φ-Divergences Measures. *J. Stat. Plan. Inference.* **2006**, *136*, 982–1006. [CrossRef]
48. Pardo, J.A.; Pardo, M.C. Minimum φ-Divergence Estimator and φ-Divergence Statistics in Generalized Linear Models with Binary Data. *Methodol. Comput. Appl. Probab.* **2008**, *10*, 357–379. [CrossRef]
49. Simpson, D.G. Minimum Hellinger Distance Estimation for the Analysis of Count Data. *J. Am. Stat. Assoc.* **1987**, *82*, 802–807. [CrossRef]
50. Simpson, D.G. Hellinger Deviance Tests: Efficiency, Breakdown Points, and Examples. *J. Am. Stat. Assoc.* **1989**, *84*, 104–113. [CrossRef]
51. Markatou, M.; Basu, A.; Lindsay, B.G. Weighted Likelihood Estimating Equations: The Discrete Case with Applications to Logistic Regression. *J. Stat. Plan. Inference* **1997**, *57*, 215–232. [CrossRef]
52. Basu, A.; Basu, S. Penalized Minimum Disparity Methods for Multinomial Models. *Stat. Sin.* **1998**, *8*, 841–860.
53. Gupta, A.K.; Nguyen, T.; Pardo, L. Inference Procedures for Polytomous Logistic Regression Models Based on φ-Divergence Measures. *Math. Methods Stat.* **2006**, *15*, 269–288.
54. Martín, N.; Pardo, L. New Influence Measures in Polytomous Logistic Regression Models Based on Phi-Divergence Measures. *Commun. Stat.* **2014**, *43*, 2311–2321. [CrossRef]
55. Castilla, E.; Ghosh, A.; Martín, N.; Pardo, L. New Robust Statistical Procedures for Polytomous Logistic Regression Models. *Biometrics* **2018**, *74*, 1282–1291. [CrossRef] [PubMed]
56. Martín, N.; Pardo, L. Minimum Phi-Divergence Estimators for Loglinear Models with Linear Constraints and Multinomial Sampling. *Stat. Pap.* **2008**, *49*, 2311–2321. [CrossRef]
57. Pardo, L.; Martín, N. Minimum Phi Divergence Estimators and Phi-Divergence Test for Statistics in Contingency Tables with Symmetric Structure: An Overview. *Symmetry* **2010**, *2*, 1108–1120. [CrossRef]
58. Pardo, L.; Pardo, M.C. Minimum Power-Divergence Estimator in Three-Way Contingency Tables. *J. Stat. Comput. Simul.* **2003**, *73*, 819–831. [CrossRef]
59. Pardo, L.; Pardo, M.C.; Zografos, K. Minimum φ-Divergence Estimator for Homogeneity in Multinomial Populations. *Sankhya Indian J. Stat. Ser. A* **2001**, *63*, 72–92.
60. Basu, A.; Harris, I.A.; Hjort, N.L.; Jones, M.C. Robust and Efficient Estimation by Minimising a Density Power Divergence. *Biometrika* **1998**, *85*, 549–559. [CrossRef]
61. Vidyashankar, A.N.; Collamore, J.F. Event Analysis for Minimum Hellinger Distance Estimators via Large Deviation Theory. *Entropy* **2021**, *23*, 386. [CrossRef]

Article

Some Dissimilarity Measures of Branching Processes and Optimal Decision Making in the Presence of Potential Pandemics

Niels B. Kammerer [1] and Wolfgang Stummer [2],*

[1]   Königinstrasse 75, 80539 Munich, Germany; nielskammerer@gmx.de
[2]   Department of Mathematics, University of Erlangen–Nürnberg, Cauerstrasse 11, 91058 Erlangen, Germany
*   Correspondence: stummer@math.fau.de

Received: 26 June 2020; Accepted: 28 July 2020; Published: 8 August 2020

Abstract: We compute exact values respectively bounds of dissimilarity/distinguishability measures–in the sense of the Kullback-Leibler information distance (relative entropy) and some transforms of more general power divergences and Renyi divergences–between two competing discrete-time *Galton-Watson branching processes with immigration* GWI for which the offspring as well as the immigration (importation) is arbitrarily Poisson-distributed; especially, we allow for arbitrary type of extinction-concerning criticality and thus for non-stationarity. We apply this to optimal decision making in the context of the spread of potentially pandemic infectious diseases (such as e.g., the current COVID-19 pandemic), e.g., covering different levels of dangerousness and different kinds of intervention/mitigation strategies. Asymptotic distinguishability behaviour and diffusion limits are investigated, too.

Keywords: Galton-Watson branching processes with immigration; Hellinger integrals; power divergences; Kullback-Leibler information distance/divergence; relative entropy; Renyi divergences; epidemiology; COVID-19 pandemic; Bayesian decision making; INARCH(1) model; GLM model; Bhattacharyya coefficient/distance

---

Contents

1. Introduction

(This paper is a thoroughly revised, extended and retitled version of the preprint arXiv:1005.3758v1 of both authors) Over the past twenty years, *density-based divergences* $D(P,Q)$ –also known as (dis)similarity measures, directed distances, disparities, distinguishability measures, proximity measures–between probability distributions P and Q, have turned out to be of substantial importance for decisive statistical tasks such as parameter estimation, testing for goodness-of-fit, Bayesian decision procedures, change-point detection, clustering, as well as for other research fields such as information theory, artificial intelligence, machine learning, signal processing (including image and speech processing), pattern recognition, econometrics, and statistical physics. For some comprehensive overviews on the divergence approach to statistics and probability, the reader is referred to the insightful books of e.g., Liese & Vajda [1], Read & Cressie [2], Vajda [3], Csiszár & Shields [4], Stummer [5], Pardo [6], Liese & Miescke [7], Basu et al. [8], Voinov et al. [9], the survey articles of e.g., Liese & Vajda [10], Vajda & van der Meulen [11], the structure-building papers of Stummer & Vajda [12], Kißlinger & Stummer [13] and Broniatowski & Stummer [14], and the references therein. Divergence-based bounds of minimal mean decision risks (e.g., Bayes risks in finance) can be found e.g., in Stummer & Vajda [15] and Stummer & Lao [16].

Amongst the above-mentioned dissimilarity measures, an important omnipresent subclass are the so-called $f-$divergences of Csiszar [17], Ali & Silvey [18] and Morimoto [19]; important special cases thereof are the total variation distance and the very frequently used $\lambda-$*order power divergences* $I_\lambda(P,Q)$ (also known as alpha-entropies, Cressie-Read measures, Tsallis cross-entropies) with $\lambda \in \mathbb{R}$. The latter cover e.g., the very prominent Kullback-Leibler information divergence $I_1(P,Q)$ (also called relative entropy), the (squared) Hellinger distance $I_{1/2}(P,Q)$, as well as the Pearson chi-square divergence $I_2(P,Q)$. It is well known that the power divergences can be build with the help of the $\lambda-$*order Hellinger integrals* $H_\lambda(P,Q)$ (where e.g., the case $\lambda = 1/2$ corresponds to the well-known Bhattacharyya coefficient), which are information measures of interest by their own and which are also the crucial ingredients of $\lambda-$*order Renyi divergences* $R_\lambda(P,Q)$ (see e.g., Liese & Vajda [1], van Erven & Harremoes [20]); the case $R_{1/2}(P,Q)$ corresponds to the well-known Bhattacharyya distance.

The above-mentioned information/dissimilarity measures have been also investigated in non-static, time-dynamic frameworks such as for various different contexts of *stochastic processes* like *processes with independent increments* (see e.g., Newman [21], Liese [22], Memin & Shiryaev [23], Jacod & Shiryaev [24], Liese & Vajda [1], Linkov & Shevlyakov [25]), *Poisson point processes* (see e.g., Liese [26], Jacod & Shiryaev [24], Liese & Vajda [1]), *diffusion prcoesses and solutions of stochastic differential equations with continuous paths* (see e.g., Kabanov et al. [27], Liese [28], Jacod & Shiryaev [24], Liese & Vajda [1], Vajda [29], Stummer [30–32], Stummer & Vajda [15]), and *generalized binomial processes* (see e.g., Stummer & Lao [16]); further related literature can be found e.g., in references of the aforementioned papers and books.

Another important class of time-dynamic models is given by *discrete-time integer-valued branching processes*, in particular *(Bienaymé-)Galton-Watson processes without immigration* GW respectively *with immigration (resp. importation, invasion)* GWI, which have numerous applications in biotechnology, population genetics, internet traffic research, clinical trials, asset price modelling, derivative pricing, and many others. As far as important terminology is concerned, we abbreviatingly subsume both models as GW(I) and, simply as GWI in case that GW appears as a parameter-special-case of GWI; recall that a GW(I) is called *subcritical* respectively *critical* respectively *supercritical* if its offspring mean is less than 1 respectively equal to 1 respectively larger than 1.

For applications of GW(I) in *epidemiology*, see e.g., the works of Bartoszynski [33], Ludwig [34], Becker [35,36], Metz [37], Heyde [38], von Bahr & Martin-Löf [39], Ball [40], Jacob [41], Barbour & Reinert [42], Section 1.2 of Britton & Pardoux [43]); for more details see Section 2.3 below.

For connections of GW(I) to *time series of counts* including GLM models, see e.g., Dion, Gauthier & Latour [44], Grunwald et al. [45], Kedem & Fokianos [46], Held, Höhle & Hofmann [47], and Weiß [48]; a more comprehensive discussion can be found in Section 2.2 below.

As far as the combined study of information measures and GW processes is concerned, let us first mention that (transforms of) power divergences have been used for supercritical Galton-Watson processes without immigration for instance as follows: Feigin & Passy [49] study the problem to find an offspring distribution which is closest (in terms of relative entropy type distance) to the original offspring distribution and under which ultimate extinction is certain. Furthermore, Mordecki [50] gives an equivalent characterization for the stable convergence of the corresponding log-likelihood process to a mixed Gaussian limit, in terms of conditions on Hellinger integrals of the involved offspring laws. Moreover, Sriram & Vidyashankar [51] study the properties of offspring-distribution-parameters · which minimize the squared Hellinger distance between the model offspring distribution and the corresponding non-parametric maximum likelihood estimator of Guttorp [52]. For the setup of GWI with Poisson offspring and nonstochastic immigration of constant value 1, Linkov & Lunyova [53] investigate the asymptotics of Hellinger integrals in order to deduce large deviation assertions in hypotheses testing problems.

In contrast to the above-mentioned contexts, this paper pursues the following main goals:

(MG1) for any time horizon and any criticality scenario (allowing for non-stationarities), to compute lower and upper bounds–and sometimes even exact values–of the Hellinger integrals $H_\lambda (P_\mathcal{A}||P_\mathcal{H})$, power divergences $I_\lambda (P_\mathcal{A}||P_\mathcal{H})$ and Renyi divergences $R_\lambda (P_\mathcal{A}||P_\mathcal{H})$ of two alternative Galton-Watson branching processes $P_\mathcal{A}$ and $P_\mathcal{H}$ (on path/scenario space), where (i) $P_\mathcal{A}$ has Poisson($\beta_\mathcal{A}$) distributed offspring as well as Poisson($\alpha_\mathcal{A}$) distributed immigration, and (ii) $P_\mathcal{H}$ has Poisson($\beta_\mathcal{H}$) distributed offspring as well as Poisson($\alpha_\mathcal{H}$) distributed immigration; the non-immigration cases are covered as $\alpha_\mathcal{A} = \alpha_\mathcal{H} = 0$; as a side effect, we also aim for corresponding asymptotic distinguishability results;

(MG2) to compute the corresponding limit quantities for the context in which (a proper rescaling of) the two alternative Galton-Watson processes with immigration converge to *Feller*-type branching diffusion processes, as the time-lags between the generation-size observations tend to zero;

(MG3) as an exemplary field of application, to indicate how to use the results of (MG1) for Bayesian decision making in the epidemiological context of an infectious-disease pandemic (e.g., the current COVID-19), where e.g., potential state-budgetary losses can be controlled by alternative public policies (such as e.g., different degrees of lockdown) for mitigations of the time-evolution of the number of infectious persons (being quantified by a GW(I)). Corresponding Neyman-Pearson testing will be treated, too.

Because of the involved Poisson distributions, these goals can be tackled with a high degree of tractability, which is worked out in detail with the following structure (see also the full table of contents after this paragraph): in Section 2, we first introduce (i) the basic ingredients of Galton-Watson processes together with their interpretations in the above-mentioned pandemic setup where it is essential to study *all* types of criticality (being connected with levels of reproduction numbers), (ii) the employed fundamental information measures such as Hellinger integrals, power divergences and Renyi divergences, (iii) the underlying decision-making framework, as well as (iv) connections to time series of counts and asymptotical distinguishability. Thereafter, we start our detailed technical analyses by giving *recursive* exact values respectively *recursive* bounds–as well as their applications–of Hellinger integrals $H_\lambda \left(P_\mathcal{A}||P_\mathcal{H}\right)$ (see Section 3), power divergences $I_\lambda \left(P_\mathcal{A}||P_\mathcal{H}\right)$ and Renyi divergences $R_\lambda \left(P_\mathcal{A}||P_\mathcal{H}\right)$ (see Sections 4 and 5). *Explicit closed-form* bounds of Hellinger integrals $H_\lambda \left(P_\mathcal{A}||P_\mathcal{H}\right)$ will be worked out in Section 6, whereas Section 7 deals with Hellinger integrals and power divergences of the above-mentioned Galton-Watson type diffusion approximations.

2. The Framework and Application Setups

2.1. Process Setup

We investigate dissimilarity measures and apply them to decisions, in the following context. Let the integer-valued random variable X_n ($n \in \mathbb{N}_0$) denote the size of the nth generation of a population (of persons, organisms, spreading news, other kind of objects, etc.) with specified characteristics, and suppose that for the modelling of the time-evolution $n \mapsto X_n$ we have the choice between the following two (e.g., alternative, competing) models $(\mathcal{H})$ and $(\mathcal{A})$:

$(\mathcal{H})$ a discrete-time homogeneous *Galton-Watson process with immigration GWI*, given by the recursive description

$$X_0 \in \mathbb{N}; \qquad \mathbb{N}_0 \ni X_n = \sum_{k=1}^{X_{n-1}} Y_{n-1,k} + \widetilde{Y}_n, \qquad n \in \mathbb{N}, \tag{1}$$

where $Y_{n-1,k}$ is the number of offspring of the kth object (e.g., organism, person) within the $(n-1)$th generation, and $\widetilde{Y}_n$ denotes the number of immigrating objects in the nth generation. Notice that we employ an arbitrary *deterministic* (i.e., degenerate random) initial generation size X_0. We always assume that under the corresponding dynamics-governing law $P_\mathcal{H}$

(GWI1) the collection $Y := \left\{Y_{n-1,k}, \, n \in \mathbb{N}, k \in \mathbb{N}\right\}$ consists of independent and identically distributed (i.i.d.) random variables which are Poisson distributed with parameter $\beta_\mathcal{H} > 0$,

(GWI2) the collection $\widetilde{Y} := \left\{\widetilde{Y}_n, \, n \in \mathbb{N}\right\}$ consists of i.i.d. random variables which are Poisson distributed with parameter $\alpha_\mathcal{H} \geq 0$ (where $\alpha_\mathcal{H} = 0$ stands for the degenerate case of having no immigration),

(GWI3) Y and $\widetilde{Y}$ are independent.

$(\mathcal{A})$ a discrete-time homogeneous *Galton-Watson process with immigration GWI* given by the same recursive description (1), but with different dynamics-governing law $P_\mathcal{A}$ under which (GWI1) holds with parameter $\beta_\mathcal{A} > 0$ (instead of $\beta_\mathcal{H} > 0$), (GWI2) holds with $\alpha_\mathcal{A} \geq 0$ (instead of $\alpha_\mathcal{H} \geq 0$), and (GWI3) holds. As a side remark, in some contexts the two models $(\mathcal{H})$ and $(\mathcal{A})$ may function as a "sandwich" of a more complicated not fully known model.

Basic and advanced facts on general GWI (introduced by Heathcote [54]) can be found e.g., in the monographs of Athreya & Ney [55], Jagers [56], Asmussen & Hering [57], Haccou [58]; see also e.g., Heyde & Seneta [59], Basawa & Rao [60], Basawa & Scott [61], Sankaranarayanan [62], Wei & Winnicki [63], Winnicki [64], Guttorp [52] as well as Yanev [65] (and also the references therein all those) for adjacent fundamental statistical issues including the involved technical and conceptual challenges.

For the sake of brevity, wherever we introduce or discuss corresponding quantities *simultaneously* for both models $\mathcal{H}$ and $\mathcal{A}$, we will use the subscript $\bullet$ as a synonym for either the symbol $\mathcal{H}$ or $\mathcal{A}$.

For illustration, recall the well-known fact that the corresponding conditional probabilities $P_\bullet(X_n = \cdot \mid X_{n-1} = k)$ are again Poisson-distributed, with parameter $\beta_\bullet \cdot k + \alpha_\bullet$.

In oder to achieve a transparently representable structure of our results, we subsume the involved parameters as follows:

(PS1) $\mathcal{P}_{\mathrm{SP}}$ is the set of all constellations $(\beta_\mathcal{A}, \beta_\mathcal{H}, \alpha_\mathcal{A}, \alpha_\mathcal{H})$ of real-valued parameters $\beta_\mathcal{A} > 0$, $\beta_\mathcal{H} > 0$, $\alpha_\mathcal{A} > 0$, $\alpha_\mathcal{H} > 0$, such that $\beta_\mathcal{A} \neq \beta_\mathcal{H}$ or $\alpha_\mathcal{A} \neq \alpha_\mathcal{H}$ (or both); in other words, both models are non-identical and have non-vanishing immigration;

(PS2) $\mathcal{P}_{\mathrm{NI}}$ is the set of all $(\beta_\mathcal{A}, \beta_\mathcal{H}, \alpha_\mathcal{A}, \alpha_\mathcal{H})$ of real-valued parameters $\beta_\mathcal{A} > 0$, $\beta_\mathcal{H} > 0$, $\alpha_\mathcal{A} = \alpha_\mathcal{H} = 0$, such that $\beta_\mathcal{A} \neq \beta_\mathcal{H}$; this corresponds to the important special case that both models have no immigration and are non-identical;

(PS3) the resulting disjoint union will be denoted by $\mathcal{P} = \mathcal{P}_{\mathrm{SP}} \cup \mathcal{P}_{\mathrm{NI}}$.

Notice that for (unbridgeable) technical reasons, we *do not allow for* "crossovers" between "immigration and no-immigration" (i.e., $\alpha_\mathcal{A} = 0$ and $\alpha_\mathcal{H} \neq 0$, respectively, $\alpha_\mathcal{A} \neq 0$ and $\alpha_\mathcal{H} = 0$). For practice, this is not a strong restriction, since one may take e.g., $\alpha_\mathcal{A} = 10^{-12}$ and $\alpha_\mathcal{H} = 1$.

For the non-immigration case $\alpha_\bullet = 0$ one has the following *extinction properties* (see e.g., Harris [66], Athreya & Ney [55]). As usual, let us define the extinction time $\tau := \min\{i \in \mathbb{N} : X_\ell = 0 \text{ for all integers } \ell \geq i\}$ if this minimum exists, and $\tau := \infty$ else. Correspondingly, let $\mathcal{B} := \{\tau < \infty\}$ be the extinction set. If the *offspring mean* $\beta_\bullet$ satisfies $\beta_\bullet < 1$—which is called the *subcritical* case– or $\beta_\bullet = 1$—which is known as the *critical* case–then extinction is certain, i.e., there holds $P(\mathcal{B} \mid X_0 = 1) = 1$. However, if the offspring mean satisfies $\beta_\bullet > 1$—which is called the *supercritical* case–then there is a probability greater than zero, that the population never dies out, i.e., $P(\mathcal{B} \mid X_0 = 1) \in]0, 1[$. In the latter case, X_n explodes (a.s.) to infinity as $n \to \infty$.

In contrast, for the (nondegenerate, nonvanishing) immigration case $\alpha_\bullet \neq 0$ there is *no extinction*, viz. $P(\mathcal{B} \mid X_0 = 1) = 0$, although there may be zero population $X_{\ell_0} = 0$ for some intermediate time $\ell_0 \in \mathbb{N}$; but due to the immigration, with probability one there is always a later time $\ell_1 > \ell_0$, such that $X_{\ell_1} > 0$. Nevertheless, also for the setup $\alpha_\bullet \neq 0$ it is important to know whether $\beta_\bullet \gtreqless 1$—which is still called (super-, sub-)criticality–since e.g., in the case $\beta_\bullet < 1$ the population size X_n converges (as $n \to \infty$) to a stationary distribution on $\mathbb{N}$ whereas for $\beta_\bullet > 1$ the behaviour is non-stationary (non-ergodic), see e.g., Athreya & Ney [55].

At this point, let us emphasize that in our investigations (both for $\alpha_\bullet = 0$ and for $\alpha_\bullet \neq 0$) we *do allow for* "crossovers" between "different criticalities", i.e., we deal with all cases $\beta_\mathcal{A} \gtreqless 1$ versus all cases $\beta_\mathcal{H} \gtreqless 1$; as will be explained in the following, this unifying flexibility is especially important for corresponding epidemiological-model comparisons (e.g., for the sake of decision making).

One of our main goals is to quantitatively compare (the time-evolution of) two competing GWI models $\mathcal{H}$ and $\mathcal{A}$ with respective parameter sets $(\beta_\mathcal{H}, \alpha_\mathcal{H})$ and $(\beta_\mathcal{A}, \alpha_\mathcal{A})$, in terms of the information measures $H_\lambda(P_\mathcal{A} || P_\mathcal{H})$ (Hellinger intergrals), $I_\lambda(P_\mathcal{A} || P_\mathcal{H})$ (power divergences), $R_\lambda(P_\mathcal{A} || P_\mathcal{H})$ (Renyi divergences). The latter two express a distance (degree of dissimilarity) between $\mathcal{H}$ and $\mathcal{A}$. From this, we shall particularly derive applications for decision making under uncertainty (including tests).

2.2. Connections to Time Series of Counts

It is well known that a Galton-Watson process with Poisson offspring (with parameter $\beta_\bullet$) and Poisson immigration (with parameter $\alpha_\bullet$) is "distributionally" equal to each of the following models (listed in "tree-type" chronological order):

(M1) a Poissonian *Generalized Integer-valued Autoregressive process* GINAR(1) in the sense of Gauthier & Latour [67] (see also Dion, Gauthier & Latour [44], Latour [68], as well as Grunwald et al. [45]), that is, a first-order autoregressive times series with Poissonian thinning (with parameter $\beta_\bullet$) and Poissonian innovations (with parameter $\alpha_\bullet$);

(M2) Poissonian *first order Conditional Linear Autoregressive model* (Poissonian CLAR(1)) in the sense of Grunwald et al. [45] (and earlier preprints thereof) (since the conditional expectation is $EP_\bullet[X_n|\mathcal{F}_{n-1}] = \alpha_\bullet + \beta_\bullet \cdot X_{n-1}$); this can be equally seen as Poissonian autoregressive *Generalized Linear Model* GLM with identity link function (cf. [45] as well as Chapter 4 of Kedem & Fokianos [46]), that is, an autoregressive GLM with Poisson distribution as random component and the identity link as systematic component; the same model was used (and generalized)

    (M2i) under the name BIN(1) by Rydberg & Shephard [69] for the description of the number X_n of stock transactions/trades recorded up to time n;

    (M2ii) under the name *Poisson autoregressive model* PAR(1) by Brandt & Williams [70] for the description of event counts in political and other social science applications;

    (M2iii) under the name *Autoregressive Conditional Poisson model* ACP(1,0) by Heinen [71];

    (M2iv) by Held, Höhle & Hofmann [47] as well as Held et al. [72], as a description of the time-evolution of counts from infectious disease surveillance databases, where $\beta_\bullet$ (respectively, $\alpha_\bullet$) is interpreted as driving parameter of epidemic (respectively, endemic) component; in principle, this type of modelling can be also implicitly recovered as a special case of the epidemics-treating work of Finkenstädt, Bjornstad & Grenfell [73], by assuming trend- and season-neglecting (e.g., intra-year) measles data in urban areas of about 10 million people (provided that their population size approximation extends linearly);

    (M2v) under the name *integer-valued Generalized Autoregressive Conditional Heteroscedastic model* INGARCH(1,0) by Ferland, Latour & Oraichi [74] (since the conditional variance is $VarP_\bullet[X_n|\mathcal{F}_{n-1}] = \alpha_\bullet + \beta_\bullet \cdot X_{n-1}$), see also Weiß [75]; this has been refinely named as INARCH(1) model by Weiß [76,77], and frequently applied thereafter; for an "overlapping-generation type" interpretation of the INARCH(1) model, which is an adequate description for the time-evolution of overdispersed counts with an autoregressive serial dependence structure, see Weiß & Testik [78]; for a corresponding comprehensive recent survey (also to more general count time series), the reader is referred to the book of Weiß [48];

Moreover, according to the general considerations of Grunwald et al. [45], the Poissonian Galton-Watson model with immigration may possibly be "distributionally equal" to an integer-valued autoregressive model with random coefficient (thinning).

Nowadays, besides the name *homogeneous Galton-Watson model with immigration GWI*, the name *INARCH(1)* seems to be the most used one, and we follow this terminology (with emphasis on GWI). Typical features of the above-mentioned models (M1) to (M2v), are the use of $\mathbb{Z}$ as the set of times, and the assumptions $\alpha_\bullet > 0$ as well as $\beta_\bullet \in]0,1[$, which guarantee stationarity and ergodicity (see above). In contrast, we employ $\mathbb{N}_0$ as the set of times, degenerate (and thus, non-equilibrium) starting distribution, and arbitrary $\alpha_\bullet \geq 0$ as well as $\beta_\bullet > 0$. For such a situation, as explained above, we quantitatively compare two competing GWI models $\mathcal{H}$ and $\mathcal{A}$ with respective parameter sets $(\beta_\mathcal{H}, \alpha_\mathcal{H})$ and $(\beta_\mathcal{A}, \alpha_\mathcal{A})$. Since—as can be seen e.g., in (29) below—we basically employ only (conditionally) distributional ingredients, such as the corresponding likelihood ratio (see e.g., (13) to (15), (27) to (29) below), *all the results of the Sections 3–6 can be immediately carried over to the above-mentioned time-series contexts* (where we even allow for non-stationarities, in fact we start with a one-point/Dirac distribution); for the sake of brevity, in the rest of the paper this will not be mentioned explicitly anymore.

Notice that a Poissonian GWI as well as all models (M1) and (M2) are–despite of their *conditional* Poisson law– typically overdispersed since

$$EP_\bullet[X_n] = \alpha_\bullet + \beta_\bullet \cdot EP_\bullet[X_{n-1}] \leq \alpha_\bullet + \beta_\bullet \cdot EP_\bullet[X_{n-1}] + \beta_\bullet^2 \cdot VarP_\bullet[X_{n-1}] = VarP_\bullet[X_n], \quad n \in \mathbb{N}\setminus\{1\},$$

with equality iff (i.e., if and only if) $\alpha_\bullet = 0$ (NI) and $X_{n-2} = 0$ (extinction at $n - 2$ with $n \geq 3$).

2.3. Applicability to Epidemiology

The above-mentioned framework can be used for any of the numerous fields of applications of discrete-time branching processes, and of the closely related INARCH(1) models. For the sake of brevity, we explain this—as a kind of running-example—in detail for the currently highly important context of the epidemiology of infectious diseases. For insightful non-mathematical introductions to the latter, see e.g., Kaslow & Evans [79], Osterholm & Hedberg [80]; for a first entry as well as overviews on modelling, the reader is referred to e.g., Grassly & Fraser [81], Keeling & Rohani [82], Yan [83,84], Britton [85], Diekmann, Heesterbeek & Britton [86], Cummings & Lessler [87], Just et al. [88], Britton & Giardina [89], Britton & Pardoux [43]. A survey on the particular role of branching processes in epidemiology can be found e.g., in Jacob [41].

Undoubtedly, by nature, the spreading of an infectious disease through a (human, animal, plant) population is a branching process with possible immigration. Indeed, typically one has the following mechanism:

(D1)   at some time t_k^E–called the time of exposure (moment of infection)—an individual k of a specified population is infected in a wide sense, i.e., entered/invaded/colonized by a number of transmissible disease-causative pathogens (etiologic agents such as viruses, bacteria, protozoans and other parasites, subviruses (e.g., prions and plant viroids), etc.); the individual is then a *host* (of pathogens);

(D2)   depending on the level of immunity and some other factors, these pathogens may multiply/replicate within the host to an extent (over a threshold number) such that at time t_k^I some of the pathogens start to leave their host (*shedding of pathogens*); in other words, the individual k becomes *infectious* at the time t_k^I of *onset of infectiousness*. Ex post, one can then say that the individual became infected in the narrow sense at earlier time t_k^E and call it a *primary case*. The time interval $[t_k^E, t_k^I[$ is called the *latent/latency/pre-infectious period* of k, and $t_k^I - t_k^E$ its duration (in some literature, there is no verbal distinction between them); notice that t_k^I may differ from the time t_k^{OS} of *onset (first appearance) of symptoms*, which leads to the so-called *incubation period* $[t_k^E, t_k^{OS}[$; if $t_k^I < t_k^{OS}$ then $[t_k^I, t_k^{OS}[$ is called the *pre-symptomatic period*;

(D3)   as long as the individual k stays infectious, by shedding of pathogens it may infect in a narrow sense a random number $Y_k \in \mathbb{N}_0$ of other individuals which are *susceptible* (i.e., neither immune nor already infected in a narrow sense), where the distribution of Y_k depends on the individual's (natural, voluntary, forced) behaviour, its environment, as well as some other factors e.g., connected with the type of pathogen transmission; the newly infected individuals are called *offspring of k*, and *secondary cases* if they are from the same specified population or *exportations* if they are from a different population; from the view of the latter, these infections are *imported cases* and thus can be viewed as *immigrants*;

(D4)   at the time t_k^R of *cessation of infectiousness*, the individual stops being infectious (e.g., because of recovery, death, or total isolation); the time interval $[t_k^I, t_k^R[$ is called the *period of infectiousness (also period of communicability, infectious/infective/shedding/contagious period)* of k, and $t_k^R - t_k^I$ its duration (in some literature, there is no verbal distinction between them); notice that t_k^R may differ from the time t_k^{CS} of *cessation (last appearance) of symptoms* which leads to the so-called *sickness period* $[t_k^{OS}, t_k^{CS}[$;

(D5)   this branching mechanism continues within the specified population until there are no infectious individuals and also no importations anymore (eradication, full extinction, total elimination)– up to a specified final time (which may be large or even infinite);

All the above-mentioned times $t_k^\cdot$ and time intervals are random, by nature. Two further connected quantities are also important for modelling (see e.g., Yan & Chowell [84] (p. 241ff), including a history of corresponding terminology). Firstly, the *generation interval* (generation time, transmission interval) is the time interval from the onset of infectiousness in a primary case (called the infector) to the onset of infectiousness in a secondary case (called the infectee) infected by the primary case; clearly, the generation interval is random, and so is its duration (often, the (population-)mean of the latter is also called generation interval). Typically, generation intervals are important ingredients of branching process models of infectious diseases. Secondly, the *serial interval* describes time interval from the onset of symptoms in a primary case to the onset of symptoms in a secondary case infected by the primary case. By nature, the serial interval is random, and so is its duration (often, the (population-)mean of the latter is also called serial interval). Typically, the serial interval is easier to observe than the generation interval, and thus, the latter is often approximately estimated from data of the former. For further investigations on generation and serial intervals, the reader is referred to e.g., Fine [90], Svensson [91,92], Wallinga & Lipsitch [93], Forsberg White & Pagano [94], Nishiura [95], Scalia Tomba et al. [96], Trichereau et al. [97], Vink, Bootsma & Wallinga [98], Champredon & Dushoff [99], Just et al. [88], and–especially for the novel COVID-19 pandemics—An der Heiden & Hamouda [100], Ferretti et al. [101], Ganyani et al. [102], Li et al. [103], Nishiura, Linton & Akhmetzhanov [104], Park et al. [105].

With the help of the above-mentioned *individual* ingredients, one can aggregatedly build numerous different *population-wide* models of infectious diseases in discrete time as well as in continuous time; the latter are typically observed only in discrete-time steps (discrete-time sampling), and hence in the following we concentrate on discrete-time modelling (of the real or the observational process). In fact, we confine ourselves to the important task of modelling the evolution $n \mapsto X_n$ of the number of *incidences* at "stage" n, where *incidence* refers to the number of *new* infected/infectious individuals. Here, n may be a generation number where, inductively, $n = 0$ refers to the generation of the first appearing primary cases in the population (also called *initial importations*), and n refers to the generation of offsprings of all individuals of generation $n - 1$. Alternatively, n may be the index of a physical ("calender") point of time t_n, which may be deterministic or random; e.g., $(t_n)_{n \in \mathbb{N}}$ may be a strictly increasing series of (i) equidistant deterministic time points (and thus, one can identify $t_n = n$ in appropriate time units such as days, weeks, bi-weeks, months), or (ii) non-equidistant deterministic time points, or (iii) random time points (as a side remark, let us mention that in some situations, X_n may alternatively denote the number of *prevalences* at "stage" n, where *prevalence* refers to the total number of infected/infectious individuals (e.g., through some methodical tricks like "self-infection")).

In the light of this, one can loosely define an *epidemic* as the rapid spread of an infectious disease within a specified population, where the numbers X_n of incidences are high (or much higher than expected) for that kind of population. A *pandemic* is a geographically large-scale (e.g., multicontinental or worldwide) epidemic. An *outbreak/onset* of an epidemic in the narrow sense is the (time of) change where an infectious disease turns into an epidemic, which is typically quantified by exceedance over an threshold; analogously, an *outbreak/onset* of a pandemic is the (time of) change where the epidemic turns into a pandemic. Of course, one goal of infectious-disease modelling is to quantify "early enough" the potential danger of an emerging outbreak of an epidemic or a pandemic.

Returning to possible models of the incidence-evolution $n \mapsto X_n$, its description may be theoretically derived from more detailed, time-finer, highly sophisticated, individual-based "mechanistic" infectious-disease models such as e.g., continuous-time suscetible-exposed-infectious-recovered (SEIR) models (see the above-mentioned introductory texts); however, as e.g., pointed out in Held et al. [72], the estimation of the correspondingly involved numerous parameters may be too ambitious for routinely collected, non-detailed disease data, such as e.g., daily/weekly counts X_n of incidences–especially in decisive emerging/early phases of a novel disease (such as the current COVID-19 pandemic). Accordingly, in the following we assume that X_n can be approximately described by a Poissonian Galton-Watson process with immigration respectively a ("distributionally equal") Poissonian autoregressive Generalized Linear Model in the sense of (M2). Depending on the situation, this can be quite reasonable, for the following arguments (apart from the usual "if the data say so"). Firstly, it is well known (see e.g., Bartoszynski [33], Ludwig [34],

Becker [35,36], Metz [37], Heyde [38], von Bahr & Martin-Löf [39], Ball [40], Jacob [41], Barbour & Reinert [42], Section 1.2 of Britton & Pardoux [43]) that in populations with a relatively high number of susceptible individuals and a relatively low number of infectious individuals (e.g., in a large population and in decisive emerging/early phases of the disease spreading), the incidence-evolution $n \mapsto X_n$ can be well approximated by a (e.g., Poissonian) Galton-Watson process with possible immigration where n plays the role of a *generation number*. If the above-mentioned generation interval is "nearly" deterministic (leading to nearly synchronous, non-overlapping generations)—which is the case e.g., for (phases of) Influenza A(H1N1)pdm09, Influenza A(H3N2), Rubella (cf. Vink, Bootsma & Wallinga [98]), and COVID-19 (cf. Ferretti et al. [101])—and the length of the generation interval is approximated by its mean length and the latter is tuned to be equal to the unit time between consecutive observations, then n plays the role of an *observation (surveillance) time*. This effect is even more realistic if the period of infectiousness is nearly deterministic and relatively short. Secondly, as already mentioned above, the spreading of an infectious disease is intrinsically a (not necessarily Poissonian Galton-Watson) branching mechanism, which may be blurred by other effects in a way that a Poissonian autoregressive Generalized Linear Model is still a reasonably fitting model for the observational process in disease surveillance. The latter have been used e.g., by Finkenstädt, Bjornstad & Grenfell [73], Held, Höhle & Hofmann [47], and Held et al. [72]; they all use non-constant parameters (e.g., to describe seasonal effects, which are however unknown in early phases of a novel infectious disease such as COVID-19). In contrast, we employ different new–namely divergence-based–statistical techniques, for which we assume constant parameters but also indicate procedures for the detection of changes; the extension to non-constant parameters is straightforward.

Returning to Galton-Watson processes, let us mention as a *side remark* that they can be also used to model the above-mentioned within-host replication dynamics (D2) (e.g., in the time-interval $[t_k^E, t_k^I[$ and beyond) on a sub-cellular level, see e.g., Spouge [106], as well as Taneyhill, Dunn & Hatcher [107] for parasitic pathogens; on the other hand, one can also employ Galton-Watson processes for quantifying snowball-effect (avalanche-effect, cascade-effect) type, economic-crisis triggered consequences of large epidemics and pandemics, such as e.g., the potential spread of transmissible (i) foreclosures of homes (cf. Parnes [108]), or clearly also (ii) company insolvencies, downsizings and credit-risk downgradings; moreover, the time-evolution of integer-valued indicators concerning the spread of (rational or unwarranted) fears resp. perceived threats may be modelled, too.

Summing up things, we model the evolution $n \mapsto X_n$ of the number of incidences at stage n by a Poissonian Galton Watson process with immigration GWI

$$X_0 \in \mathbb{N}; \qquad \mathbb{N}_0 \ni X_n = \sum_{k=1}^{X_{n-1}} Y_{n-1,k} + \widetilde{Y}_n, \qquad n \in \mathbb{N}, \qquad \text{cf. (1), (GWI1)–(GWI3) with law } P_{\bullet},$$

(where $Y_{n-1,k}$ corresponds to the Y_k of (D3), equipped with an additional stage-index $n-1$), respectively by a corresponding "distributionally equal"–possibly non-stationary– Poissonian autoregressive Generalized Linear Model in the sense of (M2); depending on the situation, we may also fix a (deterministic or random) upper time horizon other than infinity. Recall that both models are overdispersed, which is consistent with the current debate on overdispersion in connection with the current COVID-19 pandemic. In infectious-disease language, the sum $\sum_{k=1}^{X_{n-1}} Y_{n-1,k}$ can also be loosely interpreted as *epidemic component* (in a narrow sense) driven by the parameter $\beta_{\bullet}$, and $\widetilde{Y}_n$ as *endemic component* driven by the parameter $\alpha_{\bullet}$. In fact, the offspring mean (here, $\beta_{\bullet}$) is called *reproduction number* and plays a major role–also e.g., in the current public debate about the COVID-19 pandemic–because it crucially determines the rapidity of the spread of the disease and—as already indicated above in the second and third paragraph after (PS3)–also the probability that the epidemic/pandemic becomes (maybe temporally) extinct or at least stationary at a low level (that is, *endemic*). For this to happen, $\beta_{\bullet}$ should be subcritical, i.e., $\beta_{\bullet} < 1$, and even better, close to zero. Of course, the size of the *importation mean* $\alpha_{\bullet} \geq 0$ matters, too, in a secondary order.

Keeping this in mind, let us discuss on which factors the reproduction number $\beta_\bullet$ and the *importation mean* $\alpha_\bullet$ depend upon, and how they can be influenced/controlled. To begin with, by recalling the above-mentioned points (D1) to (D5) and by adapting the considerations of e.g., Grassly & Fraser [81] to our model, one encounters the fact that the distribution of the offspring $Y_{n-1,k}$—here driven by the reproduction number (offspring mean) $\beta_\bullet$—depends on the following factors:

(B1) the *degree of infectiousness* of the individual k, with three major components:

    (B1a) degree of *biological* infectiousness; this reflects the within-host dynamics (D2) of the "representative" individual k, in particular the duration and amount of the corresponding replication and shedding/excretion of the infectious pathogens; this degree depends thus on (i) the number of host-invading pathogens (called the *initial infectious dose*), (ii) the type of the pathogen with respect to e.g., its principal capabilities of replication speed, range of spread and drug-sensitivity, (iii) features of the immune system of the host k including the level of innate or acquired immunity, and (iv) the interaction between the genetic determinants of disease progression in both the pathogen and the host;

    (B1b) degree of *behavioural* infectiousness; this depends on the contact patterns of an infected/infectious individual (and, if relevant, the contact patterns of intermediate hosts or vectors), in relation to the disease-specific type of route(s) of transmission of the infectious pathogens (for an overview of the latter, see e.g., Table 3 of Kaslow & Evans [79]); a long-distance-travel behaviour may also lead to the disease exportation to another, outside population (and thus, for the latter to a disease importation);

    (B1c) degree of *environmental* infectiousness; this depends on the location and environment of the host k, which influences the duration of outside-host survival of the pathogens (and, if relevant, of the intermediate hosts or vectors) as well as the speed and range of their outside-host spread; for instance, high temperature may kill the pathogens, high airflow or rainfall dynamics may ease their spread, etc.

(B2) the *degree of susceptibility* of uninfected individuals who have contact with k, with the following three major components (with similar background as their infectiousness counterparts):

    (B2a) degree of *biological* susceptibility;
    (B2b) degree of *behavioural* susceptibility;
    (B2c) degree of *environmental* susceptibility.

All these factors (B1a) to (B2c) can be principally influenced/controlled to a certain–respective–extent. Let us briefly discuss this for *human* infectious diseases, where one major goal of epidemic risk management is to operate countermeasures/interventions in order to slow down the disease transmission (e.g., by reducing the reproduction number $\beta_\bullet$ to less than 1) and eventually even break the chain of transmission, for the sake of containment or mitigation; preparedness and preparation are motives, too, for instance as a part of governmental pandemic risk management.

For instance, (B1a) can be reduced or even erased through pharmaceutical interventions such as medication (if available), and preventive strengthening of the immune system through non-extreme sports activities and healthy food.

Moreover, the following exemplary control measures for (B2) can be either put into action by common-sense self-behaviour, or by large-scale public recommendations (e.g., through mass media), or by rules/requirements from authorities:

(i) personal preventive measures such as frequent washing and disinfecting of hands; keeping hands away from face; covering coughs; avoidance of handshakes and hugs with non-family-members; maintaining physical distance (e.g., of two meters) from non-family-members; wearing a face-mask of respective security degree (such as homemade cloth face mask, particulate-filtering face-piece respirator, medical (non-surgical) mask, surgical mask); self-quarantine;

(ii)   environmental measures, such as e.g., cleaning of surfaces;

(iii)   community measures aimed at mild or stringent social distancing, such as e.g., prohibiting/cancelling/banning gatherings of more than z non-family members (e.g., $z = 2, 5, 10, 100, 1000$ in various different phases and countries during the current COVID-19 pandemic); mask-wearing (see above); closing of schools, universities, some or even all nonessential ("system-irrelevant") businesses and venues; home-officing/work ban; home isolation of disease cases; isolation of homes for the elderly/aged (nursing homes); stay-at-home orders with exemptions, household or even general quarantine; testing & tracing; lockdown of entire cities and beyond; restricting the degrees of travel freedom/allowed mobility (e.g., local, union-state, national, international including border and airport closure). The latter also affects the mean importation rate $\alpha_\bullet$, which can be controlled by vaccination programs in "outside populations", too.

As far as the degree of *biological* susceptibility (B2a) is concerned, one obvious therapeutic countermeasure is a mass vaccination program/campaign (if available).

In case of *highly virulent* infectious diseases causing epidemics and pandemics with substantial *fatality rates*, some of the above-mentioned control strategies and countermeasures may (have to) be "drastic" (e.g., lockdown), and thus imply considerable social and economic costs, with a huge impact and potential danger of triggering severe social, economic and political disruptions.

In order to prepare corresponding suggestions for decisions about appropriate control measures (e.g., public policies), it is therefore important–especially for a novel infectious disease such as the current COVID-19 pandemic–to have a model for the time-evolution of the incidences in (i) a natural (basically uncontrolled) set-up, as well as in (ii) the control set-ups under consideration. As already mentioned above, we assume that all these situations can be distilled into an incidence evolution $n \mapsto X_n$ which follows a Poissonian Galton-Watson process with respectively different parameter pairs $(\beta_\bullet, \alpha_\bullet)$. Correspondingly, we always compare two alternative models $(\mathcal{H})$ and $(\mathcal{A})$ with parameter pairs $(\beta_{\mathcal{H}}, \alpha_{\mathcal{H}})$ and $(\beta_{\mathcal{A}}, \alpha_{\mathcal{A}})$ which reflect either a "pure" statistical uncertainty (under the *same* uncontrolled or controlled set-up), or the uncertainty between two *different* potential control set-ups (for the sake of assessing the potential impact/efficiency of some planned interventions, compared with alternative ones); the economic impact can be also taken into account, within a Bayesian decision framework discussed in Section 2.5 below. As will be explained in the next subsections, we achieve such comparisons by means of density-based dissimilarity distances/divergences and related quantities thereof.

From the above-mentioned detailed explanations, it is immediately clear that for the described epidemiological context one should investigate *all* types of criticality and importation means for the therein involved two Poissonian Galton-Watson processes with/without immigration (respectively the equally distributed INARCH(1) models); in particular, this motivates (or even "justifies") the necessity of the very lengthy detailed studies in the Sections 3–7 below.

2.4. Information Measures

Having two competing models $(\mathcal{H})$ and $(\mathcal{A})$ at stake, it makes sense to study questions such as "how far are they apart?" and thus "how dissimilar are they?". This can be quantified in terms of divergences in the sense of directed (i.e., not necessarily symmetric) distances, where usually the triangular inequality fails. Let us first discuss our employed divergence subclasses in a *general* set-up of two *equivalent* probability measures $P_{\mathcal{H}}, P_{\mathcal{A}}$ on a measurable space $(\Omega, \mathcal{F})$. In terms of the parameter

$\lambda \in \mathbb{R}$, the *power divergences*—also known as Cressie-Read divergences, relative Tsallis entropies, or generalized cross-entropy family– are defined as (see e.g., Liese & Vajda [1,10])

$$0 \leq I_\lambda \left(P_\mathcal{A}||P_\mathcal{H}\right) := \begin{cases} I\left(P_\mathcal{A}||P_\mathcal{H}\right), & \text{if } \lambda = 1, \\ \frac{1}{\lambda(\lambda-1)}\left(H_\lambda\left(P_\mathcal{A}||P_\mathcal{H}\right) - 1\right), & \text{if } \lambda \in \mathbb{R}\backslash\{0,1\}, \\ I\left(P_\mathcal{H}||P_\mathcal{A}\right), & \text{if } \lambda = 0, \end{cases} \tag{2}$$

where

$$I\left(P_\mathcal{A}||P_\mathcal{H}\right) := \int p_\mathcal{A} \log \frac{p_\mathcal{A}}{p_\mathcal{H}} \, d\mu \geq 0 \tag{3}$$

is the *Kullback-Leibler information divergence* (also known as *relative entropy*) and

$$H_\lambda\left(P_\mathcal{A}||P_\mathcal{H}\right) := \int_\Omega p_\mathcal{A}^\lambda p_\mathcal{H}^{1-\lambda} \, d\mu \geq 0 \tag{4}$$

is the *Hellinger integral of order* $\lambda \in \mathbb{R}\backslash\{0,1\}$; for this, we assume as usual without loss of generality that the probability measures $P_\mathcal{H}$, $P_\mathcal{A}$ are dominated by some $\sigma-$finite measure μ, with densities

$$p_\mathcal{A} = \frac{dP_\mathcal{A}}{d\mu} \quad \text{and} \quad p_\mathcal{H} = \frac{dP_\mathcal{H}}{d\mu} \tag{5}$$

defined on Ω (the zeros of $p_\mathcal{H}$, $p_\mathcal{A}$ are handled in (3) and (4) with the usual conventions). Clearly, for $\lambda \in \{0,1\}$ one trivially gets

$$H_0\left(P_\mathcal{A}||P_\mathcal{H}\right) = H_1\left(P_\mathcal{A}||P_\mathcal{H}\right) = 1 .$$

The Kullback-Leibler information divergences (relative entropies) in (2) and (3) can alternatively be expressed as (see, e.g., Liese & Vajda [1])

$$I\left(P_\mathcal{A}||P_\mathcal{H}\right) = \lim_{\lambda \nearrow 1} \frac{1 - H_\lambda\left(P_\mathcal{A}||P_\mathcal{H}\right)}{\lambda(1-\lambda)}, \qquad I\left(P_\mathcal{H}||P_\mathcal{A}\right) = \lim_{\lambda \searrow 0} \frac{1 - H_\lambda\left(P_\mathcal{A}||P_\mathcal{H}\right)}{\lambda(1-\lambda)} . \tag{6}$$

Apart from the Kullback-Leibler information divergence (relative entropy), other prominent examples of power divergences are the squared Hellinger distance $\frac{1}{2} I_{1/2}\left(P_\mathcal{A}||P_\mathcal{H}\right)$ and Pearson's χ^2-divergence $2\, I_2\left(P_\mathcal{A}||P_\mathcal{H}\right)$; the Hellinger integral $H_{1/2}\left(P_\mathcal{A}||P_\mathcal{H}\right)$ is also known as (multiple of) the *Bhattacharyya coefficent*. Extensive studies about basic and advanced general facts on power divergences, Hellinger integrals and the related Renyi divergences of order $\lambda \in \mathbb{R}\backslash\{0,1\}$

$$0 \leq R_\lambda\left(P_\mathcal{A}||P_\mathcal{H}\right) := \frac{1}{\lambda(\lambda-1)} \log H_\lambda\left(P_\mathcal{A}||P_\mathcal{H}\right) , \qquad \text{with } \log 0 = -\infty, \tag{7}$$

can be found e.g., in Liese & Vajda [1,10], Jacod & Shiryaev [24], van Erven & Harremoes [20] (as a side remark, $R_{1/2}\left(P_\mathcal{A}||P_\mathcal{H}\right)$ is also known as (multiple of) *Bhattacharyya distance*). For instance, the integrals in (3) and (4) do not depend on the choice of μ. Furthermore, one has the skew symmetries

$$H_\lambda\left(P_\mathcal{A}||P_\mathcal{H}\right) = H_{1-\lambda}\left(P_\mathcal{H}||P_\mathcal{A}\right) , \qquad \text{as well as} \qquad I_\lambda\left(P_\mathcal{A}||P_\mathcal{H}\right) = I_{1-\lambda}\left(P_\mathcal{H}||P_\mathcal{A}\right), \tag{8}$$

for all $\lambda \in \mathbb{R}$ (see e.g., Liese & Vajda [1]). As far as finiteness is concerned, for $\lambda \in]0,1[$ one gets the rudimentary bounds

$$0 < H_\lambda\left(P_\mathcal{A}||P_\mathcal{H}\right) \leq 1 , \qquad \text{and equivalently,} \tag{9}$$

$$0 \leq I_\lambda\left(P_\mathcal{A}||P_\mathcal{H}\right) = \frac{1 - H_\lambda\left(P_\mathcal{A}||P_\mathcal{H}\right)}{\lambda(1-\lambda)} < \frac{1}{\lambda(1-\lambda)} , \tag{10}$$

where the lower bound in (10) (upper bound in (9)) is achieved iff $P_{\mathcal{A}} = P_{\mathcal{H}}$. For $\lambda \in \mathbb{R}\backslash]0,1[$, one gets the bounds

$$0 \leq I_\lambda\left(P_{\mathcal{A}}||P_{\mathcal{H}}\right) \leq \infty, \qquad \text{and equivalently,} \qquad 1 \leq H_\lambda\left(P_{\mathcal{A}}||P_{\mathcal{H}}\right) \leq \infty, \tag{11}$$

where, in contrast to above, both the lower bound of $H_\lambda\left(P_{\mathcal{A}}||P_{\mathcal{H}}\right)$ and the lower bound of $I_\lambda\left(P_{\mathcal{A}}||P_{\mathcal{H}}\right)$ is achieved iff $P_{\mathcal{A}} = P_{\mathcal{H}}$; however, the power divergence $I_\lambda\left(P_{\mathcal{A}}||P_{\mathcal{H}}\right)$ and Hellinger integral $H_\lambda\left(P_{\mathcal{A}}||P_{\mathcal{H}}\right)$ might be infinite, depending on the particular setup.

The Hellinger integrals can be also used for bounds of the well-known *total variation*

$$0 \leq V(P_{\mathcal{A}}||P_{\mathcal{H}}) := 2 \sup_{A \in \mathcal{F}} \left\{ P_{\mathcal{A}}(A) - P_{\mathcal{H}}(A) \right\} = \int_\Omega |p_{\mathcal{A}} - p_{\mathcal{H}}| \, d\mu,$$

with $p_{\mathcal{A}}$ and $p_{\mathcal{H}}$ defined in (5). Certainly, the total variation is one of the best known statistical distances, see e.g., Le Cam [109]. For arbitrary $\lambda \in]0,1[$ there holds (cf. Liese & Vajda [1])

$$1 - \frac{V(P_{\mathcal{A}}||P_{\mathcal{H}})}{2} \leq H_\lambda(P_{\mathcal{A}}||P_{\mathcal{H}}) \leq \left(1 + \frac{V(P_{\mathcal{A}}||P_{\mathcal{H}})}{2}\right)^{\max\{\lambda, 1-\lambda\}} \left(1 - \frac{V(P_{\mathcal{A}}||P_{\mathcal{H}})}{2}\right)^{\min\{\lambda, 1-\lambda\}}.$$

From this together with the particular choice $\lambda = \frac{1}{2}$, we can derive the fundamental universal bounds

$$2\left(1 - H_{\frac{1}{2}}(P_{\mathcal{A}}||P_{\mathcal{H}})\right) \leq V(P_{\mathcal{A}}||P_{\mathcal{H}}) \leq 2\sqrt{1 - \left(H_{\frac{1}{2}}(P_{\mathcal{A}}||P_{\mathcal{H}})\right)^2}. \tag{12}$$

We apply these concepts to our setup of Section 2.1 with two competing models $(\mathcal{H})$ and $(\mathcal{A})$ of Galton-Watson processes with immigration, where one can take $\Omega \subset \mathbb{N}_0^{\mathbb{N}_0}$ to be the space of all paths of $(X_n)_{n\in\mathbb{N}}$. More detailed, in terms of the extinction set $\mathcal{B} := \{\tau < \infty\}$ and the parameter-set notation (PS1) to (PS3), it is known that for $\mathcal{P}_{\mathrm{SP}}$ the two laws $P_{\mathcal{H}}$ and $P_{\mathcal{A}}$ are equivalent, whereas for $\mathcal{P}_{\mathrm{NI}}$ the two restrictions $P_{\mathcal{H}}|_{\mathcal{B}}$ and $P_{\mathcal{A}}|_{\mathcal{B}}$ are equivalent (see e.g., Lemma 1.1.3 of Guttorp [52]); with a slight abuse of notation we shall henceforth omit $|_{\mathcal{B}}$. Consistently, for fixed time $n \in \mathbb{N}_0$ we introduce $P_{\mathcal{A},n} := P_{\mathcal{A}}|_{\mathcal{F}_n}$ and $P_{\mathcal{H},n} := P_{\mathcal{H}}|_{\mathcal{F}_n}$ as well as the corresponding Radon-Nikodym-derivative (likelihood ratio)

$$Z_n := \frac{dP_{\mathcal{A},n}}{dP_{\mathcal{H},n}}, \tag{13}$$

where $(\mathcal{F}_n)_{n\in\mathbb{N}}$ denotes the corresponding canonical filtration generated by $X := (X_n)_{n\in\mathbb{N}}$; in other words, $\mathcal{F}_n$ reflects the "process-intrinsic" information known at stage n. Clearly, $Z_0 = 1$. By choosing the reference measure $\mu = P_{\mathcal{H},n}$ one obtains from (4) the Hellinger integral $H_\lambda\left(P_{\mathcal{A},0}||P_{\mathcal{H},0}\right) = 1$, as well as and for all $n \in \mathbb{N}$

$$H_\lambda\left(P_{\mathcal{A},n}||P_{\mathcal{H},n}\right) = EP_{\mathcal{H},n}\left[(Z_n)^\lambda\right], \tag{14}$$

$$I\left(P_{\mathcal{A},n}||P_{\mathcal{H},n}\right) = EP_{\mathcal{A},n}\left[\log Z_n\right], \tag{15}$$

from which one can immediately build $I_\lambda\left(P_{\mathcal{A},n}||P_{\mathcal{H},n}\right)$ $(\lambda \in \mathbb{R})$ respectively $R_\lambda\left(P_{\mathcal{A},n}||P_{\mathcal{H},n}\right)$ $(\lambda \in \mathbb{R}\backslash\{0,1\})$ respectively bounds of $V\left(P_{\mathcal{A},n}||P_{\mathcal{H},n}\right)$ via (2) respectively (7) respectively (12).

The outcoming values (respectively bounds) of $H_\lambda\left(P_{\mathcal{A},n}||P_{\mathcal{H},n}\right)$ are quite diverse and depend on the choice of the involved parameter pairs $(\beta_{\mathcal{H}}, \alpha_{\mathcal{H}})$, $(\beta_{\mathcal{A}}, \alpha_{\mathcal{A}})$ as well as λ; the exact details will be given in the Sections 3 and 6 below.

Before we achieve this, in the following we explain how the outcoming dissimilarity results can be applied to Bayesian testing and more general Bayesian decision making, as well as to Neyman-Pearson testing.

2.5. Decision Making under Uncertainty

Within the above-mentioned context of two competing models $(\mathcal{H})$ and $(\mathcal{A})$ of Galton-Watson processes with immigration, let us briefly discuss how knowledge about the time-evolution of the Hellinger integrals $H_\lambda\left(P_{\mathcal{A},n}||P_{\mathcal{H},n}\right)$–or equivalently, of the power divergences $I_\lambda\left(P_{\mathcal{A},n}||P_{\mathcal{H},n}\right)$, cf. (2)—can be used in order to take decisions under uncertainty, within a framework of Bayesian decision making BDM, or alternatively, of Neyman-Pearson testing NPT.

In our context of BDM, we decide between an action $d_{\mathcal{H}}$ "associated with" the (say) hypothesis law $P_{\mathcal{H}}$ and an action $d_{\mathcal{A}}$ "associated with" the (say) alternative law $P_{\mathcal{A}}$, based on the sample path observation $\mathcal{X}_n := \{X_l : l \in \{0, 1, \ldots, n\}\}$ of the GWI-generation-sizes (e.g., infectious-disease incidences, cf. Section 2.3) up to observation horizon $n \in \mathbb{N}$. Following the lines of Stummer & Vajda [15] (adapted to our branching process context), for our BDM let us consider as admissible decision rules $\delta_n : \Omega_n \mapsto \{d_{\mathcal{H}}, d_{\mathcal{A}}\}$ the ones generated by all path sets $G_n \in \Omega_n$ (where Ω_n denotes the space of all possible paths of $(X_k)_{k\in\{1,\ldots,n\}}$) through

$$\delta_n(\mathcal{X}_n) := \delta_{G_n}(\mathcal{X}_n) := \begin{cases} d_{\mathcal{A}}, & \text{if } \mathcal{X}_n \in G_n, \\ d_{\mathcal{H}}, & \text{if } \mathcal{X}_n \notin G_n, \end{cases}$$

as well as loss functions of the form

$$\begin{pmatrix} L(d_{\mathcal{H}}, \mathcal{H}) & L(d_{\mathcal{H}}, \mathcal{A}) \\ L(d_{\mathcal{A}}, \mathcal{H}) & L(d_{\mathcal{A}}, \mathcal{A}) \end{pmatrix} := \begin{pmatrix} 0 & L_{\mathcal{A}} \\ L_{\mathcal{H}} & 0 \end{pmatrix} \tag{16}$$

with pregiven constants $L_{\mathcal{A}} > 0$, $L_{\mathcal{H}} > 0$ (e.g., arising as bounds from quantities in worst-case scenarios); notice that in (16), $d_{\mathcal{H}}$ is assumed to be a zero-loss action under $\mathcal{H}$ and $d_{\mathcal{A}}$ a zero-loss action under $\mathcal{A}$. Per definition, the *Bayes decision rule* $\delta_{G_{n,\min}}$ minimizes–over G_n—the *mean decision loss*

$$\begin{aligned} \mathcal{L}(\delta_{G_n}) &:= p_{\mathcal{H}}^{\text{prior}} \cdot L_{\mathcal{H}} \cdot Pr\left(\delta_{G_n}(\mathcal{X}_n) = d_{\mathcal{A}}\big|\mathcal{H}\right) + p_{\mathcal{A}}^{\text{prior}} \cdot L_{\mathcal{A}} \cdot Pr\left(\delta_{G_n}(\mathcal{X}_n) = d_{\mathcal{H}}\big|\mathcal{A}\right) \\ &= p_{\mathcal{H}}^{\text{prior}} \cdot L_{\mathcal{H}} \cdot P_{\mathcal{H},n}(G_n) + p_{\mathcal{A}}^{\text{prior}} \cdot L_{\mathcal{A}} \cdot P_{\mathcal{A},n}(\Omega_n - G_n) \end{aligned} \tag{17}$$

for given prior probabilities $p_{\mathcal{H}}^{\text{prior}} = Pr(\mathcal{H}) \in]0, 1[$ for $\mathcal{H}$ and $p_{\mathcal{A}}^{\text{prior}} := Pr(\mathcal{A}) = 1 - p_{\mathcal{H}}^{\text{prior}}$ for $\mathcal{A}$. As a side remark let us mention that, in a certain sense, the involved model (parameter) uncertainty expressed by the "superordinate" Bernoulli-type law $Pr = Bin(1, p_{\mathcal{H}}^{\text{prior}})$ can also be reinterpreted as a rudimentary static random environment caused e.g., by a random Bernoulli-type external static force. By straightforward calculations, one gets with (13) the minimizing path set $G_{n,\min} = \left\{ Z_n \geq \frac{p_{\mathcal{H}}^{\text{prior}} L_{\mathcal{H}}}{p_{\mathcal{A}}^{\text{prior}} L_{\mathcal{A}}} \right\}$ leading to the *minimal mean decision loss*, i.e., the *Bayes risk*,

$$\mathcal{R}_n := \min_{G_n} \mathcal{L}(\delta_{G_n}) = \mathcal{L}(\delta_{G_{n,\min}}) = \int_{\Omega_n} \min\left\{ p_{\mathcal{H}}^{\text{prior}} L_{\mathcal{H}}, p_{\mathcal{A}}^{\text{prior}} L_{\mathcal{A}} Z_n \right\} dP_{\mathcal{H},n}. \tag{18}$$

Notice that—by straightforward standard arguments—the *alternative* decision procedure

$$\text{take action } d_{\mathcal{A}} \text{ (resp. } d_{\mathcal{H}}) \quad \text{if } \quad L_{\mathcal{H}} \cdot p_{\mathcal{H}}^{\text{post}}(\mathcal{X}_n) \leq (\text{resp. } >) \ L_{\mathcal{A}} \cdot p_{\mathcal{A}}^{\text{post}}(\mathcal{X}_n)$$

with posterior probabilities $p_{\mathcal{H}}^{\text{post}}(\mathcal{X}_n) := \frac{p_{\mathcal{H}}^{\text{prior}}}{(1-p_{\mathcal{H}}^{\text{prior}}) \cdot Z_n(\mathcal{X}_n) + p_{\mathcal{H}}^{\text{prior}}} =: 1 - p_{\mathcal{A}}^{\text{post}}(\mathcal{X}_n)$, leads exactly to the same actions as $\delta_{G_{n,\min}}$. By adapting the Lemma 6.5 of Stummer & Vajda [15]—which on general probability spaces gives *fundamental universal* inequalities relating Hellinger integrals (or equivalently, power divergences) and Bayes risks—one gets for all $L_{\mathcal{H}} > 0$, $L_{\mathcal{A}} > 0$, $p_{\mathcal{H}}^{\text{prior}} \in]0, 1[$, $\lambda \in]0, 1[$ and $n \in \mathbb{N}$ the upper bound

$$\mathcal{R}_n \leq \Lambda_{\mathcal{A}}^\lambda \Lambda_{\mathcal{H}}^{1-\lambda} H_\lambda\left(P_{\mathcal{A},n}||P_{\mathcal{H},n}\right), \quad \text{with } \Lambda_{\mathcal{H}} := p_{\mathcal{H}}^{\text{prior}} L_{\mathcal{H}}, \ \Lambda_{\mathcal{A}} := (1 - p_{\mathcal{H}}^{\text{prior}}) L_{\mathcal{A}}, \tag{19}$$

as well as the lower bound

$$\left(\mathcal{R}_n\right)^{\min\{\lambda,1-\lambda\}} \cdot \left(\Lambda_\mathcal{H} + \Lambda_\mathcal{A} - \mathcal{R}_n\right)^{\max\{\lambda,1-\lambda\}} \geq \Lambda_\mathcal{A}^\lambda\, \Lambda_\mathcal{H}^{1-\lambda}\, H_\lambda\left(P_{\mathcal{A},n}||P_{\mathcal{H},n}\right)$$

which implies in particular the "direct" lower bound

$$\mathcal{R}_n \geq \frac{\Lambda_\mathcal{A}^{\max\{1,\frac{\lambda}{1-\lambda}\}}\, \Lambda_\mathcal{H}^{\max\{1,\frac{1-\lambda}{\lambda}\}}}{\left(\Lambda_\mathcal{A} + \Lambda_\mathcal{H}\right)^{\max\{\frac{\lambda}{1-\lambda},\frac{1-\lambda}{\lambda}\}}} \cdot \left(H_\lambda\left(P_{\mathcal{A},n}||P_{\mathcal{H},n}\right)\right)^{\max\{\frac{1}{\lambda},\frac{1}{1-\lambda}\}}. \tag{20}$$

By using (19) (respectively (20)) together with the exact values and the upper (respectively lower) bounds of the Hellinger integrals $H_\lambda\left(P_{\mathcal{A},n}||P_{\mathcal{H},n}\right)$ derived in the following sections, we end up with upper (respectively lower) bounds of the Bayes risk $\mathcal{R}_n$. Of course, with the help of (2) the bounds (19) and (20) can be (i) immediately rewritten in terms of the power divergences $I_\lambda\left(P_{\mathcal{A},n}||P_{\mathcal{H},n}\right)$ and (ii) thus be *directly* interpreted in terms of dissimilarity-size arguments. As a side-remark, in such a Bayesian context the $\lambda-$order Hellinger integral $H_\lambda\left(P_{\mathcal{A},n}||P_{\mathcal{H},n}\right) = EP_{\mathcal{H},n}\left[(Z_n)^\lambda\right]$ (cf. (14)) can be also interpreted as $\lambda-$order Bayes-factor moment (with respect to $P_{\mathcal{H},n}$), since $Z_n = Z_n(\mathcal{X}_n) = \frac{p_\mathcal{A}^{\text{post}}(\mathcal{X}_n)}{p_\mathcal{H}^{\text{post}}(\mathcal{X}_n)} \Big/ \frac{p_\mathcal{A}^{\text{prior}}}{p_\mathcal{H}^{\text{prior}}}$ is the Bayes factor (i.e., the posterior odds ratio of $(\mathcal{A})$ to $(\mathcal{H})$, divided by the prior odds ratio of $(\mathcal{A})$ to $(\mathcal{H})$).

At this point, the potential applicant should be warned about the *usual way of* asynchronous decision making, where one first *tests* $(\mathcal{A})$ versus $(\mathcal{H})$ (i.e., $L_\mathcal{A} = L_\mathcal{H} = 1$ which leads to 0–1 losses in (16)) and afterwards, based on the outcoming result (e.g., in favour of $(\mathcal{A})$), takes the attached economic decision (e.g., $d_\mathcal{A}$); this can lead to distortions compared with synchronous decision making with "full" monetary losses $L_\mathcal{A}$ and $L_\mathcal{H}$, as is shown in Stummer & Lao [16] within an economic context in connection with discrete approximations of financial diffusion processes (they call this distortion effect a *non-commutativity between Bayesian statistical and investment decisions*).

For different types of–mainly parameter estimation (squared-error type loss function) concerning—Bayesian analyses based on GW(I) generation size observations, see e.g., Jagers [56], Heyde [38], Heyde & Johnstone [110], Johnson et al. [111], Basawa & Rao [60], Basawa & Scott [61], Scott [112], Guttorp [52], Yanev & Tsokos [113], Mendoza & Gutierrez-Pena [114], and the references therein.

Within our running-example epidemiological context of Section 2.3, let us briefly discuss the role of the above-mentioned losses $L_\mathcal{A}$ and $L_\mathcal{H}$. To begin with, as mentioned above the *unit-free* choice $L_\mathcal{A} = L_\mathcal{H} = 1$ corresponds to *Bayesian testing*. Recall that this concerns with two alternative infectious-disease models $(\mathcal{H})$ and $(\mathcal{A})$ with parameter pairs (recall the interpretation of $\beta_\bullet$ as reproduction number and $\alpha_\bullet$ as importation mean) $(\beta_\mathcal{H}, \alpha_\mathcal{H})$ and $(\beta_\mathcal{A}, \alpha_\mathcal{A})$ which reflect either a "pure" statistical uncertainty (under the *same* uncontrolled or controlled set-up), or the uncertainty between two *different* potential control set-ups (for the sake of assessing the potential impact/efficiency of some planned interventions, compared with alternative ones). As far as *non-unit-free*–e.g., macroeconomic or monetary–losses is concerned, recall that some of the above-mentioned control strategies (countermeasures, public policies, governmental pandemic risk management plans) may imply considerable social and economic costs, with a huge impact and potential danger of triggering severe social, economic and political disruptions; a corresponding tradeoff between health and economic issues can be incorporated by choosing $L_\mathcal{A}$ and $L_\mathcal{H}$ to be (e.g., monetary) values which reflect estimates or upper bounds of losses due to wrong decisions, e.g., if at stage n due to the observed data one erroneously thinks (reinforced by fear) that a novel infectious disease (e.g., COVID-19) will lead (or re-emerge) to a severe pandemic and consequently decides for a lockdown with drastic future economic consequences, versus, if one erroneously thinks (reinforced by carelessness) that the infectious disease is (or stays) non-severe and consequently eases some/all control measures which will lead to extremely devastating future economic consequences. For the estimates/bounds of $L_\mathcal{A}$ and $L_\mathcal{H}$, one can e.g., employ (i) the comprehensive stochastic studies of Feicht & Stummer [115] on the quantitative degree of elasticity and speed of recovery of economies after a sudden macroeconomic disaster, or (ii) the more short-term, German-specific, scenario-type (basically non-stochastic) studies of Dorn et al. [116,117] in connection with the current COVID-19 pandemic.

Of course, the above-mentioned Bayesian decision procedure can be also operated in *sequential way*. For instance, suppose that we are encountered with a novel infectious disease (e.g., COVID-19) of non-negligible fatality rate and let $(\mathcal{A})$ reflect a "potentially dangerous" infectious-disease-transmission situation (e.g., a reproduction number of substantially supercritical case $\beta_{\mathcal{A}} = 2$, and an importation mean of $\alpha_{\mathcal{A}} = 10$, for *weekly* appearing new incidence-generations) whereas $(\mathcal{H})$ describes a "relatively harmless/mild" situation (e.g., a substantially subcritical $\beta_{\mathcal{H}} = 0.5$, $\alpha_{\mathcal{H}} = 0.2$). Moreover, let $d_{\mathcal{A}}$ respectively $d_{\mathcal{H}}$ denote (non-quantitatively) the decision/action to accept $(\mathcal{A})$ respectively $(\mathcal{H})$. It can then be reasonable to decide to stop the observation process $n \mapsto \mathcal{X}_n$ (also called *surveillance* or *online-monitoring*) of incidence numbers at the first time at which $n \mapsto Z_n = Z_n(\mathcal{X}_n)$ exceeds the threshold $p_{\mathcal{H}}^{\text{prior}}/p_{\mathcal{A}}^{\text{prior}}$; if this happens, one takes $d_{\mathcal{A}}$ as decision (and e.g., declare the situation as *occurrence of an epidemic outbreak* and start with control/intervention measures (however, as explained above, one should synchronously involve also the potential economic losses)) whereas as long as this does not happen, one continues the observation (and implicitly takes $d_{\mathcal{H}}$ as decision). This can be modelled in terms of the pair $(\tilde{\tau}, d_{\mathcal{A}})$ with (random) stopping time $\tilde{\tau} := \inf\left\{n \in \mathcal{N} : Z_n \geq \frac{p_{\mathcal{H}}^{\text{prior}}}{p_{\mathcal{A}}^{\text{prior}}}\right\}$ (with the usual convention that the infimum of the empty set is infinity), and the corresponding decision $d_{\mathcal{A}}$. After the time $\tilde{\tau} < \infty$ and e.g., immediate subsequent employment of some control/counter measures, one can e.g., take the old model $(\mathcal{A})$ as new $(\mathcal{H})$, declare a new target $(\mathcal{A})$ for the desired quantification of the effectiveness of the employed control measures (e.g., a mitigation to a slightly subcritical case of $\beta_{\mathcal{A}} = 0.95$, $\alpha_{\mathcal{H}} = 0.8$), and starts to observe the new incidence numbers until the new target $(\mathcal{A})$ has been reached. This can be interpreted as online-detection of a distributional change; a related comprehensive new framework for the use of divergences (even much beyond power divergences) for distributional change detection can be found e.g., in the recent work of Kißlinger & Stummer [118]. A completely different, SIR-model based, approach for the detection of change points in the spread of COVID-19 is given in Dehning et al. [119]. Moreover, other different surveillance methods can be also found e.g., in the corresponding overview of Frisen [120] and the Swedish epidemics outbreak investigations of Friesen & Andersson & Schiöler [121].

One can refine the above-mentioned sequential procedure via two (instead of one) appropriate thresholds $c_1 < c_2$ and the pair $(\check{\tau}, \delta_{\check{\tau}})$, with the stopping time $\check{\tau} := \inf\left\{n \in \mathcal{N} : Z_n \notin [c_1, c_2]\right\}$ as well as corresponding decision rule

$$
\delta_{\check{\tau}} \quad := \quad \begin{cases} d_{\mathcal{A}}, & \text{if } Z_{\check{\tau}} > c_2, \\ d_{\mathcal{H}}, & \text{if } Z_{\check{\tau}} < c_1. \end{cases}
$$

An exact optimized treatment on the two above-mentioned sequential procedures, and their connection to Hellinger integrals (and power divergences) of Galton-Watson processes with immigration, is beyond the scope of this paper.

As a side remark, let us mention that our above-mentioned suggested method of Bayesian decision making with Hellinger integrals of GWIs differs completely from the very recent work of Brauner et al. [122] who use a Bayesian hierarchical model for the concrete, very comprehensive study on the effectiveness and burden of non-pharmaceutical interventions against COVID-19 transmission.

The power divergences $I_\lambda\left(P_{\mathcal{A},n}||P_{\mathcal{H},n}\right)$ ($\lambda \in \mathbb{R}$) can be employed also in other ways within Bayesian decision making, of statistical nature. Namely, by adapting the general lines of Österreicher & Vajda [123] (see also Liese & Vajda [10], as well as diffusion-process applications in Stummer [5,31,32]) to our context of Galton-Watson processes with immigration, we can proceed as follows. For the sake of comfortable notations, we first attach the value $\theta := 1$ to the GWI model $(\mathcal{A})$ (which has prior probability $p_{\mathcal{A}}^{\text{prior}} \in\,]0, 1[$) and $\theta := 0$ to $(\mathcal{H})$ (which has prior probability $1 - p_{\mathcal{A}}^{\text{prior}}$). Suppose we want to decide, in an optimal Bayesian way, which *degree of evidence* $\mathrm{deg} \in [0,1]$ we should attribute (according to a pregiven *loss function* $\mathcal{LO}$) to the model $(\mathcal{A})$. In order to achieve this goal, we choose a nonnegatively-valued loss function $\mathcal{LO}(\theta, \mathrm{deg})$ defined on $\{0,1\} \times [0,1]$, of two types which will be specified below. The risk at stage 0 (i.e., prior to the GWI-path observations $\mathcal{X}_n$), from the optimal decision about the degree of evidence deg concerning the decision parameter θ, is defined as

$$\mathcal{BR}_{\mathcal{LO}}\left(p_{\mathcal{A}}^{\mathrm{prior}}\right) := \min_{\mathrm{deg} \in [0,1]} \left\{ (1 - p_{\mathcal{A}}^{\mathrm{prior}}) \cdot \mathcal{LO}(0, \mathrm{deg}) + p_{\mathcal{A}}^{\mathrm{prior}} \cdot \mathcal{LO}(1, \mathrm{deg}) \right\},$$

which can be thus interpreted as a *minimal prior expected loss* (the minimum will always exist). The corresponding risk *posterior* to the GWI-path observations $\mathcal{X}_n$, from the optimal decision about the degree of evidence deg concerning the parameter θ, is given by

$$\mathcal{BR}_{\mathcal{LO}}^{\mathrm{post}}\left(p_{\mathcal{A}}^{\mathrm{prior}}\right) := \int_{\Omega_n} \mathcal{BR}_{\mathcal{LO}}\left(p_{\mathcal{A}}^{\mathrm{post}}(\mathcal{X}_n)\right) \left(p_{\mathcal{A}}^{\mathrm{prior}} \, dP_{\mathcal{A},n} + (1 - p_{\mathcal{A}}^{\mathrm{prior}}) \, dP_{\mathcal{H},n}\right),$$

which is achieved by the optimal decision rule (about the degree of evidence)

$$\mathfrak{D}^{*}(\mathcal{X}_n) := \arg \min_{\mathrm{deg} \in [0,1]} \left\{ \left(1 - p_{\mathcal{A}}^{\mathrm{post}}(\mathcal{X}_n)\right) \cdot \mathcal{LO}(0, \mathrm{deg}) + p_{\mathcal{A}}^{\mathrm{post}}(\mathcal{X}_n) \cdot \mathcal{LO}(1, \mathrm{deg}) \right\}.$$

The corresponding *statistical information measure* (in the sense of De Groot [124])

$$\Delta \mathcal{BR}_{\mathcal{LO}}(p_{\mathcal{A}}^{\mathrm{prior}}) := \mathcal{BR}_{\mathcal{LO}}\left(p_{\mathcal{A}}^{\mathrm{prior}}\right) - \mathcal{BR}_{\mathcal{LO}}^{\mathrm{post}}\left(p_{\mathcal{A}}^{\mathrm{prior}}\right) \geq 0$$

represents the *reduction of the decision risk* about the degree of evidence deg concerning the parameter θ, that can be attained by observing the GWI-path $\mathcal{X}_n$ until stage n. For the first-type loss function $\widetilde{\mathcal{LO}}(\theta, \mathrm{deg}) := \mathrm{deg} - (2\,\mathrm{deg} - 1) \cdot \mathbf{1}_{\{1\}}(\theta)$, defined on $\{0,1\} \times [0,1]$ with the help of the indicator function $\mathbf{1}_A(.)$ on the set A, one can show that

$$\mathfrak{D}^{*}(\mathcal{X}_n) := \begin{cases} 0, & \text{if } p_{\mathcal{A}}^{\mathrm{post}}(\mathcal{X}_n) \in [0, \tfrac{1}{2}[, \\ 1, & \text{if } p_{\mathcal{A}}^{\mathrm{post}}(\mathcal{X}_n) \in]\tfrac{1}{2}, 1[, \\ \text{any number in } [0,1], & \text{if } p_{\mathcal{A}}^{\mathrm{post}}(\mathcal{X}_n) = \tfrac{1}{2}, \end{cases}$$

as well as the representation formula

$$I_{\lambda}\left(P_{\mathcal{A},n} || P_{\mathcal{H},n}\right) = \int_{0}^{1} \Delta \mathcal{BR}_{\widetilde{\mathcal{LO}}}(p_{\mathcal{A}}^{\mathrm{prior}}) \cdot \left(1 - p_{\mathcal{A}}^{\mathrm{prior}}\right)^{\lambda - 2} \cdot \left(p_{\mathcal{A}}^{\mathrm{prior}}\right)^{-1 - \lambda} \, dp_{\mathcal{A}}^{\mathrm{prior}}, \qquad \lambda \in \mathbb{R}, \qquad (21)$$

(cf. Österreicher & Vajda [123], Liese & Vajda [10], adapted to our GWI context); in other words, the power divergence $I_{\lambda}\left(P_{\mathcal{A},n} || P_{\mathcal{H},n}\right)$ can be regarded as a *weighted-average statistical information measure* (*weighted-average decision risk reduction*). One can also use other weights of $p_{\mathcal{A}}^{\mathrm{prior}}$ in order to get bounds of $I_{\lambda}\left(P_{\mathcal{A},n} || P_{\mathcal{H},n}\right)$ (analogously to Stummer [5]).

For the second-type loss function $\mathcal{LO}_{\lambda,\chi}(\theta, \mathrm{deg}) := \dfrac{\lambda^{\theta - 1} \, \mathrm{deg}^{\lambda - \theta}}{\chi^{\lambda} \, (1-\chi)^{1-\lambda} \, (1-\lambda)^{\theta} \, (1-\mathrm{deg})^{\lambda - \theta}}$ defined on $\{0,1\} \times [0,1]$ with parameters $\lambda \in]0,1[$ and $\chi \in]0,1[$, one can derive the optimal decision rule

$$\mathfrak{D}^{*}(\mathcal{X}_n) = p_{\mathcal{A}}^{\mathrm{post}}(\mathcal{X}_n)$$

as well as the representation formula as a *limit statistical information measure* (*limit decision risk reduction*)

$$I_{\lambda}\left(P_{\mathcal{A},n} || P_{\mathcal{H},n}\right) = \lim_{\chi \to p_{\mathcal{A}}^{\mathrm{prior}}} \Delta \mathcal{BR}_{\mathcal{LO}_{\lambda,\chi}}\left(p_{\mathcal{A}}^{\mathrm{prior}}\right) =: \Delta \mathcal{BR}_{\mathcal{LO}_{\lambda, p_{\mathcal{A}}^{\mathrm{prior}}}}\left(p_{\mathcal{A}}^{\mathrm{prior}}\right) \qquad (22)$$

(cf. Österreicher & Vajda [123], Stummer [5], adapted to our GWI context).

As an alternative to the above-mentioned Bayesian-decision-making applications of Hellinger integrals $H_\lambda\left(P_{A,n}||P_{\mathcal{H},n}\right)$, let us now briefly discuss the use of the latter for the corresponding *Neyman-Pearson* (NPT) framework with randomized tests $\mathcal{T}_n : \Omega_n \mapsto [0,1]$ of the hypothesis $P_{\mathcal{H}}$ against the alternative P_A, based on the GWI-generation-size sample path observations $\mathcal{X}_n := \{X_l : l \in \{0,1,\ldots,n\}\}$. In contrast to (17) and (18) a Neyman-Pearson test minimizes—over $\mathcal{T}_n$–the type II error probability $\int_{\Omega_n}(1 - \mathcal{T}_n)\,\mathrm{d}P_{A,n}$ in the class of the tests for which the type I error probability $\int_{\Omega_n}\mathcal{T}_n\,\mathrm{d}P_{\mathcal{H},n}$ is at most $\varsigma \in\,]0,1[$. The corresponding minimal type II error probability

$$\mathcal{E}_\varsigma\left(P_{A,i}||P_{\mathcal{H},i}\right) := \inf_{\mathcal{T}_i:\int_{\Omega_i}\mathcal{T}_i\,\mathrm{d}P_{\mathcal{H},i}\leq\varsigma}\int_{\Omega_i}(1 - \mathcal{T}_i)\,\mathrm{d}P_{A,i}$$

can for all $\varsigma \in\,]0,1[$, $\lambda \in\,]0,1[$, $i \in \mathcal{I}$ be bounded from above by

$$\mathcal{E}_\varsigma\left(P_{A,i}||P_{\mathcal{H},i}\right) \leq \mathcal{E}_\varsigma^U\left(P_{A,i}||P_{\mathcal{H},i}\right) := \min\left\{(1-\lambda)\cdot\left(\frac{\lambda}{\varsigma}\right)^{\lambda/(1-\lambda)}\cdot\left(H_\lambda\left(P_{A,i}||P_{\mathcal{H},i}\right)\right)^{1/(1-\lambda)}, 1\right\}, \quad (23)$$

and for all $\lambda > 1$, $i \in \mathcal{I}$ it can be bounded from below by

$$\mathcal{E}_\varsigma\left(P_{A,i}||P_{\mathcal{H},i}\right) > \mathcal{E}_\varsigma^L\left(P_{A,i}||P_{\mathcal{H},i}\right) := (1-\varsigma)^{\lambda/(\lambda-1)}\cdot\left(H_\lambda\left(P_{A,i}||P_{\mathcal{H},i}\right)\right)^{1/(1-\lambda)}, \quad (24)$$

which is an adaption of a general result of Krafft & Plachky [125], see also Liese & Vajda [1] as well as Stummer & Vajda [15]. Hence, by combining (23) and (24) with the exact values respectively upper bounds of the Hellinger integrals $H_{1-\lambda}\left(P_{A,n}||P_{\mathcal{H},n}\right)$ from the following sections, we obtain for our context of Galton-Watson processes with Poisson offspring and Poisson immigration (including the non-immigration case) some upper bounds of $\mathcal{E}_\varsigma\left(P_{A,n}||P_{\mathcal{H},n}\right)$, which can also be immediately rewritten as lower bounds for the power $1 - \mathcal{E}_\varsigma\left(P_{A,n}||P_{\mathcal{H},n}\right)$ of a most powerful test at level ς. In contrast to such finite-time-horizon results, for the (to our context) incompatible setup of Galton-Watson processes with Poisson offspring but nonstochastic immigration of constant value 1, the asymptotic rates of decrease as $n \to \infty$ of the unconstrained type II error probabilities as well as the type I error probabilites were studied in Linkov & Lunyova [53] by a different approach employing also Hellinger integrals. Some other types of Galton-Watson-process concerning Neyman-Pearson testing investigations different to ours can be found e.g., in Basawa & Scott [126], Feigin [127], Sweeting [128], Basawa & Scott [61], and the references therein.

2.6. Asymptotical Distinguishability

The next two concepts deal with two general families $(P_{A,i})_{i\in\mathcal{I}}$ and $(P_{\mathcal{H},i})_{i\in\mathcal{I}}$ of probability measures on the measurable spaces $(\Omega_i, \mathcal{F}_i)_{i\in\mathcal{I}}$, where the index set $\mathcal{I}$ is either $\mathbb{N}_0$ or $\mathbb{R}_+$. For them, the following two general types of asymptotical distinguishability are well known (see e.g., LeCam [109], Liese & Vajda [1], Jacod & Shiryaev [24], Linkov [129], and the references therein).

Definition 1. *The family $(P_{A,i})_{i\in\mathcal{I}}$ is contiguous to the family $(P_{\mathcal{H},i})_{i\in\mathcal{I}}$ – in symbols, $(P_{A,i}) \triangleleft (P_{\mathcal{H},i})$– if for all sets $A_i \in \mathcal{F}_i$ with $\lim_{i\to\infty} P_{\mathcal{H},i}(A_i) = 0$ there holds $\lim_{i\to\infty} P_{A,i}(A_i) = 0$.*

Definition 2. *Families of measures $(P_{A,i})_{i\in\mathcal{I}}$ and $(P_{\mathcal{H},i})_{i\in\mathcal{I}}$ are called entirely separated (completely asymptotically distinguishable)—in symbols, $(P_{A,i}) \triangle (P_{\mathcal{H},i})$–if there exist a sequence $i_m \uparrow \infty$ as $m \uparrow \infty$ and for each $m \in \mathbb{N}_0$ an $A_{i_m} \in \mathcal{F}_{i_m}$ such that $\lim_{m\to\infty} P_{A,i_m}(A_{i_m}) = 1$ and $\lim_{m\to\infty} P_{\mathcal{H},i_m}(A_{i_m}) = 0$.*

It is clear that the notion of contiguity is the attempt to carry the concept of absolute continuity over to families of measures. Loosely speaking, $(P_{A,i})$ is contiguous to $(P_{\mathcal{H},i})$, if the limit $\lim_{i\to\infty}(P_{A,i})$ (existence preconditioned) is absolute continuous to the limit $\lim_{i\to\infty}(P_{\mathcal{H},i})$. However, for the definition of contiguity, we do not need to require the probability measures to converge to limiting probability measures. On the other hand, entire separation is the generalization of singularity to families of measures.

The corresponding negations will be denoted by $\overline{\lhd}$ and $\overline{\triangle}$. One can easily check that a family $(P_{\mathcal{A},i})$ cannot be both contiguous and entirely separated to a family $(P_{\mathcal{H},i})$. In fact, as shown in Linkov [129], the relation between the families $(P_{\mathcal{A},i})$ and $(P_{\mathcal{H},i})$ can be uniquely classified into the following *distinguishability types*:

(a) $(P_{\mathcal{A},i}) \lhd\rhd (P_{\mathcal{H},i})$;

(b) $(P_{\mathcal{A},i}) \lhd (P_{\mathcal{H},i})$, $(P_{\mathcal{H},i}) \overline{\lhd} (P_{\mathcal{A},i})$;

(c) $(P_{\mathcal{A},i}) \overline{\lhd} (P_{\mathcal{H},i})$, $(P_{\mathcal{H},i}) \lhd (P_{\mathcal{A},i})$;

(d) $(P_{\mathcal{A},i}) \overline{\lhd}\,\overline{\rhd} (P_{\mathcal{H},i})$, $(P_{\mathcal{A},i}) \overline{\triangle} (P_{\mathcal{H},i})$;

(e) $(P_{\mathcal{A},i}) \triangle (P_{\mathcal{H},i})$.

As demonstrated in the above-mentioned references for a general context, one can conclude the type of distinguishability from the time-evolution of Hellinger integrals. Indeed, the following assertions can be found e.g., in Linkov [129], where part (c) was established in Liese & Vajda [1] and (f), (g) in Vajda [3].

Proposition 1. *The following assertions are equivalent:*

(a) $(P_{\mathcal{A},i}) \triangle (P_{\mathcal{H},i})$,

(b) $\liminf\limits_{i\to\infty} H_\lambda(P_{\mathcal{A},i}\|P_{\mathcal{H},i}) = 0 \quad \text{for all } \lambda \in\,]0,1[$,

(c) *there exists a* $\lambda \in\,]0,1[:\quad \liminf\limits_{i\to\infty} H_\lambda(P_{\mathcal{A},i}\|P_{\mathcal{H},i}) = 0$,　　　　　(25)

(d) *there exists a* $\pi \in\,]0,1[:\quad \liminf\limits_{i\to\infty} e_\pi(P_{\mathcal{A},i}\|P_{\mathcal{H},i}) = 0$,

(e) $\limsup\limits_{i\to\infty} V(P_{\mathcal{A},i}\|P_{\mathcal{H},i}) = 2$,

(f) *there exists a* $\lambda \in\,]0,1[:\quad \limsup\limits_{i\to\infty} I_\lambda(P_{\mathcal{A},i}\|P_{\mathcal{H},i}) = \dfrac{1}{\lambda\cdot(1-\lambda)}$,

(g) $\limsup\limits_{i\to\infty} I_\lambda(P_{\mathcal{A},i}\|P_{\mathcal{H},i}) = \dfrac{1}{\lambda\cdot(1-\lambda)}$, *for all* $\lambda \in\,]0,1[$.

In combination with the discussion after Definition 2, one can thus interpret the $\lambda-$order Hellinger integral $H_\lambda(P_{\mathcal{A},i}\|P_{\mathcal{H},i})$ as a "measure" for the distinctness of the two families $P_{\mathcal{A},i}$ and $P_{\mathcal{H},i}$ up to a fixed finite time horizon $i \in \mathcal{I}$.

Furthermore, for the contiguity we obtain the equivalence (see e.g., Liese & Vajda [1], Linkov [129])

$$(P_{\mathcal{A},i}) \lhd (P_{\mathcal{H},i}) \qquad \Longleftrightarrow \qquad \liminf_{\lambda\nearrow 1}\left\{\liminf_{i\to\infty} H_\lambda(P_{\mathcal{A},i}\|P_{\mathcal{H},i})\right\} = 1 \qquad (26)$$

$$\Longleftrightarrow \qquad \limsup_{\lambda\nearrow 1}\left\{\limsup_{i\to\infty} \lambda\cdot(1-\lambda)\cdot I_\lambda(P_{\mathcal{A},i}\|P_{\mathcal{H},i})\right\} = 0.$$

All the above-mentioned general results can be applied to our context of two competing Poissonian Galton-Watson processes with immigration (GWI) $(\mathcal{H})$ and $(\mathcal{A})$ (reflected by the two different laws $P_\mathcal{H}$ resp. $P_\mathcal{A}$ with parameter pairs $(\beta_\mathcal{H}, \alpha_\mathcal{H})$ resp. $(\beta_\mathcal{A}, \alpha_\mathcal{A})$), by taking $P_{\mathcal{A},i} := P_\mathcal{A}|_{\mathcal{F}_i}$ and $P_{\mathcal{H},i} := P_\mathcal{H}|_{\mathcal{F}_i}$. Recall from the preceding subsections (by identifying i with n) that the latter two describe the stochastic dynamics of the respective GWI within the restricted time-/stage-frame $\{0, 1, \ldots, i\}$.

In the following, we study in detail the evolution of Hellinger integrals between two competing models of Galton-Watson processes with immigration, which turns out to be quite extensive.

3. Detailed Recursive Analyses of Hellinger Integrals

3.1. A First Basic Result

In terms of our notations (PS1) to (PS3), a typical situation for applications in our mind is that one particular constellation $(\beta_A, \beta_H, \alpha_A, \alpha_H) \in \mathcal{P}$ (e.g., obtained from theoretical or previous statistical investigations) is fixed, whereas–in contrast–the parameter $\lambda \in \mathbb{R}\backslash\{0,1\}$ for the Hellinger integral or the power divergence might be chosen freely, e.g., depending on which (transform of a) dissimilarity measure one decides to choose for further analysis. At this point, let us emphasize that *in general* we will not make assumptions of the form $\beta_\bullet \gtreqless 1$, i.e., upon the type of criticality.

To start with our investigations, in order to justify for all $n \in \mathbb{N}_0$

$$Z_n := \frac{\mathrm{d}P_{A,n}}{\mathrm{d}P_{H,n}} \qquad \text{(cf. (13))},$$

(14) and (15) (as well as $I_\lambda\left(P_{A,n}||P_{H,n}\right)$ for $\lambda \in \mathbb{R}$ respectively $R_\lambda\left(P_{A,n}||P_{H,n}\right)$ for $\lambda \in \mathbb{R}\backslash\{0,1\}$), we first mention the following straightforward facts: (i) if $(\beta_A, \beta_H, \alpha_A, \alpha_H) \in \mathcal{P}_{\mathrm{NI}}$, then $P_{A,n}$ and $P_{H,n}$ are equivalent (i.e., $P_{A,n} \sim P_{H,n}$), as well as (ii) if $(\beta_A, \beta_H, \alpha_A, \alpha_H) \in \mathcal{P}_{\mathrm{SP}}$, then $P_{A,n}$ and $P_{H,n}$ are equivalent (i.e., $P_{A,n} \sim P_{H,n}$). Moreover, by recalling $Z_0 = 1$ and using the "rate functions" $f_\bullet(x) = \beta_\bullet x + \alpha_\bullet$ ($x \in [0,\infty)$), a version of (13) can be easily determined by calculating for each $\vec{x} :- (x_0, x_1, x_2, \cdots) \in \Omega := \mathbb{N} \times \mathbb{N}_0 \times \mathbb{N}_0 \times \cdots$

$$Z_n(\vec{x}) = \prod_{k=1}^{n} Z_{n,k}(\vec{x}) \qquad \text{with } Z_{n,k}(\vec{x}) := \exp\left\{-\left(f_A(x_{k-1}) - f_H(x_{k-1})\right)\right\} \left[\frac{f_A(x_{k-1})}{f_H(x_{k-1})}\right]^{x_k},$$

where for the last term we use the convention $\left(\frac{0}{0}\right)^x = 1$ for all $x \in \mathbb{N}_0$. Furthermore, we define for each $\vec{x} \in \Omega$

$$Z_{n,k}^{(\lambda)}(\vec{x}) := \exp\left\{-\left(\lambda f_A(x_{k-1}) + (1-\lambda)f_H(x_{k-1})\right)\right\} \frac{\left[(f_A(x_{k-1}))^\lambda \, (f_H(x_{k-1}))^{1-\lambda}\right]^{x_k}}{x_k!} \tag{27}$$

with the convention $\frac{(0)^0}{0!} = 1$ for the last term. Accordingly, one obtains from (14) the Hellinger integral $H_\lambda\left(P_{A,0}||P_{H,0}\right) = 1$, as well as for all $(\beta_A, \beta_H, \alpha_A, \alpha_H, \lambda) \in \mathcal{P} \times (\mathbb{R}\backslash\{0,1\})$

$$H_\lambda\left(P_{A,1}||P_{H,1}\right) = \exp\left\{(f_A(x_0))^\lambda \, (f_H(x_0))^{(1-\lambda)} - (\lambda f_A(x_0) + (1-\lambda)f_H(x_0))\right\} \tag{28}$$

for $x_0 = X_0 \in \mathbb{N}$, and for all $n \in \mathbb{N}\backslash\{1\}$

$$\begin{aligned}
H_\lambda\left(P_{A,n}||P_{H,n}\right) &= EP_{H,n}\left[(Z_n)^\lambda\right] = \sum_{x_1=0}^{\infty} \cdots \sum_{x_n=0}^{\infty} \prod_{k=1}^{n} Z_{n,k}^{(\lambda)}(\vec{x}) \\
&= \sum_{x_1=0}^{\infty} \cdots \sum_{x_{n-1}=0}^{\infty} \prod_{k=1}^{n-1} Z_{n,k}^{(\lambda)}(\vec{x}) \cdot e^{-(\lambda f_A(x_{n-1})+(1-\lambda)f_H(x_{n-1}))} \sum_{x_n=0}^{\infty} \frac{\left[(f_A(x_{n-1}))^\lambda \, (f_H(x_{n-1}))^{1-\lambda}\right]^{x_n}}{x_n!} \\
&= \sum_{x_1=0}^{\infty} \cdots \sum_{x_{n-1}=0}^{\infty} \prod_{k=1}^{n-1} Z_{n,k}^{(\lambda)}(\vec{x}) \cdot \exp\{(f_A(x_{n-1}))^\lambda \, (f_H(x_{n-1}))^{1-\lambda} - (\lambda f_A(x_{n-1}) + (1-\lambda)f_H(x_{n-1}))\}. \tag{29}
\end{aligned}$$

From (29), one can see that a crucial role for the exact calculation (respectively the derivation of bounds) of the Hellinger integral is played by the functions defined for $x \in [0, \infty[$

$$\phi_\lambda(x) := \phi(x, \beta_A, \beta_H, \alpha_A, \alpha_H, \lambda) := \varphi_\lambda(x) - f_\lambda(x), \qquad \text{with} \tag{30}$$

$$\varphi_\lambda(x) := \varphi(x, \beta_A, \beta_H, \alpha_A, \alpha_H, \lambda) := (f_A(x))^\lambda \, (f_H(x))^{1-\lambda} \qquad \text{and} \tag{31}$$

$$f_\lambda(x) := f(x, \beta_{\mathcal{A}}, \beta_{\mathcal{H}}, \alpha_{\mathcal{A}}, \alpha_{\mathcal{H}}, \lambda) := \lambda f_{\mathcal{A}}(x) + (1-\lambda)f_{\mathcal{H}}(x) = \alpha_\lambda + \beta_\lambda x, \tag{32}$$

where we have used the λ-*weighted-averages*

$$\alpha_\lambda := \alpha(\alpha_{\mathcal{A}}, \alpha_{\mathcal{H}}, \lambda) := \lambda \cdot \alpha_{\mathcal{A}} + (1-\lambda) \cdot \alpha_{\mathcal{H}} \quad \text{and} \quad \beta_\lambda := \beta(\beta_{\mathcal{A}}, \beta_{\mathcal{H}}, \lambda) := \lambda \cdot \beta_{\mathcal{A}} + (1-\lambda) \cdot \beta_{\mathcal{H}}.$$

Since λ plays a special role, henceforth we typically use it as index and often omit $(\beta_{\mathcal{A}}, \beta_{\mathcal{H}}, \alpha_{\mathcal{A}}, \alpha_{\mathcal{H}})$. According to Lemma A1 in the Appendix A.1, it follows that for $\lambda \in]0, 1[$ (respectively $\lambda \in \mathbb{R}\backslash[0, 1]$) one gets $\phi_\lambda(x) \leq 0$ (respectively $\phi_\lambda(x) \geq 0$) for all $x \in [0, \infty[$. Furthermore, in both cases there holds $\phi_\lambda(x) = 0$ iff $f_{\mathcal{A}}(x) = f_{\mathcal{H}}(x)$, i.e., for $x = x^* := \frac{\alpha_{\mathcal{A}} - \alpha_{\mathcal{H}}}{\beta_{\mathcal{H}} - \beta_{\mathcal{A}}} \geq 0$. This is consistent with the corresponding generally valid upper and lower bounds (cf. (9) and (11)) $0 < H_\lambda(P_{\mathcal{A},n}||P_{\mathcal{H},n}) \leq 1$, for $\lambda \in]0,1[$, $1 \leq H_\lambda(P_{\mathcal{A},n}||P_{\mathcal{H},n}) \leq \infty$, for $\lambda \in \mathbb{R}\backslash[0, 1]$.

As a first indication for our proposed method, let us start by illuminating the simplest case $\lambda \in \mathbb{R}\backslash\{0, 1\}$ and $\gamma := \alpha_{\mathcal{H}}\beta_{\mathcal{A}} - \alpha_{\mathcal{A}}\beta_{\mathcal{H}} = 0$. This means that $(\beta_{\mathcal{A}}, \beta_{\mathcal{H}}, \alpha_{\mathcal{A}}, \alpha_{\mathcal{H}}) \in \mathcal{P}_{\text{NI}} \cup \mathcal{P}_{\text{SP},1}$, where $\mathcal{P}_{\text{SP},1}$ is the set of all (componentwise) strictly positive $(\beta_{\mathcal{A}}, \beta_{\mathcal{H}}, \alpha_{\mathcal{A}}, \alpha_{\mathcal{H}})$ with $\beta_{\mathcal{A}} \neq \beta_{\mathcal{H}}$, $\alpha_{\mathcal{A}} \neq \alpha_{\mathcal{H}}$ and $\frac{\beta_{\mathcal{A}}}{\beta_{\mathcal{H}}} = \frac{\alpha_{\mathcal{A}}}{\alpha_{\mathcal{H}}} \neq 1$ ("the equal-fraction-case"). In this situation, *all* the three functions (30) to (32) are linear. Indeed,

$$\varphi_\lambda(x) = p_\lambda^E + q_\lambda^E x \tag{33}$$

with $p_\lambda^E := \alpha_{\mathcal{A}}^\lambda \alpha_{\mathcal{H}}^{1-\lambda}$ and $q_\lambda^E := \beta_{\mathcal{A}}^\lambda \beta_{\mathcal{H}}^{1-\lambda}$ (where the index E stands for $\underline{e}$xact linearity). Clearly, $q_\lambda^E > 0$ on $\mathcal{P}_{\text{NI}} \cup \mathcal{P}_{\text{SP},1}$, as well as $p_\lambda^E > 0$ on $\mathcal{P}_{\text{SP},1}$ and $p_\lambda^E = 0$ on $\mathcal{P}_{\text{NI}}$. Furthermore,

$$\phi_\lambda(x) = r_\lambda^E + s_\lambda^E x$$

with $r_\lambda^E := p_\lambda^E - \alpha_\lambda = \alpha_{\mathcal{A}}^\lambda \alpha_{\mathcal{H}}^{1-\lambda} - (\lambda\alpha_{\mathcal{A}} + (1-\lambda)\alpha_{\mathcal{H}})$ and $s_\lambda^E := q_\lambda^E - \beta_\lambda = \beta_{\mathcal{A}}^\lambda \beta_{\mathcal{H}}^{1-\lambda} - (\lambda\beta_{\mathcal{A}} + (1-\lambda)\beta_{\mathcal{H}})$. Due to Lemma A1 one knows that on $\mathcal{P}_{\text{NI}} \cup \mathcal{P}_{\text{SP},1}$ one gets $s_\lambda^E < 0$ for $\lambda \in]0, 1[$ and $s_\lambda^E > 0$ for $\lambda \in \mathbb{R}\backslash[0, 1]$. Furthermore, on $\mathcal{P}_{\text{SP},1}$ one gets $r_\lambda^E < 0$ (resp. $r_\lambda^E > 0$) for $\lambda \in]0, 1[$ (resp. $\lambda \in \mathbb{R}\backslash[0, 1]$), whereas on $\mathcal{P}_{\text{NI}}$, the no-immigration setup, we get for all $\lambda \in \mathbb{R}\backslash\{0, 1\}$ $r_\lambda^E = 0$.

As it will be seen later on, such kind of linearity properties are useful for the recursive handling of the Hellinger integrals. However, only on the parameter set $\mathcal{P}_{\text{NI}} \cup \mathcal{P}_{\text{SP},1}$ the functions φ_λ and ϕ_λ are linear. Hence, in the general case $(\beta_{\mathcal{A}}, \beta_{\mathcal{H}}, \alpha_{\mathcal{A}}, \alpha_{\mathcal{H}}, \lambda) \in \mathcal{P} \times \mathbb{R}\backslash\{0, 1\}$ we aim for linear lower and upper bounds

$$\varphi_\lambda^L(x) := p_\lambda^L + q_\lambda^L x \leq \varphi_\lambda(x) \leq \varphi_\lambda^U(x) := p_\lambda^U + q_\lambda^U x, \tag{34}$$

$x \in [0, \infty[$ (ultimately, $x \in \mathbb{N}_0$), which by (30) and (31) leads to

$$\phi_\lambda(x) \begin{cases} \leq \phi_\lambda^U(x) := r_\lambda^U + s_\lambda^U \cdot x := (p_\lambda^U - \alpha_\lambda) + (q_\lambda^U - \beta_\lambda) \cdot x, \\[2mm] \geq \phi_\lambda^L(x) := r_\lambda^L + s_\lambda^L \cdot x := (p_\lambda^L - \alpha_\lambda) + (q_\lambda^L - \beta_\lambda) \cdot x, \end{cases} \tag{35}$$

$x \in [0, \infty[$ (ultimately, $x \in \mathbb{N}_0$). Of course, the involved slopes and intercepts should satisfy reasonable restrictions. Later on, we shall impose further restrictions on the involved slopes and intercepts, in order to guarantee nice properties of the general Hellinger integral bounds given in Theorem 1 below (for instance, in consistency with the nonnegativity of φ_λ we could require $p_\lambda^U \geq p_\lambda^L \geq 0$, $q_\lambda^U \geq q_\lambda^L \geq 0$ which nontrivially implies that these bounds possess certain monotonicity properties). For the formulation of our first assertions on Hellinger integrals, we make use of the following notation:

Definition 3. *For all* $(\beta_{\mathcal{A}}, \beta_{\mathcal{H}}, \alpha_{\mathcal{A}}, \alpha_{\mathcal{H}}, \lambda) \in \mathcal{P} \times \mathbb{R}\backslash\{0,1\}$ *and all* $p, q \in \mathbb{R}$ *let us define the sequences* $\left(a_n^{(q)}\right)_{n\in\mathbb{N}_0}$ *and* $\left(b_n^{(p,q)}\right)_{n\in\mathbb{N}_0}$ *recursively by*

$$a_0^{(q)} := 0 \quad ; \qquad a_n^{(q)} := \xi_\lambda^{(q)}\left(a_{n-1}^{(q)}\right) := q \cdot e^{a_{n-1}^{(q)}} - \beta_\lambda, \ n \in \mathbb{N}, \tag{36}$$

$$b_0^{(p,q)} := 0 \quad ; \qquad b_n^{(p,q)} := p \cdot e^{a_{n-1}^{(q)}} - \alpha_\lambda, \ n \in \mathbb{N}. \tag{37}$$

Notice the interrelation $a_1^{(q_\lambda^A)} = s_\lambda^A$ and $b_1^{(p_\lambda^A, q_\lambda^A)} = r_\lambda^A$ for $A \in \{E, L, U\}$. Clearly, for all $q \in \mathbb{R}\backslash\{0\}$ and $p \in \mathbb{R}$ one has the linear interrelation

$$b_n^{(p,q)} = \frac{p}{q} a_n^{(q)} + \frac{p}{q} \beta_\lambda - \alpha_\lambda, \ n \in \mathbb{N}. \tag{38}$$

Accordingly, we obtain fundamental Hellinger integral evaluations:

Theorem 1.

(a) *For all* $(\beta_{\mathcal{A}}, \beta_{\mathcal{H}}, \alpha_{\mathcal{A}}, \alpha_{\mathcal{H}}, \lambda) \in (\mathcal{P}_{NI} \cup \mathcal{P}_{SP,1}) \times \mathbb{R}\backslash\{0,1\}$, *all initial population sizes* $X_0 \in \mathbb{N}$ *and all observation horizons* $n \in \mathbb{N}$ *one can recursively compute the **exact value***

$$H_\lambda(P_{\mathcal{A},n}||P_{\mathcal{H},n}) = \exp\left\{a_n^{(q_\lambda^E)} X_0 + \frac{\alpha_{\mathcal{A}}}{\beta_{\mathcal{A}}} \sum_{k=1}^{n} a_k^{(q_\lambda^E)}\right\} =: V_{\lambda, X_0, n}, \tag{39}$$

where $\frac{\alpha_{\mathcal{A}}}{\beta_{\mathcal{A}}}$ *can be equivalently replaced by* $\frac{\alpha_{\mathcal{H}}}{\beta_{\mathcal{H}}}$. *Recall that* $q_\lambda^E := \beta_{\mathcal{A}}^\lambda \beta_{\mathcal{H}}^{1-\lambda}$. *Notice that on* $\mathcal{P}_{NI} \times (\mathbb{R}\backslash\{0,1\})$ *the formula* (39) *simplifies significantly, since* $\alpha_{\mathcal{A}} = \alpha_{\mathcal{H}} = 0$.

(b) *For all* $(\beta_{\mathcal{A}}, \beta_{\mathcal{H}}, \alpha_{\mathcal{A}}, \alpha_{\mathcal{H}}, \lambda) \in (\mathcal{P}_{SP}\backslash\mathcal{P}_{SP,1}) \times (\mathbb{R}\backslash\{0,1\})$, *all coefficients* p_λ^L, p_λ^U, q_λ^L, $q_\lambda^U \in \mathbb{R}$ *which satisfy* (35) *for all* $x \in \mathbb{N}_0$ *(and thus in particular* $p_\lambda^L \le p_\lambda^U$, $q_\lambda^L \le q_\lambda^U$), *all initial population sizes* $X_0 \in \mathbb{N}$ *and all observation horizons* $n \in \mathbb{N}$ *one gets the following **recursive** (i.e., recursively computable) **bounds** for the Hellinger integrals:*

$$\text{for } \lambda \in]0,1[: \quad B_{\lambda,X_0,n}^L := \widetilde{B}_{\lambda,X_0,n}^{(p_\lambda^L, q_\lambda^L)} < H_\lambda(P_{\mathcal{A},n}||P_{\mathcal{H},n}) \le \min\left\{\widetilde{B}_{\lambda,X_0,n}^{(p_\lambda^U, q_\lambda^U)}, 1\right\} =: B_{\lambda,X_0,n}^U, \tag{40}$$

$$\text{for } \lambda \in \mathbb{R}\backslash[0,1]: \quad B_{\lambda,X_0,n}^L := \max\left\{\widetilde{B}_{\lambda,X_0,n}^{(p_\lambda^L, q_\lambda^L)}, 1\right\} \le H_\lambda(P_{\mathcal{A},n}||P_{\mathcal{H},n}) < \widetilde{B}_{\lambda,X_0,n}^{(p_\lambda^U, q_\lambda^U)} =: B_{\lambda,X_0,n}^U, \tag{41}$$

where for general $\lambda \in \mathbb{R}\backslash\{0,1\}$, $p \in \mathbb{R}$, $q \in \mathbb{R}\backslash\{0\}$ *we use the definitions*

$$\widetilde{B}_{\lambda,X_0,n}^{(p,q)} := \exp\left\{a_n^{(q)} \cdot X_0 + \sum_{k=1}^{n} b_k^{(p,q)}\right\} = \exp\left\{a_n^{(q)} \cdot X_0 + \frac{p}{q}\sum_{k=1}^{n} a_k^{(q)} + n \cdot \left(\frac{p}{q}\beta_\lambda - \alpha_\lambda\right)\right\}, \tag{42}$$

as well as

$$\widetilde{B}_{\lambda,X_0,n}^{(p,0)} := \exp\left\{-\beta_\lambda \cdot X_0 + \left(p \cdot e^{-\beta_\lambda} - \alpha_\lambda\right) \cdot n\right\}.$$

Remark 1.

(a) *Notice that the expression* $\widetilde{B}_{\lambda,X_0,n}^{(p,q)}$ *can analogously be defined on the parameter set* $\mathcal{P}_{NI} \cup \mathcal{P}_{SP,1}$. *For the choices* $q_\lambda^E := \beta_{\mathcal{A}}^\lambda \beta_{\mathcal{H}}^{1-\lambda} > 0$ *and* $p_\lambda^E := \alpha_{\mathcal{A}}^\lambda \alpha_{\mathcal{H}}^{1-\lambda} = q_\lambda^E \cdot \frac{\alpha_{\mathcal{A}}}{\beta_{\mathcal{A}}} = q_\lambda^E \cdot \frac{\alpha_{\mathcal{H}}}{\beta_{\mathcal{H}}} \ge 0$ *one gets* $(p_\lambda^E/q_\lambda^E) \cdot \beta_\lambda - \alpha_\lambda = 0$, *and thus the characterization* $\widetilde{B}_{\lambda,X_0,n}^{(p_\lambda^E, q_\lambda^E)} = V_{\lambda,X_0,n}$ *as the exact value (rather than a lower/upper bound (component))*.

(b) *In the case* $q = \beta_\lambda$ *one gets the explicit representation* $\widetilde{B}_{\lambda,X_0,n}^{(p,q)} = \exp\left\{(p - \alpha_\lambda) \cdot n\right\}$.

(c) *Using the skew symmetry* (8), *one can derive alternative bounds of the Hellinger integral by switching to the transformed parameter setup* $(\overleftrightarrow{\beta}_{\mathcal{A}}, \overleftrightarrow{\beta}_{\mathcal{H}}, \overleftrightarrow{\alpha}_{\mathcal{A}}, \overleftrightarrow{\alpha}_{\mathcal{H}}, \overleftrightarrow{\lambda}) := (\beta_{\mathcal{H}}, \beta_{\mathcal{A}}, \alpha_{\mathcal{H}}, \alpha_{\mathcal{A}}, 1 - \lambda)$. *However, this does not lead to different bounds: define* $\overleftrightarrow{\phi}_{\overleftrightarrow{\lambda}}$, $\overleftrightarrow{\varphi}_{\overleftrightarrow{\lambda}}$ *and* $\overleftrightarrow{f}_{\overleftrightarrow{\lambda}}$ *analogously to* (30), (31) *and* (32) *by* replacing the parameters $(\beta_{\mathcal{A}}, \beta_{\mathcal{H}}, \alpha_{\mathcal{A}}, \alpha_{\mathcal{H}}, \lambda)$ with $(\overleftrightarrow{\beta_{\mathcal{A}}}, \overleftrightarrow{\beta_{\mathcal{H}}}, \overleftrightarrow{\alpha_{\mathcal{A}}}, \overleftrightarrow{\alpha_{\mathcal{H}}}, \overleftrightarrow{\lambda})$. Then, there holds $\overleftrightarrow{f}_{\overleftrightarrow{\lambda}}(x) = f_\lambda(x)$, $\overleftrightarrow{\varphi}_{\overleftrightarrow{\lambda}}(x) = \varphi_\lambda(x)$ and $\overleftrightarrow{\phi}_{\overleftrightarrow{\lambda}}(x) = \phi_\lambda(x)$, and the set of (lower and upper bound) parameters $p_\lambda^L, q_\lambda^L, p_\lambda^U, q_\lambda^U$ satisfying (35) does not change under this transformation.

(d)   If there are no other restrictions on p_λ^L, p_λ^U, q_λ^L, q_λ^U than (35), the bounds in (40) and (41) can have some inconvenient features, e.g., being 1 for all (large enough) $n \in \mathbb{N}$, having oscillating n-behaviour, being suboptimal in certain (other) senses. For a detailed discussion, the reader is referred to Section 3.16 ff. below.

(e)   For the (to our context) incompatible setup of GWI with Poisson offspring but nonstochastic immigration of constant value 1, the exact values of the corresponding Hellinger integrals (i.e., an "analogue" of part (a)) was established in Linkov & Lunyova [53].

Proof of Theorem 1. Let us fix $(\beta_{\mathcal{A}}, \beta_{\mathcal{H}}, \alpha_{\mathcal{A}}, \alpha_{\mathcal{H}}) \in \mathcal{P}$ as well as $x_0 := X_0 \in \mathbb{N}$, and start with arbitrary $\lambda \in\,]0,1[$. We first prove the upper bound $B_{\lambda,X_0,n}^U$ of part (b). Correspondingly, we suppose that the coefficients p_λ^U, q_λ^U satisfy (35) for all $x \in \mathbb{N}_0$. From (28), (30), (31), (32) and (35) one gets immediately $B_{\lambda,X_0,1}^U$ in terms of the first sequence-element $a_1^{(q_\lambda^U)}$ (cf. (36)). With the help of (29) for all observation horizons $n \in \mathbb{N}\backslash\{1\}$ we get (with the obvious shortcut for $n = 2$)

$$
\begin{aligned}
H_\lambda\left(P_{\mathcal{A},n} \| P_{\mathcal{H},n}\right) &= \sum_{x_1=0}^{\infty} \cdots \sum_{x_{n-1}=0}^{\infty} \prod_{k=1}^{n-1} Z_{n,k}^{(\lambda)}(\vec{x}) \cdot \exp\left\{\varphi_\lambda(x_{n-1}) - f_\lambda(x_{n-1})\right\} \\
&< \sum_{x_1=0}^{\infty} \cdots \sum_{x_{n-1}=0}^{\infty} \prod_{k=1}^{n-1} Z_{n,k}^{(\lambda)}(\vec{x}) \cdot \exp\left\{(p_\lambda^U - \alpha_\lambda) + (q_\lambda^U - \beta_\lambda) x_{n-1}\right\} \\
&= \sum_{x_1=0}^{\infty} \cdots \sum_{x_{n-1}=0}^{\infty} \prod_{k=1}^{n-1} Z_{n,k}^{(\lambda)}(\vec{x}) \cdot \exp\left\{b_1^{(p_\lambda^U, q_\lambda^U)} + a_1^{(q_\lambda^U)} x_{n-1}\right\} \\
&= \exp\left\{b_1^{(p_\lambda^U, q_\lambda^U)}\right\} \sum_{x_1=0}^{\infty} \cdots \sum_{x_{n-2}=0}^{\infty} \prod_{k=1}^{n-2} Z_{n,k}^{(\lambda)}(\vec{x}) \cdot \exp\left\{\exp\left\{a_1^{(q_\lambda^U)}\right\} \varphi_\lambda(x_{n-2}) - f_\lambda(x_{n-2})\right\} \\
&< \exp\left\{b_1^{(p_\lambda^U, q_\lambda^U)}\right\} \sum_{x_1=0}^{\infty} \cdots \sum_{x_{n-2}=0}^{\infty} \prod_{k=1}^{n-2} Z_{n,k}^{(\lambda)}(\vec{x}) \\
&\quad \cdot \exp\left\{\left(\exp\left\{a_1^{(q_\lambda^U)}\right\} p_\lambda^U - \alpha_\lambda\right) + \left(\exp\left\{a_1^{(q_\lambda^U)}\right\} q_\lambda^U - \beta_\lambda\right) \cdot x_{n-2}\right\} \\
&< \exp\left\{b_1^{(p_\lambda^U, q_\lambda^U)}\right\} \sum_{x_1=0}^{\infty} \cdots \sum_{x_{n-2}=0}^{\infty} \prod_{k=1}^{n-2} Z_{n,k}^{(\lambda)}(\vec{x}) \cdot \exp\left\{b_2^{(p_\lambda^U, q_\lambda^U)} + a_2^{(q_\lambda^U)} x_{n-2}\right\} \\
&< \cdots < \exp\left\{a_n^{(q_\lambda^U)} x_0 + \sum_{k=1}^{n} b_k^{(p_\lambda^U, q_\lambda^U)}\right\}.
\end{aligned}
\tag{43}
$$

Notice that for the strictness of the above inequalities we have used the fact that $\phi_\lambda(x) < \phi_\lambda^U(x)$ for some (in fact, all but at most two) $x \in \mathbb{N}_0$ (cf. Properties 3(P19) below). Since for some admissible choices of p_λ^U, q_λ^U and some $n \in \mathbb{N}$ the last term in (43) can become larger than 1, one needs to take into account the cutoff-point 1 arising from (9). The lower bound $B_{\lambda,X_0,n}^L$ of part (b), as well as the exact value of part (a) follow from (29) in an analogous manner by employing p_λ^L, q_λ^L and p_λ^E, q_λ^E respectively. Furthermore, we use the fact that for $(\beta_{\mathcal{A}}, \beta_{\mathcal{H}}, \alpha_{\mathcal{A}}, \alpha_{\mathcal{H}}, \lambda) \in (\mathcal{P}_{\mathrm{NI}} \cup \mathcal{P}_{\mathrm{SP},1}) \times\,]0,1[$ one gets from (38) the relation $b_n^{(p_\lambda^E, q_\lambda^E)} = \frac{\alpha_{\mathcal{A}}}{\beta_{\mathcal{A}}} a_n^{(q_\lambda^E)}$. For the sake of brevity, the corresponding straightforward details are omitted here. Although we take the minimum of the upper bound derived in (43) and 1, the inequality $B_{\lambda,X_0,n}^L < B_{\lambda,X_0,n}^U$ is nevertheless valid: the reason is that for constituting a lower bound, the parameters p_λ^L, q_λ^L must fulfill either the conditions $[p_\lambda^L - \alpha_\lambda < 0$ and $q_\lambda^L - \beta_\lambda \leq 0]$ or $[p_\lambda^L - \alpha_\lambda \leq 0$ and $q_\lambda^L - \beta_\lambda < 0]$ (or both), which guarantees that $B_{\lambda,X_0,n}^L < 1$. The proof for all $\lambda \in \mathbb{R}\backslash[0,1]$ works out completely analogous, by taking into account the generally valid lower bound $H_\lambda(P_{A,n}||P_{\mathcal{H},n}) \geq 1$ (cf. (11)). $\quad\square$

3.2. Some Useful Facts for Deeper Analyses

Theorem 1(b) and Remark 1(a) indicate the crucial role of the expression $\widetilde{B}_{\lambda,X_0,n}^{(p,q)}$ and that the choice of the quantities p,q depends on the underlying (e.g., fixed) offspring-immigration parameter constellation $(\beta_A, \beta_{\mathcal{H}}, \alpha_A, \alpha_{\mathcal{H}})$ as well as on the (e.g., selectable) value of λ, i.e., $p_\lambda^A = p^A(\beta_A, \beta_{\mathcal{H}}, \alpha_A, \alpha_{\mathcal{H}}, \lambda)$ and $q_\lambda^A = q^A(\beta_A, \beta_{\mathcal{H}}, \alpha_A, \alpha_{\mathcal{H}}, \lambda)$ with $A \in \{E, L, U\}$. In order to study the desired time-behaviour $n \mapsto \widetilde{B}_{\lambda,X_0,n}^{(\cdot,\cdot)}$ of the Hellinger integral bounds resp. exact values, one therefore faces a six-dimensional (and thus highly non-obvious) detailed analysis, including the search for criteria (in addition to (35)) on good/optimal choices of $p_\lambda^L, q_\lambda^L, p_\lambda^U, q_\lambda^U$. Since these criteria will (almost) always imply the nonnegativity of p_λ^A, q_λ^A ($A \in \{L, U\}$) and $p_\lambda^E \geq 0$, $q_\lambda^E > 0$ (cf. Remark 1(a)), let us first present some fundamental properties of the underlying crucial sequences $\left(a_n^{(q)}\right)_{n\in\mathbb{N}}$ and $\left(b_n^{(p,q)}\right)_{n\in\mathbb{N}}$ for general $p \geq 0$, $q \geq 0$.

Properties 1. *For all $\lambda \in \mathbb{R}$ the following holds:*

(P1) *If $0 < q < \beta_\lambda$, then the sequence $\left(a_n^{(q)}\right)_{n\in\mathbb{N}}$ is strictly negative, strictly decreasing and converges to the unique negative solution $x_0^{(q)} \in\,]-\beta_\lambda, q - \beta_\lambda[$ of the equation*

$$\xi_\lambda^{(q)}(x) = q \cdot e^x - \beta_\lambda = x. \tag{44}$$

(P2) *If $0 < q = \beta_\lambda$, then $a_n^{(q)} \equiv 0$.*

(P3) *If $q > \max\{0, \beta_\lambda\}$, then the sequence $\left(a_n^{(q)}\right)_{n\in\mathbb{N}}$ is strictly positive and strictly increasing. Notice that in this setup, $q = 1$ implies $\min\{1, e^{\beta_\lambda - 1}\} = e^{\beta_\lambda - 1} < q$.*

    (P3a) *If additionally $q \leq \min\{1, e^{\beta_\lambda - 1}\}$, then the sequence $\left(a_n^{(q)}\right)_{n\in\mathbb{N}}$ converges to the smallest positive solution $x_0^{(q)} \in\,]0, -\log q]$ of the Equation (44).*

    (P3b) *If additionally $q > \min\{1, e^{\beta_\lambda - 1}\}$, then the sequence $\left(a_n^{(q)}\right)_{n\in\mathbb{N}}$ diverges to ∞, faster than exponentially (i.e., there do not exist constants $c_1, c_2 \in \mathbb{R}$ such that $a_n^{(q)} \leq e^{c_1 + c_2 n}$ for all $n \in \mathbb{N}$).*

(P4) *If $q = 0$, then one gets $a_n^{(0)} \equiv -\beta_\lambda$.*

    Due to the linear interrelation (38), these results directly carry over to the behaviour of the sequence $\left(b_n^{(p,q)}\right)_{n\in\mathbb{N}}$:

(P5) *If $p > 0$ and $0 < q < \beta_\lambda$, then the sequence $\left(b_n^{(p,q)}\right)_{n\in\mathbb{N}}$ is strictly decreasing and converges to $p \cdot e^{x_0^{(q)}} - \alpha_\lambda$. Trivially, $b_1^{(p,q)} = p - \alpha_\lambda$.*

    (P5a) *If additionally $p < \alpha_\lambda$, then $\left(b_n^{(p,q)}\right)_{n\in\mathbb{N}}$ is strictly negative for all $n \in \mathbb{N}$.*

    (P5b) *If additionally $p = \alpha_\lambda$, then $\left(b_n^{(p,q)}\right)_{n\in\mathbb{N}}$ is strictly negative for all $n \in \mathbb{N}\backslash\{1\}$.*

    (P5c) *If additionally $p > \alpha_\lambda$, then $\left(b_n^{(p,q)}\right)_{n\in\mathbb{N}}$ is strictly positive for some (and possibly for all) $n \in \mathbb{N}$.*

(P6) *If $0 < q = \beta_\lambda$, then $b_n^{(p,q)} \equiv p - \alpha_\lambda$.*

(P7) *If $p > 0$ and $q > \max\{0, \beta_\lambda\}$, then the sequence $\left(b_n^{(p,q)}\right)_{n\in\mathbb{N}}$ is strictly increasing.*

(P7a)   *If additionally $q \leq \min\{1, e^{\beta_\lambda - 1}\}$, then the sequence $\left(b_n^{(p,q)}\right)_{n\in\mathbb{N}}$ converges to $p \cdot e^{x_0^{(q)}} - \alpha_\lambda \in$
$]p - \alpha_\lambda, p/q - \alpha_\lambda]$; this limit can take any sign, depending on the parameter constellation.*

(P7b)   *If additionally $q > \min\{1, e^{\beta_\lambda - 1}\}$, then the sequence $\left(b_n^{(p,q)}\right)_{n\in\mathbb{N}}$ diverges to ∞, faster than exponentially.*

(P8)   *For the remaining cases we get: $b_n^{(0,q)} \equiv -\alpha_\lambda$ and $b_n^{(p,0)} \equiv p \cdot e^{-\beta_\lambda} - \alpha_\lambda$ ($p \in \mathbb{R}$, $q \in \mathbb{R}$).*

Moreover, in our investigations we will repeatedly make use of the function $\xi_\lambda^{(q)}(\cdot)$ from the definition (36) of $a_n^{(q)}$ (see also (44)), which has the following properties:

(P9)   *For $q \in]0, \infty[$ and all $\lambda \in \mathbb{R}\setminus\{0,1\}$ the function $\xi_\lambda^{(q)}(\cdot)$ is strictly increasing, strictly convex and smooth, and there holds*

(P9a)
$$\xi_\lambda^{(q)}(0) \begin{cases} < 0, & \text{if } q < \beta_\lambda, \\ = 0, & \text{if } q = \beta_\lambda, \\ > 0, & \text{if } q > \beta_\lambda. \end{cases}$$

(P9b)
$$\lim_{x \to -\infty} \xi_\lambda^{(q)}(x) = -\beta_\lambda, \qquad \text{and} \qquad \lim_{x \to \infty} \xi_\lambda^{(q)}(x) = \infty.$$

The proof of these properties is provided in Appendix A.1. From Properties 1 (P1) to (P4) we can see, that the behaviour of the sequence $\left(a_n^{(q)}\right)_{n\in\mathbb{N}}$ can be classified basically into four different types; besides the case (P2) where $a_n^{(q)}$ is *constant*, the sequence can be either (i) *strictly decreasing and convergent* (e.g., for the NI case $(\beta_\mathcal{A}, \beta_\mathcal{H}, \alpha_\mathcal{A}, \alpha_\mathcal{H}, \lambda) = (0.5, 2, 0, 0, 0.5)$ leading to $\beta_\lambda = \lambda\beta_\mathcal{A} + (1 - \lambda)\beta_\mathcal{H} = 1.25$ and to $q := q_\lambda^E = \beta_\mathcal{A}^\lambda \beta_\mathcal{H}^{1-\lambda} = 1$, cf. (33) resp. Theorem 1(a)), or (ii) *strictly increasing and convergent* (e.g., for $(\beta_\mathcal{A}, \beta_\mathcal{H}, \alpha_\mathcal{A}, \alpha_\mathcal{H}, \lambda) = (0.5, 2, 0, 0, 1.5)$ leading to $\beta_\lambda = -0.25$, $q := q_\lambda^E = 0.25$), or (iii) *strictly increasing and divergent* (e.g., for $(\beta_\mathcal{A}, \beta_\mathcal{H}, \alpha_\mathcal{A}, \alpha_\mathcal{H}, \lambda) = (0.5, 2, 0, 0, 2.7)$ leading to $\beta_\lambda = -2.05$, $q := q_\lambda^E \approx 0.047366$). Within our running-example epidemiological context of Section 2.3, this corresponds to a "potentially dangerous" infectious-disease-transmission situation $(\mathcal{H})$ (with supercritical reproduction number $\beta_\mathcal{H} = 2$), whereas $(\mathcal{A})$ describes a "mild" situation (with "low" subcritical $\beta_\mathcal{A} = 0.5$).

As already mentioned before, the sequences $\left(a_n^{(q)}\right)_{n\in\mathbb{N}}$ and $\left(b_n^{(p,q)}\right)_{n\in\mathbb{N}}$—whose behaviours for general $p \geq 0$ and $q \geq 0$ were described by the Properties 1—have to be evaluated at setup-dependent choices $p = p_\lambda = p(\beta_\mathcal{A}, \beta_\mathcal{H}, \alpha_\mathcal{A}, \alpha_\mathcal{H}, \lambda)$ and $q = q_\lambda = q(\beta_\mathcal{A}, \beta_\mathcal{H}, \alpha_\mathcal{A}, \alpha_\mathcal{H}, \lambda)$. Hence, for fixed $(\beta_\mathcal{A}, \beta_\mathcal{H}, \alpha_\mathcal{A}, \alpha_\mathcal{H})$, one of the questions—which arises in the course of the desired investigations of the time-behaviour of the Hellinger integral bounds (resp. exact values)—is for which $\lambda \in \mathbb{R}$ the sequence $\left(a_n^{(q_\lambda)}\right)_{n\in\mathbb{N}}$ converges. In the following, we illuminate this for the important special case $q_\lambda = \beta_\mathcal{A}^\lambda \beta_\mathcal{H}^{1-\lambda}$. Suppose at first that $\beta_\mathcal{A} \neq \beta_\mathcal{H}$. Properties 1 (P1) implies that for $\lambda \in]0, 1[$ one has $\lim_{n\to\infty} a_n^{(q_\lambda)} = x_0^{(q_\lambda)} \in]-\beta_\lambda, q_\lambda - \beta_\lambda[$, and Lemma A1 states that $q_\lambda - \beta_\lambda < 0$. For $\lambda \in \mathbb{R}\setminus[0,1]$, there holds $q_\lambda > \max\{0, \beta_\lambda\}$, and from (P3) one can see that $\left(a_n^{(q_\lambda)}\right)_{n\in\mathbb{N}}$ does not converge to $x_0^{(q_\lambda)}$ in general, but for $q_\lambda \leq \min\{1, e^{\beta_\lambda - 1}\}$ which constitutes an implicit condition on λ. This can be made explicit, with the help of the auxiliary variables

$$\lambda_- := \lambda_-(\beta_\mathcal{A}, \beta_\mathcal{H}) := \begin{cases} \inf\left\{\lambda \leq 0 : \beta_\mathcal{A}^\lambda \beta_\mathcal{H}^{1-\lambda} \leq \min\{1, \exp\{\lambda\beta_\mathcal{A} + (1-\lambda)\beta_\mathcal{H} - 1\}\}\right\}, \\ \qquad\qquad \text{in case that the set is nonempty,} \\ 0, \qquad\qquad \text{else,} \end{cases}$$

$$\lambda_+ := \lambda_+(\beta_\mathcal{A}, \beta_\mathcal{H}) := \begin{cases} \sup\left\{\lambda \geq 1 : \beta_\mathcal{A}^\lambda \beta_\mathcal{H}^{1-\lambda} \leq \min\{1, \exp\{\lambda\beta_\mathcal{A} + (1-\lambda)\beta_\mathcal{H} - 1\}\}\right\}, \\ \qquad\qquad \text{in case that the set is nonempty,} \\ 1, \qquad\qquad \text{else.} \end{cases}$$

For the constellation $\beta_{\mathcal{A}} = \beta_{\mathcal{H}} > 0$ we clearly obtain $q_{\lambda} = \beta_{\mathcal{A}}^{\lambda}\beta_{\mathcal{H}}^{1-\lambda} = \beta_{\mathcal{A}} = \beta_{\mathcal{H}} = \beta_{\lambda}$. Hence, (P2) implies that the sequence $\left(a_n^{(q_{\lambda})}\right)_{n\in\mathbb{N}}$ converges *for all* $\lambda \in \mathbb{R}\backslash\{0,1\}$ and we can set $\lambda_- := -\infty$ as well as $\lambda_+ := \infty$. Incorporating this and by adapting a result of Linkov & Lunyova [53] on $\lambda_-(v_1, v_2)$, $\lambda_+(v_1, v_2)$ for $\beta_{\mathcal{A}} \neq \beta_{\mathcal{H}}$, we end up with

Lemma 1. *(a) For all $\beta_{\mathcal{A}} > 0$, $\beta_{\mathcal{H}} > 0$ with $\beta_{\mathcal{A}} \neq \beta_{\mathcal{H}}$ there holds*

$$
\lambda_- = \lambda_-(\beta_{\mathcal{A}}, \beta_{\mathcal{H}}) = \begin{cases} 0, & \text{if } \beta_{\mathcal{H}} \geq 1, \\ \check{\lambda}, & \text{if } \beta_{\mathcal{H}} < 1 \text{ and } \beta_{\mathcal{A}} \notin [\beta_{\mathcal{H}}, \beta_{\mathcal{H}} \, z(\beta_{\mathcal{H}})], \\ -\infty, & \text{if } \beta_{\mathcal{H}} < 1 \text{ and } \beta_{\mathcal{A}} \in]\beta_{\mathcal{H}}, \beta_{\mathcal{H}} \, z(\beta_{\mathcal{H}})], \end{cases}
$$

$$
\lambda_+ = \lambda_+(\beta_{\mathcal{A}}, \beta_{\mathcal{H}}) = \begin{cases} 1, & \text{if } \beta_{\mathcal{A}} \geq 1, \\ \check{\lambda}, & \text{if } \beta_{\mathcal{A}} < 1 \text{ and } \beta_{\mathcal{H}} \notin [\beta_{\mathcal{A}}, \beta_{\mathcal{A}} \, z(\beta_{\mathcal{A}})], \\ \infty, & \text{if } \beta_{\mathcal{A}} < 1 \text{ and } \beta_{\mathcal{H}} \in]\beta_{\mathcal{A}}, \beta_{\mathcal{A}} \, z(\beta_{\mathcal{A}})], \end{cases}
$$

where

$$
\check{\lambda} := \check{\lambda}(\beta_{\mathcal{A}}, \beta_{\mathcal{H}}) := \frac{\beta_{\mathcal{H}} - 1 - \log(\beta_{\mathcal{H}})}{\beta_{\mathcal{H}} - \beta_{\mathcal{A}} + \log\left(\frac{\beta_{\mathcal{A}}}{\beta_{\mathcal{H}}}\right)} \begin{cases} < 0, & \text{if } \beta_{\mathcal{H}} < 1 \text{ and } \beta_{\mathcal{A}} \notin [\beta_{\mathcal{H}}, \beta_{\mathcal{H}} \, z(\beta_{\mathcal{H}})], \\ > 1, & \text{if } \beta_{\mathcal{A}} < 1 \text{ and } \beta_{\mathcal{H}} \notin [\beta_{\mathcal{A}}, \beta_{\mathcal{A}} \, z(\beta_{\mathcal{A}})]. \end{cases}
$$

Here, for fixed $\beta \in]0, \infty[\backslash\{1\}$ we denote by $z(\beta)$ the unique solution of the equation $\log(x) - \beta(x - 1) = 0$, $x \in]0, \infty[\backslash\{1\}$. For $\beta = 1$, $z(\beta) = 1$ denotes the unique solution of $\log(x) - (x - 1) = 0$, $x \in]0, \infty[$. (b) For all $\beta_{\mathcal{A}} = \beta_{\mathcal{H}} > 0$ one gets $\lambda_- = \lambda_-(\beta_{\mathcal{A}}, \beta_{\mathcal{H}}) = -\infty$ as well as $\lambda_+ = \lambda_+(\beta_{\mathcal{A}}, \beta_{\mathcal{H}}) = \infty$. Notice that the relationship $\check{\lambda}(\beta_{\mathcal{A}}, \beta_{\mathcal{H}}) = 1 - \check{\lambda}(\beta_{\mathcal{H}}, \beta_{\mathcal{A}})$ is consistent with the skew symmetry (8).

A corresponding proof is given in Appendix A.1.

With these auxiliary basic facts in hand, let us now work out our detailed investigations of the time-behaviour $n \mapsto H_{\lambda}(P_{\mathcal{A},n}||P_{\mathcal{H},n})$, where we start with the exactly treatable case (a) in Theorem 1.

3.3. Detailed Analyses of the Exact Recursive Values, i.e., for the Cases $(\beta_{\mathcal{A}}, \beta_{\mathcal{H}}, \alpha_{\mathcal{A}}, \alpha_{\mathcal{H}}) \in \mathcal{P}_{NI} \cup \mathcal{P}_{SP,1}$

In the no-immigration-case $(\beta_{\mathcal{A}}, \beta_{\mathcal{H}}, \alpha_{\mathcal{A}}, \alpha_{\mathcal{H}}) \in \mathcal{P}_{NI}$ and in the equal-fraction-case $(\beta_{\mathcal{A}}, \beta_{\mathcal{H}}, \alpha_{\mathcal{A}}, \alpha_{\mathcal{H}}) \in \mathcal{P}_{SP,1}$, the Hellinger integral can be calculated exactly in terms of $H_{\lambda}(P_{\mathcal{A},n}||P_{\mathcal{H},n}) = V_{\lambda, X_0, n}$ (cf. (39)), as proposed in part (a) of Theorem 1. This quantity depends on the behaviour of the sequence $\left(a_n^{(q_{\lambda}^E)}\right)_{n\in\mathbb{N}}$, with $q_{\lambda}^E := \beta_{\mathcal{A}}^{\lambda}\beta_{\mathcal{H}}^{1-\lambda} > 0$, and of the sum $\left(\frac{\alpha_{\mathcal{A}}}{\beta_{\mathcal{A}}}\sum_{k=1}^{n} a_k^{(q_{\lambda}^E)}\right)_{n\in\mathbb{N}}$. The last expression is equal to zero on $\mathcal{P}_{NI}$. On $\mathcal{P}_{SP,1}$, this sum is unequal to zero. Using Lemma A1 we conclude that $q_{\lambda}^E < \beta_{\lambda}$ (resp. $q_{\lambda}^E > \beta_{\lambda}$) iff $\lambda \in]0, 1[$ (resp. $\lambda \in \mathbb{R}\backslash[0, 1]$), since on $\mathcal{P}_{NI} \cup \mathcal{P}_{SP,1}$ there holds $\beta_{\mathcal{A}} \neq \beta_{\mathcal{H}}$. Thus, from Properties 1 (P1) we can see that the sequence $\left(a_n^{(q_{\lambda}^E)}\right)_{n\in\mathbb{N}}$ is strictly negative, strictly decreasing and it converges to the unique solution $x_0^{(q_{\lambda}^E)} \in]-\beta_{\lambda}, q_{\lambda}^E - \beta_{\lambda}[$ of the Equation (44) if $\lambda \in]0, 1[$. For $\lambda \in \mathbb{R}\backslash[0, 1]$, (P3) implies that the sequence $\left(a_n^{(q_{\lambda}^E)}\right)_{n\in\mathbb{N}}$ is strictly positive, strictly increasing and converges to the smallest positive solution $x_0^{(q_{\lambda}^E)} \in]0, -\log(q_{\lambda}^E)]$ of the Equation (44) in case that (P3a) is satisfied, otherwise it diverges to ∞. Thus, we have shown the following detailed behaviour of Hellinger integrals:

Proposition 2. *For all* $(\beta_A, \beta_H, \alpha_A, \alpha_H, \lambda) \in \mathcal{P}_{NI} \times\,]0,1[$ *and all initial population sizes* $X_0 \in \mathbb{N}$ *there holds*

(a) $\quad H_\lambda(P_{A,1}||P_{H,1}) = \exp\left\{ \left(\beta_A^\lambda \beta_H^{1-\lambda} - \lambda\beta_A - (1-\lambda)\beta_H \right) X_0 \right\} < 1,$

(b) $\quad$ *the sequence* $(H_\lambda(P_{A,n}||P_{H,n}))_{n\in\mathbb{N}}$ *given by*

$$H_\lambda(P_{A,n}||P_{H,n}) = \exp\left\{ a_n^{(q_\lambda^E)} X_0 \right\} =: V_{\lambda,X_0,n}$$

$\quad$ *is strictly decreasing,*

(c) $\quad \lim\limits_{n\to\infty} H_\lambda(P_{A,n}||P_{H,n}) = \exp\left\{ x_0^{(q_\lambda^E)} X_0 \right\} \in\,]0,1[,$

(d) $\quad \lim\limits_{n\to\infty} \dfrac{1}{n} \log H_\lambda(P_{A,n}||P_{H,n}) = 0$

(e) $\quad$ *the map* $X_0 \mapsto V_{\lambda,X_0,n}$ *is strictly decreasing.*

Proposition 3. *For all* $(\beta_A, \beta_H, \alpha_A, \alpha_H, \lambda) \in \mathcal{P}_{NI} \times (\mathbb{R}\backslash[0,1])$ *and all initial population sizes* $X_0 \in \mathbb{N}$ *there holds with* $q_\lambda^E := \beta_A^\lambda \beta_H^{1-\lambda}$

(a) $\quad H_\lambda(P_{A,1}||P_{H,1}) = \exp\left\{ \left(\beta_A^\lambda \beta_H^{1-\lambda} - \beta_\lambda \right) \cdot X_0 \right\} > 1,$

(b) $\quad$ *the sequence* $(H_\lambda(P_{A,n}||P_{H,n}))_{n\in\mathbb{N}}$ *given by*

$$H_\lambda(P_{A,n}||P_{H,n}) = \exp\left\{ a_n^{(q_\lambda^E)} \cdot X_0 \right\} =: V_{\lambda,X_0,n}$$

$\quad$ *is strictly increasing,*

(c) $\quad \lim\limits_{n\to\infty} H_\lambda(P_{A,n}||P_{H,n}) = \begin{cases} \exp\left\{ x_0^{(q_\lambda^E)} \cdot X_0 \right\} > 1, & \text{if } \lambda \in [\lambda_-, \lambda_+] \backslash [0,1], \\ \infty, & \text{if } \lambda \in\,]-\infty, \lambda_-[\, \cup\,]\lambda_+, \infty[, \end{cases}$

(d) $\quad \lim\limits_{n\to\infty} \dfrac{1}{n} \log H_\lambda(P_{A,n}||P_{H,n}) = \begin{cases} 0, & \text{if } \lambda \in [\lambda_-, \lambda_+] \backslash [0,1], \\ \infty, & \text{if } \lambda \in\,]-\infty, \lambda_-[\, \cup\,]\lambda_+, \infty[, \end{cases}$

(e) $\quad$ *the map* $X_0 \mapsto V_{\lambda,X_0,n}$ *is strictly increasing.*

In the case $(\beta_A, \beta_H, \alpha_A, \alpha_H) \in \mathcal{P}_{SP,1}$, the sequence $\left(a_n^{(q_\lambda^E)} \right)_{n\in\mathbb{N}}$ under consideration is formally the same, with the parameter $q_\lambda^E := \beta_A^\lambda \beta_H^{1-\lambda} > 0$. However, in contrast to the case $\mathcal{P}_{NI}$, on $\mathcal{P}_{SP,1}$ both the sequence $\left(a_n^{(q_\lambda^E)} \right)_{n\in\mathbb{N}}$ and the sum $\left(\frac{\alpha_A}{\beta_A} \sum_{k=1}^n a_k^{(q_\lambda^E)} \right)_{n\in\mathbb{N}}$ are strictly decreasing in case that $\lambda \in\,]0,1[$, and strictly increasing in case that $\lambda \in \mathbb{R}\backslash[0,1]$. The respective convergence behaviours are given in Properties 1 (P1) and (P3). We thus obtain

Proposition 4. *For all* $(\beta_A, \beta_H, \alpha_A, \alpha_H, \lambda) \in \mathcal{P}_{SP,1} \times\,]0,1[$ *and all initial population sizes* $X_0 \in \mathbb{N}$ *there holds with* $q_\lambda^E := \beta_A^\lambda \beta_H^{1-\lambda}$

(a) $\quad H_\lambda(P_{A,1}||P_{H,1}) = \exp\left\{ \left(\beta_A^\lambda \beta_H^{1-\lambda} - \beta_\lambda \right) \cdot \left(X_0 + \dfrac{\alpha_A}{\beta_A} \right) \right\} < 1,$

(b) $\quad$ *the sequence* $(H_\lambda(P_{A,n}||P_{H,n}))_{n\in\mathbb{N}}$ *given by*

$$H_\lambda(P_{A,n}||P_{H,n}) = \exp\left\{ a_n^{(q_\lambda^E)} \cdot X_0 + \dfrac{\alpha_A}{\beta_A} \sum_{k=1}^n a_k^{(q_\lambda^E)} \right\} =: V_{\lambda,X_0,n}$$

$\quad$ *is strictly decreasing,*

(c) $\quad \lim\limits_{n\to\infty} H_\lambda(P_{A,n}||P_{H,n}) = 0,$

(d) $\quad \lim\limits_{n\to\infty} \dfrac{1}{n} \log H_\lambda(P_{A,n}||P_{H,n}) = \dfrac{\alpha_A}{\beta_A} \cdot x_0^{(q_\lambda^E)} < 0,$

(e) $\quad$ *the map* $X_0 \mapsto V_{\lambda,X_0,n}$ *is strictly decreasing.*

Proposition 5. *For all* $(\beta_{\mathcal{A}}, \beta_{\mathcal{H}}, \alpha_{\mathcal{A}}, \alpha_{\mathcal{H}}, \lambda) \in \mathcal{P}_{SP,1} \times (\mathbb{R} \backslash [0,1])$ *and all initial population sizes* $X_0 \in \mathbb{N}$ *there holds with* $q_{\lambda}^{E} := \beta_{\mathcal{A}}^{\lambda} \beta_{\mathcal{H}}^{1-\lambda}$

$(a) \qquad H_{\lambda}(P_{\mathcal{A},1} || P_{\mathcal{H},1}) \; = \; \exp\left\{ \left(\beta_{\mathcal{A}}^{\lambda} \beta_{\mathcal{H}}^{1-\lambda} - \beta_{\lambda} \right) \cdot \left(X_0 + \frac{\alpha_{\mathcal{A}}}{\beta_{\mathcal{A}}} \right) \right\} > 1,$

$(b) \qquad$ *the sequence* $\left(H_{\lambda}(P_{\mathcal{A},n} || P_{\mathcal{H},n}) \right)_{n \in \mathbb{N}}$ *given by*

$$H_{\lambda}(P_{\mathcal{A},n} || P_{\mathcal{H},n}) \; = \; \exp\left\{ a_n^{(q_{\lambda}^{E})} \cdot X_0 + \frac{\alpha_{\mathcal{A}}}{\beta_{\mathcal{A}}} \sum_{k=1}^{n} a_k^{(q_{\lambda}^{E})} \right\} \; =: \; V_{\lambda, X_0, n}$$

$\qquad$ *is strictly increasing,*

$(c) \qquad \lim\limits_{n \to \infty} H_{\lambda}(P_{\mathcal{A},n} || P_{\mathcal{H},n}) \; = \; \infty,$

$(d) \qquad \lim\limits_{n \to \infty} \dfrac{1}{n} \log H_{\lambda}(P_{\mathcal{A},n} || P_{\mathcal{H},n}) \; = \; \begin{cases} \frac{\alpha_{\mathcal{A}}}{\beta_{\mathcal{A}}} \cdot x_0^{(q_{\lambda}^{E})} > 0, & \text{if } \; \lambda \in [\lambda_-, \lambda_+] \backslash [0,1], \\ \infty, & \text{if } \lambda \in \,] - \infty, \lambda_-[\, \cup \,]\lambda_+, \infty[, \end{cases}$

$(e) \qquad$ *the map* $\quad X_0 \mapsto V_{\lambda, X_0, n} \quad$ *is strictly increasing.*

Due to the nature of the equal-fraction-case $\mathcal{P}_{SP,1}$, in the assertions (a), (b), (d) of the Propositions 4 and 5, the fraction $\alpha_{\mathcal{A}}/\beta_{\mathcal{A}}$ can be equivalently replaced by $\alpha_{\mathcal{H}}/\beta_{\mathcal{H}}$.

Remark 2. *For the (to our context) incompatible setup of GWI with Poisson offspring but nonstochastic immigration of constant value 1, an "analogue" of part (d) of the Propositions 4 resp. 5 was established in Linkov & Lunyova* [53].

3.4. Some Preparatory Basic Facts for the Remaining Cases $(\beta_{\mathcal{A}}, \beta_{\mathcal{H}}, \alpha_{\mathcal{A}}, \alpha_{\mathcal{H}}) \in \mathcal{P}_{SP} \backslash \mathcal{P}_{SP,1}$

The bounds $B_{\lambda, X_0, n}^{L}$, $B_{\lambda, X_0, n}^{U}$ for the Hellinger integral introduced in formula (40) in Theorem 1 can be chosen arbitrarily from a $(p_{\lambda}^{L}, q_{\lambda}^{L}, p_{\lambda}^{U}, q_{\lambda}^{U})$-indexed set of context-specific parameters satisfying (34), or equivalently (35).

In order to derive bounds which are optimal, with respect to goals that will be discussed later, the following monotonicity properties of the sequences $\left(a_n^{(q)} \right)_{n \in \mathbb{N}}$ and $\left(b_n^{(p,q)} \right)_{n \in \mathbb{N}}$ (cf. (36), (37)) for general, context-independent parameters q and p, will turn out to be very useful:

Properties 2.

(P10) *For* $0 \leq q_1 < q_2 < \infty$ *there holds* $a_n^{(q_1)} < a_n^{(q_2)}$ *for all* $n \in \mathbb{N}$.

(P11) *For each fixed* $q \geq 0$ *and* $0 \leq p_1 < p_2 < \infty$ *there holds* $b_n^{(p_1, q)} < b_n^{(p_2, q)}$, *for all* $n \in \mathbb{N}$.

(P12) *For fixed* $p > 0$ *and* $0 \leq q_1 < q_2$ *it follows* $b_n^{(p, q_1)} < b_n^{(p, q_2)}$ *for all* $n \in \mathbb{N}$.

(P13) *Suppose that* $0 \leq p_1 < p_2$ *and* $0 \leq q_2 < q_1$. *For fixed* $n \in \mathbb{N}$, *no dominance assertion can be conjectured for* $b_n^{(p_1, q_1)}, b_n^{(p_2, q_2)}$. *As an example, consider the setup* $(\beta_{\mathcal{A}}, \beta_{\mathcal{H}}, \alpha_{\mathcal{A}}, \alpha_{\mathcal{H}}, \lambda) = (0.4, 0.8, 5, 3, 0.5)$; *within our running-example epidemiological context of Section 2.3, this corresponds to a "nearly dangerous" infectious-disease-transmission situation* $(\mathcal{H})$ *(with nearly critical reproduction number* $\beta_{\mathcal{H}} = 0.8$ *and importation mean of* $\alpha_{\mathcal{H}} = 3$), *whereas* $(\mathcal{A})$ *describes a "mild" situation (with "low" subcritical* $\beta_{\mathcal{A}} = 0.4$ *and* $\alpha_{\mathcal{A}} = 5$). *On the nonnegative real line, the function* $\phi_{\lambda}(x)$ *can be bounded from above by the linear functions* $\phi_{\lambda}^{U,1}(x) := p_1 + q_1 x := 4.040 + 0.593 \cdot x$ *as well as by* $\phi_{\lambda}^{U,2}(x) := p_2 + q_2 x := 4.110 + 0.584 \cdot x$. *Clearly,* $p_1 < p_2$ *and* $q_1 > q_2$. *Let us show the first eight elements and the respective limits of the corresponding sequences* $b_n^{(p_1, q_1)}, b_n^{(p_2, q_2)}$:

n	1	2	3	4	5	6	7	8	$\cdots$	∞
$b_n^{(p_1, q_1)}$	0.040	0.011	-0.005	-0.015	-0.021	-0.024	-0.026	-0.028	$\cdots$	-0.029
$b_n^{(p_2, q_2)}$	0.110	0.045	0.007	-0.014	-0.026	-0.033	-0.036	-0.039	$\cdots$	-0.041

(P14)   *For arbitrary $0 < p_1, p_2$ and $0 \le q_1, q_2 \le \min\{1, e^{\beta_\lambda - 1}\}$ suppose that $\log(p_1) + x_0^{(q_1)} < \log(p_2) + x_0^{(q_2)}$. Then there holds*

$$p_1 \cdot e^{x_0^{(q_1)}} - \alpha_\lambda \; = \; \lim_{n \to \infty} \frac{1}{n} \sum_{k=1}^{n} b_k^{(p_1, q_1)} \; < \; \lim_{n \to \infty} \frac{1}{n} \sum_{k=1}^{n} b_k^{(p_2, q_2)} \; = \; p_2 \cdot e^{x_0^{(q_2)}} - \alpha_\lambda \,.$$

From (P10) to (P12) one deduces that both sequences $\left(a_n^{(q)} \right)_{n \in \mathbb{N}}$ and $\left(b_n^{(p,q)} \right)_{n \in \mathbb{N}}$ are monotone in the general parameters $p, q \ge 0$. Thus, for the upper bound of the Hellinger integral $B_{\lambda, X_0, n}^{U}$ we should use nonnegative context-specific parameters $p_\lambda^{U} = p^{U}(\beta_\mathcal{A}, \beta_\mathcal{H}, \alpha_\mathcal{A}, \alpha_\mathcal{H}, \lambda)$ and $q_\lambda^{U} = q^{U}(\beta_\mathcal{A}, \beta_\mathcal{H}, \alpha_\mathcal{A}, \alpha_\mathcal{H}, \lambda)$ which are as small as possible, and for the lower bound $B_{\lambda, X_0, n}^{L}$ we should use nonnegative context-specific parameters $p_\lambda^{L} = p^{L}(\beta_\mathcal{A}, \beta_\mathcal{H}, \alpha_\mathcal{A}, \alpha_\mathcal{H}, \lambda)$ and $q_\lambda^{L} = q^{L}(\beta_\mathcal{A}, \beta_\mathcal{H}, \alpha_\mathcal{A}, \alpha_\mathcal{H}, \lambda)$ which are as large as possible, of course, subject to the (equivalent) restrictions (34) and (35).

To find "optimal" parameter pairs, we have to study the following properties of the function $\phi_\lambda(\cdot) = \phi(\cdot, \beta_\mathcal{A}, \beta_\mathcal{H}, \alpha_\mathcal{A}, \alpha_\mathcal{H}, \lambda)$ defined on $[0, \infty[$ in (30) (which are also valid for the previous parameter context $(\beta_\mathcal{A}, \beta_\mathcal{H}, \alpha_\mathcal{A}, \alpha_\mathcal{H}) \in (\mathcal{P}_{\mathrm{NI}} \cup \mathcal{P}_{\mathrm{SP,1}})$):

Properties 3.

(P15)   *One has*

$$\phi_\lambda(x) = (\alpha_\mathcal{A} + \beta_\mathcal{A} x)^\lambda (\alpha_\mathcal{H} + \beta_\mathcal{H} x)^{1-\lambda} - \lambda(\alpha_\mathcal{A} + \beta_\mathcal{A} x) + (1-\lambda)(\alpha_\mathcal{H} + \beta_\mathcal{H} x) \begin{cases} \le 0, & \textit{if } \lambda \in]0, 1[, \\ \ge 0, & \textit{if } \lambda \in \mathbb{R} \backslash [0, 1], \end{cases}$$

where equality holds iff $f_\mathcal{A}(x) = f_\mathcal{H}(x)$ for some $x \in [0, \infty[$ iff $x = x^* := \frac{\alpha_\mathcal{A} - \alpha_\mathcal{H}}{\beta_\mathcal{H} - \beta_\mathcal{A}} \in [0, \infty[$.

(P16)   *There holds*

$$\phi_\lambda(0) = \alpha_\mathcal{A}^\lambda \alpha_\mathcal{H}^{1-\lambda} - \alpha_\lambda \begin{cases} \le 0, & \textit{if } \lambda \in]0, 1[, \\ \ge 0, & \textit{if } \lambda \in \mathbb{R} \backslash [0, 1], \end{cases}$$

with equality iff $\alpha_\mathcal{A} = \alpha_\mathcal{H}$ together with $\beta_\mathcal{A} \ne \beta_\mathcal{H}$ (cf. Lemma A1).

(P17)   *For all $\lambda \in \mathbb{R} \backslash \{0, 1\}$ one gets*

$$\phi_\lambda'(x) \; = \; \lambda \beta_\mathcal{A} \left(f_\mathcal{A}(x) \right)^{\lambda - 1} \left(f_\mathcal{H}(x) \right)^{1-\lambda} + (1-\lambda) \beta_\mathcal{H} \left(f_\mathcal{A}(x) \right)^{\lambda} \left(f_\mathcal{H}(x) \right)^{-\lambda} - \beta_\lambda \,.$$

(P18)   *There holds*

$$\lim_{x \to \infty} \phi_\lambda'(x) = \beta_\mathcal{A}^\lambda \beta_\mathcal{H}^{1-\lambda} - \beta_\lambda \begin{cases} \le 0, & \textit{if } \lambda \in]0, 1[, \\ \ge 0, & \textit{if } \lambda \in \mathbb{R} \backslash [0, 1], \end{cases}$$

with equality iff $\beta_\mathcal{A} = \beta_\mathcal{H}$ together with $\alpha_\mathcal{A} \ne \alpha_\mathcal{H}$ (cf. Lemma A1).

(P19)   *There holds*

$$\phi_\lambda''(x) \; = \; -\lambda(1-\lambda) \left(f_\mathcal{A}(x) \right)^{\lambda - 2} \left(f_\mathcal{H}(x) \right)^{-\lambda - 1} \left(\alpha_\mathcal{A} \beta_\mathcal{H} - \alpha_\mathcal{H} \beta_\mathcal{A} \right)^2 \begin{cases} \le 0, & \textit{if } \lambda \in]0, 1[, \\ \ge 0, & \textit{if } \lambda \in \mathbb{R} \backslash [0, 1], \end{cases}$$

with equality iff $(\beta_\mathcal{A}, \beta_\mathcal{H}, \alpha_\mathcal{A}, \alpha_\mathcal{H}) \in (\mathcal{P}_{\mathrm{NI}} \cup \mathcal{P}_{\mathrm{SP,1}})$. Hence, for $(\beta_\mathcal{A}, \beta_\mathcal{H}, \alpha_\mathcal{A}, \alpha_\mathcal{H}) \in \mathcal{P}_{\mathrm{SP}} \backslash \mathcal{P}_{\mathrm{SP,1}}$, the function ϕ_λ is strictly concave (convex) for $\lambda \in]0, 1[$ ($\lambda \in \mathbb{R} \backslash [0, 1]$). Notice that $\phi_\lambda'(0) = \lambda \beta_\mathcal{A} \left(\frac{\alpha_\mathcal{A}}{\alpha_\mathcal{H}} \right)^{\lambda - 1} + (1-\lambda) \beta_\mathcal{H} \left(\frac{\alpha_\mathcal{A}}{\alpha_\mathcal{H}} \right)^{\lambda} - \beta_\lambda$ can be either negative (e.g., for the setup $(\beta_\mathcal{A}, \beta_\mathcal{H}, \alpha_\mathcal{A}, \alpha_\mathcal{H}, \lambda) \in \{(4, 2, 3, 1, 0.5), (4, 2, 5, 1, 2)\}$, or zero (e.g., for $(\beta_\mathcal{A}, \beta_\mathcal{H}, \alpha_\mathcal{A}, \alpha_\mathcal{H}, \lambda) \in \{(4, 2, 4, 1, 0.5), (4, 2, 3, 1, 2)\}$), or positive (e.g., for $(\beta_\mathcal{A}, \beta_\mathcal{H}, \alpha_\mathcal{A}, \alpha_\mathcal{H}, \lambda) \in \{(4, 2, 5, 1, 0.5), (4, 2, 2, 1, 2)\}$), where the exemplary parameter constellations have concrete interpretations in our running-example epidemiological context of Section 2.3. Accordingly, for $\lambda \in]0,1[$, due to concavity and (P17), the function $\phi_\lambda(\cdot)$ can be either strictly decreasing, or can obtain its global maximum in $]0,\infty[$, or—only in the case $\beta_\mathcal{A} = \beta_\mathcal{H}$—can be strictly increasing. Analogously, for $\lambda \in \mathbb{R}\backslash[0,1]$, the function $\phi_\lambda(\cdot)$ can be either strictly increasing, or can obtain its global minimum in $]0,\infty[$, or—only in the case $\beta_\mathcal{A} = \beta_\mathcal{H}$—can be strictly decreasing.

(P20) For all $\lambda \in \mathbb{R}\backslash\{0,1\}$ one has

$$\lim_{x\to\infty}\left(\phi_\lambda(x) - (\widetilde{r_\lambda} + \widetilde{s_\lambda}\, x)\right) = 0,$$

$$\text{for}\quad \widetilde{r_\lambda} := \widetilde{p_\lambda} - \alpha_\lambda := \lambda\alpha_\mathcal{A}\left(\frac{\beta_\mathcal{A}}{\beta_\mathcal{H}}\right)^{\lambda-1} + (1-\lambda)\alpha_\mathcal{H}\left(\frac{\beta_\mathcal{A}}{\beta_\mathcal{H}}\right)^{\lambda} - \alpha_\lambda$$

$$\text{and}\quad \widetilde{s_\lambda} := \widetilde{q_\lambda} - \beta_\lambda := \beta_\mathcal{A}^\lambda\beta_\mathcal{H}^{1-\lambda} - \beta_\lambda.$$

The linear function $\widetilde{\phi_\lambda}(x) := \widetilde{r_\lambda} + \widetilde{s_\lambda} \cdot x$ constitutes the asymptote of $\phi_\lambda(\cdot)$. Notice that if $\beta_\mathcal{A} = \beta_\mathcal{H}$ one has $\widetilde{s_\lambda} = 0 = \widetilde{r_\lambda}$; if $\beta_\mathcal{A} \neq \beta_\mathcal{H}$ we have $\widetilde{s_\lambda} < 0$ in the case $\lambda \in]0,1[$ and $\widetilde{s_\lambda} > 0$ if $\lambda \in \mathbb{R}\backslash[0,1]$. Furthermore, $\phi_\lambda(0) < \widetilde{r_\lambda}$ if $\lambda \in]0,1[$ and $\phi_\lambda(0) > \widetilde{r_\lambda}$ if $\lambda \in \mathbb{R}\backslash[0,1]$, (cf. Lemma A1(c1) and (c2)). If $\alpha_\mathcal{A} = \alpha_\mathcal{H}$ (and thus $\beta_\mathcal{A} \neq \beta_\mathcal{H}$), then the intercept $\widetilde{r_\lambda}$ is strictly positive if $\lambda \in]0,1[$ resp. strictly negative if $\lambda \in \mathbb{R}\backslash[0,1]$. In contrast, for the case $\alpha_\mathcal{A} \neq \alpha_\mathcal{H}$, the intercept $\widetilde{r_\lambda}$ can assume any sign, take e.g., $(\beta_\mathcal{A}, \beta_\mathcal{H}, \alpha_\mathcal{A}, \alpha_\mathcal{H}, \lambda) \in \{(3.7, 0.9, 2.0, 1.0, 0.5), (4, 2, 1.6, 1, 2)\}$ for $\widetilde{r_\lambda} > 0$, $(\beta_\mathcal{A}, \beta_\mathcal{H}, \alpha_\mathcal{A}, \alpha_\mathcal{H}, \lambda) \in \{(3.6, 0.9, 2.0, 1.0, 0.5), (4, 2, 1.5, 1, 2)\}$ for $\widetilde{r_\lambda} = 0$, and $(\beta_\mathcal{A}, \beta_\mathcal{H}, \alpha_\mathcal{A}, \alpha_\mathcal{H}, \lambda) \in \{(3.5, 0.9, 2.0, 1.0, 0.5), (4, 2, 1.4, 1, 2)\}$ for $\widetilde{r_\lambda} < 0$; again, the exemplary parameter constellations have concrete interpretations in our running-example epidemiological context of Section 2.3.

The properties (P15) to (P20) above describe in detail the characteristics of the function $\phi_\lambda(\cdot) = \phi(\cdot, \beta_\mathcal{A}, \beta_\mathcal{H}, \alpha_\mathcal{A}, \alpha_\mathcal{H}, \lambda)$. In the previous parameter setup $\mathcal{P}_\mathrm{NI} \cup \mathcal{P}_\mathrm{SP,1}$, this function is linear, which can be seen from (P19). In the current parameter setup $\mathcal{P}_\mathrm{SP}\backslash\mathcal{P}_\mathrm{SP,1}$, this function can basically be classified into four different types. From (P16) to (P20) it is easy to see that for all current parameter constellations the particular choices

$$p_\lambda^A := \alpha_\mathcal{A}^\lambda\alpha_\mathcal{H}^{1-\lambda} > 0, \qquad q_\lambda^A := \beta_\mathcal{A}^\lambda\beta_\mathcal{H}^{1-\lambda} > 0, \tag{45}$$

which correspond to the following choices in (35)

$$r_\lambda^A := \alpha_\mathcal{A}^\lambda\alpha_\mathcal{H}^{1-\lambda} - \alpha_\lambda \leq 0 \quad (\text{resp.} \geq 0), \qquad s_\lambda^A := \beta_\mathcal{A}^\lambda\beta_\mathcal{H}^{1-\lambda} - \beta_\lambda \leq 0 \quad (\text{resp.} \geq 0),$$

– where $A = L$ (resp. $A = U$)–lead to the tightest lower bound $B_{\lambda,X_0,n}^L$ (resp. upper bound $B_{\lambda,X_0,n}^U$) for $H_\lambda(P_{\mathcal{A},n}||P_{\mathcal{H},n})$ in (40) in the case $\lambda \in]0,1[$ (resp. $\lambda \in \mathbb{R}\backslash[0,1])$. Notice that for the previous parameter setup $(\beta_\mathcal{A}, \beta_\mathcal{H}, \alpha_\mathcal{A}, \alpha_\mathcal{H}) \in (\mathcal{P}_\mathrm{NI} \cup \mathcal{P}_\mathrm{SP,1})$ these choices led to the exact values of the Hellinger integral and to the simplification $(p_\lambda^E/q_\lambda^E) \cdot \beta_\lambda - \alpha_\lambda = 0$, which implies $b_n^{(p_\lambda^E, q_\lambda^E)} = (\alpha_\mathcal{A}/\beta_\mathcal{A}) \cdot a_n^{(q_\lambda^E)}$. In contrast, in the current parameter setup $(\beta_\mathcal{A}, \beta_\mathcal{H}, \alpha_\mathcal{A}, \alpha_\mathcal{H}) \in \mathcal{P}_\mathrm{SP}\backslash\mathcal{P}_\mathrm{SP,1}$ we only derive the *optimal* lower (resp. upper) bound for $\lambda \in]0,1[$ (resp. $\lambda \in \mathbb{R}\backslash[0,1])$ by using the parameters p_λ^A, q_λ^A for $A = L$ (resp. $A = U$) and $(p_\lambda^A/q_\lambda^A) \cdot \beta_\lambda - \alpha_\lambda \neq 0$. For a better distinguishability and easier reference we thus stick to the $L-$notation (resp. $U-$notation) here.

3.5. Lower Bounds for the Cases $(\beta_\mathcal{A}, \beta_\mathcal{H}, \alpha_\mathcal{A}, \alpha_\mathcal{H}, \lambda) \in (\mathcal{P}_\mathrm{SP}\backslash\mathcal{P}_\mathrm{SP,1})\times]0,1[$

The discussion above implies that the lower bound $B_{\lambda,X_0,n}^L$ for the Hellinger integral $H_\lambda(P_{\mathcal{A},n}||P_{\mathcal{H},n})$ in (40) is optimal for the choices p_λ^L, $q_\lambda^L > 0$ defined in (45). If $\beta_\mathcal{A} \neq \beta_\mathcal{H}$, due to Properties 1 (P1) and Lemma A1, the sequence $\left(a_n^{(q_\lambda^L)}\right)_{n\in\mathbb{N}}$ is strictly negative and strictly decreasing and converges to the unique negative solution of the Equation (44). Furthermore, due to (P5), the sequence $\left(b_n^{(p_\lambda^L, q_\lambda^L)} \right)_{n \in \mathbb{N}}$, as defined in (37), is strictly decreasing. Since $b_1^{(p_\lambda^L, q_\lambda^L)} = p_\lambda^L - \alpha_\lambda \leq 0$ by Lemma A1, with equality iff $\alpha_\mathcal{A} = \alpha_\mathcal{H}$, the sequence $\left(b_n^{(p_\lambda^L, q_\lambda^L)} \right)_{n \in \mathbb{N}}$ is also strictly negative (with the exception $b_1^{(p_\lambda^L, q_\lambda^L)} = 0$ for $\alpha_\mathcal{A} = \alpha_\mathcal{H}$) and strictly decreasing. If $\beta_\mathcal{A} = \beta_\mathcal{H}$ and thus $\alpha_\mathcal{A} \neq \alpha_\mathcal{H}$, due to (P2), (P6) and Lemma A1, there holds $a_n^{(q_\lambda^L)} \equiv 0$ and $b_n^{(q_\lambda^L)} \equiv p_\lambda^L - \alpha_\lambda < 0$. Thus, analogously to the cases $\mathcal{P}_{\mathrm{NI}} \cup \mathcal{P}_{\mathrm{SP,1}}$ we obtain

Proposition 6. *For all* $(\beta_\mathcal{A}, \beta_\mathcal{H}, \alpha_\mathcal{A}, \alpha_\mathcal{H}, \lambda) \in (\mathcal{P}_{SP} \backslash \mathcal{P}_{SP,1}) \times]0,1[$ *and all initial population sizes* $X_0 \in \mathbb{N}$ *there holds with* $p_\lambda^L := \alpha_\mathcal{A}^\lambda \alpha_\mathcal{H}^{1-\lambda}$, $q_\lambda^L := \beta_\mathcal{A}^\lambda \beta_\mathcal{H}^{1-\lambda}$

$$(a) \qquad B_{\lambda, X_0, 1}^L = \exp\left\{ \left(\beta_\mathcal{A}^\lambda \beta_\mathcal{H}^{1-\lambda} - \beta_\lambda \right) \cdot X_0 + \alpha_\mathcal{A}^\lambda \alpha_\mathcal{H}^{1-\lambda} - \alpha_\lambda \right\} < 1,$$

(b) *the sequence of lower bounds* $\left(B_{\lambda, X_0, n}^L \right)_{n \in \mathbb{N}}$ *for* $H_\lambda(P_{\mathcal{A},n} || P_{\mathcal{H},n})$ *given by*

$$B_{\lambda, X_0, n}^L = \exp\left\{ a_n^{(q_\lambda^L)} \cdot X_0 + \frac{p_\lambda^L}{q_\lambda^L} \sum_{k=1}^n a_k^{(q_\lambda^L)} + n \cdot \left(\frac{p_\lambda^L}{q_\lambda^L} \cdot \beta_\lambda - \alpha_\lambda \right) \right\} \quad \text{is strictly decreasing,}$$

$$(c) \qquad \lim_{n \to \infty} B_{\lambda, X_0, n}^L = 0,$$

$$(d) \qquad \lim_{n \to \infty} \frac{1}{n} \log B_{\lambda, X_0, n}^L = \frac{p_\lambda^L}{q_\lambda^L} \cdot \left(x_0^{(q_\lambda^L)} + \beta_\lambda \right) - \alpha_\lambda = p_\lambda^L \cdot e^{x_0^{(q_\lambda^L)}} - \alpha_\lambda < 0.$$

$(e) \qquad$ *the map* $X_0 \mapsto B_{\lambda, X_0, n}^L$ *is strictly decreasing.*

3.6. Goals for Upper Bounds for the Cases $(\beta_\mathcal{A}, \beta_\mathcal{H}, \alpha_\mathcal{A}, \alpha_\mathcal{H}, \lambda) \in (\mathcal{P}_{SP} \backslash \mathcal{P}_{SP,1}) \times]0,1[$

For parameter constellations $(\beta_\mathcal{A}, \beta_\mathcal{H}, \alpha_\mathcal{A}, \alpha_\mathcal{H}, \lambda) \in (\mathcal{P}_{SP} \backslash \mathcal{P}_{SP,1}) \times]0,1[$, in contrast to the treatment of the lower bounds (cf. the previous Section 3.5), the fine-tuning of the *upper bounds* of the Hellinger integrals $H_\lambda(P_{\mathcal{A},n} || P_{\mathcal{H},n})$ is much more involved. To begin with, let us mention that the monotonicity-concerning Properties 2 (P10) to (P12) imply that for a tight upper bound $B_{\lambda, X_0, n}^U$ (cf. (40)) one should choose parameters $p_\lambda^U \geq p_\lambda^L > 0$, $q_\lambda^U \geq q_\lambda^L > 0$ as small as possible. Due to the concavity (cf. Properties 3 (P19)) of the function $\phi_\lambda(\cdot)$, the linear upper bound $\phi_\lambda^U(\cdot)$ (on the ultimately relevant subdomain $\mathbb{N}_0$) thus must hit the function $\phi_\lambda(\cdot)$ in at least one point $x \in \mathbb{N}_0$, which corresponds to some "discrete tangent line" of $\phi_\lambda(\cdot)$ in x, or in at most two points $x, x+1 \in \mathbb{N}_0$, which corresponds to the secant line of $\phi_\lambda(\cdot)$ across its arguments x and $x+1$. Accordingly, there is in general *no overall best* upper bound; of course, one way to obtain "good" upper bounds for $H_\lambda(P_{\mathcal{A},n} || P_{\mathcal{H},n})$ is to solve the optimization problem

$$\left(\overline{p_\lambda^U}, \overline{q_\lambda^U} \right) := \underset{(p_\lambda^U, q_\lambda^U)}{\arg\min} \left\{ \exp\left\{ a_n^{(q_\lambda^U)} \cdot X_0 + \sum_{k=1}^n b_k^{(p_\lambda^U, q_\lambda^U)} \right\} \right\}, \tag{46}$$

subject to the constraint (35). However, the corresponding result generally depends on the particular choice of the initial population $X_0 \in \mathbb{N}$ and on the observation time horizon $n \in \mathbb{N}$. Hence, there is in general no overall optimal choice of p_λ^U, q_λ^U without the incorporation of further goal-dependent constraints such as $\lim_{n \to \infty} B_{\lambda, X_0, n}^U = 0$ in case of $\lim_{n \to \infty} H_\lambda(P_{\mathcal{A},n} || P_{\mathcal{H},n}) = 0$. By the way, mainly because of the non-explicitness of the sequence $\left(a_n^{(q_\lambda^U)} \right)_{n \in \mathbb{N}}$ (due to the generally not explicitly solvable recursion (36)) and the discreteness of the constraint (35), this optimization problem seems to be not straightforward to solve, anyway. The choice of parameters p_λ^U, q_λ^U for the upper bound $B_{\lambda, X_0, n}^U \geq H_\lambda(P_{\mathcal{A},n} || P_{\mathcal{H},n})$ can be made according to different, partially incompatible ("optimality-" resp. "goodness-") criteria and goals, such as:

(G1) the validity of $B^U_{\lambda,X_0,n} < 1$ *simultaneously* for all initial configurations $X_0 \in \mathbb{N}$, all observation horizons $n \in \mathbb{N}$ and all $\lambda \in\,]0,1[$, which leads to a *strict* improvement of the general upper bound $H_\lambda(P_{\mathcal{A},n}||P_{\mathcal{H},n}) < 1$ (cf. (9));

(G2) the determination of the long-term-limits $\lim_{n\to\infty} H_\lambda(P_{\mathcal{A},n}||P_{\mathcal{H},n})$ respectively $\lim_{n\to\infty} B^U_{\lambda,X_0,n}$ for all $X_0 \in \mathbb{N}$ and all $\lambda \in\,]0,1[$; in particular, one would like to check whether $\lim_{n\to\infty} H_\lambda(P_{\mathcal{A},n}||P_{\mathcal{H},n}) = 0$, which implies that the families of probability distributions $(P_{\mathcal{A},n})_{n\in\mathbb{N}}$ and $(P_{\mathcal{H},n})_{n\in\mathbb{N}}$ are *asymptotically distinguishable* (entirely separated), cf. (25);

(G3) the determination of the time-asymptotical growth rates $\lim_{n\to\infty} \frac{1}{n} \log\left(H_\lambda(P_{\mathcal{A},n}||P_{\mathcal{H},n})\right)$ resp. $\lim_{n\to\infty} \frac{1}{n} \log\left(B^U_{\lambda,X_0,n}\right)$ for all $X_0 \in \mathbb{N}$ and all $\lambda \in\,]0,1[$.

Further goals–with which we do not deal here for the sake of brevity–are for instance (i) a very good tightness of the upper bound $B^U_{\lambda,X_0,n}$ for $n \geq N$ for some fixed large $N \in \mathbb{N}$, or (ii) the criterion (G1) with *fixed* (rather than arbitrary) initial population size $X_0 \in \mathbb{N}$.

Let us briefly discuss the three Goals (G1) to (G3) and their challenges: due to Theorem 1, Goal (G1) can only be achieved if the sequence $\left(a_n^{(q^U_\lambda)}\right)_{n\in\mathbb{N}}$ is non-increasing, since otherwise, for each fixed observation horizon $n \in \mathbb{N}$ there is a large enough initial population size X_0 such that the upper bound component $\widetilde{B}^{(p^U_\lambda,q^U_\lambda)}_{\lambda,X_0,n}$ becomes larger than 1, and thus $B^U_{\lambda,X_0,n} = 1$ (cf. (40)). Hence, Properties 1 (P1) and (P2) imply that one should have $q^U_\lambda \leq \beta_\lambda$. Then, the sequence $\left(b_n^{(p^U_\lambda,q^U_\lambda)}\right)_{n\in\mathbb{N}}$ is also non-increasing. However, since $b_n^{(p^U_\lambda,q^U_\lambda)}$ might be positive for some (even all) $n \in \mathbb{N}$, the sum $\left(\sum_{k=1}^n b_k^{(p^U_\lambda,q^U_\lambda)}\right)_{n\in\mathbb{N}}$ is not necessarily decreasing. Nevertheless, the restriction

$$q^U_\lambda - \beta_\lambda \leq 0 \quad \text{and} \quad p^U_\lambda - \alpha_\lambda \leq 0, \qquad \text{where at least one of the inequalities is strict,} \qquad (47)$$

ensures that both sequences $\left(a_n^{(q^U_\lambda)}\right)_{n\in\mathbb{N}}$ and $\left(b_n^{(p^L_\lambda,q^U_\lambda)}\right)_{n\in\mathbb{N}}$ are nonpositive and decreasing, where at least one sequence is strictly negative, implying that the sum $\left(\sum_{k=1}^n b_k^{(p^U_\lambda,q^U_\lambda)}\right)_{n\in\mathbb{N}}$ is strictly negative for $n \geq 2$ and strictly decreasing. To see this, suppose that (47) is satisfied with two strict inequalities. Then, $\left(a_n^{(q^U_\lambda)}\right)_{n\in\mathbb{N}}$ as well as $\left(b_n^{(p^L_\lambda,q^U_\lambda)}\right)_{n\in\mathbb{N}}$ are strictly negative and strictly decreasing. If $q^U_\lambda = \beta_\lambda$ and $p^U_\lambda < \alpha_\lambda$, we see from (P2) and (P6) that $a_n^{(q^U_\lambda)} \equiv 0$ and that $b_n^{(p^U_\lambda,q^U_\lambda)} \equiv p^U_\lambda - \alpha_\lambda < 0$ (notice that $\alpha_\lambda = 0$ is not possible in the current setup $\mathcal{P}_{SP}\backslash\mathcal{P}_{SP,1}$ and for $\lambda \in\,]0,1[$). In the last case $q^U_\lambda < \beta_\lambda$ and $p^U_\lambda = \alpha_\lambda$, from (P1) and (P5) it follows that $\left(a_n^{(q^U_\lambda)}\right)_{n\in\mathbb{N}}$ is strictly negative and strictly decreasing, as well as that $b_1^{(p^U_\lambda,q^U_\lambda)} = 0$ and $\left(b_n^{(p^L_\lambda,q^U_\lambda)}\right)_{n\in\mathbb{N}}$ is strictly decreasing and strictly negative for $n \geq 2$. Thus, whenever (47) is satisfied, the sum $\left(\sum_{k=1}^n b_k^{(p^U_\lambda,q^U_\lambda)}\right)_{n\in\mathbb{N}}$ is strictly negative for $n \geq 2$ and strictly decreasing.

To achieve Goal (G2), we have to require that the sequence $\left(a_n^{(q^U_\lambda)}\right)_{n\in\mathbb{N}}$ converges, which is the case if either $q^U_\lambda \leq \beta_\lambda$ or $\beta_\lambda < q^U_\lambda \leq \min\{1, e^{\beta_\lambda - 1}\}$ (cf. Properties 1 (P1) to (P3)). From the upper bound component $\widetilde{B}^{(p^U_\lambda,q^U_\lambda)}_{\lambda,X_0,n}$ (42) we conclude that Goal (G2) is met if the sequence $\left(b_n^{(p^U_\lambda,q^U_\lambda)}\right)_{n\in\mathbb{N}}$ converges to a negative limit, i.e., $\lim_{n\to\infty} b_n^{(p^U_\lambda,q^U_\lambda)} = p^U_\lambda \cdot e^{x_0^{(q^U_\lambda)}} - \alpha_\lambda < 0$. Notice that this condition holds true if (47) is satisfied: suppose that $q^U_\lambda < \beta_\lambda$, then $x_0^{(q^U_\lambda)} < 0$ and $p^U_\lambda \cdot e^{x_0^{(q^U_\lambda)}} - \alpha_\lambda < p^U_\lambda - \alpha_\lambda \leq 0$. On the other hand, if $p^U_\lambda - \alpha_\lambda < 0$, one obtains $x_0^{(q^U_\lambda)} \leq 0$ leading to $p^U_\lambda \cdot e^{x_0^{(q^U_\lambda)}} - \alpha_\lambda \leq p^U_\lambda - \alpha_\lambda < 0$.

The examination of Goal (G2) above enters into the discussion of Goal (G3): if the sequence $\left(a_n^{(q_\lambda^U)} \right)_{n \in \mathbb{N}}$ converges and $\lim_{n \to \infty} B_{\lambda, X_0, n}^U = 0$, then there holds

$$\lim_{n \to \infty} \frac{1}{n} \log \left(B_{\lambda, X_0, n}^U \right) = \lim_{n \to \infty} \frac{1}{n} \log \left(\widetilde{B}_{\lambda, X_0, n}^{(p_\lambda^U, q_\lambda^U)} \right) = p_\lambda^U \cdot e^{x_0^{(q_\lambda^U)}} - \alpha_\lambda . \tag{48}$$

For the case $(\beta_{\mathcal{A}}, \beta_{\mathcal{H}}, \alpha_{\mathcal{A}}, \alpha_{\mathcal{H}}, \lambda) \in (\mathcal{P}_{SP} \backslash \mathcal{P}_{SP,1}) \times]0, 1[$, let us now start with our comprehensive investigations of the upper bounds, where we focus on fulfilling the condition (47) which tackles Goals (G1) and (G2) simultaneously; then, the Goal (G3) can be achieved by (48). As indicated above, various different parameter subcases can lead to different Hellinger-integral-upper-bound details, which we work out in the following. For better transparency, we employ the following notations (where the first four are just reminders of sets which were already introduced above)

$$\mathcal{P}_{NI} := \left\{ (\beta_{\mathcal{A}}, \beta_{\mathcal{H}}, \alpha_{\mathcal{A}}, \alpha_{\mathcal{H}}) \in [0, \infty[^4 : \alpha_{\mathcal{A}} = \alpha_{\mathcal{H}} = 0; \ \beta_{\mathcal{A}} > 0; \ \beta_{\mathcal{H}} > 0; \ \beta_{\mathcal{A}} \neq \beta_{\mathcal{H}} \right\},$$

$$\mathcal{P}_{SP} := \left\{ (\beta_{\mathcal{A}}, \beta_{\mathcal{H}}, \alpha_{\mathcal{A}}, \alpha_{\mathcal{H}}) \in]0, \infty[^4 : (\alpha_{\mathcal{A}} \neq \alpha_{\mathcal{H}}) \text{ or } (\beta_{\mathcal{A}} \neq \beta_{\mathcal{H}}) \text{ or both} \right\},$$

$$\mathcal{P} := \mathcal{P}_{NI} \cup \mathcal{P}_{SP},$$

$$\mathcal{P}_{SP,1} := \left\{ (\beta_{\mathcal{A}}, \beta_{\mathcal{H}}, \alpha_{\mathcal{A}}, \alpha_{\mathcal{H}}) \in \mathcal{P}_{SP} : \alpha_{\mathcal{A}} \neq \alpha_{\mathcal{H}}, \ \beta_{\mathcal{A}} \neq \beta_{\mathcal{H}}, \ \frac{\alpha_{\mathcal{A}}}{\beta_{\mathcal{A}}} = \frac{\alpha_{\mathcal{H}}}{\beta_{\mathcal{H}}} \right\},$$

$$\mathcal{P}_{SP,2} := \left\{ (\beta_{\mathcal{A}}, \beta_{\mathcal{H}}, \alpha_{\mathcal{A}}, \alpha_{\mathcal{H}}) \in \mathcal{P}_{SP} : \alpha_{\mathcal{A}} = \alpha_{\mathcal{H}}, \ \beta_{\mathcal{A}} \neq \beta_{\mathcal{H}} \right\},$$

$$\mathcal{P}_{SP,3} := \left\{ (\beta_{\mathcal{A}}, \beta_{\mathcal{H}}, \alpha_{\mathcal{A}}, \alpha_{\mathcal{H}}) \in \mathcal{P}_{SP} : \alpha_{\mathcal{A}} \neq \alpha_{\mathcal{H}}, \ \beta_{\mathcal{A}} \neq \beta_{\mathcal{H}}, \ \frac{\alpha_{\mathcal{A}}}{\beta_{\mathcal{A}}} \neq \frac{\alpha_{\mathcal{H}}}{\beta_{\mathcal{H}}} \right\} = \mathcal{P}_{SP,3a} \cup \mathcal{P}_{SP,3b} \cup \mathcal{P}_{SP,3c} ,$$

$$\mathcal{P}_{SP,3a} := \left\{ (\beta_{\mathcal{A}}, \beta_{\mathcal{H}}, \alpha_{\mathcal{A}}, \alpha_{\mathcal{H}}) \in \mathcal{P}_{SP} : \alpha_{\mathcal{A}} \neq \alpha_{\mathcal{H}}, \ \beta_{\mathcal{A}} \neq \beta_{\mathcal{H}}, \ \frac{\alpha_{\mathcal{A}}}{\beta_{\mathcal{A}}} \neq \frac{\alpha_{\mathcal{H}}}{\beta_{\mathcal{H}}}, \ \frac{\alpha_{\mathcal{A}} - \alpha_{\mathcal{H}}}{\beta_{\mathcal{H}} - \beta_{\mathcal{A}}} \in]-\infty, 0[\right\},$$

$$\mathcal{P}_{SP,3b} := \left\{ (\beta_{\mathcal{A}}, \beta_{\mathcal{H}}, \alpha_{\mathcal{A}}, \alpha_{\mathcal{H}}) \in \mathcal{P}_{SP} : \alpha_{\mathcal{A}} \neq \alpha_{\mathcal{H}}, \ \beta_{\mathcal{A}} \neq \beta_{\mathcal{H}}, \ \frac{\alpha_{\mathcal{A}}}{\beta_{\mathcal{A}}} \neq \frac{\alpha_{\mathcal{H}}}{\beta_{\mathcal{H}}}, \ \frac{\alpha_{\mathcal{A}} - \alpha_{\mathcal{H}}}{\beta_{\mathcal{H}} - \beta_{\mathcal{A}}} \in]0, \infty[\backslash \mathbb{N} \right\},$$

$$\mathcal{P}_{SP,3c} := \left\{ (\beta_{\mathcal{A}}, \beta_{\mathcal{H}}, \alpha_{\mathcal{A}}, \alpha_{\mathcal{H}}) \in \mathcal{P}_{SP} : \alpha_{\mathcal{A}} \neq \alpha_{\mathcal{H}}, \ \beta_{\mathcal{A}} \neq \beta_{\mathcal{H}}, \ \frac{\alpha_{\mathcal{A}}}{\beta_{\mathcal{A}}} \neq \frac{\alpha_{\mathcal{H}}}{\beta_{\mathcal{H}}}, \ \frac{\alpha_{\mathcal{A}} - \alpha_{\mathcal{H}}}{\beta_{\mathcal{H}} - \beta_{\mathcal{A}}} \in \mathbb{N} \right\},$$

$$\mathcal{P}_{SP,4} := \left\{ (\beta_{\mathcal{A}}, \beta_{\mathcal{H}}, \alpha_{\mathcal{A}}, \alpha_{\mathcal{H}}) \in \mathcal{P}_{SP} : \alpha_{\mathcal{A}} \neq \alpha_{\mathcal{H}} > 0, \ \beta_{\mathcal{A}} = \beta_{\mathcal{H}} \right\} = \mathcal{P}_{SP,4a} \cup \mathcal{P}_{SP,4b} ,$$

$$\mathcal{P}_{SP,4a} := \left\{ (\beta_{\mathcal{A}}, \beta_{\mathcal{H}}, \alpha_{\mathcal{A}}, \alpha_{\mathcal{H}}) \in \mathcal{P}_{SP} : \alpha_{\mathcal{A}} \neq \alpha_{\mathcal{H}} > 0, \ \beta_{\mathcal{A}} = \beta_{\mathcal{H}} \in]0, 1[\right\},$$

$$\mathcal{P}_{SP,4b} := \left\{ (\beta_{\mathcal{A}}, \beta_{\mathcal{H}}, \alpha_{\mathcal{A}}, \alpha_{\mathcal{H}}) \in \mathcal{P}_{SP} : \alpha_{\mathcal{A}} \neq \alpha_{\mathcal{H}} > 0, \ \beta_{\mathcal{A}} = \beta_{\mathcal{H}} \in [1, \infty[\right\}; \tag{49}$$

notice that because of Lemma A1 and of the Properties 3 (P15) one gets on the domain $]0, \infty[$ the relation $\phi_\lambda(x) = 0$ iff $f_{\mathcal{A}}(x) = f_{\mathcal{H}}(x)$ iff $x = x^* := \frac{\alpha_{\mathcal{H}} - \alpha_{\mathcal{A}}}{\beta_{\mathcal{A}} - \beta_{\mathcal{H}}} \in]0, \infty[$.

3.7. Upper Bounds for the Cases $(\beta_{\mathcal{A}}, \beta_{\mathcal{H}}, \alpha_{\mathcal{A}}, \alpha_{\mathcal{H}}, \lambda) \in \mathcal{P}_{SP,2} \times]0, 1[$

For this parameter constellation, one has $\phi_\lambda(0) = 0$ and $\phi_\lambda'(0) = 0$ (cf. Properties 3 (P16), (P17)). Thus, the only admissible intercept choice satisfying (47) is $r_\lambda^U = 0 = p_\lambda^U - \alpha_\lambda$ (i.e., $p_\lambda^U = p^U(\beta_{\mathcal{A}}, \beta_{\mathcal{H}}, \alpha_{\mathcal{A}}, \alpha_{\mathcal{H}}, \lambda) = \alpha_\lambda = \alpha > 0$), and the minimal admissible slope which implies (35) for all $x \in \mathbb{N}_0$ is given by $s_\lambda^U = \frac{\phi_\lambda(1) - \phi_\lambda(0)}{1 - 0} = q_\lambda^U - \beta_\lambda = a_1^{(q_\lambda^U)} < 0$ (i.e., $q_\lambda^U = q^U(\beta_{\mathcal{A}}, \beta_{\mathcal{H}}, \alpha_{\mathcal{A}}, \alpha_{\mathcal{H}}, \lambda) = (\alpha + \beta_{\mathcal{A}})^\lambda (\alpha + \beta_{\mathcal{H}})^{1 - \lambda} - \alpha > 0$). Analogously to the investigation for $\mathcal{P}_{SP,1}$ in the above-mentioned Section 3.3, one can derive that $\left(a_n^{(q_\lambda^U)} \right)_{n \in \mathbb{N}}$ is strictly negative, strictly decreasing, and converges to $x_0^{(q_\lambda^U)} \in]-\beta_\lambda, q_\lambda^U - \beta_\lambda[$ as indicated in Properties 1 (P1). Moreover, in the same manner as for the case $\mathcal{P}_{SP,1}$ this leads to

Proposition 7. *For all* $(\beta_{\mathcal{A}}, \beta_{\mathcal{H}}, \alpha_{\mathcal{A}}, \alpha_{\mathcal{H}}, \lambda) \in \mathcal{P}_{SP,2} \times]0,1[$ *and all initial population sizes* $X_0 \in \mathbb{N}$ *there holds with* $p_\lambda^U = \alpha$, $q_\lambda^U = (\alpha + \beta_{\mathcal{A}})^\lambda (\alpha + \beta_{\mathcal{H}})^{1-\lambda} - \alpha$

$$(a) \qquad B_{\lambda,X_0,1}^U = \exp\left\{ \left(q_\lambda^U - \beta_\lambda \right) \cdot X_0 \right\} < 1,$$

(b) *the sequence* $\left(B_{\lambda,X_0,n}^U \right)_{n \in \mathbb{N}}$ *of upper bounds for* $H_\lambda(P_{\mathcal{A},n} || P_{\mathcal{H},n})$ *given by*

$$B_{\lambda,X_0,n}^U = \exp\left\{ a_n^{(q_\lambda^U)} \cdot X_0 + \sum_{k=1}^n b_k^{(p_\lambda^U, q_\lambda^U)} \right\}$$

is strictly decreasing,

$$(c) \qquad \lim_{n \to \infty} B_{\lambda,X_0,n}^U = 0 = \lim_{n \to \infty} H_\lambda(P_{\mathcal{A},n} || P_{\mathcal{H},n}),$$

$$(d) \qquad \lim_{n \to \infty} \frac{1}{n} \log B_{\lambda,X_0,n}^U = p_\lambda^U \cdot e^{x_0^{(q_\lambda^U)}} - \alpha_\lambda = \alpha \left(e^{x_0^{(q_\lambda^U)}} - 1 \right) < 0.$$

(e) *the map* $X_0 \mapsto B_{\lambda,X_0,n}^U$ *is strictly decreasing.*

3.8. Upper Bounds for the Cases $(\beta_{\mathcal{A}}, \beta_{\mathcal{H}}, \alpha_{\mathcal{A}}, \alpha_{\mathcal{H}}, \lambda) \in \mathcal{P}_{SP,3a} \times]0,1[$

From Properties 3 (P16) one gets $\phi_\lambda(0) < 0$, whereas $\phi_\lambda'(0)$ can assume any sign, take e.g., the parameters $(\beta_{\mathcal{A}}, \beta_{\mathcal{H}}, \alpha_{\mathcal{A}}, \alpha_{\mathcal{H}}, \lambda) = (1.8, 0.9, 2.7, 0.7, 0.5)$ for $\phi_\lambda'(0) < 0$, $(\beta_{\mathcal{A}}, \beta_{\mathcal{H}}, \alpha_{\mathcal{A}}, \alpha_{\mathcal{H}}, \lambda) = (1.8, 0.9, 2.8, 0.7, 0.5)$ for $\phi_\lambda'(0) = 0$ and $(\beta_{\mathcal{A}}, \beta_{\mathcal{H}}, \alpha_{\mathcal{A}}, \alpha_{\mathcal{H}}, \lambda) = (1.8, 0.9, 2.9, 0.7, 0.5)$ for $\phi_\lambda'(0) > 0$; within our running-example epidemiological context of Section 2.3, this corresponds to a "nearly dangerous" infectious-disease-transmission situation $(\mathcal{H})$ (with nearly critical reproduction number $\beta_{\mathcal{H}} = 0.9$ and importation mean of $\alpha_{\mathcal{H}} = 0.7$), whereas $(\mathcal{A})$ describes a "dangerous" situation (with supercritical $\beta_{\mathcal{A}} = 1.8$ and $\alpha_{\mathcal{A}} = 2.7, 2.8, 2.9$). However, in all three subcases there holds $\max_{x \in \mathbb{N}_0} \phi_\lambda(x) \leq \max_{x \in [0,\infty[} \phi_\lambda(x) < 0$. Thus, there clearly exist parameters $p_\lambda^U = p^U(\beta_{\mathcal{A}}, \beta_{\mathcal{H}}, \alpha_{\mathcal{A}}, \alpha_{\mathcal{H}}, \lambda)$, $q_\lambda^U = q^U(\beta_{\mathcal{A}}, \beta_{\mathcal{H}}, \alpha_{\mathcal{A}}, \alpha_{\mathcal{H}}, \lambda)$ with $p_\lambda^U \in [\alpha_{\mathcal{A}}^\lambda \alpha_{\mathcal{H}}^{1-\lambda}, \alpha_\lambda[$ and $q_\lambda^U \in [\beta_{\mathcal{A}}^\lambda \beta_{\mathcal{H}}^{1-\lambda}, \beta_\lambda[$ (implying (47)) such that (35) is satisfied. As explained above, we get the following

Proposition 8. *For all* $(\beta_{\mathcal{A}}, \beta_{\mathcal{H}}, \alpha_{\mathcal{A}}, \alpha_{\mathcal{H}}, \lambda) \in \mathcal{P}_{SP,3a} \times]0,1[$ *there exist parameters* p_λ^U, q_λ^U *which satisfy* $p_\lambda^U \in [\alpha_{\mathcal{A}}^\lambda \alpha_{\mathcal{H}}^{1-\lambda}, \alpha_\lambda[$ *and* $q_\lambda^U \in [\beta_{\mathcal{A}}^\lambda \beta_{\mathcal{H}}^{1-\lambda}, \beta_\lambda[$ *as well as* (35) *for all* $x \in \mathbb{N}_0$, *and for all such pairs* $(p_\lambda^U, q_\lambda^U)$ *and all initial population sizes* $X_0 \in \mathbb{N}$ *there holds*

$$(a) \qquad B_{\lambda,X_0,1}^U = \exp\left\{ \left(q_\lambda^U - \beta_\lambda \right) \cdot X_0 + p_\lambda^U - \alpha_\lambda \right\} < 1,$$

(b) *the sequence* $\left(B_{\lambda,X_0,n}^U \right)_{n \in \mathbb{N}}$ *of upper bounds for* $H_\lambda(P_{\mathcal{A},n} || P_{\mathcal{H},n})$ *given by*

$$B_{\lambda,X_0,n}^U = \exp\left\{ a_n^{(q_\lambda^U)} X_0 + \sum_{k=1}^n b_k^{(p_\lambda^U, q_\lambda^U)} \right\}$$

is strictly decreasing,

$$(c) \qquad \lim_{n \to \infty} B_{\lambda,X_0,n}^U = 0 = \lim_{n \to \infty} H_\lambda(P_{\mathcal{A},n} || P_{\mathcal{H},n}),$$

$$(d) \qquad \lim_{n \to \infty} \frac{1}{n} \log B_{\lambda,X_0,n}^U = p_\lambda^U \cdot e^{x_0^{(q_\lambda^U)}} - \alpha_\lambda < 0,$$

(e) *the map* $X_0 \mapsto B_{\lambda,X_0,n}^U$ *is strictly decreasing.*

Notice that all parts of this proposition also hold true for parameter pairs $(p_\lambda^U, q_\lambda^U)$ satisfying (35) and additionally either $p_\lambda^U = \alpha_\lambda$, $q_\lambda^U < \beta_\lambda$ or $p_\lambda^U < \alpha_\lambda$, $q_\lambda^U = \beta_\lambda$.

Let us briefly illuminate the above-mentioned possible parameter choices, where we begin with the case of $\phi_\lambda'(0) \leq 0$, which corresponds to $\lambda \beta_{\mathcal{A}} (\alpha_{\mathcal{A}}/\alpha_{\mathcal{H}})^{\lambda-1} + (1-\lambda)\beta_{\mathcal{H}} (\alpha_{\mathcal{A}}/\alpha_{\mathcal{H}})^\lambda - \beta_\lambda \leq 0$ (cf. (P17)); then, the function $\phi_\lambda(\cdot)$ is strictly negative, strictly decreasing, and–due to (P19)–strictly concave (and thus, the assumption $\frac{\alpha_{\mathcal{H}} - \alpha_{\mathcal{A}}}{\beta_{\mathcal{A}} - \beta_{\mathcal{H}}} < 0$ is superfluous here). One pragmatic but yet reasonable parameter choice is the following: take any intercept $p_\lambda^U \in [\alpha_\mathcal{A}^\lambda \alpha_\mathcal{H}^{1-\lambda}, \alpha_\lambda]$ such that $(p_\lambda^U - \alpha_\lambda) + 2(\phi_\lambda(1) - (p_\lambda^U - \alpha_\lambda)) \geq \phi_\lambda(2)$ $\left(\text{i.e., } 2(\alpha_\mathcal{A} + \beta_\mathcal{A})^\lambda (\alpha_\mathcal{H} + \beta_\mathcal{H})^{1-\lambda} - p_\lambda^U + \alpha_\lambda \geq (\alpha_\mathcal{A} + 2\beta_\mathcal{A})^\lambda (\alpha_\mathcal{H} + 2\beta_\mathcal{H})^{1-\lambda} \right)$ and $q_\lambda^U := \phi_\lambda(1) - (p_\lambda^U - \alpha_\lambda) + \beta_\lambda = (\alpha_\mathcal{A} + \beta_\mathcal{A})^\lambda (\alpha_\mathcal{H} + \beta_\mathcal{H})^{1-\lambda} - p_\lambda^U$, which corresponds to a linear function ϕ_λ^U which is (i) nonpositive on $\mathbb{N}_0$ and strictly negative on $\mathbb{N}$, and (ii) larger than or equal to ϕ_λ on $\mathbb{N}_0$, strictly larger than ϕ_λ on $\mathbb{N}\backslash\{1,2\}$, and equal to ϕ_λ at the point $x = 1$ ("discrete tangent or secant line through $x = 1$"). One can easily see that (due to the restriction (34)) not all $p_\lambda^U \in [\alpha_\mathcal{A}^\lambda \alpha_\mathcal{H}^{1-\lambda}, \alpha_\lambda]$ might qualify for the current purpose. For the particular choice $p_\lambda^U = \alpha_\mathcal{A}^\lambda \alpha_\mathcal{H}^{1-\lambda}$ and $q_\lambda^U = (\alpha_\mathcal{A} + \beta_\mathcal{A})^\lambda (\alpha_\mathcal{H} + \beta_\mathcal{H})^{1-\lambda} - \alpha_\mathcal{A}^\lambda \alpha_\mathcal{H}^{1-\lambda}$ one obtains $r_\lambda^U = p_\lambda^U - \alpha_\lambda = b_1^{(p_\lambda^U, q_\lambda^U)} < 0$ (cf. Lemma A1) and $s_\lambda^U = q_\lambda^U - \beta_\lambda = \phi_\lambda(1) - \phi_\lambda(0) = a_1^{(q_\lambda^U)} < 0$ (secant line through $\phi_\lambda(0)$ and $\phi_\lambda(1)$).

For the remaining case $\phi_\lambda'(0) > 0$, which corresponds to $\lambda\beta_\mathcal{A} (\alpha_\mathcal{A}/\alpha_\mathcal{H})^{\lambda-1} + (1 - \lambda)\beta_\mathcal{H} (\alpha_\mathcal{A}/\alpha_\mathcal{H})^\lambda - \beta_\lambda > 0$, the function $\phi_\lambda(\cdot)$ is strictly negative, strictly concave and hump-shaped (cf. (P18)). For the derivation of the parameter choices, we employ $x_{\max} := \mathrm{argmax}_{x\in]0,\infty[}\phi_\lambda(x)$ which is the unique solution of

$$\lambda\beta_\mathcal{A} \left[\left(\frac{f_\mathcal{A}(x)}{f_\mathcal{H}(x)} \right)^{\lambda-1} - 1 \right] + (1 - \lambda)\beta_\mathcal{H} \left[\left(\frac{f_\mathcal{A}(x)}{f_\mathcal{H}(x)} \right)^\lambda - 1 \right] = 0, \qquad x \in]0, \infty[, \tag{50}$$

(cf. (P17), (P19)); notice that $x = x^* := \frac{\alpha_\mathcal{H} - \alpha_\mathcal{A}}{\beta_\mathcal{A} - \beta_\mathcal{H}} \in]0, \infty[$ formally satisfies the Equation (50) but does not qualify because of the current restriction $x^* < 0$.

Let us first inspect the case $\phi_\lambda(\lfloor x_{\max}\rfloor) > \phi_\lambda(\lfloor x_{\max}\rfloor + 1)$, where $\lfloor x \rfloor$ denotes the integer part of x. Consider the subcase $\phi_\lambda(\lfloor x_{\max}\rfloor) + \lfloor x_{\max}\rfloor (\phi_\lambda(\lfloor x_{\max}\rfloor) - \phi_\lambda(\lfloor x_{\max}\rfloor + 1)) \leq 0$, which means that the secant line through $\phi_\lambda(\lfloor x_{\max}\rfloor)$ and $\phi_\lambda(\lfloor x_{\max}\rfloor + 1)$ possesses a non-positive intercept. In this situation it is reasonable to choose as *intercept* any $p_\lambda^U - \alpha_\lambda = b_1^{(p_\lambda^U, q_\lambda^U)} = r_\lambda^U \in [\phi_\lambda(\lfloor x_{\max}\rfloor), \phi_\lambda(\lfloor x_{\max}\rfloor) + \lfloor x_{\max}\rfloor (\phi_\lambda(\lfloor x_{\max}\rfloor) - \phi_\lambda(\lfloor x_{\max}\rfloor + 1))]$, and as corresponding *slope* $q_\lambda^U - \alpha_\lambda = a_1^{(q_\lambda^U)} = s_\lambda^U = \frac{\phi_\lambda(\lfloor x_{\max}\rfloor) - r_\lambda^U}{(\lfloor x_{\max}\rfloor) - 0} \leq 0$. A larger intercept would lead to a linear function ϕ_λ^U for which (35) is not valid at $\lfloor x_{\max}\rfloor + 1$. In the other subcase $\phi_\lambda(\lfloor x_{\max}\rfloor) + x_{\max} (\phi_\lambda(\lfloor x_{\max}\rfloor) - \phi_\lambda(\lfloor x_{\max}\rfloor + 1)) > 0$, one can choose any intercept $p_\lambda^U - \alpha_\lambda = b_1^{(p_\lambda^U, q_\lambda^U)} = r_\lambda^U \in [\phi_\lambda(\lfloor x_{\max}\rfloor), 0]$ and as corresponding slope $q_\lambda^U - \alpha_\lambda = a_1^{(q_\lambda^U)} = s_\lambda^U = \frac{\phi_\lambda(\lfloor x_{\max}\rfloor) - r_\lambda^U}{(\lfloor x_{\max}\rfloor) - 0} \leq 0$ (notice that the corresponding line ϕ_λ^U is on $]\lfloor x_{\max}\rfloor, \infty[$ strictly larger than the secant line through $\phi_\lambda(\lfloor x_{\max}\rfloor)$ and $\phi_\lambda(\lfloor x_{\max}\rfloor + 1)$).

If $\phi_\lambda(\lfloor x_{\max}\rfloor) \leq \phi_\lambda(\lfloor x_{\max}\rfloor + 1)$, one can proceed as above by substituting the crucial pair of points $(\lfloor x_{\max}\rfloor, \lfloor x_{\max}\rfloor + 1)$ with $(\lfloor x_{\max}\rfloor + 1, \lfloor x_{\max}\rfloor + 2)$ and examining the analogous two subcases.

3.9. Upper Bounds for the Cases $(\beta_\mathcal{A}, \beta_\mathcal{H}, \alpha_\mathcal{A}, \alpha_\mathcal{H}, \lambda) \in \mathcal{P}_{SP,3b}\times]0, 1[$

The only difference to the preceding Section 3.8 is that–due to Properties 3 (P15)–the maximum value of $\phi_\lambda(\cdot)$ now achieves 0, at the positive *non-integer* point $x_{\max} = x^* = \frac{\alpha_\mathcal{H} - \alpha_\mathcal{A}}{\beta_\mathcal{A} - \beta_\mathcal{H}} \in]0, \infty[\backslash\mathbb{N}$ (take e.g., $(\beta_\mathcal{A}, \beta_\mathcal{H}, \alpha_\mathcal{A}, \alpha_\mathcal{H}, \lambda) = (1.8, 0.9, 1.1, 3.0, 0.5)$ as an example, which within our running-example epidemiological context of Section 2.3 corresponds to a "nearly dangerous" infectious-disease-transmission situation $(\mathcal{H})$ (with nearly critical reproduction number $\beta_\mathcal{H} = 0.9$ and importation mean of $\alpha_\mathcal{H} = 3$), whereas $(\mathcal{A})$ describes a "dangerous" situation (with supercritical $\beta_\mathcal{A} = 1.8$ and $\alpha_\mathcal{A} = 1.1$)); this implies that $\phi_\lambda(x) < 0$ for all x on the relevant subdomain $\mathbb{N}_0$. Due to (P16), (P17) and (P19) one gets automatically $\lambda\beta_\mathcal{A} (\alpha_\mathcal{A}/\alpha_\mathcal{H})^{\lambda-1} + (1 - \lambda)\beta_\mathcal{H} (\alpha_\mathcal{A}/\alpha_\mathcal{H})^\lambda - \beta_\lambda > 0$ for all $\lambda \in]0, 1[$. Analogously to Section 3.8, there exist parameter $p_\lambda^U \in [\alpha_\mathcal{A}^\lambda \alpha_\mathcal{H}^{1-\lambda}, \alpha_\lambda]$ and $q_\lambda^U \in [\beta_\mathcal{A}^\lambda \beta_\mathcal{H}^{1-\lambda}, \beta_\lambda]$ such that (47) and (35) are satisfied. Thus, all the assertions (a) to (e) of Proposition 8 also hold true for the current parameter constellations.

3.10. Upper Bounds for the Cases $(\beta_{\mathcal{A}}, \beta_{\mathcal{H}}, \alpha_{\mathcal{A}}, \alpha_{\mathcal{H}}, \lambda) \in \mathcal{P}_{SP,3c} \times\,]0, 1[$

The only difference to the preceding Section 3.9 is that the maximum value of $\phi_\lambda(\cdot)$ now achieves 0 at the *integer* point $x_{\max} = x^* = \frac{\alpha_{\mathcal{H}} - \alpha_{\mathcal{A}}}{\beta_{\mathcal{A}} - \beta_{\mathcal{H}}} \in \mathbb{N}$ (take e.g., $(\beta_{\mathcal{A}}, \beta_{\mathcal{H}}, \alpha_{\mathcal{A}}, \alpha_{\mathcal{H}}, \lambda) = (1.8, 0.9, 1.2, 3.0, 0.5)$ as an example). Accordingly, there do not exist parameters p_λ^U, q_λ^U, such that (35) and (47) are satisfied simultaneously. The only parameter pair that ensures $\exp\left\{a_n^{(q_\lambda^U)} \cdot X_0 + \sum_{k=1}^n b_k^{(p_\lambda^U, q_\lambda^U)}\right\} \leq 1$ for all $n \in \mathbb{N}$ and all $X_0 \in \mathbb{N}$ without further investigations, leads to the choices $p_\lambda^U = \alpha_\lambda$ as well as $q_\lambda^U = \beta_\lambda$. Consequently, $B_{\lambda, X_0, n}^U \equiv 1$, which coincides with the general upper bound (9), but violates the above-mentioned desired Goal (G1). However, there might exist parameters $p_\lambda^U < \alpha_\lambda$, $q_\lambda^U > \beta_\lambda$ or $p_\lambda^U > \alpha_\lambda$, $q_\lambda^U < \beta_\lambda$, such that at least the parts (c) and (d) of Proposition 8 are satisfied. Nevertheless, by using a conceptually different method we can prove

$$H_\lambda(P_{\mathcal{A},n} \| P_{\mathcal{H},n}) < 1 \quad \forall\, n \in \mathbb{N}\backslash\{1\} \qquad \text{as well as the convergence} \quad \lim_{n\to\infty} H_\lambda(P_{\mathcal{A},n} \| P_{\mathcal{H},n}) = 0 \quad (51)$$

which will be used for the study of complete asymptotical distinguishability (entire separation) below. This proof is provided in Appendix A.1.

3.11. Upper Bounds for the Cases $(\beta_{\mathcal{A}}, \beta_{\mathcal{H}}, \alpha_{\mathcal{A}}, \alpha_{\mathcal{H}}, \lambda) \in \mathcal{P}_{SP,4a} \times\,]0, 1[$

This setup and the remaining setup $(\beta_{\mathcal{A}}, \beta_{\mathcal{H}}, \alpha_{\mathcal{A}}, \alpha_{\mathcal{H}}, \lambda) \in \mathcal{P}_{SP,4b} \times\,]0, 1[$ (see the next Section 3.12) are the only constellations where $\phi_\lambda(\cdot)$ is strictly negative and strictly increasing, with $\lim_{x\to\infty} \phi_\lambda(x) = \lim_{x\to\infty} \phi_\lambda'(x) = 0$, leading to the choices $p_\lambda^U = \alpha_\lambda$ as well as $q_\lambda^U = \beta_\lambda = \beta$ under the restriction that $\exp\left\{a_n^{(q_\lambda^U)} \cdot X_0 + \sum_{k=1}^n b_k^{(p_\lambda^U, q_\lambda^U)}\right\} \leq 1$ for all $n \in \mathbb{N}$ and all $X_0 \in \mathbb{N}$. Consequently, one has $B_{\lambda, X_0, n}^U \equiv 1$, which is consistent with the general upper bound (9) but violates the above-mentioned desired Goal (G1). Unfortunately, the proof method of (51) (cf. Appendix A.1) can't be carried over to the current setup. The following proposition states two of the above-mentioned desired assertions which can be verified by a completely different proof method, which is also given in Appendix A.1.

Proposition 9. *For all* $(\beta_{\mathcal{A}}, \beta_{\mathcal{H}}, \alpha_{\mathcal{A}}, \alpha_{\mathcal{H}}, \lambda) \in \mathcal{P}_{SP,4a} \times\,]0, 1[$ *there exist parameters* $p_\lambda^U < \alpha_\lambda$, $1 > q_\lambda^U > \beta_\lambda = \beta$ *such that (35) is satisfied for all* $x \in [0, \infty[$ *and such that for all initial population sizes* $X_0 \in \mathbb{N}$ *the parts (c) and (d) of Proposition 8 hold true.*

3.12. Upper Bounds for the Cases $(\beta_{\mathcal{A}}, \beta_{\mathcal{H}}, \alpha_{\mathcal{A}}, \alpha_{\mathcal{H}}, \lambda) \in \mathcal{P}_{SP,4b} \times\,]0, 1[$

The assertions preceding Proposition 9 remain valid. However, any linear upper bound of the function $\phi_\lambda(\cdot)$ on the domain $\mathbb{N}_0$ possesses the slope $q_\lambda^U - \beta_\lambda \geq 0$. If $q_\lambda^U = \beta_\lambda$, then the intercept is $p_\lambda^U - \alpha_\lambda = 0$ leading to $B_{\lambda, X_0, n}^U \equiv 1$ and thus Goal (G1) is violated. If we use a slope $q_\lambda^U - \beta_\lambda > 0$, then both the sequences $\left(a_n^{(q_\lambda^U)}\right)_{n \in \mathbb{N}}$ and $\left(b_n^{(p_\lambda^U, q_\lambda^U)}\right)_{n \in \mathbb{N}}$ are strictly increasing and diverge to ∞. This comes from Properties 1 (P3b) and (P7b) since $q_\lambda^U > \beta_\lambda = \beta \geq 1$. Altogether, this implies that the corresponding upper bound component $\widetilde{B}_{\lambda, X_0, n}^{(p_\lambda^U, q_\lambda^U)}$ (cf. (42)) diverges to ∞ as well. This leads to

Proposition 10. *For all* $(\beta_{\mathcal{A}}, \beta_{\mathcal{H}}, \alpha_{\mathcal{A}}, \alpha_{\mathcal{H}}, \lambda) \in \mathcal{P}_{SP,4b} \times\,]0, 1[$ *and all initial population sizes* $X_0 \in \mathbb{N}$ *there do not exist parameters* $p_\lambda^U \geq 0$, $q_\lambda^U \geq 0$ *such that (35) is satisfied and such that the parts (c) and (d) of Proposition 8 hold true.*

3.13. Concluding Remarks on Alternative Upper Bounds for all Cases $(\beta_{\mathcal{A}}, \beta_{\mathcal{H}}, \alpha_{\mathcal{A}}, \alpha_{\mathcal{H}}, \lambda) \in (\mathcal{P}_{SP} \backslash \mathcal{P}_{SP,1}) \times\,]0, 1[$

As mentioned earlier on, starting from Section 3.6 we have principally focused on constructing upper bounds $B_{\lambda, X_0, n}^U$ of the Hellinger integrals, starting from p_λ^U, q_λ^U which fulfill (35) as well as further constraints depending on the Goals (G1) and (G2). For the setups in the Sections 3.7–3.9, we have proved the existence of *special parameter choices* p_λ^U, q_λ^U which were consistent with (G1) and (G2). Furthermore, for the constellation in the Section 3.11 we have found parameters such that at least (G2) is satisfied. In contrast, for the setup of Section 3.12 we have not found any choices which are consistent with (G1) and (G2), leading to the "cut-off bound" $B_{\lambda,X_0,n}^U \equiv 1$ which gives no improvement over the generally valid upper bound (9).

In the following, we present some *alternative choices* of p_λ^U, q_λ^U which–depending on the parameter constellation $(\beta_\mathcal{A}, \beta_\mathcal{H}, \alpha_\mathcal{A}, \alpha_\mathcal{H}, \lambda) \in (\mathcal{P}_{SP} \backslash \mathcal{P}_{SP,1}) \times]0,1[$–may or may not lead to upper bounds $B_{\lambda,X_0,n}^U$ which are consistent with Goal (G1) or with (G2) (and which are maybe weaker or better than resp. incomparable with the previous upper bounds when dealing with some relaxations of (G1), such as e.g., $H_\lambda(P_{\mathcal{A},n} || P_{\mathcal{H},n}) < 1$ for all but finitely many $n \in \mathbb{N}$).

As a first alternative choice for a linear upper bound of $\phi_\lambda(\cdot)$ (cf. (35)) one could use the asymptote $\widetilde{\phi_\lambda}(\cdot)$ (cf. Properties 3 (P20)) with the parameters $p_\lambda^U := \widetilde{p_\lambda} = \lambda \alpha_\mathcal{A} (\beta_\mathcal{A}/\beta_\mathcal{H})^{\lambda-1} + (1 - \lambda)\alpha_\mathcal{H} (\beta_\mathcal{A}/\beta_\mathcal{H})^\lambda$ and $q_\lambda^U := \widetilde{q_\lambda} = \beta_\mathcal{A}^\lambda \beta_\mathcal{H}^{1-\lambda}$. Another important linear upper bound of $\phi_\lambda(\cdot)$ is the tangent line $\phi_{\lambda,y}^{\tan}(\cdot)$ on $\phi_\lambda(\cdot)$ at an arbitrarily fixed point $y \in [0, \infty[$, which amounts to

$$\phi_{\lambda,y}^{\tan}(x) := r_{\lambda,y}^{\tan} + s_{\lambda,y}^{\tan} \cdot x := \left(p_{\lambda,y}^{\tan} - \alpha_\lambda\right) + \left(q_{\lambda,y}^{\tan} - \beta_\lambda\right) \cdot x := \left(\phi_\lambda(y) - y \cdot \phi_\lambda'(y)\right) + \phi_\lambda'(y) \cdot x, \quad (52)$$

where $\phi_\lambda'(\cdot)$ is given by (P17). Notice that this upper bound is for $y \in]0, \infty[\backslash \mathbb{N}$ "not tight" in the sense that $\phi_{\lambda,y}^{\tan}(\cdot)$ does not hit the function $\phi_\lambda(\cdot)$ on $\mathbb{N}_0$ (where the generation sizes "live"); moreover, $\phi_{\lambda,y}^{\tan}(x)$ might take on strictly positive values for large enough points x which is counter-productive for Goal (G1). Another alternative choice of a linear upper bound for $\phi_\lambda(\cdot)$, which in contrast to the tangent line is "tight" (but not necessarily avoiding the strict positivity), is the secant line $\phi_{\lambda,k}^{sec}(\cdot)$ across its arguments k and $k+1$, given by

$$\begin{aligned}
\phi_{\lambda,k}^{sec}(x) &:= r_{\lambda,k}^{sec} + s_{\lambda,k}^{sec} \cdot x := \left(p_{\lambda,k}^{sec} - \alpha_\lambda\right) + \left(q_{\lambda,k}^{sec} - \beta_\lambda\right) \cdot x \\
&:= \left[\phi_\lambda(k) - k \cdot \left(\phi_\lambda(k+1) - \phi_\lambda(k)\right)\right] + \left(\phi_\lambda(k+1) - \phi_\lambda(k)\right) \cdot x.
\end{aligned} \quad (53)$$

Another alternative choice is the horizontal line

$$\phi_\lambda^{hor}(x) \equiv \max\left\{\phi_\lambda(y), \, y \in \mathbb{N}_0\right\}. \quad (54)$$

For $p_\lambda^U \in \left\{\widetilde{p_\lambda}, p_{\lambda,y}^{\tan}, p_{\lambda,y}^{sec}\right\}$ and $q_\lambda^U \in \left\{q_{\lambda,y}^{\tan}, q_{\lambda,y}^{sec}\right\}$ it is possible that in some parameter cases $(\beta_\mathcal{A}, \beta_\mathcal{H}, \alpha_\mathcal{A}, \alpha_\mathcal{H})$ either the intercept $r_\lambda^U = p_\lambda^U - \alpha_\lambda$ is strictly larger than zero or the slope $s_\lambda^U = q_\lambda^U - \beta_\lambda$ is strictly larger than zero. Thus, it can happen that $\widetilde{B}_{\lambda,X_0,n}^{(p_\lambda^U, q_\lambda^U)} > 1$ for some (and even for all) $n \in \mathbb{N}$, such that the corresponding upper bound $B_{\lambda,X_0,n}^U$ for the Hellinger integral $H_\lambda(P_{\mathcal{A},n} || P_{\mathcal{H},n})$ amounts to the cut-off at 1. However, due to Properties 1 (P5) and (P7a), the sequence $\left(\widetilde{B}_{\lambda,X_0,n}^{(p_\lambda^U, q_\lambda^U)}\right)_{n \in \mathbb{N}}$ may become smaller than 1 and may finally converge to zero. Due to Properties 2 (P14), this upper bound can even be tighter (smaller) than those bounds derived from parameters p_λ^U, q_λ^U fulfilling (47).

As far as our desired Hellinger integral bounds are concerned, in the setup of Section 3.11 —where $\lim_{y\to\infty} \phi_{\lambda,y}^{\tan}(\cdot) \equiv 0$–for the proof of Proposition 9 in Appendix A.1 we shall employ the mappings $y \mapsto \phi_{\lambda,y}^{\tan}$ resp. $y \mapsto p_{\lambda,y}^{\tan}$ resp. $y \mapsto q_{\lambda,y}^{\tan}$. These will also be used for the proof of the below-mentioned Theorem 4.

3.14. Intermezzo 1: Application to Asymptotical Distinguishability

The above-mentioned investigations can be applied to the context of Section 2.6 on asymptotical distinguishability. Indeed, with the help of the Definitions 1 and 2 as well as the equivalence relations (25) and (26) we obtain the following

Corollary 1.

(a) For all $(\beta_{\mathcal{A}}, \beta_{\mathcal{H}}, \alpha_{\mathcal{A}}, \alpha_{\mathcal{H}}) \in \mathcal{P}_{SP} \backslash \mathcal{P}_{SP,4b}$ and all initial population sizes $X_0 \in \mathbb{N}$, the corresponding sequences $(P_{\mathcal{A},n})_{n \in \mathbb{N}_0}$ and $(P_{\mathcal{H},n})_{n \in \mathbb{N}_0}$ are entirely separated (completely asymptotically distinguishable).

(b) For all $(\beta_{\mathcal{A}}, \beta_{\mathcal{H}}, \alpha_{\mathcal{A}}, \alpha_{\mathcal{H}}) \in \mathcal{P}_{NI}$ with $\beta_{\mathcal{A}} \leq 1$ and all initial population sizes $X_0 \in \mathbb{N}$, the sequence $(P_{\mathcal{A},n})_{n \in \mathbb{N}_0}$ is contiguous to $(P_{\mathcal{H},n})_{n \in \mathbb{N}_0}$.

(c) For all $(\beta_{\mathcal{A}}, \beta_{\mathcal{H}}, \alpha_{\mathcal{A}}, \alpha_{\mathcal{H}}) \in \mathcal{P}_{NI}$ with $\beta_{\mathcal{A}} > 1$ and all initial population sizes $X_0 \in \mathbb{N}$, the sequence $(P_{\mathcal{A},n})_{n \in \mathbb{N}_0}$ is neither contiguous to nor entirely separated to $(P_{\mathcal{H},n})_{n \in \mathbb{N}_0}$.

The proof of Corollary 1 will be given in Appendix A.1.

Remark 3.

(a) Assertion (c) of Corollary 1 contrasts the case of Gaussian processes with independent increments where one gets either entire separation or mutual contiguity (see e.g., Liese & Vajda [1]).

(b) By putting Corollary 1(b) and (c) together, we obtain for different "criticality pairs" in the non-immigration case $\mathcal{P}_{NI}$ the following asymptotical distinguishability types:
$(P_{\mathcal{A},n}) \vartriangleleft\vartriangleright (P_{\mathcal{H},n})$ if $\beta_{\mathcal{A}} \leq 1, \beta_{\mathcal{H}} \leq 1$; $\qquad$ $(P_{\mathcal{A},n}) \vartriangleleft\overline{\vartriangleright} (P_{\mathcal{H},n})$ if $\beta_{\mathcal{A}} \leq 1, \beta_{\mathcal{H}} > 1$;
$(P_{\mathcal{A},n}) \vartriangleleft\vartriangleright (P_{\mathcal{H},n})$ if $\beta_{\mathcal{A}} > 1, \beta_{\mathcal{H}} \leq 1$; $\qquad$ $(P_{\mathcal{A},n}) \overline{\vartriangleleft}\overline{\vartriangleright} (P_{\mathcal{H},n})$ and $(P_{\mathcal{A},n}) \overline{\triangle} (P_{\mathcal{H},n})$ if $\beta_{\mathcal{A}} > 1, \beta_{\mathcal{H}} > 1$;
in particular, for $\mathcal{P}_{NI}$ the sequences $(P_{\mathcal{A},n})_{n \in \mathbb{N}_0}$ and $(P_{\mathcal{H},n})_{n \in \mathbb{N}_0}$ are not completely asymptotically inseparable (indistinguishable).

(c) In the light of the above-mentioned characterizations of contiguity resp. entire separation by means of Hellinger integral limits, the finite-time-horizon results on Hellinger integrals given in the "$\lambda \in]0,1[$ parts" of Theorem 1, the Sections 3.3–3.13 and also in the below-mentioned Section 6 can loosely be interpreted as "finite-sample (rather than asymptotical) distinguishability" assertions.

3.15. Intermezzo 2: Application to Decision Making under Uncertainty

3.15.1. Bayesian Decision Making

The above-mentioned investigations can be applied to the context of Section 2.5 on *dichotomous* Bayesian decision making on the space of all possible path scenarios (path space) of Poissonian Galton-Watson processes without/with immigration GW(I) (e.g., in combination with our running-example epidemiological context of Section 2.3). More detailed, for the minimal mean decision loss (Bayes risk) $\mathcal{R}_n$ defined by (18) we can derive upper (respectively lower) bounds by using (19) respectively (20) together with the exact values or the upper (respectively lower) bounds of the Hellinger integrals $H_\lambda(P_{\mathcal{A},n} || P_{\mathcal{H},n})$ derived in the "$\lambda \in]0,1[$ parts" of Theorem 1, the Sections 3.3–3.13 (and also in the below-mentioned Section 6); instead of providing the corresponding outcoming formulas–which is merely repetitive–we give the illustrative

Example 1. *Based on a sample path observation* $\mathcal{X}_n := \{X_\ell : \ell = 1, ..., n\}$ *of a GWI, which is either governed by a hypothesis law* $P_{\mathcal{H}}$ *or an alternative law* $P_{\mathcal{A}}$, *we want to make a dichotomous optimal Bayesian decision described in Section 2.5, namely, decide between an action* $d_{\mathcal{H}}$ *"associated with"* $P_{\mathcal{H}}$ *and an action* $d_{\mathcal{A}}$ *"associated with"* $P_{\mathcal{A}}$, *with pregiven loss function (16) involving constants* $L_{\mathcal{A}} > 0, L_{\mathcal{H}} > 0$ *which e.g., arise as bounds from quantities in worst-case scenarios.*

For this, let us exemplarily deal with initial population $X_0 = 5$ *as well as parameter setup* $(\beta_{\mathcal{A}}, \beta_{\mathcal{H}}, \alpha_{\mathcal{A}}, \alpha_{\mathcal{H}}) = (1.2, 0.9, 4, 3) \in \mathcal{P}_{SP,1}$; *within our running-example epidemiological context of Section 2.3, this corresponds e.g., to a setup where one is encountered with a novel infectious disease (such as COVID-19) of non-negligible fatality rate, and* $(\mathcal{A})$ *reflects a "potentially dangerous" infectious-disease-transmission situation (with supercritical reproduction number* $\beta_{\mathcal{A}} = 1.2$ *and importation mean of* $\alpha_{\mathcal{A}} = 4$, *for weekly appearing new incidence-generations) whereas* $(\mathcal{H})$ *describes a "milder" situation (with subcritical* $\beta_{\mathcal{H}} = 0.9$*

and $\alpha_{\mathcal{H}} = 3$). *Moreover, let $d_{\mathcal{H}}$ and $d_{\mathcal{A}}$ reflect two possible sets of interventions (control measures) in the course of pandemic risk management, with respective "worst-case type" decision losses $L_{\mathcal{A}} = 600$ and $L_{\mathcal{H}} = 300$ (e.g., in units of billion Euros or U.S. Dollars). Additionally we assume the prior probabilities $\pi = Pr(\mathcal{H}) = 1 - Pr(\mathcal{A}) = 0.5$, which results in the prior-loss constants $\mathfrak{L}_{\mathcal{A}} = 300$ and $\mathfrak{L}_{\mathcal{H}} = 150$. In order to obtain bounds for the corresponding minimal mean decision loss (Bayes Risk) $\mathcal{R}_n$ defined in* (18) *we can employ the general Stummer-Vajda bounds (cf. [15])* (19) *and* (20) *in terms of the Hellinger integral $H_\lambda(P_{\mathcal{A},n}||P_{\mathcal{H},n})$ (with arbitrary $\lambda \in\ ]0,1[$), and combine this with the appropriate detailed results on the latter from the preceding subsections. To demonstrate this, let us choose $\lambda = 0.5$ (for which $H_{1/2}(P_{\mathcal{A},n}||P_{\mathcal{H},n})$ can be interpreted as a multiple of the Bhattacharyya coefficient between the two competing GWI) respectively $\lambda = 0.9$, leading to the parameters $p_{0.5}^E = 3.464$, $q_{0.5}^E = 1.039$ respectively $p_{0.9}^E = 3.887$, $q_{0.9}^E = 1.166$ (cf.* (33)*). Combining* (19) *and* (20) *with Theorem* 1 *(a)– which provides us with the exact recursive values of $H_\lambda(P_{\mathcal{A},n}||P_{\mathcal{H},n})$ in terms of the sequence $a_n^{(q_\lambda^E)}$ (cf.* (36)*)– we obtain for $\lambda = 0.5$ the bounds*

$$\mathcal{R}_n \ \leq\ \mathcal{R}_n^U := 2.121 \cdot 10^2 \cdot \exp\left\{5 \cdot a_n^{(1.039)} + \frac{10}{3} \cdot \sum_{k=1}^{n} a_k^{(1.039)}\right\},$$

$$\mathcal{R}_n \ \geq\ \mathcal{R}_n^L := 100 \cdot \exp\left\{10 \cdot a_n^{(1.039)} + \frac{20}{3} \cdot \sum_{k=1}^{n} a_k^{(1.039)}\right\},$$

whereas for $\lambda = 0.9$ we get

$$\mathcal{R}_n \ \leq\ \mathcal{R}_n^U := 2.799 \cdot 10^2 \cdot \exp\left\{5 \cdot a_n^{(1.166)} + \frac{10}{3} \cdot \sum_{k=1}^{n} a_k^{(1.166)}\right\},$$

$$\mathcal{R}_n \ \geq\ \mathcal{R}_n^L := 3.902 \cdot \exp\left\{50 \cdot a_n^{(1.166)} + \frac{100}{3} \cdot \sum_{k=1}^{n} a_k^{(1.166)}\right\}.$$

Figure 1 *illustrates the lower (orange resp. cyan) and upper (red resp. blue) bounds $\mathcal{R}_n^L$ resp. $\mathcal{R}_n^U$ of the Bayes Risk $\mathcal{R}_n$ employing $\lambda = 0.5$ resp. $\lambda = 0.9$ on both a unit scale (left graph) and a logarithmic scale (right graph). The lightgrey/grey/black curves correspond to the* (18)*-based empirical evaluation of the Bayes risk sequence $\left(\mathcal{R}_n^{sample}\right)_{n=1,\dots,50}$ from three independent Monte Carlo simulations of 10000 GWI sample paths (each) up to time horizon 50.*

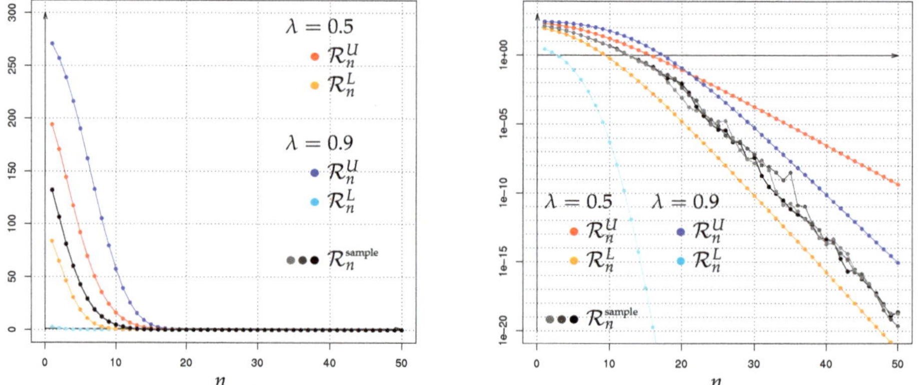

Figure 1. Bayes risk bounds (using $\lambda = 0.5$ (red/orange) resp. $\lambda = 0.9$ (blue/cyan)) and Bayes risk simulations (lightgrey/grey/black) on a unit (**left graph**) and logarithmic (**right graph**) scale in the parameter setup $(\beta_{\mathcal{A}}, \beta_{\mathcal{H}}, \alpha_{\mathcal{A}}, \alpha_{\mathcal{H}}) = (1.2, 0.9, 4, 3) \in \mathcal{P}_{SP,1}$, with initial population $X_0 = 5$ and prior-loss constants $\mathfrak{L}_{\mathcal{A}} = 300$ and $\mathfrak{L}_{\mathcal{H}} = 150$.

3.15.2. Neyman-Pearson Testing

By combining (23) with the exact values resp. upper bounds of the Hellinger integrals $H_\lambda\left(P_{\mathcal{A},n}||P_{\mathcal{H},n}\right)$ from the preceding subsections, we obtain for our context of GW(I) with Poisson offspring and Poisson immigration (including the non-immigration case) some upper bounds of the *minimal* type II error probability $\mathcal{E}_\varsigma\left(P_{\mathcal{A},n}||P_{\mathcal{H},n}\right)$ in the class of the tests for which the type I error probability is at most $\varsigma \in]0,1[$, which can also be immediately rewritten as lower bounds for the power $1 - \mathcal{E}_\varsigma\left(P_{\mathcal{A},n}||P_{\mathcal{H},n}\right)$ of a most powerful test at level ς. As for the Bayesian context of Section 3.15.1, instead of providing the–merely repetitive–outcoming formulas for the bounds of $\mathcal{E}_\varsigma\left(P_{\mathcal{A},n}||P_{\mathcal{H},n}\right)$ we give the illustrative

Example 2. *Consider the Figures 2 and 3 which deal with initial population $X_0 = 5$ and the parameter setup $(\beta_{\mathcal{A}}, \beta_{\mathcal{H}}, \alpha_{\mathcal{A}}, \alpha_{\mathcal{H}}) = (0.3, 1.2, 1, 4) \in \mathcal{P}_{SP,1}$; within our running-example epidemiological context of Section 2.3, this corresponds to a "potentially dangerous" infectious-disease-transmission situation $(\mathcal{H})$ (with supercritical reproduction number $\beta_{\mathcal{H}} = 1.2$ and importation mean of $\alpha_{\mathcal{H}} = 4$), whereas $(\mathcal{A})$ describes a "very mild" situation (with "low" subcritical $\beta_{\mathcal{A}} = 0.3$ and $\alpha_{\mathcal{A}} = 1$). Figure 2 shows the lower and upper bounds of $\mathcal{E}_\varsigma\left(P_{\mathcal{A},n}||P_{\mathcal{H},n}\right)$ with $\varsigma = 0.05$, evaluated from the Formulas (23) and (24), together with the exact values of the Hellinger integral $H_\lambda\left(P_{\mathcal{A},n}||P_{\mathcal{H},n}\right)$, cf. Theorem 1 (recall that we are in the setup $\mathcal{P}_{SP,1}$) on both a unit scale (left graph) and a logarithmic scale (right graph). The orange resp. red resp. purple curves correspond to the outcoming upper bounds $\mathcal{E}_n^U := \mathcal{E}_n^U\left(P_{\mathcal{A},n}||P_{\mathcal{H},n}\right)$ (cf. (23)) with parameters $\lambda = 0.3$ resp. $\lambda = 0.5$ resp. $\lambda = 0.7$. The green resp. cyan resp. blue curves correspond to the lower bounds $\mathcal{E}_n^L := \mathcal{E}_n^L\left(P_{\mathcal{A},n}||P_{\mathcal{H},n}\right)$ (cf. (24)) with parameters $\lambda = 2$ resp. $\lambda = 1.5$ resp. $\lambda = 1.1$. Notice the different λ-ranges in (23) and (24). In contrast, Figure 3 compares the lower bound $\mathcal{E}_n^L$ (for fixed $\lambda = 1.1$) with the upper bound $\mathcal{E}_n^U$ (for fixed $\lambda = 0.5$) of the minimal type II error probability $\mathcal{E}_\varsigma(P_{\mathcal{A},n}||P_{\mathcal{H},n})$ for different levels $\varsigma = 0.1$ (orange for the lower and cyan for the upper bound), $\varsigma = 0.05$ (green and magenta) and $\varsigma = 0.01$ (blue and purple) on both a unit scale (left graph) and a logarithmic scale (right graph).*

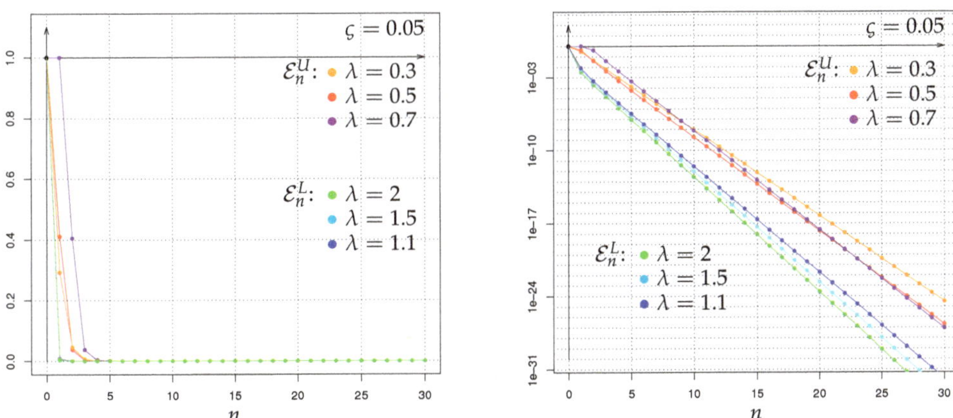

Figure 2. Different lower bounds $\mathcal{E}_n^L$ (using $\lambda \in \{1.1, 1.5, 2\}$) and upper bounds $\mathcal{E}_n^U$ (using $\lambda \in \{0.3, 0.5, 0.7\}$) of the minimal type II error probability $\mathcal{E}_\varsigma\left(P_{\mathcal{A},n}||P_{\mathcal{H},n}\right)$ for fixed level $\varsigma = 0.05$ in the parameter setup $(\beta_{\mathcal{A}}, \beta_{\mathcal{H}}, \alpha_{\mathcal{A}}, \alpha_{\mathcal{H}}) = (0.3, 1.2, 1, 4) \in \mathcal{P}_{SP,1}$ together with initial population $X_0 = 5$ on both a unit scale (**left graph**) and a logarithmic scale (**right graph**).

3.16. Goals for Lower Bounds for the Cases $(\beta_{\mathcal{A}}, \beta_{\mathcal{H}}, \alpha_{\mathcal{A}}, \alpha_{\mathcal{H}}, \lambda) \in (\mathcal{P}_{SP} \backslash \mathcal{P}_{SP,1}) \times (\mathbb{R} \backslash [0, 1])$

Recall from (49) the set $\mathcal{P}_{SP} := \left\{ (\beta_{\mathcal{A}}, \beta_{\mathcal{H}}, \alpha_{\mathcal{A}}, \alpha_{\mathcal{H}}) \in]0, \infty[^4 : (\alpha_{\mathcal{A}} \neq \alpha_{\mathcal{H}}) \text{ or } (\beta_{\mathcal{A}} \neq \beta_{\mathcal{H}}) \text{ or both} \right\}$ and the "equal-fraction-case" set $\mathcal{P}_{SP,1} := \left\{ (\beta_{\mathcal{A}}, \beta_{\mathcal{H}}, \alpha_{\mathcal{A}}, \alpha_{\mathcal{H}}) \in \mathcal{P}_{SP} : \alpha_{\mathcal{A}} \neq \alpha_{\mathcal{H}}, \beta_{\mathcal{A}} \neq \beta_{\mathcal{H}}, \frac{\alpha_{\mathcal{A}}}{\beta_{\mathcal{A}}} = \frac{\alpha_{\mathcal{H}}}{\beta_{\mathcal{H}}} \right\}$, where for the latter we have derived in Theorem 1(a) and in Proposition 5 the *exact* recursive values for the time-behaviour of the Hellinger integrals $H_\lambda(P_{\mathcal{A},1}||P_{\mathcal{H},1})$ of order $\lambda \in \mathbb{R}\backslash[0,1]$. Moreover, recall that for the case $(\beta_\mathcal{A}, \beta_\mathcal{H}, \alpha_\mathcal{A}, \alpha_\mathcal{H}, \lambda) \in (\mathcal{P}_{SP}\backslash\mathcal{P}_{SP,1})\times]0,1[$ we have obtained in the Sections 3.4 and 3.5 some "optimal" linear lower bounds $\phi_\lambda^L(\cdot)$ for the strictly concave function $\phi_\lambda(x) := \phi(x, \beta_\mathcal{A}, \beta_\mathcal{H}, \alpha_\mathcal{A}, \alpha_\mathcal{H}, \lambda)$ on the domain $x \in [0, \infty[$; due to the monotonicity Properties 2 (P10) to (P12) of the sequences $\left(a_n^{(q_\lambda^L)}\right)_{n\in\mathbb{N}}$ and $\left(b_n^{(p_\lambda^L, q_\lambda^L)}\right)_{n\in\mathbb{N}}$, these bounds have led to the "optimal" recursive lower bound $B_{\lambda,X_0,n}^L$ of the Hellinger integral $H_\lambda(P_{\mathcal{A},n}||P_{\mathcal{H},n})$ in (40) of Theorem 1(b)).

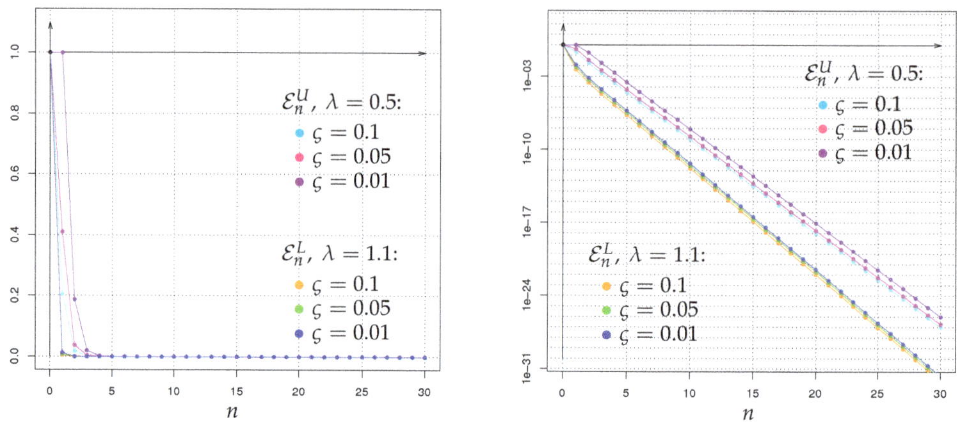

Figure 3. The lower bound $\mathcal{E}_n^L$ (using $\lambda = 1.1$) and the upper bound $\mathcal{E}_n^U$ (using $\lambda = 0.5$) of the minimal type II error probability $\mathcal{E}_\varsigma(P_{\mathcal{A},n}||P_{\mathcal{H},n})$ for different levels $\varsigma \in \{0.01, 0.05, 0.1\}$ in the parameter setup $(\beta_\mathcal{A}, \beta_\mathcal{H}, \alpha_\mathcal{A}, \alpha_\mathcal{H}) = (0.3, 1.2, 1, 4) \in \mathcal{P}_{SP,1}$ together with initial population $X_0 = 5$ on both a unit scale (**left graph**) and a logarithmic scale (**right graph**).

In contrast, the strict *convexity* of the function $\phi_\lambda(\cdot)$ in the case $(\beta_\mathcal{A}, \beta_\mathcal{H}, \alpha_\mathcal{A}, \alpha_\mathcal{H}, \lambda) \in (\mathcal{P}_{SP}\backslash\mathcal{P}_{SP,1}) \times (\mathbb{R}\backslash[0,1])$ implies that we cannot maximize both parameters p_λ^L, $q_\lambda^L \in \mathbb{R}$ *simultaneously* subject to the constraint (35). This effect carries over to the lower bounds $B_{\lambda,X_0,n}^L$ of the Hellinger integrals $H_\lambda(P_{\mathcal{A},n}||P_{\mathcal{H},n})$ (cf. (41)); in general, these bounds cannot be maximized *simultaneously* for all initial population sizes $X_0 \in \mathbb{N}$ and all observation horizons $n \in \mathbb{N}$.

Analogously to (46), one way to obtain "good" recursive lower bounds for $H_\lambda(P_{\mathcal{A},n}||P_{\mathcal{H},n})$ from (41) in Theorem 1 (b) is to solve the optimization problem,

$$\left(\overline{p_\lambda^L}, \overline{q_\lambda^L}\right) := \underset{(p_\lambda^L, q_\lambda^L)\in\mathbb{R}^2}{\arg\max}\left\{\exp\left\{a_n^{(q_\lambda^L)}\cdot X_0 + \sum_{k=1}^n b_k^{(p_\lambda^L, q_\lambda^L)}\right\}\right\} \qquad \text{such that (35) is satisfied,} \qquad (55)$$

for each fixed initial population size $X_0 \in \mathbb{N}$ and observation horizon $n \in \mathbb{N}$. But due to the same reasons as explained right after (46), the optimization problem (55) seems to be not straightforward to solve explicitly. In a congeneric way as in the discussion of the upper bounds for the case $\lambda \in]0,1[$ above, we now have to look for suitable parameters p_λ^L, q_λ^L for the lower bound $B_{\lambda,X_0,n}^L \leq H_\lambda(P_{\mathcal{A},n}||P_{\mathcal{H},n})$ that fulfill (35) and that guarantee certain reasonable criteria and goals; these are similar to the goals (G1) to (G3) from Section 3.6, and are therefore supplemented by an additional " $'$ ":

(G1′) the validity of $B_{\lambda,X_0,n}^L > 1$ *simultaneously* for all initial configurations $X_0 \in \mathbb{N}$, all observation horizons $n \in \mathbb{N}$ and all $\lambda \in \mathbb{R}\backslash[0,1]$, which leads to a *strict* improvement of the general upper bound $H_\lambda(P_{\mathcal{A},n}||P_{\mathcal{H},n}) > 1$ (cf. (11));

(G2′) the determination of the long-term-limits $\lim_{n\to\infty} H_\lambda(P_{A,n}||P_{\mathcal{H},n})$ respectively $\lim_{n\to\infty} B^L_{\lambda,X_0,n}$ for all $X_0 \in \mathbb{N}$ and all $\lambda \in \mathbb{R}\backslash[0,1]$; in particular, one would like to check whether $\lim_{n\to\infty} H_\lambda(P_{A,n}||P_{\mathcal{H},n}) = \infty$;

(G3′) the determination of the time-asymptotical growth rates $\lim_{n\to\infty} \frac{1}{n}\log\left(H_\lambda(P_{A,n}||P_{\mathcal{H},n})\right)$ resp. $\lim_{n\to\infty} \frac{1}{n}\log\left(B^L_{\lambda,X_0,n}\right)$ for all $X_0 \in \mathbb{N}$ and all $\lambda \in \mathbb{R}\backslash[0,1]$.

In the following, let us briefly discuss how these three goals can be achieved in principle, where we confine ourselves to parameters p^L_λ, q^L_λ which–in addition to (35)–fulfill the requirement

$$\left\{ q^L_\lambda \geq \max\{0,\beta_\lambda\} \quad \wedge \quad p^L_\lambda > \max\{0,\alpha_\lambda\} \right\} \quad \vee \quad \left\{ q^L_\lambda > \max\{0,\beta_\lambda\} \quad \wedge \quad p^L_\lambda \geq \max\{0,\alpha_\lambda\} \right\}, \quad (56)$$

where $\wedge$ is the logical "AND" and $\vee$ the logical "OR" operator. This is sufficient to tackle all three Goals (G1′) to (G3′). To see this, assume that p^L_λ, q^L_λ satisfy (35). Let us begin with the two "extremal" cases in (56), i.e., with (i) $q^L_\lambda = \max\{0,\beta_\lambda\}$, $p^L_\lambda > \max\{0,\alpha_\lambda\}$, respectively (ii) $q^L_\lambda > \max\{0,\beta_\lambda\}$, $p^L_\lambda = \max\{0,\alpha_\lambda\}$.

Suppose in the first extremal case (i) that $\beta_\lambda \leq 0$. Then, $q^L_\lambda = 0$ and Properties 1 (P4) implies that $a_n^{(q^L_\lambda)} = -\beta_\lambda \geq 0$ and hence $b_n^{(p^L_\lambda,q^L_\lambda)} = p^L_\lambda e^{-\beta_\lambda} - \alpha_\lambda \geq p^L_\lambda - \alpha_\lambda > 0$ for all $n \in \mathbb{N}$. This enters into (41) as follows: the Hellinger integral lower bound becomes $B^L_{\lambda,X_0,n} \geq \widetilde{B}^{(p^L_\lambda,q^L_\lambda)}_{\lambda,X_0,n} = \exp\{-\beta_\lambda \cdot X_0 + (p^L_\lambda e^{-\beta_\lambda} - \alpha_\lambda)\cdot n\} > 1$. Furthermore, one clearly has $\lim_{n\to\infty} B^L_{\lambda,X_0,n} = \infty$ as well as $\lim_{n\to\infty} \frac{1}{n}\log\left(B^L_{\lambda,X_0,n}\right) = p^L_\lambda e^{-\beta_\lambda} - \alpha_\lambda > 0$. Assume now that $\beta_\lambda > 0$. Then, $q^L_\lambda = \beta_\lambda > 0$, $a_n^{(q^L_\lambda)} = 0$ (cf. (P2)), $b_n^{(p^L_\lambda,q^L_\lambda)} = p^L_\lambda - \alpha_\lambda > 0$ and thus $B^L_{\lambda,X_0,n} = \exp\{(p^L_\lambda - \alpha_\lambda)\cdot n\} > 1$ for all $n \in \mathbb{N}$. Furthermore, one gets $\lim_{n\to\infty} B^L_{\lambda,X_0,n} = \infty$ as well as $\lim_{n\to\infty} \frac{1}{n}\log\left(B^L_{\lambda,X_0,n}\right) = p^L_\lambda - \alpha_\lambda > 0$.

Let us consider the other above-mentioned extremal case (ii). Suppose that $q^L_\lambda > \max\{0,\beta_\lambda\}$ together with $q^L_\lambda > \min\{1,e^{\beta_\lambda-1}\}$ which implies that the sequence $\left(a_n^{(q^L_\lambda)}\right)_{n\in\mathbb{N}}$ is strictly positive, strictly increasing and grows to infinity faster than exponentially, cf. (P3b). Hence, $B^L_{\lambda,X_0,n} \geq \exp\{a_n^{(q^L_\lambda)}\cdot X_0\} > 1$, $\lim_{n\to\infty} B^L_{\lambda,X_0,n} = \infty$ as well as $\lim_{n\to\infty} \frac{1}{n}\log\left(B^L_{\lambda,X_0,n}\right) = \infty$. If $\max\{0,\beta_\lambda\} < q^L_\lambda \leq \min\{1,e^{\beta_\lambda-1}\}$, then $\left(a_n^{(q^L_\lambda)}\right)_{n\in\mathbb{N}}$ is strictly positive, strictly increasing and converges to $x_0^{(q_\lambda)} \in]0,-\log(q^L_\lambda)]$ (cf. (P3a)). This carries over to the sequence $\left(b_n^{(p^L_\lambda,q^L_\lambda)}\right)_{n\in\mathbb{N}}$: one gets $b_1^{(p^L_\lambda,q^L_\lambda)} = p^L_\lambda - \alpha_\lambda \geq 0$ and $b_n^{(p^L_\lambda,q^L_\lambda)} > 0$ for all $n \geq 2$. Furthermore, $b_n^{(p^L_\lambda,q^L_\lambda)}$ is strictly increasing and converges to $p^L_\lambda \cdot e^{x_0^{(q^L_\lambda)}} - \alpha_\lambda > 0$, leading to $B^L_{\lambda,X_0,n} > 1$ for all $n \in \mathbb{N}$, to $\lim_{n\to\infty} B^L_{\lambda,X_0,n} = \infty$ as well as to $\lim_{n\to\infty} \frac{1}{n}\log\left(B^L_{\lambda,X_0,n}\right) = p^L_\lambda \cdot e^{x_0^{(q^L_\lambda)}} - \alpha_\lambda > 0$.

It remains to look at the cases where p^L_λ, q^L_λ satisfy (35), and (56) with two strict inequalities. For this situation, one gets

- $\left(a_n^{(q^L_\lambda)}\right)_{n\in\mathbb{N}}$ is strictly positive, strictly increasing and–iff $q^L_\lambda \leq \min\{1,e^{\beta_\lambda-1}\}$–convergent (namely to the smallest positive solution $x_0^{(q^L_\lambda)} \in]0,-\log(q^L_\lambda)]$ of (44)), cf. (P3);

- $\left(b_n^{(p^L_\lambda,q^L_\lambda)}\right)_{n\in\mathbb{N}}$ is strictly increasing, strictly positive (since $b_1^{(p^L_\lambda,q^L_\lambda)} = p^L_\lambda - \alpha_\lambda > 0$) and–iff $q^L_\lambda \leq \min\{1,e^{\beta_\lambda-1}\}$–convergent (namely to $p^L_\lambda e^{x_0^{(q^L_\lambda)}} - \alpha_\lambda \in [p^L_\lambda - \alpha_\lambda, p^L_\lambda/q^L_\lambda - \alpha_\lambda])$, cf (P7).

Hence, under the assumptions (35) and $(p^L_\lambda > \max\{0,\alpha_\lambda\}) \wedge (q^L_\lambda > \max\{0,\beta_\lambda\})$ the corresponding lower bounds $B^L_{\lambda,X_0,n}$ of the Hellinger integral $H_\lambda(P_{A,n}||P_{\mathcal{H},n})$ fulfill for all $X_0 \in \mathbb{N}$

- $B^L_{\lambda,X_0,n} > 1$ for all $n \in \mathbb{N}$,

- $\lim_{n \to \infty} B^{L}_{\lambda, X_0, n} = \infty$,

- $\lim_{n \to \infty} \frac{1}{n} \log \left(B^{L}_{\lambda, X_0, n} \right) = p^{L}_{\lambda} e^{x_0^{(q^L_\lambda)}} - \alpha_\lambda > 0$ for the case $q^{L}_{\lambda} \in \,] \max\{0, \beta_\lambda\}, \min\{1, e^{\beta_\lambda - 1}\}]$, respectively $\lim_{n \to \infty} \frac{1}{n} \log \left(B^{L}_{\lambda, X_0, n} \right) = \infty$ for the remaining case $q^{L}_{\lambda} > \min\{1, e^{\beta_\lambda - 1}\}$.

Putting these considerations together we conclude that the constraints (35) and (56) are sufficient to achieve the Goals (G1′) to (G3′). Hence, for fixed parameter constellation $(\beta_\mathcal{A}, \beta_\mathcal{H}, \alpha_\mathcal{A}, \alpha_\mathcal{H}, \lambda)$, we aim for finding $p^{L}_{\lambda} = p^{L}(\beta_\mathcal{A}, \beta_\mathcal{H}, \alpha_\mathcal{A}, \alpha_\mathcal{H}, \lambda)$ and $q^{L}_{\lambda} = q^{L}(\beta_\mathcal{A}, \beta_\mathcal{H}, \alpha_\mathcal{A}, \alpha_\mathcal{H}, \lambda)$ which satisfy (35) and (56). This can be achieved mostly, but not always, as we shall show below. As an auxiliary step for further investigations, it is useful to examine the set of all $\lambda \in \mathbb{R} \backslash [0, 1]$ for which $\alpha_\lambda \leq 0$ or $\beta_\lambda \leq 0$ (or both). By straightforward calculations, we see that

$$\alpha_\lambda \leq 0 \iff \lambda \begin{cases} \leq \frac{-\alpha_\mathcal{H}}{\alpha_\mathcal{A} - \alpha_\mathcal{H}}, & \text{if } \alpha_\mathcal{A} > \alpha_\mathcal{H}, \\[2mm] \geq \frac{\alpha_\mathcal{H}}{\alpha_\mathcal{H} - \alpha_\mathcal{A}}, & \text{if } \alpha_\mathcal{A} < \alpha_\mathcal{H}, \end{cases} \quad \text{and} \quad \beta_\lambda \leq 0 \iff \lambda \begin{cases} \leq \frac{-\beta_\mathcal{H}}{\beta_\mathcal{A} - \beta_\mathcal{H}}, & \text{if } \beta_\mathcal{A} > \beta_\mathcal{H}, \\[2mm] \geq \frac{\beta_\mathcal{H}}{\beta_\mathcal{H} - \beta_\mathcal{A}}, & \text{if } \beta_\mathcal{A} < \beta_\mathcal{H}. \end{cases} \tag{57}$$

Furthermore, recall that (35) implies the general bounds $p^{L}_{\lambda} \leq \alpha^{\lambda}_{\mathcal{A}} \alpha^{1-\lambda}_{\mathcal{H}} = \varphi_\lambda(0)$ (being equivalent to the requirement $\phi^{L}_{\lambda}(0) = \phi_\lambda(0)$) and $q^{L}_{\lambda} \leq \beta^{\lambda}_{\mathcal{A}} \beta^{1-\lambda}_{\mathcal{H}} = \tilde{q}_\lambda$ (the latter being the maximal slope due to Properties 3 (P19), (P20)).

Let us now undertake the desired *detailed* investigations on lower and upper bounds of the Hellinger integrals $H_\lambda(P_{\mathcal{A}, n} || P_{\mathcal{H}, n})$ of order $\lambda \in \mathbb{R} \backslash [0, 1]$, for the various different subclasses of $\mathcal{P}_{\text{SP}} \backslash \mathcal{P}_{\text{SP,1}}$.

3.17. Lower Bounds for the Cases $(\beta_\mathcal{A}, \beta_\mathcal{H}, \alpha_\mathcal{A}, \alpha_\mathcal{H}, \lambda) \in \mathcal{P}_{\text{SP,2}} \times (\mathbb{R} \backslash [0, 1])$

In such a constellation, where $\mathcal{P}_{\text{SP,2}} := \{ (\beta_\mathcal{A}, \beta_\mathcal{H}, \alpha_\mathcal{A}, \alpha_\mathcal{H}) \in \mathcal{P}_{\text{SP}} : \alpha_\mathcal{A} = \alpha_\mathcal{H}, \beta_\mathcal{A} \neq \beta_\mathcal{H} \}$ (cf. (49)), one gets $\phi_\lambda(0) = 0$ (cf. Properties 3 (P16)), $\phi'_\lambda(0) = 0$ (cf. (P17)). Thus, the only choice for the intercept and the slope of the linear lower bound $\phi^{L}_{\lambda}(\cdot)$ for $\phi_\lambda(\cdot)$, which satisfies (35) for all $x \in \mathbb{N}$ and (potentially) (56), is $r^{L}_{\lambda} = 0 = p^{L}_{\lambda} - \alpha_\lambda$ (i.e., $p^{L}_{\lambda} = \alpha_\lambda = \alpha > 0$) and $s^{L}_{\lambda} = \frac{\phi_\lambda(1) - \phi_\lambda(0)}{1 - 0} = q^{L}_{\lambda} - \beta_\lambda = a_1^{(q^L_\lambda)} > 0$ (i.e., $q^{L}_{\lambda} = (\alpha + \beta_\mathcal{A})^{\lambda} (\alpha + \beta_\mathcal{H})^{1-\lambda} - \alpha$). However, since $p^{L}_{\lambda} = \alpha_\lambda = \alpha > 0$, the restriction (56) is fulfilled iff $q^{L}_{\lambda} > 0$, which is equivalent to

$$\lambda \in \mathcal{I}_{\text{SP,2}} := \begin{cases} \left] \dfrac{\log\left(\frac{\alpha}{\alpha + \beta_\mathcal{H}}\right)}{\log\left(\frac{\alpha + \beta_\mathcal{A}}{\alpha + \beta_\mathcal{H}}\right)}, 0 \right[\cup \,]1, \infty[, & \text{if } \beta_\mathcal{A} > \beta_\mathcal{H}, \\[6mm]]-\infty, 0[\, \cup \left] 1, \dfrac{\log\left(\frac{\alpha}{\alpha + \beta_\mathcal{H}}\right)}{\log\left(\frac{\alpha + \beta_\mathcal{A}}{\alpha + \beta_\mathcal{H}}\right)} \right[, & \text{if } \beta_\mathcal{A} < \beta_\mathcal{H}. \end{cases} \tag{58}$$

Suppose that $\lambda \in \mathcal{I}_{\text{SP,2}}$. As we have seen above, from Properties 1 (P3a) and (P3b) one can derive that $\left(a_n^{(q^L_\lambda)} \right)_{n \in \mathbb{N}}$ is strictly positive, strictly increasing, and converges to $x_0^{(q^L_\lambda)} \in]0, -\log(q^{L}_{\lambda})]$ iff $q^{L}_{\lambda} \leq \min\{1, e^{\beta_\lambda - 1}\}$, and otherwise it diverges to ∞. Notice that both cases can occur: consider the parameter setup $(\beta_\mathcal{A}, \beta_\mathcal{H}, \alpha_\mathcal{A}, \alpha_\mathcal{H}) = (1.5, 0.5, 0.5, 0.5) \in \mathcal{P}_{\text{SP,2}}$, which leads to $\mathcal{I}_{\text{SP,2}} =]-1, 0[\cup]1, \infty[$; within our running-example epidemiological context of Section 2.3, this corresponds to a "mild" infectious-disease-transmission situation $(\mathcal{H})$ (with "low" reproduction number $\beta_\mathcal{H} = 0.5$ and importation mean of $\alpha_\mathcal{H} = 0.5$), whereas $(\mathcal{A})$ describes a "dangerous" situation (with supercritical $\beta_\mathcal{A} = 1.5$ and $\alpha_\mathcal{A} = 0.5$). For $\lambda = -0.5 \in \mathcal{I}_{\text{SP,2}}$ one obtains $q^{L}_{\lambda} \approx 0.207 \leq \min\{1, e^{\beta_\lambda - 1}\} \approx 0.368$, whereas for $\lambda = 2 \in \mathcal{I}_{\text{SP,2}}$ one gets $q^{L}_{\lambda} = 3.5 > \min\{1, e^{\beta_\lambda - 1}\} = 1$. Altogether, this leads to

Proposition 11. *For all* $(\beta_\mathcal{A}, \beta_\mathcal{H}, \alpha_\mathcal{A}, \alpha_\mathcal{H}, \lambda) \in \mathcal{P}_{SP,2} \times \mathcal{I}_{SP,2}$ *and all initial population sizes* $X_0 \in \mathbb{N}$ *there holds with* $p_\lambda^L = \alpha_\mathcal{A} = \alpha_\mathcal{H} = \alpha$, $q_\lambda^L = (\alpha + \beta_\mathcal{A})^\lambda (\alpha + \beta_\mathcal{H})^{1-\lambda} - \alpha$

$$(a) \qquad B_{\lambda,X_0,1}^L = \widetilde{B}_{\lambda,X_0,1}^{(p_\lambda^L, q_\lambda^L)} = \exp\left\{ \left(q_\lambda^L - \beta_\lambda \right) \cdot X_0 \right\} > 1,$$

(b) *the sequence* $\left(B_{\lambda,X_0,n}^L \right)_{n \in \mathbb{N}}$ *of lower bounds for* $H_\lambda(P_{\mathcal{A},n} || P_{\mathcal{H},n})$ *given by*

$$B_{\lambda,X_0,n}^L = \widetilde{B}_{\lambda,X_0,n}^{(p_\lambda^L, q_\lambda^L)} = \exp\left\{ a_n^{(q_\lambda^L)} \cdot X_0 + \sum_{k=1}^{n} b_k^{(p_\lambda^L, q_\lambda^L)} \right\}$$

is strictly increasing,

$$(c) \qquad \lim_{n \to \infty} B_{\lambda,X_0,n}^L = \infty = \lim_{n \to \infty} H_\lambda(P_{\mathcal{A},n} || P_{\mathcal{H},n}),$$

$$(d) \quad \lim_{n \to \infty} \frac{1}{n} \log B_{\lambda,X_0,n}^L = \begin{cases} p_\lambda^L \cdot \exp\left\{ x_0^{(q_\lambda^L)} \right\} - \alpha > 0, & \text{if } q_\lambda^L \leq \min\left\{1, e^{\beta_\lambda - 1}\right\}, \\[2mm] \infty, & \text{if } q_\lambda^L > \min\left\{1, e^{\beta_\lambda - 1}\right\}, \end{cases}$$

$(e) \qquad$ *the map* $\quad X_0 \mapsto B_{\lambda,X_0,n}^L = \widetilde{B}_{\lambda,X_0,n}^{(p_\lambda^L, q_\lambda^L)} \quad$ *is strictly increasing.*

Nevertheless, for the remaining constellations $(\beta_\mathcal{A}, \beta_\mathcal{H}, \alpha_\mathcal{A}, \alpha_\mathcal{H}, \lambda) \in \mathcal{P}_{SP,2} \times \mathbb{R} \setminus (\mathcal{I}_{SP,2} \cup [0,1])$, all observation time horizons $n \in \mathbb{N}$ and all initial population sizes $X_0 \in \mathbb{N}$ one can still prove

$$1 < H_\lambda(P_{\mathcal{A},n} || P_{\mathcal{H},n}) \qquad \text{and} \qquad \lim_{n \to \infty} H_\lambda(P_{\mathcal{A},n} || P_{\mathcal{H},n}) = \infty, \tag{59}$$

(i.e., the achievement of the Goals (G1′), (G2′)), which is done by a conceptually different method (without involving p_λ^L, q_λ^L) in Appendix A.1.

3.18. Lower Bounds for the Cases $(\beta_\mathcal{A}, \beta_\mathcal{H}, \alpha_\mathcal{A}, \alpha_\mathcal{H}, \lambda) \in \mathcal{P}_{SP,3a} \times (\mathbb{R} \setminus [0,1])$

In the current setup, where $\mathcal{P}_{SP,3a} := \left\{ (\beta_\mathcal{A}, \beta_\mathcal{H}, \alpha_\mathcal{A}, \alpha_\mathcal{H}) \in \mathcal{P}_{SP} : \alpha_\mathcal{A} \neq \alpha_\mathcal{H}, \beta_\mathcal{A} \neq \beta_\mathcal{H}, \frac{\alpha_\mathcal{A}}{\beta_\mathcal{A}} \neq \frac{\alpha_\mathcal{H}}{\beta_\mathcal{H}}, \frac{\alpha_\mathcal{A} - \alpha_\mathcal{H}}{\beta_\mathcal{H} - \beta_\mathcal{A}} \in\ ] -\infty, 0[\right\}$ (cf. (49)), we *always* have either $(\alpha_\mathcal{A} > \alpha_\mathcal{H}) \wedge (\beta_\mathcal{A} > \beta_\mathcal{H})$ or $(\alpha_\mathcal{A} < \alpha_\mathcal{H}) \wedge (\beta_\mathcal{A} < \beta_\mathcal{H})$. Furthermore, from Properties 3 (P16) we obtain $\phi_\lambda(0) > 0$. As in the case $\lambda \in\ ]0,1[$, the derivative $\phi_\lambda'(0)$ can assume any sign on $\mathcal{P}_{SP,3a}$, take e.g., $(\beta_\mathcal{A}, \beta_\mathcal{H}, \alpha_\mathcal{A}, \alpha_\mathcal{H}, \lambda) = (2.2, 4.5, 1, 3, 2)$ for $\phi_\lambda'(0) < 0$, $(\beta_\mathcal{A}, \beta_\mathcal{H}, \alpha_\mathcal{A}, \alpha_\mathcal{H}, \lambda) = (2.25, 4.5, 1, 3, 2)$ for $\phi_\lambda'(0) = 0$ and $(\beta_\mathcal{A}, \beta_\mathcal{H}, \alpha_\mathcal{A}, \alpha_\mathcal{H}, \lambda) = (2.3, 4.5, 1, 3, 2)$ for $\phi_\lambda'(0) > 0$ (these parameter constellations reflect "dangerous" ($\mathcal{A}$) versus "highly dangerous" ($\mathcal{H}$) situations within our running-example epidemiological context of Section 2.3). Nevertheless, in all three subcases one gets $\min_{x \in \mathbb{N}_0} \phi_\lambda(x) \geq \min_{x \geq 0} \phi_\lambda(x) > 0$. Thus, there exist parameters $p_\lambda^L \in\ ]\alpha_\lambda, \alpha_\mathcal{A}^\lambda \alpha_\mathcal{H}^{1-\lambda}]$ and $q_\lambda^L \in\ ]\beta_\lambda, \beta_\mathcal{A}^\lambda \beta_\mathcal{H}^{1-\lambda}]$ which satisfy (35) (in particular, $p_\lambda^L - \alpha_\lambda > 0$, $q_\lambda^L - \beta_\lambda > 0$). We now have to look for a condition which guarantees that these parameters *additionally* fulfill (56); such a condition is clearly that both $\alpha_\lambda \geq 0$ and $\beta_\lambda \geq 0$ hold, which is equivalent (cf. (57)) with

$$\lambda \in \mathcal{I}_{SP,3a}^{(\geq)} := \begin{cases} \left[\max\left\{ \frac{-\alpha_\mathcal{H}}{\alpha_\mathcal{A} - \alpha_\mathcal{H}}, \frac{-\beta_\mathcal{H}}{\beta_\mathcal{A} - \beta_\mathcal{H}} \right\}, 0 \right[\cup\ ]1, \infty[, & \text{if } (\alpha_\mathcal{A} > \alpha_\mathcal{H}) \wedge (\beta_\mathcal{A} > \beta_\mathcal{H}), \\[3mm] \left] -\infty, 0 \right[\cup\ \left]1, \min\left\{ \frac{\alpha_\mathcal{H}}{\alpha_\mathcal{H} - \alpha_\mathcal{A}}, \frac{\beta_\mathcal{H}}{\beta_\mathcal{H} - \beta_\mathcal{A}} \right\} \right], & \text{if } (\alpha_\mathcal{A} < \alpha_\mathcal{H}) \wedge (\beta_\mathcal{A} < \beta_\mathcal{H}); \end{cases}$$

recall that $\alpha_\lambda = 0$ and $\beta_\lambda = 0$ cannot occur simultaneously in the current setup. If $\alpha_\lambda \leq 0$ and $\beta_\lambda \leq 0$, i.e., if

$$\lambda \in \mathcal{I}_{SP,3a}^{(<)} := \begin{cases} \left] -\infty, \min\left\{ \frac{-\alpha_\mathcal{H}}{\alpha_\mathcal{A} - \alpha_\mathcal{H}}; \frac{-\beta_\mathcal{H}}{\beta_\mathcal{A} - \beta_\mathcal{H}} \right\} \right], & \text{if } (\alpha_\mathcal{A} > \alpha_\mathcal{H}) \wedge (\beta_\mathcal{A} > \beta_\mathcal{H}), \\[3mm] \left[\max\left\{ \frac{\alpha_\mathcal{H}}{\alpha_\mathcal{H} - \alpha_\mathcal{A}}; \frac{\beta_\mathcal{H}}{\beta_\mathcal{H} - \beta_\mathcal{A}} \right\}, \infty \right[, & \text{if } (\alpha_\mathcal{A} < \alpha_\mathcal{H}) \wedge (\beta_\mathcal{A} < \beta_\mathcal{H}), \end{cases}$$

then–due to the strict positivity of the function $\varphi_\lambda(\cdot)$ (cf. (31))–there exist parameters $p_\lambda^L > 0 = \max\{0, \alpha_\lambda\}$ and $q_\lambda^L > 0 = \max\{0, \beta_\lambda\}$ which satisfy (56) and (34) (where the latter implies (35) and thus $p_\lambda^L \le \alpha_{\mathcal{A}}^\lambda \alpha_{\mathcal{H}}^{1-\lambda}$, $q_\lambda^L \le \beta_{\mathcal{A}}^\lambda \beta_{\mathcal{H}}^{1-\lambda}$). With

$$\mathcal{I}_{\mathrm{SP,3a}} := \mathcal{I}_{\mathrm{SP,3a}}^{(\ge)} \cup \mathcal{I}_{\mathrm{SP,3a}}^{(<)} \tag{60}$$

and with the discussion below (56), we thus derive the following

Proposition 12. *For all* $(\beta_{\mathcal{A}}, \beta_{\mathcal{H}}, \alpha_{\mathcal{A}}, \alpha_{\mathcal{H}}, \lambda) \in \mathcal{P}_{\mathrm{SP,3a}} \times \mathcal{I}_{\mathrm{SP,3a}}$ *there exist parameters* p_λ^L, q_λ^L *which satisfy* $\max\{0, \alpha_\lambda\} < p_\lambda^L \le \alpha_{\mathcal{A}}^\lambda \alpha_{\mathcal{H}}^{1-\lambda}$, $\max\{0, \beta_\lambda\} < q_\lambda^L \le \beta_{\mathcal{A}}^\lambda \beta_{\mathcal{H}}^{1-\lambda}$ *as well as* (35) *for all* $x \in \mathbb{N}_0$, *and for all such pairs* $(p_\lambda^L, q_\lambda^L)$ *and all initial population sizes* $X_0 \in \mathbb{N}$ *one gets*

(a) $\quad B_{\lambda,X_0,1}^L = \widetilde{B}_{\lambda,X_0,1}^{(p_\lambda^L, q_\lambda^L)} = \exp\left\{ \left(q_\lambda^L - \beta_\lambda\right) \cdot X_0 + p_\lambda^L - \alpha_\lambda \right\} > 1,$

(b) $\quad$ *the sequence* $\left(B_{\lambda,X_0,n}^L \right)_{n \in \mathbb{N}}$ *of lower bounds for* $H_\lambda(P_{\mathcal{A},n} \| P_{\mathcal{H},n})$ *given by*

$$B_{\lambda,X_0,n}^L = \widetilde{B}_{\lambda,X_0,n}^{(p_\lambda^L, q_\lambda^L)} = \exp\left\{ a_n^{(q_\lambda^L)} \cdot X_0 + \sum_{k=1}^n b_k^{(p_\lambda^L, q_\lambda^L)} \right\}$$

$\quad$ *is strictly increasing,*

(c) $\quad \lim\limits_{n \to \infty} B_{\lambda,X_0,n}^L = \infty = \lim\limits_{n \to \infty} H_\lambda(P_{\mathcal{A},n} \| P_{\mathcal{H},n}),$

(d) $\quad \lim\limits_{n \to \infty} \dfrac{1}{n} \log B_{\lambda,X_0,n}^L = \begin{cases} p_\lambda^L \cdot \exp\left\{ x_0^{(q_\lambda^L)} \right\} - \alpha_\lambda > 0, & \text{if } q_\lambda^L \le \min\left\{1, e^{\beta_\lambda - 1}\right\}, \\ \infty, & \text{if } q_\lambda^L > \min\left\{1, e^{\beta_\lambda - 1}\right\}, \end{cases}$

(e) $\quad$ *the map* $X_0 \mapsto B_{\lambda,X_0,n}^L = \widetilde{B}_{\lambda,X_0,n}^{(p_\lambda^L, q_\lambda^L)}$ *is strictly increasing.*

Notice that the assertions (a) to (e) of Proposition 12 hold true for parameter pairs $(p_\lambda^L, q_\lambda^L)$ *whenever* they satisfy (35) and (56); in particular, we may allow either $p_\lambda^L = \max\{0, \alpha_\lambda\}$ or $q_\lambda^L = \max\{0, \beta_\lambda\}$. Let us furthermore mention that in part (d) both asymptotical behaviours can occur: consider e.g., the parameter setup $(\beta_{\mathcal{A}}, \beta_{\mathcal{H}}, \alpha_{\mathcal{A}}, \alpha_{\mathcal{H}}) = (0.3, 0.2, 4, 3) \in \mathcal{P}_{\mathrm{SP,3a}}$, leading to $]1, \infty[\subsetneq \mathcal{I}_{\mathrm{SP,3a}}^{(\ge)} \subsetneq \mathcal{I}_{\mathrm{SP,3a}}$. For $\lambda = 2 \in \mathcal{I}_{\mathrm{SP,3a}}$, the parameters $p_\lambda^L := \widetilde{p}_\lambda := 5.25$, $q_\lambda^L := \widetilde{q}_\lambda := 0.45$ (corresponding to the asymptote $\widetilde{\phi}_\lambda(\cdot)$, cf. (P20)) fulfill (35), (56) and additionally $q_\lambda^L = 0.45 < \min\{1, e^{\beta_\lambda - 1}\} \approx 0.549$. Analogously, in the setup $(\beta_{\mathcal{A}}, \beta_{\mathcal{H}}, \alpha_{\mathcal{A}}, \alpha_{\mathcal{H}}, \lambda) = (3, 2, 4, 3, 2) \in \mathcal{P}_{\mathrm{SP,3a}} \times \mathcal{I}_{\mathrm{SP,3a}}$, the choices $p_\lambda^L := \widetilde{p}_\lambda := 5.25$, $q_\lambda^L := \widetilde{q}_\lambda := 4.5$ satisfy (35), (56) and there holds $q_\lambda^L = 4.5 > \min\{1, e^{\beta_\lambda - 1}\} = 1$.

For the remaining two cases $(\alpha_\lambda \le 0) \wedge (\beta_\lambda > 0)$ (e.g., $(\beta_{\mathcal{A}}, \beta_{\mathcal{H}}, \alpha_{\mathcal{A}}, \alpha_{\mathcal{H}}, \lambda) = (6, 5, 3, 2, -3)$) and $(\alpha_\lambda > 0) \wedge (\beta_\lambda \le 0)$ (e.g., $(\beta_{\mathcal{A}}, \beta_{\mathcal{H}}, \alpha_{\mathcal{A}}, \alpha_{\mathcal{H}}, \lambda) = (3, 2, 6, 5, -3)$), one has to proceed differently. Indeed, for all parameter constellations $(\beta_{\mathcal{A}}, \beta_{\mathcal{H}}, \alpha_{\mathcal{A}}, \alpha_{\mathcal{H}}, \lambda) \in \mathcal{P}_{\mathrm{SP,3a}} \times \mathbb{R} \setminus (\mathcal{I}_{\mathrm{SP,3a}} \cup [0, 1])$, all observation time horizons $n \in \mathbb{N}$ and all initial population sizes $X_0 \in \mathbb{N}$ one can still prove

$$1 < H_\lambda\left(P_{\mathcal{A},n} \| P_{\mathcal{H},n}\right), \qquad \text{and} \qquad \lim\limits_{n \to \infty} H_\lambda\left(P_{\mathcal{A},n} \| P_{\mathcal{H},n}\right) = \infty, \tag{61}$$

which is done in Appendix A.1, using a similar method as in the proof of assertion (59).

3.19. Lower Bounds for the Cases $(\beta_{\mathcal{A}}, \beta_{\mathcal{H}}, \alpha_{\mathcal{A}}, \alpha_{\mathcal{H}}, \lambda) \in \mathcal{P}_{\mathrm{SP,3b}} \times (\mathbb{R} \setminus [0, 1])$

Within such a constellation, where $\mathcal{P}_{\mathrm{SP,3b}} := \Big\{ (\beta_{\mathcal{A}}, \beta_{\mathcal{H}}, \alpha_{\mathcal{A}}, \alpha_{\mathcal{H}}) \in \mathcal{P}_{\mathrm{SP}} : \alpha_{\mathcal{A}} \ne \alpha_{\mathcal{H}}, \beta_{\mathcal{A}} \ne \beta_{\mathcal{H}},$ $\frac{\alpha_{\mathcal{A}}}{\beta_{\mathcal{A}}} \ne \frac{\alpha_{\mathcal{H}}}{\beta_{\mathcal{H}}}, \frac{\alpha_{\mathcal{A}} - \alpha_{\mathcal{H}}}{\beta_{\mathcal{H}} - \beta_{\mathcal{A}}} \in \,]0, \infty[\setminus \mathbb{N} \Big\}$ (cf. (49)), one *always* has either $(\alpha_{\mathcal{A}} < \alpha_{\mathcal{H}}) \wedge (\beta_{\mathcal{A}} > \beta_{\mathcal{H}})$ or $(\alpha_{\mathcal{A}} > \alpha_{\mathcal{H}}) \wedge (\beta_{\mathcal{A}} < \beta_{\mathcal{H}})$. Moreover, from Properties 3 (P15) one can see that $\phi_\lambda(x) = 0$ for $x = x^* = \frac{\alpha_{\mathcal{H}} - \alpha_{\mathcal{A}}}{\beta_{\mathcal{A}} - \beta_{\mathcal{H}}} > 0$. However, $x^* \notin \mathbb{N}_0$, which implies $\phi_\lambda(x) > 0$ for all x on the relevant subdomain $\mathbb{N}_0$.

Again, we incorporate (57) and consider the set of all $\lambda \in \mathbb{R}\backslash[0,1]$ such that $\alpha_\lambda \geq 0$ and $\beta_\lambda \geq 0$ (where $\alpha_\lambda = 0 \wedge \beta_\lambda = 0$ cannot appear), i.e.,

$$
\lambda \in \mathcal{I}_{\mathrm{SP,3b}}^{(\geq)} := \begin{cases}
\left[\frac{-\beta_\mathcal{H}}{\beta_\mathcal{A}-\beta_\mathcal{H}}, 0\right[\cup \left]1, \frac{\alpha_\mathcal{H}}{\alpha_\mathcal{H}-\alpha_\mathcal{A}}\right], & \text{if } (\alpha_\mathcal{A} < \alpha_\mathcal{H}) \wedge (\beta_\mathcal{A} > \beta_\mathcal{H}), \\[2mm]
\left[\frac{-\alpha_\mathcal{H}}{\alpha_\mathcal{A}-\alpha_\mathcal{H}}, 0\right[\cup \left]1, \frac{\beta_\mathcal{H}}{\beta_\mathcal{H}-\beta_\mathcal{A}}\right], & \text{if } (\alpha_\mathcal{A} > \alpha_\mathcal{H}) \wedge (\beta_\mathcal{A} < \beta_\mathcal{H}).
\end{cases}
\tag{62}
$$

As above in Section 3.18, if $\lambda \in \mathcal{I}_{\mathrm{SP,3b}}^{(\geq)}$ then there exist parameters $p_\lambda^L \in \,]\alpha_\lambda, \alpha_\mathcal{A}^\lambda \alpha_\mathcal{H}^{1-\lambda}]$, $q_\lambda^L \in \,]\beta_\lambda, \beta_\mathcal{A}^\lambda \beta_\mathcal{H}^{1-\lambda}]$ (which thus fulfill (56)) such that (35) is satisfied for all $x \in \mathbb{N}_0$. Hence, for all $\lambda \in \mathcal{I}_{\mathrm{SP,3b}} := \mathcal{I}_{\mathrm{SP,3b}}^{(\geq)}$, all assertions (a) to (e) of Proposition 12 hold true. Notice that for the current setup $\mathcal{P}_{\mathrm{SP,3b}}$ one cannot have $\alpha_\lambda \leq 0$ and $\beta_\lambda \leq 0$ simultaneously. Furthermore, in each of the two remaining cases $(\alpha_\lambda < 0) \wedge (\beta_\lambda > 0)$ respectively $(\alpha_\lambda > 0) \wedge (\beta_\lambda < 0)$ it can happen that there do not exist parameters $p_\lambda^L, q_\lambda^L > 0$ which satisfy both (35) and (56). However, as in the case $\mathcal{P}_{\mathrm{SP,3a}}$ above, for all $\lambda \notin \mathcal{I}_{\mathrm{SP,3b}}$ we prove in Appendix A.1 (by a method without p_λ^L, q_λ^L) that for all observation times $n \in \mathbb{N}$ and all initial population sizes $X_0 \in \mathbb{N}$ there holds

$$
1 < H_\lambda\left(P_{\mathcal{A},n} || P_{\mathcal{H},n}\right) \qquad \text{and} \qquad \lim_{n\to\infty} H_\lambda\left(P_{\mathcal{A},n} || P_{\mathcal{H},n}\right) = \infty.
\tag{63}
$$

3.20. Lower Bounds for the Cases $(\beta_\mathcal{A}, \beta_\mathcal{H}, \alpha_\mathcal{A}, \alpha_\mathcal{H}, \lambda) \in \mathcal{P}_{\mathrm{SP,3c}} \times (\mathbb{R}\backslash[0,1])$

Since in this subcase one has $\mathcal{P}_{\mathrm{SP,3c}} := \Big\{ (\beta_\mathcal{A}, \beta_\mathcal{H}, \alpha_\mathcal{A}, \alpha_\mathcal{H}) \in \mathcal{P}_{\mathrm{SP}} : \alpha_\mathcal{A} \neq \alpha_\mathcal{H}, \beta_\mathcal{A} \neq \beta_\mathcal{H},$ $\frac{\alpha_\mathcal{A}}{\beta_\mathcal{A}} \neq \frac{\alpha_\mathcal{H}}{\beta_\mathcal{H}}, \frac{\alpha_\mathcal{A}-\alpha_\mathcal{H}}{\beta_\mathcal{H}-\beta_\mathcal{A}} \in \mathbb{N} \Big\}$ (cf. (49)) and thus $\phi_\lambda(x^*) = 0$ for $x^* \in \mathbb{N}$, there do not exist parameters p_λ^L, q_λ^L such that (35) and (56) are satisfied. The only parameter pair that ensures $\exp\left\{ a_n^{(q_\lambda^L)} \cdot X_0 + \sum_{k=1}^{n} b_k^{(p_\lambda^L, q_\lambda^L)} \right\} \geq 1$ for all $n \in \mathbb{N}$ and all $X_0 \in \mathbb{N}$ within our proposed method, is the choice $p_\lambda^L = \alpha_\lambda$, $q_\lambda^L = \beta_\lambda$. Consequently, $B_{\lambda,X_0,n}^L \equiv 1$, which coincides with the general lower bound (11) but violates the above-mentioned desired Goal (G1'). However, in some constellations there exist *nonnegative* parameters $p_\lambda^L < \alpha_\lambda, q_\lambda^L > \beta_\lambda$ or $p_\lambda^L > \alpha_\lambda, q_\lambda^L < \beta_\lambda$, such that at least the parts (c) and (d) of Proposition 12 are satisfied. As in Section 3.19 above, by using a conceptually different method (without p_λ^L, q_λ^L) we prove in Appendix A.1 that for all $\lambda \in \mathbb{R}\backslash[0,1]$, all observation times $n \in \mathbb{N}$ and all initial population sizes $X_0 \in \mathbb{N}$ there holds

$$
1 < H_\lambda\left(P_{\mathcal{A},n} || P_{\mathcal{H},n}\right) \qquad \text{and} \qquad \lim_{n\to\infty} H_\lambda\left(P_{\mathcal{A},n} || P_{\mathcal{H},n}\right) = \infty.
\tag{64}
$$

3.21. Lower Bounds for the Cases $(\beta_\mathcal{A}, \beta_\mathcal{H}, \alpha_\mathcal{A}, \alpha_\mathcal{H}, \lambda) \in \mathcal{P}_{\mathrm{SP,4a}} \times (\mathbb{R}\backslash[0,1])$

In the current setup, where $\mathcal{P}_{\mathrm{SP,4a}} := \{ (\beta_\mathcal{A}, \beta_\mathcal{H}, \alpha_\mathcal{A}, \alpha_\mathcal{H}) \in \mathcal{P}_{\mathrm{SP}} : \alpha_\mathcal{A} \neq \alpha_\mathcal{H} > 0, \beta_\mathcal{A} = \beta_\mathcal{H} \in\,]0,1[\}$ (cf. (49)), the function $\phi_\lambda(\cdot)$ is strictly positive and strictly decreasing, with $\lim_{x\to\infty} \phi_\lambda(x) = \lim_{x\to\infty} \phi_\lambda'(x) = 0$. The only choice of parameters p_λ^L, q_λ^L which fulfill (35) and $\exp\left\{ a_n^{(q_\lambda^L)} \cdot X_0 + \sum_{k=1}^{n} b_k^{(p_\lambda^L, q_\lambda^L)} \right\} \geq 1$ for all $n \in \mathbb{N}$ and all $X_0 \in \mathbb{N}$, is the choice $p_\lambda^L = \alpha_\lambda$ as well as $q_\lambda^L = \beta_\lambda = \beta_\bullet$, where $\beta_\bullet$ stands for both (equal) $\beta_\mathcal{H}$ and $\beta_\mathcal{A}$. Of course, this leads to $B_{\lambda,X_0,n}^L \equiv 1$, which is consistent with the general lower bound (11), but violates the above-mentioned desired Goal (G1'). Nevertheless, in Appendix A.1 we prove the following

Proposition 13. *For all* $(\beta_\mathcal{A}, \beta_\mathcal{H}, \alpha_\mathcal{A}, \alpha_\mathcal{H}, \lambda) \in \mathcal{P}_{\mathrm{SP,4a}} \times \mathbb{R}\backslash[0,1]$ *there exist parameters* $p_\lambda^L > \alpha_\lambda$ *(not necessarily satisfying* $p_\lambda^L \geq 0$*) and* $0 < q_\lambda^L < \beta_\lambda = \beta_\bullet < \min\{1, e^{\beta_\bullet - 1}\} = e^{\beta_\bullet - 1}$ *such that (35) holds for all* $x \in [0, \infty[$ *and such that for all initial population sizes* $X_0 \in \mathbb{N}$ *the parts (c) and (d) of Proposition 12 hold true.*

3.22. Lower Bounds for the Cases $(\beta_{\mathcal{A}}, \beta_{\mathcal{H}}, \alpha_{\mathcal{A}}, \alpha_{\mathcal{H}}, \lambda) \in \mathcal{P}_{SP,4b} \times (\mathbb{R} \backslash [0,1])$

By recalling $\mathcal{P}_{SP,4b} := \{ (\beta_{\mathcal{A}}, \beta_{\mathcal{H}}, \alpha_{\mathcal{A}}, \alpha_{\mathcal{H}}) \in \mathcal{P}_{SP} : \alpha_{\mathcal{A}} \neq \alpha_{\mathcal{H}} > 0, \beta_{\mathcal{A}} = \beta_{\mathcal{H}} \in [1, \infty[\}$ (cf.(49)), the assertions preceding Proposition 13 remain valid. However, the proof of Proposition 13 in Appendix A.1 contains details which explain why it cannot be carried over to the current case $\mathcal{P}_{SP,4b}$. Thus, the generally valid lower bound $B^L_{\lambda, X_0, n} \equiv 1$ cannot be improved with our methods.

3.23. Concluding Remarks on Alternative Lower Bounds for all Cases
$(\beta_{\mathcal{A}}, \beta_{\mathcal{H}}, \alpha_{\mathcal{A}}, \alpha_{\mathcal{H}}, \lambda) \in (\mathcal{P}_{SP} \backslash \mathcal{P}_{SP,1}) \times (\mathbb{R} \backslash [0,1])$

To achieve the Goals (G1′) to (G3′), in the above-mentioned investigations about lower bounds of the Hellinger integral $H_\lambda(P_{\mathcal{A},n} || P_{\mathcal{H},n})$, $\lambda \in \mathbb{R} \backslash [0,1]$, we have mainly focused on parameters p_λ^L, q_λ^L which satisfy (35) and additionally (56). Nevertheless, Theorem 1 (b) gives lower bounds $B^L_{\lambda, X_0, n}$ *whenever* (35) is fulfilled. However, this lower bound can be the trivial one, $B^L_{\lambda, X_0, n} \equiv 1$. Let us remark here that for the parameter constellations $(\beta_{\mathcal{A}}, \beta_{\mathcal{H}}, \alpha_{\mathcal{A}}, \alpha_{\mathcal{H}}, \lambda) \in \left(\mathcal{P}_{SP,2} \times \mathbb{R} \backslash ([0,1] \cup \mathcal{I}_{SP,2}) \right) \cup$
$\left(\mathcal{P}_{SP,3a} \times \mathbb{R} \backslash ([0,1] \cup \mathcal{I}_{SP,3a}) \right) \cup \left(\mathcal{P}_{SP,3b} \times \mathbb{R} \backslash ([0,1] \cup \mathcal{I}_{SP,3b}) \right)$ one can prove that there exist p_λ^L, q_λ^L which satisfy (35) for all $x \in \mathbb{N}_0$ as well as the condition (generalizing (56))

$$ p_\lambda^L \geq \alpha_\lambda, \qquad q_\lambda^L \geq \beta_\lambda, \qquad \text{(where at least one of the inequalities is strict)}, $$

and that for such p_λ^L, q_λ^L one gets the validity of $H_\lambda(P_{\mathcal{A},n} || P_{\mathcal{H},n}) \geq B^L_{\lambda, X_0, n} = \widetilde{B}^{(p_\lambda^L, q_\lambda^L)}_{\lambda, X_0, n} > 1$ for all $X_0 \in \mathbb{N}$ and all $n \in \mathbb{N}$; consequently, Goal (G1′) is achieved. However, in these parameter constellations it can unpleasantly happen that $n \mapsto B^L_{\lambda, X_0, n}$ is oscillating (in contrast to the monotone behaviour in the Propositions 11 (b), 12 (b)).

As a final general remark, let us mention that the functions $\phi_{\lambda,y}^{\tan}(\cdot)$, $\phi_{\lambda,k}^{\sec}(\cdot)$, $\phi_\lambda^{\hor}(\cdot)$, $\widetilde{\phi}_\lambda(\cdot)$ –defined in (52)–(54) and Properties 3 (P20)–constitute linear lower bounds for $\phi_\lambda(\cdot)$ on the domain $\mathbb{N}_0$ in the case $\lambda \in \mathbb{R} \backslash [0,1]$. Their parameters $p_\lambda^L \in \left\{ p_{\lambda,y}^{\tan}, p_{\lambda,y}^{\sec}, p_{\lambda,y}^{\hor}, \widetilde{p}_\lambda \right\}$ and $q_\lambda^L \in \left\{ q_{\lambda,y}^{\tan}, q_{\lambda,y}^{\sec}, q_{\lambda,y}^{\hor}, \widetilde{q}_\lambda \right\}$ lead to lower bounds $B^L_{\lambda, X_0, n}$ of the Hellinger integrals that may or may not be consistent with Goals (G1′) to (G3′), and which may be possibly better respectively weaker respectively incomparable with the previous lower bounds when adding some relaxation of (G1′), such as e.g., the validity of $H_\lambda(P_{\mathcal{A},n} || P_{\mathcal{H},n}) > 1$ for all but finitely many $n \in \mathbb{N}$.

3.24. Upper Bounds for the Cases $(\beta_{\mathcal{A}}, \beta_{\mathcal{H}}, \alpha_{\mathcal{A}}, \alpha_{\mathcal{H}}, \lambda) \in (\mathcal{P}_{SP} \backslash \mathcal{P}_{SP,1}) \times (\mathbb{R} \backslash [0,1])$

For the cases $\lambda \in \mathbb{R} \backslash [0,1]$, the investigation of upper bounds for the Hellinger integral $H_\lambda(P_{\mathcal{A},n} || P_{\mathcal{H},n})$ is much easier than the above-mentioned derivations of lower bounds. In fact, we face a situation which is similar to the lower-bounds-studies for the cases $\lambda \in]0,1[$: due to Properties 3 (P19), the function $\phi_\lambda(\cdot)$ is strictly convex on the nonnegative real line. Furthermore, it is asymptotically linear, as stated in (P20). The monotonicity Properties 2 (P10) to (P12) imply that for the tightest upper bound (within our framework) one should use the parameters $p_\lambda^U := \alpha_{\mathcal{A}}^\lambda \alpha_{\mathcal{H}}^{1-\lambda} > 0$ and $q_\lambda^U := \beta_{\mathcal{A}}^\lambda \beta_{\mathcal{H}}^{1-\lambda} > 0$. Lemma A1 states that $p_\lambda^U \geq \alpha_\lambda$ resp. $q_\lambda^U \geq \beta_\lambda$, with equality iff $\alpha_{\mathcal{A}} = \alpha_{\mathcal{H}}$ resp. iff $\beta_{\mathcal{A}} = \beta_{\mathcal{H}}$. From Properties 1 (P3a) we see that for $\beta_{\mathcal{A}} \neq \beta_{\mathcal{H}}$ the corresponding sequence $\left(a_n^{(q_\lambda^U)} \right)_{n \in \mathbb{N}}$ is convergent to $x_0^{(q_\lambda^U)} \in]0, -\log(q_\lambda^U)]$ if $q_\lambda^U \leq \min\{1, e^{\beta_{\mathcal{A}}-1}\}$ (i.e., if $\lambda \in [\lambda_-, \lambda_+]$, cf. Lemma 1 (a)), and otherwise it diverges to ∞ faster than exponentially (cf. (P3b)). If $\beta_{\mathcal{A}} = \beta_{\mathcal{H}}$ (i.e., if $(\beta_{\mathcal{A}}, \beta_{\mathcal{H}}, \alpha_{\mathcal{A}}, \alpha_{\mathcal{H}}) \in \mathcal{P}_{SP,4} = \mathcal{P}_{SP,4a} \cup \mathcal{P}_{SP,4b}$), then one gets $q_\lambda^U = \beta_\lambda$ and $a_n^{(q_\lambda^U)} = 0 = x_0^{(q_\lambda^U)}$ for all $n \in \mathbb{N}$ (cf. (P2)). Altogether, this leads to

Proposition 14. *For all* $(\beta_{\mathcal{A}}, \beta_{\mathcal{H}}, \alpha_{\mathcal{A}}, \alpha_{\mathcal{H}}, \lambda) \in (\mathcal{P}_{SP} \backslash \mathcal{P}_{SP,1}) \times (\mathbb{R}\backslash[0,1])$ *and all initial population sizes* $X_0 \in \mathbb{N}$ *there holds with* $p_{\lambda}^{U} := \alpha_{\mathcal{A}}^{\lambda}\alpha_{\mathcal{H}}^{1-\lambda}$, $q_{\lambda}^{U} := \beta_{\mathcal{A}}^{\lambda}\beta_{\mathcal{H}}^{1-\lambda}$

$(a) \qquad B_{\lambda,X_0,1}^{U} \;=\; \widetilde{B}_{\lambda,X_0,1}^{(p_{\lambda}^{U},q_{\lambda}^{U})} \;=\; \exp\left\{ \left(\beta_{\mathcal{A}}^{\lambda}\beta_{\mathcal{H}}^{1-\lambda} - \beta_{\lambda}\right) \cdot X_0 \,+\, \alpha_{\mathcal{A}}^{\lambda}\alpha_{\mathcal{H}}^{1-\lambda} - \alpha_{\lambda} \right\} \;>\; 1,$

$(b) \qquad$ *the sequence* $\left(B_{\lambda,X_0,n}^{U}\right)_{n\in\mathbb{N}}$ *of upper bounds for* $H_{\lambda}(P_{\mathcal{A},n}||P_{\mathcal{H},n})$ *given by*

$$B_{\lambda,X_0,n}^{U} \;=\; \widetilde{B}_{\lambda,X_0,n}^{(p_{\lambda}^{U},q_{\lambda}^{U})} \;=\; \exp\left\{ a_n^{(q_{\lambda}^{U})} \cdot X_0 \,+\, \sum_{k=1}^{n} b_k^{(p_{\lambda}^{U},q_{\lambda}^{U})} \right\}$$

is strictly increasing,

$(c) \qquad \displaystyle\lim_{n\to\infty} B_{\lambda,X_0,n}^{U} \;=\; \infty \,,$

$(d) \quad \displaystyle\lim_{n\to\infty} \frac{1}{n} \log B_{\lambda,X_0,n}^{U} \;=\; \begin{cases} p_{\lambda}^{U} \cdot \exp\left\{ x_0^{(q_{\lambda}^{U})} \right\} - \alpha_{\lambda} \;>\; 0, & \text{if } \lambda \in [\lambda_-,\lambda_+] \setminus [0,1]\,, \\[2mm] \infty, & \text{if } \lambda \in\,]-\infty,\lambda_-[\,\cup\,]\lambda_+,\infty[\,, \end{cases}$

$(e) \qquad$ *the map* $\quad X_0 \mapsto B_{\lambda,X_0,n}^{U} = \widetilde{B}_{\lambda,X_0,n}^{(p_{\lambda}^{U},q_{\lambda}^{U})} \quad$ *is strictly increasing.*

4. Power Divergences of Non-Kullback-Leibler-Information-Divergence Type

4.1. A First Basic Result

For orders $\lambda \in \mathbb{R}\backslash\{0,1\}$, all the results of the previous Section 3 carry correspondingly over from the Hellinger integrals $H_{\lambda}(\cdot||\cdot)$ to the total variation distance $V(\cdot||\cdot)$, by virtue of the relation (cf. (12))

$$2\left(1 - H_{\frac{1}{2}}(P_{\mathcal{A},n}||P_{\mathcal{H},n})\right) \;\leq\; V(P_{\mathcal{A},n}||P_{\mathcal{H},n}) \;\leq\; 2\sqrt{1 - \left(H_{\frac{1}{2}}(P_{\mathcal{A},n}||P_{\mathcal{H},n})\right)^2}\,,$$

to the Renyi divergences $R_{\lambda}(\cdot||\cdot)$, by virtue of the relation (cf. (7))

$$0 \leq R_{\lambda}(P_{\mathcal{A},n}||P_{\mathcal{H},n}) \;=\; \frac{1}{\lambda(\lambda-1)}\, \log H_{\lambda}(P_{\mathcal{A},n}||P_{\mathcal{H},n})\,, \qquad \text{with } \log 0 := -\infty,$$

as well as to the power divergences $I_{\lambda}(\cdot||\cdot)$, by virtue of the relation (cf. (2))

$$I_{\lambda}(P_{\mathcal{A},n}||P_{\mathcal{H},n}) \;=\; \frac{1 - H_{\lambda}(P_{\mathcal{A},n}||P_{\mathcal{H},n})}{\lambda\cdot(1-\lambda)}\,, \qquad n \in \mathbb{N};$$

in the following, we concentrate on the latter. In particular, the above-mentioned carrying-over procedure leads to bounds on $I_{\lambda}(P_{\mathcal{A}}||P_{\mathcal{H}})$ which are tighter than the general rudimentary bounds (cf. (10) and (11))

$$0 \leq I_{\lambda}(P_{\mathcal{A},n}||P_{\mathcal{H},n}) \;<\; \frac{1}{\lambda(1-\lambda)}, \quad \text{for } \lambda \in\,]0,1[\,, \qquad 0 \leq I_{\lambda}(P_{\mathcal{A},n}||P_{\mathcal{H},n}) \leq \infty, \quad \text{for } \lambda \in \mathbb{R}\backslash[0,1]\,.$$

Because power divergences have a *very insightful interpretation* as "directed distances" between two probability distributions (e.g., within our running-example epidemiological context), and function as important tools in statistics, information theory, machine learning, and artificial intelligence, we present explicitly the outcoming exact values respectively bounds of $I_{\lambda}(P_{\mathcal{A}}||P_{\mathcal{H}})$ ($\lambda \in \mathbb{R}\backslash\{0,1\}$, $n \in \mathbb{N}$), in the current and the following subsections. For this, recall the case-dependent parameters $p^A = p_{\lambda}^A = p^A(\beta_{\mathcal{A}}, \beta_{\mathcal{H}}, \alpha_{\mathcal{A}}, \alpha_{\mathcal{H}}, \lambda)$ and $q^A = q_{\lambda}^A = q^A(\beta_{\mathcal{A}}, \beta_{\mathcal{H}}, \alpha_{\mathcal{A}}, \alpha_{\mathcal{H}}, \lambda)$ ($A \in \{E, L, U\}$). To begin with, we can deduce from Theorem 1

Theorem 2.

(a) For all $(\beta_{\mathcal{A}}, \beta_{\mathcal{H}}, \alpha_{\mathcal{A}}, \alpha_{\mathcal{H}}) \in (\mathcal{P}_{NI} \cup \mathcal{P}_{SP,1})$, all initial population sizes $X_0 \in \mathbb{N}_0$, all observation horizons $n \in \mathbb{N}$ and all $\lambda \in \mathbb{R} \backslash \{0, 1\}$ one can recursively compute the **exact value**

$$I_\lambda(P_{\mathcal{A},n} || P_{\mathcal{H},n}) \;=\; \frac{1}{\lambda(\lambda-1)} \cdot \left[\exp\left\{ a_n^{(q_\lambda^E)} \cdot X_0 + \frac{\alpha_{\mathcal{A}}}{\beta_{\mathcal{A}}} \sum_{k=1}^{n} a_k^{(q_\lambda^E)} \right\} - 1 \right] \;=:\; V_{\lambda, X_0, n}^I \ , \tag{65}$$

where $\frac{\alpha_{\mathcal{A}}}{\beta_{\mathcal{A}}}$ can be equivalently replaced by $\frac{\alpha_{\mathcal{H}}}{\beta_{\mathcal{H}}}$ and $q_\lambda^E := \beta_{\mathcal{A}}^\lambda \beta_{\mathcal{H}}^{1-\lambda}$. Notice that on $\mathcal{P}_{NI}$ the formula (65) simplifies significantly, since $\alpha_{\mathcal{A}} = \alpha_{\mathcal{H}} = 0$.

(b) For general parameters $p \in \mathbb{R}$, $q \neq 0$ recall the general expression (cf. (42))

$$\widetilde{B}_{\lambda, X_0, n}^{(p,q)} \;:=\; \exp\left\{ a_n^{(q)} \cdot X_0 + \frac{p}{q} \sum_{k=1}^{n} a_k^{(q)} + n \cdot \left(\frac{p}{q} \beta_\lambda - \alpha_\lambda \right) \right\}$$

as well as

$$\widetilde{B}_{\lambda, X_0, n}^{(p,0)} \;:=\; \exp\left\{ -\beta_\lambda \cdot X_0 + \left(p \cdot e^{-\beta_\lambda} - \alpha_\lambda \right) \cdot n \right\} .$$

Then, for all $(\beta_{\mathcal{A}}, \beta_{\mathcal{H}}, \alpha_{\mathcal{A}}, \alpha_{\mathcal{H}}) \in \mathcal{P}_{SP} \backslash \mathcal{P}_{SP,1}$, all $\lambda \in \mathbb{R} \backslash \{0, 1\}$, all coefficients p_λ^L, p_λ^U, q_λ^L, $q_\lambda^U \in \mathbb{R}$ which satisfy (35) for all $x \in \mathbb{N}_0$, all initial population sizes $X_0 \in \mathbb{N}$ and all observation horizons $n \in \mathbb{N}$ one gets the following recursive bounds for the power divergences: for $\lambda \in \,]0, 1[$ there holds

$$I_\lambda(P_{\mathcal{A},n} || P_{\mathcal{H},n}) \begin{cases} < & \frac{1}{\lambda(1-\lambda)} \cdot \left(1 - B_{\lambda, X_0, n}^L \right) = \frac{1}{\lambda(1-\lambda)} \cdot \left(1 - \widetilde{B}_{\lambda, X_0, n}^{(p_\lambda^L, q_\lambda^L)} \right) =: B_{\lambda, X_0, n}^{I,U} \ , \\[2ex] \geq & \frac{1}{\lambda(1-\lambda)} \cdot \left(1 - B_{\lambda, X_0, n}^U \right) = \frac{1}{\lambda(1-\lambda)} \cdot \left(1 - \min\left\{ \widetilde{B}_{\lambda, X_0, n}^{(p_\lambda^U, q_\lambda^U)}, 1 \right\} \right) =: B_{\lambda, X_0, n}^{I,L} \ , \end{cases}$$

whereas for $\lambda \in \mathbb{R} \backslash [0, 1]$ there holds

$$I_\lambda(P_{\mathcal{A},n} || P_{\mathcal{H},n}) \begin{cases} < & \frac{1}{\lambda(\lambda-1)} \cdot \left(B_{\lambda, X_0, n}^U - 1 \right) = \frac{1}{\lambda(\lambda-1)} \cdot \left(\widetilde{B}_{\lambda, X_0, n}^{(p_\lambda^U, q_\lambda^U)} - 1 \right) =: B_{\lambda, X_0, n}^{I,U} \ , \\[2ex] \geq & \frac{1}{\lambda(\lambda-1)} \cdot \left(B_{\lambda, X_0, n}^L - 1 \right) = \frac{1}{\lambda(\lambda-1)} \cdot \left(\max\left\{ \widetilde{B}_{\lambda, X_0, n}^{(p_\lambda^L, q_\lambda^L)}, 1 \right\} - 1 \right) =: B_{\lambda, X_0, n}^{I,L} \ . \end{cases}$$

In order to deduce the subsequent *detailed* recursive analyses of power divergences, we also employ the obvious relations

$$\lim_{n \to \infty} \frac{1}{n} \log \left(\frac{1}{\lambda(1-\lambda)} - I_\lambda(P_{\mathcal{A},n} || P_{\mathcal{H},n}) \right) = \lim_{n \to \infty} \frac{1}{n} \left[-\log\left(\lambda(1-\lambda) \right) + \log\left(H_\lambda(P_{\mathcal{A},n} || P_{\mathcal{H},n}) \right) \right]$$
$$= \lim_{n \to \infty} \frac{1}{n} \log \left(H_\lambda(P_{\mathcal{A},n} || P_{\mathcal{H},n}) \right) , \qquad \text{for } \lambda \in \,]0, 1[\ , \tag{66}$$

as well as

$$\lim_{n \to \infty} \frac{1}{n} \log \left(I_\lambda(P_{\mathcal{A},n} || P_{\mathcal{H},n}) \right) = \lim_{n \to \infty} \frac{1}{n} \left[-\log\left(\lambda(\lambda-1) \right) + \log\left(H_\lambda(P_{\mathcal{A},n} || P_{\mathcal{H},n}) - 1 \right) \right]$$
$$= \lim_{n \to \infty} \frac{1}{n} \left[\log \left(1 - \frac{1}{H_\lambda(P_{\mathcal{A},n} || P_{\mathcal{H},n})} \right) + \log\left(H_\lambda(P_{\mathcal{A},n} || P_{\mathcal{H},n}) \right) \right] = \lim_{n \to \infty} \frac{1}{n} \log \left(H_\lambda(P_{\mathcal{A},n} || P_{\mathcal{H},n}) \right) , \tag{67}$$

for $\lambda \in \mathbb{R} \backslash [0, 1]$ (provided that $\liminf_{n \to \infty} H_\lambda(P_{\mathcal{A},n} || P_{\mathcal{H},n}) > 1$).

4.2. Detailed Analyses of the Exact Recursive Values of $I_\lambda(\cdot||\cdot)$, i.e., for the Cases $(\beta_\mathcal{A}, \beta_\mathcal{H}, \alpha_\mathcal{A}, \alpha_\mathcal{H}, \lambda) \in (\mathcal{P}_{NI} \cup \mathcal{P}_{SP,1}) \times (\mathbb{R}\backslash\{0,1\})$

Corollary 2. *For all $(\beta_\mathcal{A}, \beta_\mathcal{H}, \alpha_\mathcal{A}, \alpha_\mathcal{H}, \lambda) \in \mathcal{P}_{NI} \times\,]0,1[$ and all initial population sizes $X_0 \in \mathbb{N}$ there holds with $q_\lambda^E := \beta_\mathcal{A}^\lambda \beta_\mathcal{H}^{1-\lambda}$*

$(a) \qquad I_\lambda(P_{\mathcal{A},1}||P_{\mathcal{H},1}) \; = \; \dfrac{1}{\lambda(1-\lambda)} \cdot \left(1 - \exp\left\{\left(\beta_\mathcal{A}^\lambda \beta_\mathcal{H}^{1-\lambda} - \beta_\lambda\right) \cdot X_0\right\}\right) \; > \; 0,$

$(b) \qquad$ *the sequence $(I_\lambda(P_{\mathcal{A},n}||P_{\mathcal{H},n}))_{n\in\mathbb{N}}$ given by*

$$I_\lambda(P_{\mathcal{A},n}||P_{\mathcal{H},n}) \; = \; \frac{1}{\lambda(1-\lambda)} \cdot \left(1 - \exp\left\{a_n^{(q_\lambda^E)} \cdot X_0\right\}\right) \; =: \; V_{\lambda,X_0,n}^I$$

   is strictly increasing,

$(c) \qquad \lim\limits_{n\to\infty} I_\lambda(P_{\mathcal{A},n}||P_{\mathcal{H},n}) \; = \; \dfrac{1}{\lambda(1-\lambda)} \cdot \left(1 - \exp\left\{x_0^{(q_\lambda^E)} \cdot X_0\right\}\right) \; \in \; \left]0, \dfrac{1}{\lambda(1-\lambda)}\right[,$

$(d) \qquad \lim\limits_{n\to\infty} \dfrac{1}{n} \log\left(\dfrac{1}{\lambda(1-\lambda)} - I_\lambda(P_{\mathcal{A},n}||P_{\mathcal{H},n})\right) \; = \; \lim\limits_{n\to\infty} \dfrac{1}{n} \log H_\lambda(P_{\mathcal{A},n}||P_{\mathcal{H},n}) \; = \; 0,$

$(e) \qquad$ *the map $\quad X_0 \mapsto V_{\lambda,X_0,n}^I \quad$ is strictly increasing.*

Corollary 3. *For all $(\beta_\mathcal{A}, \beta_\mathcal{H}, \alpha_\mathcal{A}, \alpha_\mathcal{H}, \lambda) \in \mathcal{P}_{NI} \times (\mathbb{R}\backslash[0,1])$ and all initial population sizes $X_0 \in \mathbb{N}$ there holds with $q_\lambda^E := \beta_\mathcal{A}^\lambda \beta_\mathcal{H}^{1-\lambda}$*

$(a) \qquad I_\lambda(P_{\mathcal{A},1}||P_{\mathcal{H},1}) \; = \; \dfrac{1}{\lambda(\lambda-1)} \cdot \left(\exp\left\{\left(\beta_\mathcal{A}^\lambda \beta_\mathcal{H}^{1-\lambda} - \beta_\lambda\right) \cdot X_0\right\} - 1\right) \; > \; 0,$

$(b) \qquad$ *the sequence $(I_\lambda(P_{\mathcal{A},n}||P_{\mathcal{H},n}))_{n\in\mathbb{N}}$ given by*

$$I_\lambda(P_{\mathcal{A},n}||P_{\mathcal{H},n}) \; = \; \frac{1}{\lambda(\lambda-1)} \cdot \left(\exp\left\{a_n^{(q_\lambda^E)} \cdot X_0\right\} - 1\right) \; =: \; V_{\lambda,X_0,n}^I$$

   is strictly increasing,

$(c) \qquad \lim\limits_{n\to\infty} I_\lambda(P_{\mathcal{A},n}||P_{\mathcal{H},n}) \; = \; \begin{cases} \dfrac{1}{\lambda(\lambda-1)} \cdot \left(\exp\left\{x_0^{(q_\lambda^E)} \cdot X_0\right\} - 1\right) > 0, & \text{if } \lambda \in [\lambda_-, \lambda_+] \backslash [0,1], \\ \infty, & \text{if } \lambda \in\,]-\infty, \lambda_-[\, \cup\,]\lambda_+, \infty[, \end{cases}$

$(d) \qquad \lim\limits_{n\to\infty} \dfrac{1}{n} \log I_\lambda(P_{\mathcal{A},n}||P_{\mathcal{H},n}) \; = \; \begin{cases} 0, & \text{if } \lambda \in [\lambda_-, \lambda_+] \backslash [0,1], \\ \infty, & \text{if } \lambda \in\,]-\infty, \lambda_-[\, \cup\,]\lambda_+, \infty[, \end{cases}$

$(e) \qquad$ *the map $\quad X_0 \mapsto V_{\lambda,X_0,n}^I \quad$ is strictly increasing.*

Corollary 4. *For all $(\beta_\mathcal{A}, \beta_\mathcal{H}, \alpha_\mathcal{A}, \alpha_\mathcal{H}, \lambda) \in \mathcal{P}_{SP,1} \times\,]0,1[$ and all initial population sizes $X_0 \in \mathbb{N}$ there holds with $q_\lambda^E := \beta_\mathcal{A}^\lambda \beta_\mathcal{H}^{1-\lambda}$*

$(a) \qquad I_\lambda(P_{\mathcal{A},1}||P_{\mathcal{H},1}) \; = \; \dfrac{1}{\lambda(1-\lambda)} \cdot \left(1 - \exp\left\{\left(\beta_\mathcal{A}^\lambda \beta_\mathcal{H}^{1-\lambda} - \beta_\lambda\right) \cdot \left(X_0 + \dfrac{\alpha_\mathcal{A}}{\beta_\mathcal{A}}\right)\right\}\right) \; > \; 0,$

$(b) \qquad$ *the sequence $(I_\lambda(P_{\mathcal{A},n}||P_{\mathcal{H},n}))_{n\in\mathbb{N}}$ given by*

$$I_\lambda(P_{\mathcal{A},n}||P_{\mathcal{H},n}) \; = \; \frac{1}{\lambda(1-\lambda)} \cdot \left(1 - \exp\left\{a_n^{(q_\lambda^E)} \cdot X_0 + \frac{\alpha_\mathcal{A}}{\beta_\mathcal{A}} \sum_{k=1}^{n} a_k^{(q_\lambda^E)}\right\}\right) \; =: \; V_{\lambda,X_0,n}^I$$

   is strictly increasing,

$(c) \qquad \lim\limits_{n\to\infty} I_\lambda(P_{\mathcal{A},n}||P_{\mathcal{H},n}) \; = \; \dfrac{1}{\lambda(1-\lambda)},$

$(d) \qquad \lim\limits_{n\to\infty} \dfrac{1}{n} \log\left(\dfrac{1}{\lambda(1-\lambda)} - I_\lambda(P_{\mathcal{A},n}||P_{\mathcal{H},n})\right) \; = \; \dfrac{\alpha_\mathcal{A}}{\beta_\mathcal{A}} \cdot x_0^{(q_\lambda^E)} \; < \; 0,$

$(e) \qquad$ *the map $\quad X_0 \mapsto V_{\lambda,X_0,n}^I \quad$ is strictly increasing.*

Corollary 5. *For all* $(\beta_{\mathcal{A}}, \beta_{\mathcal{H}}, \alpha_{\mathcal{A}}, \alpha_{\mathcal{H}}, \lambda) \in \mathcal{P}_{SP,1} \times (\mathbb{R} \setminus [0,1])$ *and all initial population sizes* $X_0 \in \mathbb{N}$ *there holds with* $q_\lambda^E := \beta_{\mathcal{A}}^\lambda \beta_{\mathcal{H}}^{1-\lambda}$

(a) $\quad I_\lambda(P_{\mathcal{A},1} || P_{\mathcal{H},1}) = \dfrac{1}{\lambda(\lambda-1)} \cdot \left(\exp\left\{ \left(\beta_{\mathcal{A}}^\lambda \beta_{\mathcal{H}}^{1-\lambda} - \beta_\lambda \right) \cdot \left(X_0 + \dfrac{\alpha_{\mathcal{A}}}{\beta_{\mathcal{A}}} \right) \right\} - 1 \right) > 0,$

(b) $\quad$ *the sequence* $\left(I_\lambda(P_{\mathcal{A},n} || P_{\mathcal{H},n}) \right)_{n \in \mathbb{N}}$ *given by*

$$I_\lambda(P_{\mathcal{A},n} || P_{\mathcal{H},n}) = \dfrac{1}{\lambda(\lambda-1)} \cdot \left(\exp\left\{ a_n^{(q_\lambda^E)} \cdot X_0 + \dfrac{\alpha_{\mathcal{A}}}{\beta_{\mathcal{A}}} \sum_{k=1}^{n} a_k^{(q_\lambda^E)} \right\} - 1 \right) =: V_{\lambda,X_0,n}^I$$

$\qquad$ *is strictly increasing,*

(c) $\quad \displaystyle\lim_{n \to \infty} I_\lambda(P_{\mathcal{A},n} || P_{\mathcal{H},n}) = \infty,$

(d) $\quad \displaystyle\lim_{n \to \infty} \dfrac{1}{n} \log I_\lambda(P_{\mathcal{A},n} || P_{\mathcal{H},n}) = \begin{cases} \dfrac{\alpha_{\mathcal{A}}}{\beta_{\mathcal{A}}} \cdot x_0^{(q_\lambda^E)} > 0, & \text{if} \quad \lambda \in [\lambda_-, \lambda_+] \setminus [0,1], \\ \infty, & \text{if} \quad \lambda \in]-\infty, \lambda_-[\,\cup\,]\lambda_+, \infty[, \end{cases}$

(e) $\quad$ *the map* $\quad X_0 \mapsto V_{\lambda,X_0,n}^I \quad$ *is strictly increasing.*

In the assertions (a), (b), (d) of the Corollaries 4 and 5 the fraction $\alpha_{\mathcal{A}}/\beta_{\mathcal{A}}$ can be equivalently replaced by $\alpha_{\mathcal{H}}/\beta_{\mathcal{H}}$.

Let us now derive the corresponding detailed results for the bounds of the power divergences for the parameter cases $\mathcal{P}_{SP} \setminus \mathcal{P}_{SP,1}$, where the Hellinger integral, and thus $I_\lambda(P_{\mathcal{A},n} || P_{\mathcal{H},n})$, cannot be determined exactly. The extensive discussion on the Hellinger-integral bounds in the Sections 3.4–3.13, as well as in the Sections 3.16–3.24 can be carried over directly to obtain power-divergence bounds. In the following, we summarize the outcoming key results, referring a detailed discussion on the possible choices of $p_\lambda^A = p^A(\beta_{\mathcal{A}}, \beta_{\mathcal{H}}, \alpha_{\mathcal{A}}, \alpha_{\mathcal{H}}, \lambda)$ and $q_\lambda^A = q^A(\beta_{\mathcal{A}}, \beta_{\mathcal{H}}, \alpha_{\mathcal{A}}, \alpha_{\mathcal{H}}, \lambda)$ $(A \in \{L, U\})$ to the corresponding above-mentioned subsections.

4.3. Lower Bounds of $I_\lambda(\cdot || \cdot)$ for the Cases $(\beta_{\mathcal{A}}, \beta_{\mathcal{H}}, \alpha_{\mathcal{A}}, \alpha_{\mathcal{H}}, \lambda) \in (\mathcal{P}_{SP} \setminus \mathcal{P}_{SP,1}) \times \,]0,1[$

Corollary 6. *For all* $(\beta_{\mathcal{A}}, \beta_{\mathcal{H}}, \alpha_{\mathcal{A}}, \alpha_{\mathcal{H}}, \lambda) \in (\mathcal{P}_{SP,2} \cup \mathcal{P}_{SP,3a} \cup \mathcal{P}_{SP,3b}) \times \,]0,1[$ *there exist parameters* p_λ^U, q_λ^U *which satisfy* $p_\lambda^U \in [\alpha_{\mathcal{A}}^\lambda \alpha_{\mathcal{H}}^{1-\lambda}, \alpha_\lambda]$ *and* $q_\lambda^U \in [\beta_{\mathcal{A}}^\lambda \beta_{\mathcal{H}}^{1-\lambda}, \beta_\lambda[$ *as well as* (35) *for all* $x \in \mathbb{N}_0$, *and for all such pairs* $(p_\lambda^U, q_\lambda^U)$ *and all initial population sizes* $X_0 \in \mathbb{N}$ *there holds*

(a) $\quad B_{\lambda,X_0,1}^{I,L} = \dfrac{1}{\lambda(1-\lambda)} \cdot \left(1 - \exp\left\{ \left(q_\lambda^U - \beta_\lambda \right) \cdot X_0 + p_\lambda^U - \alpha_\lambda \right\} \right) > 0,$

(b) $\quad$ *the sequence* $\left(B_{\lambda,X_0,n}^{I,L} \right)_{n \in \mathbb{N}}$ *of lower bounds for* $I_\lambda(P_{\mathcal{A},n} || P_{\mathcal{H},n})$ *given by*

$$B_{\lambda,X_0,n}^{I,L} = \dfrac{1}{\lambda(1-\lambda)} \cdot \left(1 - \exp\left\{ a_n^{(q_\lambda^U)} \cdot X_0 + \sum_{k=1}^{n} b_k^{(p_\lambda^U, q_\lambda^U)} \right\} \right)$$

$\qquad$ *is strictly increasing,*

(c) $\quad \displaystyle\lim_{n \to \infty} B_{\lambda,X_0,n}^{I,L} = \lim_{n \to \infty} I_\lambda(P_{\mathcal{A},n} || P_{\mathcal{H},n}) = \dfrac{1}{\lambda(1-\lambda)},$

(d) $\quad \displaystyle\lim_{n \to \infty} \dfrac{1}{n} \log\left(\dfrac{1}{\lambda(1-\lambda)} - B_{\lambda,X_0,n}^{I,L} \right) = p_\lambda^U \cdot e^{x_0^{(q_\lambda^U)}} - \alpha_\lambda < 0,$

(e) $\quad$ *the map* $\quad X_0 \mapsto B_{\lambda,X_0,n}^{I,L} \quad$ *is strictly increasing.*

Remark 4.

(a) $\quad$ Notice that in the case $(\beta_{\mathcal{A}}, \beta_{\mathcal{H}}, \alpha_{\mathcal{A}}, \alpha_{\mathcal{H}}, \lambda) \in \mathcal{P}_{SP,2} \times \,]0,1[$—where $\alpha_{\mathcal{A}}^\lambda \alpha_{\mathcal{H}}^{1-\lambda} = \alpha_\lambda = \alpha_{\mathcal{A}} = \alpha_{\mathcal{H}} = \alpha$—we get the special choice $p_\lambda^U = \alpha$ and $q_\lambda^U = (\alpha + \beta_{\mathcal{A}})^\lambda (\alpha + \beta_{\mathcal{H}})^{1-\lambda} - \alpha$ (cf. Section 3.7). For the constellations $(\beta_{\mathcal{A}}, \beta_{\mathcal{H}}, \alpha_{\mathcal{A}}, \alpha_{\mathcal{H}}, \lambda) \in (\mathcal{P}_{SP,3a} \cup \mathcal{P}_{SP,3b}) \times \,]0,1[$ there exist parameters

$$p_\lambda^U \in [\alpha_{\mathcal{A}}^\lambda \alpha_{\mathcal{H}}^{1-\lambda}, \alpha_\lambda[\ , q_\lambda^U \in [\beta_{\mathcal{A}}^\lambda \beta_{\mathcal{H}}^{1-\lambda}, \beta_\lambda[\ \textit{which satisfy (35) for all } x \in \mathbb{N}_0.$$

(b)   *For the parameter setups* $(\beta_{\mathcal{A}}, \beta_{\mathcal{H}}, \alpha_{\mathcal{A}}, \alpha_{\mathcal{H}}, \lambda) \in (\mathcal{P}_{SP,2} \cup \mathcal{P}_{SP,3a} \cup \mathcal{P}_{SP,3b}) \times]0, 1[$ *there might exist parameter pairs* $(p_\lambda^U, q_\lambda^U)$ *satisfying (35) and either* $p_\lambda^U = \alpha_\lambda$ *or* $q_\lambda^U = \beta_\lambda$, *for which all assertions of Corollary 6 still hold true.*

(c)   *Following the discussion in Section 3.10 for all* $(\beta_{\mathcal{A}}, \beta_{\mathcal{H}}, \alpha_{\mathcal{A}}, \alpha_{\mathcal{H}}, \lambda) \in \mathcal{P}_{SP,3c} \times]0, 1[$ *at least part (c) still holds true.*

Corollary 7. *For all* $(\beta_{\mathcal{A}}, \beta_{\mathcal{H}}, \alpha_{\mathcal{A}}, \alpha_{\mathcal{H}}, \lambda) \in \mathcal{P}_{SP,4a} \times]0, 1[$ *there exist parameters* $p_\lambda^U < \alpha_\lambda, 1 > q_\lambda^U > \beta_\lambda = \beta$ *such that (35) is satisfied for all* $x \in [0, \infty[$ *and such that for all initial population sizes* $X_0 \in \mathbb{N}$ *at least the parts (c) and (d) of Corollary 6 hold true.*

As in Section 3.12, for the parameter setup $(\beta_{\mathcal{A}}, \beta_{\mathcal{H}}, \alpha_{\mathcal{A}}, \alpha_{\mathcal{H}}, \lambda) \in \mathcal{P}_{SP,4b} \times]0, 1[$ we cannot derive a lower bound for the power divergences which improves the generally valid lower bound $I_\lambda(P_{\mathcal{A},n} || P_{\mathcal{H},n}) \geq 0$ (cf. (10)) by employing our proposed $(p_\lambda^U, q_\lambda^U)$-method.

4.4. Upper Bounds of $I_\lambda(\cdot || \cdot)$ *for the Cases* $(\beta_{\mathcal{A}}, \beta_{\mathcal{H}}, \alpha_{\mathcal{A}}, \alpha_{\mathcal{H}}, \lambda) \in (\mathcal{P}_{SP} \backslash \mathcal{P}_{SP,1}) \times]0, 1[$

Since in this setup the upper bounds of the power divergences can be derived from the lower bounds of the Hellinger integrals, we here appropriately adapt the results of Proposition 6.

Corollary 8.   *For all* $(\beta_{\mathcal{A}}, \beta_{\mathcal{H}}, \alpha_{\mathcal{A}}, \alpha_{\mathcal{H}}, \lambda) \in (\mathcal{P}_{SP} \backslash \mathcal{P}_{SP,1}) \times]0, 1[$ *and all initial population sizes* $X_0 \in \mathbb{N}$ *there holds with* $p_\lambda^L := \alpha_{\mathcal{A}}^\lambda \alpha_{\mathcal{H}}^{1-\lambda}$ *and* $q_\lambda^L := \beta_{\mathcal{A}}^\lambda \beta_{\mathcal{H}}^{1-\lambda}$

(a) $$B_{\lambda, X_0, 1}^{I,U} = \frac{1}{\lambda(1-\lambda)} \cdot \left(1 - \exp\left\{\left(\beta_{\mathcal{A}}^\lambda \beta_{\mathcal{H}}^{1-\lambda} - \beta_\lambda\right) \cdot X_0 + \alpha_{\mathcal{A}}^\lambda \alpha_{\mathcal{H}}^{1-\lambda} - \alpha_\lambda\right\}\right) > 0,$$

(b)   *the sequence of upper bounds* $\left(B_{\lambda, X_0, n}^{I,U}\right)_{n \in \mathbb{N}}$ *for* $I_\lambda(P_{\mathcal{A},n} || P_{\mathcal{H},n})$ *given by*

$$B_{\lambda, X_0, n}^{I,U} = \frac{1}{\lambda(1-\lambda)} \cdot \left(1 - \exp\left\{a_n^{(q_\lambda^L)} \cdot X_0 + \frac{p_\lambda^L}{q_\lambda^L} \sum_{k=1}^n a_k^{(q_\lambda^L)} + n \cdot \left(\frac{p_\lambda^L}{q_\lambda^L} \cdot \beta_\lambda - \alpha_\lambda\right)\right\}\right)$$

     is strictly increasing,

(c) $$\lim_{n \to \infty} B_{\lambda, X_0, n}^{I,U} = \frac{1}{\lambda(1-\lambda)},$$

(d) $$\lim_{n \to \infty} \frac{1}{n} \log\left(\frac{1}{\lambda(1-\lambda)} - B_{\lambda, X_0, n}^{I,U}\right) = \frac{p_\lambda^L}{q_\lambda^L} \cdot \left(x_0^{(q_\lambda^L)} + \beta_\lambda\right) - \alpha_\lambda = p_\lambda^L \cdot e^{x_0^{(q_\lambda^L)}} - \alpha_\lambda < 0,$$

(e)   *the map* $X_0 \mapsto B_{\lambda, X_0, n}^{I,U}$ *is strictly increasing.*

4.5. Lower Bounds of $I_\lambda(\cdot || \cdot)$ *for the Cases* $(\beta_{\mathcal{A}}, \beta_{\mathcal{H}}, \alpha_{\mathcal{A}}, \alpha_{\mathcal{H}}, \lambda) \in (\mathcal{P}_{SP} \backslash \mathcal{P}_{SP,1}) \times (\mathbb{R} \backslash [0,1])$

In order to derive detailed results on lower bounds of the power divergences in the case $\lambda \in \mathbb{R} \backslash [0, 1]$, we have to subsume and adapt the Hellinger-integral concerning lower-bounds investigations from the Sections 3.16–3.23. Recall the λ-sets $\mathcal{I}_{SP,2}, \mathcal{I}_{SP,3a}, \mathcal{I}_{SP,3b}$ (cf. (58), (60), (62)). For the constellations $\mathcal{P}_{SP,2} \times \mathcal{I}_{SP,2}$ we employ the special choice $p_\lambda^L = \alpha_{\mathcal{A}}^\lambda \alpha_{\mathcal{H}}^{1-\lambda} = \alpha_\lambda = \alpha_{\mathcal{A}} = \alpha_{\mathcal{H}} = \alpha$ together with $q_\lambda^L = (\alpha + \beta_{\mathcal{A}})^\lambda (\alpha + \beta_{\mathcal{H}})^{1-\lambda} - \alpha > \max\{0, \beta_\lambda\}$ (cf. (58)) which satisfy (35) for all $x \in \mathbb{N}_0$ and (56), whereas for the constellations $(\mathcal{P}_{SP,3a} \times \mathcal{I}_{SP,3a}) \cup (\mathcal{P}_{SP,3b} \times \mathcal{I}_{SP,3b})$ we have proved the existence of parameters p_λ^L, q_λ^L satisfying both (35) for all $x \in \mathbb{N}_0$ and (56) with two strict inequalities. Subsuming this, we obtain

Corollary 9. *For all* $(\beta_{\mathcal{A}}, \beta_{\mathcal{H}}, \alpha_{\mathcal{A}}, \alpha_{\mathcal{H}}, \lambda) \in (\mathcal{P}_{SP,2} \times \mathcal{I}_{SP,2}) \cup (\mathcal{P}_{SP,3a} \times \mathcal{I}_{SP,3a}) \cup (\mathcal{P}_{SP,3b} \times \mathcal{I}_{SP,3b})$ *there exist parameters* p_λ^L, q_λ^L *which satisfy* $\max\{0, \alpha_\lambda\} \leq p_\lambda^L \leq \alpha_{\mathcal{A}}^\lambda \alpha_{\mathcal{H}}^{1-\lambda}$, $\max\{0, \beta_\lambda\} < q_\lambda^L \leq \beta_{\mathcal{A}}^\lambda \beta_{\mathcal{H}}^{1-\lambda}$ *as well as* (35) *for all* $x \in \mathbb{N}_0$, *and for all such pairs* $(p_\lambda^L, q_\lambda^L)$ *and all initial population sizes* $X_0 \in \mathbb{N}$ *one gets*

$$(a) \qquad B_{\lambda, X_0, 1}^{I,L} = \frac{1}{\lambda(\lambda-1)} \cdot \left(\exp\left\{ \left(q_\lambda^L - \beta_\lambda \right) \cdot X_0 + p_\lambda^L - \alpha_\lambda \right\} - 1 \right) > 0,$$

$(b) \qquad$ *the sequence* $\left(B_{\lambda, X_0, n}^{I,L} \right)_{n \in \mathbb{N}}$ *of lower bounds for* $I_\lambda(P_{\mathcal{A},n} || P_{\mathcal{H},n})$ *given by*

$$B_{\lambda, X_0, n}^{I,L} = \frac{1}{\lambda(\lambda-1)} \cdot \left(\exp\left\{ a_n^{(q_\lambda^L)} \cdot X_0 + \sum_{k=1}^{n} b_k^{(p_\lambda^L, q_\lambda^L)} \right\} - 1 \right)$$

$\qquad$ *is strictly increasing,*

$$(c) \qquad \lim_{n \to \infty} B_{\lambda, X_0, n}^{I,L} = \lim_{n \to \infty} I_\lambda(P_{\mathcal{A},n} || P_{\mathcal{H},n}) = \infty,$$

$$(d) \qquad \lim_{n \to \infty} \frac{1}{n} \log B_{\lambda, X_0, n}^{I,L} = \begin{cases} p_\lambda^L \cdot \exp\left\{ x_0^{(q_\lambda^L)} \right\} - \alpha_\lambda > 0, & \text{if } q_\lambda^L \leq \min\left\{ 1; e^{\beta_\lambda - 1} \right\}, \\ \infty, & \text{if } q_\lambda^L > \min\left\{ 1; e^{\beta_\lambda - 1} \right\}, \end{cases}$$

$(e) \qquad$ *the map* $X_0 \mapsto B_{\lambda, X_0, n}^{I,L}$ *is strictly increasing.*

Analogously to the discussions in the Sections 3.17–3.20, for the parameter setups $\Big(\mathcal{P}_{SP,2} \times \mathbb{R} \backslash (\mathcal{I}_{SP,2} \cup [0,1]) \Big) \cup \Big(\mathcal{P}_{SP,3a} \times \mathbb{R} \backslash (\mathcal{I}_{SP,3a} \cup [0,1]) \Big) \cup \Big(\mathcal{P}_{SP,3b} \times \mathbb{R} \backslash (\mathcal{I}_{SP,3b} \cup [0,1]) \Big) \cup \Big(\mathcal{P}_{SP,3c} \times \mathbb{R} \backslash [0,1] \Big)$ and for all initial population sizes $X_0 \in \mathbb{N}$ one can still show

$$0 < I_\lambda(P_{\mathcal{A},n} || P_{\mathcal{H},n}), \qquad \text{and} \qquad \lim_{n \to \infty} I_\lambda(P_{\mathcal{A},n} || P_{\mathcal{H},n}) = \infty.$$

For the penultimate case we obtain

Corollary 10. *For all* $(\beta_{\mathcal{A}}, \beta_{\mathcal{H}}, \alpha_{\mathcal{A}}, \alpha_{\mathcal{H}}, \lambda) \in \mathcal{P}_{SP,4a} \times (\mathbb{R} \backslash [0,1])$ *there exist parameters* $p_\lambda^L > \alpha_\lambda$ *(where not necessarily* $p_\lambda^L \geq 0$*) and* $0 < q_\lambda^L < \beta_\lambda = \beta_\bullet < \min\{1, e^{\beta_\bullet - 1}\} = e^{\beta_\bullet - 1}$ *such that* (35) *is satisfied for all* $x \in [0, \infty[$ *and such that for all initial population sizes* $X_0 \in \mathbb{N}$ *at least the parts (c) and (d) of Corollary* 9 *hold true.*

Notice that for the last case $(\beta_{\mathcal{A}}, \beta_{\mathcal{H}}, \alpha_{\mathcal{A}}, \alpha_{\mathcal{H}}, \lambda) \in \mathcal{P}_{SP,4b} \times \mathbb{R} \backslash [0,1]$ (where $(\beta_{\mathcal{A}} = \beta_{\mathcal{H}} \geq 1)$ we cannot derive lower bounds of the power divergences which improve the generally valid lower bound $I_\lambda(P_{\mathcal{A},n} || P_{\mathcal{H},n}) \geq 0$ (cf. (11)) by employing our proposed $(p_\lambda^U, q_\lambda^U)$-method.

4.6. Upper Bounds of $I_\lambda(\cdot || \cdot)$ for the Cases $(\beta_{\mathcal{A}}, \beta_{\mathcal{H}}, \alpha_{\mathcal{A}}, \alpha_{\mathcal{H}}, \lambda) \in (\mathcal{P}_{SP} \backslash \mathcal{P}_{SP,1}) \times (\mathbb{R} \backslash [0,1])$

For these constellations we adapt Proposition 14, which after modulation becomes

Corollary 11. *For all* $(\beta_{\mathcal{A}}, \beta_{\mathcal{H}}, \alpha_{\mathcal{A}}, \alpha_{\mathcal{H}}, \lambda) \in (\mathcal{P}_{SP} \backslash \mathcal{P}_{SP,1}) \times (\mathbb{R} \backslash [0,1])$ *and all initial population sizes* $X_0 \in \mathbb{N}$ *there holds with* $p_\lambda^U := \alpha_{\mathcal{A}}^\lambda \alpha_{\mathcal{H}}^{1-\lambda}$ *and* $q_\lambda^U := \beta_{\mathcal{A}}^\lambda \beta_{\mathcal{H}}^{1-\lambda}$

(a) $\quad B_{\lambda,X_0,1}^{I,U} \;=\; \dfrac{1}{\lambda(\lambda-1)} \cdot \left(\exp\left\{ \left(\beta_{\mathcal{A}}^\lambda \beta_{\mathcal{H}}^{1-\lambda} - \beta_\lambda \right) \cdot X_0 + \alpha_{\mathcal{A}}^\lambda \alpha_{\mathcal{H}}^{1-\lambda} - \alpha_\lambda \right\} - 1 \right) \;>\; 0,$

(b) $\quad$ *the sequence* $\left(B_{\lambda,X_0,n}^{I,U} \right)_{n \in \mathbb{N}}$ *of upper bounds for* $I_\lambda(P_{\mathcal{A},n} \| P_{\mathcal{H},n})$ *given by*

$$B_{\lambda,X_0,n}^{I,U} \;=\; \dfrac{1}{\lambda(\lambda-1)} \cdot \left(\exp\left\{ a_n^{(q_\lambda^U)} \cdot X_0 + \sum_{k=1}^{n} b_k^{(p_\lambda^U, q_\lambda^U)} \right\} - 1 \right)$$

$\quad$ *is strictly increasing,*

(c) $\quad \lim\limits_{n \to \infty} B_{\lambda,X_0,n}^{I,U} \;=\; \infty,$

(d) $\quad \lim\limits_{n \to \infty} \dfrac{1}{n} \log B_{\lambda,X_0,n}^{I,U} \;=\; \begin{cases} p_\lambda^U \cdot \exp\left\{ x_0^{(q_\lambda^U)} \right\} - \alpha_\lambda \;>\; 0, & \text{if } \lambda \in [\lambda_-, \lambda_+] \backslash [0,1], \\[2mm] \infty, & \text{if } \lambda \in]-\infty, \lambda_-[\cup]\lambda_+, \infty[, \end{cases}$

(e) $\quad$ *the map* $\quad X_0 \mapsto B_{\lambda,X_0,n}^{I,U} \quad$ *is strictly increasing.*

4.7. Applications to Bayesian Decision Making

As explained in Section 2.5, the power divergences fulfill

$$I_\lambda\left(P_{\mathcal{A},n} \| P_{\mathcal{H},n}\right) \;=\; \int_0^1 \Delta\mathcal{BR}_{\widehat{\mathcal{LO}}}(p_{\mathcal{A}}^{\text{prior}}) \cdot \left(1 - p_{\mathcal{A}}^{\text{prior}}\right)^{\lambda-2} \cdot \left(p_{\mathcal{A}}^{\text{prior}}\right)^{-1-\lambda} \, dp_{\mathcal{A}}^{\text{prior}}, \qquad \lambda \in \mathbb{R}, \qquad \text{(cf. (21))},$$

and

$$I_\lambda\left(P_{\mathcal{A},n} \| P_{\mathcal{H},n}\right) \;=\; \lim_{\chi \to p_{\mathcal{A}}^{\text{prior}}} \Delta\mathcal{BR}_{\mathcal{LO}_{\lambda,\chi}}\left(p_{\mathcal{A}}^{\text{prior}}\right), \qquad \lambda \in]0,1[, \qquad \text{(cf. (22))},$$

and thus can be interpreted as (i) *weighted-average* decision risk reduction (weighted-average statistical information measure) about the degree of evidence ᴅᴇɢ concerning the parameter θ that can be attained by observing the GWI-path $\mathcal{X}_n$ until stage n, and as (ii) *limit* decision risk reduction (limit statistical information measure). Hence, by combining (21) and (22) with the investigations in the previous Sections 4.1–4.6, we obtain exact recursive values respectively recursive bounds of the above-mentioned decision risk reductions. For the sake of brevity, we omit the details here.

5. Kullback-Leibler Information Divergence (Relative Entropy)

5.1. Exact Values Respectively Upper Bounds of $I(\cdot \| \cdot)$

From (2), (3) and (6) in Section 2.4, one can immediately see that the Kullback-Leibler information divergence (relative entropy) between two competing Galton-Watson processes without/with immigration can be obtained by the limit

$$I(P_{\mathcal{A},n} \| P_{\mathcal{H},n}) \;=\; \lim_{\lambda \nearrow 1} I_\lambda\left(P_{\mathcal{A},n} \| P_{\mathcal{H},n}\right), \tag{68}$$

and the reverse Kullback-Leibler information divergence (reverse relative entropy) by $I\left(P_{\mathcal{H},n} \| P_{\mathcal{A},n}\right) = \lim_{\lambda \searrow 0} I_\lambda\left(P_{\mathcal{A},n} \| P_{\mathcal{H},n}\right)$. Hence, in the following we concentrate only on (68), the reverse case works analogously. Accordingly, we can use (68) in appropriate combination with the $\lambda \in]0,1[$-parts of the previous Section 4 (respectively, the corresponding parts of Section 3) in order to obtain detailed analyses for $I\left(P_{\mathcal{H},n} \| P_{\mathcal{A},n}\right)$. Let us start with the following assertions on exact values respectively upper bounds, which will be proved in Appendix A.2:

Theorem 3.

(a)   For all $(\beta_{\mathcal{A}}, \beta_{\mathcal{H}}, \alpha_{\mathcal{A}}, \alpha_{\mathcal{H}}) \in (\mathcal{P}_{NI} \cup \mathcal{P}_{SP,1})$, all initial population sizes $X_0 \in \mathbb{N}$ and all observation horizons $n \in \mathbb{N}$ the Kullback-Leibler information divergence (relative entropy) is given by

$$
I(P_{\mathcal{A},n} || P_{\mathcal{H},n}) = I_{X_0,n} :=
\begin{cases}
\dfrac{\beta_{\mathcal{A}} \cdot \left(\log\left(\frac{\beta_{\mathcal{A}}}{\beta_{\mathcal{H}}}\right) - 1\right) + \beta_{\mathcal{H}}}{1 - \beta_{\mathcal{A}}} \cdot \left[X_0 - \frac{\alpha_{\mathcal{A}}}{1 - \beta_{\mathcal{A}}}\right] \cdot (1 - (\beta_{\mathcal{A}})^n) & \\[2mm]
\quad + \dfrac{\alpha_{\mathcal{A}} \cdot \left[\beta_{\mathcal{A}} \cdot \left(\log\left(\frac{\beta_{\mathcal{A}}}{\beta_{\mathcal{H}}}\right) - 1\right) + \beta_{\mathcal{H}}\right]}{\beta_{\mathcal{A}}(1 - \beta_{\mathcal{A}})} \cdot n, & \text{if } \beta_{\mathcal{A}} \neq 1, \\[4mm]
[\beta_{\mathcal{H}} - \log \beta_{\mathcal{H}} - 1] \cdot \left[\frac{\alpha_{\mathcal{A}}}{2} \cdot n^2 + \left(X_0 + \frac{\alpha_{\mathcal{A}}}{2}\right) \cdot n\right], & \text{if } \beta_{\mathcal{A}} = 1.
\end{cases}
$$

(69)

(b)   For all $(\beta_{\mathcal{A}}, \beta_{\mathcal{H}}, \alpha_{\mathcal{A}}, \alpha_{\mathcal{H}}) \in \mathcal{P}_{SP} \backslash \mathcal{P}_{SP,1}$, all initial population sizes $X_0 \in \mathbb{N}$ and all observation horizons $n \in \mathbb{N}$ there holds $I(P_{\mathcal{A},n} || P_{\mathcal{H},n}) \leq E^U_{X_0,n}$, where

$$
E^U_{X_0,n} :=
\begin{cases}
\dfrac{\beta_{\mathcal{A}} \cdot \left(\log\left(\frac{\beta_{\mathcal{A}}}{\beta_{\mathcal{H}}}\right) - 1\right) + \beta_{\mathcal{H}}}{1 - \beta_{\mathcal{A}}} \cdot \left[X_0 - \frac{\alpha_{\mathcal{A}}}{1 - \beta_{\mathcal{A}}}\right] \cdot (1 - (\beta_{\mathcal{A}})^n) & \\[2mm]
\quad + \left[\dfrac{\alpha_{\mathcal{A}} \cdot \left[\beta_{\mathcal{A}} \cdot \left(\log\left(\frac{\beta_{\mathcal{A}}}{\beta_{\mathcal{H}}}\right) - 1\right) + \beta_{\mathcal{H}}\right]}{\beta_{\mathcal{A}}(1 - \beta_{\mathcal{A}})} + \alpha_{\mathcal{A}}\left[\log\left(\frac{\alpha_{\mathcal{A}}\beta_{\mathcal{H}}}{\alpha_{\mathcal{H}}\beta_{\mathcal{A}}}\right) - \frac{\beta_{\mathcal{H}}}{\beta_{\mathcal{A}}}\right] + \alpha_{\mathcal{H}}\right] \cdot n, & \text{if } \beta_{\mathcal{A}} \neq 1, \\[4mm]
[\beta_{\mathcal{H}} - \log \beta_{\mathcal{H}} - 1] \cdot \left[\frac{\alpha_{\mathcal{A}}}{2} \cdot n^2 + \left(X_0 + \frac{\alpha_{\mathcal{A}}}{2}\right) \cdot n\right] & \\[2mm]
\quad + \left[\alpha_{\mathcal{A}}\left[\log\left(\frac{\alpha_{\mathcal{A}}\beta_{\mathcal{H}}}{\alpha_{\mathcal{H}}}\right) - \beta_{\mathcal{H}}\right] + \alpha_{\mathcal{H}}\right] \cdot n, & \text{if } \beta_{\mathcal{A}} = 1.
\end{cases}
$$

(70)

Remark 5.

(i) Notice that the exact values respectively upper bounds are in closed form (rather than in recursive form).

(ii) The n−behaviour of (the bounds of) the Kullback-Leibler information divergence/relative entropy $I(P_{\mathcal{A},n} || P_{\mathcal{H},n})$ in Theorem 3 is influenced by the following facts:

(a) $\beta_{\mathcal{A}} \cdot \left(\log\left(\frac{\beta_{\mathcal{A}}}{\beta_{\mathcal{H}}}\right) - 1\right) + \beta_{\mathcal{H}} \geq 0$ with equality iff $\beta_{\mathcal{A}} = \beta_{\mathcal{H}}$.

(b)   In the case $\beta_{\mathcal{A}} \neq 1$ of (70), there holds $\dfrac{\alpha_{\mathcal{A}} \cdot \left[\beta_{\mathcal{A}} \cdot \left(\log\left(\frac{\beta_{\mathcal{A}}}{\beta_{\mathcal{H}}}\right) - 1\right) + \beta_{\mathcal{H}}\right]}{\beta_{\mathcal{A}}(1 - \beta_{\mathcal{A}})} + \alpha_{\mathcal{A}}\left[\log\left(\frac{\alpha_{\mathcal{A}}\beta_{\mathcal{H}}}{\alpha_{\mathcal{H}}\beta_{\mathcal{A}}}\right) - \frac{\beta_{\mathcal{H}}}{\beta_{\mathcal{A}}}\right] + \alpha_{\mathcal{H}} \geq 0$,
with equality iff $\alpha_{\mathcal{A}} = \alpha_{\mathcal{H}}$ and $\beta_{\mathcal{A}} = \beta_{\mathcal{H}}$.

5.2. Lower Bounds of $I(\cdot || \cdot)$ for the Cases $(\beta_{\mathcal{A}}, \beta_{\mathcal{H}}, \alpha_{\mathcal{A}}, \alpha_{\mathcal{H}}) \in (\mathcal{P}_{SP} \backslash \mathcal{P}_{SP,1})$

Again by using (68) in appropriate combination with the "$\lambda \in]0, 1[$-parts" of the previous Section 4 (respectively, the corresponding parts of Section 3), we obtain the following *(semi-)closed-form* lower bounds of $I(P_{\mathcal{H},n} || P_{\mathcal{A},n})$:

Theorem 4. For all $(\beta_{\mathcal{A}}, \beta_{\mathcal{H}}, \alpha_{\mathcal{A}}, \alpha_{\mathcal{H}}) \in \mathcal{P}_{SP} \backslash \mathcal{P}_{SP,1}$, all initial population sizes $X_0 \in \mathbb{N}$ and all observation horizons $n \in \mathbb{N}$

$$
I(P_{\mathcal{A},n} || P_{\mathcal{H},n}) \geq E^L_{X_0,n} := \sup_{k \in \mathbb{N}_0, \, y \in [0,\infty[} \left\{E^{L,tan}_{y,X_0,n}, \, E^{L,sec}_{k,X_0,n}, \, E^{L,hor}_{X_0,n}\right\} \in [0, \infty[,
$$

(71)

where for all $y \in [0, \infty[$ we define the – possibly negatively valued– finite bound component

$$E_{y,X_0,n}^{L,tan} := \begin{cases} \left[\beta_{\mathcal{A}} \log\left(\frac{\alpha_{\mathcal{A}}+\beta_{\mathcal{A}}y}{\alpha_{\mathcal{H}}+\beta_{\mathcal{H}}y}\right) + \beta_{\mathcal{H}}\left(1 - \frac{\alpha_{\mathcal{A}}+\beta_{\mathcal{A}}y}{\alpha_{\mathcal{H}}+\beta_{\mathcal{H}}y}\right)\right] \cdot \frac{1-(\beta_{\mathcal{A}})^n}{1-\beta_{\mathcal{A}}} \cdot \left[X_0 - \frac{\alpha_{\mathcal{A}}}{1-\beta_{\mathcal{A}}}\right] \\ \quad + \left[\frac{\alpha_{\mathcal{A}}}{\beta_{\mathcal{A}}(1-\beta_{\mathcal{A}})}\left[\beta_{\mathcal{A}} \log\left(\frac{\alpha_{\mathcal{A}}+\beta_{\mathcal{A}}y}{\alpha_{\mathcal{H}}+\beta_{\mathcal{H}}y}\right) + \beta_{\mathcal{H}}\left(1 - \frac{\alpha_{\mathcal{A}}+\beta_{\mathcal{A}}y}{\alpha_{\mathcal{H}}+\beta_{\mathcal{H}}y}\right)\right] \right. \\ \qquad \left. + \left(\alpha_{\mathcal{H}} - \alpha_{\mathcal{A}}\frac{\beta_{\mathcal{H}}}{\beta_{\mathcal{A}}}\right)\left(1 - \frac{\alpha_{\mathcal{A}}+\beta_{\mathcal{A}}y}{\alpha_{\mathcal{H}}+\beta_{\mathcal{H}}y}\right)\right] \cdot n, & \text{if } \beta_{\mathcal{A}} \neq 1, \\[2ex] \left[\log\left(\frac{\alpha_{\mathcal{A}}+y}{\alpha_{\mathcal{H}}+\beta_{\mathcal{H}}y}\right) + \beta_{\mathcal{H}}\left(1 - \frac{\alpha_{\mathcal{A}}+y}{\alpha_{\mathcal{H}}+\beta_{\mathcal{H}}y}\right)\right] \cdot \left[\frac{\alpha_{\mathcal{A}}}{2}\cdot n^2 + \left(X_0 + \frac{\alpha_{\mathcal{A}}}{2}\right)\cdot n\right] \\ \quad + \left(\alpha_{\mathcal{H}} - \alpha_{\mathcal{A}}\beta_{\mathcal{H}}\right)\left(1 - \frac{\alpha_{\mathcal{A}}+y}{\alpha_{\mathcal{H}}+\beta_{\mathcal{H}}y}\right)\cdot n, & \text{if } \beta_{\mathcal{A}} = 1, \end{cases} \tag{72}$$

and for all $k \in \mathbb{N}_0$ the – possibly negatively valued– finite bound component

$$E_{k,X_0,n}^{L,sec} := \begin{cases} \left[f_{\mathcal{A}}(k+1)\log\left(\frac{f_{\mathcal{A}}(k+1)}{f_{\mathcal{H}}(k+1)}\right) - f_{\mathcal{A}}(k)\log\left(\frac{f_{\mathcal{A}}(k)}{f_{\mathcal{H}}(k)}\right) + \beta_{\mathcal{H}} - \beta_{\mathcal{A}}\right] \cdot \frac{1-(\beta_{\mathcal{A}})^n}{1-\beta_{\mathcal{A}}} \cdot \left[X_0 - \frac{\alpha_{\mathcal{A}}}{1-\beta_{\mathcal{A}}}\right] \\ \quad + \left[\frac{\alpha_{\mathcal{A}}}{\beta_{\mathcal{A}}(1-\beta_{\mathcal{A}})}\left(f_{\mathcal{A}}(k+1)\log\left(\frac{f_{\mathcal{A}}(k+1)}{f_{\mathcal{H}}(k+1)}\right) - f_{\mathcal{A}}(k)\log\left(\frac{f_{\mathcal{A}}(k)}{f_{\mathcal{H}}(k)}\right) + \beta_{\mathcal{H}} - \beta_{\mathcal{A}}\right)\right. \\ \qquad - \left(f_{\mathcal{A}}(k+1)\log\left(\frac{f_{\mathcal{A}}(k+1)}{f_{\mathcal{H}}(k+1)}\right) - f_{\mathcal{A}}(k)\log\left(\frac{f_{\mathcal{A}}(k)}{f_{\mathcal{H}}(k)}\right)\right)\cdot\left(k + \frac{\alpha_{\mathcal{A}}}{\beta_{\mathcal{A}}}\right) \\ \qquad \left. + f_{\mathcal{A}}(k)\log\left(\frac{f_{\mathcal{A}}(k)}{f_{\mathcal{H}}(k)}\right) - \frac{\alpha_{\mathcal{A}}\beta_{\mathcal{H}}}{\beta_{\mathcal{A}}} + \alpha_{\mathcal{H}}\right]\cdot n, & \text{if } \beta_{\mathcal{A}} \neq 1, \\[2ex] \left[f_{\mathcal{A}}(k+1)\log\left(\frac{f_{\mathcal{A}}(k+1)}{f_{\mathcal{H}}(k+1)}\right) - f_{\mathcal{A}}(k)\log\left(\frac{f_{\mathcal{A}}(k)}{f_{\mathcal{H}}(k)}\right) + \beta_{\mathcal{H}} - 1\right]\cdot\left[\frac{\alpha_{\mathcal{A}}}{2}\cdot n^2 + \left(X_0 + \frac{\alpha_{\mathcal{A}}}{2}\right)\cdot n\right] \\ \quad - \left[\left(f_{\mathcal{A}}(k+1)\log\left(\frac{f_{\mathcal{A}}(k+1)}{f_{\mathcal{H}}(k+1)}\right) - f_{\mathcal{A}}(k)\log\left(\frac{f_{\mathcal{A}}(k)}{f_{\mathcal{H}}(k)}\right)\right)(k + \alpha_{\mathcal{A}})\right. \\ \qquad \left. - f_{\mathcal{A}}(k)\log\left(\frac{f_{\mathcal{A}}(k)}{f_{\mathcal{H}}(k)}\right) + \alpha_{\mathcal{A}}\beta_{\mathcal{H}} - \alpha_{\mathcal{H}}\right]\cdot n, & \text{if } \beta_{\mathcal{A}} = 1. \end{cases} \tag{73}$$

Furthermore, on $\mathcal{P}_{SP,4}$ we set $E_{X_0,n}^{L,hor} := 0$ for all $n \in \mathbb{N}$ whereas on $\mathcal{P}_{SP}\backslash(\mathcal{P}_{SP,1} \cup \mathcal{P}_{SP,4})$ we define

$$E_{X_0,n}^{L,hor} := \left[(\alpha_{\mathcal{A}} + \beta_{\mathcal{A}}z^*)\cdot\left[\log\left(\frac{\alpha_{\mathcal{A}} + \beta_{\mathcal{A}}z^*}{\alpha_{\mathcal{H}} + \beta_{\mathcal{H}}z^*}\right) - 1\right] + \alpha_{\mathcal{H}} + \beta_{\mathcal{H}}z^*\right]\cdot n, \qquad , n \in \mathbb{N}, \tag{74}$$

with $z^* := \arg\max_{x \in \mathbb{N}_0}\left\{(\alpha_{\mathcal{A}} + \beta_{\mathcal{A}}x)\left[-\log\left(\frac{\alpha_{\mathcal{A}}+\beta_{\mathcal{A}}x}{\alpha_{\mathcal{H}}+\beta_{\mathcal{H}}x}\right) + 1\right] - (\alpha_{\mathcal{H}} + \beta_{\mathcal{H}}x)\right\}$.
On $\mathcal{P}_{SP}\backslash(\mathcal{P}_{SP,1} \cup \mathcal{P}_{SP,3c})$ one even gets $E_{X_0,n}^{L} > 0$ for all $X_0 \in \mathbb{N}$ and all $n \in \mathbb{N}$.
For the subcase $\mathcal{P}_{SP,3c}$, one obtains for each fixed $n \in \mathbb{N}$ and each fixed $X_0 \in \mathbb{N}$ the strict positivity $E_{X_0,n}^{L} > 0$ if
$\left(\frac{\partial}{\partial y}E_{y,n}^{L,tan}\right)(y^*) \neq 0$, where $y^* := \frac{\alpha_{\mathcal{A}}-\alpha_{\mathcal{H}}}{\beta_{\mathcal{H}}-\beta_{\mathcal{A}}} \in \mathbb{N}$ and hence

$$\left(\frac{\partial}{\partial y}E_{y,X_0,n}^{L,tan}\right)(y^*) \tag{75}$$

$$= \begin{cases} -\frac{(\beta_{\mathcal{A}}-\beta_{\mathcal{H}})^3}{\alpha_{\mathcal{A}}\beta_{\mathcal{H}}-\alpha_{\mathcal{H}}\beta_{\mathcal{A}}} \cdot \frac{1-(\beta_{\mathcal{A}})^n}{1-\beta_{\mathcal{A}}} \cdot \left[X_0 - \frac{\alpha_{\mathcal{A}}}{1-\beta_{\mathcal{A}}}\right] - \frac{(\beta_{\mathcal{A}}-\beta_{\mathcal{H}})^2}{\beta_{\mathcal{A}}}\left(1 + \frac{\alpha_{\mathcal{A}}(\beta_{\mathcal{A}}-\beta_{\mathcal{H}})}{(1-\beta_{\mathcal{A}})(\alpha_{\mathcal{A}}\beta_{\mathcal{H}}-\alpha_{\mathcal{H}}\beta_{\mathcal{A}})}\right)\cdot n, & \text{if } \beta_{\mathcal{A}} \neq 1, \\[2ex] -\frac{(1-\beta_{\mathcal{H}})^3}{\alpha_{\mathcal{A}}\beta_{\mathcal{H}}-\alpha_{\mathcal{H}}} \cdot \left[\frac{\alpha_{\mathcal{A}}}{2}\cdot n^2 + \left(X_0 + \frac{\alpha_{\mathcal{A}}}{2}\right)\cdot n\right] - (1-\beta_{\mathcal{H}})^2\cdot n, & \text{if } \beta_{\mathcal{A}} = 1. \end{cases}$$

A proof of this theorem is given in in Appendix A.2.

Remark 6. *Consider the exemplary parameter setup $(\beta_{\mathcal{A}}, \beta_{\mathcal{H}}, \alpha_{\mathcal{A}}, \alpha_{\mathcal{H}}) = \left(\frac{1}{3}, \frac{2}{3}, 2, 1\right) \in \mathcal{P}_{SP,3c}$; within our running-example epidemiological context of Section 2.3, this corresponds to a "semi-mild" infectious-disease-transmission situation ($\mathcal{H}$) (with subcritical reproduction number $\beta_{\mathcal{H}} = \frac{2}{3}$ and importation mean of $\alpha_{\mathcal{H}} = 1$), whereas ($\mathcal{A}$) describes a "mild" situation (with "low" subcritical $\beta_{\mathcal{A}} = \frac{1}{3}$ and $\alpha_{\mathcal{A}} = 2$). In the case of $X_0 = 3$ there holds $\left(\frac{\partial}{\partial y}E_{y,X_0,n}^{L,tan}\right)(y^*) = 0$ for all $n \in \mathbb{N}$, whereas for $X_0 \neq 3$ one obtains $\left(\frac{\partial}{\partial y}E_{y,X_0,n}^{L,tan}\right)(y^*) \neq 0$ for all $n \in \mathbb{N}$.*

It seems that the optimization problem in (71) admits in general only an implicitly representable solution, and thus we have used the prefix "(semi-)" above. Of course, as a less tight but less involved *explicit* lower bound of the Kullback-Leibler information divergence (relative entropy) $I(P_{\mathcal{A},n}||P_{\mathcal{H},n})$ one can use any term of the form $\max\left\{ E_{y,X_0,n}^{L,tan}, E_{k,X_0,n}^{L,sec}, E_{X_0,n}^{L,hor}\right\}$ ($y \in [0,\infty[, k \in \mathbb{N}_0$), as well as the following

Corollary 12. *(a) For all $(\beta_{\mathcal{A}}, \beta_{\mathcal{H}}, \alpha_{\mathcal{A}}, \alpha_{\mathcal{H}}) \in \mathcal{P}_{SP}\backslash\mathcal{P}_{SP,1}$, all initial population sizes $X_0 \in \mathbb{N}$ and all observation horizons $n \in \mathbb{N}$*

$$I(P_{\mathcal{A},n}||P_{\mathcal{H},n}) \geq E_{X_0,n}^{L} \geq \widetilde{E}_{X_0,n}^{L} := \max\left\{ E_{\infty,X_0,n}^{L,tan}, E_{0,X_0,n}^{L,sec}, E_{X_0,n}^{L,hor}\right\} \in [0,\infty[\, ,$$

with $E_{X_0,n}^{L,hor}$ defined by (74), with – possibly negatively valued– finite bound component $E_{\infty,X_0,n}^{L,tan} := \lim_{y\to\infty} E_{y,X_0,n}^{L,tan}$, where

$$E_{\infty,X_0,n}^{L,tan} := \begin{cases} \frac{\beta_{\mathcal{A}}\cdot\left(\log\left(\frac{\beta_{\mathcal{A}}}{\beta_{\mathcal{H}}}\right)-1\right)+\beta_{\mathcal{H}}}{1-\beta_{\mathcal{A}}} \cdot \left[X_0 - \frac{\alpha_{\mathcal{A}}}{1-\beta_{\mathcal{A}}}\right] \cdot \left(1-(\beta_{\mathcal{A}})^n\right) \\ + \left[\frac{\alpha_{\mathcal{A}}\cdot\left[\beta_{\mathcal{A}}\cdot\left(\log\left(\frac{\beta_{\mathcal{A}}}{\beta_{\mathcal{H}}}\right)-1\right)+\beta_{\mathcal{H}}\right]}{\beta_{\mathcal{A}}(1-\beta_{\mathcal{A}})} + \alpha_{\mathcal{A}}\left(1-\frac{\beta_{\mathcal{H}}}{\beta_{\mathcal{A}}}\right) + \alpha_{\mathcal{H}}\left(1-\frac{\beta_{\mathcal{A}}}{\beta_{\mathcal{H}}}\right)\right]\cdot n, & \text{if } \beta_{\mathcal{A}} \neq 1, \\[2ex] \left[\beta_{\mathcal{H}} - \log\beta_{\mathcal{H}} - 1\right] \cdot \left[\frac{\alpha_{\mathcal{A}}}{2}\cdot n^2 + \left(X_0 + \frac{\alpha_{\mathcal{A}}}{2}\right)\cdot n\right] \\ + \left[\alpha_{\mathcal{A}}(1-\beta_{\mathcal{H}}) + \alpha_{\mathcal{H}}\left(1-\frac{1}{\beta_{\mathcal{H}}}\right)\right]\cdot n, & \text{if } \beta_{\mathcal{A}} = 1, \end{cases}$$

and –possibly negatively valued–finite bound component

$$E_{0,X_0,n}^{L,sec} = \begin{cases} \left[(\alpha_{\mathcal{A}}+\beta_{\mathcal{A}})\cdot\log\left(\frac{\alpha_{\mathcal{A}}+\beta_{\mathcal{A}}}{\alpha_{\mathcal{H}}+\beta_{\mathcal{H}}}\right) - \alpha_{\mathcal{A}}\cdot\log\left(\frac{\alpha_{\mathcal{A}}}{\alpha_{\mathcal{H}}}\right) + \beta_{\mathcal{H}} - \beta_{\mathcal{A}}\right]\cdot\frac{1-(\beta_{\mathcal{A}})^n}{1-\beta_{\mathcal{A}}}\cdot\left[X_0 - \frac{\alpha_{\mathcal{A}}}{1-\beta_{\mathcal{A}}}\right] \\ + \left\{\frac{\alpha_{\mathcal{A}}}{\beta_{\mathcal{A}}(1-\beta_{\mathcal{A}})}\left((\alpha_{\mathcal{A}}+\beta_{\mathcal{A}})\cdot\log\left(\frac{\alpha_{\mathcal{A}}+\beta_{\mathcal{A}}}{\alpha_{\mathcal{H}}+\beta_{\mathcal{H}}}\right) - \alpha_{\mathcal{A}}\cdot\log\left(\frac{\alpha_{\mathcal{A}}}{\alpha_{\mathcal{H}}}\right)\right) - \frac{\alpha_{\mathcal{A}}}{1-\beta_{\mathcal{A}}}(1-\beta_{\mathcal{H}})\right. \\ \left. -\alpha_{\mathcal{A}}\left(1+\frac{\alpha_{\mathcal{A}}}{\beta_{\mathcal{A}}}\right)\cdot\log\left(\frac{\alpha_{\mathcal{H}}(\alpha_{\mathcal{A}}+\beta_{\mathcal{A}})}{\alpha_{\mathcal{A}}(\alpha_{\mathcal{H}}+\beta_{\mathcal{H}})}\right) + \alpha_{\mathcal{H}}\right\}\cdot n, & \text{if } \beta_{\mathcal{A}} \neq 1, \\[2ex] \left[(\alpha_{\mathcal{A}}+1)\cdot\log\left(\frac{\alpha_{\mathcal{A}}+1}{\alpha_{\mathcal{H}}+\beta_{\mathcal{H}}}\right) - \alpha_{\mathcal{A}}\cdot\log\left(\frac{\alpha_{\mathcal{A}}}{\alpha_{\mathcal{H}}}\right) + \beta_{\mathcal{H}} - 1\right]\cdot\left[n\cdot X_0 + \frac{\alpha_{\mathcal{A}}}{2}\cdot n^2\right] \\ + \left\{\frac{\alpha_{\mathcal{A}}}{2}\left[(\alpha_{\mathcal{A}}+1)\cdot\log\left(\frac{\alpha_{\mathcal{A}}+1}{\alpha_{\mathcal{H}}+\beta_{\mathcal{H}}}\right) - \alpha_{\mathcal{A}}\cdot\log\left(\frac{\alpha_{\mathcal{A}}}{\alpha_{\mathcal{H}}}\right) - \beta_{\mathcal{H}} - 1\right]\right. \\ \left. -\alpha_{\mathcal{A}}(1+\alpha_{\mathcal{A}})\cdot\log\left(\frac{\alpha_{\mathcal{H}}(\alpha_{\mathcal{A}}+1)}{\alpha_{\mathcal{A}}(\alpha_{\mathcal{H}}+\beta_{\mathcal{H}})}\right) + \alpha_{\mathcal{H}}\right\}\cdot n, & \text{if } \beta_{\mathcal{A}} = 1. \end{cases}$$

For the cases $\mathcal{P}_{SP,2} \cup \mathcal{P}_{SP,3a} \cup \mathcal{P}_{SP,3b}$ one gets even $\widetilde{E}_{X_0,n}^{L} > 0$ for all $X_0 \in \mathbb{N}$ and all $n \in \mathbb{N}$.

5.3. Applications to Bayesian Decision Making

As explained in Section 2.5, the Kullback-Leibler information divergence fulfills

$$I(P_{\mathcal{A},n}||P_{\mathcal{H},n}) = \int_0^1 \Delta\mathcal{BR}_{\widetilde{LO}}(p_{\mathcal{A}}^{\text{prior}}) \cdot \left(1-p_{\mathcal{A}}^{\text{prior}}\right)^{-1} \cdot \left(p_{\mathcal{A}}^{\text{prior}}\right)^{-2} \, dp_{\mathcal{A}}^{\text{prior}}, \qquad \text{(cf. (21) with } \lambda = 1\text{),}$$

and thus can be interpreted as *weighted-average* decision risk reduction (weighted-average statistical information measure) about the degree of evidence deg concerning the parameter θ that can be attained by observing the GWI-path $\mathcal{X}_n$ until stage n. Hence, by combining (21) with the investigations in the previous Sections 5.1 and 5.2, we obtain exact values respectively bounds of the above-mentioned decision risk reductions. For the sake of brevity, we omit the details here.

6. Explicit Closed-Form Bounds of Hellinger Integrals

6.1. Principal Approach

Depending on the parameter constellation $(\beta_A, \beta_H, \alpha_A, \alpha_H, \lambda) \in \mathcal{P} \times (\mathbb{R} \backslash \{0, 1\})$, for the Hellinger integrals $H_\lambda (P_{A,n} || P_{H,n})$ we have derived in Section 3 corresponding lower/upper bounds respectively exact values–of recursive nature– which can be obtained by choosing appropriate $p = p_\lambda^A = p^A (\beta_A, \beta_H, \alpha_A, \alpha_H, \lambda)$, $q = q_\lambda^A = q^A (\beta_A, \beta_H, \alpha_A, \alpha_H, \lambda)$ $(A \in \{E, L, U\})$ and by using those together with the recursion $\left(a_n^{(q)} \right)_{n \in \mathbb{N}}$ defined by (36) as well as the sequence $\left(b_n^{(p,q)} \right)_{n \in \mathbb{N}}$ obtained from $\left(a_n^{(q)} \right)_{n \in \mathbb{N}}$ by the linear transformation (38). Both sequences are "stepwise fully evaluable" but generally seem not to admit a closed-form representation in the observation horizons n; consequently, the time-evolution $n \mapsto H_\lambda (P_{A,n} || P_{H,n})$–respectively the time-evolution of the corresponding recursive bounds– can generally *not be seen explicitly*. On order to avoid this *intransparency* (at the expense of losing some precision) one can approximate (36) by a recursion that allows for a closed-form representation; by the way, this will also turn out to be useful for investigations concerning diffusion limits (cf. the next Section 7).

To explain the basic underlying principle, let us first assume some *general* $q \in]0, \beta_\lambda[$ and $\lambda \in]0, 1[$. With Properties 1 (P1) we see that the sequence $\left(a_n^{(q)} \right)_{n \in \mathbb{N}}$ is strictly negative, strictly decreasing and converges to $x_0^{(q)} \in] - \beta_\lambda, q - \beta_\lambda [$. Recall that this sequence is obtained by the recursive application of the function $\xi_\lambda^{(q)}(x) := q \cdot e^x - \beta_\lambda$, through $a_1^{(q)} = \xi_\lambda^{(q)}(0) = q - \beta_\lambda < 0$, $a_n^{(q)} = \xi_\lambda^{(q)} \left(a_{n-1}^{(q)} \right) = q e^{a_{n-1}^{(q)}} - \beta_\lambda$ (cf. (36)). As a first step, we want to approximate $\xi_\lambda^{(q)}(\cdot)$ by a linear function on the interval $\left[x_0^{(q)}, 0 \right]$. Due to convexity (P9), this is done by using the tangent line of $\xi_\lambda^{(q)}(\cdot)$ at $x_0^{(q)}$

$$\xi_\lambda^{(q),T}(x) := c^{(q),T} + d^{(q),T} \cdot x := x_0^{(q)} \left(1 - q \cdot e^{x_0^{(q)}} \right) + q \cdot e^{x_0^{(q)}} \cdot x , \tag{76}$$

as a linear lower bound, and the secant line of $\xi_\lambda^{(q)}(\cdot)$ across its arguments 0 and $x_0^{(q)}$

$$\xi_\lambda^{(q),S}(x) := c^{(q),S} + d^{(q),S} \cdot x := q - \beta_\lambda + \frac{x_0^{(q)} - (q - \beta_\lambda)}{x_0^{(q)}} \cdot x , \tag{77}$$

as a linear upper bound. With the help of these functions, we can define the *linear* recursions

$$a_0^{(q),T} := 0 , \qquad a_n^{(q),T} := \xi_\lambda^{(q),T} \left(a_{n-1}^{(q),T} \right) , \; n \in \mathbb{N} , \tag{78}$$

as well as
$$a_0^{(q),S} := 0 , \qquad a_n^{(q),S} := \xi_\lambda^{(q),S} \left(a_{n-1}^{(q),S} \right) , \; n \in \mathbb{N} . \tag{79}$$

In the following, we will refer to these sequences as the *rudimentary closed-form sequence-bounds*. Clearly, both sequences are strictly negative (on $\mathbb{N}$), strictly decreasing, and one gets the sandwiching

$$a_n^{(q),T} < a_n^{(q)} \leq a_n^{(q),S} \tag{80}$$

for all $n \in \mathbb{N}$, with equality on the right side iff $n = 1$ (where $a_1^{(q)} = q - \beta_\lambda < 0$); moreover,

$$\lim_{n \to \infty} a_n^{(q),T} = \lim_{n \to \infty} a_n^{(q),S} = \lim_{n \to \infty} a_n^{(q)} = x_0^{(q)} . \tag{81}$$

Furthermore, such linear recursions allow for a closed-form representation, namely

$$a_n^{(q),*} = \frac{c^{(q),*}}{1 - d^{(q),*}} \cdot \left(1 - \left(d^{(q),*}\right)^n\right) = x_0^{(q)} \cdot \left(1 - \left(d^{(q),*}\right)^n\right), \tag{82}$$

where the " * " stands for either S or T. Notice that this representation is valid due to $d^{(q),T}, d^{(q),S} \in]0,1[$. So far, we have considered the case $q \in]0, \beta_\lambda[$. If $q = \beta_\lambda$, then one can see from Properties 1 (P2) that $a_n^{(q)} \equiv 0$, which is also an explicitly given (though trivial) sequence. For the remaining case, where $q > \beta_\lambda$ and thus $\xi_\lambda^{(q)}(0) = a_1^{(q)} = q - \beta_\lambda > 0$), we want to exclude $q \geq \min\left\{1, e^{\beta_\lambda - 1}\right\}$ for the following reasons. Firstly, if $q > \min\left\{1, e^{\beta_\lambda - 1}\right\}$, then from (P3) we see that the sequence $\left(a_n^{(q)}\right)_{n \in \mathbb{N}}$ is strictly increasing and divergent to ∞, at a rate faster than exponentially (P3b); but a linear recursion is too weak to approximate such a growth pattern. Secondly, if $q = \min\left\{1, e^{\beta_\lambda - 1}\right\}$, then one necessarily gets $q = e^{\beta_\lambda - 1} < 1$ (since we have required $q > \beta_\lambda$, and otherwise one obtains the contradiction $\beta_\lambda < q = 1 \leq e^{\beta_\lambda - 1}$). This means that the function $\xi_\lambda^{(q)}(\cdot)$ now touches the straight line $id(\cdot)$ in the point $-\log(q)$, i.e., $\xi_\lambda^{(q)}\left(-\log(q)\right) = -\log(q)$. Our above-proposed method, namely to use the tangent line of $\xi_\lambda^{(q)}(\cdot)$ at $x = x_0^{(q)} = -\log(q)$ as a linear lower bound for $\xi_\lambda^{(q)}(\cdot)$, leads then to the recursion $a_n^{(q),T} \equiv 0$ (cf. (78)). This is due to the fact that the tangent line $\xi_\lambda^{(q),T}(\cdot)$ is in the current case equivalent with the straight line $id(\cdot)$. Consequently, (81) would not be satisfied.

Notice that in the case $\beta_\lambda < q < \min\left\{1, e^{\beta_\lambda - 1}\right\}$, the above-introduced functions $\xi_\lambda^{(q),T}(\cdot)$, $\xi_\lambda^{(q),S}(\cdot)$ constitute again linear lower and upper bounds for $\xi_\lambda^{(q)}(\cdot)$, however, this time on the interval $\left[0, x_0^{(q)}\right]$. The sequences defined in (78) and (79) still fulfill the assertions (80) and (81), and additionally allow for the closed-form representation (82). Furthermore, let us mention that these rudimentary closed-form sequence-bounds can be defined analogously for $\lambda \in \mathbb{R} \backslash [0,1]$ and either $0 < q < \beta_\lambda$, or $q = \beta_\lambda$, or $\max\{0, \beta_\lambda\} < q < \min\{1, e^{\beta_\lambda - 1}\}$.

In a second step, we want to *improve* the above-mentioned linear (lower and upper) approximations of the sequence $a_n^{(q)}$ by reducing the faced error within each iteration. To do so, in both cases of lower and upper approximates we shall employ context-adapted linear *inhomogeneous difference equations* of the form

$$\tilde{a}_0 := 0 \quad ; \qquad \tilde{a}_n := \tilde{\xi}\left(\tilde{a}_{n-1}\right) + \rho_{n-1}, \quad n \in \mathbb{N}, \tag{83}$$

with

$$\tilde{\xi}(x) := c + d \cdot x, \qquad\qquad x \in \mathbb{R}, \tag{84}$$

$$\rho_{n-1} := K_1 \cdot \varkappa^{n-1} + K_2 \cdot v^{n-1}, \qquad n \in \mathbb{N}, \tag{85}$$

for some constants $c \in \mathbb{R}$, $d \in]0,1[$, $K_1, K_2, \varkappa, v \in \mathbb{R}$ with $0 \leq v < \varkappa \leq d$. This will be applied to $c := c^{(q),S}$, $c := c^{(q),T}$, $d := d^{(q),S}$ and $d := d^{(q),T}$ later on. Meanwhile, let us first present some facts and expressions which are insightful for further formulations and analyses.

Lemma 2. *Consider the sequence* $(\tilde{a}_n)_{n \in \mathbb{N}_0}$ *defined in* (83) *to* (85). *If* $0 \leq v < \varkappa < d$, *then one gets the closed-form representation*

$$\tilde{a}_n = \tilde{a}_n^{hom} + \tilde{c}_n \quad \text{with} \quad \tilde{a}_n^{hom} = c \cdot \frac{1 - d^n}{1 - d} \quad \text{and} \quad \tilde{c}_n = K_1 \cdot \frac{d^n - \varkappa^n}{d - \varkappa} + K_2 \cdot \frac{d^n - v^n}{d - v}, \tag{86}$$

which leads for all $n \in \mathbb{N}$ *to*

$$\sum_{k=1}^{n} \tilde{a}_k = \left(\frac{K_1}{d - \varkappa} + \frac{K_2}{d - v} - \frac{c}{1 - d}\right) \cdot \frac{d \cdot (1 - d^n)}{1 - d} - \frac{K_1 \cdot \varkappa \cdot (1 - \varkappa^n)}{(d - \varkappa)(1 - \varkappa)} - \frac{K_2 \cdot v \cdot (1 - v^n)}{(d - v)(1 - v)} + \frac{c}{1 - d} \cdot n. \tag{87}$$

If $0 \leq v < \varkappa = d$, then one gets the closed-form representation

$$\tilde{a}_n = \tilde{a}_n^{hom} + \tilde{c}_n \quad \text{with} \quad \tilde{a}_n^{hom} = c \cdot \frac{1 - d^n}{1 - d} \quad \text{and} \quad \tilde{c}_n = K_1 \cdot n \cdot d^{n-1} + K_2 \cdot \frac{d^n - v^n}{d - v}, \tag{88}$$

which leads for all $n \in \mathbb{N}$ to

$$\sum_{k=1}^{n} \tilde{a}_k = \left(\frac{K_1}{d(1-d)} + \frac{K_2}{d-v} - \frac{c}{1-d} \right) \cdot \frac{d \cdot (1 - d^n)}{1 - d} - \frac{K_2 \cdot v \cdot (1 - v^n)}{(d-v)(1-v)} + \left(\frac{c}{1-d} - \frac{K_1 \cdot d^n}{1-d} \right) \cdot n. \tag{89}$$

Lemma 2 will be proved in Appendix A.3. Notice that (88) is consistent with taking the limit $\varkappa \nearrow d$ in (86). Furthermore, for the special case $K_2 = -K_1 > 0$ one has from (85) for all integers $n \geq 2$ the relation $\rho_{n-1} < 0$ and thus $\tilde{a}_n - \tilde{a}_n^{hom} < 0$, leading to

$$\tilde{c}_n < 0 \quad \text{and} \quad \sum_{k=1}^{n} \tilde{c}_n < 0. \tag{90}$$

Lemma 2 gives explicit expressions for a linear inhomogeneous recursion of the form (83) possessing the extra term given by (85). Therefrom we derive lower and upper bounds for the sequence $\left(a_n^{(q)}\right)_{n \in \mathbb{N}}$ by employing $a_n^{(q),T}$ resp. $a_n^{(q),S}$ as the homogeneous solution of (83), i.e., by setting $\tilde{a}_n^{hom} := a_n^{(q),T}$ resp. $\tilde{a}_n^{hom} := a_n^{(q),S}$. Moreover, our concrete approximation-error-reducing "correction terms" ρ_n will have different form, depending on whether $0 < q < \beta_\lambda$ or $q > \max\{0, \beta_\lambda\}$. In both cases, we express ρ_n by means of the slopes $d^{(q),T} = q e^{x_0^{(q)}}$ resp. $d^{(q),S} = \frac{x_0^{(q)} - (q - \beta_\lambda)}{x_0^{(q)}}$ of the tangent line $\xi_\lambda^{(q),T}(\cdot)$ (cf. (76)) resp. the secant line $\xi_\lambda^{(q),S}(\cdot)$ (cf. (77)), as well as in terms of the parameters

$$\Gamma_<^{(q)} := \frac{1}{2} \cdot \left(x_0^{(q)}\right)^2 \cdot q \cdot e^{x_0^{(q)}}, \quad \text{for } 0 < q < \beta_\lambda, \quad \text{and} \quad \Gamma_>^{(q)} := \frac{q}{2} \cdot \left(x_0^{(q)}\right)^2, \quad \text{for } q > \max\{0, \beta_\lambda\}. \tag{91}$$

In detail, let us first define the lower approximate by

$$\underline{a}_0^{(q)} := 0, \qquad \underline{a}_n^{(q)} := \xi_\lambda^{(q),T}\left(\underline{a}_{n-1}^{(q)}\right) + \underline{\rho}_{n-1}^{(q)}, \quad n \in \mathbb{N}, \tag{92}$$

where

$$\underline{\rho}_{n-1}^{(q)} := \begin{cases} \Gamma_<^{(q)} \cdot \left(d^{(q),T}\right)^{2(n-1)}, & \text{if } 0 < q < \beta_\lambda, \\[2ex] \Gamma_>^{(q)} \cdot \left(d^{(q),S}\right)^{2(n-1)}, & \text{if } \max\{0, \beta_\lambda\} < q < \min\{1, e^{\beta_\lambda - 1}\}. \end{cases} \tag{93}$$

The upper approximate is defined by

$$\overline{a}_0^{(q)} := 0, \qquad \overline{a}_n^{(q)} := \xi_\lambda^{(q),S}\left(\overline{a}_{n-1}^{(q)}\right) + \overline{\rho}_{n-1}^{(q)}, \quad n \in \mathbb{N}, \tag{94}$$

where

$$\overline{\rho}_{n-1}^{(q)} := \begin{cases} -\Gamma_<^{(q)} \cdot \left(d^{(q),T}\right)^{n-1} \cdot \left[1 - \left(d^{(q),S}\right)^{n-1}\right], & \text{if } 0 < q < \beta_\lambda, \\[2ex] -\Gamma_>^{(q)} \cdot \left(d^{(q),S}\right)^{n-1} \cdot \left[1 - \left(d^{(q),T}\right)^{n-1}\right], & \text{if } \max\{0, \beta_\lambda\} < q < \min\{1, e^{\beta_\lambda - 1}\}. \end{cases} \tag{95}$$

In terms of (85), we use for $\underline{\rho}_n^{(q)}$ the constants $K_2 = v = 0$ as well as $K_1 = \Gamma_<^{(q)}$, $\varkappa = \left(d^{(q),T}\right)^2$ for $0 < q < \beta_\lambda$ respectively $K_1 = \Gamma_>^{(q)}$, $\varkappa = \left(d^{(q),S}\right)^2$ for $\max\{0, \beta_\lambda\} < q < \min\{1, e^{\beta_\lambda - 1}\}$. For $\overline{\rho}_n^{(q)}$ we shall employ the constants $-K_1 = K_2 = \Gamma_{<}^{(q)}$, $\varkappa = d^{(q),T}$, $v = d^{(q),S}d^{(q),T}$ for $0 < q < \beta_\lambda$, and $-K_1 = K_2 = \Gamma_{>}^{(q)}$, $\varkappa = d^{(q),S}$, $v = d^{(q),S}d^{(q),T}$ for $\max\{0,\beta_\lambda\} < q < \min\{1, e^{\beta_\lambda - 1}\}$. Recall from (76) the constants $c^{(q),T} := x_0^{(q)}(1 - qe^{x_0^{(q)}})$, $d^{(q),T} := qe^{x_0^{(q)}}$ and from (77) $c^{(q),S} := q - \beta_\lambda$, $d^{(q),S} := \frac{x_0^{(q)} - (q - \beta_\lambda)}{x_0^{(q)}}$.

In the following, we will refer to the sequences $\underline{a}_n^{(q)}$ resp. $\bar{a}_n^{(q)}$ as the *improved closed-form sequence-bounds*. Putting all ingredients together, we arrive at the

Lemma 3. *For all* $(\beta_{\mathcal{A}}, \beta_{\mathcal{H}}, \alpha_{\mathcal{A}}, \alpha_{\mathcal{H}}) \in \mathcal{P}$ *there holds with* $d^{(q),T} = qe^{x_0^{(q)}}$ *and* $d^{(q),S} = \frac{x_0^{(q)} - (q - \beta_\lambda)}{x_0^{(q)}}$

(a) in the case $0 < q < \beta_\lambda$:

(i)

$$\underline{a}_n^{(q)} < a_n^{(q)} \leq \bar{a}_n^{(q)} \qquad \text{for all } n \in \mathbb{N},$$

with equality on the right-hand side iff $n = 1$, *where*

$$\underline{a}_n^{(q)} = x_0^{(q)} \cdot \left(1 - \left(d^{(q),T}\right)^n\right) + \Gamma_{<}^{(q)} \cdot \frac{\left(d^{(q),T}\right)^{n-1}}{1 - d^{(q),T}} \cdot \left(1 - \left(d^{(q),T}\right)^n\right) > a_n^{(q),T}, \quad \text{and}$$

$$\bar{a}_n^{(q)} = x_0^{(q)} \cdot \left(1 - \left(d^{(q),S}\right)^n\right) - \Gamma_{<}^{(q)} \cdot \left[\frac{\left(d^{(q),S}\right)^n - \left(d^{(q),T}\right)^n}{d^{(q),S} - d^{(q),T}} - \left(d^{(q),S}\right)^{n-1}\frac{1 - \left(d^{(q),T}\right)^n}{1 - d^{(q),T}}\right] \leq a_n^{(q),S},$$

with $a_n^{(q),T}$ *and* $a_n^{(q),S}$ *defined by* (78) *and* (79).

(ii) Both sequences $\left(\underline{a}_n^{(q)}\right)_{n \in \mathbb{N}}$ *and* $\left(\bar{a}_n^{(q)}\right)_{n \in \mathbb{N}}$ *are strictly decreasing.*

(iii)

$$\lim_{n \to \infty} \underline{a}_n^{(q)} = \lim_{n \to \infty} \bar{a}_n^{(q)} = \lim_{n \to \infty} a_n^{(q)} = x_0^{(q)} \in\,]-\beta_\lambda, q - \beta_\lambda[.$$

(b) in the case $\max\{0, \beta_\lambda\} < q < \min\left\{1, e^{\beta_\lambda - 1}\right\}$:

(i)

$$\underline{a}_n^{(q)} < a_n^{(q)} \leq \bar{a}_n^{(q)}, \qquad \text{for all } n \in \mathbb{N},$$

with equality on the right-hand side iff $n = 1$, *where*

$$\underline{a}_n^{(q)} = x_0^{(q)} \cdot \left(1 - \left(d^{(q),T}\right)^n\right) + \Gamma_{>}^{(q)} \cdot \frac{\left(d^{(q),T}\right)^n - \left(d^{(q),S}\right)^{2n}}{d^{(q),T} - \left(d^{(q),S}\right)^2} > a_n^{(q),T} \quad \text{and}$$

$$\bar{a}_n^{(q)} = x_0^{(q)} \cdot \left(1 - \left(d^{(q),S}\right)^n\right) - \Gamma_{>}^{(q)} \cdot \left(d^{(q),S}\right)^{n-1}\left[n - \frac{1 - \left(d^{(q),T}\right)^n}{1 - d^{(q),T}}\right] \leq a_n^{(q),S},$$

with $a_n^{(q),T}$ *and* $a_n^{(q),S}$ *defined by* (78) *and* (79).

(ii) Both sequences $\left(\underline{a}_n^{(q)}\right)_{n \in \mathbb{N}}$ *and* $\left(\bar{a}_n^{(q)}\right)_{n \in \mathbb{N}}$ *are strictly increasing.*

(iii)

$$\lim_{n \to \infty} \underline{a}_n^{(q)} = \lim_{n \to \infty} \bar{a}_n^{(q)} = \lim_{n \to \infty} a_n^{(q)} = x_0^{(q)} \in\,]q - \beta_\lambda, -\log(q)[.$$

A detailed proof of Lemma 3 is provided in Appendix A.3. In the following, we employ the above-mentioned investigations in order to derive the desired closed-form bounds of the Hellinger integrals $H_\lambda(P_{\mathcal{A},n} || P_{\mathcal{H},n})$.

6.2. Explicit Closed-Form Bounds for the Cases $(\beta_\mathcal{A}, \beta_\mathcal{H}, \alpha_\mathcal{A}, \alpha_\mathcal{H}, \lambda) \in (\mathcal{P}_{NI} \cup \mathcal{P}_{SP,1}) \times (\mathbb{R}\backslash\{0,1\})$

Recall that in this setup, we have obtained the recursive, non-explicit *exact* values $V_{\lambda,X_0,n} = H_\lambda(P_{\mathcal{A},n}||P_{\mathcal{H},n})$ given in (39) of Theorem 1, where we used $q = q_\lambda^E = q^E(\beta_\mathcal{A}, \beta_\mathcal{H}, \lambda) = \beta_\mathcal{A}^\lambda \beta_\mathcal{H}^{1-\lambda} \in]0, \beta_\lambda[$ in the case $\lambda \in]0,1[$ respectively $q = q_\lambda^E = \beta_\mathcal{A}^\lambda \beta_\mathcal{H}^{1-\lambda} > \max\{0, \beta_\lambda\}$ in the case $\lambda \in \mathbb{R}\backslash[0,1]$. For the latter, Lemma 1 implies that $q_\lambda^E < \min\{1, e^{\beta_\lambda - 1}\}$ iff $\lambda \in]\lambda_-, \lambda_+[\, \backslash [0,1]$. This—together with (39) from Theorem 1, Lemma 2 and with the quantities $d^{(q),T}$, $d^{(q),S}$, $\Gamma_<^{(q)}$ and $\Gamma_>^{(q)}$ as defined in (76) and (77) resp. (91) –leads to

Theorem 5. *Let $p_\lambda^E := \alpha_\mathcal{A}^\lambda \alpha_\mathcal{H}^{1-\lambda}$ and $q_\lambda^E := \beta_\mathcal{A}^\lambda \beta_\mathcal{H}^{1-\lambda}$. For all $(\beta_\mathcal{A}, \beta_\mathcal{H}, \alpha_\mathcal{A}, \alpha_\mathcal{H}, \lambda) \in (\mathcal{P}_{NI} \cup \mathcal{P}_{SP,1}) \times \left( ]\lambda_-, \lambda_+[\, \backslash \{0,1\} \right)$, all initial population sizes $X_0 \in \mathbb{N}$ and for all observation horizons $n \in \mathbb{N}$ the following assertions hold true:*

(a) the Hellinger integral can be bounded by the closed-form lower and upper bounds

$$C_{\lambda,X_0,n}^{(p_\lambda^E, q_\lambda^E),T} \leq C_{\lambda,X_0,n}^{(p_\lambda^E, q_\lambda^E),L} \leq V_{\lambda,X_0,n} = H_\lambda(P_{\mathcal{A},n}||P_{\mathcal{H},n}) \leq C_{\lambda,X_0,n}^{(p_\lambda^E, q_\lambda^E),U} \leq C_{\lambda,X_0,n}^{(p_\lambda^E, q_\lambda^E),S} \, ,$$

(b)

$$\lim_{n\to\infty} \frac{1}{n} \log\left(V_{\lambda,X_0,n}\right) = \lim_{n\to\infty} \frac{1}{n} \log\left(C_{\lambda,X_0,n}^{(p_\lambda^E, q_\lambda^E),L}\right) = \lim_{n\to\infty} \frac{1}{n} \log\left(C_{\lambda,X_0,n}^{(p_\lambda^E, q_\lambda^E),U}\right)$$

$$= \lim_{n\to\infty} \frac{1}{n} \log\left(C_{\lambda,X_0,n}^{(p_\lambda^E, q_\lambda^E),T}\right) = \lim_{n\to\infty} \frac{1}{n} \log\left(C_{\lambda,X_0,n}^{(p_\lambda^E, q_\lambda^E),S}\right) = \frac{\alpha_\mathcal{A}}{\beta_\mathcal{A}} \cdot x_0^{(q_\lambda^E)} \, ,$$

where the involved closed-form lower bounds are defined by

$$C_{\lambda,X_0,n}^{(p_\lambda^E, q_\lambda^E),L} := C_{\lambda,X_0,n}^{(p_\lambda^E, q_\lambda^E),T} \cdot \exp\left\{\underline{\zeta}_n^{(q_\lambda^E)} \cdot X_0 + \frac{\alpha_\mathcal{A}}{\beta_\mathcal{A}} \cdot \underline{\vartheta}_n^{(q_\lambda^E)}\right\}, \qquad with \tag{96}$$

$$C_{\lambda,X_0,n}^{(p_\lambda^E, q_\lambda^E),T} := \exp\left\{x_0^{(q_\lambda^E)} \cdot \left[X_0 - \frac{\alpha_\mathcal{A}}{\beta_\mathcal{A}} \cdot \frac{d^{(q_\lambda^E),T}}{1 - d^{(q_\lambda^E),T}}\right] \cdot \left(1 - \left(d^{(q_\lambda^E),T}\right)^n\right) + \frac{\alpha_\mathcal{A}}{\beta_\mathcal{A}} x_0^{(q_\lambda^E)} \cdot n\right\},$$

and the closed-form upper bounds are defined by

$$C_{\lambda,X_0,n}^{(p_\lambda^E, q_\lambda^E),U} := C_{\lambda,X_0,n}^{(p_\lambda^E, q_\lambda^E),S} \cdot \exp\left\{- \overline{\zeta}_n^{(q_\lambda^E)} \cdot X_0 - \frac{\alpha_\mathcal{A}}{\beta_\mathcal{A}} \cdot \overline{\vartheta}_n^{(q_\lambda^E)}\right\}, \qquad with \tag{97}$$

$$C_{\lambda,X_0,n}^{(p_\lambda^E, q_\lambda^E),S} := \exp\left\{x_0^{(q_\lambda^E)} \cdot \left[X_0 - \frac{\alpha_\mathcal{A}}{\beta_\mathcal{A}} \cdot \frac{d^{(q_\lambda^E),S}}{1 - d^{(q_\lambda^E),S}}\right] \cdot \left(1 - \left(d^{(q_\lambda^E),S}\right)^n\right) + \frac{\alpha_\mathcal{A}}{\beta_\mathcal{A}} x_0^{(q_\lambda^E)} \cdot n\right\},$$

where in the case $\lambda \in]0,1[$

$$\underline{\zeta}_n^{(q_\lambda^E)} := \Gamma_<^{(q_\lambda^E)} \cdot \frac{\left(d^{(q_\lambda^E),T}\right)^{n-1}}{1 - d^{(q_\lambda^E),T}} \cdot \left(1 - \left(d^{(q_\lambda^E),T}\right)^n\right) > 0 \, , \tag{98}$$

$$\underline{\vartheta}_n^{(q_\lambda^E)} := \Gamma_<^{(q_\lambda^E)} \cdot \frac{1 - \left(d^{(q_\lambda^E),T}\right)^n}{\left(1 - d^{(q_\lambda^E),T}\right)^2} \cdot \left[1 - \frac{d^{(q_\lambda^E),T}\left(1 + \left(d^{(q_\lambda^E),T}\right)^n\right)}{1 + d^{(q_\lambda^E),T}}\right] > 0 \, , \tag{99}$$

$$\overline{\zeta}_n^{(q_\lambda^E)} := \Gamma_<^{(q_\lambda^E)} \cdot \left[\frac{\left(d^{(q_\lambda^E),S}\right)^n - \left(d^{(q_\lambda^E),T}\right)^n}{d^{(q_\lambda^E),S} - d^{(q_\lambda^E),T}} - \left(d^{(q_\lambda^E),S}\right)^{n-1} \cdot \frac{1 - \left(d^{(q_\lambda^E),T}\right)^n}{1 - d^{(q_\lambda^E),T}}\right] > 0 \, , \tag{100}$$

$$\overline{\vartheta}_n^{(q_\lambda^E)} := \Gamma_<^{(q_\lambda^E)} \cdot \frac{d^{(q_\lambda^E),T}}{1 - d^{(q_\lambda^E),T}} \cdot \left[\frac{1 - \left(d^{(q_\lambda^E),S} d^{(q_\lambda^E),T}\right)^n}{1 - d^{(q_\lambda^E),S} d^{(q_\lambda^E),T}} - \frac{\left(d^{(q_\lambda^E),S}\right)^n - \left(d^{(q_\lambda^E),T}\right)^n}{d^{(q_\lambda^E),S} - d^{(q_\lambda^E),T}}\right] > 0 \, , \tag{101}$$

and where in the case $\lambda \in \,]\lambda_-, \lambda_+[\, \backslash [0,1]$

$$\underline{\zeta}_n^{(q_\lambda^E)} \; := \; \Gamma_>^{(q_\lambda^E)} \cdot \frac{\left(d^{(q_\lambda^E),T}\right)^n - \left(d^{(q_\lambda^E),S}\right)^{2n}}{d^{(q_\lambda^E),T} - \left(d^{(q_\lambda^E),S}\right)^2} \; > \; 0 \, , \tag{102}$$

$$\underline{\vartheta}_n^{(q_\lambda^E)} \; := \; \frac{\Gamma_>^{(q_\lambda^E)}}{d^{(q_\lambda^E),T} - \left(d^{(q_\lambda^E),S}\right)^2} \left[\frac{d^{(q_\lambda^E),T}\left(1 - \left(d^{(q_\lambda^E),T}\right)^n\right)}{1 - d^{(q_\lambda^E),T}} - \frac{\left(d^{(q_\lambda^E),S}\right)^2\left(1 - \left(d^{(q_\lambda^E),S}\right)^{2n}\right)}{1 - \left(d^{(q_\lambda^E),S}\right)^2} \right]$$

$$> \; 0 \, , \tag{103}$$

$$\overline{\zeta}_n^{(q_\lambda^E)} \; := \; \Gamma_>^{(q_\lambda^E)} \cdot \left(d^{(q_\lambda^E),S}\right)^{n-1} \cdot \left[n - \frac{1 - \left(d^{(q_\lambda^E),T}\right)^n}{1 - d^{(q_\lambda^E),T}} \right] \; > \; 0 \, , \tag{104}$$

$$\overline{\vartheta}_n^{(q_\lambda^E)} \; := \; \Gamma_>^{(q_\lambda^E)} \cdot \left[\frac{d^{(q_\lambda^E),S} - d^{(q_\lambda^E),T}}{\left(1 - d^{(q_\lambda^E),S}\right)^2 \left(1 - d^{(q_\lambda^E),T}\right)} \cdot \left(1 - \left(d^{(q_\lambda^E),S}\right)^n\right) \right.$$

$$\left. + \frac{d^{(q_\lambda^E),T}\left(1 - \left(d^{(q_\lambda^E),S}d^{(q_\lambda^E),T}\right)^n\right)}{\left(1 - d^{(q_\lambda^E),T}\right)\left(1 - d^{(q_\lambda^E),S}d^{(q_\lambda^E),T}\right)} - \frac{\left(d^{(q_\lambda^E),S}\right)^n}{1 - d^{(q_\lambda^E),S}} \cdot n \right] \; > \; 0 \, . \tag{105}$$

Notice that $\frac{\alpha_{\mathcal{A}}}{\beta_{\mathcal{A}}}$ *can be equivalently be replaced by* $\frac{\alpha_{\mathcal{H}}}{\beta_{\mathcal{H}}}$ *in* (96) *and in* (97).

A proof of Theorem 5 is given in Appendix A.3.

6.3. Explicit Closed-Form Bounds for the Cases $(\beta_{\mathcal{A}}, \beta_{\mathcal{H}}, \alpha_{\mathcal{A}}, \alpha_{\mathcal{H}}, \lambda) \in (\mathcal{P}_{SP} \backslash \mathcal{P}_{SP,1}) \times]0,1[$

To derive (explicit) closed-form lower bounds of the (nonexplicit) recursive lower bounds $B_{\lambda, X_0, n}^L$ for the Hellinger integral $H_\lambda(P_{\mathcal{A},n} \| P_{\mathcal{H},n})$ respectively closed-form upper bounds of the recursive upper bounds $B_{\lambda, X_0, n}^U$ for all parameters cases $(\beta_{\mathcal{A}}, \beta_{\mathcal{H}}, \alpha_{\mathcal{A}}, \alpha_{\mathcal{H}}, \lambda) \in (\mathcal{P}_{SP} \backslash \mathcal{P}_{SP,1}) \times (\mathbb{R} \backslash \{0,1\})$, we combine part (b) of Theorem 1, Lemma 2, Lemma 3 together with appropriate parameters $p_\lambda^L = p^L(\beta_{\mathcal{A}}, \beta_{\mathcal{H}}, \alpha_{\mathcal{A}}, \alpha_{\mathcal{H}}, \lambda)$, $p_\lambda^U = p^U(\beta_{\mathcal{A}}, \beta_{\mathcal{H}}, \alpha_{\mathcal{A}}, \alpha_{\mathcal{H}}, \lambda) \geq 0$ and $q_\lambda^L = q^L(\beta_{\mathcal{A}}, \beta_{\mathcal{H}}, \alpha_{\mathcal{A}}, \alpha_{\mathcal{H}}, \lambda)$, $q_\lambda^U = q^U(\beta_{\mathcal{A}}, \beta_{\mathcal{H}}, \alpha_{\mathcal{A}}, \alpha_{\mathcal{H}}, \lambda) > 0$ satisfying (35). Notice that the representations of the lower and upper closed-form sequence-bounds depend on whether $0 < q_\lambda^A < \beta_\lambda$, $0 < q_\lambda^A = \beta_\lambda$ or $\max\{0, \beta_\lambda\} < q_\lambda^A < \min\{1, e^{\beta_\lambda - 1}\}$ $(A \in \{L, U\})$.

Let us start with closed-form *lower* bounds for the case $\lambda \in]0,1[$; recall that the choice $p_\lambda^L = \alpha_{\mathcal{A}}^\lambda \alpha_{\mathcal{H}}^{1-\lambda}$, $q_\lambda^L = \beta_{\mathcal{A}}^\lambda \beta_{\mathcal{H}}^{1-\lambda}$ led to the optimal recursive lower bounds $B_{\lambda, X_0, n}^L$ of the Hellinger integral (cf. Theorem 1(b) and Section 3.5). Correspondingly, we can derive

Theorem 6. *Let* $p_\lambda^L = \alpha_{\mathcal{A}}^\lambda \alpha_{\mathcal{H}}^{1-\lambda}$ *and* $q_\lambda^L = \beta_{\mathcal{A}}^\lambda \beta_{\mathcal{H}}^{1-\lambda}$. *Then, the following assertions hold true:*

(a) For all $(\beta_{\mathcal{A}}, \beta_{\mathcal{H}}, \alpha_{\mathcal{A}}, \alpha_{\mathcal{H}}, \lambda) \in (\mathcal{P}_{SP,2} \cup \mathcal{P}_{SP,3a} \cup \mathcal{P}_{SP,3b} \cup \mathcal{P}_{SP,3c}) \times]0,1[$ *(for which particularly $0 < q_\lambda^L < \beta_\lambda$, $\beta_{\mathcal{A}} \neq \beta_{\mathcal{H}}$), all initial population sizes $X_0 \in \mathbb{N}$ and all observation horizons $n \in \mathbb{N}$ there holds*

$$C_{\lambda,X_0,n}^{(p_\lambda^L,q_\lambda^L),T} \quad \leq \quad C_{\lambda,X_0,n}^{(p_\lambda^L,q_\lambda^L),L} \leq B_{\lambda,X_0,n}^L < 1,$$

where
$$C_{\lambda,X_0,n}^{(p_\lambda^L,q_\lambda^L),L} := C_{\lambda,X_0,n}^{(p_\lambda^L,q_\lambda^L),T} \cdot \exp\left\{ \underline{\zeta}_n^{(q_\lambda^L)} \cdot X_0 + \frac{p_\lambda^L}{q_\lambda^L} \cdot \underline{\vartheta}_n^{(q_\lambda^L)} \right\} \tag{106}$$

with
$$C_{\lambda,X_0,n}^{(p_\lambda^L,q_\lambda^L),T} := \exp\left\{ x_0^{(q_\lambda^L)} \cdot \left[X_0 - \frac{p_\lambda^L}{q_\lambda^L} \cdot \frac{d^{(q_\lambda^L),T}}{1 - d^{(q_\lambda^L),T}} \right] \cdot \left(1 - \left(d^{(q_\lambda^L),T} \right)^n \right) \right.$$
$$\left. + \left(\frac{p_\lambda^L}{q_\lambda^L} \cdot \left(\beta_\lambda + x_0^{(q_\lambda^L)} \right) - \alpha_\lambda \right) \cdot n \right\},$$

and with
$$\underline{\zeta}_n^{(q_\lambda^L)} := \Gamma_<^{(q_\lambda^L)} \cdot \frac{\left(d^{(q_\lambda^L),T} \right)^{n-1}}{1 - d^{(q_\lambda^L),T}} \cdot \left(1 - \left(d^{(q_\lambda^L),T} \right)^n \right) > 0, \tag{107}$$

$$\underline{\vartheta}_n^{(q_\lambda^L)} := \Gamma_<^{(q_\lambda^L)} \cdot \frac{1 - \left(d^{(q_\lambda^L),T} \right)^n}{\left(1 - d^{(q_\lambda^L),T} \right)^2} \cdot \left[1 - \frac{d^{(q_\lambda^L),T} \left(1 + \left(d^{(q_\lambda^L),T} \right)^n \right)}{1 + d^{(q_\lambda^L),T}} \right] > 0. \tag{108}$$

(b) For all $(\beta_{\mathcal{A}}, \beta_{\mathcal{H}}, \alpha_{\mathcal{A}}, \alpha_{\mathcal{H}}, \lambda) \in (\mathcal{P}_{SP,4a} \cup \mathcal{P}_{SP,4b}) \times]0,1[$ *(for which particularly $0 < q_\lambda^L = \beta_\lambda$, $\beta_{\mathcal{A}} = \beta_{\mathcal{H}}$), all initial population sizes $X_0 \in \mathbb{N}$ and all observation horizons $n \in \mathbb{N}$ there holds*

$$C_{\lambda,X_0,n}^{(p_\lambda^L,q_\lambda^L),L} := C_{\lambda,X_0,n}^{(p_\lambda^L,q_\lambda^L),T} := B_{\lambda,X_0,n}^L = \exp\left\{ \left(p_\lambda^L - \alpha_\lambda \right) \cdot n \right\} < 1.$$

(c) For all $(\beta_{\mathcal{A}}, \beta_{\mathcal{H}}, \alpha_{\mathcal{A}}, \alpha_{\mathcal{H}}, \lambda) \in (\mathcal{P}_{SP} \backslash \mathcal{P}_{SP,1}) \times]0,1[$ *and all initial population sizes $X_0 \in \mathbb{N}$ one gets*

$$\lim_{n \to \infty} \frac{1}{n} \log\left(C_{\lambda,X_0,n}^{(p_\lambda^L,q_\lambda^L),T} \right) = \lim_{n \to \infty} \frac{1}{n} \log\left(C_{\lambda,X_0,n}^{(p_\lambda^L,q_\lambda^L),L} \right) = \lim_{n \to \infty} \frac{1}{n} \log\left(B_{\lambda,X_0,n}^L \right)$$
$$= \frac{p_\lambda^L}{q_\lambda^L} \cdot \left(\beta_\lambda + x_0^{(q_\lambda^L)} \right) - \alpha_\lambda < 0,$$

where in the case $\beta_{\mathcal{A}} = \beta_{\mathcal{H}}$ there holds $q_\lambda^L = \beta_\lambda$ and $x_0^{(q_\lambda^L)} = 0$.

The proof will be provided in Appendix A.3.

In order to deduce closed-form *upper* bounds for the case $\lambda \in]0,1[$, we first recall from the Sections 3.6–3.13, that we have to employ suitable parameters $p_\lambda^U = p^U(\beta_{\mathcal{A}}, \beta_{\mathcal{H}}, \alpha_{\mathcal{A}}, \alpha_{\mathcal{H}}, \lambda)$, $q_\lambda^U = q^U(\beta_{\mathcal{A}}, \beta_{\mathcal{H}}, \alpha_{\mathcal{A}}, \alpha_{\mathcal{H}}, \lambda)$ satisfying (35). Notice that we automatically obtain $p_\lambda^U \geq p_\lambda^L = \alpha_{\mathcal{A}}^\lambda \alpha_{\mathcal{H}}^{1-\lambda} > 0$. Correspondingly, we obtain

Theorem 7. *For all $(\beta_{\mathcal{A}}, \beta_{\mathcal{H}}, \alpha_{\mathcal{A}}, \alpha_{\mathcal{H}}, \lambda) \in (\mathcal{P}_{SP} \backslash \mathcal{P}_{SP,1}) \times]0,1[$, all coefficients p_λ^U, q_λ^U which satisfy (35) for all $x \in \mathbb{N}_0$ and additionally either $0 < q_\lambda^U \leq \beta_\lambda$ or $\beta_\lambda < q_\lambda^U < \min\{1, e^{\beta_\lambda - 1}\}$, all initial population sizes $X_0 \in \mathbb{N}$ and all observation horizons $n \in \mathbb{N}$ the following assertions hold true:*

$$C_{\lambda,X_0,n}^{(p_\lambda^U,q_\lambda^U),S} \geq C_{\lambda,X_0,n}^{(p_\lambda^U,q_\lambda^U),U} \geq \widetilde{B}_{\lambda,X_0,n}^{(p_\lambda^U,q_\lambda^U)} \geq B_{\lambda,X_0,n}^U, \qquad where \tag{109}$$

(a) in the case $0 < q_\lambda^U < \beta_\lambda$ one has

$$C_{\lambda,X_0,n}^{(p_\lambda^U,q_\lambda^U),U} := C_{\lambda,X_0,n}^{(p_\lambda^U,q_\lambda^U),S} \cdot \exp\left\{ -\overline{\zeta}_n^{(q_\lambda^U)} \cdot X_0 - \frac{p_\lambda^U}{q_\lambda^U} \cdot \overline{\vartheta}_n^{(q_\lambda^U)} \right\} \tag{110}$$

with $\quad C_{\lambda,X_0,n}^{(p_\lambda^U,q_\lambda^U),S} := \exp\left\{ x_0^{(q_\lambda^U)} \cdot \left[X_0 - \frac{p_\lambda^U}{q_\lambda^U} \cdot \frac{d^{(q_\lambda^U),S}}{1 - d^{(q_\lambda^U),S}} \right] \cdot \left(1 - \left(d^{(q_\lambda^U),S} \right)^n \right) \right.$

$$\left. + \left(\frac{p_\lambda^U}{q_\lambda^U} \cdot \left(\beta_\lambda + x_0^{(q_\lambda^U)} \right) - \alpha_\lambda \right) \cdot n \right\},$$

$$\overline{\zeta}_n^{(q_\lambda^U)} := \Gamma_<^{(q_\lambda^U)} \cdot \left[\frac{\left(d^{(q_\lambda^U),S} \right)^n - \left(d^{(q_\lambda^U),T} \right)^n}{d^{(q_\lambda^U),S} - d^{(q_\lambda^U),T}} - \left(d^{(q_\lambda^U),S} \right)^{n-1} \cdot \frac{1 - \left(d^{(q_\lambda^U),T} \right)^n}{1 - d^{(q_\lambda^U),T}} \right] > 0, \tag{111}$$

$$\overline{\vartheta}_n^{(q_\lambda^U)} := \Gamma_<^{(q_\lambda^U)} \cdot \frac{d^{(q_\lambda^U),T}}{1 - d^{(q_\lambda^U),T}} \cdot \left[\frac{1 - \left(d^{(q_\lambda^U),S} d^{(q_\lambda^U),T} \right)^n}{1 - d^{(q_\lambda^U),S} d^{(q_\lambda^U),T}} - \frac{\left(d^{(q_\lambda^U),S} \right)^n - \left(d^{(q_\lambda^U),T} \right)^n}{d^{(q_\lambda^U),S} - d^{(q_\lambda^U),T}} \right] > 0; \tag{112}$$

furthermore, whenever p_λ^U, q_λ^U satisfy additionally (47) (such parameters exist particularly in the setups $\mathcal{P}_{SP,2} \cup \mathcal{P}_{SP,3a} \cup \mathcal{P}_{SP,3b}$, cf. Sections 3.7–3.9), then

$$1 > C_{\lambda,X_0,n}^{(p_\lambda^U,q_\lambda^U),S} \quad \text{and} \quad \widetilde{B}_{\lambda,X_0,n}^{(p_\lambda^U,q_\lambda^U)} = B_{\lambda,X_0,n}^U \quad \forall\, n \in \mathbb{N};$$

(b) in the case $0 < q_\lambda^U = \beta_\lambda$ one has

$$C_{\lambda,X_0,n}^{(p_\lambda^U,q_\lambda^U),U} := C_{\lambda,X_0,n}^{(p_\lambda^U,q_\lambda^U),S} := \widetilde{B}_{\lambda,X_0,n}^{(p_\lambda^U,q_\lambda^U)} = \exp\left\{ \left(p_\lambda^U - \alpha_\lambda \right) \cdot n \right\};$$

(c) in the case $\beta_\lambda < q_\lambda^U < \min\left\{ 1, e^{\beta_\lambda - 1} \right\}$ the formulas (109) and (110) remain valid, but with

$$\overline{\zeta}_n^{(q_\lambda^U)} := \Gamma_>^{(q_\lambda^U)} \cdot \left(d^{(q_\lambda^U),S} \right)^{n-1} \cdot \left[n - \frac{1 - \left(d^{(q_\lambda^U),T} \right)^n}{1 - d^{(q_\lambda^U),T}} \right] > 0, \tag{113}$$

$$\overline{\vartheta}_n^{(q_\lambda^U)} := \Gamma_>^{(q_\lambda^U)} \cdot \left[\frac{d^{(q_\lambda^U),S} - d^{(q_\lambda^U),T}}{\left(1 - d^{(q_\lambda^U),S} \right)^2 \left(1 - d^{(q_\lambda^U),T} \right)} \cdot \left(1 - \left(d^{(q_\lambda^U),S} \right)^n \right) \right.$$

$$\left. + \frac{d^{(q_\lambda^U),T} \left(1 - \left(d^{(q_\lambda^U),S} d^{(q_\lambda^U),T} \right)^n \right)}{\left(1 - d^{(q_\lambda^U),T} \right) \left(1 - d^{(q_\lambda^U),S} d^{(q_\lambda^U),T} \right)} - \frac{\left(d^{(q_\lambda^U),S} \right)^n}{1 - d^{(q_\lambda^U),S}} \cdot n \right] > 0; \tag{114}$$

(d) for all cases (a) to (c) one gets

$$\lim_{n\to\infty} \frac{1}{n} \log\left(C_{\lambda,X_0,n}^{(p_\lambda^U,q_\lambda^U),S} \right) = \lim_{n\to\infty} \frac{1}{n} \log\left(C_{\lambda,X_0,n}^{(p_\lambda^U,q_\lambda^U),U} \right) = \lim_{n\to\infty} \frac{1}{n} \log\left(\widetilde{B}_{\lambda,X_0,n}^{(p_\lambda^U,q_\lambda^U)} \right)$$

$$= \frac{p_\lambda^U}{q_\lambda^U} \cdot \left(\beta_\lambda + x_0^{(q_\lambda^U)} \right) - \alpha_\lambda,$$

where in the case $q_\lambda^U = \beta_\lambda$ there holds $x_0^{(q_\lambda^U)} = 0$.

This Theorem 7 will be proved in Appendix A.3. Notice that for an inadequate choice of p_λ^U, q_λ^U it may hold that $\frac{p_\lambda^U}{q_\lambda^U}(\beta_\lambda + x_0^{(q_\lambda^U)}) - \alpha_\lambda > 0$ in part (d) of Theorem 7.

6.4. Explicit Closed-Form Bounds for the Cases $(\beta_\mathcal{A}, \beta_\mathcal{H}, \alpha_\mathcal{A}, \alpha_\mathcal{H}, \lambda) \in (\mathcal{P}_{SP} \backslash \mathcal{P}_{SP,1}) \times (\mathbb{R} \backslash [0,1])$

For $\lambda \in \mathbb{R} \backslash [0,1]$, let us now construct closed-form *lower* bounds of the recursive lower bound components $\widetilde{B}_{\lambda,X_0,n}^{(p_\lambda^L,q_\lambda^L)}$, for suitable parameters $p_\lambda^L \geq 0$ and either $0 < q_\lambda^L \leq \beta_\lambda$ or $\max\{0, \beta_\lambda\} < q_\lambda^L < \min\{1, e^{\beta_\lambda - 1}\}$ satisfying (35).

Theorem 8. *For all $(\beta_\mathcal{A}, \beta_\mathcal{H}, \alpha_\mathcal{A}, \alpha_\mathcal{H}, \lambda) \in (\mathcal{P}_{SP} \backslash \mathcal{P}_{SP,1}) \times (\mathbb{R} \backslash [0,1])$, all coefficients $p_\lambda^L \geq 0$, $q_\lambda^L > 0$ which satisfy (35) for all $x \in \mathbb{N}_0$ and either $0 < q_\lambda^L \leq \beta_\lambda$ or $\max\{0, \beta_\lambda\} < q_\lambda^L < \min\{1, e^{\beta_\lambda - 1}\}$, all initial population sizes $X_0 \in \mathbb{N}$ and all observation horizons $n \in \mathbb{N}$ the following assertions hold true:*

$$C_{\lambda,X_0,n}^{(p_\lambda^L,q_\lambda^L),T} \leq C_{\lambda,X_0,n}^{(p_\lambda^L,q_\lambda^L),L} \leq \widetilde{B}_{\lambda,X_0,n}^{(p_\lambda^L,q_\lambda^L)} \leq B_{\lambda,X_0,n}^L, \qquad \text{where} \tag{115}$$

(a) in the case $0 < q_\lambda^L < \beta_\lambda$ one has

$$C_{\lambda,X_0,n}^{(p_\lambda^L,q_\lambda^L),L} := C_{\lambda,X_0,n}^{(p_\lambda^L,q_\lambda^L),T} \cdot \exp\left\{\underline{\zeta}_n^{(q_\lambda^L)} \cdot X_0 + \frac{p_\lambda^L}{q_\lambda^L} \cdot \underline{\vartheta}_n^{(q_\lambda^L)}\right\}, \tag{116}$$

$$\text{with} \qquad C_{\lambda,X_0,n}^{(p_\lambda^L,q_\lambda^L),T} := \exp\left\{x_0^{(q_\lambda^L)} \cdot \left[X_0 - \frac{p_\lambda^L}{q_\lambda^L} \cdot \frac{d^{(q_\lambda^L),T}}{1 - d^{(q_\lambda^L),T}}\right] \cdot \left(1 - \left(d^{(q_\lambda^L),T}\right)^n\right)\right.$$

$$\left. + \left(\frac{p_\lambda^L}{q_\lambda^L} \cdot \left(\beta_\lambda + x_0^{(q_\lambda^L)}\right) - \alpha_\lambda\right) \cdot n\right\}$$

$$\underline{\zeta}_n^{(q_\lambda^L)} := \Gamma_<^{(q_\lambda^L)} \cdot \frac{\left(d^{(q_\lambda^L),T}\right)^{n-1}}{1 - d^{(q_\lambda^L),T}} \cdot \left(1 - \left(d^{(q_\lambda^L),T}\right)^n\right) > 0, \tag{117}$$

$$\underline{\vartheta}_n^{(q_\lambda^L)} := \Gamma_<^{(q_\lambda^L)} \cdot \frac{1 - \left(d^{(q_\lambda^L),T}\right)^n}{\left(1 - d^{(q_\lambda^L),T}\right)^2} \cdot \left[1 - \frac{d^{(q_\lambda^L),T}\left(1 + \left(d^{(q_\lambda^L),T}\right)^n\right)}{1 + d^{(q_\lambda^L),T}}\right] > 0; \tag{118}$$

furthermore, whenever p_λ^L, q_λ^L satisfy additionally (56) (such parameters exist particularly in the setups $\mathcal{P}_{SP,2} \cup \mathcal{P}_{SP,3a} \cup \mathcal{P}_{SP,3b}$, cf. Sections 3.17–3.19), then

$$1 < C_{\lambda,X_0,n}^{(p_\lambda^L,q_\lambda^L),T} \qquad \text{and} \qquad \widetilde{B}_{\lambda,X_0,n}^{(p_\lambda^L,q_\lambda^L)} = B_{\lambda,X_0,n}^L \quad \forall\, n \in \mathbb{N};$$

(b) in the case $0 < q_\lambda^L = \beta_\lambda$ one has

$$C_{\lambda,X_0,n}^{(p_\lambda^L,q_\lambda^L),L} := C_{\lambda,X_0,n}^{(p_\lambda^L,q_\lambda^L),T} = \widetilde{B}_{\lambda,X_0,n}^{(p_\lambda^L,q_\lambda^L)} = \exp\left\{\left(p_\lambda^L - \alpha_\lambda\right) \cdot n\right\};$$

(c) in the case $\max\{0, \beta_\lambda\} < q_\lambda^L < \min\left\{1, e^{\beta_\lambda - 1}\right\}$ the formulas (115) and (116) remain valid, but with

$$\underline{\zeta}_n^{(q_\lambda^L)} := \Gamma_>^{(q_\lambda^L)} \cdot \frac{\left(d^{(q_\lambda^L),T}\right)^n - \left(d^{(q_\lambda^L),S}\right)^{2n}}{d^{(q_\lambda^L),T} - \left(d^{(q_\lambda^L),S}\right)^2} > 0, \tag{119}$$

$$\underline{\vartheta}_n^{(q_\lambda^L)} := \frac{\Gamma_>^{(q_\lambda^L)}}{d^{(q_\lambda^L),T} - \left(d^{(q_\lambda^L),S}\right)^2} \cdot \left[\frac{d^{(q_\lambda^L),T} \cdot \left(1 - \left(d^{(q_\lambda^L),T}\right)^n\right)}{1 - d^{(q_\lambda^L),T}} - \frac{\left(d^{(q_\lambda^L),S}\right)^2 \cdot \left(1 - \left(d^{(q_\lambda^L),S}\right)^{2n}\right)}{1 - \left(d^{(q_\lambda^L),S}\right)^2}\right] > 0; \tag{120}$$

(d)   for all cases (a) to (c) one gets

$$\lim_{n\to\infty} \frac{1}{n} \log\left(C^{(p_\lambda^L,q_\lambda^L),T}_{\lambda,X_0,n} \right) \;=\; \lim_{n\to\infty} \frac{1}{n} \log\left(C^{(p_\lambda^L,q_\lambda^L),L}_{\lambda,X_0,n} \right) \;=\; \lim_{n\to\infty} \frac{1}{n} \log\left(\widetilde{B}^{(p_\lambda^L,q_\lambda^L)}_{\lambda,X_0,n} \right)$$

$$= \;\; \frac{p_\lambda^L}{q_\lambda^L} \cdot \left(\beta_\lambda + x_0^{(q_\lambda^L)} \right) - \alpha_\lambda \,,$$

where in the case $q_\lambda^L = \beta_\lambda$ there holds $x_0^{(q_\lambda^L)} = 0$.

For the proof of Theorem 8, see Appendix A.3. Notice that for an inadequate choice of p_λ^L, q_λ^L it may hold that $\frac{p_\lambda^L}{q_\lambda^L}(\beta_\lambda + x_0^{(q_\lambda^U)}) - \alpha_\lambda < 0$ in the last assertion of Theorem 8.

To derive closed-form *upper* bounds of the recursive upper bounds $B^U_{\lambda,X_0,n}$ of the Hellinger integral in the case $\lambda \in \mathbb{R}\backslash[0,1]$, let us first recall from Section 3.24 that we have to use the parameters $p_\lambda^U = \alpha_{\mathcal{A}}^\lambda \alpha_{\mathcal{H}}^{1-\lambda} > 0$ and $q_\lambda^U = \beta_{\mathcal{A}}^\lambda \beta_{\mathcal{H}}^{1-\lambda} > 0$. Furthermore, in the case $\beta_{\mathcal{A}} \neq \beta_{\mathcal{H}}$ we obtain from Lemma 1 (setting $q_\lambda = q_\lambda^U$) the assertion that $\max\{0, \beta_\lambda\} < q_\lambda^U < \min\{1, e^{\beta_\lambda - 1}\}$ iff $\lambda \in]\lambda_-, \lambda_+[\,\backslash\, [0,1]$ (implying that the sequence $(a_n^{(q_\lambda^U)})_{n\in\mathbb{N}}$ converges). In the case $\beta_{\mathcal{A}} = \beta_{\mathcal{H}}$ on gets $q_\lambda^U = \beta_{\mathcal{A}}^\lambda \beta_{\mathcal{H}}^{1-\lambda} = \beta_{\mathcal{A}} = \beta_{\mathcal{H}} = \beta_\lambda$ and therefore (cf. (P2)) $a_n^{(q_\lambda^U)} = 0$ for all $n \in \mathbb{N}$ and for all $\lambda \in \mathbb{R}\backslash[0,1]$. Correspondingly, we deduce

Theorem 9. *Let $p_\lambda^U = \alpha_{\mathcal{A}}^\lambda \alpha_{\mathcal{H}}^{1-\lambda}$ and $q_\lambda^U = \beta_{\mathcal{A}}^\lambda \beta_{\mathcal{H}}^{1-\lambda}$. Then, the following assertions hold true:*

(a)   For all $(\beta_{\mathcal{A}}, \beta_{\mathcal{H}}, \alpha_{\mathcal{A}}, \alpha_{\mathcal{H}}, \lambda) \in (\mathcal{P}_{SP,2} \cup \mathcal{P}_{SP,3a} \cup \mathcal{P}_{SP,3b} \cup \mathcal{P}_{SP,3c}) \times (\,]\lambda_-, \lambda_+[\,\backslash\, [0,1] \,)$ (in particular for $\beta_{\mathcal{A}} \neq \beta_{\mathcal{H}}$), all initial population sizes $X_0 \in \mathbb{N}$ and all observation horizons $n \in \mathbb{N}$ there holds

$$\infty \;>\; C^{(p_\lambda^U,q_\lambda^U),S}_{\lambda,X_0,n} \;\geq\; C^{(p_\lambda^U,q_\lambda^U),U}_{\lambda,X_0,n} \;\geq\; B^U_{\lambda,X_0,n} \;>\; 1\,,$$

where
$$C^{(p_\lambda^U,q_\lambda^U),U}_{\lambda,X_0,n} := C^{(p_\lambda^U,q_\lambda^U),S}_{\lambda,X_0,n} \cdot \exp\left\{ -\overline{\zeta}_n^{(q_\lambda^U)} \cdot X_0 - \frac{p_\lambda^U}{q_\lambda^U} \cdot \overline{\vartheta}_n^{(q_\lambda^U)} \right\} \tag{121}$$

with
$$C^{(p_\lambda^U,q_\lambda^U),S}_{\lambda,X_0,n} := \exp\left\{ x_0^{(q_\lambda^U)} \cdot \left[X_0 - \frac{p_\lambda^U}{q_\lambda^U} \cdot \frac{d^{(q_\lambda^U),T}}{1 - d^{(q_\lambda^U),T}} \right] \cdot \left(1 - \left(d^{(q_\lambda^U),T} \right)^n \right) \right.$$
$$\left. + \left(\frac{p_\lambda^U}{q_\lambda^U} \cdot \left(\beta_\lambda + x_0^{(q_\lambda^U)} \right) - \alpha_\lambda \right) \cdot n \right\}\,,$$

$$\overline{\zeta}_n^{(q_\lambda^U)} := \Gamma_>^{(q_\lambda^U)} \cdot \left(d^{(q_\lambda^U),S} \right)^{n-1} \cdot \left[n - \frac{1 - \left(d^{(q_\lambda^U),T} \right)^n}{1 - d^{(q_\lambda^U),T}} \right] > 0\,, \tag{122}$$

$$\overline{\vartheta}_n^{(q_\lambda^U)} := \Gamma_>^{(q_\lambda^U)} \cdot \left[\frac{d^{(q_\lambda^U),S} - d^{(q_\lambda^U),T}}{\left(1 - d^{(q_\lambda^U),S}\right)^2 \left(1 - d^{(q_\lambda^U),T}\right)} \cdot \left(1 - \left(d^{(q_\lambda^U),S} \right)^n \right) \right.$$
$$\left. + \frac{d^{(q_\lambda^U),T}\left(1 - \left(d^{(q_\lambda^U),S} d^{(q_\lambda^U),T} \right)^n \right)}{\left(1 - d^{(q_\lambda^U),T}\right)\left(1 - d^{(q_\lambda^U),S} d^{(q_\lambda^U),T}\right)} - \frac{\left(d^{(q_\lambda^U),S} \right)^n}{1 - d^{(q_\lambda^U),S}} \cdot n \right] > 0\,. \tag{123}$$

(b)   For all $(\beta_{\mathcal{A}}, \beta_{\mathcal{H}}, \alpha_{\mathcal{A}}, \alpha_{\mathcal{H}}, \lambda) \in (\mathcal{P}_{SP,4a} \cup \mathcal{P}_{SP,4b}) \times (\mathbb{R}\backslash[0,1] \,)$ (for which particularly $0 < q_\lambda^U = \beta_\lambda$, $\beta_{\mathcal{A}} = \beta_{\mathcal{H}}$), all initial population sizes $X_0 \in \mathbb{N}$ and all observation horizons $n \in \mathbb{N}$ there holds

$$C^{(p_\lambda^U,q_\lambda^U),U}_{\lambda,X_0,n} := C^{(p_\lambda^U,q_\lambda^U),S}_{\lambda,X_0,n} := B^U_{\lambda,X_0,n} = \exp\left\{ \left(p_\lambda^U - \alpha_\lambda \right) \cdot n \right\} > 1\,.$$

(c) For all $(\beta_\mathcal{A}, \beta_\mathcal{H}, \alpha_\mathcal{A}, \alpha_\mathcal{H}, \lambda) \in (\mathcal{P}_{SP} \backslash \mathcal{P}_{SP,1}) \times (\,]\lambda_-, \lambda_+[\,\backslash [0,1]\,)$ and all initial population sizes $X_0 \in \mathbb{N}$ one gets

$$\lim_{n\to\infty} \frac{1}{n} \log\left(C_{\lambda,X_0,n}^{(p_\lambda^U, q_\lambda^U), S}\right) \;=\; \lim_{n\to\infty} \frac{1}{n} \log\left(C_{\lambda,X_0,n}^{(p_\lambda^U, q_\lambda^U), U}\right) \;=\; \lim_{n\to\infty} \frac{1}{n} \log\left(B_{\lambda,X_0,n}^U\right)$$

$$= \;\frac{p_\lambda^U}{q_\lambda^U} \cdot \left(\beta_\lambda + x_0^{(q_\lambda^U)}\right) - \alpha_\lambda \;>\; 0\,,$$

where in the case $\beta_\mathcal{A} = \beta_\mathcal{H}$ there holds $q_\lambda^U = \beta_\lambda$ and $x_0^{(q_\lambda^U)} = 0$.

A proof of Theorem 9 is provided in Appendix A.3.

Remark 7. *Substituting* $a_n^{(q)}$ *by* $a_n^{(q),T}$ *resp.* $a_n^{(q),S}$ *(cf. (78) resp. (79)) in* $\widetilde{B}_{\lambda,X_0,n}^{(p,q)}$ *from (42) leads to the "rudimentary" closed-form bounds* $C_{\lambda,X_0,n}^{(p,q),T}$ *resp.* $C_{\lambda,X_0,n}^{(p,q),S}$, *whereas substituting* $a_n^{(q)}$ *by* $\underline{a}_n^{(q)}$ *resp.* $\overline{a}_n^{(q)}$ *(cf. (92) resp. (94)) in* $\widetilde{B}_{\lambda,X_0,n}^{(p,q)}$ *from (42) leads to the "improved" closed-form bounds* $C_{\lambda,X_0,n}^{(p,q),L}$ *resp.* $C_{\lambda,X_0,n}^{(p,q),U}$ *in all the Theorems 5–9.*

6.5. Totally Explicit Closed-Form Bounds

The above-mentioned results give closed-form lower bounds $C_{\lambda,X_0,n}^{(p,q),L}$, $C_{\lambda,X_0,n}^{(p,q),T}$ resp. closed-form upper bounds $C_{\lambda,X_0,n}^{(p,q),U}$, $C_{\lambda,X_0,n}^{(p,q),S}$ of the Hellinger integrals $H_\lambda(P_{\mathcal{A},n} || P_{\mathcal{H},n})$ for case-dependent choices of p, q. However, these bounds still involve the fixed point $x_0^{(q)}$ which in general has to be calculated implicitly. In order to get "totally" explicit but "slightly" less tight closed-form bounds of $H_\lambda(P_{\mathcal{A},n} || P_{\mathcal{H},n})$, one can proceed as follows:

1. in all the closed-form lower bound formulas of the Theorems 5, 6 and 8–including the definitions (76), (77) and (91)–replace the implicit $x_0^{(q)}$ by a close explicitly known point $\underline{x}_0^{(q)} < x_0^{(q)}$;
2. in all closed-form upper bound formulas of the Theorems 5, 7 and 9–including (76), (77) and (91)–replace $x_0^{(q)}$ by a close explicitly known point $\overline{x}_0^{(q)} > x_0^{(q)}$.

For instance, one can use the following choices which will be also employed as an auxiliary tool for the diffusion-limit-concerning proof of Lemma A6 in Appendix A.4:

$$\underline{x}_0^{(q)} \;:=\; \begin{cases} q^{-1} \cdot e^{-\underline{x}_0^{(q)}} \cdot \left[(1-q) - \sqrt{(1-q)^2 - 2 \cdot q \cdot e^{\underline{x}_0^{(q)}} \cdot (q - \beta_\lambda)}\right], & \text{if } \; q \in]0, \beta_\lambda[, \\[2ex] q^{-1} \cdot \left[(1-q) - \sqrt{(1-q)^2 - 2 \cdot q \cdot (q - \beta_\lambda)}\right], & \text{if } \; \max\{0, \beta_\lambda\} < q < \min\{1, e^{\beta_\lambda - 1}\}, \end{cases} \tag{124}$$

$$\text{where} \quad \underline{\underline{x}}_0^{(q)} \;:=\; \begin{cases} \max\left\{-\beta_\lambda, \, \frac{q - \beta_\lambda}{1-q}\right\}, & \text{if } q \in]0, 1[, \\ -\beta_\lambda, & \text{if } q \geq 1, \end{cases} \tag{125}$$

$$\overline{x}_0^{(q)} \;:=\; \begin{cases} q^{-1} \cdot \left[(1-q) - \sqrt{(1-q)^2 - 2 \cdot q \cdot (q - \beta_\lambda)}\right], & \text{if } \; q \in]0, \beta_\lambda[, \\[2ex] (1-q) - \sqrt{(1-q)^2 - 2 \cdot (q - \beta_\lambda)}, & \text{if } \; \max\{0, \beta_\lambda\} < q < \min\{1, e^{\beta_\lambda - 1}\} \\ & \quad \text{and } (1-q)^2 - 2 \cdot q \cdot (q - \beta_\lambda) \geq 0, \\[2ex] \overline{\overline{x}}_0^{(q)} := -\log(q), & \text{if } \; \max\{0, \beta_\lambda\} < q < \min\{1, e^{\beta_\lambda - 1}\} \\ & \quad \text{and } (1-q)^2 - 2 \cdot q \cdot (q - \beta_\lambda) < 0. \end{cases} \tag{126}$$

Behind this choice "lies" the idea that–in contrast to the solution $x_0^{(q)}$ of $\xi_\lambda^{(q)}{}'(x) := q e^x - \beta_\lambda = x$–the point $\underline{x}_0^{(q)}$ is a solution of (the obviously explicitly solvable) $\underline{Q}_\lambda^{(q)}{}'(x) := \underline{a}_\lambda^{(q)} x^2 + \underline{b}_\lambda^{(q)} x + \underline{c}_\lambda^{(q)} = x$ in both cases $0 < q < \beta_\lambda$ and $\max\{0, \beta_\lambda\} < q < \min\{1, e^{\beta_\lambda - 1}\}$, whereas the point $\overline{x}_0^{(q)}$ is a solution of $\overline{Q}_\lambda^{(q)}{}'(x) := \overline{a}_\lambda^{(q)} x^2 + \overline{b}_\lambda^{(q)} x + \overline{c}_\lambda^{(q)} = x$ in the case $0 < q < \beta_\lambda$ and in the case $\max\{0, \beta_\lambda\} < q < \min\{1, e^{\beta_\lambda - 1}\}$ together with $(1-q)^2 - 2 \cdot q \cdot (q - \beta_\lambda) \geq 0$. Thereby, $\underline{Q}_\lambda^{(q)}(\cdot)$ and $\overline{Q}_\lambda^{(q)}(\cdot)$ are the lower resp. upper quadratic approximates of $\xi_\lambda^{(q)}(\cdot)$ satisfying the following constraints:

- for $q \in]0, \beta_\lambda[$ (mostly but not only for $\lambda \in]0, 1[$) (lower bound):

$$\underline{Q}_\lambda^{(q)}(0) = \xi_\lambda^{(q)}(0) = q - \beta_\lambda, \qquad \underline{Q}_\lambda^{(q)}{}'(0) = \xi_\lambda^{(q)}{}'(0) = q, \qquad \underline{Q}_\lambda^{(q)}{}''(x) = \xi_\lambda^{(q)}{}''(y) = q e^y, \quad x \in \mathbb{R},$$

for some explicitly known approximate $y < x_0^{(q)}$ (leading to the (tighter) explicit lower approximate $\underline{x}_0^{(q)} \in]y, x_0^{(q)}[$); here, we choose

$$y := \underline{x}_0^{(q)} := \begin{cases} \max\left\{-\beta_\lambda, \frac{q - \beta_\lambda}{1 - q}\right\}, & \text{if } q < 1, \\ -\beta_\lambda, & \text{if } q \geq 1; \end{cases}$$

- for $q \in]0, \beta_\lambda[$ (mostly but not only for $\lambda \in]0, 1[$) (upper bound):

$$\overline{Q}_\lambda^{(q)}(0) = \xi_\lambda^{(q)}(0) = q - \beta_\lambda, \qquad \overline{Q}_\lambda^{(q)}{}'(0) = \xi_\lambda^{(q)}{}'(0) = q, \qquad \overline{Q}_\lambda^{(q)}{}''(x) = \xi_\lambda^{(q)}{}''(0) = q, \quad x \in \mathbb{R};$$

- for $\max\{0, \beta_\lambda\} < q < \min\{1, e^{\beta_\lambda - 1}\}$ (mostly but not only for $\lambda \in \mathbb{R} \setminus [0, 1]$) (lower bound):

$$\underline{Q}_\lambda^{(q)}(0) = \xi_\lambda^{(q)}(0) = q - \beta_\lambda, \qquad \underline{Q}_\lambda^{(q)}{}'(0) = \xi_\lambda^{(q)}{}'(0) = q, \qquad \underline{Q}_\lambda^{(q)}{}''(x) = \xi_\lambda^{(q)}{}''(0) = q, \quad x \in \mathbb{R};$$

- for $\max\{0, \beta_\lambda\} < q < \min\{1, e^{\beta_\lambda - 1}\}$ in combination with $(1-q)^2 - 2 \cdot q \cdot (q - \beta_\lambda) \geq 0$ (mostly but not only for $\lambda \in \mathbb{R} \setminus [0, 1]$) (upper bound):

$$\overline{Q}_\lambda^{(q)}(0) = \xi_\lambda^{(q)}(0) = q - \beta_\lambda, \quad \overline{Q}_\lambda^{(q)}{}'(0) = \xi_\lambda^{(q)}{}'(0) = q, \quad \overline{Q}_\lambda^{(q)}{}''(x) = \xi_\lambda^{(q)}{}''(-\log(q)) = 1, \quad x \in \mathbb{R}.$$

If $\max\{0, \beta_\lambda\} < q < \min\{1, e^{\beta_\lambda - 1}\}$ and $(1-q)^2 - 2 \cdot q \cdot (q - \beta_\lambda) < 0$, then a real-valued solution $\overline{Q}_\lambda^{(q)}{}'(x) = x$ does not exist and we set $\overline{x}_0^{(q)} := \overline{\overline{x}}_0^{(q)} := -\log(q)$, with $\xi_\lambda^{(q)}{}'(\overline{\overline{x}}_0^{(q)}) = 1$. The above considerations lead to corresponding unique choices of constants $\underline{a}_\lambda^{(q)}$, $\underline{b}_\lambda^{(q)}$, $\underline{c}_\lambda^{(q)}$, $\overline{a}_\lambda^{(q)}$, $\overline{b}_\lambda^{(q)}$, $\overline{c}_\lambda^{(q)}$ culminating in

$$\underline{Q}_\lambda^{(q)}(x) := \begin{cases} \frac{q}{2} \cdot e^{\underline{x}_0^{(q)}} \cdot x^2 + q \cdot x + q - \beta_\lambda, & \text{if } 0 < q < \beta_\lambda, \\[2mm] \frac{q}{2} \cdot x^2 + q \cdot x + q - \beta_\lambda, & \text{if } \max\{0, \beta_\lambda\} < q < \min\{1, e^{\beta_\lambda - 1}\}, \end{cases} \tag{127}$$

$$\overline{Q}_\lambda^{(q)}(x) := \begin{cases} \frac{q}{2} \cdot x^2 + q \cdot x + q - \beta_\lambda, & \text{if } 0 < q < \beta_\lambda, \\[2mm] \frac{1}{2} \cdot x^2 + q \cdot x + q - \beta_\lambda, & \text{if } \max\{0, \beta_\lambda\} < q < \min\{1, e^{\beta_\lambda - 1}\}. \end{cases} \tag{128}$$

6.6. Closed-Form Bounds for Power Divergences of Non-Kullback-Leibler-Information-Divergence Type

Analogously to Section 4 (see especially Section 4.1), for orders $\lambda \in \mathbb{R} \setminus \{0, 1\}$ all the results of the previous Sections 6.1–6.5 carry correspondingly over from closed-form bounds of the Hellinger integrals $H_\lambda(\cdot||\cdot)$ to closed-form bounds of the total variation distance $V(\cdot||\cdot)$, by virtue of the relation (cf. (12))

$$2\left(1 - H_{\frac{1}{2}}(P_{\mathcal{A},n}||P_{\mathcal{H},n})\right) \leq V(P_{\mathcal{A},n}||P_{\mathcal{H},n}) \leq 2\sqrt{1 - \left(H_{\frac{1}{2}}(P_{\mathcal{A},n}||P_{\mathcal{H},n})\right)^2},$$

to closed-form bounds of the Renyi divergences $R_\lambda(\cdot||\cdot)$, by virtue of the relation (cf. (7))

$$0 \leq R_\lambda\left(P_{\mathcal{A},n}||P_{\mathcal{H},n}\right) = \frac{1}{\lambda(\lambda-1)} \log H_\lambda\left(P_{\mathcal{A},n}||P_{\mathcal{H},n}\right), \qquad \text{with } \log 0 := -\infty,$$

as well as to closed-form bounds of the power divergences $I_\lambda(\cdot||\cdot)$, by virtue of the relation (cf. (2))

$$I_\lambda\left(P_{\mathcal{A},n}||P_{\mathcal{H},n}\right) = \frac{1 - H_\lambda(P_{\mathcal{A},n}||P_{\mathcal{H},n})}{\lambda\cdot(1-\lambda)}, \qquad n \in \mathbb{N}.$$

For the sake of brevity, the–merely repetitive–exact details are omitted.

6.7. Applications to Decision Making

The above-mentioned investigations of the Sections 6.1 to 6.6 can be applied to the context of Section 2.5 on *dichotomous* decision making on the space of all possible path scenarios (path space) of Poissonian Galton-Watson processes without (with) immigration GW(I) (e.g., in combination with our running-example epidemiological context of Section 2.3). More detailed, for the minimal mean decision loss (Bayes risk) $\mathcal{R}_n$ defined by (18) we can derive explicit closed-form upper (respectively lower) bounds by using (19) respectively (20) together with the results of the Sections 6.1–6.5 concerning Hellinger integrals of order $\lambda \in\ ]0,1[$; we can proceed analogously in the Neyman-Pearson context in order to deduce closed-form bounds of type II error probabilities, by means of (23) and (24). Moreover, in an analogous way we can employ the investigations of Section 6.6 on power divergences in order to obtain closed-form bounds of (i) the corresponding (cf. (21)) *weighted-average* decision risk reduction (weighted-average statistical information measure) about the degree of evidence $\mathfrak{deg}$ concerning the parameter θ that can be attained by observing the GW(I)-path $\mathcal{X}_n$ until stage n, as well as (ii) the corresponding (cf. (22)) *limit* decision risk reduction (limit statistical information measure). For the sake of brevity, the–merely repetitive–exact details are omitted.

7. Hellinger Integrals and Power Divergences of Galton-Watson Type Diffusion Approximations

7.1. Branching-Type Diffusion Approximations

One can show that a properly rescaled Galton-Watson process without (respectively with) immigration GW(I) converges weakly to a diffusion process $\widetilde{X} := \left\{\widetilde{X}_s, s \in [0,\infty[\right\}$ which is the unique, strong, nonnegative – and in case of $\frac{\eta}{\sigma^2} \geq \frac{1}{2}$ strictly positive– solution of the stochastic differential equation (SDE) of the form

$$d\widetilde{X}_s = \left(\eta - \kappa\,\widetilde{X}_s\right)ds + \sigma\sqrt{\widetilde{X}_s}\,dW_s, \qquad s \in [0,\infty[, \qquad \widetilde{X}_0 \in\ ]0,\infty[\text{ given}, \tag{129}$$

where $\eta \in [0,\infty[$, $\kappa \in [0,\infty[$, $\sigma \in\ ]0,\infty[$ are constants and $\{W_s,\ s \in [0,\infty[\}$ denotes a standard Brownian motion with respect to the underlying probability measure P; see e.g., Feller [130], Jirina [131], Lamperti [132,133], Lindvall [134,135], Grimvall [136], Jagers [56], Borovkov [137], Ethier & Kurtz [138], Durrett [139] for the non-immigration case corresponding to $\eta = 0$, $\kappa \geq 0$, Kawazu & Watanabe [140], Wei & Winnicki [141], Winnicki [64] for the immigration case corresponding to $\eta \neq 0$, $\kappa = 0$, as well as Sriram [142] for the general case $\eta \in [0,\infty[$, $\kappa \in \mathbb{R}$. Feller-type branching processes of the form (129), which are special cases of continuous state branching processes with immigration (see e.g., Kawazu & Watanabe [140], Li [143], as well as Dawson & Li [144] for imbeddings to affine processes) play for instance an important role in the modelling of the term structure of interest rates, cf. the seminal Cox-Ingersoll-Ross CIR model [145] and the vast follow-up literature thereof. Furthermore, (129) is also prominently used as (a special case of) Cox & Ross's [146] constant elasticity of variance CEV asset price process, as (part of) Heston's [147] stochastic asset-volatility framework, as a model of neuron activity (see e.g., Lansky & Lanska [148], Giorno et al. [149], Lanska et al. [150], Lansky et al [151], Ditlevsen & Lansky [152], Höpfner [153], Lansky & Ditlevsen [154]), as a time-dynamic description of the nitrous oxide emission rate from the soil surface (see e.g., Pedersen [155]), as well as a model for the individual hazard rate in a survival analysis context (see e.g., Aalen & Gjessing [156]).

Along these lines of branching-type diffusion limits, it makes sense to consider the solutions of two SDEs (129) with different fixed parameter sets $(\eta, \kappa_\mathcal{A}, \sigma)$ and $(\eta, \kappa_\mathcal{H}, \sigma)$, determine for each of them a corresponding approximating GW(I), investigate the Hellinger integral between the laws of these two GW(I), and finally calculate the limit of the Hellinger integral (bounds) as the GW(I) approach their SDE solutions. Notice that for technicality reasons (which will be explained below), the constants η and σ ought to be independent of $\mathcal{A}$, $\mathcal{H}$ in our current context.

In order to make the above-mentioned limit procedure rigorous, it is reasonable to work with appropriate approximations such that in each convergence step m one faces the setup $\mathcal{P}_{\mathrm{NI}} \cup \mathcal{P}_{\mathrm{SP},1}$ (i.e., the non-immigration or the equal-fraction case), where the corresponding Hellinger integral can be calculated exactly in a recursive way, as stated in Theorem 1. Let us explain the details in the following.

Consider a sequence of GW(I) $\left(X^{(m)}\right)_{m \in \mathbb{N}}$ with probability laws $P_\bullet^{(m)}$ on a measurable space $(\Omega, \mathcal{F})$, where as above the subscript $\bullet$ stands for either the hypothesis $\mathcal{H}$ or the alternative $\mathcal{A}$. Analogously to (1), we use for each fixed step $m \in \mathbb{N}$ the representation $X^{(m)} := \left\{ X_\ell^{(m)}, \ell \in \mathbb{N} \right\}$ with

$$X_\ell^{(m)} := \sum_{j=1}^{X_{\ell-1}^{(m)}} Y_{\ell-1,j}^{(m)} + \widetilde{Y}_\ell^{(m)}, \qquad \ell \in \mathbb{N}, \qquad X_0^{(m)} \in \mathbb{N} \text{ given}, \tag{130}$$

where under the law $P_\bullet^{(m)}$

- the collection $Y^{(m)} := \left\{ Y_{i,j}^{(m)}, i \in \mathbb{N}_0, j \in \mathbb{N} \right\}$ consists of i.i.d. random variables which are Poisson distributed with parameter $\beta_\bullet^{(m)} > 0$,
- the collection $\widetilde{Y}^{(m)} := \left\{ \widetilde{Y}_i^{(m)}, i \in \mathbb{N} \right\}$ consists of i.i.d. random variables which are Poisson distributed with parameter $\alpha_\bullet^{(m)} \geq 0$,
- $Y^{(m)}$ and $\widetilde{Y}^{(m)}$ are independent.

From arbitrary drift-parameters $\eta \in [0, \infty[$, $\kappa_\bullet \in [0, \infty[$, and diffusion-term-parameter $\sigma > 0$, we construct the offspring-distribution-parameter and the immigration-distribution parameter of the sequence $\left(X_\ell^{(m)}\right)_{\ell \in \mathbb{N}}$ by

$$\beta_\bullet^{(m)} := 1 - \frac{\kappa_\bullet}{\sigma^2 m} \qquad \text{and} \qquad \alpha_\bullet^{(m)} := \beta_\bullet^{(m)} \cdot \frac{\eta}{\sigma^2}. \tag{131}$$

Here and henceforth, we always assume that the approximation step m is large enough to ensure that $\beta_\bullet^{(m)} \in]0, 1]$ and at least one of $\beta_\mathcal{A}^{(m)}$, $\beta_\mathcal{H}^{(m)}$ is strictly less than 1; this will be abbreviated by $m \in \overline{\mathbb{N}}$. Let us point out that – as mentioned above–our choice entails the best-to-handle setup $\mathcal{P}_{\mathrm{NI}} \cup \mathcal{P}_{\mathrm{SP},1}$ (which does not happen if instead of η one uses $\eta_\bullet$ with $\eta_\mathcal{A} \neq \eta_\mathcal{H}$). Based on the GW(I) $X^{(m)}$, let us construct the *continuous-time* branching process $\widetilde{X}^{(m)} := \left\{ \widetilde{X}_s^{(m)}, s \in [0, \infty[\right\}$ by

$$\widetilde{X}_s^{(m)} := \frac{1}{m} X_{\lfloor \sigma^2 m s \rfloor}^{(m)}, \tag{132}$$

living on the state space $E^{(m)} := \frac{1}{m}\mathbb{N}_0$. Notice that $\widetilde{X}^{(m)}$ is constant on each time-interval $\left[\frac{k}{\sigma^2 m}, \frac{k+1}{\sigma^2 m}\right[$ and takes at $s = \frac{k}{\sigma^2 m}$ the value $\frac{1}{m}X_k^{(m)}$ of the k-th GW(I) generation size, divided by m, i.e., it "jumps" with the jump-size $\frac{1}{m}\left(X_k^{(m)} - X_{k-1}^{(m)}\right)$ which is equal to the $\frac{1}{m}$-fold difference to the previous generation size. From (132) one can immediately see the necessity of having σ to be independent of $\mathcal{A}, \mathcal{H}$ because for the required law-equivalence in (the corresponding version of) (13) both models at stake have to "live" on the same time-scale $\tau_s^{(m)} := \lfloor \sigma^2 ms \rfloor$. For this setup, one obtains the following convergenc result:

Theorem 10. *Let $\eta \in [0,\infty[$, $\kappa_\bullet \in [0,\infty[$, $\sigma \in]0,\infty[$ and $\widetilde{X}^{(m)}$ be as defined in (130) to (132). Furthermore, let us suppose that $\lim_{m\to\infty} \frac{1}{m} X_0^{(m)} = \widetilde{X}_0 > 0$ and denote by $D([0,\infty[, [0,\infty[)$ the space of right-continuous functions $f : [0,\infty[\mapsto [0,\infty[$ with left limits. Then the sequence of processes $\left(\widetilde{X}^{(m)}\right)_{m\in\overline{\mathbb{N}}}$ convergences in distribution in $D([0,\infty[, [0,\infty[)$ to a diffusion process $\widetilde{X}$ which is the unique strong, nonnegative–and in case of $\frac{\eta}{\sigma^2} \geq \frac{1}{2}$ strictly positive–solution of the SDE*

$$d\widetilde{X}_s = \left(\eta - \kappa_\bullet \widetilde{X}_s\right) ds + \sigma\sqrt{\widetilde{X}_s}\, dW_s^\bullet, \quad s \in [0,\infty[, \qquad \widetilde{X}_0 \in]0,\infty[\text{ given}, \tag{133}$$

where $\{W_\sigma^\bullet, s \in [0,\infty[\}$ denotes a standard Brownian motion with respect to the limit probability measure $\widetilde{P}_\bullet$.

Remark 8. *Notice that the condition $\frac{\eta}{\sigma^2} \geq \frac{1}{2}$ can be interpreted in our approximation setup (131) as $\alpha_\bullet^{(m)} \geq \beta_\bullet^{(m)}/2$, which quantifies the intuitively reasonable indication that if the probability $P_\bullet[\widetilde{Y}_\ell^{(m)} = 0] = e^{-\alpha_\bullet^{(m)}}$ of having no immigration is small enough relative to the probability $P_\bullet[Y_{\ell-1,k}^{(m)} = 0] = e^{-\beta_\bullet^{(m)}}$ of having no offspring ($m \in \overline{\mathbb{N}}$), then the limiting diffusion $\widetilde{X}$ never hits zero almost surely.*

The corresponding proof of Theorem 10–which is outlined in Appendix A.4–is an adaption of the proof of Theorem 9.1.3 in Ethier & Kurtz [138] which deals with drift-parameters $\eta = 0$, $\kappa_\bullet = 0$ in the SDE (133) whose solution is approached on a σ–independent time scale by a sequence of (critical) Galton-Watson processes without immigration but with general offspring distribution with mean 1 and variance σ. Notice that due to (131) the latter is inconsistent with our Poissonian setup, but this is compensated by our chosen σ–dependent time scale. Other limit investigations for (133) involving offspring/immigration distributions and parametrizations which are also incompatible to ours, are e.g., treated in Sriram [142].

As illustration of our proposed approach, let us give the following

Example 3. *Consider the parameter setup $(\eta, \kappa_\bullet, \sigma) = (5, 2, 0.4)$ and initial generation size $\widetilde{X}_0 = 3$. Figure 4 shows the diffusion-approximation $\widetilde{X}_s^{(m)}$ (blue) of the corresponding solution $\widetilde{X}_s$ of the SDE (133) up to the time horizon $T = 10$, for the approximation steps $m \in \{13, 50, 200, 1000\}$. Notice that in this setup there holds $\overline{\mathbb{N}} = \{k \in \mathbb{N} : k \geq 13\}$ (recall that $\overline{\mathbb{N}}$ is the subset of the positive integers such that $\beta_\bullet^{(m)} = 1 - \frac{\kappa_\bullet}{\sigma^2 \cdot m} > 0$). The "long-term mean" of the limit process $\widetilde{X}_s$ is $\frac{\eta}{\kappa_\bullet} = 2.5$ and is indicated as red line. The "long-term mean" of the approximations $\widetilde{X}_s^{(m)}$ is equal to $\frac{\alpha_\bullet^{(m)}}{1-\beta_\bullet^{(m)}} = \frac{\eta}{\kappa_\bullet} - \frac{\eta}{\sigma^2 \cdot m} = 2.5 - 31.25/m$ and is displayed as green line.*

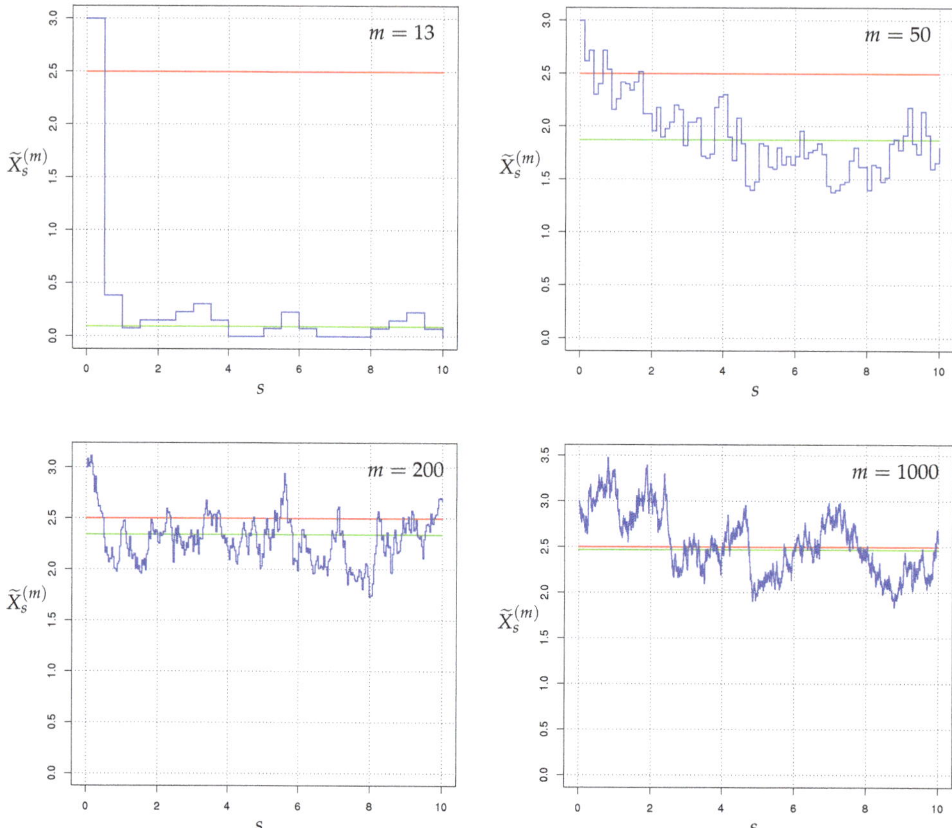

Figure 4. Simulation of the process $\widetilde{X}_s^{(m)}$ for the approximation steps $m \in \{13, 50, 200, 1000\}$ in the parameter setup $(\eta, \kappa_\bullet, \sigma) = (5, 2, 0.4)$ and with initial starting value $\widetilde{X}_0 = 3$.

7.2. Bounds of Hellinger Integrals for Diffusion Approximations

For each approximation step m and each observation horizon $t \in [0, \infty[$, let us now investigate the behaviour of the Hellinger integrals $H_\lambda \left(P_{\mathcal{A},t}^{(m),CDA} \middle\| P_{\mathcal{H},t}^{(m),CDA} \right)$, where $P_{\bullet,t}^{(m),CDA}$ denotes the canonical law (under $\mathcal{H}$ resp. $\mathcal{A}$) of the <u>c</u>ontinuous-time <u>d</u>iffusion <u>a</u>pproximation $\widetilde{X}^{(m)}$ (cf. (132)), restricted to $[0, t]$. It is easy to see that $H_\lambda \left(P_{\mathcal{A},t}^{(m),CDA} \middle\| P_{\mathcal{H},t}^{(m),CDA} \right)$ coincides with $H_\lambda \left(P_{\mathcal{A},\lfloor \sigma^2 mt \rfloor}^{(m)} \middle\| P_{\mathcal{H},\lfloor \sigma^2 mt \rfloor}^{(m)} \right)$ of the law restrictions of the GW(I) generations sizes $\left(X_\ell^{(m)} \right)_{\ell \in \{0,\dots,\lfloor \sigma^2 mt \rfloor\}}$, where $\frac{\lfloor \sigma^2 mt \rfloor}{\sigma^2 m}$ can be interpreted as the last "jump-time" of $\widetilde{X}^{(m)}$ before t. These Hellinger integrals obey the results of

- the Propositions 2 and 3 (for $\eta = 0$) respectively the Propositions 4 and 5 (for $\eta \in]0, \infty[$), as far as recursively computable exact values are concerned,
- Theorem 5 as far as closed-form bounds are concerned; recall that the current setup is of type $\mathcal{P}_{NI} \cup \mathcal{P}_{SP,1}$, and thus we can use the simplifications proposed in the Remark 7(a).

In order to obtain the desired Hellinger integral limits $\lim_{m \to \infty} H_\lambda \left(P_{\mathcal{A},\lfloor \sigma^2 mt \rfloor}^{(m)} \middle\| P_{\mathcal{H},\lfloor \sigma^2 mt \rfloor}^{(m)} \right)$, one faces several technical problems which will be described in the following. To begin with, for fixed

$m \in \overline{\mathbb{N}}$ we apply the Propositions 2(b), 3(b), 4(b), 5(b) to the current setup $\left(\beta_{\mathcal{A}}^{(m)}, \beta_{\mathcal{H}}^{(m)}, \alpha_{\mathcal{A}}^{(m)}, \alpha_{\mathcal{H}}^{(m)}\right) \in \mathcal{P}_{NI} \cup \mathcal{P}_{SP,1}$ with

$$\beta_{\bullet}^{(m)} := \beta_{\bullet}(m, \kappa_{\bullet}, \sigma^2) := 1 - \frac{\kappa_{\bullet}}{\sigma^2 m} \quad \text{and} \quad \alpha_{\bullet}^{(m)} := \alpha_{\bullet}(m, \kappa_{\bullet}, \sigma^2, \eta) := \beta_{\bullet}^{(m)} \cdot \frac{\eta}{\sigma^2} \quad \text{(cf. (131)).}$$

Notice that $\eta = 0$ corresponds to the no-immigration (NI) case and that $\frac{\alpha_{\bullet}^{(m)}}{\beta_{\bullet}^{(m)}} = \frac{\eta}{\sigma^2}$. Accordingly, we set $\alpha_{\lambda}^{(m)} := \lambda \cdot \alpha_{\mathcal{A}}^{(m)} + (1 - \lambda) \cdot \alpha_{\mathcal{H}}^{(m)}$, $\beta_{\lambda}^{(m)} := \lambda \cdot \beta_{\mathcal{A}}^{(m)} + (1 - \lambda) \cdot \beta_{\mathcal{H}}^{(m)}$. By using

$$q_{\lambda}^{(m)} := q(m, \kappa_{\bullet}, \sigma^2, \lambda) := \left(\beta_{\mathcal{A}}^{(m)}\right)^{\lambda} \left(\beta_{\mathcal{H}}^{(m)}\right)^{1-\lambda}, \qquad \lambda \in \mathbb{R}\backslash\{0, 1\}, \tag{134}$$

as well as the connected sequence $\left(a_n^{(m)}\right)_{n \in \mathbb{N}} := \left(a_n^{\left(q_{\lambda}^{(m)}\right)}\right)_{n \in \mathbb{N}}$ we arrive at the

Corollary 13. *For all* $\left(\beta_{\mathcal{A}}^{(m)}, \beta_{\mathcal{H}}^{(m)}, \alpha_{\mathcal{A}}^{(m)}, \alpha_{\mathcal{H}}^{(m)}, \lambda\right) \in (\mathcal{P}_{NI} \cup \mathcal{P}_{SP,1}) \times (\mathbb{R}\backslash\{0, 1\})$ *and all population sizes* $X_0^{(m)} \in \mathbb{N}$ *there holds*

$$h_{\lambda}^{(m)} := H_{\lambda}\left(P_{\mathcal{A}, \lfloor \sigma^2 mt \rfloor}^{(m)} \middle|\middle| P_{\mathcal{H}, \lfloor \sigma^2 mt \rfloor}^{(m)}\right) = \exp\left\{a_{\lfloor \sigma^2 mt \rfloor}^{\left(q_{\lambda}^{(m)}\right)} \cdot X_0^{(m)} + \frac{\eta}{\sigma^2} \sum_{k=1}^{\lfloor \sigma^2 mt \rfloor} a_k^{\left(q_{\lambda}^{(m)}\right)}\right\} \tag{135}$$

with $\eta = 0$ *in the NI case.*

In the following, we employ the SDE-parameter constellations (which are consistent with (131) in combination with our requirement to work here only on $(\mathcal{P}_{NI} \cup \mathcal{P}_{SP,1})$)

$$\widetilde{\mathcal{P}}_{NI} := \left\{(\kappa_{\mathcal{A}}, \kappa_{\mathcal{H}}, \eta), \eta = 0, \kappa_{\mathcal{A}} \in [0, \infty[, \kappa_{\mathcal{H}} \in [0, \infty[, \kappa_{\mathcal{A}} \neq \kappa_{\mathcal{H}}\right\}, \tag{136}$$

$$\widetilde{\mathcal{P}}_{SP,1} := \left\{(\kappa_{\mathcal{A}}, \kappa_{\mathcal{H}}, \eta), \eta > 0, \kappa_{\mathcal{A}} \in [0, \infty[, \kappa_{\mathcal{H}} \in [0, \infty[, \kappa_{\mathcal{A}} \neq \kappa_{\mathcal{H}}\right\}. \tag{137}$$

Due to the–not in closed-form representable–recursive nature of the sequences $\left(a_n^{(q)}\right)_{n \in \mathbb{N}}$ defined by (36), the calculation of $\lim_{m \to \infty} h_{\lambda}^{(m)}$ in (135) seems to be not (straightforwardly) tractable; after all, one "has to move along" a *sequence* of recursions (roughly speaking) since $\lfloor \sigma^2 mt \rfloor \to \infty$ as m tends to infinity. One way to "circumvent" such technical problems is to compute instead of the limit $\lim_{m \to \infty} h_{\lambda}^{(m)}$ of the (exact values of the) Hellinger integrals $h_{\lambda}^{(m)}$, the limits of the corresponding (explicit) closed-form lower resp. upper bounds adapted from Theorem 5. In order to achieve this, one first needs a preparatory step, due to the fact that the sequence $\left(a_{\lfloor \sigma^2 mt \rfloor}^{\left(q_{\lambda}^{(m)}\right)}\right)_{m \in \overline{\mathbb{N}}}$ (and hence its bounds leading to closed-form expressions) does not necessarily converge for all $\lambda \in \mathbb{R}\backslash[0, 1]$; roughly, this can be conjectured from the Propositions 3(c) and 5(c) in combination with $\lfloor \sigma^2 mt \rfloor \to \infty$. Correspondingly, for our "sequence-of-recursions" context equipped with the diffusion-limit's drift-parameter constellations $(\kappa_{\mathcal{A}}, \kappa_{\mathcal{H}}, \eta)$ we have to derive a "convergence interval" $[\widetilde{\lambda}_-, \widetilde{\lambda}_+]\backslash[0, 1]$ which replaces the single-recursion-concerning $[\lambda_-, \lambda_+]\backslash[0, 1]$ (cf. Lemma 1). This amounts to

Proposition 15. *For all* $(\kappa_{\mathcal{A}}, \kappa_{\mathcal{H}}, \eta) \in \widetilde{\mathcal{P}}_{NI} \cup \widetilde{\mathcal{P}}_{SP,1}$ *define*

$$0 > \widetilde{\lambda}_- := \begin{cases} -\infty, & \text{if } \kappa_{\mathcal{A}} < \kappa_{\mathcal{H}}, \\ -\frac{\kappa_{\mathcal{H}}^2}{\kappa_{\mathcal{A}}^2 - \kappa_{\mathcal{H}}^2}, & \text{if } \kappa_{\mathcal{A}} > \kappa_{\mathcal{H}}, \end{cases} \quad \text{and} \quad 1 < \widetilde{\lambda}_+ := \begin{cases} \frac{\kappa_{\mathcal{H}}^2}{\kappa_{\mathcal{H}}^2 - \kappa_{\mathcal{A}}^2}, & \text{if } \kappa_{\mathcal{A}} < \kappa_{\mathcal{H}}, \\ \infty, & \text{if } \kappa_{\mathcal{A}} > \kappa_{\mathcal{H}}. \end{cases} \tag{138}$$

Then, for all $(\kappa_{\mathcal{A}}, \kappa_{\mathcal{H}}, \eta, \lambda) \in (\widetilde{\mathcal{P}}_{NI} \cup \widetilde{\mathcal{P}}_{SP,1}) \times]\tilde{\lambda}_-, \tilde{\lambda}_+[\setminus [0,1]$ there holds for all sufficiently large $m \in \overline{\mathbb{N}}$

$$q_\lambda^{(m)} := \left(1 - \frac{\kappa_{\mathcal{A}}}{\sigma^2 m}\right)^\lambda \left(1 - \frac{\kappa_{\mathcal{H}}}{\sigma^2 m}\right)^{1-\lambda} < \min\left\{1, e^{\beta_\lambda^{(m)} - 1}\right\}, \tag{139}$$

and thus the sequence $\left(a_n^{(q_\lambda^{(m)})}\right)_{n \in \mathbb{N}}$ converges to the fixed point $x_0^{(m)} \in \left]0, -\log\left(q_\lambda^{(m)}\right)\right[$.

This will be proved in Appendix A.4.

We are now in the position to determine bounds of the Hellinger integral limits $\lim_{m \to \infty} H_\lambda\left(P_{\mathcal{A}, \lfloor \sigma^2 mt \rfloor}^{(m)} \middle\| P_{\mathcal{H}, \lfloor \sigma^2 mt \rfloor}^{(m)}\right)$ in form of m-limits of appropriate versions of closed-form bounds

from Section 6. For the sake of brevity, let us henceforth use the abbreviations $x_0^{(m)} := x_0^{(q_\lambda^{(m)})}$, $\Gamma_<^{(m)} := \Gamma_<^{(q_\lambda^{(m)})} = \frac{q_\lambda^{(m)}}{2} \cdot e^{x_0^{(m)}} \cdot \left(x_0^{(m)}\right)^2$, $\Gamma_>^{(m)} := \Gamma_>^{(q_\lambda^{(m)})} = \frac{q_\lambda^{(m)}}{2} \cdot \left(x_0^{(m)}\right)^2$, $d^{(m),S} := d^{(q_\lambda^{(m)}),S} = \frac{x_0^{(m)} - (q_\lambda^{(m)} - \beta_\lambda^{(m)})}{x_0^{(m)}}$

and $d^{(m),T} := d^{(q_\lambda^{(m)}),T} = q_\lambda^{(m)} \cdot e^{x_0^{(m)}}$. By the above considerations, the Theorem 5 (together with Remark 7(a)) adapts to the current setup as follows:

Corollary 14. *(a) For all $(\kappa_{\mathcal{A}}, \kappa_{\mathcal{H}}, \eta, \lambda) \in (\widetilde{\mathcal{P}}_{NI} \cup \widetilde{\mathcal{P}}_{SP,1}) \times]0,1[$, all $t \in [0, \infty[$, all approximation steps $m \in \overline{\mathbb{N}}$ and all initial population sizes $X_0^{(m)} \in \mathbb{N}$ the Hellinger integral can be bounded by*

$$\begin{aligned}
C_{\lambda, X_0^{(m)}, t}^{(m), L} &:= \exp\left\{x_0^{(m)} \cdot \left[X_0^{(m)} - \frac{\eta}{\sigma^2} \frac{d^{(m),T}}{1 - d^{(m),T}}\right]\left(1 - \left(d^{(m),T}\right)^{\lfloor \sigma^2 mt \rfloor}\right) + x_0^{(m)} \frac{\eta}{\sigma^2} \cdot \lfloor \sigma^2 mt \rfloor \right. \\
&\qquad \left. + \underline{\zeta}_{\lfloor \sigma^2 mt \rfloor}^{(m)} \cdot X_0^{(m)} + \frac{\eta}{\sigma^2} \cdot \underline{\vartheta}_{\lfloor \sigma^2 mt \rfloor}^{(m)}\right\} \tag{140} \\[1ex]
&\leq H_\lambda\left(P_{\mathcal{A}, \lfloor \sigma^2 mt \rfloor}^{(m)} \middle\| P_{\mathcal{H}, \lfloor \sigma^2 mt \rfloor}^{(m)}\right) \\[1ex]
&\leq \exp\left\{x_0^{(m)} \cdot \left[X_0^{(m)} - \frac{\eta}{\sigma^2} \frac{d^{(m),S}}{1 - d^{(m),S}}\right]\left(1 - \left(d^{(m),S}\right)^{\lfloor \sigma^2 mt \rfloor}\right) + x_0^{(m)} \frac{\eta}{\sigma^2} \cdot \lfloor \sigma^2 mt \rfloor \right. \\
&\qquad \left. - \overline{\zeta}_{\lfloor \sigma^2 mt \rfloor}^{(m)} \cdot X_0^{(m)} - \frac{\eta}{\sigma^2} \cdot \overline{\vartheta}_{\lfloor \sigma^2 mt \rfloor}^{(m)}\right\} =: C_{\lambda, X_0^{(m)}, t}^{(m), U}, \tag{141}
\end{aligned}$$

where we define analogously to (98) to (101)

$$\underline{\zeta}_n^{(m)} := \Gamma_<^{(m)} \cdot \frac{\left(d^{(m),T}\right)^{n-1}}{1 - d^{(m),T}} \cdot \left(1 - \left(d^{(m),T}\right)^n\right) > 0, \tag{142}$$

$$\underline{\vartheta}_n^{(m)} := \Gamma_<^{(m)} \cdot \frac{1 - \left(d^{(m),T}\right)^n}{\left(1 - d^{(m),T}\right)^2} \cdot \left[1 - \frac{d^{(m),T}\left(1 + \left(d^{(m),T}\right)^n\right)}{1 + d^{(m),T}}\right] > 0, \tag{143}$$

$$\overline{\zeta}_n^{(m)} := \Gamma_<^{(m)} \cdot \left[\frac{\left(d^{(m),S}\right)^n - \left(d^{(m),T}\right)^n}{d^{(m),S} - d^{(m),T}} - \left(d^{(m),S}\right)^{n-1} \cdot \frac{1 - \left(d^{(m),T}\right)^n}{1 - d^{(m),T}}\right] > 0, \tag{144}$$

$$\overline{\vartheta}_n^{(m)} := \Gamma_<^{(m)} \cdot \frac{d^{(m),T}}{1 - d^{(m),T}} \cdot \left[\frac{1 - \left(d^{(m),S} d^{(m),T}\right)^n}{1 - d^{(m),S} d^{(m),T}} - \frac{\left(d^{(m),S}\right)^n - \left(d^{(m),T}\right)^n}{d^{(m),S} - d^{(m),T}}\right] > 0. \tag{145}$$

Notice that (140) and (141) simplify significantly for $(\kappa_{\mathcal{A}}, \kappa_{\mathcal{H}}, \eta, \lambda) \in \widetilde{\mathcal{P}}_{NI} \times]0,1[$ for which $\eta = 0$ holds.
(b) For all $(\kappa_{\mathcal{A}}, \kappa_{\mathcal{H}}, \eta, \lambda) \in (\widetilde{\mathcal{P}}_{NI} \cup \widetilde{\mathcal{P}}_{SP,1}) \times]\tilde{\lambda}_-, \tilde{\lambda}_+[\setminus [0,1]$ and all initial population sizes $X_0^{(m)} \in \mathbb{N}$ the

Hellinger integral bounds (140) *and* (141) *are valid for all sufficiently large* $m \in \overline{\mathbb{N}}$, *where the expressions* (142) *to* (145) *have to be replaced by*

$$\underline{\zeta}_n^{(m)} \;:=\; \Gamma_>^{(m)} \cdot \frac{\left(d^{(m),T}\right)^n - \left(d^{(m),S}\right)^{2n}}{d^{(m),T} - \left(d^{(m),S}\right)^2} \;>\; 0\,, \tag{146}$$

$$\underline{\vartheta}_n^{(m)} \;:=\; \frac{\Gamma_>^{(m)}}{d^{(m),T} - \left(d^{(m),S}\right)^2} \cdot \left[\frac{d^{(m),T} \cdot \left(1 - \left(d^{(m),T}\right)^n\right)}{1 - d^{(m),T}} - \frac{\left(d^{(m),S}\right)^2 \cdot \left(1 - \left(d^{(m),S}\right)^{2n}\right)}{1 - \left(d^{(m),S}\right)^2}\right] \;>\; 0\,,$$

$$\overline{\zeta}_n^{(m)} \;:=\; \Gamma_>^{(m)} \cdot \left(d^{(m),S}\right)^{n-1} \cdot \left[n - \frac{1 - \left(d^{(m),T}\right)^n}{1 - d^{(m),T}}\right] \;>\; 0\,, \tag{147}$$

$$\overline{\vartheta}_n^{(m)} \;:=\; \Gamma_>^{(m)} \cdot \left[\frac{d^{(m),S} - d^{(m),T}}{(1 - d^{(m),S})^2\,(1 - d^{(m),T})} \cdot \left(1 - \left(d^{(m),S}\right)^n\right)\right. \tag{148}$$

$$\left. + \frac{d^{(m),T}\left(1 - \left(d^{(m),S}d^{(m),T}\right)^n\right)}{(1 - d^{(m),T})\,(1 - d^{(m),S}d^{(m),T})} - \frac{\left(d^{(m),S}\right)^n}{1 - d^{(m),S}} \cdot n\right]\,. \tag{149}$$

Let us finally present the desired assertions on the limits of the bounds given in Corollary 14 as the approximation step m tends to infinity, by employing for $\lambda \in\,]\widetilde{\lambda}_-, \widetilde{\lambda}_+[\,\supsetneq\, [0,1]$ the quantities

$$\kappa_\lambda := \lambda \kappa_{\mathcal{A}} + (1-\lambda)\kappa_{\mathcal{H}} \qquad \text{as well as} \qquad \Lambda_\lambda := \sqrt{\lambda \kappa_{\mathcal{A}}^2 + (1-\lambda)\kappa_{\mathcal{H}}^2}\,, \tag{150}$$

for which the following relations hold:

$$\Lambda_\lambda > \kappa_\lambda > 0, \qquad \text{for} \quad \lambda \in\,]0,1[, \tag{151}$$

$$0 < \Lambda_\lambda < \kappa_\lambda, \qquad \text{for} \quad \lambda \in\,]\widetilde{\lambda}_-, \widetilde{\lambda}_+[\setminus [0,1]\,. \tag{152}$$

Theorem 11. *Let the initial SDE-value* $\widetilde{X}_0 \in]0, \infty[$ *be arbitrary but fixed, and suppose that* $\lim_{m \to \infty} \frac{1}{m} X_0^{(m)} = \widetilde{X}_0$. *Then, for all* $(\kappa_{\mathcal{A}}, \kappa_{\mathcal{H}}, \eta, \lambda) \in (\widetilde{\mathcal{P}}_{NI} \cup \widetilde{\mathcal{P}}_{SP,1}) \times\,]\widetilde{\lambda}_-, \widetilde{\lambda}_+[\setminus \{0,1\}$ *and all* $t \in [0, \infty[$ *the Hellinger integral limit can be bounded by*

$$
\begin{aligned}
D_{\lambda,\widetilde{X}_0,t}^L \;:=\; \exp&\left\{ -\frac{\Lambda_\lambda - \kappa_\lambda}{\sigma^2}\left[\widetilde{X}_0 - \frac{\eta}{\Lambda_\lambda}\right]\left(1 - e^{-\Lambda_\lambda \cdot t}\right) - \frac{\eta}{\sigma^2}\left(\Lambda_\lambda - \kappa_\lambda\right) \cdot t \right. \\
&\left. + L_\lambda^{(1)}(t) \cdot \widetilde{X}_0 + \frac{\eta}{\sigma^2} \cdot L_\lambda^{(2)}(t) \right\} \tag{153} \\
\leq\; &\lim_{m \to \infty} H_\lambda\left(P_{\mathcal{A},\lfloor \sigma^2 mt\rfloor}^{(m)} \,\big\|\, P_{\mathcal{H},\lfloor \sigma^2 mt\rfloor}^{(m)}\right) \\
\leq\; \exp&\left\{ -\frac{\Lambda_\lambda - \kappa_\lambda}{\sigma^2}\left[\widetilde{X}_0 - \frac{\eta}{\frac{1}{2}(\Lambda_\lambda + \kappa_\lambda)}\right]\left(1 - e^{-\frac{1}{2}(\Lambda_\lambda + \kappa_\lambda)\cdot t}\right) - \frac{\eta}{\sigma^2}\left(\Lambda_\lambda - \kappa_\lambda\right) \cdot t \right. \\
&\left. - U_\lambda^{(1)}(t) \cdot \widetilde{X}_0 - \frac{\eta}{\sigma^2} \cdot U_\lambda^{(2)}(t) \right\} \;=:\; D_{\lambda,\widetilde{X}_0,t}^U\,, \tag{154}
\end{aligned}
$$

where for the (sub)case of all $\lambda \in\,]0,1[$ and all $t \geq 0$

$$L_\lambda^{(1)}(t) \;:=\; \frac{(\Lambda_\lambda - \kappa_\lambda)^2}{2\sigma^2 \cdot \Lambda_\lambda} \cdot e^{-\Lambda_\lambda \cdot t} \cdot \left(1 - e^{-\Lambda_\lambda \cdot t}\right), \tag{155}$$

$$L_\lambda^{(2)}(t) \;:=\; \frac{1}{4} \cdot \left(\frac{\Lambda_\lambda - \kappa_\lambda}{\Lambda_\lambda}\right)^2 \cdot \left(1 - e^{-\Lambda_\lambda \cdot t}\right)^2, \tag{156}$$

$$U_\lambda^{(1)}(t) \;:=\; \frac{(\Lambda_\lambda - \kappa_\lambda)^2}{\sigma^2} \cdot \left[\frac{e^{-\frac{1}{2}(\Lambda_\lambda + \kappa_\lambda)\cdot t} - e^{-\Lambda_\lambda \cdot t}}{\Lambda_\lambda - \kappa_\lambda} - \frac{e^{-\frac{1}{2}(\Lambda_\lambda + \kappa_\lambda)\cdot t}\left(1 - e^{-\Lambda_\lambda \cdot t}\right)}{2 \cdot \Lambda_\lambda}\right], \tag{157}$$

$$U_\lambda^{(2)}(t) \;:=\; \frac{(\Lambda_\lambda - \kappa_\lambda)^2}{\Lambda_\lambda} \cdot \left[\frac{1 - e^{-\frac{1}{2}(3\Lambda_\lambda + \kappa_\lambda)\cdot t}}{3\Lambda_\lambda + \kappa_\lambda} + \frac{e^{-\Lambda_\lambda \cdot t} - e^{-\frac{1}{2}(\Lambda_\lambda + \kappa_\lambda)\cdot t}}{\Lambda_\lambda - \kappa_\lambda}\right], \tag{158}$$

and for the remaining (sub)case of all $\lambda \in\,]\widetilde{\lambda}_-, \widetilde{\lambda}_+[\,\setminus\,[0,1]$ and all $t \geq 0$

$$L_\lambda^{(1)}(t) \;:=\; \frac{(\Lambda_\lambda - \kappa_\lambda)^2}{2\sigma^2 \cdot \kappa_\lambda} \cdot e^{-\Lambda_\lambda \cdot t} \cdot \left(1 - e^{-\kappa_\lambda \cdot t}\right), \tag{159}$$

$$L_\lambda^{(2)}(t) \;:=\; \frac{(\Lambda_\lambda - \kappa_\lambda)^2}{2 \cdot \kappa_\lambda} \cdot \left[\frac{1 - e^{-\Lambda_\lambda \cdot t}}{\Lambda_\lambda} - \frac{1 - e^{-(\Lambda_\lambda + \kappa_\lambda)\cdot t}}{\Lambda_\lambda + \kappa_\lambda}\right], \tag{160}$$

$$U_\lambda^{(1)}(t) \;:=\; \frac{(\Lambda_\lambda - \kappa_\lambda)^2}{2 \cdot \sigma^2} \cdot e^{-\frac{1}{2}(\Lambda_\lambda + \kappa_\lambda)\cdot t} \cdot \left[t - \frac{1 - e^{-\Lambda_\lambda \cdot t}}{\Lambda_\lambda}\right], \tag{161}$$

$$U_\lambda^{(2)}(t) \;:=\; (\Lambda_\lambda - \kappa_\lambda)^2 \cdot \left[\frac{(\Lambda_\lambda - \kappa_\lambda)\left(1 - e^{-\frac{1}{2}(\Lambda_\lambda + \kappa_\lambda)\cdot t}\right)}{\Lambda_\lambda \cdot (\Lambda_\lambda + \kappa_\lambda)^2} + \frac{1 - e^{-\frac{1}{2}(3\Lambda_\lambda + \kappa_\lambda)\cdot t}}{\Lambda_\lambda \cdot (3\Lambda_\lambda + \kappa_\lambda)} - \frac{e^{-\frac{1}{2}(\Lambda_\lambda + \kappa_\lambda)\cdot t}}{\Lambda_\lambda + \kappa_\lambda} \cdot t\right]. \tag{162}$$

Notice that the components $L_\lambda^{(i)}(t)$ and $U_\lambda^{(i)}(t)$ (for $i = 1,2$ and in both cases $\lambda \in\,]0,1[$ and $\lambda \in\,]\widetilde{\lambda}_-, \widetilde{\lambda}_+[\,\setminus\,[0,1]$) are strictly positive for $t > 0$ and do not depend on the parameter η. Furthermore, the bounds $D^L_{\lambda, \widetilde{X}_0, t}$ and $D^U_{\lambda, \widetilde{X}_0, t}$ simplify significantly in the case $(\kappa_\mathcal{A}, \kappa_\mathcal{H}, \eta) \in \widetilde{\mathcal{P}}_{NI}$, for which $\eta = 0$ holds.

This will be proved in Appendix A.4. For the time-asymptotics, we obtain the

Corollary 15. *Let the initial SDE-value $\widetilde{X}_0 \in\,]0,\infty[$ be arbitrary but fixed, and suppose that $\lim_{m\to\infty} \frac{1}{m} X_0^{(m)} = \widetilde{X}_0$. Then:*

(a) For all $(\kappa_\mathcal{A}, \kappa_\mathcal{H}, \eta, \lambda) \in \widetilde{\mathcal{P}}_{NI} \times\,]\widetilde{\lambda}_-, \widetilde{\lambda}_+[\,\setminus\{0,1\}$ the Hellinger integral limit converges to

$$\lim_{t\to\infty} \lim_{m\to\infty} \log\left(H_\lambda\left(P^{(m)}_{\mathcal{A},\lfloor \sigma^2 mt\rfloor} \Big\| P^{(m)}_{\mathcal{H},\lfloor \sigma^2 mt\rfloor}\right)\right) = -\frac{\widetilde{X}_0}{\sigma^2} \cdot (\Lambda_\lambda - \kappa_\lambda) \begin{cases} < 0, & \text{for } \lambda \in\,]0,1[\,, \\[2ex] > 0, & \text{for } \lambda \in\,]\widetilde{\lambda}_-, \widetilde{\lambda}_+[\,\setminus\,[0,1]\,. \end{cases}$$

(b) For all $(\kappa_\mathcal{A}, \kappa_\mathcal{H}, \eta, \lambda) \in \widetilde{\mathcal{P}}_{SP,1} \times\,]\widetilde{\lambda}_-, \widetilde{\lambda}_+[\,\setminus\{0,1\}$ the Hellinger integral limit possesses the asymptotical behaviour

$$\lim_{t\to\infty} \frac{1}{t} \log\left(\lim_{m\to\infty} H_\lambda\left(P^{(m)}_{\mathcal{A},\lfloor \sigma^2 mt\rfloor} \Big\| P^{(m)}_{\mathcal{H},\lfloor \sigma^2 mt\rfloor}\right)\right) = -\frac{\eta}{\sigma^2} \cdot (\Lambda_\lambda - \kappa_\lambda) \begin{cases} < 0, & \text{for } \lambda \in\,]0,1[\,, \\[2ex] > 0, & \text{for } \lambda \in\,]\widetilde{\lambda}_-, \widetilde{\lambda}_+[\,\setminus\,[0,1]\,. \end{cases}$$

The assertions of Corollary 15 follow immediately by inspecting the expressions in the exponential of (153) and (154) in combination with (155) to (162).

7.3. Bounds of Power Divergences for Diffusion Approximations

Analogously to Section 4 (see especially Section 4.1), for orders $\lambda \in \mathbb{R}\backslash\{0,1\}$ all the results of the previous Section 7.2 carry correspondingly over from (limits of) bounds of the Hellinger integrals $H_\lambda\left(P^{(m)}_{\mathcal{A},\lfloor\sigma^2 mt\rfloor}\big|\big|P^{(m)}_{\mathcal{H},\lfloor\sigma^2 mt\rfloor}\right)$ to (limits of) bounds of the total variation distance $V\left(P^{(m)}_{\mathcal{A},\lfloor\sigma^2 mt\rfloor}\big|\big|P^{(m)}_{\mathcal{H},\lfloor\sigma^2 mt\rfloor}\right)$ (by virtue of (12)), to (limits of) bounds of the Renyi divergences $R_\lambda\left(P^{(m)}_{\mathcal{A},\lfloor\sigma^2 mt\rfloor}\big|\big|P^{(m)}_{\mathcal{H},\lfloor\sigma^2 mt\rfloor}\right)$ (by virtue of (7)) as well as to (limits of) bounds of the power divergences $I_\lambda\left(P^{(m)}_{\mathcal{A},\lfloor\sigma^2 mt\rfloor}\big|\big|P^{(m)}_{\mathcal{H},\lfloor\sigma^2 mt\rfloor}\right)$ (by virtue of (2)). For the sake of brevity, the–merely repetitive–exact details are omitted. Moreover, by combining the outcoming results on the above-mentioned power divergences with parts of the Bayesian-decision-making context of Section 2.5, we obtain corresponding assertions on (i) the (cf. (21)) *weighted-average* decision risk reduction (weighted-average statistical information measure) about the degree of evidence $\mathfrak{deg}$ concerning the parameter θ that can be attained by observing the GWI-path $\mathcal{X}_n$ until stage n, as well as (ii) the (cf. (22)) *limit* decision risk reduction (limit statistical information measure).

In the following, let us concentrate on the derivation of the Kullback-Leibler information divergence KL (relative entropy) within the current diffusion-limit framework. Notice that altogether we face two limit procedures simultaneously: by the first limit $\lim_{\lambda\uparrow 1} I_\lambda\left(P^{(m)}_{\mathcal{A},\lfloor\sigma^2 mt\rfloor}\big|\big|P^{(m)}_{\mathcal{H},\lfloor\sigma^2 mt\rfloor}\right)$ we obtain the KL $I\left(P^{(m)}_{\mathcal{A},\lfloor\sigma^2 mt\rfloor}\big|\big|P^{(m)}_{\mathcal{H},\lfloor\sigma^2 mt\rfloor}\right)$ for every fixed approximation step $m \in \overline{\mathbb{N}}$; on the other hand, for each fixed $\lambda \in]0,1[$, the second limit $\lim_{m\to\infty} I_\lambda\left(P^{(m)}_{\mathcal{A},\lfloor\sigma^2 mt\rfloor}\big|\big|P^{(m)}_{\mathcal{H},\lfloor\sigma^2 mt\rfloor}\right)$ describes the limit of the power divergence – as the sequence of rescaled and continuously interpolated GW(I)'s $\left(\left(\widetilde{X}^{(m)}_s\right)_{s\in[0,\infty[}\right)_{m\in\overline{\mathbb{N}}}$ (equipped with probability law $P^{(m)}_{\mathcal{A},\lfloor\sigma^2 mt\rfloor}$ resp. $P^{(m)}_{\mathcal{H},\lfloor\sigma^2 mt\rfloor}$ up to time $\lfloor\sigma^2 mt\rfloor$) converges weakly to the continuous-time CIR-type diffusion process $\left(\widetilde{X}_s\right)_{s\in[0,\infty[}$ (with probability law $\widetilde{P}_{\mathcal{A},t}$ resp. $\widetilde{P}_{\mathcal{H},t}$ up to time t). In Appendix A.4 we shall prove that these two limits can be interchanged:

Theorem 12. *Let the initial SDE-value $\widetilde{X}_0 \in]0,\infty[$ be arbitrary but fixed, and suppose that $\lim_{m\to\infty}\frac{1}{m}X^{(m)}_0 = \widetilde{X}_0$. Then, for all $(\kappa_{\mathcal{A}},\kappa_{\mathcal{H}},\eta) \in \widetilde{\mathcal{P}}_{NI}\cup\widetilde{\mathcal{P}}_{SP,1}$ and all $t \in [0,\infty[$, one gets the Kullback-Leibler information divergence (relative entropy) convergences*

$$\lim_{m\to\infty} I\left(P^{(m)}_{\mathcal{A},\lfloor\sigma^2 mt\rfloor}\big|\big|P^{(m)}_{\mathcal{H},\lfloor\sigma^2 mt\rfloor}\right) = \lim_{m\to\infty}\lim_{\lambda\nearrow 1} I_\lambda\left(P^{(m)}_{\mathcal{A},\lfloor\sigma^2 mt\rfloor}\big|\big|P^{(m)}_{\mathcal{H},\lfloor\sigma^2 mt\rfloor}\right)$$

$$= \begin{cases} \frac{(\kappa_{\mathcal{A}}-\kappa_{\mathcal{H}})^2}{2\sigma^2\cdot\kappa_{\mathcal{A}}}\cdot\left[\left(\widetilde{X}_0-\frac{\eta}{\kappa_{\mathcal{A}}}\right)\cdot(1-e^{-\kappa_{\mathcal{A}}\cdot t})+\eta\cdot t\right], & \text{if } \kappa_{\mathcal{A}} > 0, \\[2ex] \frac{\kappa_{\mathcal{H}}^2}{2\sigma^2}\cdot\left[\frac{\eta}{2}\cdot t^2+\widetilde{X}_0\cdot t\right], & \text{if } \kappa_{\mathcal{A}} = 0, \end{cases}$$

$$= \lim_{\lambda\nearrow 1}\lim_{m\to\infty} I_\lambda\left(P^{(m)}_{\mathcal{A},\lfloor\sigma^2 mt\rfloor}\big|\big|P^{(m)}_{\mathcal{H},\lfloor\sigma^2 mt\rfloor}\right). \tag{163}$$

This immediately leads to the following

Corollary 16. *Let the initial SDE-value $\widetilde{X}_0 \in]0,\infty[$ be arbitrary but fixed, and suppose that $\lim_{m\to\infty} \frac{1}{m} X_0^{(m)} = \widetilde{X}_0$. Then, the KL limit (163) possesses the following time-asymptotical behaviour:*
(a) For all $(\kappa_{\mathcal{A}}, \kappa_{\mathcal{H}}, \eta) \in \widetilde{\mathcal{P}}_{NI}$ (i.e., $\eta = 0$) one gets

(i) *in the case $\kappa_{\mathcal{A}} > 0$* $\displaystyle \lim_{t\to\infty} \lim_{m\to\infty} I\left(P^{(m)}_{\mathcal{A}, \lfloor \sigma^2 mt \rfloor} \middle|\middle| P^{(m)}_{\mathcal{H}, \lfloor \sigma^2 mt \rfloor} \right) = \frac{\widetilde{X}_0 \cdot (\kappa_{\mathcal{A}} - \kappa_{\mathcal{H}})^2}{2\sigma^2 \cdot \kappa_{\mathcal{A}}}$,

(ii) *in the case $\kappa_{\mathcal{A}} = 0$* $\displaystyle \lim_{t\to\infty} \lim_{m\to\infty} \frac{1}{t} \cdot I\left(P^{(m)}_{\mathcal{A}, \lfloor \sigma^2 mt \rfloor} \middle|\middle| P^{(m)}_{\mathcal{H}, \lfloor \sigma^2 mt \rfloor} \right) = \frac{\widetilde{X}_0 \cdot \kappa_{\mathcal{H}}^2}{4\sigma^2}$.

(b) For all $(\kappa_{\mathcal{A}}, \kappa_{\mathcal{H}}, \eta) \in \widetilde{\mathcal{P}}_{SP,1}$ (i.e., $\eta > 0$) one gets

(i) *in the case $\kappa_{\mathcal{A}} > 0$* $\displaystyle \lim_{t\to\infty} \lim_{m\to\infty} \frac{1}{t} \cdot I\left(P^{(m)}_{\mathcal{A}, \lfloor \sigma^2 mt \rfloor} \middle|\middle| P^{(m)}_{\mathcal{H}, \lfloor \sigma^2 mt \rfloor} \right) = \frac{\eta \cdot (\kappa_{\mathcal{A}} - \kappa_{\mathcal{H}})^2}{2\sigma^2 \cdot \kappa_{\mathcal{A}}}$,

(ii) *in the case $\kappa_{\mathcal{A}} = 0$* $\displaystyle \lim_{t\to\infty} \lim_{m\to\infty} \frac{1}{t^2} \cdot I\left(P^{(m)}_{\mathcal{A}, \lfloor \sigma^2 mt \rfloor} \middle|\middle| P^{(m)}_{\mathcal{H}, \lfloor \sigma^2 mt \rfloor} \right) = \frac{\eta \cdot \kappa_{\mathcal{H}}^2}{4\sigma^2}$.

Remark 9. *In Appendix A.4 we shall see that the proof of the last (limit-interchange concerning) equality in (163) relies heavily on the use of the extra terms $L_\lambda^{(1)}(t)$, $L_\lambda^{(2)}(t)$, $U_\lambda^{(1)}(t)$, $U_\lambda^{(2)}(t)$ in (153) and (154). Recall that these terms ultimately stem from (manipulations of) the corresponding parts of the "improved closed-form bounds" in Theorem 5, which were derived by using the linear inhomogeneous difference equations $\underline{a}_n^{(q)}$ resp. $\overline{a}_n^{(q)}$ (cf. (92) resp. (94)) instead of the linear homogeneous difference equations $a_n^{(q),T}$ resp. $a_n^{(q),S}$ (cf. (78) resp. (79)) as explicit approximates of the sequence $a_n^{(q)}$. Not only this fact shows the importance of this more tedious approach.*

Interesting comparisons of the above-mentioned results in Sections 7.2 and 7.3 with corresponding information measures of the solutions of the SDE (129) themselves (rather their branching approximations), can be found in Kammerer [157].

7.4. Applications to Decision Making

Analogously to Section 6.7, the above-mentioned investigations of the Sections 7.1–7.3 can be applied to the context of Section 2.5 on *dichotomous* decision making about GW(I)-type diffusion approximations of solutions of the stochastic differential Equation (129). For the sake of brevity, the–merely repetitive–exact details are omitted.

Author Contributions: Conceptualization, N.B.K. and W.S.; Formal analysis, N.B.K. and W.S.; Methodology, N.B.K. and W.S.; Visualization, N.B.K.; Writing, N.B.K. and W.S. All authors have read and agreed to the published version of the manuscript.

Funding: Niels B. Kammerer received a scholarship of the "Studienstiftung des Deutschen Volkes" for his PhD Thesis.

Acknowledgments: We are very grateful to the referees for their patience to review this long manuscript, and for their helpful suggestions. Moreover, we would like to thank Andreas Greven for some useful remarks.

Conflicts of Interest: The authors declare no conflict of interest.

Appendix A. Proofs and Auxiliary Lemmas

Appendix A.1. Proofs and Auxiliary Lemmas for Section 3

Lemma A1. *For all real numbers $x, y, z > 0$ and all $\lambda \in \mathbb{R}$ one has*

$$x^\lambda y^{1-\lambda} - \left(\lambda\, x\, z^{\lambda-1} + (1-\lambda)\, y\, z^\lambda \right) \begin{cases} \leq 0, & \text{for } \lambda \in]0,1[\,, \\ = 0, & \text{for } \lambda \in \{0,1\}\,, \\ \geq 0, & \text{for } \lambda \in \mathbb{R}\backslash[0,1]\,, \end{cases}$$

with equality in the cases $\lambda \in \mathbb{R}\backslash\{0,1\}$ iff $\frac{x}{y} = z$.

Proof of Lemma A1. For fixed $\tilde{x} := x z^{\lambda-1} > 0$, $\tilde{y} := y z^\lambda > 0$ with $\tilde{x} \neq \tilde{y}$ we inspect the function g on $\mathbb{R}$ defined by $g(\lambda) := \tilde{x}^\lambda \tilde{y}^{1-\lambda} - (\lambda \tilde{x} + (1-\lambda)\tilde{y})$ which satisfies $g(0) = g(1) = 0$, $g'(0) = \tilde{y} \log(\tilde{x}/\tilde{y}) - (\tilde{x} - \tilde{y}) < \tilde{y}((\tilde{x}/\tilde{y}) - 1) - (\tilde{x} - \tilde{y}) = 0$ and which is strictly convex. Thus, the assertion follows immediately by taking into account the obvious case $\tilde{x} = \tilde{y}$. $\square$

Proof of Properties 1. Property (P9) is trivially valid. To show (P1) we assume $0 < q < \beta_\lambda$, which implies $a_1^{(q)} = \xi_\lambda^{(q)}(0) = q - \beta_\lambda < 0$. By induction, $(a_n)_{n \in \mathbb{N}}$ is strictly negative and strictly decreasing. As stated in (P9), the function $\xi_\lambda^{(q)}$ is strictly increasing, strictly convex and converges to $-\beta_\lambda$ for $x \to -\infty$. Thus, it hits the straight line $id(x) = x$ once and only once on the negative real line at $x_0^{(q)} \in\,]-\beta_\lambda, 0[$ (cf. (44)). This implies that the sequence $\left(a_n^{(q)} \right)_{n \in \mathbb{N}}$ converges to $x_0^{(q)} \in\,]-\beta_\lambda, q - \beta_\lambda[$. Property (P2) follows immediately. In order to prove (P3), let us fix $q > \max\{0, \beta_\lambda\}$, implying $a_1^{(q)} = \xi_\lambda^{(q)}(0) = q - \beta_\lambda > 0$; notice that in this setup, the special choice $q = 1$ implies $\min\{1, e^{\beta_\lambda - 1}\} = e^{\beta_\lambda - 1} < q$. By induction, $\left(a_n^{(q)} \right)_{n \in \mathbb{N}}$ is strictly positive and strictly increasing. Since $\lim_{x \to \infty} \xi_\lambda^{(q)}(x) = \infty$, the function $\xi_\lambda^{(q)}$ does not necessarily hit the straight line $id(x) = x$ on the positive real line. In fact, due to strict convexity (cf. (P9)), this is excluded if $\xi_\lambda^{(q)\prime}(0) = q \geq 1$. Suppose that $q < 1$. To prove that there exists a positive solution of the equation $\xi_\lambda^{(q)}(x) = x$ it is sufficient to show that the unique global minimum of the strict convex function $h_\lambda^{(q)}(x) := \xi_\lambda^{(q)}(x) - x$ is taken at some point $x_0 \in\,]0, \infty[$ and that $h_\lambda^{(q)}(x_0) \leq 0$. It holds $h_\lambda^{(q)\prime}(x) = q \cdot e^x - 1$, and therefore $h_\lambda^{(q)\prime}(x) = 0$ iff $x = x_0 = -\log q$. We have $h_\lambda^{(q)}(-\log q) = 1 - \beta_\lambda + \log q$, which is less or equal to zero iff $q \leq e^{\beta_\lambda - 1}$. It remains to show that for $q > \beta_\lambda$ and $q > \min\left\{1, e^{\beta_\lambda - 1}\right\}$ the sequence $\left(a_n^{(q)} \right)_{n \in \mathbb{N}}$ grows faster than exponentially, i.e., there do not exist constants $c_1, c_2 \in \mathbb{R}$ such that $a_n^{(q)} \leq e^{c_1 + c_2 n}$ for all $n \in \mathbb{N}$. We already know that (in the current case) $a_n^{(q)} \overset{n \to \infty}{\longrightarrow} \infty$. Notice that it is sufficient to verify $\limsup_{n \to \infty} \left(\log(a_{n+1}^{(q)}) - \log(a_n^{(q)}) \right) = \infty$. For the case $\beta_\lambda \geq 0$ the latter is obtained by

$$\log\left(a_{n+1}^{(q)} \right) - \log\left(a_n^{(q)} \right) = \log\left((q - \beta_\lambda)e^{a_n^{(q)}} + \beta_\lambda(e^{a_n^{(q)}} - 1) \right) - \log\left(q e^{a_{n-1}^{(q)}} - \beta_\lambda \right)$$

$$\geq \left(\log(q - \beta_\lambda) - \log(q) \right) + \left(q e^{a_{n-1}^{(q)}} - \beta_\lambda - a_{n-1}^{(q)} \right) \overset{a_{n-1}^{(q)} \to \infty}{\longrightarrow} \infty\,.$$

An analogous consideration works out for the case $\beta_\lambda < 0$. Property (P4) is trivial, and (P5) to (P8) are direct implications of the already proven properties (P1) to (P4). $\square$

Proof of Lemma 1. (a) Let $\beta_{\mathcal{A}} > 0$, $\beta_{\mathcal{H}} > 0$ with $\beta_{\mathcal{A}} \neq \beta_{\mathcal{H}}$, $\lambda \in \mathbb{R}\backslash\,]0,1[$, $\beta_\lambda := \lambda\beta_{\mathcal{A}} + (1-\lambda)\beta_{\mathcal{H}}$ and $q_\lambda := \beta_{\mathcal{A}}^\lambda \beta_{\mathcal{H}}^{1-\lambda} > \max\{0, \beta_\lambda\}$ (cf. Lemma A1). Below, we follow the lines of Linkov & Lunyova [53], appropriately adapted to our context. We have to find those $\lambda \in \mathbb{R}\backslash\,]0,1[$ for which the following two conditions hold:

(i) $q_\lambda \leq 1$, i.e., $\xi_\lambda^{(q_\lambda)\,\prime}(0) \leq 1$,

(ii) $q_\lambda \leq e^{\beta_\lambda - 1}$ (cf.(P3a)), which is equivalent with the existence of a–positive, if (i) is satisfied,–solution of the equation $\xi_\lambda^{(q_\lambda)}(x) = x$.

Notice that the case $q_\lambda = 1$, $\lambda \in \mathbb{R}\backslash[0,1]$, cannot appear in (i), provided that (ii) holds (since due to Lemma A1 $e^{\beta_\lambda - 1} < e^{q_\lambda - 1} = 1$). For (i), it is easy to check that we have to require

$$\lambda \begin{cases} < \dfrac{\log(\beta_{\mathcal{H}})}{\log(\beta_{\mathcal{H}}/\beta_{\mathcal{A}})}, & \text{if } \beta_{\mathcal{A}} > \beta_{\mathcal{H}}, \\[2ex] > \dfrac{\log(\beta_{\mathcal{H}})}{\log(\beta_{\mathcal{H}}/\beta_{\mathcal{A}})}, & \text{if } \beta_{\mathcal{A}} < \beta_{\mathcal{H}}. \end{cases} \tag{A1}$$

To proceed, straightforward analysis leads to $-\log(q_\lambda) = \arg\min_{x\in\mathbb{R}}\{\xi_\lambda^{(q_\lambda)}(x) - x\}$. To check (ii), we first notice that $q_\lambda \leq e^{\beta_\lambda - 1}$ iff $\xi_\lambda^{(q_\lambda)}(x) - x \leq 0$ for some $x \in \mathbb{R}$. Hence, we calculate

$$\xi_\lambda^{(q_\lambda)}\big(-\log(q_\lambda)\big) + \log(q_\lambda) \leq 0 \quad\Longleftrightarrow\quad 1 - \lambda(\beta_{\mathcal{A}} - \beta_{\mathcal{H}}) - \beta_{\mathcal{H}} + \lambda\log\left(\frac{\beta_{\mathcal{A}}}{\beta_{\mathcal{H}}}\right) + \log(\beta_{\mathcal{H}}) \leq 0$$

$$\Longleftrightarrow\quad \lambda\cdot\left[\beta_{\mathcal{H}}\left(1 - \frac{\beta_{\mathcal{A}}}{\beta_{\mathcal{H}}}\right) + \log\left(\frac{\beta_{\mathcal{A}}}{\beta_{\mathcal{H}}}\right)\right] \leq \beta_{\mathcal{H}} - 1 - \log(\beta_{\mathcal{H}}). \tag{A2}$$

In order to isolate λ in (A2), one has to find out for which $(\beta_{\mathcal{A}}, \beta_{\mathcal{H}})$ the term in the square bracket is positive resp. zero resp. negative. To achieve this, we aim for the substitutions $x := \beta_{\mathcal{A}}/\beta_{\mathcal{H}}$, $\beta = \beta_{\mathcal{H}}$ and thus study first the auxiliary function $h_\beta(x) := \log(x) - \beta(x-1)$, $x > 0$, with fixed parameters $\beta > 0$. Straightforwardly, we obtain $h'_\beta(x) = x^{-1} - \beta$ and $h''_\beta(x) = -x^{-2}$. Thus, the function $h_\beta(\cdot)$ is strictly concave and attains a maximum at $x = \beta^{-1}$. Since additionally $h_\beta(1) = 0$ and $h'_\beta(1) = 1 - \beta$, there exists a second solution $z(\beta) \neq 1$ of the equation $h_\beta(x) = 0$ iff $\beta \neq 1$. Thus, one gets

- for $\beta = 1$: for all $x > 0$ there holds $h_\beta(x) \leq 0$, with equality iff $x = \beta^{-1}$,
- for $\beta < 1$: $h_\beta(x) \geq 0$ iff $x \in [1, z(\beta)]$, with equality iff $x \in \{1, z(\beta)\}$ (notice that $z(\beta) > 1$),
- for $\beta > 1$: $h_\beta(x) \geq 0$ iff $x \in [z(\beta), 1]$, with equality iff $x \in \{z(\beta), 1\}$ (notice that $z(\beta) < 1$).

Suppose that $\lambda < 0$.

Case 1: If $\beta_{\mathcal{H}} = 1$, then condition (ii) is not satisfied whenever $\beta_{\mathcal{A}} \neq \beta_{\mathcal{H}}$, since the right side of (A2) is equal to zero and the left side is strictly greater than zero. Hence, $\lambda_- = 0$.

Case 2: Let $\beta_{\mathcal{H}} > 1$. If $\beta_{\mathcal{A}} < \beta_{\mathcal{H}}$, then condition (i) is not satisfied and hence $\lambda_- = 0$. If $\beta_{\mathcal{A}} > \beta_{\mathcal{H}}$, then condition (i) is satisfied iff $\lambda < \breve{\lambda} := \breve{\lambda}(\beta_{\mathcal{A}}, \beta_{\mathcal{H}}) := \dfrac{\log(\beta_{\mathcal{H}})}{\log(\beta_{\mathcal{H}}/\beta_{\mathcal{A}})} < 0$. On the other hand, incorporating the discussion of the function $h_\beta(\cdot)$, we see that $h_{\beta_{\mathcal{H}}}\left(\frac{\beta_{\mathcal{A}}}{\beta_{\mathcal{H}}}\right) < 0$. Thus, (A2) implies that condition (ii) is satisfied when $\lambda \geq \check{\lambda} := \check{\lambda}(\beta_{\mathcal{A}}, \beta_{\mathcal{H}}) := \dfrac{\beta_{\mathcal{H}} - 1 - \log(\beta_{\mathcal{H}})}{\beta_{\mathcal{H}} - \beta_{\mathcal{A}} + \log\left(\frac{\beta_{\mathcal{A}}}{\beta_{\mathcal{H}}}\right)}$. We claim that $\breve{\lambda} < \check{\lambda}$ and conclude that the conditions (i) and (ii) are not fulfilled jointly, which leads to $\lambda_- = 0$. To see this, we notice that due to $1 < \beta_{\mathcal{H}} < \beta_{\mathcal{A}}$ we get $\log(\beta_{\mathcal{A}})/(\beta_{\mathcal{A}} - 1) < \log(\beta_{\mathcal{H}})/(\beta_{\mathcal{H}} - 1)$ and thus

$$\log(\beta_{\mathcal{A}})(\beta_{\mathcal{H}} - 1) \; < \; \log(\beta_{\mathcal{H}})(\beta_{\mathcal{A}} - 1)$$

$$\iff \quad \beta_{\mathcal{H}}\log(\beta_{\mathcal{H}}) - \beta_{\mathcal{A}}\log(\beta_{\mathcal{H}}) \; < \; \beta_{\mathcal{H}}\log(\beta_{\mathcal{H}}) - \beta_{\mathcal{H}}\log(\beta_{\mathcal{A}}) - \log(\beta_{\mathcal{H}}) + \log(\beta_{\mathcal{A}})$$

$$\iff \quad \log(\beta_{\mathcal{H}})(\beta_{\mathcal{H}} - \beta_{\mathcal{A}}) + \log(\beta_{\mathcal{H}})\log\left(\frac{\beta_{\mathcal{A}}}{\beta_{\mathcal{H}}}\right) \; < \; \log\left(\frac{\beta_{\mathcal{H}}}{\beta_{\mathcal{A}}}\right)(\beta_{\mathcal{H}} - 1) + \log(\beta_{\mathcal{H}})\log\left(\frac{\beta_{\mathcal{A}}}{\beta_{\mathcal{H}}}\right)$$

$$\iff \quad \frac{\log(\beta_{\mathcal{H}})}{\log\left(\frac{\beta_{\mathcal{H}}}{\beta_{\mathcal{A}}}\right)} \; < \; \frac{\beta_{\mathcal{H}} - 1 - \log(\beta_{\mathcal{H}})}{\beta_{\mathcal{H}} - \beta_{\mathcal{A}} + \log\left(\frac{\beta_{\mathcal{A}}}{\beta_{\mathcal{H}}}\right)} \qquad \iff \quad \check{\lambda} < \bar{\lambda}. \tag{A3}$$

Case 3: Let $\beta_{\mathcal{H}} < 1$. For this, one gets $h_{\beta_{\mathcal{H}}}\left(\frac{\beta_{\mathcal{A}}}{\beta_{\mathcal{H}}}\right) \geq 0$ for $\beta_{\mathcal{A}} \in \,]\beta_{\mathcal{H}}, \beta_{\mathcal{H}} z(\beta_{\mathcal{H}})]$. Hence, condition (ii) is satisfied if either $\beta_{\mathcal{A}} \in \,]\beta_{\mathcal{H}}, \beta_{\mathcal{H}} z(\beta_{\mathcal{H}})]$, or $\beta_{\mathcal{A}} \notin \,]\beta_{\mathcal{H}}, \beta_{\mathcal{H}} z(\beta_{\mathcal{H}})]$ and $\lambda \geq \bar{\lambda}$. If $\beta_{\mathcal{A}} > \beta_{\mathcal{H}} z(\beta_{\mathcal{H}})$, then condition (i) is trivially satisfied for all $\lambda < 0$. In the case $\beta_{\mathcal{A}} < \beta_{\mathcal{H}}$, condition (i) is satisfied whenever $\lambda > \check{\lambda}$. Notice that since $0 < \beta_{\mathcal{A}} < \beta_{\mathcal{H}} < 1$, an analogous consideration as in (A3) leads to $\check{\lambda} < \bar{\lambda}$. This implies that $\lambda_{-} = \bar{\lambda}$. The last case $\beta_{\mathcal{A}} \in \,]\beta_{\mathcal{H}}, \beta_{\mathcal{H}} z(\beta_{\mathcal{H}})]$ is easy to handle: since $\frac{\log(\beta_{\mathcal{H}})}{\log(\beta_{\mathcal{H}}/\beta_{\mathcal{A}})} > 0$ as well as $z_{\beta_{\mathcal{H}}}\left(\frac{\beta_{\mathcal{A}}}{\beta_{\mathcal{H}}}\right) > 0$, both conditions (i) and (ii) hold trivially.

The representation of λ_{+} follows straightforwardly from the λ_{-}-result and the skew symmetry (8), by employing $1 - \check{\lambda}(\beta_{\mathcal{H}}, \beta_{\mathcal{A}}) = \check{\lambda}(\beta_{\mathcal{A}}, \beta_{\mathcal{H}})$. Alternatively, one can proceed analogously to the λ_{-}-case.

Part (b) is much easier to prove: if $\beta_{\bullet} := \beta_{\mathcal{A}} = \beta_{\mathcal{H}} > 0$, then for all $\lambda \in \mathbb{R}\backslash[0,1]$ one gets $q_{\lambda} = \beta_{\mathcal{A}}^{\lambda}\beta_{\mathcal{H}}^{1-\lambda} = \beta_{\bullet}$ as well as $\beta_{\lambda} = \beta_{\bullet}$. Hence, Properties 1 (P2) implies that $a_{n}^{(q_{\lambda})} \equiv 0$ and thus it is convergent, independently of the choice $\lambda \in \mathbb{R}\backslash[0,1]$. $\square$

Proof of Formula (51). For the parameter constellation in Section 3.10, we employ as upper bound for $\phi_{\lambda}(x)$ $(x \in \mathbb{N}_0)$ the function

$$\overline{\phi_{\lambda}}(x) := \begin{cases} \phi_{\lambda}(0), & \text{if } x = 0, \\ 0, & \text{if } x > 0. \end{cases}$$

Notice that this method is rather crude, and gives in the other cases treated in the Sections 3.7–3.9 worse bounds than those derived there. Since $\lambda \in \,]0,1[$ and $\alpha_{\mathcal{A}} \neq \alpha_{\mathcal{H}}$, one has $\phi_{\lambda}(0) < 0$. In order to derive an upper bound of the Hellinger integral, we first set $\bar{\epsilon} := 1 - e^{\phi_{\lambda}(0)} \in \,]0,1[$. Hence, for all $n \in \mathbb{N}\backslash\{1\}$ we obtain the auxiliary expression

$$\sum_{x_{n-1}=0}^{\infty} \frac{[\varphi_{\lambda}(x_{n-2})]^{x_{n-1}}}{x_{n-1}!} \cdot \exp\left\{\phi_{\lambda}(x_{n-1})\right\} \leq \sum_{x_{n-1}=0}^{\infty} \frac{[\varphi_{\lambda}(x_{n-2})]^{x_{n-1}}}{x_{n-1}!} \cdot \exp\left\{\overline{\phi_{\lambda}}(x_{n-1})\right\}$$

$$= \exp\left\{\varphi_{\lambda}(x_{n-2})\right\} - \bar{\epsilon} = \exp\left\{\varphi_{\lambda}(x_{n-2})\right\} \cdot \left[1 - \bar{\epsilon} \cdot \exp\left\{-\varphi_{\lambda}(x_{n-2})\right\}\right].$$

Moreover, since $\beta_{\mathcal{A}} \neq \beta_{\mathcal{H}}$, one gets $\lim_{x\to\infty} \phi_{\lambda}(x) = -\infty$ (cf. Properties 3 (P20) and Lemma A1). This–together with the nonnegativity of $\varphi_{\lambda}(\cdot)$–implies

$$\sup_{x\in\mathbb{N}_0} \left\{\exp\left\{\phi_{\lambda}(x)\right\} \cdot \left[1 - \bar{\epsilon} \cdot \exp\left\{-\varphi_{\lambda}(x)\right\}\right]\right\} =: \bar{\delta} \in \,]0,1[.$$

Incorporating these considerations as well as the formulas (27) to (32), we get for $n = 1$ the relation $H_\lambda\left(P_{\mathcal{A},n}||P_{\mathcal{H},n}\right) = \exp\{\phi_\lambda(x_0)\} \leq 1$ (with equality iff $x_0 = x^* = \frac{\alpha_\mathcal{A} - \alpha_\mathcal{H}}{\beta_\mathcal{H} - \beta_\mathcal{A}}$), and–as a continuation of formula (29)– for all $n \in \mathbb{N}\backslash\{1\}$ (recall that $\vec{x} := (x_0, x_1, \dots) \in \Omega$)

$$
\begin{aligned}
H_\lambda\left(P_{\mathcal{A},n}||P_{\mathcal{H},n}\right) &= \sum_{x_1=0}^{\infty} \cdots \sum_{x_n=0}^{\infty} \prod_{k=1}^{n} Z_{n,k}^{(\lambda)}(\vec{x}) \\
&= \sum_{x_1=0}^{\infty} \cdots \sum_{x_{n-1}=0}^{\infty} \prod_{k=1}^{n-1} Z_{n,k}^{(\lambda)}(\vec{x}) \\
&\quad \cdot \exp\left\{ (f_\mathcal{A}(x_{n-1}))^\lambda \, (f_\mathcal{H}(x_{n-1}))^{(1-\lambda)} - (\lambda f_\mathcal{A}(x_{n-1}) + (1-\lambda)f_\mathcal{H}(x_{n-1})) \right\} \\
&= \sum_{x_1=0}^{\infty} \cdots \sum_{x_{n-2}=0}^{\infty} \prod_{k=1}^{n-2} Z_{n,k}^{(\lambda)}(\vec{x}) \cdot \exp\left\{-f_\lambda(x_{n-2})\right\} \sum_{x_{n-1}=0}^{\infty} \frac{[\varphi_\lambda(x_{n-2})]^{x_{n-1}}}{x_{n-1}!} \cdot \exp\{\phi_\lambda(x_{n-1})\} \\
&\leq \sum_{x_1=0}^{\infty} \cdots \sum_{x_{n-2}=0}^{\infty} \prod_{k=1}^{n-2} Z_{n,k}^{(\lambda)}(\vec{x}) \cdot \exp\left\{\phi_\lambda(x_{n-2})\right\} \cdot \left[1 - \bar{\epsilon} \cdot \exp\left\{-\varphi_\lambda(x_{n-2})\right\}\right] \\
&\leq \bar{\delta} \cdot \sum_{x_1=0}^{\infty} \cdots \sum_{x_{n-2}=0}^{\infty} \prod_{k=1}^{n-2} Z_{n,k}^{(\lambda)}(\vec{x}) \leq \cdots \leq \bar{\delta}^{\lfloor n/2 \rfloor} .
\end{aligned}
\tag{A4}
$$

Hence, $H_\lambda\left(P_{\mathcal{A},n}||P_{\mathcal{H},n}\right) < 1$ for (at least) all $n \in \mathbb{N}\backslash\{1\}$, and $\lim_{n \to \infty} H_\lambda\left(P_{\mathcal{A},n}||P_{\mathcal{H},n}\right) = 0$. $\square$

Notice that the above proof method of formula (51) does not work for the parameter setup in Section 3.11, because there one gets $\bar{\delta} = \sup_{x \in \mathbb{N}_0} \left\{ \exp\left\{\phi_\lambda(x)\right\} \cdot \left[1 - \bar{\epsilon} \cdot \exp\left\{-\varphi_\lambda(x)\right\}\right] \right\} = 1$.

Proof of Proposition 9. In the setup $(\beta_\mathcal{A}, \beta_\mathcal{H}, \alpha_\mathcal{A}, \alpha_\mathcal{H}, \lambda) \in \mathcal{P}_{\mathrm{SP4a}} \times \,]0, 1[$ we require $\beta_\bullet := \beta_\mathcal{A} = \beta_\mathcal{H} < 1$. As a linear upper bound for $\phi_\lambda(\cdot)$, we employ the tangent line at $y \geq 0$ (cf. (52))

$$
\phi_{\lambda,y}^{\tan}(x) := (p_y - \alpha_\lambda) + (q_y - \beta_\bullet) \cdot x := (p_{\lambda,y}^{\tan} - \alpha_\lambda) + (q_{\lambda,y}^{\tan} - \beta_\lambda) \cdot x := \left(\phi_\lambda(y) - y \cdot \phi_\lambda'(y)\right) + \phi_\lambda'(y) \cdot x . \tag{A5}
$$

Since in the current setup $\mathcal{P}_{\mathrm{SP4a}}$ the function $\phi_\lambda(\cdot)$ is strictly increasing, the slope $\phi_\lambda'(y)$ of the tangent line at y is positive. Thus we have $q_y > \beta_\lambda$ and Properties 1 (P3) implies that the sequence $\left(a_n^{(q_y)}\right)_{n \in \mathbb{N}}$ is strictly increasing and converges to $x_0^{(q_y)} \in \,]0, -\log(q_y)]$ iff $q_y \leq \min\{1, e^{\beta_\bullet - 1}\} = e^{\beta_\bullet - 1} < 1$ (cf. (P3a)), where $x_0^{(q_y)}$ is the smallest solution of the equation $\xi_\lambda^{(q_y)}(x) = q_y \cdot e^x - \beta_\bullet = x$. Since $q_y \searrow \beta_\bullet$ for $y \to \infty$ (cf. Properties 3 (P18)) and additionally $e^{\beta_\bullet - 1} > \beta_\bullet$, there exists a large enough $y \geq 0$ such that the sequence $\left(a_n^{(q_y)}\right)_{n \in \mathbb{N}}$ converges. If this y is also large enough to additionally guarantee $h(y) < 0$ for

$$
h(y) := \lim_{n \to \infty} \frac{1}{n} \log\left(\widetilde{B}_{\lambda, X_0, n}^{(p_y, q_y)}\right) = p_y \cdot e^{x_0^{(q_y)}} - \alpha_\lambda ,
$$

then one can conclude that $\lim_{n \to \infty} H_\lambda(P_{\mathcal{A},n}||P_{\mathcal{H},n}) = 0$. As a first step, for verifying $h(y) < 0$ we look for an upper bound $\bar{x}_0^{(q_y)}$ for the fixed point $x_0^{(q_y)}$ where the latter exists for $y \geq y_1$ (say). Notice that

$$
\overline{Q}_\lambda^{(q_y)}(x) := \frac{1}{2}x^2 + q_y x + q_y - \beta_\bullet \geq q_y \cdot e^x - \beta_\bullet = \xi_\lambda^{(q_y)}(x) , \tag{A6}
$$

since $\overline{Q}_\lambda^{(q_y)}(0) = \xi_\lambda^{(q_y)}(0)$, $\overline{Q}_\lambda^{(q_y)\prime}(0) = \xi_\lambda^{(q_y)\prime}(0)$ and $\overline{Q}_\lambda^{(q_y)\prime\prime}(x) \geq \xi_\lambda^{(q_y)\prime\prime}(x)$ for $x \in [0, -\log(q_y)]$. For sufficiently large $y \geq y_2 \geq y_1$ (say), we easily obtain the smaller solution of $\overline{Q}_\lambda^{(q_y)}(x) = x$ as

$$
\bar{x}_0^{(q_y)} = (1 - q_y) - \sqrt{(1 - q_y)^2 - 2(q_y - \beta_\bullet)} = (1 - \phi_\lambda'(y) - \beta_\bullet) - \sqrt{(1 - \phi_\lambda'(y) - \beta_\bullet)^2 - 2\phi_\lambda'(y)} \geq x_0^{(q_y)} \tag{A7}
$$

where the expression in the root is positive since $q_y \searrow \beta_\bullet$ for $y \to \infty$. We now have

$$h(y) = p_y \cdot e^{\overline{x}_0^{(qy)}} - \alpha_\lambda \leq p_y \cdot e^{\overline{x}_0^{(qy)}} - \alpha_\lambda =: \overline{h}(y), \qquad \forall\, y \geq y_2. \tag{A8}$$

Hence, it suffices to show that $\overline{h}(y) < 0$ for some $y \geq y_2$. We recall from Properties 3 (P15), (P17) and (P19) that

$$\phi_\lambda(y) = (\alpha_\mathcal{A} + \beta_\bullet \cdot y)^\lambda (\alpha_\mathcal{H} + \beta_\bullet \cdot y)^{1-\lambda} - \lambda\,(\alpha_\mathcal{A} + \beta_\bullet \cdot y) - (1-\lambda)\,(\alpha_\mathcal{H} + \beta_\bullet \cdot y) < 0,$$

$$\phi_\lambda'(y) = \lambda \cdot \beta_\bullet \cdot \left(\frac{\alpha_\mathcal{A} + \beta_\bullet \cdot y}{\alpha_\mathcal{H} + \beta_\bullet \cdot y}\right)^{\lambda-1} + (1-\lambda) \cdot \beta_\bullet \cdot \left(\frac{\alpha_\mathcal{A} + \beta_\bullet \cdot y}{\alpha_\mathcal{H} + \beta_\bullet \cdot y}\right)^{\lambda} - \beta_\bullet > 0 \qquad \text{and that}$$

$$\phi_\lambda''(y) = -\left(\frac{\alpha_\mathcal{A} + \beta_\bullet \cdot y}{\alpha_\mathcal{H} + \beta_\bullet \cdot y}\right)^\lambda \cdot \frac{\lambda(1-\lambda) \cdot \beta_\bullet^2 \cdot (\alpha_\mathcal{A} - \alpha_\mathcal{H})^2}{(\alpha_\mathcal{A} + \beta_\bullet \cdot y)^2 (\alpha_\mathcal{H} + \beta_\bullet \cdot y)} < 0, \tag{A9}$$

which immediately implies $\lim_{y\to\infty} \phi_\lambda(y) = \lim_{y\to\infty} \phi_\lambda'(y) = \lim_{y\to\infty} \phi_\lambda''(y) = 0$ and with l'Hospital's rule

$$\lim_{y\to\infty} y \cdot \phi_\lambda(y) = \lim_{y\to\infty} -y^2 \cdot \phi_\lambda'(y) = \lim_{y\to\infty} \frac{y^3}{2} \cdot \phi_\lambda''(y) \tag{A10}$$

$$= -\frac{1}{2} \lim_{y\to\infty} \left(\frac{\alpha_\mathcal{A} + \beta_\bullet \cdot y}{\alpha_\mathcal{H} + \beta_\bullet \cdot y}\right)^\lambda \cdot \frac{\lambda(1-\lambda) \cdot \beta_\bullet^2 \cdot (\alpha_\mathcal{A} - \alpha_\mathcal{H})^2}{(\alpha_\mathcal{A}/y + \beta_\bullet)^2 (\alpha_\mathcal{H}/y + \beta_\bullet)} = -\frac{1}{2}\lambda(1-\lambda) \frac{(\alpha_\mathcal{A} - \alpha_\mathcal{H})^2}{\beta_\bullet}.$$

The formulas (A5), (A7) and (A9) imply the limits $\lim_{y\to\infty} p_y = \alpha_\lambda$, $\lim_{y\to\infty} q_y = \beta_\bullet$, $\lim_{y\to\infty} \overline{x}_0^{(qy)} = 0$. Notice that $p_y < \alpha_\lambda$ holds trivially for all $y \geq 0$ since the intercept $(p_y - \alpha_\lambda)$ of the tangent line $\phi_{\lambda,y}^{\tan}(\cdot)$ is negative. Incorporating (A8) we therefore obtain $\lim_{y\to\infty} h(y) \leq \lim_{y\to\infty} \overline{h}(y) = 0$. As mentioned before, for the proof it is sufficient to show that $\overline{h}(y) < 0$ for some $y \geq y_2$. This holds true if $\lim_{y\to\infty} y \cdot \overline{h}(y) < 0$. To verify this, notice first that from (A5), (A7) and (A8) we get

$$\overline{h}'(y) = -p_y \cdot e^{\overline{x}_0^{(qy)}} \cdot \phi_\lambda''(y) \cdot \left[1 - \frac{2 - \phi_\lambda'(y) - \beta_\bullet}{\sqrt{(1-q_y)^2 - 2(q_y - \beta_\bullet)}}\right] - y \cdot \phi_\lambda''(y) \cdot e^{\overline{x}_0^{(qy)}} \overset{y\to\infty}{\longrightarrow} 0. \tag{A11}$$

Finally we obtain with (A10)

$$\lim_{y\to\infty} y \cdot \overline{h}(y) = -\lim_{y\to\infty} y^2 \cdot \overline{h}'(y)$$

$$= \lim_{y\to\infty} p_y \cdot e^{\overline{x}_0^{(qy)}} \cdot y^2 \cdot \phi_\lambda''(y) \cdot \left[1 - \frac{2 - \phi_\lambda'(y) - \beta_\bullet}{\sqrt{(1-q_y)^2 - 2(q_y - \beta_\bullet)}}\right] + y^3 \cdot \phi_\lambda''(y) \cdot e^{\overline{x}_0^{(qy)}}$$

$$= 0 - \lambda(1-\lambda) \cdot \frac{(\alpha_\mathcal{A} - \alpha_\mathcal{H})^2}{\beta_\bullet} < 0. \qquad \square$$

Proof of Corollary 1. Part (a) follows directly from Proposition 1 (a),(b) and the limit $\lim_{n\to\infty} H_\lambda(P_{\mathcal{A},n}||P_{\mathcal{H},n}) = 0$ in the respective part (c) of the Propositions 7, 8, 9 as well as from (51). To prove part (b), according to (26) we have to verify $\liminf_{\lambda\nearrow 1}\{\liminf_{n\to\infty} H_\lambda(P_{\mathcal{A},n}||P_{\mathcal{H},n})\} = 1$. From part (c) of Proposition 2 we see that this is satisfied iff $\lim_{\lambda\uparrow 1} x_0^{(q_\lambda^E)} = 0$. Recall that for fixed $\lambda \in\,]0,1[$ we have $\beta_\lambda = \lambda\beta_\mathcal{A} + (1-\lambda)\beta_\mathcal{H} > 0$, $q_\lambda^E = \beta_\mathcal{A}^\lambda \beta_\mathcal{H}^{1-\lambda} < \beta_\lambda$ (cf. Lemma A1) and from Properties 1 (P1) the unique negative solution $x_0^{(q_\lambda^E)} \in\,]-\beta_\lambda, q_\lambda^E - \beta_\lambda[$ of $\zeta_\lambda^{(q_\lambda^E)}(x) = q_\lambda^E e^x - \beta_\lambda = x$ (cf. (44)). Due to the continuity and boundedness of the map $\lambda \mapsto x_0^{(q_\lambda^E)}$ (for $\lambda \in [0,1]$) one gets that $\lim_{\lambda\nearrow 1} x_0^{(q_\lambda^E)}$ exists and is the smallest nonpositive solution of $\beta_\mathcal{A} e^x - \beta_\mathcal{A} = x$. From this, the part (b) as well as the non-contiguity in part (c) follow immediately. The other part of (c) is a direct consequence of

Proposition 1 (a),(b) and Proposition 2 (c). $\square$

Proof of Formula (59) . One can proceed similarly to the proof of formula (51) above. Recall $H_\lambda(P_{\mathcal{A},1}||P_{\mathcal{H},1}) = \exp\{\phi_\lambda(X_0)\} > 1$ for $X_0 \in \mathbb{N}$ (cf. (28), Lemma A1 and $f_{\mathcal{A}}(X_0) \neq f_{\mathcal{H}}(X_0)$ for all $X_0 \in \mathbb{N}$). For $(\beta_{\mathcal{A}}, \beta_{\mathcal{H}}, \alpha_{\mathcal{A}}, \alpha_{\mathcal{H}}, \lambda) \in \mathcal{P}_{\mathrm{SP},2} \times (\mathbb{R} \setminus [0,1])$ one gets $\phi_\lambda(0) = 0$, $\phi_\lambda(1) > 0$, and we define for $x \geq 0$

$$\underline{\phi_\lambda}(x) := \left\{ \begin{array}{ll} \phi_\lambda(1), & \text{if } x = 1, \\ 0, & \text{if } x \neq 1. \end{array} \right.$$

By means of the choice $\underline{\epsilon} := \varphi_\lambda(1) \cdot \left(e^{\phi_\lambda(1)} - 1 \right) > 0$, we obtain for all $n \in \mathbb{N} \setminus \{1\}$

$$\sum_{x_{n-1}=0}^{\infty} \frac{[\varphi_\lambda(x_{n-2})]^{x_{n-1}}}{x_{n-1}!} \cdot \exp\{\phi_\lambda(x_{n-1})\} \geq \sum_{x_{n-1}=0}^{\infty} \frac{[\varphi_\lambda(x_{n-2})]^{x_{n-1}}}{x_{n-1}!} \cdot \exp\{\underline{\phi_\lambda}(x_{n-1})\}$$

$$= \exp\{\varphi_\lambda(x_{n-2})\} + \underline{\epsilon} = \exp\{\varphi_\lambda(x_{n-2})\} \cdot \left[1 + \underline{\epsilon} \cdot \exp\{-\varphi_\lambda(x_{n-2})\}\right].$$

Incorporating

$$\inf_{x \in \mathbb{N}_0} \left\{ \exp\{\phi_\lambda(x)\} \cdot \left[1 + \underline{\epsilon} \cdot \exp\{-\varphi_\lambda(x)\}\right] \right\} =: \underline{\delta} > 1,$$

one can show analogously to (A4) that

$$H_\lambda(P_{\mathcal{A},n}||P_{\mathcal{H},n}) \geq \cdots \geq \underline{\delta}^{\lfloor n/2 \rfloor} \stackrel{n \to \infty}{\longrightarrow} \infty. \qquad \square$$

Proof of the Formulas (61), (63) **and** (64). In the following, we slightly adapt the above-mentioned proof of formula (59). Let us define

$$\underline{\phi_\lambda}(x) := \left\{ \begin{array}{ll} \phi_\lambda(0), & \text{if } x = 0, \\ 0, & \text{if } x > 0. \end{array} \right.$$

In all respective subcases one clearly has $\underline{\phi_\lambda}(0) = \phi_\lambda(0) > 0$. With $\underline{\epsilon} := e^{\phi_\lambda(0)} - 1 > 0$ we obtain for all $n \in \mathbb{N} \setminus \{1\}$

$$\sum_{x_{n-1}=0}^{\infty} \frac{[\varphi_\lambda(x_{n-2})]^{x_{n-1}}}{x_{n-1}!} \cdot \exp\{\phi_\lambda(x_{n-1})\} \geq \sum_{x_{n-1}=0}^{\infty} \frac{[\varphi_\lambda(x_{n-2})]^{x_{n-1}}}{x_{n-1}!} \cdot \exp\{\underline{\phi_\lambda}(x_{n-1})\}$$

$$= \exp\{\varphi_\lambda(x_{n-2})\} + \underline{\epsilon} = \exp\{\varphi_\lambda(x_{n-2})\} \cdot \left[1 + \underline{\epsilon} \cdot \exp\{-\varphi_\lambda(x_{n-2})\}\right].$$

By employing

$$\inf_{x \in \mathbb{N}_0} \left\{ \exp\{\phi_\lambda(x)\} \cdot \left[1 + \underline{\epsilon} \cdot \exp\{-\varphi_\lambda(x)\}\right] \right\} =: \underline{\delta} > 1, \qquad (A12)$$

one can show analogously to (A4) that

$$H_\lambda(P_{\mathcal{A},n}||P_{\mathcal{H},n}) \geq \cdots \geq \underline{\delta}^{\lfloor n/2 \rfloor} \stackrel{n \to \infty}{\longrightarrow} \infty.$$

Notice that this method does not work for the parameter cases $\mathcal{P}_{\mathrm{SP},4a} \cup \mathcal{P}_{\mathrm{SP},4b}$, since there the infimum in (A12) is equal to one. $\square$

Proof of Proposition 13. In the setup $(\beta_{\mathcal{A}}, \beta_{\mathcal{H}}, \alpha_{\mathcal{A}}, \alpha_{\mathcal{H}}, \lambda) \in \mathcal{P}_{SP4a} \times (\mathbb{R} \backslash [0,1])$ we require $\beta_\bullet := \beta_{\mathcal{A}} = \beta_{\mathcal{H}} < 1$. As in the proof of Proposition 9, we stick to the tangent line $\phi_{\lambda,y}^{\tan}(\cdot)$ at $y \geq 0$ (cf. (52)) as a linear lower bound for $\phi_\lambda(\cdot)$, i.e., we use the function

$$\phi_{\lambda,y}^{\tan}(x) := (p_y - \alpha_\lambda) + (q_y - \beta_\bullet) \cdot x := \left(p_{\lambda,y}^{\tan} - \alpha_\lambda\right) + \left(q_{\lambda,y}^{\tan} - \beta_\lambda\right) \cdot x := \left(\phi_\lambda(y) - y \cdot \phi_\lambda'(y)\right) + \phi_\lambda'(y) \cdot x \,. \quad \text{(A13)}$$

As already mentioned in Section 3.21, on $\mathcal{P}_{SP4a}$ the function $\phi_\lambda(\cdot)$ is strictly decreasing and converges to 0. Thus, for all $y \geq 0$ the slope $\phi_\lambda'(y)$ of the tangent line at y is negative, which implies that $q_y < \beta_\lambda = \beta_\bullet$. For $\lambda \in \mathbb{R} \backslash [0,1]$ there clearly may hold $q_y < 0$ for some $y \in \mathbb{R}$. However, there exists a sufficiently large $y_1 > 0$ such that $q_y > 0$ for all $y > y_1$, since $\lim_{y \to \infty} \phi_\lambda'(y) = 0$ and hence $q_y \nearrow \beta_\bullet > 0$ for $y \to \infty$. Thus, let us suppose that $y > y_1$. Then, the sequence $\left(a_n^{(q_y)}\right)_{n \in \mathbb{N}}$ is strictly negative, strictly decreasing and converges to $x_0^{(q_y)} \in]-\beta_\bullet, q_y - \beta_\bullet[$ (cf. Properties 1 (P1)). If there is some $y \geq y_1$ such that $h(y) > 0$ with

$$h(y) := \lim_{n \to \infty} \frac{1}{n} \log \left(\widetilde{B}_{\lambda, X_0, n}^{(p_y, q_y)}\right) = p_y \cdot e^{x_0^{(q_y)}} - \alpha_\lambda \,,$$

then one can conclude that $\lim_{n \to \infty} H_\lambda(P_{\mathcal{A}, n} || P_{\mathcal{H}, n}) = \infty$. Let us at first consider the case $\alpha_\lambda \geq 0$. By employing $p_y \searrow \alpha_\lambda$ for $y \to \infty$, one gets $p_y > 0$ for all $y \geq 0$. Analogously to the proof of Proposition 9, we now look for a lower bound $\underline{x}_0^{(q_y)}$ of the fixed point $x_0^{(q_y)}$. Notice that $x_0^{(q_y)} > -\beta_\bullet$ implies

$$\underline{Q}_\lambda^{(q_y)}(x) := \frac{e^{-\beta_\bullet}}{2} \cdot q_y \cdot x^2 + q_y \cdot x + q_y - \beta_\bullet \leq q_y \cdot e^x - \beta_\bullet = \xi_\lambda^{(q_y)}(x) \,, \quad \text{(A14)}$$

since $\underline{Q}_\lambda^{(q_y)}(0) = \xi_\lambda^{(q_y)}(0) < 0$, $\underline{Q}_\lambda^{(q_y)\,'}(0) = \xi_\lambda^{(q_y)\,'}(0) > 0$ and $0 < \underline{Q}_\lambda^{(q_y)\,''}(x) < \xi_\lambda^{(q_y)\,''}(x)$ for $x \in]-\beta_\bullet, 0]$. Thus, the negative solution $\underline{x}_0^{(q_y)}$ of the equation $\underline{Q}_\lambda^{(q_y)}(x) = x$ (which definitely exists) implies that there holds $\underline{x}_0^{(q_y)} \leq x_0^{(q_y)}$. We easily obtain

$$\begin{aligned} \underline{x}_0^{(q_y)} &= \frac{e^{\beta_\bullet}}{q_y} \left[(1 - q_y) - \sqrt{(1 - q_y)^2 - 2e^{-\beta_\bullet} q_y (q_y - \beta_\bullet)}\right] \\ &= \frac{e^{\beta_\bullet}}{\phi_\lambda'(y) + \beta_\bullet} \left[(1 - \phi_\lambda'(y) - \beta_\bullet) - \sqrt{(1 - \phi_\lambda'(y) - \beta_\bullet)^2 - 2 \cdot e^{-\beta_\bullet} q_y \cdot \phi_\lambda'(y)}\right] < 0. \quad \text{(A15)} \end{aligned}$$

Since

$$h(y) = p_y \cdot e^{x_0^{(q_y)}} - \alpha_\lambda \geq p_y \cdot e^{\underline{x}_0^{(q_y)}} - \alpha_\lambda =: \underline{h}(y) \,, \quad \text{(A16)}$$

it is sufficient to show $\underline{h}(y) > 0$ for some $y > y_1$. We recall from Properties 3 (P15), (P17) and (P19) that

$$\begin{aligned} \phi_\lambda(y) &= (\alpha_{\mathcal{A}} + \beta_\bullet \cdot y)^\lambda (\alpha_{\mathcal{H}} + \beta_\bullet \cdot y)^{1-\lambda} - \lambda (\alpha_{\mathcal{A}} + \beta_\bullet \cdot y) - (1 - \lambda)(\alpha_{\mathcal{H}} + \beta_\bullet \cdot y) > 0, \\ \phi_\lambda'(y) &= \lambda \cdot \beta_\bullet \cdot \left(\frac{\alpha_{\mathcal{A}} + \beta_\bullet \cdot y}{\alpha_{\mathcal{H}} + \beta_\bullet \cdot y}\right)^{\lambda - 1} + (1 - \lambda) \cdot \beta_\bullet \cdot \left(\frac{\alpha_{\mathcal{A}} + \beta_\bullet \cdot y}{\alpha_{\mathcal{H}} + \beta_\bullet \cdot y}\right)^\lambda - \beta_\bullet < 0 \quad \text{and} \\ \phi_\lambda''(y) &= -\left(\frac{\alpha_{\mathcal{A}} + \beta_\bullet \cdot y}{\alpha_{\mathcal{H}} + \beta_\bullet \cdot y}\right)^\lambda \cdot \frac{\lambda(1 - \lambda) \cdot \beta_\bullet^2 \cdot (\alpha_{\mathcal{A}} - \alpha_{\mathcal{H}})^2}{(\alpha_{\mathcal{A}} + \beta_\bullet \cdot y)^2 (\alpha_{\mathcal{H}} + \beta_\bullet \cdot y)} > 0, \quad \text{(A17)} \end{aligned}$$

which immediately implies $\lim_{y \to \infty} \phi_\lambda(y) = \lim_{y \to \infty} \phi_\lambda'(y) = \lim_{y \to \infty} \phi_\lambda''(y) = 0$, and by means of l'Hospital's rule

$$\lim_{y \to \infty} y \cdot \phi_\lambda(y) = \lim_{y \to \infty} -y^2 \cdot \phi_\lambda'(y) = \lim_{y \to \infty} \frac{y^3}{2} \cdot \phi_\lambda''(y) \quad \text{(A18)}$$

$$= -\frac{1}{2} \lim_{y \to \infty} \left(\frac{\alpha_{\mathcal{A}} + \beta_\bullet \cdot y}{\alpha_{\mathcal{H}} + \beta_\bullet \cdot y}\right)^\lambda \cdot \frac{\lambda(1 - \lambda) \cdot \beta_\bullet^2 \cdot (\alpha_{\mathcal{A}} - \alpha_{\mathcal{H}})^2}{(\alpha_{\mathcal{A}}/y + \beta_\bullet)^2 (\alpha_{\mathcal{H}}/y + \beta_\bullet)} = -\frac{1}{2} \lambda(1 - \lambda) \cdot \frac{(\alpha_{\mathcal{A}} - \alpha_{\mathcal{H}})^2}{\beta_\bullet} \,.$$

The Formulas (A13), (A15), (A17) imply the limits $\lim_{y\to\infty} p_y = \alpha_\lambda$, $\lim_{y\to\infty} q_y = \beta_\bullet$ and $\lim_{y\to\infty} \underline{x}_0^{(q_y)} = 0$ iff $\beta_\bullet \leq 1$. The latter is due to the fact that for $\beta_\bullet > 1$ one gets with (A15) $\lim_{y\to\infty} \underline{x}_0^{(q_y)} = \frac{e^{\beta_\bullet}}{\beta_\bullet}\left[(1-\beta_\bullet) - \sqrt{(1-\beta_\bullet)^2}\right] = \frac{e^{\beta_\bullet}}{\beta_\bullet}\left[2 - 2\beta_\bullet\right] \neq 0$. In the following, let us assume $\beta_\bullet < 1$ (the reason why we exclude the case $\beta_\bullet = 1$ is explained below). One gets $\lim_{y\to\infty} h(y) \geq \lim_{y\to\infty} \underline{h}(y) = 0$. Since we have to prove that $\underline{h}(y) > 0$ for some $y > y_1$, it is sufficient to show that $\lim_{y\to\infty} y \cdot \underline{h}(y) > 0$. To verify the latter, we first derive with l'Hospital's rule and with (A17), (A18)

$$
\begin{aligned}
\lim_{y\to\infty} y \cdot \left(1 - e^{\underline{x}_0^{(q_y)}}\right) &= \lim_{y\to\infty} y^2 \cdot e^{\underline{x}_0^{(q_y)}} \cdot \left(\frac{\partial}{\partial y}\underline{x}_0^{(q_y)}\right) \\
&= \lim_{y\to\infty} \left\{ y^2 \cdot \frac{-e^{\beta_\bullet} \cdot \phi''_\lambda(y)}{(\phi'_\lambda(y) + \beta_\bullet)^2} \cdot \left[(1-q_y) - \sqrt{(1-q_y)^2 - 2e^{-\beta_\bullet} q_y(q_y - \beta_\bullet)}\right] \right. \\
&\quad \left. + \frac{e^{\beta_\bullet}}{q_y} \cdot \left[-y^2 \cdot \phi''_\lambda(y) - \frac{-2y^2\phi''_\lambda(y)(1-q_y) - 2y^2\phi''_\lambda(y)e^{-\beta_\bullet}q_y - 2y^2\phi''_\lambda(y)e^{-\beta_\bullet}\phi'_\lambda(y)}{2 \cdot \sqrt{(1-q_y)^2 - 2e^{-\beta_\bullet}q_y(q_y - \beta_\bullet)}}\right] \right\} \\
&= 0 .
\end{aligned}
\tag{A19}
$$

Notice that without further examination this limit would not necessarily hold for $\beta_\bullet = 1$, since then the denominator in (A19) converges to zero. With (A13), (A16), (A18) and (A19) we finally obtain

$$
\begin{aligned}
\lim_{y\to\infty} y \cdot \underline{h}(y) &= \lim_{y\to\infty} \left\{ \left(y \cdot \phi_\lambda(y) - y^2 \cdot \phi'_\lambda(y)\right) \cdot e^{\underline{x}_0^{(q_y)}} - y \cdot \left(1 - e^{\underline{x}_0^{(q_y)}}\right) \alpha_\lambda \right\} \\
&= -\lambda(1-\lambda)\frac{(\alpha_\mathcal{A} - \alpha_\mathcal{H})^2}{\beta_\bullet} > 0 .
\end{aligned}
\tag{A20}
$$

Let us now consider the case $\alpha_\lambda < 0$. The proof works out almost completely analogous to the case $\alpha_\lambda \geq 0$. We indicate the main differences. Since $p_y \searrow \alpha_\lambda < 0$ and $q_y \nearrow \beta_\bullet \in\,]0,1[$ for $y \to \infty$, there is a sufficiently large $y_2 > y_1$, such that $p_y < 0$ and $q_y > 0$. Thus,

$$
\overline{Q}_\lambda^{(q_y)}(x) := \frac{q_y}{2} \cdot x^2 + q_y \cdot x + q_y - \beta_\bullet \geq \zeta_\lambda^{(q_y)}(x) = q_y e^x - \beta_\bullet \qquad \text{for } x \in\,]-\infty, 0].
$$

The corresponding (existing) smaller solution of $\overline{Q}_\lambda^{(q_y)}(x) = x$ is

$$
\overline{x}_0^{(q_y)} = \frac{1}{q_y}\left[(1-q_y) - \sqrt{(1-q_y)^2 - 2q_y(q_y - \beta_\bullet)}\right] ,
$$

having the same form as the solution (A15) with $e^{-\beta_\bullet}$ substituted by 1. Notice that there clearly holds $\underline{x}_0^{(q_y)} < \overline{x}_0^{(q_y)} < 0$. However, since $p_y < 0$, we now get $h(y) = p_y \cdot e^{\underline{x}_0^{(q_y)}} - \alpha_\lambda \geq p_y \cdot e^{\overline{x}_0^{(q_y)}} - \alpha_\lambda =: \underline{h}(y)$, as in (A16). Since all calculations (A17) to (A20) remain valid (with $e^{-\beta_\bullet}$ substituted by 1), this proof is finished. $\square$

Appendix A.2. Proofs and Auxiliary Lemmas for Section 5

We start with two lemmas which will be useful for the proof of Theorem 3. They deal with the sequence $\left(a_n^{(q_\lambda)}\right)_{n\in\mathbb{N}}$ from (36).

Lemma A2. *For arbitrarily fixed parameter constellation $(\beta_\mathcal{A}, \beta_\mathcal{H}, \alpha_\mathcal{A}, \alpha_\mathcal{H}, \lambda) \in \mathcal{P}\times]0,1[$, suppose that $q_\lambda > 0$ and $\lim_{\lambda\nearrow 1} q_\lambda = \beta_\mathcal{A}$ holds. Then one gets the limit*

$$
\forall\, n \in \mathbb{N}: \quad \lim_{\lambda\nearrow 1} a_n^{(q_\lambda)} = 0.
\tag{A21}
$$

Proof. This can be easily seen by induction: for $n = 1$ there clearly holds

$$\lim_{\lambda \nearrow 1} a_1^{(q_\lambda)} = \lim_{\lambda \nearrow 1} (q_\lambda - \beta_\lambda) = \beta_\mathcal{A} - \beta_\mathcal{A} = 0.$$

Assume now that $\lim_{\lambda \nearrow 1} a_k^{(q_\lambda)} = 0$ holds for all $k \in \mathbb{N}, k \le n-1$, then

$$\lim_{\lambda \nearrow 1} a_n^{(q_\lambda)} = \lim_{\lambda \nearrow 1} (q_\lambda \cdot e^{a_{n-1}^{(q_\lambda)}} - \beta_\lambda) = \beta_\mathcal{A} \cdot 1 - \beta_\mathcal{A} = 0. \qquad \square$$

Lemma A3. *In addition to the assumptions of Lemma A2, suppose that $\lambda \mapsto q_\lambda$ is continuously differentiable on $]0, 1[$ and that the limit $l := \lim_{\lambda \nearrow 1} \frac{\partial q_\lambda}{\partial \lambda}$ is finite. Then, for all $n \in \mathbb{N}$ one obtains*

$$\lim_{\lambda \nearrow 1} \frac{\partial a_n^{(q_\lambda)}}{\partial \lambda} = u_n := \begin{cases} \frac{l + \beta_\mathcal{H} - \beta_\mathcal{A}}{1 - \beta_\mathcal{A}} \cdot \left(1 - (\beta_\mathcal{A})^n\right), & \text{if } \beta_\mathcal{A} \ne 1, \\[2mm] n \cdot (l + \beta_\mathcal{H} - 1), & \text{if } \beta_\mathcal{A} = 1, \end{cases} \tag{A22}$$

which is the unique solution of the linear recursion equation

$$u_n = l + \beta_\mathcal{H} - \beta_\mathcal{A} + \beta_\mathcal{A} \cdot u_{n-1}, \qquad u_0 = 0. \tag{A23}$$

Furthermore, for all $n \in \mathbb{N}$ there holds

$$\sum_{k=1}^{n} \lim_{\lambda \nearrow 1} \frac{\partial a_k^{(q_\lambda)}}{\partial \lambda} = \sum_{k=1}^{n} u_k = \begin{cases} \frac{l + \beta_\mathcal{H} - \beta_\mathcal{A}}{1 - \beta_\mathcal{A}} \cdot \left[n - \frac{\beta_\mathcal{A}}{1 - \beta_\mathcal{A}} \left(1 - (\beta_\mathcal{A})^n\right)\right], & \text{if } \beta_\mathcal{A} \ne 1, \\[2mm] \frac{n \cdot (n+1)}{2} \cdot (l + \beta_\mathcal{H} - 1), & \text{if } \beta_\mathcal{A} = 1. \end{cases}$$

Proof. Clearly, u_n defined by (A22) is the unique solution of (A23). We prove by induction that $\lim_{\lambda \nearrow 1} \frac{\partial a_n^{(q_\lambda)}}{\partial \lambda} = u_n$ holds. For $n = 1$ one gets

$$\lim_{\lambda \nearrow 1} \frac{\partial a_1^{(q_\lambda)}}{\partial \lambda} = \lim_{\lambda \nearrow 1} \frac{\partial (q_\lambda - \beta_\lambda)}{\partial \lambda} = l - (\beta_\mathcal{A} - \beta_\mathcal{H}) = u_1.$$

Suppose now that (A22) holds for all $k \in \mathbb{N}, k \le n-1$. Then, by incorporating (A21) we obtain

$$\lim_{\lambda \nearrow 1} \frac{\partial a_n^{(q_\lambda)}}{\partial \lambda} = \lim_{\lambda \nearrow 1} \frac{\partial}{\partial \lambda} \left(q_\lambda \cdot e^{a_{n-1}^{(q_\lambda)}} - \beta_\lambda\right) = \lim_{\lambda \nearrow 1} e^{a_{n-1}^{(q_\lambda)}} \cdot \left(\frac{\partial q_\lambda}{\partial \lambda} + q_\lambda \frac{\partial a_{n-1}^{(q_\lambda)}}{\partial \lambda}\right) - (\beta_\mathcal{A} - \beta_\mathcal{H})$$

$$= l - (\beta_\mathcal{A} - \beta_\mathcal{H}) + \beta_\mathcal{A} \cdot u_{n-1} = u_n.$$

The remaining assertions follow immediately. $\square$

We are now ready to give the

Proof of Theorem 3. (a) Recall that for the setup $(\beta_\mathcal{A}, \beta_\mathcal{H}, \alpha_\mathcal{A}, \alpha_\mathcal{H}) \in (\mathcal{P}_{\mathrm{NI}} \cup \mathcal{P}_{\mathrm{SP},1})$ we chose the intercept as $p_\lambda := p_\lambda^E := \alpha_\mathcal{A}^\lambda \alpha_\mathcal{H}^{1-\lambda}$ and the slope as $q_\lambda := q_\lambda^E := \beta_\mathcal{A}^\lambda \beta_\mathcal{H}^{1-\lambda}$, which in (39) lead to the exact value $V_{\lambda,X_0,n}$ of the Hellinger integral. Because of $\frac{p_\lambda}{q_\lambda}\beta_\lambda - \alpha_\lambda = 0$ as well as $\lim_{\lambda\nearrow 1} q_\lambda = \beta_\mathcal{A}$, we obtain by using (38) and Lemma A2 for all $X_0 \in \mathbb{N}$ and for all $n \in \mathbb{N}$

$$\lim_{\lambda\nearrow 1} V_{\lambda,X_0,n} := \lim_{\lambda\nearrow 1} \exp\left\{ a_n^{(q_\lambda)} \cdot X_0 + \sum_{k=1}^{n} b_k^{(p_\lambda,q_\lambda)} \right\} = \lim_{\lambda\nearrow 1} \exp\left\{ a_n^{(q_\lambda)} \cdot X_0 + \frac{\alpha_\mathcal{A}}{\beta_\mathcal{A}} \sum_{k=1}^{n} a_k^{(q_\lambda)} \right\} = 1,$$

which leads by (68) to

$$
\begin{aligned}
I(P_{\mathcal{A},n}||P_{\mathcal{H},n}) &= \lim_{\lambda\nearrow 1} \frac{1 - H_\lambda(P_{\mathcal{A},n}||P_{\mathcal{H},n})}{\lambda\cdot(1-\lambda)} = \lim_{\lambda\nearrow 1} \frac{1 - V_{\lambda,X_0,n}}{\lambda\cdot(1-\lambda)} \\[2mm]
&= \lim_{\lambda\nearrow 1} \frac{-V_{\lambda,X_0,n}}{1-2\lambda} \cdot \frac{\partial}{\partial\lambda}\left[a_n^{(q_\lambda)} \cdot X_0 + \frac{p_\lambda}{q_\lambda} \sum_{k=1}^{n} a_k^{(q_\lambda)} \right] \\[2mm]
&= \lim_{\lambda\nearrow 1} \left[\frac{\partial a_n^{(q_\lambda)}}{\partial\lambda} \cdot X_0 + \left(\frac{\partial}{\partial\lambda}\frac{p_\lambda}{q_\lambda}\right) \cdot \sum_{k=1}^{n} a_k^{(q_\lambda)} + \frac{p_\lambda}{q_\lambda} \cdot \sum_{k=1}^{n} \frac{\partial a_k^{(q_\lambda)}}{\partial\lambda} \right]. \quad (A24)
\end{aligned}
$$

For further analysis, we use the obvious derivatives

$$\frac{\partial p_\lambda}{\partial\lambda} = p_\lambda \log\left(\frac{\alpha_\mathcal{A}}{\alpha_\mathcal{H}}\right), \qquad \frac{\partial}{\partial\lambda}\frac{p_\lambda}{q_\lambda} = \frac{p_\lambda}{q_\lambda} \log\left(\frac{\alpha_\mathcal{A}\beta_\mathcal{H}}{\alpha_\mathcal{H}\beta_\mathcal{A}}\right), \qquad \frac{\partial q_\lambda}{\partial\lambda} = q_\lambda \log\left(\frac{\beta_\mathcal{A}}{\beta_\mathcal{H}}\right), \quad (A25)$$

where the subcase $(\beta_\mathcal{A},\beta_\mathcal{H},\alpha_\mathcal{A},\alpha_\mathcal{H}) \in \mathcal{P}_{\mathrm{NI}}$ (with $p_\lambda \equiv 0$) is consistently covered. From (A25) and Lemma A3 we deduce

$$\lim_{\lambda\nearrow 1} \frac{\partial a_n^{(q_\lambda)}}{\partial\lambda} \cdot X_0 = \begin{cases} \left(\beta_\mathcal{A} \log\left(\frac{\beta_\mathcal{A}}{\beta_\mathcal{H}}\right) - (\beta_\mathcal{A} - \beta_\mathcal{H})\right) \cdot \frac{1-(\beta_\mathcal{A})^n}{1-\beta_\mathcal{A}} \cdot X_0, & \text{if } \beta_\mathcal{A} \neq 1, \\[2mm] n \cdot \left(\beta_\mathcal{A} \log\left(\frac{\beta_\mathcal{A}}{\beta_\mathcal{H}}\right) - (\beta_\mathcal{A} - \beta_\mathcal{H})\right) \cdot X_0, & \text{if } \beta_\mathcal{A} = 1, \end{cases}$$

and by means of (A21)

$$\forall\, n \in \mathbb{N}: \quad \lim_{\lambda\nearrow 1} \left[\left(\frac{\partial}{\partial\lambda}\frac{p_\lambda}{q_\lambda}\right) \cdot \sum_{k=1}^{n} a_k^{(q_\lambda)} \right] = 0.$$

For the last expression in (A24) we again apply Lemma A3 to end up with

$$\lim_{\lambda\nearrow 1} \frac{p_\lambda}{q_\lambda} \cdot \sum_{k=1}^{n} \frac{\partial}{\partial\lambda} a_k^{(q_\lambda)} = \begin{cases} \frac{\alpha_\mathcal{A}\cdot\left[\beta_\mathcal{A} \log\left(\frac{\beta_\mathcal{A}}{\beta_\mathcal{H}}\right) - (\beta_\mathcal{A}-\beta_\mathcal{H})\right]}{\beta_\mathcal{A}(1-\beta_\mathcal{A})} \cdot \left[n - \frac{\beta_\mathcal{A}}{1-\beta_\mathcal{A}}\left(1 - (\beta_\mathcal{A})^n\right)\right], & \text{if } \beta_\mathcal{A} \neq 1, \\[2mm] n \cdot (n+1)\frac{\alpha_\mathcal{A}}{2\beta_\mathcal{A}} \cdot \left[\beta_\mathcal{A} \log\left(\frac{\beta_\mathcal{A}}{\beta_\mathcal{H}}\right) - (\beta_\mathcal{A}-\beta_\mathcal{H})\right], & \text{if } \beta_\mathcal{A} = 1, \end{cases} \quad (A26)$$

which finishes the proof of part (a). To show part (b), for the corresponding setup $(\beta_\mathcal{A},\beta_\mathcal{H},\alpha_\mathcal{A},\alpha_\mathcal{H}) \in \mathcal{P}_{\mathrm{SP}}\backslash\mathcal{P}_{\mathrm{SP},1}$ let us first choose – according to (45) in Section 3.4—the intercept as $p_\lambda := p_\lambda^L := \alpha_\mathcal{A}^\lambda \alpha_\mathcal{H}^{1-\lambda}$ and the slope as $q_\lambda := q_\lambda^L := \beta_\mathcal{A}^\lambda \beta_\mathcal{H}^{1-\lambda}$, which in part (b) of Proposition 6 lead to the lower bounds $B_{\lambda,X_0,n}^L$ of the Hellinger integral. This is formally the same choice as in part (a) satisfying $\lim_{\lambda\nearrow 1} p_\lambda = \alpha_\mathcal{A}$, $\lim_{\lambda\nearrow 1} q_\lambda = \beta_\mathcal{A}$ but in contrast to (a) we now have $\frac{p_\lambda}{q_\lambda}\beta_\lambda - \alpha_\lambda \neq 0$ but nevertheless

$$\lim_{\lambda\nearrow 1} \frac{p_\lambda}{q_\lambda}\beta_\lambda - \alpha_\lambda = 0.$$

From this, (38), part (b) of Proposition 6 and Lemma A2 we obtain

$$\lim_{\lambda\nearrow 1} B_{\lambda,X_0,n}^L = \lim_{\lambda\nearrow 1} \exp\left\{ a_n^{(q_\lambda)} \cdot X_0 + \frac{p_\lambda}{q_\lambda} \sum_{k=1}^{n} a_k^{(q_\lambda)} + n \cdot \left(\frac{p_\lambda}{q_\lambda}\beta_\lambda - \alpha_\lambda\right) \right\} = 1 \quad (A27)$$

and hence

$$I(P_{\mathcal{A},n}\|P_{\mathcal{H},n}) \;\leq\; \lim_{\lambda\nearrow 1}\frac{1-B^{L}_{\lambda,X_0,n}}{\lambda\cdot(1-\lambda)} = \lim_{\lambda\nearrow 1}\frac{-B^{L}_{\lambda,X_0,n}}{1-2\lambda}\cdot\frac{\partial}{\partial\lambda}\left[a_n^{(q_\lambda)}X_0 + \frac{p_\lambda}{q_\lambda}\sum_{k=1}^{n}a_k^{(q_\lambda)} + n\left(\frac{p_\lambda}{q_\lambda}\beta_\lambda - \alpha_\lambda\right)\right]$$

$$= \lim_{\lambda\nearrow 1}\left[\frac{\partial a_n^{(q_\lambda)}}{\partial\lambda}X_0 + \left(\frac{\partial}{\partial\lambda}\frac{p_\lambda}{q_\lambda}\right)\sum_{k=1}^{n}a_k^{(q_\lambda)} + \frac{p_\lambda}{q_\lambda}\sum_{k=1}^{n}\frac{\partial a_k^{(q_\lambda)}}{\partial\lambda} + n\frac{\partial}{\partial\lambda}\left(\frac{p_\lambda}{q_\lambda}\beta_\lambda - \alpha_\lambda\right)\right]. \qquad (A28)$$

In the current setup, the first three expressions in (A28) can be evaluated in exactly the same way as in (A25) to (A26), and for the last expression one has the limit

$$\frac{\partial}{\partial\lambda}\left(\frac{p_\lambda}{q_\lambda}\beta_\lambda - \alpha_\lambda\right) \;=\; \frac{p_\lambda}{q_\lambda}\log\left(\frac{\alpha_{\mathcal{A}}\beta_{\mathcal{H}}}{\alpha_{\mathcal{H}}\beta_{\mathcal{A}}}\right)\cdot\beta_\lambda + \frac{p_\lambda}{q_\lambda}\cdot(\beta_{\mathcal{A}}-\beta_{\mathcal{H}}) - (\alpha_{\mathcal{A}}-\alpha_{\mathcal{H}})$$

$$\xrightarrow{\lambda\nearrow 1}\; \alpha_{\mathcal{A}}\left[\log\left(\frac{\alpha_{\mathcal{A}}\beta_{\mathcal{H}}}{\alpha_{\mathcal{H}}\beta_{\mathcal{A}}}\right) - \frac{\beta_{\mathcal{H}}}{\beta_{\mathcal{A}}}\right] + \alpha_{\mathcal{H}}\,,$$

which finishes the proof of part (b). $\square$

Proof of Theorem 4. Let us fix $(\beta_{\mathcal{A}},\beta_{\mathcal{H}},\alpha_{\mathcal{A}},\alpha_{\mathcal{H}}) \in \mathcal{P}_{\mathrm{SP}}\backslash\mathcal{P}_{\mathrm{SP},1}$, $X_0 \in \mathbb{N}$, $n \in \mathbb{N}$ and $y \in [0,\infty[$. The lower bound $E^{L,tan}_{y,X_0,n}$ of the Kullback-Leibler information divergence (relative entropy) is derived by using $\phi_\lambda^{U} \equiv \phi_{\lambda,y}^{tan}$ (cf. (52)), which corresponds to the tangent line of ψ_λ at y, as a linear upper bound for ϕ_λ ($\lambda \subset]0,1[$). More precisely, one gets $\phi_\lambda^{U}(x) := (p_\lambda^{U} - \alpha_\lambda) + (q_\lambda^{U} - \beta_\lambda)x$ ($x \in [0,\infty[$) with $p_\lambda := p_\lambda(y) := \phi_\lambda(y) - y\phi_\lambda'(y) + \alpha_\lambda$ and $q_\lambda := q_\lambda(y) := \phi_\lambda'(y) + \beta_\lambda$, implying $q_\lambda > 0$ because of Properties 3 (P17). Analogously to (A27) and (A28), we obtain from (38) and (40) the convergence $\lim_{\lambda\nearrow 1}B^{U}_{\lambda,X_0,n} = 1$ and thus

$$I(P_{\mathcal{A},n}\|P_{\mathcal{H},n}) \;\geq\; \lim_{\lambda\nearrow 1}\left[\frac{\partial a_n^{(q_\lambda)}}{\partial\lambda}X_0 + \left(\frac{\partial}{\partial\lambda}\frac{p_\lambda}{q_\lambda}\right)\sum_{k=1}^{n}a_k^{(q_\lambda)} + \frac{p_\lambda}{q_\lambda}\sum_{k=1}^{n}\frac{\partial a_k^{(q_\lambda)}}{\partial\lambda} + n\frac{\partial}{\partial\lambda}\left(\frac{p_\lambda}{q_\lambda}\beta_\lambda - \alpha_\lambda\right)\right]. \qquad (A29)$$

As before, we compute the involved derivatives. From (30) to (32) as well as (P17) we get

$$\frac{\partial p_\lambda}{\partial\lambda} = \left(\frac{f_{\mathcal{A}}(y)}{f_{\mathcal{H}}(y)}\right)^{\lambda}f_{\mathcal{H}}(y)\log\left(\frac{f_{\mathcal{A}}(y)}{f_{\mathcal{H}}(y)}\right) - \beta_{\mathcal{A}}y\left(\frac{f_{\mathcal{A}}(y)}{f_{\mathcal{H}}(y)}\right)^{\lambda-1} - \lambda\beta_{\mathcal{A}}y\left(\frac{f_{\mathcal{A}}(y)}{f_{\mathcal{H}}(y)}\right)^{\lambda-1}\log\left(\frac{f_{\mathcal{A}}(y)}{f_{\mathcal{H}}(y)}\right)$$

$$+ \beta_{\mathcal{H}}y\left(\frac{f_{\mathcal{A}}(y)}{f_{\mathcal{H}}(y)}\right)^{\lambda} - (1-\lambda)\beta_{\mathcal{H}}y\left(\frac{f_{\mathcal{A}}(y)}{f_{\mathcal{H}}(y)}\right)^{\lambda}\log\left(\frac{f_{\mathcal{A}}(y)}{f_{\mathcal{H}}(y)}\right)$$

$$\xrightarrow{\lambda\nearrow 1}\; \alpha_{\mathcal{A}}\log\left(\frac{f_{\mathcal{A}}(y)}{f_{\mathcal{H}}(y)}\right) + \frac{y\cdot(\alpha_{\mathcal{A}}\beta_{\mathcal{H}} - \alpha_{\mathcal{H}}\beta_{\mathcal{A}})}{f_{\mathcal{H}}(y)}\,, \qquad (A30)$$

and

$$\frac{\partial q_\lambda}{\partial\lambda} = \beta_{\mathcal{A}}\left(\frac{f_{\mathcal{A}}(y)}{f_{\mathcal{H}}(y)}\right)^{\lambda-1} + \lambda\beta_{\mathcal{A}}\left(\frac{f_{\mathcal{A}}(y)}{f_{\mathcal{H}}(y)}\right)^{\lambda-1}\log\left(\frac{f_{\mathcal{A}}(y)}{f_{\mathcal{H}}(y)}\right) - \beta_{\mathcal{H}}\left(\frac{f_{\mathcal{A}}(y)}{f_{\mathcal{H}}(y)}\right)^{\lambda}$$

$$+ (1-\lambda)\beta_{\mathcal{H}}\left(\frac{f_{\mathcal{A}}(y)}{f_{\mathcal{H}}(y)}\right)^{\lambda}\log\left(\frac{f_{\mathcal{A}}(y)}{f_{\mathcal{H}}(y)}\right)$$

$$\xrightarrow{\lambda\nearrow 1}\; \beta_{\mathcal{A}}\left(1 + \log\left(\frac{f_{\mathcal{A}}(y)}{f_{\mathcal{H}}(y)}\right)\right) - \beta_{\mathcal{H}}\frac{f_{\mathcal{A}}(y)}{f_{\mathcal{H}}(y)} \;=:\; l. \qquad (A31)$$

Combining these two limits we get

$$\frac{\partial}{\partial\lambda}\left(\frac{p_\lambda}{q_\lambda}\beta_\lambda - \alpha_\lambda\right) \;=\; \frac{q_\lambda\left(\frac{\partial p_\lambda}{\partial\lambda}\right) - p_\lambda\left(\frac{\partial q_\lambda}{\partial\lambda}\right)}{(q_\lambda)^2}\cdot\beta_\lambda + \frac{p_\lambda}{q_\lambda}\cdot(\beta_{\mathcal{A}}-\beta_{\mathcal{H}}) - (\alpha_{\mathcal{A}}-\alpha_{\mathcal{H}})$$

$$\xrightarrow{\lambda\nearrow 1}\; \left[\frac{y\cdot(\alpha_{\mathcal{A}}\beta_{\mathcal{H}} - \alpha_{\mathcal{H}}\beta_{\mathcal{A}})}{f_{\mathcal{H}}(y)} - \alpha_{\mathcal{A}}\left(1 - \frac{\beta_{\mathcal{H}}f_{\mathcal{A}}(y)}{\beta_{\mathcal{A}}f_{\mathcal{H}}(y)}\right)\right] + \alpha_{\mathcal{H}} - \frac{\alpha_{\mathcal{A}}\beta_{\mathcal{H}}}{\beta_{\mathcal{A}}}.$$

$$= \left(\alpha_{\mathcal{H}} - \alpha_{\mathcal{A}}\frac{\beta_{\mathcal{H}}}{\beta_{\mathcal{A}}}\right)\left(1 - \frac{f_{\mathcal{A}}(y)}{f_{\mathcal{H}}(y)}\right). \qquad (A32)$$

The above calculation also implies that $\lim_{\lambda \nearrow 1} \left(\frac{\partial}{\partial \lambda} \frac{p_\lambda}{q_\lambda} \right)$ is finite and thus $\lim_{\lambda \nearrow 1} \left(\frac{\partial}{\partial \lambda} \frac{p_\lambda}{q_\lambda} \right) \sum_{k=1}^{n} a_k^{(q_\lambda)} = 0$ by means of Lemma A2. The proof of $I(P_{\mathcal{A},n} || P_{\mathcal{H},n}) \geq E_{y,X_0,n}^{L,\tan}$ is finished by using Lemma A3 with l defined in (A31) and by plugging the limits (A30) to (A32) in (A29).

To derive the lower bound $E_{k,X_0,n}^{L,\text{sec}}$ (cf. (73)) for fixed $k \in \mathbb{N}_0$, we use as a linear upper bound ϕ_λ^U for $\phi_\lambda(\cdot)$ ($\lambda \in]0,1[$) the secant line $\phi_{\lambda,k}^{\text{sec}}$ (cf. (53)) of ϕ_λ across its arguments k and $k+1$, corresponding to the choices $p_\lambda := p_{\lambda,k}^{\text{sec}} = (k+1) \cdot \phi_\lambda(k) - k \cdot \phi_\lambda(k+1) + \alpha_\lambda$ and $q_\lambda := q_{\lambda,k}^{\text{sec}} := \phi_\lambda(k+1) - \phi_\lambda(k) + \beta_\lambda$, implying $q_\lambda > 0$ because of Properties 3 (P18). As a side remark, notice that this $\phi_\lambda^U(x)$ may become positive for some $x \in [0,\infty[$ (which is not always consistent with Goal (G1) for fixed λ, but leads to a tractable limit bound as λ tends to 1). Analogously to (A27) and (A28) we get again $\lim_{\lambda \nearrow 1} B_{\lambda,X_0,n}^U = 1$, which leads to the lower bound given in (A29) with appropriately plugged-in quantities. As in the above proof of the lower bound $E_{y,X_0,n}^{L,\tan}$, the inequality $I(P_{\mathcal{A},n} || P_{\mathcal{H},n}) \geq E_{k,X_0,n}^{L,\text{sec}}$ follows straightforwardly from Lemma A2, Lemma A3 and the three limits

$$\frac{\partial p_\lambda}{\partial \lambda} = \left(\frac{f_{\mathcal{A}}(k)}{f_{\mathcal{H}}(k)} \right)^\lambda f_{\mathcal{H}}(k) \cdot (k+1) \log \left(\frac{f_{\mathcal{A}}(k)}{f_{\mathcal{H}}(k)} \right) - \left(\frac{f_{\mathcal{A}}(k+1)}{f_{\mathcal{H}}(k+1)} \right)^\lambda f_{\mathcal{H}}(k+1) \cdot k \log \left(\frac{f_{\mathcal{A}}(k+1)}{f_{\mathcal{H}}(k+1)} \right)$$

$$\xrightarrow{\lambda \nearrow 1} f_{\mathcal{A}}(k)(k+1) \log \left(\frac{f_{\mathcal{A}}(k)}{f_{\mathcal{H}}(k)} \right) - f_{\mathcal{A}}(k+1) k \log \left(\frac{f_{\mathcal{A}}(k+1)}{f_{\mathcal{H}}(k+1)} \right),$$

$$\frac{\partial q_\lambda}{\partial \lambda} = \left(\frac{f_{\mathcal{A}}(k+1)}{f_{\mathcal{H}}(k+1)} \right)^\lambda f_{\mathcal{H}}(k+1) \log \left(\frac{f_{\mathcal{A}}(k+1)}{f_{\mathcal{H}}(k+1)} \right) - \left(\frac{f_{\mathcal{A}}(k)}{f_{\mathcal{H}}(k)} \right)^\lambda f_{\mathcal{H}}(k) \log \left(\frac{f_{\mathcal{A}}(k)}{f_{\mathcal{H}}(k)} \right)$$

$$\xrightarrow{\lambda \nearrow 1} f_{\mathcal{A}}(k+1) \log \left(\frac{f_{\mathcal{A}}(k+1)}{f_{\mathcal{H}}(k+1)} \right) - f_{\mathcal{A}}(k) \log \left(\frac{f_{\mathcal{A}}(k)}{f_{\mathcal{H}}(k)} \right) =: l, \qquad \text{and}$$

$$\frac{\partial}{\partial \lambda} \left(\frac{p_\lambda}{q_\lambda} \beta_\lambda - \alpha_\lambda \right) = \frac{q_\lambda \left(\frac{\partial p_\lambda}{\partial \lambda} \right) - p_\lambda \left(\frac{\partial q_\lambda}{\partial \lambda} \right)}{(q_\lambda)^2} \cdot \beta_\lambda + \frac{p_\lambda}{q_\lambda} \cdot (\beta_{\mathcal{A}} - \beta_{\mathcal{H}}) - (\alpha_{\mathcal{A}} - \alpha_{\mathcal{H}})$$

$$\xrightarrow{\lambda \nearrow 1} f_{\mathcal{A}}(k) \log \left(\frac{f_{\mathcal{A}}(k)}{f_{\mathcal{H}}(k)} \right) \left(k+1 + \frac{\alpha_{\mathcal{A}}}{\beta_{\mathcal{A}}} \right) - f_{\mathcal{A}}(k+1) \log \left(\frac{f_{\mathcal{A}}(k+1)}{f_{\mathcal{H}}(k+1)} \right) \left(k + \frac{\alpha_{\mathcal{A}}}{\beta_{\mathcal{A}}} \right) - \frac{\alpha_{\mathcal{A}} \beta_{\mathcal{H}}}{\beta_{\mathcal{A}}} + \alpha_{\mathcal{H}}.$$

To construct the third lower bound $E_{X_0,n}^{L,\text{hor}}$ (cf. (74)), we start by using the horizontal line $\phi_\lambda^{\text{hor}}(\cdot)$ (cf. (54)) as an upper bound of ϕ_λ. For each fixed $\lambda \in]0,1[$, it is defined by the intercept $\sup_{x \in \mathbb{N}_0} \phi_\lambda(x)$. On $\mathcal{P}_{\text{SP,3a}} \cup \mathcal{P}_{\text{SP,3b}}$, this supremum is achieved at the finite integer point $z_\lambda^* := \arg\max_{x \in \mathbb{N}_0} \phi_\lambda(x)$ (since $\lim_{x \to \infty} \phi_\lambda(x) = -\infty$) and there holds $\phi_\lambda(z_\lambda^*) < 0$ which leads with the parameters $q_\lambda = \beta_\lambda$, $p_\lambda = \phi_\lambda(z_\lambda^*) + \alpha_\lambda$ to the Hellinger integral upper bound $B_{\lambda,X_0,n}^U = \exp\{\phi_\lambda(z_\lambda^*) \cdot n\} < 1$ (cf. Remark 1 (b)). We strive for computing the limit $\lim_{\lambda \nearrow 1} \frac{1 - B_{\lambda,X_0,n}^U}{\lambda(1-\lambda)}$, which is not straightforward to solve since in general it seems to be intractable to express z_λ^* explicitly in terms of λ. To circumvent this problem, we notice that it is sufficient to determine z_λ^* in a small ϵ−environment $]1 - \epsilon, 1[$. To accomplish this, we incorporate $\lim_{\lambda \nearrow 1} \phi_\lambda(x) = 0$ for all $x \in [0,\infty[$ and calculate by using l'Hospital's rule

$$\lim_{\lambda \nearrow 1} \frac{\phi_\lambda(x)}{1 - \lambda} = (\alpha_{\mathcal{A}} + \beta_{\mathcal{A}} x) \left[-\log \left(\frac{\alpha_{\mathcal{A}} + \beta_{\mathcal{A}} x}{\alpha_{\mathcal{H}} + \beta_{\mathcal{H}} x} \right) + 1 \right] - (\alpha_{\mathcal{H}} + \beta_{\mathcal{H}} x).$$

Accordingly, let us define $z^* := \arg\max_{x \in \mathbb{N}_0} \left\{ (\alpha_{\mathcal{A}} + \beta_{\mathcal{A}} x) \left[-\log \left(\frac{\alpha_{\mathcal{A}} + \beta_{\mathcal{A}} x}{\alpha_{\mathcal{H}} + \beta_{\mathcal{H}} x} \right) + 1 \right] - (\alpha_{\mathcal{H}} + \beta_{\mathcal{H}} x) \right\}$ (note that the maximum exists since $\lim_{x \to \infty} \left\{ (\alpha_{\mathcal{A}} + \beta_{\mathcal{A}} x) \left[-\log \left(\frac{\alpha_{\mathcal{A}} + \beta_{\mathcal{A}} x}{\alpha_{\mathcal{H}} + \beta_{\mathcal{H}} x} \right) + 1 \right] - (\alpha_{\mathcal{H}} + \beta_{\mathcal{H}} x) \right\} = -\infty$). Due to continuity of the function $(\lambda, x) \mapsto \frac{\phi_\lambda(x)}{1-\lambda}$, there exists an $\epsilon > 0$ such that for all $\lambda \in]1 - \epsilon, 1[$ there holds $z_\lambda^* = z^*$. Applying these considerations, we get with l'Hospital's rule

$$I(P_{\mathcal{A},n} || P_{\mathcal{H},n}) \geq \lim_{\lambda \nearrow 1} \frac{1 - \exp\{\phi_\lambda(z^*) \cdot n\}}{\lambda(1-\lambda)} = \left[f_{\mathcal{A}}(z^*) \cdot \left[\log \left(\frac{f_{\mathcal{A}}(z^*)}{f_{\mathcal{H}}(z^*)} \right) - 1 \right] + f_{\mathcal{H}}(z^*) \right] \cdot n \geq 0. \qquad \text{(A33)}$$

In fact, for the current parameter constellation $\mathcal{P}_{\text{SP,3a}} \cup \mathcal{P}_{\text{SP,3b}}$ we have $\phi_\lambda(x) < 0$ for all $\lambda \in]0,1[$ and all $x \in \mathbb{N}_0$ which implies $f_{\mathcal{A}}(z^*) \neq f_{\mathcal{H}}(z^*)$ by Lemma A1; thus, we even get $E_{X_0,n}^{L,\text{hor}} > 0$ for all $n \in \mathbb{N}$ by virtue of the inequality $-\log \left(\frac{f_{\mathcal{H}}(z^*)}{f_{\mathcal{A}}(z^*)} \right) > -\frac{f_{\mathcal{H}}(z^*)}{f_{\mathcal{A}}(z^*)} + 1$.

For the case $\mathcal{P}_{\text{SP,2}}$, the above-mentioned procedure leads to $z_\lambda^* = 0 = z^*$ ($\lambda \in \,]0,1[$) which implies $\phi_\lambda(z_\lambda^*) = 0$, $B_{\lambda,X_0,n}^U \equiv 1$ and thus the trivial lower bound $E_{X_0,n}^{L,hor} = \lim_{\lambda \nearrow 1} \frac{1 - B_{\lambda,X_0,n}^U}{\lambda(1-\lambda)} = 0$ follows for all $n \in \mathbb{N}$. In contrast, for the case $\mathcal{P}_{\text{SP,3c}}$ one gets $z_\lambda^* = \frac{\alpha_\mathcal{A} - \alpha_\mathcal{H}}{\beta_\mathcal{H} - \beta_\mathcal{A}} = z^*$ ($\lambda \in \,]0,1[$) which nevertheless also implies $\phi_\lambda(z_\lambda^*) = 0$ and hence $E_{X_0,n}^{L,hor} \equiv 0$. On $\mathcal{P}_{\text{SP,4}}$, we have $\sup_{x \in \mathbb{N}_0} \phi_\lambda(x) = \lim_{x \to \infty} \phi_\lambda(x) = 0$ and hence we set $E_{X_0,n}^{L,hor} \equiv 0$.

To show the strict positivity $E_{X_0,n}^L > 0$ in the parameter case $\mathcal{P}_{\text{SP,2}}$, we inspect the bound $E_{0,X_0,n}^{L,sec}$. With $\alpha := \alpha_\bullet := \alpha_\mathcal{A} = \alpha_\mathcal{H}$ (the bullet will be omitted in this proof) and the auxiliary variable $x := \frac{\beta_\mathcal{H}}{\beta_\mathcal{A}} > 0$, the definition (73) respectively its special case (76) rewrites for all $n \in \mathbb{N}$ as

$$E_{0,X_0,n}^{L,sec} := E_{0,X_0,n}^{L,sec}(x) := \begin{cases} \begin{aligned}&\left[-(\alpha + \beta_\mathcal{A}) \cdot \log\left(\frac{\alpha + \beta_\mathcal{A}x}{\alpha + \beta_\mathcal{A}}\right) + \beta_\mathcal{A}(x-1) \right] \cdot \frac{1-(\beta_\mathcal{A})^n}{1-\beta_\mathcal{A}} \cdot \left[X_0 - \frac{\alpha}{1-\beta_\mathcal{A}} \right] \\ &+ \left[\frac{\alpha}{\beta_\mathcal{A}(1-\beta_\mathcal{A})} \left(-(\alpha + \beta_\mathcal{A}) \cdot \log\left(\frac{\alpha + \beta_\mathcal{A}x}{\alpha + \beta_\mathcal{A}}\right) + \beta_\mathcal{A}(x-1) \right) \right. \\ &\quad \left. + \frac{\alpha}{\beta_\mathcal{A}} (\alpha + \beta_\mathcal{A}) \cdot \log\left(\frac{\alpha + \beta_\mathcal{A}x}{\alpha + \beta_\mathcal{A}}\right) - \alpha(x-1) \right] \cdot n, \qquad \text{if } \beta_\mathcal{A} \neq 1, \\[2mm] &\left[-(\alpha+1) \cdot \log\left(\frac{\alpha+x}{\alpha+1}\right) + x - 1 \right] \cdot \left[\frac{\alpha}{2} \cdot n^2 + \left(X_0 + \frac{\alpha}{2}\right) \cdot n \right] \\ &+ \left[(\alpha+1) \cdot \log\left(\frac{\alpha+x}{\alpha+1}\right) - x + 1 \right] \cdot \alpha \cdot n, \qquad \text{if } \beta_\mathcal{A} = 1. \end{aligned} \end{cases} \quad \text{(A34)}$$

To prove that $E_{0,X_0,n}^{L,sec} > 0$ for all $X_0 \in \mathbb{N}$ and all $n \in \mathbb{N}$ it suffices to show that $E_{0,X_0,n}^{L,sec}(1) = \left(\frac{\partial}{\partial x} E_{0,X_0,n}^{L,sec} \right)(1) = 0$ and $\left(\frac{\partial^2}{\partial x^2} E_{0,X_0,n}^{L,sec} \right)(x) > 0$ for all $x \in \,]0,\infty[\,\backslash\{1\}$. The assertion $E_{0,X_0,n}^{L,sec}(1) = 0$ is trivial from (A34). Moreover, we obtain

$$\left(\frac{\partial}{\partial x} E_{0,X_0,n}^{L,sec} \right)(x) = \begin{cases} \begin{aligned}&\beta_\mathcal{A} \cdot \left[1 - \frac{\alpha + \beta_\mathcal{A}}{\alpha + \beta_\mathcal{A}x} \right] \cdot \frac{1-(\beta_\mathcal{A})^n}{1-\beta_\mathcal{A}} \cdot \left[X_0 - \frac{\alpha}{1-\beta_\mathcal{A}} \right] \\ &+ \alpha \cdot \left(1 - \frac{\alpha + \beta_\mathcal{A}}{\alpha + \beta_\mathcal{A}x} \right) \cdot \frac{\beta_\mathcal{A}}{1-\beta_\mathcal{A}} \cdot n, \qquad \text{if } \beta_\mathcal{A} \neq 1, \\[2mm] &\left[1 - \frac{\alpha+1}{\alpha+x} \right] \cdot \left[\frac{\alpha}{2} \cdot n^2 + \left(X_0 - \frac{\alpha}{2}\right) \cdot n \right], \qquad \text{if } \beta_\mathcal{A} = 1, \end{aligned} \end{cases}$$

which immediately yields $\left(\frac{\partial}{\partial x} E_{0,X_0,n}^{L,sec} \right)(1) = 0$. For the second derivative we get

$$\left(\frac{\partial^2}{\partial x^2} E_{0,X_0,n}^{L,sec} \right)(x) = \begin{cases} \begin{aligned}&\frac{(\alpha + \beta_\mathcal{A}) \cdot \beta_\mathcal{A}^2}{(\alpha + \beta_\mathcal{A}x)^2} \cdot \frac{1-(\beta_\mathcal{A})^n}{1-\beta_\mathcal{A}} \cdot \left[X_0 - \frac{\alpha}{1-\beta_\mathcal{A}} \right] \\ &+ \alpha \frac{\alpha + \beta_\mathcal{A}}{(\alpha + \beta_\mathcal{A}x)^2} \cdot \frac{\beta_\mathcal{A}^2}{1-\beta_\mathcal{A}} \cdot n > 0, \qquad \text{if } \beta_\mathcal{A} \neq 1, \\[2mm] &\frac{\alpha+1}{(\alpha+x)^2} \cdot \left[\frac{\alpha}{2} \cdot n^2 + \left(X_0 - \frac{\alpha}{2}\right) \cdot n \right] > 0, \qquad \text{if } \beta_\mathcal{A} = 1, \end{aligned} \end{cases} \quad \text{(A35)}$$

where the strict positivity of $E_{0,X_0,n}^{L,sec}$ in the case $\beta_\mathcal{A} \neq 1$ follows immediately by replacing X_0 with 0 and by using the obvious relation $\frac{1}{1-\beta_\mathcal{A}} \cdot \left[n - \frac{1-\beta_\mathcal{A}^n}{1-\beta_\mathcal{A}} \right] = \frac{1}{1-\beta_\mathcal{A}} \sum_{k=0}^{n-1} \left(1 - \beta_\mathcal{A}^k \right) > 0$. The strict positivity in the case $\beta_\mathcal{A} = 1$ is trivial by inspection.

For the constellation $\mathcal{P}_{\text{SP,4}}$ with parameters $\beta := \beta_\bullet := \beta_\mathcal{A} = \beta_\mathcal{H}$, $\alpha_\mathcal{A} \neq \alpha_\mathcal{H}$, the strict positivity of $E_{X_0,n}^L > 0$ follows by showing that $E_{y,X_0,n}^{L,tan}$ converges from above to zero as y tends to infinity. This is done by proving $\lim_{y \to \infty} y \cdot E_{y,X_0,n}^{L,tan} \in \,]0,\infty[$. To see this, let us first observe that by l'Hospital's rule we get

$$\lim_{y \to \infty} y \cdot \log\left(\frac{\alpha_\mathcal{A} + \beta y}{\alpha_\mathcal{H} + \beta y} \right) = \frac{\alpha_\mathcal{A} - \alpha_\mathcal{H}}{\beta} \qquad \text{as well as} \qquad \lim_{y \to \infty} y \cdot \left(1 - \frac{\alpha_\mathcal{A} + \beta y}{\alpha_\mathcal{H} + \beta y} \right) = -\frac{\alpha_\mathcal{A} - \alpha_\mathcal{H}}{\beta}.$$

From this and (72), we obtain $\lim_{y \to \infty} y \cdot E_{y,X_0,n}^{L,tan} = \frac{(\alpha_\mathcal{A} - \alpha_\mathcal{H})^2}{\beta} \cdot n > 0$ in both cases $\beta \neq 1$ and $\beta = 1$.

Finally, for the parameter case $\mathcal{P}_{\text{SP},3c}$ we consider the bound $E^{L,tan}_{y^*,X_0,n}$, with $y^* = \frac{\alpha_A - \alpha_{\mathcal{H}}}{\beta_{\mathcal{H}} - \beta_A}$. Since $\alpha_A + \beta_A y^* = \alpha_{\mathcal{H}} + \beta_{\mathcal{H}} y^*$, it is easy to see that $E^{L,tan}_{y^*,X_0,n} = 0$ for all $n \in \mathbb{N}$. However, the condition $\left(\frac{\partial}{\partial y} E^{L,tan}_{y,X_0,n} \right)(y^*) \neq 0$ implies that $\sup_{y \geq 0} E^{L,tan}_{y,X_0,n} > 0$. The explicit form (75) of this condition follows from

$$
\left(\frac{\partial}{\partial y} E^{L,tan}_{y,X_0,n} \right)(y) =
\begin{cases}
\frac{(\alpha_A \beta_{\mathcal{H}} - \alpha_{\mathcal{H}} \beta_A)^2}{f_A(y)(f_{\mathcal{H}}(y))^2} \cdot \frac{1-(\beta_A)^n}{1-\beta_A} \cdot \left[X_0 - \frac{\alpha_A}{1-\beta_A} \right] \\
\quad + \frac{\alpha_A \beta_{\mathcal{H}} - \alpha_{\mathcal{H}} \beta_A}{(f_{\mathcal{H}}(y))^2} \cdot \left[\frac{\alpha_A}{\beta_A(1-\beta_A)f_A(y)} - \frac{\alpha_A \beta_{\mathcal{H}} - \alpha_{\mathcal{H}} \beta_A}{\beta_A} \right] \cdot n, & \text{if } \beta_A \neq 1, \\[2ex]
\frac{(\alpha_A \beta_{\mathcal{H}} - \alpha_{\mathcal{H}})^2}{f_A(y)(f_{\mathcal{H}}(y))^2} \cdot \left[\frac{\alpha_A}{2} \cdot n^2 + \left(X_0 + \frac{\alpha_A}{2} \right) \cdot n \right] - \frac{(\alpha_A \beta_{\mathcal{H}} - \alpha_{\mathcal{H}})^2}{(f_{\mathcal{H}}(y))^2} \cdot n, & \text{if } \beta_A = 1,
\end{cases}
$$

$y \geq 0$, by using the particular choice $y = y^*$ together with $f_A(y^*) = f_{\mathcal{H}}(y^*) = -\frac{\alpha_A \beta_{\mathcal{H}} - \alpha_{\mathcal{H}} \beta_A}{\beta_A - \beta_{\mathcal{H}}}$. $\quad \square$

Appendix A.3. Proofs and Auxiliary Lemmas for Section 6

Proof of Lemma 2. A closed-form representation of a sequence $(\tilde{a}_n)_{n \in \mathbb{N}_0}$ defined in (83) to (85) is given by the formula

$$
\tilde{a}_n = \sum_{k=0}^{n-1} (c + \rho_k) \, d^{n-1-k}. \tag{A36}
$$

This can be seen by induction: from (83) we obtain with $\tilde{a}_0 = 0$ for the first element $\tilde{a}_1 = c + \rho_0 = \sum_{k=0}^{0}(c + \rho_k)d^{-k}$. Supposing that (A36) holds for the n-th element, the induction step is

$$
\tilde{a}_{n+1} = c + d \cdot \tilde{a}_n + \rho_n = c + d \cdot \sum_{k=0}^{n-1} (c + \rho_k) \, d^{n-1-k} + \rho_n = \sum_{k=0}^{n} (c + \rho_k) \, d^{n-k}.
$$

In order to obtain the explicit representation of $\tilde{a}_n$, we consider first the case $0 \leq \nu < \varkappa < d$ and $\rho_n = K_1 \cdot \varkappa^n + K_2 \cdot \nu^n$, which leads to

$$
\begin{aligned}
\tilde{a}_n &= d^{n-1} \sum_{k=0}^{n-1} \left(c \cdot d^{-k} + K_1 \cdot \left(\frac{\varkappa}{d} \right)^k + K_2 \cdot \left(\frac{\nu}{d} \right)^k \right) \\
&= d^{n-1} \cdot \left[c \cdot \frac{1 - d^{-n}}{1 - d^{-1}} + K_1 \cdot \frac{1 - \left(\frac{\varkappa}{d} \right)^n}{1 - \frac{\varkappa}{d}} + K_2 \cdot \frac{1 - \left(\frac{\nu}{d} \right)^n}{1 - \frac{\nu}{d}} \right] \\
&= \frac{c}{1 - d}(1 - d^n) + K_1 \cdot \frac{d^n - \varkappa^n}{d - \varkappa} + K_2 \cdot \frac{d^n - \nu^n}{d - \nu}.
\end{aligned} \tag{A37}
$$

Hence, for the corresponding sum we get

$$
\begin{aligned}
\sum_{k=1}^{n} \tilde{a}_k &= \sum_{k=1}^{n} \left[\frac{c}{1-d} + \left(\frac{K_1}{d-\varkappa} + \frac{K_2}{d-\nu} - \frac{c}{1-d} \right) \cdot d^k - \frac{K_1}{d-\varkappa} \cdot \varkappa^k - \frac{K_2}{d-\nu} \cdot \nu^k \right] \\
&= \frac{c}{1-d} \cdot n + \left(\frac{K_1}{d-\varkappa} + \frac{K_2}{d-\nu} - \frac{c}{1-d} \right) \cdot \frac{d \cdot (1 - d^n)}{1 - d} - \frac{K_1 \cdot \varkappa \cdot (1 - \varkappa^n)}{(d-\varkappa)(1-\varkappa)} - \frac{K_2 \cdot \nu \cdot (1 - \nu^n)}{(d-\nu)(1-\nu)}.
\end{aligned} \tag{A38}
$$

Consider now the case $0 \leq \nu < \varkappa = d$. Then some expressions in (A37) and (A38) have a zero denominator. In this case, the evaluation of (A36) becomes

$$
\begin{aligned}
\tilde{a}_n &= d^{n-1} \sum_{k=0}^{n-1} \left(c \cdot d^{-k} + K_1 + K_2 \cdot \left(\frac{\nu}{d} \right)^k \right) = d^{n-1} \cdot \left[c \cdot \frac{1 - d^{-n}}{1 - d^{-1}} + K_1 \cdot n + K_2 \cdot \frac{1 - \left(\frac{\nu}{d} \right)^n}{1 - \frac{\nu}{d}} \right] \\
&= \frac{c}{1-d}(1 - d^n) + K_1 \cdot n \cdot d^{n-1} + K_2 \cdot \frac{d^n - \nu^n}{d - \nu}.
\end{aligned} \tag{A39}
$$

Before we calculate the corresponding sum $\sum_{k=1}^{n} \widetilde{a}_k$, we notice that

$$\sum_{k=1}^{n} k \cdot d^{k-1} \;=\; \sum_{k=1}^{n} \frac{\partial}{\partial d} d^k \;=\; \frac{\partial}{\partial d} \sum_{k=1}^{n} d^k \;=\; \frac{\partial}{\partial d} \left(\frac{d \cdot (1-d^n)}{1-d} \right) \;=\; \frac{1 - n \cdot d^n (1-d) - d^n}{(1-d)^2} .$$

Using this fact, we obtain

$$
\begin{aligned}
\sum_{k=1}^{n} \widetilde{a}_k \;=\;& \sum_{k=1}^{n} \left[\frac{c}{1-d}(1 - d^k) + K_1 \cdot k \cdot d^{k-1} + K_2 \cdot \frac{d^k - v^k}{d - v} \right] \\
=\;& \frac{c}{1-d} \cdot n + \sum_{k=1}^{n} \left(\frac{K_2}{d-v} - \frac{c}{1-d} \right) d^k + K_1 \sum_{k=1}^{n} k \cdot d^{k-1} - \frac{K_2}{d-v} \sum_{k=1}^{n} v^k \\
=\;& \left(\frac{K_2}{d-v} - \frac{c}{1-d} \right) \frac{d \cdot (1 - d^n)}{1-d} + K_1 \cdot \frac{1 - n \cdot d^n (1-d) - d^n}{(1-d)^2} - \frac{K_2 \cdot v (1 - v^n)}{(d-v)(1-v)} + \frac{c}{1-d} \cdot n \\
=\;& \left(\frac{K_1}{d(1-d)} + \frac{K_2}{d-v} - \frac{c}{1-d} \right) \frac{d \cdot (1 - d^n)}{1-d} - \frac{K_2 \cdot v (1 - v^n)}{(d-v)(1-v)} + \left(\frac{c}{1-d} - \frac{K_1 \cdot d^n}{1-d} \right) \cdot n. \qquad \square
\end{aligned}
$$

Proof of Lemma 3. (a) In this case we have $0 < q < \beta_\lambda$. To prove part (i), we consider the function $\xi_\lambda^{(q)}(\cdot)$ on $[x_0^{(q)}, 0]$, the range of the sequence $\left(a_n^{(q)} \right)_{n \in \mathbb{N}}$ (recall Properties 1 (P1)). For tackling the left-hand inequality in (i), we compare $\xi_\lambda^{(q)}(x) = q \, e^x - \beta_\lambda$ with the quadratic function

$$\underline{Y}_\lambda^{(q)}(x) := \frac{q}{2} e^{x_0^{(q)}} \cdot x^2 + q e^{x_0^{(q)}} \left(1 - x_0^{(q)} \right) \cdot x + x_0^{(q)} \left(1 - q e^{x_0^{(q)}} + \frac{q}{2} e^{x_0^{(q)}} x_0^{(q)} \right). \tag{A40}$$

Clearly, one has the relations $\underline{Y}_\lambda^{(q)}(x_0^{(q)}) = x_0^{(q)} = \xi_\lambda^{(q)}(x_0^{(q)})$, $\underline{Y}_\lambda^{(q)\,\prime}(x_0^{(q)}) = q \cdot e^{x_0^{(q)}} = \xi_\lambda^{(q)\,\prime}(x_0^{(q)})$, and $\underline{Y}_\lambda^{(q)\,\prime\prime}(x) < \xi_\lambda^{(q)\,\prime\prime}(x)$ for all $x \in]x_0^{(q)}, 0]$. Hence, $\underline{Y}_\lambda^{(q)}(\cdot)$ is on $]x_0^{(q)}, 0]$ a strict lower functional bound of $\xi_\lambda^{(q)}(\cdot)$. We are now ready to prove the left-hand inequality in (i) by induction. For $n = 1$, we easily see that $\underline{a}_1^{(q)} < a_1^{(q)}$ iff $x_0^{(q)} \left(1 - q e^{x_0^{(q)}} + \frac{q}{2} e^{x_0^{(q)}} x_0^{(q)} \right) < q - \beta_\lambda$ iff $\underline{Y}_\lambda^{(q)}(0) < \xi_\lambda^{(q)}(0)$, and the latter is obviously true. Let us assume that $\underline{a}_n^{(q)} \leq a_n^{(q)}$ holds. From this, (93), (78) and (80) we obtain

$$
\begin{aligned}
0 < \underline{\rho}_n^{(q)} \;=\;& \frac{q}{2} e^{x_0^{(q)}} \left(x_0^{(q)} \cdot \left(q \cdot e^{x_0^{(q)}} \right)^n \right)^2 \;=\; \frac{q}{2} e^{x_0^{(q)}} \left(a_n^{(q),T} - x_0^{(q)} \right)^2 \\
<\;& \frac{q}{2} e^{x_0^{(q)}} \left(a_n^{(q)} - x_0^{(q)} \right)^2 \;=\; \underline{Y}_\lambda^{(q)} \left(a_n^{(q)} \right) - d^{(q),T} \cdot a_n^{(q)} - x_0^{(q)} \cdot \left(1 - d^{(q),T} \right) \\
<\;& \xi_\lambda^{(q)} \left(a_n^{(q)} \right) - d^{(q),T} \cdot a_n^{(q)} - x_0^{(q)} \cdot \left(1 - d^{(q),T} \right) \\
<\;& a_{n+1}^{(q)} - d^{(q),T} \cdot \underline{a}_n^{(q)} - x_0^{(q)} \cdot \left(1 - d^{(q),T} \right) \;=\; a_{n+1}^{(q)} - \xi_\lambda^{(q),T} \left(\underline{a}_n^{(q)} \right).
\end{aligned}
$$

Thus, there holds $\underline{a}_{n+1}^{(q)} < a_{n+1}^{(q)}$. For the right-hand inequality in (i), we proceed analogously:

$$\overline{Y}_\lambda^{(q)}(x) := \frac{q}{2} e^{x_0^{(q)}} \cdot x^2 + \left(1 - \frac{q}{2} e^{x_0^{(q)}} x_0^{(q)} - \frac{q - \beta_\lambda}{x_0^{(q)}} \right) \cdot x + q - \beta_\lambda \tag{A41}$$

satisfies $\overline{Y}_\lambda^{(q)}(x_0^{(q)}) = x_0^{(q)} = \xi_\lambda^{(q)}(x_0^{(q)})$, $\overline{Y}_\lambda^{(q)}(0) = q - \beta_\lambda = \xi_\lambda^{(q)}(0)$ as well as $\overline{Y}_\lambda^{(q)\,\prime\prime}(x) < \xi_\lambda^{(q)\,\prime\prime}(x)$ for all $x \in]x_0^{(q)}, 0]$. Hence, $\overline{Y}_\lambda^{(q)}(\cdot)$ is on $]x_0^{(q)}, 0]$ a strict upper functional bound of $\xi_\lambda^{(q)}(\cdot)$. Let us first observe the obvious relation $\bar{a}_1^{(q)} = q - \beta_\lambda = a_1^{(q)} < 0$, and assume that $\bar{a}_n^{(q)} \geq a_n^{(q)}$ ($n \in \mathbb{N}$) holds. From this, (95), (79), and (80) we obtain the desired inequality $\bar{a}_{n+1}^{(q)} > a_{n+1}^{(q)}$ by

$$
\begin{aligned}
0 > \bar{\rho}_n^{(q)} &= -\Gamma_{<}^{(q)} \left(d^{(q),T} \right)^n \cdot \frac{a_n^{(q),S}}{x_0^{(q)}} = \frac{q}{2}\, e^{x_0^{(q)}} \left(a_n^{(q),T} - x_0^{(q)} \right) \cdot a_n^{(q),S} \\
&\geq \frac{q}{2}\, e^{x_0^{(q)}} \left(a_n^{(q)} - x_0^{(q)} \right) \cdot a_n^{(q)} = \overline{Y}_\lambda^{(q)} \left(a_n^{(q)} \right) - d^{(q),S} \cdot a_n^{(q)} - (q - \beta_\lambda) \\
&> \xi_\lambda^{(q)} \left(a_n^{(q)} \right) - d^{(q),S} \cdot a_n^{(q)} - (q - \beta_\lambda) \geq a_{n+1}^{(q)} - d^{(q),S} \cdot \bar{a}_n^{(q)} - (q - \beta_\lambda) = a_{n+1}^{(q)} - \xi_\lambda^{(q),S}(\bar{a}_n^{(q)}) .
\end{aligned}
$$

The explicit representations of the sequences $\left(a_n^{(q)} \right)_{n \in \mathbb{N}}$, $\left(\underline{a}_n^{(q)} \right)_{n \in \mathbb{N}}$ and $\left(\bar{a}_n^{(q)} \right)_{n \in \mathbb{N}}$ follow from (86) by incorporating the appropriate constants mentioned in the prelude of Lemma 3. With (83) to (85) and (86) we immediately achieve $a_n^{(q)} > a_n^{(q),T}$ for all $n \in \mathbb{N}$. Analogously, for all $n \geq 2$, we get $\bar{\rho}_{n-1} < 0$, which implies that $\bar{a}_n^{(q)} < a_n^{(q),S}$ for all $n \geq 2$. For $n = 1$ one obtains $\bar{\rho}_0 = 0$ as well as $\bar{a}_1^{(q)} = a_1^{(q),S} = a_1^{(q)} = q - \beta_\lambda$.

For the second part (ii), we employ the representation (A36) which leads to

$$
a_n^{(q)} = \sum_{k=0}^{n-1} \left(d^{(q),T} \right)^{n-1-k} \cdot \left(\rho_k^{(q)} + x_0^{(q)} \cdot (1 - d^{(q),T}) \right)
$$

as well as
$$
\bar{a}_n^{(q)} = \sum_{k=0}^{n-1} \left(d^{(q),S} \right)^{n-1-k} \cdot \left(\bar{\rho}_k^{(q)} + (q - \beta_\lambda) \right) .
$$

The strict decreasingness of both sequences follows from

$$
\rho_k^{(q)} + x_0^{(q)}(1 - d^{(q),T}) = \frac{q e^{x_0^{(q)}}}{2} \left(x_0^{(q)} \right)^2 \left(d^{(q),T} \right)^{2n} + x_0^{(q)} \left(1 - d^{(q),T} \right) \leq \underline{Y}_\lambda^{(q)}(0) < \xi_\lambda^{(q)}(0) = q - \beta_\lambda < 0
$$

and from the fact that $\bar{\rho}_k^{(q)} \leq 0$ for all $k \in \mathbb{N}_0$ and $q < \beta_\lambda$. Part (iii) follows directly from (i), since $d^{(q),T}, d^{(q),S} \in\,]0,1[$.

Let us now prove part (b), where $\max\{0, \beta_\lambda\} < q < \min\left\{ 1, e^{\beta_\lambda - 1} \right\}$ is assumed. To tackle part (i), we compare $\xi_\lambda^{(q)}(x) = q \cdot e^x - \beta_\lambda$ with the quadratic function

$$
\underline{v}_\lambda^{(q)}(x) := \frac{q}{2} \cdot x^2 + q \cdot \left(e^{x_0^{(q)}} - x_0^{(q)} \right) \cdot x + x_0^{(q)} \left(1 - q e^{x_0^{(q)}} + \frac{q}{2} x_0^{(q)} \right) > 0 \tag{A42}
$$

on the interval $[0, x_0^{(q)}]$. Clearly, we have $\underline{v}_\lambda^{(q)} \left(x_0^{(q)} \right) = \xi_\lambda^{(q)}(x_0^{(q)}) = x_0^{(q)}$, $\underline{v}_\lambda^{(q)\,\prime}(x_0^{(q)}) = \xi_\lambda^{(q)\,\prime}(x_0^{(q)}) = q e^{x_0^{(q)}}$ and $0 < \underline{v}_\lambda^{(q)\,\prime\prime}(x) < \xi_\lambda^{(q)\,\prime\prime}(x)$ for all $x \in\,]0, x_0^{(q)}]$. Thus, $\underline{v}_\lambda^{(q)}(\cdot)$ constitutes a positive functional lower bound for $\xi_\lambda^{(q)}(\cdot)$ on $[0, x_0^{(q)}]$. Let us now prove the left-hand inequality of (i) by induction: for $n = 1$ we get $\underline{a}_1^{(q)} = \underline{v}_\lambda^{(q)}(0) < \xi_\lambda^{(q)}(0) = a_1^{(q)}$. Moreover, by assuming $\underline{a}_n^{(q)} \leq a_n^{(q)}$ for $n \in \mathbb{N}$, we obtain with the above-mentioned considerations and (93), (80) and (82)

$$
\begin{aligned}
0 < \underline{\rho}_n^{(q)} &= \Gamma_{>}^{(q)} \left(d^{(q),S} \right)^{2n} = \frac{q}{2} \cdot \left(a_n^{(q),S} - x_0^{(q)} \right)^2 < \frac{q}{2} \cdot \left(a_n^{(q)} - x_0^{(q)} \right)^2 \\
&= \frac{q}{2} \left(a_n^{(q)} \right)^2 + q \cdot \left(e^{x_0^{(q)}} - x_0^{(q)} \right) \cdot a_n^{(q)} + x_0^{(q)} \cdot \left(1 - q e^{x_0^{(q)}} + \frac{q}{2} x_0^{(q)} \right) - d^{(q),T} a_n^{(q)} - c^{(q),T} \\
&= \underline{v}_\lambda^{(q)}(a_n^{(q)}) - d^{(q),T} a_n^{(q)} - c^{(q),T} < \xi_\lambda^{(q)}(a_n^{(q)}) - d^{(q),T} a_n^{(q)} - c^{(q),T} \\
&< a_{n+1}^{(q)} - d^{(q),T} \underline{a}_n^{(q)} - c^{(q),T} = a_{n+1}^{(q)} - \xi_\lambda^{(q),T}(\underline{a}_n^{(q)}) .
\end{aligned}
$$

Hence, $\underline{a}_{n+1}^{(q)} < a_{n+1}^{(q)}$. For the right-hand inequality in part (i), we define the quadratic function

$$\overline{\overline{v}}_\lambda^{(q)}(x) \ := \ \frac{q}{2} \cdot x^2 + \left(1 - \frac{q}{2} x_0^{(q)} - \frac{q - \beta_\lambda}{x_0^{(q)}} \right) \cdot x + q - \beta_\lambda \, , \tag{A43}$$

which is a functional upper bound for $\xi_\lambda^{(q)}(\cdot)$ on the interval $[0, x_0^{(q)}]$ since there holds $\overline{\overline{v}}_\lambda^{(q)}(0) = \xi_\lambda^{(q)}(0) = q - \beta_\lambda$, $\overline{\overline{v}}_\lambda^{(q)}(x_0^{(q)}) = \xi_\lambda^{(q)}(x_0^{(q)}) = x_0^{(q)}$ and additionally $\overline{\overline{v}}_\lambda^{(q)\,\prime\prime}(x) = q < qe^x = \xi_\lambda^{(q)\,\prime\prime}(x)$ on $]0, x_0^{(q)}[$. Obviously, $\overline{a}_1^{(q)} = q - \beta_\lambda = a_1^{(q)}$. By assuming $\overline{a}_n^{(q)} \geq a_n^{(q)}$ for $n \in \mathbb{N}$, we obtain with (80), (82) and (95)

$$
\begin{aligned}
0 > \overline{\rho}_n^{(q)} \ &= \ -\Gamma_>^{(q)} \cdot \left(d^{(q),S} \right)^n \cdot \left(1 - \left(d^{(q),T} \right)^n \right) \ = \ -\frac{q}{2} \cdot \left(x_0 - a_n^{(q),S} \right) \cdot a_n^{(q),T} \\
&> \ -\frac{q}{2} \cdot \left(x_0 - a_n^{(q)} \right) \cdot a_n^{(q)} \ = \ \overline{\overline{v}}_\lambda^{(q)}(a_n^{(q)}) - \frac{x_0^{(q)} - (q - \beta_\lambda)}{x_0^{(q)}} \cdot a_n^{(q)} - (q - \beta_\lambda) \\
&> \ \xi_\lambda^{(q)}(a_n^{(q)}) - d^{(q),S} a_n^{(q)} - c^{(q),S} \ > \ \xi_\lambda^{(q)}(a_n^{(q)}) - d^{(q),S} \overline{a}_n^{(q),S} - c^{(q),S} \ = \ a_{n+1}^{(q)} - \xi_\lambda^{(q),S}(\overline{a}_n^{(q)}) \, ,
\end{aligned}
\tag{A44}
$$

which implies $\overline{a}_{n+1}^{(q)} > a_{n+1}^{(q)}$. The explicit representations of the sequences $\left(a_n^{(q)} \right)_{n \in \mathbb{N}}$ and $\left(\overline{a}_n^{(q)} \right)_{n \in \mathbb{N}}$ follow from (86) by employing the appropriate constants mentioned in the prelude of Lemma 3. By means of (83) to (85) and (86), we directly get $a_n^{(q)} > a_n^{(q),T}$ for all $n \in \mathbb{N}$, whereas $\overline{a}_n^{(q)} < a_n^{(q),S}$ holds only for all $n \geq 2$, since $\overline{\rho}_0 = 0$ implies that $\overline{a}_1^{(q)} = a_1^{(q),S} = a_1^{(q)} = q - \beta_\lambda$.

The second part (ii) can be proved in the same way as part (ii) of (a), by employing the representation (A36). For the lower bound one has

$$\underline{a}_n^{(q)} \ = \ \sum_{k=0}^{n-1} \left(d^{(q),T} \right)^{n-1-k} \cdot \left[c^{(q),T} + \underline{\rho}_k^{(q)} \right] \, , \qquad \text{with } c^{(q),T} > 0 \quad \text{and } \underline{\rho}_k^{(q)} > 0.$$

For the upper bound we get

$$\overline{a}_n^{(q)} \ = \ \sum_{k=0}^{n-1} \left(d^{(q),S} \right)^{n-1-k} \cdot \left[c^{(q),S} + \overline{\rho}_k^{(q)} \right] ,$$

hence it is enough to show $c^{(q),S} + \overline{\rho}_n^{(q)} > 0$ for all $n \in \mathbb{N}_0$. Considering the first two lines of calculation (A44) and incorporating $c^{(q),S} = q - \beta_\lambda$, this can be seen from

$$c^{(q),S} + \overline{\rho}_n^{(q)} \ > \ \overline{\overline{v}}_\lambda^{(q)}(a_n^{(q)}) - \frac{x_0^{(q)} - (q - \beta_\lambda)}{x_0^{(q)}} \cdot a_n^{(q)} \ = \ \overline{\overline{v}}_\lambda^{(q)}(a_n^{(q)}) - d^{(q),S} \cdot a_n^{(q)} \ > \ 0 \, ,$$

because on $[0, x_0^{(q)}]$ there holds $d^{(q),S} \cdot x < x < \overline{\overline{v}}_\lambda^{(q)}(x)$. The last part (iii) can be easily deduced from (i) together with $\lim_{n \to \infty} n \cdot \left(d^{(q),S} \right)^{n-1} = 0$. $\quad\square$

The proofs of all Theorems 5–9 are mainly based on the following

Lemma A4. *Recall the quantity $\widetilde{B}_{\lambda,X_0,n}^{(p,q)}$ from (42) for general $p \geq 0$, $q > 0$ (notice that we do not consider parameters $p < 0$, $q \leq 0$ in Section 6) as well as the constants $d^{(q),T}$, $d^{(q),S}$ and $\Gamma_<^{(q)}$, $\Gamma_>^{(q)}$ defined in (76), (77) and (91). For all $(\beta_\mathcal{A}, \beta_\mathcal{H}, \alpha_\mathcal{A}, \alpha_\mathcal{H}, \lambda) \in \mathcal{P} \times \mathbb{R} \backslash \{0,1\}$, all initial population sizes $X_0 \in \mathbb{N}$ and all observation horizons $n \in \mathbb{N}$ there holds*

(a) *in the case $p \geq 0$ and $0 < q < \beta_\lambda$*

$$\widetilde{B}^{(p,q)}_{\lambda,X_0,n} \;\geq\; \exp\left\{ x_0^{(q)} \cdot \left[X_0 - \frac{p}{q} \cdot \frac{d^{(q),T}}{1 - d^{(q),T}} \right] \cdot \left(1 - \left(d^{(q),T} \right)^n \right) + \left(\frac{p}{q} \cdot \left(\beta_\lambda + x_0^{(q)} \right) - \alpha_\lambda \right) \cdot n \right.$$

$$\left. + \underline{\zeta}^{(q)}_n \cdot X_0 + \frac{p}{q} \cdot \underline{\vartheta}^{(q)}_n \right\} \;=:\; C^{(p,q),L}_{\lambda,X_0,n} \,, \tag{A45}$$

$$\widetilde{B}^{(p,q)}_{\lambda,X_0,n} \;\leq\; \exp\left\{ x_0^{(q)} \cdot \left[X_0 - \frac{p}{q} \cdot \frac{d^{(q),S}}{1 - d^{(q),S}} \right] \cdot \left(1 - \left(d^{(q),S} \right)^n \right) + \left(\frac{p}{q} \cdot \left(\beta_\lambda + x_0^{(q)} \right) - \alpha_\lambda \right) \cdot n \right.$$

$$\left. - \overline{\zeta}^{(q)}_n \cdot X_0 - \frac{p}{q} \cdot \overline{\vartheta}^{(q)}_n \right\} \;=:\; C^{(p,q),U}_{\lambda,X_0,n} \,, \tag{A46}$$

where $\quad \underline{\zeta}^{(q)}_n := \Gamma^{(q)}_< \cdot \frac{\left(d^{(q),T} \right)^{n-1}}{1 - d^{(q),T}} \cdot \left(1 - \left(d^{(q),T} \right)^n \right) > 0,$ $\tag{A47}$

$$\underline{\vartheta}^{(q)}_n \;:=\; \Gamma^{(q)}_< \cdot \frac{1 - \left(d^{(q),T} \right)^n}{\left(1 - d^{(q),T} \right)^2} \cdot \left[1 - \frac{d^{(q),T} \left(1 + \left(d^{(q),T} \right)^n \right)}{1 + d^{(q),T}} \right] > 0, \tag{A48}$$

$$\overline{\zeta}^{(q)}_n \;:=\; \Gamma^{(q)}_< \cdot \left[\frac{\left(d^{(q),S} \right)^n - \left(d^{(q),T} \right)^n}{d^{(q),S} - d^{(q),T}} - \left(d^{(q),S} \right)^{n-1} \cdot \frac{1 - \left(d^{(q),T} \right)^n}{1 - d^{(q),T}} \right] > 0, \tag{A49}$$

$$\overline{\vartheta}^{(q)}_n \;:=\; \Gamma^{(q)}_< \cdot \frac{d^{(q),T}}{1 - d^{(q),T}} \cdot \left[\frac{1 - \left(d^{(q),S} d^{(q),T} \right)^n}{1 - d^{(q),S} d^{(q),T}} - \frac{\left(d^{(q),S} \right)^n - \left(d^{(q),T} \right)^n}{d^{(q),S} - d^{(q),T}} \right] > 0. \tag{A50}$$

(b) *in the case $p \geq 0$ and $0 < q = \beta_\lambda$*

$$\widetilde{B}^{(p,q)}_{\lambda,X_0,n} \;=\; \exp\left\{ \left(\frac{p}{q} \cdot \left(\beta_\lambda + x_0^{(q)} \right) - \alpha_\lambda \right) \cdot n \right\} \;=\; \exp\left\{ (p - \alpha_\lambda) \cdot n \right\}.$$

(c) *in the case $p \geq 0$ and $\max\{0,\, \beta_\lambda\} < q < \min\left\{ 1,\, e^{\beta_\lambda - 1} \right\}$ the bounds $C^{(p,q),L}_{\lambda,X_0,n}$ and $C^{(p,q),U}_{\lambda,X_0,n}$ from (96) and (97) remain valid, but with*

$$\underline{\zeta}^{(q)}_n \;:=\; \Gamma^{(q)}_> \cdot \frac{\left(d^{(q),T} \right)^n - \left(d^{(q),S} \right)^{2n}}{d^{(q),T} - \left(d^{(q),S} \right)^2} > 0, \tag{A51}$$

$$\underline{\vartheta}^{(q)}_n \;:=\; \frac{\Gamma^{(q)}_>}{d^{(q),T} - \left(d^{(q),S} \right)^2} \cdot \left[\frac{d^{(q),T} \cdot \left(1 - \left(d^{(q),T} \right)^n \right)}{1 - d^{(q),T}} - \frac{\left(d^{(q),S} \right)^2 \cdot \left(1 - \left(d^{(q),S} \right)^{2n} \right)}{1 - \left(d^{(q),S} \right)^2} \right] > 0,$$

$$\tag{A52}$$

$$\overline{\zeta}^{(q)}_n \;:=\; \Gamma^{(q)}_> \cdot \left(d^{(q),S} \right)^{n-1} \cdot \left[n - \frac{1 - \left(d^{(q),T} \right)^n}{1 - d^{(q),T}} \right] > 0, \tag{A53}$$

$$\overline{\vartheta}^{(q)}_n \;:=\; \Gamma^{(q)}_> \cdot \left[\frac{d^{(q),S} - d^{(q),T}}{\left(1 - d^{(q),S} \right)^2 \left(1 - d^{(q),T} \right)} \cdot \left(1 - \left(d^{(q),S} \right)^n \right) \right.$$

$$\left. + \frac{d^{(q),T} \left(1 - \left(d^{(q),S} d^{(q),T} \right)^n \right)}{\left(1 - d^{(q),T} \right) \left(1 - d^{(q),S} d^{(q),T} \right)} - \frac{\left(d^{(q),S} \right)^n}{1 - d^{(q),S}} \cdot n \right]. \tag{A54}$$

(d) *for the special choices $p := p^E_\lambda := \alpha^\lambda_\mathcal{A} \alpha^{1-\lambda}_\mathcal{H} > 0$, $q := q^E_\lambda := \beta^\lambda_\mathcal{A} \beta^{1-\lambda}_\mathcal{H} > 0$ in the parameter setup $(\beta_\mathcal{A}, \beta_\mathcal{H}, \alpha_\mathcal{A}, \alpha_\mathcal{H}, \lambda) \in (\mathcal{P}_{NI} \cup \mathcal{P}_{SP,1}) \times \,]\lambda_-, \lambda_+[\,\backslash \{0,1\}$ we obtain*

$$\lim_{n \to \infty} \frac{1}{n} \log \left(V_{\lambda,X_0,n} \right) \;=\; \lim_{n \to \infty} \frac{1}{n} \log \left(C^{(p^E_\lambda, q^E_\lambda),L}_{\lambda,X_0,n} \right) \;=\; \lim_{n \to \infty} \frac{1}{n} \log \left(C^{(p^E_\lambda, q^E_\lambda),U}_{\lambda,X_0,n} \right) \;=\; \frac{\alpha_\mathcal{A}}{\beta_\mathcal{A}} \cdot x_0^{(q^E_\lambda)}.$$

(e) for all general $p \geq 0$ with either $0 < q < \beta_\lambda$ or $\max\{0, \beta_\lambda\} < q < \min\left\{1, e^{\beta_\lambda - 1}\right\}$ we get

$$\lim_{n \to \infty} \frac{1}{n} \log\left(\widetilde{B}^{(p,q)}_{\lambda, X_0, n}\right) = \lim_{n \to \infty} \frac{1}{n} \log\left(C^{(p,q),L}_{\lambda, X_0, n}\right) = \lim_{n \to \infty} \frac{1}{n} \log\left(C^{(p,q),U}_{\lambda, X_0, n}\right) = \frac{p}{q} \cdot \left(\beta_\lambda + x_0^{(q)}\right) - \alpha_\lambda .$$

Proof of Lemma A4. The closed-form bounds $C^{(p,q),L}_{\lambda, X_0, n}$ and $C^{(p,q),U}_{\lambda, X_0, n}$ are obtained by substituting in the representation (42) (for $\widetilde{B}^{(p,q)}_{\lambda, X_0, n}$, cf. Theorem 1) the recursive sequence member $a_n^{(q)}$ by the explicit sequence member $\underline{a}_n^{(q)}$ respectively $\overline{a}_n^{(q)}$. From the definitions of these sequences (92) to (95) and from (83) to (85) one can see that we basically have to evaluate the term

$$\exp\left\{ \left(\widetilde{a}_n^{hom} + \widetilde{c}_n\right) \cdot X_0 + \frac{p}{q} \cdot \sum_{k=1}^{n} \left(\widetilde{a}_k^{hom} + \widetilde{c}_k\right) + \left(\frac{p}{q} \cdot \beta_\lambda - \alpha_\lambda\right) \cdot n \right\} , \tag{A55}$$

where $\widetilde{a}_n^{hom} + \widetilde{c}_n = \widetilde{a}_n$ is either interpreted as the lower approximate $\underline{a}_n^{(q)}$ or as the upper approximate $\overline{a}_n^{(q)}$. After rearranging and incorporating that $\frac{c^{(q),S}}{1 - d^{(q),S}} = \frac{c^{(q),T}}{1 - d^{(q),T}} = x_0^{(q)}$ in both approximate cases, we obtain with the help of (86), (87) for the expression (A55) in the case $0 \leq \nu < \varkappa < d$

$$\exp\left\{ x_0^{(q)} \cdot (1 - d^n) \cdot \left[X_0 - \frac{p}{q} \cdot \frac{d}{1 - d}\right] + \left(\frac{p}{q} \cdot \left(\beta_\lambda + x_0^{(q)}\right) - \alpha_\lambda\right) \cdot n \right.$$
$$+ \left[K_1 \cdot \frac{d^n - \varkappa^n}{d - \varkappa} + K_2 \cdot \frac{d^n - \nu^n}{d - \nu}\right] \cdot X_0$$
$$\left. + \frac{p}{q} \cdot \left[\left(\frac{K_1}{d - \varkappa} + \frac{K_2}{d - \nu}\right) \cdot \frac{d \cdot (1 - d^n)}{1 - d} - \frac{K_1 \cdot \varkappa \cdot (1 - \varkappa^n)}{(d - \varkappa)(1 - \varkappa)} - \frac{K_2 \cdot \nu \cdot (1 - \nu^n)}{(d - \nu)(1 - \nu)}\right] \right\}. \tag{A56}$$

In the other case $0 \leq \nu < \varkappa = d$, the application of (88), (89) turns (A55) into

$$\exp\left\{ x_0^{(q)} \cdot (1 - d^n) \cdot \left[X_0 - \frac{p}{q} \cdot \frac{d}{1 - d}\right] + \left(\frac{p}{q} \cdot \left(\beta_\lambda + x_0^{(q)}\right) - \alpha_\lambda\right) \cdot n \right.$$
$$+ \left[K_1 \cdot n \cdot d^{n-1} + K_2 \cdot \frac{d^n - \nu^n}{d - \nu}\right] \cdot X_0$$
$$\left. + \frac{p}{q} \cdot \left[\left(\frac{K_1}{d(1 - d)} + \frac{K_2}{d - \nu}\right) \cdot \frac{d \cdot (1 - d^n)}{1 - d} - \frac{K_2 \cdot \nu \cdot (1 - \nu^n)}{(d - \nu)(1 - \nu)} - \frac{K_1 \cdot d^n}{1 - d} \cdot n\right] \right\}. \tag{A57}$$

After these preparatory considerations let us now begin with elaboration of the details.

(a) Let $0 < q < \beta_\lambda$. We obtain a closed-form lower bound for $\widetilde{B}^{(p,q)}_{\lambda, X_0, n}$ by employing the parameters $c \cong c^{(q),T}, d \cong d^{(q),T}, K_2 = \nu = 0, K_1 = \Gamma_{<}^{(q)}$, and $\varkappa = \left(d^{(q),T}\right)^2$, cf. (93) in combination with (85). Since $\varkappa < d^{(q),T}$, we have to plug in these parameters into (A56). The representations of $\underline{\zeta}_n^{(q)}$ and $\underline{\vartheta}_n^{(q)}$ in (A47) and (A48) follow immediately. For a closed-form upper bound, we employ the parameters $c \cong c^{(q),S}, d \cong d^{(q),S}, -K_1 = K_2 = \Gamma_{<}^{(q)}, \varkappa = d^{(q),T}$ and $\nu = d^{(q),S} d^{(q),T}$ (in particular, $\varkappa < d^{(q),S}$ implying that we have to use (A56)). From this, (A49) can be deduced directly; the representation (A50) comes from the expressions in the squared brackets in the last line of (A56) and from

$$-\left(\frac{\Gamma_{<}^{(q)}}{d^{(q),S}-d^{(q),T}}-\frac{\Gamma_{<}^{(q)}}{d^{(q),S}-d^{(q),S}d^{(q),T}}\right)\cdot\frac{d^{(q),S}\cdot\left(1-\left(d^{(q),S}\right)^{n}\right)}{1-d^{(q),S}}+\frac{\Gamma_{<}^{(q)}\cdot d^{(q),T}\cdot\left(1-\left(d^{(q),T}\right)^{n}\right)}{\left(d^{(q),S}-d^{(q),T}\right)\left(1-d^{(q),T}\right)}$$

$$-\frac{\Gamma_{<}^{(q)}\cdot d^{(q),S}d^{(q),T}\cdot\left(1-\left(d^{(q),S}d^{(q),T}\right)^{n}\right)}{\left(d^{(q),S}-d^{(q),S}d^{(q),T}\right)\left(1-d^{(q),S}d^{(q),T}\right)}$$

$$=-\frac{\Gamma_{<}^{(q)}\cdot d^{(q),T}\left(1-d^{(q),S}\right)}{d^{(q),S}\left(d^{(q),S}-d^{(q),T}\right)\left(1-d^{(q),T}\right)}\cdot\frac{d^{(q),S}\cdot\left(1-\left(d^{(q),S}\right)^{n}\right)}{1-d^{(q),S}}+\frac{\Gamma_{<}^{(q)}\cdot d^{(q),T}\cdot\left(1-\left(d^{(q),T}\right)^{n}\right)}{\left(d^{(q),S}-d^{(q),T}\right)\left(1-d^{(q),T}\right)}$$

$$-\frac{\Gamma_{<}^{(q)}\cdot d^{(q),T}\cdot\left(1-\left(d^{(q),S}d^{(q),T}\right)^{n}\right)}{\left(1-d^{(q),T}\right)\left(1-d^{(q),S}d^{(q),T}\right)}$$

$$=-\frac{\Gamma_{<}^{(q)}\cdot d^{(q),T}}{1-d^{(q),T}}\cdot\left[\frac{1-\left(d^{(q),S}d^{(q),T}\right)^{n}}{1-d^{(q),S}d^{(q),T}}+\frac{1-\left(d^{(q),S}\right)^{n}}{d^{(q),S}-d^{(q),T}}-\frac{1-\left(d^{(q),T}\right)^{n}}{d^{(q),S}-d^{(q),T}}\right]$$

$$=-\frac{\Gamma_{<}^{(q)}\cdot d^{(q),T}}{1-d^{(q),T}}\cdot\left[\frac{1-\left(d^{(q),S}d^{(q),T}\right)^{n}}{1-d^{(q),S}d^{(q),T}}-\frac{\left(d^{(q),S}\right)^{n}-\left(d^{(q),T}\right)^{n}}{d^{(q),S}-d^{(q),T}}\right]=-\overline{\vartheta}_{n}^{(q)}.$$

Part (b) has already been mentioned in Remark 1 (b) and is due to the fact that for $0<q=\beta_{\lambda}$, the sequence $\left(a_{n}^{(q)}\right)_{n\in\mathbb{N}}$ is itself explicitly representable by $a_{n}^{(q)}=0$ for all $n\in\mathbb{N}$ (cf. Properties 1 (P2)). Plugging this into (42) gives the desired result.

(c) Let us now consider $\max\{0,\beta_{\lambda}\}<q<\min\{1,e^{\beta_{\lambda}-1}\}$. For a closed-form lower bound for $\widetilde{B}_{\lambda,X_{0},n}^{(p,q)}$ we have to employ the parameters $c\cong c^{(q),T}$, $d\cong d^{(q),T}$, $K_{2}=v=0$, $K_{1}=\Gamma_{>}^{(q)}$ and $\varkappa=\left(d^{(q),S}\right)^{2}$, cf. (93) in combination with (85). The representations of $\underline{\zeta}_{n}^{(q)}$ and $\underline{\vartheta}_{n}^{(q)}$ in (A51) and (A52) follow immediately from (A56). For a closed-form upper bound, we use the parameters $c\cong c^{(q),S}$, $d\cong d^{(q),S}$, $-K_{1}=K_{2}=\Gamma_{>}^{(q)}$, $\varkappa=d^{(q),S}$ and $v=d^{(q),S}d^{(q),T}$. Notice that in this case we stick to the representation (A57). The formula (104) is obviously valid, and (105) is implied by

$$\left(\frac{-\Gamma_{>}^{(q)}}{d^{(q),S}\left(1-d^{(q),S}\right)}+\frac{\Gamma_{>}^{(q)}}{d^{(q),S}-d^{(q),S}d^{(q),T}}\right)\cdot\frac{d^{(q),S}\cdot\left(1-\left(d^{(q),S}\right)^{n}\right)}{1-d^{(q),S}}$$

$$=-\Gamma_{>}^{(q)}\cdot\frac{d^{(q),S}-d^{(q),T}}{\left(1-d^{(q),S}\right)^{2}\left(1-d^{(q),T}\right)}\cdot\left(1-\left(d^{(q),S}\right)^{n}\right).$$

The parts (d) and (e) are trivial by incorporating that in all respective cases one has $d^{(q),S}\in\,]0,1[$, $d^{(q),T}\in\,]0,1[$ and $\lim_{n\to\infty}n\cdot d^{(q),S}=0$. $\quad\square$

Proof of Theorem 5. (a) For $\lambda\in\,]0,1[$, we get $0<q_{\lambda}^{E}<\beta_{\lambda}$ and the assertion follows by applying part (a) of Lemma A4. Notice that in the current subcase $\mathcal{P}_{\mathrm{NI}}\cup\mathcal{P}_{\mathrm{SP},1}$ there holds $\frac{p_{\lambda}^{E}}{q_{\lambda}^{E}}\beta_{\lambda}-\alpha_{\lambda}=0$ as well as $\frac{p_{\lambda}^{E}}{q_{\lambda}^{E}}=\frac{\alpha_{A}}{\beta_{A}}=\frac{\alpha_{H}}{\beta_{H}}$. For the case $\lambda\in\mathbb{R}\backslash[0,1]$, one gets from Lemma A1 that $\max\{0,\beta_{\lambda}\}<q_{\lambda}^{E}$, and there holds $q_{\lambda}^{E}<\min\{1,e^{\beta_{\lambda}-1}\}$ iff $\lambda\in\,]\lambda_{-},\lambda_{+}[\,\backslash[0,1]$, cf. Lemma 1. Thus, an application of part (c) of Lemma A4 proves the desired result. The assertion (b) is equivalent to part (d) of Lemma A4. $\quad\square$

Proof of Theorem 6. The assertions follow immediately from (A45), Lemma A4(b),(e), Proposition 6(d) as well as the incorporation of the fact that for $\lambda \in]0,1[$ there holds $q_\lambda^L = \beta_\mathcal{A}^\lambda \beta_\mathcal{H}^{1-\lambda} < \beta_\lambda$ in the case $(\beta_\mathcal{A}, \beta_\mathcal{H}, \alpha_\mathcal{A}, \alpha_\mathcal{H}) \in (\mathcal{P}_{SP} \backslash (\mathcal{P}_{SP,1} \cup \mathcal{P}_{SP,4}))$ (i.e., $\beta_\mathcal{A} \neq \beta_\mathcal{H}$) respectively $q_\lambda^L = \beta_\lambda$ in the case $(\beta_\mathcal{A}, \beta_\mathcal{H}, \alpha_\mathcal{A}, \alpha_\mathcal{H}) \in \mathcal{P}_{SP,4}$ (i.e., $\beta_\mathcal{A} = \beta_\mathcal{H}$). $\square$

Proof of Theorem 7. This can be deduced from (A46), from the parts (b), (c) and (e) of Lemma A4 as well as the incorporation of $p_\lambda^U \geq \alpha_\mathcal{A}^\lambda \alpha_\mathcal{H}^{1-\lambda} > 0$ for $\lambda \in]0,1[$. Notice that an inadequate choice of p_λ^U, q_λ^U may lead to $\frac{p_\lambda^U}{q_\lambda^U}(\beta_\lambda + x_0^{(q_\lambda^U)}) - \alpha_\lambda > 0$. $\square$

Proof of Theorem 8. The assertions follow immediately from (A45) and from the parts (b), (c) and (e) of Lemma A4. Notice that an inadequate choice of p_λ^L, q_λ^L may lead to $\frac{p_\lambda^L}{q_\lambda^L}(\beta_\lambda + x_0^{(q_\lambda^U)}) - \alpha_\lambda < 0$. $\square$

Proof of Theorem 9. Let $p_\lambda^U = \alpha_\mathcal{A}^\lambda \alpha_\mathcal{H}^{1-\lambda} > \max\{0, \alpha_\lambda\}$ and $q_\lambda^U = \beta_\mathcal{A}^\lambda \beta_\mathcal{H}^{1-\lambda} > \max\{0, \beta_\lambda\}$. Since $q_\lambda^U < \min\{1, e^{\beta_\lambda - 1}\}$ iff $\lambda \in]\lambda_-, \lambda_+[\backslash [0,1]$ (cf. Lemma 1 for $q_\lambda := q_\lambda^U$)), this theorem follows from (A46) of Lemma A4, from the parts (b), (e) of Lemma A4 and from part (d) of Proposition 14. $\square$

Appendix A.4. Proofs and Auxiliary Lemmas for Section 7

Proof of Theorem 10. As already mentioned above, one can adapt the proof of Theorem 9.1.3 in Ethier & Kurtz [138] who deal with drift-parameters $\eta = 0$, $\kappa_\bullet = 0$, and the different setup of σ−*independent* time-scale and a sequence of *critical* Galton-Watson processes *without immigration* with *general* offspring distribution. For the sake of brevity, we basically outline here only the main differences to their proof; for similar limit investigations involving offspring/immigration distributions and parametrizations which are incompatble to ours, see e.g., Sriram [142].

As a first step, let us define the generator

$$A_\bullet f(x) := (\eta - \kappa_\bullet \cdot x) \cdot f'(x) + \frac{\sigma^2}{2} \cdot x \cdot f''(x), \quad f \in C_c^\infty([0,\infty)),$$

which corresponds to the diffusion process $\widetilde{X}$ governed by (133). In connection with (130), we study

$$T_\bullet^{(m)} f(x) := EP_\bullet\left[f\left(\frac{1}{m}\left(\sum_{k=1}^{mx} Y_{0,k}^{(m)} + \widetilde{Y}_0^{(m)} \right) \right) \right], \quad x \in E^{(m)} := \frac{1}{m}\mathbb{N}_0, \quad f \in C_c^\infty([0,\infty),$$

where the $Y_{0,k}^{(m)}$, $\widetilde{Y}_0^{(m)}$ are independent and (Poisson-$\beta_\bullet^{(m)}$ respectively Poisson-$\alpha_\bullet^{(m)}$) distributed as the members of the collection $Y^{(m)}$ respectively $\widetilde{Y}^{(m)}$. By the Theorems 8.2.1 and 1.6.5 as well as Corollary 4.8.9 of [138] it is sufficient to show

$$\lim_{m\to\infty} \sup_{x \in E^{(m)}} \left| \sigma^2 m \left(T_\bullet^{(m)} f(x) - f(x) \right) - A_\bullet f(x) \right| = 0, \quad f \in C_c^\infty([0,\infty)). \tag{A58}$$

But (A58) follows mainly from the next

Lemma A5. *Let*

$$S_n^{(m)} := \frac{1}{\sqrt{n}} \left(\sum_{k=1}^{n} \left(Y_{0,k}^{(m)} - \beta_\bullet^{(m)} \right) + \widetilde{Y}_0^{(m)} - \alpha_\bullet^{(m)} \right) , \quad n \in \mathbb{N}, \; m \in \overline{\mathbb{N}},$$

with the usual convention $S_0^{(m)} := 0$. Then for all $m \in \overline{\mathbb{N}}$, $x \in E^{(m)}$ and all $f \in C_c^\infty([0,\infty))$

$$\epsilon^{(m)}(x) := EP_\bullet \left[\int_0^1 \left(S_{mx}^{(m)} \right)^2 x(1-v) \left(f'' \left(\beta_\bullet^{(m)} x + \frac{\alpha_\bullet^{(m)}}{m} + v\sqrt{\frac{x}{m}} S_{mx}^{(m)} \right) - f''(x) \right) dv \right]$$

$$= \frac{1}{\sigma^2} \cdot \left[\sigma^2 m \cdot \left(T_\bullet^{(m)} f(x) - f(x) \right) - A_\bullet f(x) \right] + R^{(m)}, \qquad \text{where } \lim_{m\to\infty} R^{(m)} = 0. \tag{A59}$$

Proof of Lemma A5. Let us fix $f \in C_c^\infty([0,\infty))$. From the involved Poissonian expectations it is easy to see that

$$\lim_{m\to\infty} \left| \sigma^2 m \left(T_\bullet^{(m)} f(0) - f(0) \right) - A_\bullet f(0) \right| = 0 ,$$

and thus (A59) holds for $x = 0$. Accordingly, we next consider the case $x \in E^{(m)} \setminus \{0\}$, with fixed $m \in \overline{\mathbb{N}}$. From $EP_\bullet \left[\left(S_{mx}^{(m)} \right)^2 \right] = \beta_\bullet^{(m)} + \frac{\alpha_\bullet^{(m)}}{mx}$ we obtain

$$EP_\bullet \left[\left(S_{mx}^{(m)} \right)^2 x f''(x) \int_0^1 (1-v)dv \right] = \frac{1}{2} \left(\beta_\bullet^{(m)} \cdot x + \frac{\alpha_\bullet^{(m)}}{m} \right) f''(x) =: a_{mx} \frac{f''(x)}{2} =: a \frac{f''(x)}{2} . \tag{A60}$$

Furthermore, with $b_{mx} := b := a + \sqrt{x/m} \cdot S_{mx}^{(m)} = \frac{1}{m} \left(\sum_{k=1}^{mx} Y_{0,k}^{(m)} + \widetilde{Y}_0^{(m)} \right)$ we get on $\{ S_{mx}^{(m)} \neq 0 \}$

$$\int_0^1 f'' \left(\beta_\bullet^{(m)} x + \frac{\alpha_\bullet^{(m)}}{m} + v\sqrt{\frac{x}{m}} S_{mx}^{(m)} \right) dv = \sqrt{\frac{m}{x}} \cdot \frac{1}{S_{mx}^{(m)}} \int_a^b f''(y)dy = \sqrt{\frac{m}{x}} \cdot \frac{f'(b) - f'(a)}{S_{mx}^{(m)}} \tag{A61}$$

as well as

$$\int_0^1 v f'' \left(\beta_\bullet^{(m)} x + \frac{\alpha_\bullet^{(m)}}{m} + v\sqrt{\frac{x}{m}} S_{mx}^{(m)} \right) dv = \frac{m}{x \left(S_{mx}^{(m)} \right)^2} \left[\int_a^b y f''(y) \, dy - a \int_a^b f''(y) \, dy \right]$$

$$= \sqrt{\frac{m}{x}} \cdot \frac{f'(b)}{S_{mx}^{(m)}} + \frac{m}{x} \cdot \frac{f(a) - f(b)}{\left(S_{mx}^{(m)} \right)^2} . \tag{A62}$$

With our choice $\beta_\bullet^{(m)} = 1 - \frac{\kappa_\bullet}{\sigma^2 m}$ and $\alpha_\bullet^{(m)} = \beta_\bullet^{(m)} \cdot \frac{\eta}{\sigma^2}$, a Taylor expansion of f at x gives

$$f(a) = f(x) + \frac{1}{\sigma^2 m} \cdot f'(x) \left(\beta_\bullet^{(m)} \cdot \eta - \kappa_\bullet \cdot x \right) + o\left(\frac{1}{m} \right), \tag{A63}$$

where for the case $\eta = \kappa = 0$ we use the convention $o\left(\frac{1}{m}\right) \equiv 0$. Combining (A60) to (A63) and the centering $EP_\bullet\left[S_{mx}^{(m)}\right] = 0$, the left hand side of Equation (A59) becomes

$$EP_\bullet\left[\int_0^1 \left(S_{mx}^{(m)}\right)^2 x(1-v)\left(f''\left(\beta_\bullet^{(m)}x + \frac{\alpha_\bullet^{(m)}}{m} + v\sqrt{\frac{x}{m}}\,S_{mx}^{(m)}\right) - f''(x)\right)dv\right]$$

$$= EP_\bullet\left[\sqrt{mx}\cdot S_{mx}^{(m)}\cdot\left(f'(b) - f'(a)\right)\right] - EP_\bullet\left[\sqrt{mx}\cdot S_{mx}^{(m)}\cdot f'(b) + m\cdot(f(a) - f(b))\right]$$

$$-\frac{1}{2}\left(\beta_\bullet^{(m)}\cdot x + \frac{\alpha_\bullet^{(m)}}{m}\right)\cdot f''(x)$$

$$= m\cdot\left(EP_\bullet[f(b)] - f(a)\right) - \frac{1}{2}\left(\beta_\bullet^{(m)}\cdot x + \frac{\alpha_\bullet^{(m)}}{m}\right)\cdot f''(x)$$

$$= m\cdot\left\{EP_\bullet\left[f\left(\frac{1}{m}\left(\sum_{k=1}^{mx}Y_{0,k}^{(m)} + \widetilde{Y}_0\right)\right)\right] - f(x)\right\} - \frac{1}{\sigma^2}A_\bullet f(x)$$

$$+\frac{1}{\sigma^2}\left[(\eta - \kappa_\bullet\cdot x) - \beta_\bullet^{(m)}\cdot\eta + \kappa_\bullet\cdot x\right]\cdot f'(x) + \frac{x}{2}\left[1 - \beta_\bullet^{(m)} - \frac{\alpha_\bullet^{(m)}}{m}\right]\cdot f''(x) - m\cdot o\left(\frac{1}{m}\right)$$

which immediately leads to the right hand side of (A59). $\qquad\square$

To proceed with the proof of Theorem 10, we obtain for $m \geq 2\kappa_\bullet/\sigma^2$ the inequality $\beta_\bullet^{(m)} \geq 1/2$ and accordingly for all $v \in\,]0,1[$, $x \in E^{(m)}$

$$\beta_\bullet^{(m)}x + \frac{\alpha_\bullet^{(m)}}{m} + v\sqrt{\frac{x}{m}}\,S_{mx}^{(m)} = (1-v)\cdot x\cdot\beta_\bullet^{(m)} + (1-v)\frac{\alpha_\bullet^{(m)}}{m} + v\left(\sum_{k=1}^{mx}Y_{0,k}^{(m)} + \widetilde{Y}_0\right) \geq x\cdot\frac{1-v}{2}.$$

Suppose that the support of f is contained in the interval $[0,c]$. Correspondingly, for $v \leq 1 - 2c/x$ the integrand in $\epsilon^{(m)}(x)$ is zero and hence with (A64) we obtain the bounds

$$\left|\int_0^1 \left(S_{mx}^{(m)}\right)^2 x(1-v)\left(f''\left(\beta_\bullet^{(m)}x + \frac{\alpha_\bullet^{(m)}}{m} + v\sqrt{\frac{x}{m}}\,S_{mx}^{(m)}\right) - f''(x)\right)dv\right|$$

$$\leq \int_{0\vee(1-2c/x)}^1 \left(S_{mx}^{(m)}\right)^2 x(1-v)\cdot 2\,\|f''\|_\infty\,dv \leq x\cdot\left(S_{mx}^{(m)}\right)^2\left(1\wedge\frac{2c}{x}\right)^2\|f''\|_\infty.$$

From this, one can deduce $\lim_{m\to\infty}\sup_{x\in E^{(m)}}\epsilon^{(m)}(x) = 0$–and thus (A58) – in the same manner as at the end of the proof of Theorem 9.1.3 in [138] (by means of the dominated convergence theorem). $\qquad\square$

Proof of Proposition 15. Let $(\kappa_\mathcal{A}, \kappa_\mathcal{H}, \eta) \in \widetilde{\mathcal{P}}_{NI} \cup \widetilde{\mathcal{P}}_{SP,1}$ be fixed. We have to find those orders $\lambda \in \mathbb{R}\backslash[0,1]$ which satisfy for all sufficiently large $m \in \overline{\mathbb{N}}$

$$q_\lambda^{(m)} = \left(1 - \frac{\kappa_\mathcal{A}}{\sigma^2 m}\right)^\lambda \left(1 - \frac{\kappa_\mathcal{H}}{\sigma^2 m}\right)^{1-\lambda} < \min\left\{1,\,e^{\beta_\lambda^{(m)}-1}\right\}. \tag{A64}$$

In order to achieve this, we interpret $q_\lambda^{(m)} = q_\lambda\left(\frac{1}{m}\right)$ in terms of the function

$$q_\lambda(x) := \left(1 - \frac{\kappa_\mathcal{A}}{\sigma^2}\cdot x\right)^\lambda \left(1 - \frac{\kappa_\mathcal{H}}{\sigma^2}\cdot x\right)^{1-\lambda}, \qquad x \in\,]-\epsilon,\epsilon[, \tag{A65}$$

for some small enough $\epsilon > 0$ such that (A65) is well-defined. Since $\beta_\lambda^{(m)} - 1 = -\frac{\kappa_\lambda}{\sigma^2 \cdot m} = -\frac{\kappa_\lambda}{\sigma^2} \cdot x = -\frac{\lambda \kappa_\mathcal{A} + (1-\lambda)\kappa_\mathcal{H}}{\sigma^2} \cdot x$, for the verification of (A64) it suffices to show

$$\lim_{x \searrow 0} \frac{1 - q_\lambda(x)}{x} \;>\; 0, \tag{A66}$$

$$\text{and} \quad \lim_{x \searrow 0} \frac{e^{-\frac{\kappa_\lambda}{\sigma^2} \cdot x} - q_\lambda(x)}{x^2} \;>\; 0. \tag{A67}$$

By l'Hospital's rule, one gets $\lim_{x \searrow 0} \frac{1 - q_\lambda(x)}{x} = \frac{\lambda \kappa_\mathcal{A} + (1-\lambda)\kappa_\mathcal{H}}{\sigma^2} = \frac{\kappa_\lambda}{\sigma^2}$ and hence

$$(A66) \quad \Longleftrightarrow \quad \begin{cases} \lambda < \frac{\kappa_\mathcal{H}}{\kappa_\mathcal{H} - \kappa_\mathcal{A}}, & \text{if } \kappa_\mathcal{A} < \kappa_\mathcal{H}, \\[2ex] \lambda > -\frac{\kappa_\mathcal{H}}{\kappa_\mathcal{A} - \kappa_\mathcal{H}}, & \text{if } \kappa_\mathcal{A} > \kappa_\mathcal{H}. \end{cases} \tag{A68}$$

To find a condition that guarantees (A67), we use l'Hospital's rule twice to deduce

$$\lim_{x \searrow 0} \frac{e^{-\frac{\kappa_\lambda}{\sigma^2} \cdot x} - q_\lambda(x)}{x^2} = \frac{1}{2\sigma^4}\left[\kappa_\lambda^2 - \lambda(\lambda - 1)(\kappa_\mathcal{A} - \kappa_\mathcal{H})^2\right] = \frac{1}{2\sigma^4}\left[\lambda \kappa_\mathcal{A}^2 + (1-\lambda)\kappa_\mathcal{H}^2\right]$$

and hence we obtain

$$(A67) \quad \Longleftrightarrow \quad \begin{cases} \lambda < \frac{\kappa_\mathcal{H}^2}{\kappa_\mathcal{H}^2 - \kappa_\mathcal{A}^2}, & \text{if } \kappa_\mathcal{A} < \kappa_\mathcal{H}, \\[2ex] \lambda > -\frac{\kappa_\mathcal{H}^2}{\kappa_\mathcal{A}^2 - \kappa_\mathcal{H}^2}, & \text{if } \kappa_\mathcal{A} > \kappa_\mathcal{H}. \end{cases} \tag{A69}$$

To compare both the lower and upper bounds in (A68) and (A69), let us calculate

$$\frac{\kappa_\mathcal{H}^2}{\kappa_\mathcal{H}^2 - \kappa_\mathcal{A}^2} - \frac{\kappa_\mathcal{H}}{\kappa_\mathcal{H} - \kappa_\mathcal{A}} = -\frac{\kappa_\mathcal{A} \kappa_\mathcal{H}}{(\kappa_\mathcal{H} - \kappa_\mathcal{A})(\kappa_\mathcal{H} + \kappa_\mathcal{A})} \begin{cases} < 0, & \text{if } \kappa_\mathcal{A} < \kappa_\mathcal{H}, \\[2ex] > 0, & \text{if } \kappa_\mathcal{A} > \kappa_\mathcal{H}. \end{cases} \tag{A70}$$

Incorporating this, we observe that both conditions (A66) and (A67) are satisfied simultaneously iff

$$\lambda < \min\left\{\frac{\kappa_\mathcal{H}}{\kappa_\mathcal{H} - \kappa_\mathcal{A}}, \frac{\kappa_\mathcal{H}^2}{\kappa_\mathcal{H}^2 - \kappa_\mathcal{A}^2}\right\} = \frac{\kappa_\mathcal{H}^2}{\kappa_\mathcal{H}^2 - \kappa_\mathcal{A}^2} \quad \text{if } \kappa_\mathcal{A} < \kappa_\mathcal{H},$$

$$\lambda > \max\left\{-\frac{\kappa_\mathcal{H}}{\kappa_\mathcal{A} - \kappa_\mathcal{H}}, -\frac{\kappa_\mathcal{H}^2}{\kappa_\mathcal{A}^2 - \kappa_\mathcal{H}^2}\right\} = -\frac{\kappa_\mathcal{H}^2}{\kappa_\mathcal{A}^2 - \kappa_\mathcal{H}^2} \quad \text{if } \kappa_\mathcal{A} > \kappa_\mathcal{H},$$

which finishes the proof. $\quad \square$

The following lemma is the main tool for the proof of Theorem 11.

Lemma A6. *Let $(\kappa_{\mathcal{A}}, \kappa_{\mathcal{H}}, \eta, \lambda) \in (\widetilde{\mathcal{P}}_{NI} \cup \widetilde{\mathcal{P}}_{SP,1}) \times (\,]\widetilde{\lambda}_-, \widetilde{\lambda}_+[\,\backslash\{0,1\}\,)$. By using the quantities $\kappa_\lambda := \lambda\kappa_{\mathcal{A}} + (1 - \lambda)\kappa_{\mathcal{H}}$ and $\Lambda_\lambda := \sqrt{\lambda\kappa_{\mathcal{A}}^2 + (1 - \lambda)\kappa_{\mathcal{H}}^2}$ from (150) (which is well-defined, cf. (138)), one gets for all $t > 0$*

$$(a) \qquad \lim_{m \to \infty} m \cdot \left(1 - q_\lambda^{(m)}\right) = \lim_{m \to \infty} m \cdot \left(1 - \beta_\lambda^{(m)}\right) = \frac{\kappa_\lambda}{\sigma^2}.$$

$$(b) \qquad \lim_{m \to \infty} m^2 \cdot a_1^{(m)} = \lim_{m \to \infty} m^2 \cdot \left(q_\lambda^{(m)} - \beta_\lambda^{(m)}\right) = -\frac{\lambda(1-\lambda)\,(\kappa_{\mathcal{A}} - \kappa_{\mathcal{H}})^2}{2\sigma^4} = -\frac{\Lambda_\lambda^2 - \kappa_\lambda^2}{2\sigma^4}.$$

$$(c) \qquad \lim_{m \to \infty} m \cdot x_0^{(m)} = -\frac{\Lambda_\lambda - \kappa_\lambda}{\sigma^2} \begin{cases} < 0, & \text{if } \lambda \in]0,1[, \\[2mm] > 0, & \text{if } \lambda \in]\widetilde{\lambda}_-, \widetilde{\lambda}_+[\,\backslash\,[0,1]. \end{cases}$$

$$(d) \qquad \lim_{m \to \infty} m^2 \cdot \Gamma_<^{(m)} = \lim_{m \to \infty} m^2 \cdot \Gamma_>^{(m)} = \frac{(\Lambda_\lambda - \kappa_\lambda)^2}{2\sigma^4} > 0.$$

$$(e) \qquad \lim_{m \to \infty} m \cdot (1 - d^{(m),S}) = \frac{\Lambda_\lambda + \kappa_\lambda}{2\sigma^2} > 0.$$

$$(f) \qquad \lim_{m \to \infty} m \cdot (1 - d^{(m),T}) = \frac{\Lambda_\lambda}{\sigma^2} > 0.$$

$$(g) \qquad \lim_{m \to \infty} m \cdot (1 - d^{(m),S} d^{(m),T}) = \frac{3\Lambda_\lambda + \kappa_\lambda}{2\sigma^2} > 0.$$

$$(h) \qquad \lim_{m \to \infty} \left(d^{(m),S}\right)^{\sigma^2 mt} = \exp\left\{-\frac{\Lambda_\lambda + \kappa_\lambda}{2} \cdot t\right\} < 1.$$

$$(i) \qquad \lim_{m \to \infty} \left(d^{(m),T}\right)^{\sigma^2 mt} = \exp\{-\Lambda_\lambda \cdot t\} < 1.$$

$$(j) \qquad \lim_{m \to \infty} \left(d^{(m),S} d^{(m),T}\right)^{\sigma^2 mt} = \exp\left\{-\frac{3\Lambda_\lambda + \kappa_\lambda}{2} \cdot t\right\} < 1.$$

(k) *for $\lambda \in]0,1[$, there holds for the respective quantities defined in (142) to (145)*

$$\lim_{m \to \infty} m \cdot \underline{\zeta}_{\lfloor \sigma^2 mt \rfloor}^{(m)} = \frac{(\Lambda_\lambda - \kappa_\lambda)^2}{2\sigma^2 \cdot \Lambda_\lambda} \cdot e^{-\Lambda_\lambda \cdot t} \cdot \left(1 - e^{-\Lambda_\lambda \cdot t}\right) > 0,$$

$$\lim_{m \to \infty} \underline{\vartheta}_{\lfloor \sigma^2 mt \rfloor}^{(m)} = \frac{1}{4} \cdot \left(\frac{\Lambda_\lambda - \kappa_\lambda}{\Lambda_\lambda}\right)^2 \cdot \left(1 - e^{-\Lambda_\lambda \cdot t}\right)^2 > 0,$$

$$\lim_{m \to \infty} m \cdot \overline{\zeta}_{\lfloor \sigma^2 mt \rfloor}^{(m)} = \frac{(\Lambda_\lambda - \kappa_\lambda)^2}{\sigma^2} \cdot \left[\frac{e^{-\frac{1}{2}(\Lambda_\lambda + \kappa_\lambda) \cdot t} - e^{-\Lambda_\lambda \cdot t}}{\Lambda_\lambda - \kappa_\lambda} - \frac{e^{-\frac{1}{2}(\Lambda_\lambda + \kappa_\lambda) \cdot t}\left(1 - e^{-\Lambda_\lambda \cdot t}\right)}{2 \cdot \Lambda_\lambda}\right] > 0,$$

$$\lim_{m \to \infty} \overline{\vartheta}_{\lfloor \sigma^2 mt \rfloor}^{(m)} = \frac{(\Lambda_\lambda - \kappa_\lambda)^2}{\Lambda_\lambda} \cdot \left[\frac{1 - e^{-\frac{1}{2}(3\Lambda_\lambda + \kappa_\lambda) \cdot t}}{3\Lambda_\lambda + \kappa_\lambda} + \frac{e^{-\Lambda_\lambda \cdot t} - e^{-\frac{1}{2}(\Lambda_\lambda + \kappa_\lambda) \cdot t}}{\Lambda_\lambda - \kappa_\lambda}\right] > 0.$$

(l) *for $\lambda \in]\widetilde{\lambda}_-, \widetilde{\lambda}_+[\,\backslash\,[0,1]$, there holds for the respective quantities defined in (146) to (149)*

$$\lim_{m \to \infty} m \cdot \underline{\zeta}_{\lfloor \sigma^2 mt \rfloor}^{(m)} = \frac{(\Lambda_\lambda - \kappa_\lambda)^2}{2\sigma^2 \cdot \kappa_\lambda} \cdot e^{-\Lambda_\lambda \cdot t} \cdot \left(1 - e^{-\kappa_\lambda \cdot t}\right) > 0,$$

$$\lim_{m \to \infty} \underline{\vartheta}_{\lfloor \sigma^2 mt \rfloor}^{(m)} = \frac{(\Lambda_\lambda - \kappa_\lambda)^2}{2 \cdot \kappa_\lambda} \cdot \left[\frac{1 - e^{-\Lambda_\lambda \cdot t}}{\Lambda_\lambda} - \frac{1 - e^{-(\Lambda_\lambda + \kappa_\lambda) \cdot t}}{\Lambda_\lambda + \kappa_\lambda}\right] > 0,$$

$$\lim_{m \to \infty} m \cdot \overline{\zeta}_{\lfloor \sigma^2 mt \rfloor}^{(m)} = \frac{(\Lambda_\lambda - \kappa_\lambda)^2}{2 \cdot \sigma^2} \cdot e^{-\frac{1}{2}(\Lambda_\lambda + \kappa_\lambda) \cdot t} \cdot \left[t - \frac{1 - e^{-\Lambda_\lambda \cdot t}}{\Lambda_\lambda}\right] > 0,$$

$$\lim_{m \to \infty} \overline{\vartheta}_{\lfloor \sigma^2 mt \rfloor}^{(m)} = (\Lambda_\lambda - \kappa_\lambda)^2 \cdot \left[\frac{(\Lambda_\lambda - \kappa_\lambda)\left(1 - e^{-\frac{1}{2}(\Lambda_\lambda + \kappa_\lambda) \cdot t}\right)}{\Lambda_\lambda \cdot (\Lambda_\lambda + \kappa_\lambda)^2}\right.$$

$$\left. + \frac{1 - e^{-\frac{1}{2}(3\Lambda_\lambda + \kappa_\lambda) \cdot t}}{\Lambda_\lambda \cdot (3\Lambda_\lambda + \kappa_\lambda)} - \frac{e^{-\frac{1}{2}(\Lambda_\lambda + \kappa_\lambda) \cdot t}}{\Lambda_\lambda + \kappa_\lambda} \cdot t\right] > 0.$$

Proof of Lemma A6. For each of the assertions (a) to (l), we will make use of l'Hospital's rule. To begin with, we obtain for arbitrary $\mu, \nu \in \mathbb{R}$

$$\lim_{m \to \infty} m \left[1 - (\beta_{\mathcal{A}}^{(m)})^\mu (\beta_{\mathcal{H}}^{(m)})^\nu\right] = \lim_{m \to \infty} m^2 \left[\mu \cdot (\beta_{\mathcal{A}}^{(m)})^{\mu-1} (\beta_{\mathcal{H}}^{(m)})^\nu \frac{\kappa_{\mathcal{A}}}{\sigma^2 m^2} + \nu \cdot (\beta_{\mathcal{A}}^{(m)})^\mu (\beta_{\mathcal{H}}^{(m)})^{\nu-1} \frac{\kappa_{\mathcal{H}}}{\sigma^2 m^2}\right]$$

$$= \mu \frac{\kappa_{\mathcal{A}}}{\sigma^2} + \nu \frac{\kappa_{\mathcal{H}}}{\sigma^2}. \tag{A71}$$

From this, the first part of (a) follows immediately and the second part is a direct consequence of the definition of $\beta_\lambda^{(m)}$. Part (b) can be deduced from (A71):

$$\lim_{m\to\infty} m^2 \cdot a_1^{(m)} = \lim_{m\to\infty} \frac{m}{2\sigma^2} \cdot \left[\lambda \cdot \kappa_{\mathcal{A}} \left(1 - (\beta_{\mathcal{A}}^{(m)})^{\lambda-1}(\beta_{\mathcal{H}}^{(m)})^{1-\lambda} \right) \right.$$

$$\left. + (1-\lambda) \cdot \kappa_{\mathcal{H}} \left(1 - (\beta_{\mathcal{A}}^{(m)})^{\lambda}(\beta_{\mathcal{H}}^{(m)})^{-\lambda} \right) \right] = -\frac{\lambda(1-\lambda)(\kappa_{\mathcal{A}} - \kappa_{\mathcal{H}})^2}{2\sigma^4} = -\frac{\Lambda_\lambda^2 - \kappa_\lambda^2}{2\sigma^4} .$$

For the proof of (c), we rely on the inequalities $\underline{x}_0^{(m)} \le x_0^{(m)} \le \overline{x}_0^{(m)}$ ($m \in \mathbb{N}$), where $\underline{x}_0^{(m)}$ and $\overline{x}_0^{(m)}$ are the obvious notational adaptions of (124) and (126), respectively. Notice that $\underline{x}_0^{(m)}$ and $\overline{x}_0^{(m)}$ are solutions of the (again adapted) quadratic equations $\underline{Q}_\lambda^{(m)}(x) = x$ resp. $\overline{Q}_\lambda^{(m)}(x) = x$ (cf. (127) and (128)). These solutions clearly exist in the case $\lambda \in]0,1[$. For sufficiently large approximations steps $m \in \mathbb{N}$, these solutions also exist in the case $\lambda \in]\widetilde{\lambda}_-, \widetilde{\lambda}_+[\setminus [0,1]$ since (138) together with parts (a) and (b) imply

$$\lim_{m\to\infty} \left(m \cdot (1 - q_\lambda^{(m)}) \right)^2 - 2 \cdot q_\lambda^{(m)} \cdot m^2 \cdot a_1^{(m)} = \sigma^{-2} \cdot \left[\lambda \kappa_{\mathcal{A}}^2 + (1-\lambda)\kappa_{\mathcal{H}}^2 \right] > 0, \qquad \text{for } \lambda \in]\widetilde{\lambda}_-, \widetilde{\lambda}_+[\setminus [0,1].$$

To prove part (c), we show that the limits of $\underline{x}_0^{(m)}$ and $\overline{x}_0^{(m)}$ coincide. Assume first that $\lambda \in]0,1[$. Using (a) and (b), we obtain together with the obvious limit $\lim_{m\to\infty} q_\lambda^{(m)} = 1$

$$\lim_{m\to\infty} m \cdot \overline{x}_0^{(m)} = \lim_{m\to\infty} \left(q_\lambda^{(m)} \right)^{-1} \cdot \left[m \cdot (1 - q_\lambda^{(m)}) - \sqrt{\left(m \cdot (1 - q_\lambda^{(m)}) \right)^2 - 2 \cdot q_\lambda^{(m)} \cdot m^2 \cdot a_1^{(m)}} \right]$$

$$= \frac{\kappa_\lambda}{\sigma^2} - \sqrt{\left(\frac{\kappa_\lambda}{\sigma^2} \right)^2 + \frac{\Lambda_\lambda^2 - \kappa_\lambda^2}{\sigma^4}} = -\frac{\Lambda_\lambda - \kappa_\lambda}{\sigma^2} . \tag{A72}$$

Let $\underline{x}_0^{(m)}$ be the adapted version of the auxiliary fixed-point lower bound defined in (125). By incorporating $\lim_{m\to\infty} \beta_\lambda^{(m)} = 1$ we obtain with (a) and (b)

$$\lim_{m\to\infty} \underline{x}_0^{(m)} = \lim_{m\to\infty} \max \left\{ -\beta_\lambda^{(m)} , \frac{q_\lambda^{(m)} - \beta_\lambda^{(m)}}{1 - q_\lambda^{(m)}} \right\} = \lim_{m\to\infty} \frac{1}{m} \cdot \frac{m^2 \cdot a_1^{(m)}}{m \cdot \left(1 - q_\lambda^{(m)} \right)} = 0,$$

which implies

$$\lim_{m\to\infty} m \cdot \underline{x}_0^{(m)} = \lim_{m\to\infty} \frac{e^{-\underline{x}_0^{(m)}}}{q_\lambda^{(m)}} \cdot \left[m \cdot (1 - q_\lambda^{(m)}) - \sqrt{\left(m \cdot (1 - q_\lambda^{(m)}) \right)^2 - 2 \cdot e^{\underline{x}_0^{(m)}} q_\lambda^{(m)} \cdot m^2 \cdot a_1^{(m)}} \right]$$

$$= \frac{\kappa_\lambda}{\sigma^2} - \sqrt{\left(\frac{\kappa_\lambda}{\sigma^2} \right)^2 + \frac{\Lambda_\lambda^2 - \kappa_\lambda^2}{\sigma^4}} = -\frac{\Lambda_\lambda - \kappa_\lambda}{\sigma^2} . \tag{A73}$$

Combining (A72) and (A73), the desired result (c) follows for $\lambda \in]0,1[$. Assume now that $\lambda \in]\widetilde{\lambda}_-, \widetilde{\lambda}_+[\setminus [0,1]$. In this case the approximates $\underline{x}_0^{(m)}$ and $\overline{x}_0^{(m)}$ have a different form, given in (124) and (126). However, the calculations work out in the same way: with parts (a) and (b) we get

$$\lim_{m\to\infty} m \cdot \underline{x}_0^{(m)} = \lim_{m\to\infty} \frac{1}{q_\lambda^{(m)}} \cdot \left[m \cdot \left(1 - q_\lambda^{(m)} \right) - \sqrt{\left(m \cdot (1 - q_\lambda^{(m)}) \right)^2 - 2 \cdot q_\lambda^{(m)} \cdot m^2 \cdot a_1^{(m)}} \right]$$

$$= \frac{\kappa_\lambda}{\sigma^2} - \sqrt{\left(\frac{\kappa_\lambda}{\sigma^2} \right)^2 + \frac{\Lambda_\lambda^2 - \kappa_\lambda^2}{\sigma^4}} = -\frac{\Lambda_\lambda - \kappa_\lambda}{\sigma^2} ,$$

as well as

$$\lim_{m\to\infty} m \cdot \bar{x}_0^{(m)} = \lim_{m\to\infty} m \cdot \left(1 - q_\lambda^{(m)}\right) - \sqrt{\left(m \cdot (1 - q_\lambda^{(m)})\right)^2 - 2 \cdot m^2 \cdot a_1^{(m)}}$$

$$= \frac{\kappa_\lambda}{\sigma^2} - \sqrt{\left(\frac{\kappa_\lambda}{\sigma^2}\right)^2 + \frac{\Lambda_\lambda^2 - \kappa_\lambda^2}{\sigma^4}} = -\frac{\Lambda_\lambda - \kappa_\lambda}{\sigma^2},$$

which finally finishes the proof of part (c). Assertion (d) is a direct consequence of (c). Since the representations of the parameters $c^{(m),S}$, $d^{(m),S}$, $c^{(m),T}$, $d^{(m),T}$ are the same in both cases $\lambda \in]0,1[$ and $\lambda \in]\tilde{\lambda}_-, \tilde{\lambda}_+[\setminus[0,1]$, the following considerations hold generally. Part (e) follows from (b) and (c) by

$$\lim_{m\to\infty} m \cdot (1 - d^{(m),S}) = \lim_{m\to\infty} \frac{m^2 \cdot a_1^{(m)}}{m \cdot x_0^{(m)}} = \frac{\Lambda_\lambda + \kappa_\lambda}{2\sigma^2} > 0.$$

Notice that this term is positive since on $]\tilde{\lambda}_-, \tilde{\lambda}_+[\setminus\{0,1\}$ there holds $\kappa_\lambda > 0$ as well as $\Lambda_\lambda > 0$, cf. (A70). To prove (f), we apply the general limit $\lim_{x\to 0} \frac{e^x - 1}{x} = 1$ and get with (a), (c)

$$\lim_{m\to\infty} m \cdot (1 - d^{(m),T}) = \lim_{m\to\infty} \left(m \cdot \left(1 - q_\lambda^{(m)}\right) - q_\lambda^{(m)} \cdot m \cdot x_0^{(m)} \cdot \frac{e^{x_0^{(m)}} - 1}{x_0^{(m)}}\right) = \frac{\Lambda_\lambda}{\sigma^2}.$$

The limit (g) can be obtained from (e) and (f):

$$\lim_{m\to\infty} m \cdot (1 - d^{(m),S} d^{(m),T}) = \lim_{m\to\infty} \left\{m \cdot (1 - d^{(m),S}) + d^{(m),S} \cdot m \cdot (1 - d^{(m),T})\right\} = \frac{3\Lambda_\lambda + \kappa_\lambda}{2\sigma^2}.$$

The assertions (h) resp. (i) resp. (j) follow from (e) resp. (f) resp. (g) by using the general relation $\lim_{m\to\infty} \left(1 + \frac{x_m}{m}\right)^m = \exp\{\lim_{m\to\infty} x_m\}$. To get the last two parts (k) and (l), we make repeatedly use of the results (a) to (j) and combine them with the formulas (142) to (149) of Corollary 14. More detailed, for $\lambda \in]0,1[$ (and thus $q_\lambda^{(m)} < \beta_\lambda^{(m)}$) we obtain

$$m \cdot \underline{\zeta}_{\lfloor \sigma^2 mt \rfloor}^{(m)} = m^2 \cdot \Gamma_<^{(m)} \cdot \frac{\left(d^{(m),T}\right)^{\lfloor \sigma^2 mt \rfloor - 1}}{m \cdot \left(1 - d^{(m),T}\right)} \cdot \left(1 - \left(d^{(m),T}\right)^{\lfloor \sigma^2 mt \rfloor}\right)$$

$$\xrightarrow{m\to\infty} \frac{(\Lambda_\lambda - \kappa_\lambda)^2}{2\sigma^2 \cdot \Lambda_\lambda} \cdot e^{-\Lambda_\lambda \cdot t} \cdot \left(1 - e^{-\Lambda_\lambda \cdot t}\right) > 0,$$

$$\underline{\vartheta}_{\lfloor \sigma^2 mt \rfloor}^{(m)} = m^2 \cdot \Gamma_<^{(m)} \cdot \frac{1 - \left(d^{(m),T}\right)^{\lfloor \sigma^2 mt \rfloor}}{\left(m \cdot \left(1 - d^{(m),T}\right)\right)^2} \cdot \left[1 - \frac{d^{(m),T}\left(1 + \left(d^{(m),T}\right)^{\lfloor \sigma^2 mt \rfloor}\right)}{1 + d^{(m),T}}\right]$$

$$\xrightarrow{m\to\infty} \frac{1}{4} \cdot \left(\frac{\Lambda_\lambda - \kappa_\lambda}{\Lambda_\lambda}\right)^2 \cdot \left(1 - e^{-\Lambda_\lambda \cdot t}\right)^2 > 0,$$

$$m \cdot \overline{\zeta}_{\lfloor \sigma^2 mt \rfloor}^{(m)} = m^2 \cdot \Gamma_<^{(m)} \cdot \left[\frac{\left(d^{(m),S}\right)^{\lfloor \sigma^2 mt \rfloor} - \left(d^{(m),T}\right)^{\lfloor \sigma^2 mt \rfloor}}{m \cdot \left(1 - d^{(m),T}\right) - m \cdot \left(1 - d^{(m),S}\right)}\right.$$

$$\left. - \left(d^{(m),S}\right)^{\lfloor \sigma^2 mt \rfloor - 1} \cdot \frac{1 - \left(d^{(m),T}\right)^{\lfloor \sigma^2 mt \rfloor}}{m \cdot \left(1 - d^{(m),T}\right)}\right]$$

$$\xrightarrow{m\to\infty} \frac{(\Lambda_\lambda - \kappa_\lambda)^2}{\sigma^2} \cdot \left[\frac{e^{-\frac{1}{2}(\Lambda_\lambda + \kappa_\lambda)\cdot t} - e^{-\Lambda_\lambda \cdot t}}{\Lambda_\lambda - \kappa_\lambda} - \frac{e^{-\frac{1}{2}(\Lambda_\lambda + \kappa_\lambda)\cdot t}\left(1 - e^{-\Lambda_\lambda \cdot t}\right)}{2 \cdot \Lambda_\lambda}\right] > 0,$$

$$\overline{\vartheta}^{(m)}_{\lfloor\sigma^2 mt\rfloor} \;=\; \frac{m^2\cdot\Gamma^{(m)}_{<}\cdot d^{(m),T}}{m\cdot\left(1-d^{(m),T}\right)}\cdot\left[\frac{1-\left(d^{(m),S}d^{(m),T}\right)^{\lfloor\sigma^2 mt\rfloor}}{m\cdot\left(1-d^{(m),S}d^{(m),T}\right)}-\frac{\left(d^{(m),S}\right)^{\lfloor\sigma^2 mt\rfloor}-\left(d^{(m),T}\right)^{\lfloor\sigma^2 mt\rfloor}}{m\cdot\left(1-d^{(m),T}\right)-m\cdot\left(1-d^{(m),S}\right)}\right]$$

$$\xrightarrow{m\to\infty}\;\frac{(\Lambda_\lambda-\kappa_\lambda)^2}{\Lambda_\lambda}\cdot\left[\frac{1-e^{-\frac{1}{2}(3\Lambda_\lambda+\kappa_\lambda)\cdot t}}{3\Lambda_\lambda+\kappa_\lambda}+\frac{e^{-\Lambda_\lambda\cdot t}-e^{-\frac{1}{2}(\Lambda_\lambda+\kappa_\lambda)\cdot t}}{\Lambda_\lambda-\kappa_\lambda}\right]\;>\;0\,.$$

For $\lambda\in\,]\widetilde{\lambda}_-,\widetilde{\lambda}_+[\,\backslash[0,1]$ (and thus $q^{(m)}_\lambda>\beta^{(m)}_\lambda$) we get

$$m\cdot\underline{\zeta}^{(m)}_{\lfloor\sigma^2 mt\rfloor}\;=\;m^2\cdot\Gamma^{(m)}_{>}\cdot\frac{\left(d^{(m),T}\right)^{\lfloor\sigma^2 mt\rfloor}-\left(d^{(m),S}\right)^{2\cdot\lfloor\sigma^2 mt\rfloor}}{m\cdot\left(1-d^{(m),S}\right)\left(1+d^{(m),S}\right)-m\cdot\left(1-d^{(m),T}\right)}$$

$$\xrightarrow{m\to\infty}\;\frac{(\Lambda_\lambda-\kappa_\lambda)^2}{2\sigma^2\cdot\kappa_\lambda}\cdot e^{-\Lambda_\lambda\cdot t}\cdot\left(1-e^{-\kappa_\lambda\cdot t}\right)\;>\;0\,,$$

$$\underline{\vartheta}^{(m)}_{\lfloor\sigma^2 mt\rfloor}\;=\;\frac{m^2\cdot\Gamma^{(m)}_{>}}{m\cdot\left(1-d^{(m),S}\right)\left(1+d^{(m),S}\right)-m\cdot\left(1-d^{(m),T}\right)}$$

$$\cdot\left[\frac{d^{(m),T}\cdot\left(1-\left(d^{(m),T}\right)^{\lfloor\sigma^2 mt\rfloor}\right)}{m\cdot\left(1-d^{(m),T}\right)}-\frac{\left(d^{(m),S}\right)^2\cdot\left(1-\left(d^{(m),S}\right)^{2\cdot\lfloor\sigma^2 mt\rfloor}\right)}{m\cdot\left(1-d^{(m),S}\right)\left(1+d^{(m),S}\right)}\right]$$

$$\xrightarrow{m\to\infty}\;\frac{(\Lambda_\lambda-\kappa_\lambda)^2}{2\cdot\kappa_\lambda}\cdot\left[\frac{1-e^{-\Lambda_\lambda\cdot t}}{\Lambda_\lambda}-\frac{1-e^{-(\Lambda_\lambda+\kappa_\lambda)\cdot t}}{\Lambda_\lambda+\kappa_\lambda}\right]\;>\;0\,,$$

$$m\cdot\overline{\zeta}^{(m)}_{\lfloor\sigma^2 mt\rfloor}\;=\;m^2\cdot\Gamma^{(m)}_{>}\cdot\left(d^{(m),S}\right)^{\lfloor\sigma^2 mt\rfloor-1}\cdot\left[\frac{\lfloor\sigma^2 mt\rfloor}{m}-\frac{1-\left(d^{(m),T}\right)^{\lfloor\sigma^2 mt\rfloor}}{m\cdot\left(1-d^{(m),T}\right)}\right]$$

$$\xrightarrow{m\to\infty}\;\frac{(\Lambda_\lambda-\kappa_\lambda)^2}{2\cdot\sigma^2}\cdot e^{-\frac{1}{2}(\Lambda_\lambda+\kappa_\lambda)\cdot t}\cdot\left[t-\frac{1-e^{-\Lambda_\lambda\cdot t}}{\Lambda_\lambda}\right]\;>\;0\,,$$

$$\overline{\vartheta}^{(m)}_{\lfloor\sigma^2 mt\rfloor}\;=\;m^2\cdot\Gamma^{(m)}_{>}\cdot\left[\frac{m\cdot\left(1-d^{(m),T}\right)-m\cdot\left(1-d^{(m),S}\right)}{m^2\cdot\left(1-d^{(m),S}\right)^2\cdot m\cdot\left(1-d^{(m),T}\right)}\cdot\left(1-\left(d^{(m),S}\right)^{\lfloor\sigma^2 mt\rfloor}\right)\right.$$

$$\left.+\frac{d^{(m),T}\left(1-\left(d^{(m),S}d^{(m),T}\right)^{\lfloor\sigma^2 mt\rfloor}\right)}{m\cdot\left(1-d^{(m),T}\right)\cdot m\cdot\left(1-d^{(m),S}d^{(m),T}\right)}-\frac{\left(d^{(m),S}\right)^{\lfloor\sigma^2 mt\rfloor}}{m\cdot\left(1-d^{(m),S}\right)}\cdot\frac{\lfloor\sigma^2 mt\rfloor}{m}\right]$$

$$\xrightarrow{m\to\infty}\;(\Lambda_\lambda-\kappa_\lambda)^2\cdot\left[\frac{(\Lambda_\lambda-\kappa_\lambda)\left(1-e^{-\frac{1}{2}(\Lambda_\lambda+\kappa_\lambda)\cdot t}\right)}{\Lambda_\lambda\cdot(\Lambda_\lambda+\kappa_\lambda)^2}+\frac{1-e^{-\frac{1}{2}(3\Lambda_\lambda+\kappa_\lambda)\cdot t}}{\Lambda_\lambda\cdot(3\Lambda_\lambda+\kappa_\lambda)}-\frac{e^{-\frac{1}{2}(\Lambda_\lambda+\kappa_\lambda)\cdot t}}{\Lambda_\lambda+\kappa_\lambda}\cdot t\right]\;>\;0.\quad\square$$

Proof of Theorem 11. It suffices to compute the limits of the bounds given in Corollary 14 as m tends to infinity. This is done by applying Lemma A6 which provides corresponding limits of all quantities of interest. Accordingly, for all $t > 0$ the lower bound (153) in the case $\lambda \in]0,1[$ can be obtained from (140), (142) and (143) by

$$
\lim_{m \to \infty} \exp \left\{ x_0^{(m)} \cdot \left[X_0^{(m)} - \frac{\eta}{\sigma^2} \cdot \frac{d^{(m),T}}{1 - d^{(m),T}} \right] \left(1 - \left(d^{(m),T} \right)^{\lfloor \sigma^2 mt \rfloor} \right) \right.
$$

$$
\left. + x_0^{(m)} \frac{\eta}{\sigma^2} \cdot \left\lfloor \sigma^2 mt \right\rfloor + \underline{\zeta}_{\lfloor \sigma^2 mt \rfloor}^{(m)} \cdot X_0^{(m)} + \underline{\vartheta}_{\lfloor \sigma^2 mt \rfloor}^{(m)} \right\}
$$

$$
= \lim_{m \to \infty} \exp \left\{ m \cdot x_0^{(m)} \cdot \left[\frac{X_0^{(m)}}{m} - \frac{\eta}{\sigma^2} \cdot \frac{d^{(m),T}}{m \cdot \left(1 - d^{(m),T} \right)} \right] \left(1 - \left(d^{(m),T} \right)^{\lfloor \sigma^2 mt \rfloor} \right) \right.
$$

$$
\left. + m \cdot x_0^{(m)} \frac{\eta}{\sigma^2} \cdot \frac{\left\lfloor \sigma^2 mt \right\rfloor}{m} + m \cdot \underline{\zeta}_{\lfloor \sigma^2 mt \rfloor}^{(m)} \cdot \frac{X_0^{(m)}}{m} + \underline{\vartheta}_{\lfloor \sigma^2 mt \rfloor}^{(m)} \right\}
$$

$$
= \exp \left\{ -\frac{\Lambda_\lambda - \kappa_\lambda}{\sigma^2} \cdot \left[\tilde{X}_0 - \frac{\eta}{\sigma^2} \cdot \frac{\sigma^2}{\Lambda_\lambda} \right] \left(1 - e^{-\Lambda_\lambda t} \right) - \frac{\Lambda_\lambda - \kappa_\lambda}{\sigma^2} \cdot \frac{\eta}{\sigma^2} \cdot \sigma^2 t \right.
$$

$$
\left. + \frac{(\Lambda_\lambda - \kappa_\lambda)^2}{2\sigma^2 \cdot \Lambda_\lambda} \cdot e^{-\Lambda_\lambda \cdot t} \cdot \left(1 - e^{-\Lambda_\lambda \cdot t} \right) \cdot \tilde{X}_0 + \frac{\eta}{4\sigma^2} \cdot \left(\frac{\Lambda_\lambda - \kappa_\lambda}{\Lambda_\lambda} \right)^2 \cdot \left(1 - e^{-\Lambda_\lambda \cdot t} \right)^2 \right\}
$$

$$
= \exp \left\{ -\frac{\Lambda_\lambda - \kappa_\lambda}{\sigma^2} \left[\tilde{X}_0 - \frac{\eta}{\Lambda_\lambda} \right] \left(1 - e^{-\Lambda_\lambda \cdot t} \right) - \frac{\eta}{\sigma^2} (\Lambda_\lambda - \kappa_\lambda) \cdot t + L_\lambda^{(1)}(t) \cdot \tilde{X}_0 + \frac{\eta}{\sigma^2} \cdot L_\lambda^{(2)}(t) \right\}.
$$

For all $t > 0$, the upper bound (154) in the case $\lambda \in]0,1[$ follows analogously from (141), (144), (145) by

$$
\lim_{m \to \infty} \exp \left\{ x_0^{(m)} \cdot \left[X_0^{(m)} - \frac{\eta}{\sigma^2} \cdot \frac{d^{(m),S}}{1 - d^{(m),S}} \right] \left(1 - \left(d^{(m),S} \right)^{\lfloor \sigma^2 mt \rfloor} \right) \right.
$$

$$
\left. + x_0^{(m)} \frac{\eta}{\sigma^2} \cdot \left\lfloor \sigma^2 mt \right\rfloor - \overline{\zeta}_{\lfloor \sigma^2 mt \rfloor}^{(m)} \cdot X_0^{(m)} - \overline{\vartheta}_{\lfloor \sigma^2 mt \rfloor}^{(m)} \right\}
$$

$$
= \lim_{m \to \infty} \exp \left\{ m \cdot x_0^{(m)} \cdot \left[\frac{X_0^{(m)}}{m} - \frac{\eta}{\sigma^2} \cdot \frac{d^{(m),S}}{m \cdot \left(1 - d^{(m),S} \right)} \right] \left(1 - \left(d^{(m),S} \right)^{\lfloor \sigma^2 mt \rfloor} \right) \right.
$$

$$
\left. + m \cdot x_0^{(m)} \frac{\eta}{\sigma^2} \cdot \frac{\left\lfloor \sigma^2 mt \right\rfloor}{m} - m \cdot \overline{\zeta}_{\lfloor \sigma^2 mt \rfloor}^{(m)} \cdot \frac{X_0^{(m)}}{m} - \overline{\vartheta}_{\lfloor \sigma^2 mt \rfloor}^{(m)} \right\}
$$

$$
= \exp \left\{ -\frac{\Lambda_\lambda - \kappa_\lambda}{\sigma^2} \left[\tilde{X}_0 - \frac{\eta}{\sigma^2} \cdot \frac{2\sigma^2}{\Lambda_\lambda + \kappa_\lambda} \right] \left(1 - \left(e^{-\frac{1}{2}(\Lambda_\lambda + \kappa_\lambda)t} \right) \right) - \frac{\Lambda_\lambda - \kappa_\lambda}{\sigma^2} \cdot \frac{\eta}{\sigma^2} \cdot \sigma^2 t \right.
$$

$$
- \frac{(\Lambda_\lambda - \kappa_\lambda)^2}{\sigma^2} \cdot \left[\frac{e^{-\frac{1}{2}(\Lambda_\lambda + \kappa_\lambda) \cdot t} - e^{-\Lambda_\lambda \cdot t}}{\Lambda_\lambda - \kappa_\lambda} - \frac{e^{-\frac{1}{2}(\Lambda_\lambda + \kappa_\lambda) \cdot t} \left(1 - e^{-\Lambda_\lambda \cdot t} \right)}{2 \cdot \Lambda_\lambda} \right] \cdot \tilde{X}_0
$$

$$
\left. - \frac{\eta}{\sigma^2} \frac{(\Lambda_\lambda - \kappa_\lambda)^2}{\Lambda_\lambda} \cdot \left[\frac{1 - e^{-\frac{1}{2}(3\Lambda_\lambda + \kappa_\lambda) \cdot t}}{3\Lambda_\lambda + \kappa_\lambda} + \frac{e^{-\Lambda_\lambda \cdot t} - e^{-\frac{1}{2}(\Lambda_\lambda + \kappa_\lambda) \cdot t}}{\Lambda_\lambda - \kappa_\lambda} \right] \right\}
$$

$$
= \exp \left\{ -\frac{\Lambda_\lambda - \kappa_\lambda}{\sigma^2} \left[\tilde{X}_0 - \frac{\eta}{\frac{1}{2}(\Lambda_\lambda + \kappa_\lambda)} \right] \left(1 - e^{-\frac{1}{2}(\Lambda_\lambda + \kappa_\lambda) \cdot t} \right) - \frac{\eta}{\sigma^2} (\Lambda_\lambda - \kappa_\lambda) \cdot t \right.
$$

$$
\left. - U_\lambda^{(1)}(t) \cdot \tilde{X}_0 - \frac{\eta}{\sigma^2} \cdot U_\lambda^{(2)}(t) \right\}.
$$

In the case $\lambda \in]\tilde{\lambda}_-, \tilde{\lambda}_+[\setminus [0,1]$, the lower bound as well as the upper bound of the Hellinger integral limit is obtained analogously, by taking into account that the quantities $\underline{\zeta}_n^{(m)}, \underline{\vartheta}_n^{(m)}, \overline{\zeta}_n^{(m)}, \overline{\vartheta}_n^{(m)}$ now have the form (146) to (149) instead of (142) to (145). Thus, the functions $L_\lambda^{(1)}(t), U_\lambda^{(1)}(t), L_\lambda^{(2)}(t), U_\lambda^{(2)}(t)$ are obtained by employing the limits of part (l) of Lemma A6 instead of part (k). $\square$

The next Lemma (and parts of its proof) will be useful for the verification of Theorem 12:

Lemma A7. *Recall the bounds on the Hellinger integral $m-$limit given in* (153) *and* (154) *of Theorem* 11, *in terms of* $L_\lambda^{(i)}(t)$ *and* $U_\lambda^{(i)}(t)$ *($i = 1, 2$) defined by* (155) *to* (158). *Correspondingly, one gets the following $\lambda-$limits for all $t \in [0, \infty[$:*

(a) *for all $\kappa_\mathcal{A} \in]0, \infty[$ and all $\kappa_\mathcal{H} \in [0, \infty[$ with $\kappa_\mathcal{A} \neq \kappa_\mathcal{H}$*

$$\lim_{\lambda \nearrow 1} \frac{\partial L_\lambda^{(1)}(t)}{\partial \lambda} = \lim_{\lambda \nearrow 1} \frac{\partial L_\lambda^{(2)}(t)}{\partial \lambda} = \lim_{\lambda \nearrow 1} \frac{\partial U_\lambda^{(1)}(t)}{\partial \lambda} = \lim_{\lambda \nearrow 1} \frac{\partial U_\lambda^{(2)}(t)}{\partial \lambda} = 0. \tag{A74}$$

(b) *for $\kappa_\mathcal{A} = 0$ and all $\kappa_\mathcal{H} \in]0, \infty[$*

$$\lim_{\lambda \nearrow 1} \frac{\partial L_\lambda^{(1)}(t)}{\partial \lambda} = -\frac{\kappa_\mathcal{H}^2 \cdot t}{2\sigma^2}, \tag{A75}$$

$$\lim_{\lambda \nearrow 1} \frac{\partial L_\lambda^{(2)}(t)}{\partial \lambda} = -\frac{\kappa_\mathcal{H}^2 \cdot t^2}{4}, \tag{A76}$$

$$\lim_{\lambda \nearrow 1} \frac{\partial U_\lambda^{(1)}(t)}{\partial \lambda} = \lim_{\lambda \nearrow 1} \frac{\partial U_\lambda^{(2)}(t)}{\partial \lambda} = 0. \tag{A77}$$

Proof of Lemma A7. For all $\kappa_\mathcal{A}, \kappa_\mathcal{H} \in [0, \infty[$ with $\kappa_\mathcal{A} \neq \kappa_\mathcal{H}$ one can deduce from (150) as well as (155) to (158) the following derivatives:

$$\frac{\partial L_\lambda^{(1)}(t)}{\partial \lambda} = \frac{1}{2\sigma^2}\left\{ \frac{t}{2}\left(\frac{\Lambda_\lambda - \kappa_\lambda}{\Lambda_\lambda}\right)^2 \left(\kappa_\mathcal{A}^2 - \kappa_\mathcal{H}^2\right)\left[2e^{-2\Lambda_\lambda t} - e^{-\Lambda_\lambda t}\right] \right.$$
$$\left. + e^{-\Lambda_\lambda t}\frac{1 - e^{-\Lambda_\lambda t}}{\Lambda_\lambda}\left[\frac{\Lambda_\lambda - \kappa_\lambda}{\Lambda_\lambda}\left(\kappa_\mathcal{A}^2 - \kappa_\mathcal{H}^2 - 2\Lambda_\lambda(\kappa_\mathcal{A} - \kappa_\mathcal{H})\right) - \left(\frac{\Lambda_\lambda - \kappa_\lambda}{\Lambda_\lambda}\right)^2 \frac{\kappa_\mathcal{A}^2 - \kappa_\mathcal{H}^2}{2}\right] \right\}, \tag{A78}$$

$$\frac{\partial L_\lambda^{(2)}(t)}{\partial \lambda} = \frac{1}{4}\left\{ \frac{\Lambda_\lambda - \kappa_\lambda}{\Lambda_\lambda} \cdot \left(\frac{1 - e^{-\Lambda_\lambda t}}{\Lambda_\lambda}\right)^2 \cdot \left(\kappa_\mathcal{A}^2 - \kappa_\mathcal{H}^2 - 2\Lambda_\lambda(\kappa_\mathcal{A} - \kappa_\mathcal{H}) - \frac{\Lambda_\lambda - \kappa_\lambda}{\Lambda_\lambda}\left(\kappa_\mathcal{A}^2 - \kappa_\mathcal{H}^2\right)\right) \right.$$
$$\left. + t \cdot e^{-\Lambda_\lambda t} \cdot \left(\frac{\Lambda_\lambda - \kappa_\lambda}{\Lambda_\lambda}\right)^2 \cdot \frac{1 - e^{-\Lambda_\lambda t}}{\Lambda_\lambda} \cdot \left(\kappa_\mathcal{A}^2 - \kappa_\mathcal{H}^2\right) \right\}, \tag{A79}$$

$$\frac{\partial U_\lambda^{(1)}(t)}{\partial \lambda} = \frac{1}{\sigma^2}\left\{ \frac{\Lambda_\lambda - \kappa_\lambda}{2\Lambda_\lambda}\left[t e^{-\Lambda_\lambda t}\left(\kappa_\mathcal{A}^2 - \kappa_\mathcal{H}^2\right) - \frac{t}{2}e^{-\frac{1}{2}(\Lambda_\lambda + \kappa_\lambda)t}\left(\kappa_\mathcal{A}^2 - \kappa_\mathcal{H}^2 + 2\Lambda_\lambda(\kappa_\mathcal{A} - \kappa_\mathcal{H})\right)\right] \right.$$
$$- \frac{e^{-\frac{1}{2}(\Lambda_\lambda + \kappa_\lambda)t} - e^{-\Lambda_\lambda t}}{2\Lambda_\lambda} \cdot \left(\kappa_\mathcal{A}^2 - \kappa_\mathcal{H}^2 - 2\Lambda_\lambda(\kappa_\mathcal{A} - \kappa_\mathcal{H})\right)$$
$$+ \left(\frac{\Lambda_\lambda - \kappa_\lambda}{2\Lambda_\lambda}\right)^2 \left[\frac{t}{2}e^{-\frac{1}{2}(\Lambda_\lambda + \kappa_\lambda)t}\left(\kappa_\mathcal{A}^2 - \kappa_\mathcal{H}^2 + 2\Lambda_\lambda(\kappa_\mathcal{A} - \kappa_\mathcal{H})\right)\right.$$
$$- \frac{t}{2}e^{-\frac{1}{2}(3\Lambda_\lambda + \kappa_\lambda)t}\left(3\left(\kappa_\mathcal{A}^2 - \kappa_\mathcal{H}^2\right) + 2\Lambda_\lambda(\kappa_\mathcal{A} - \kappa_\mathcal{H})\right)$$
$$\left. + e^{-\frac{1}{2}(\Lambda_\lambda + \kappa_\lambda)t} \cdot \frac{1 - e^{-\Lambda_\lambda t}}{\Lambda_\lambda} \cdot \left(\kappa_\mathcal{A}^2 - \kappa_\mathcal{H}^2\right)\right]$$
$$\left. + \frac{\Lambda_\lambda - \kappa_\lambda}{\Lambda_\lambda}\left(\kappa_\mathcal{A}^2 - \kappa_\mathcal{H}^2 - 2\Lambda_\lambda(\kappa_\mathcal{A} - \kappa_\mathcal{H})\right)\left[\frac{e^{-\frac{1}{2}(\Lambda_\lambda + \kappa_\lambda)t} - e^{-\Lambda_\lambda t}}{\Lambda_\lambda - \kappa_\lambda} - \frac{e^{-\frac{1}{2}(\Lambda_\lambda + \kappa_\lambda)t}\left(1 - e^{-\Lambda_\lambda t}\right)}{2\Lambda_\lambda}\right] \right\}, \tag{A80}$$

$$\frac{\partial U_\lambda^{(2)}(t)}{\partial \lambda} = \frac{(\Lambda_\lambda - \kappa_\lambda)^2}{\Lambda_\lambda(3\Lambda_\lambda + \kappa_\lambda)}\left[\frac{t}{2}\,e^{-\frac{1}{2}(3\Lambda_\lambda+\kappa_\lambda)t}\left(3\frac{\kappa_\mathcal{A}^2 - \kappa_\mathcal{H}^2}{2\Lambda_\lambda} + \kappa_\mathcal{A} - \kappa_\mathcal{H}\right)\right.$$
$$\left. - \frac{1 - e^{-\frac{1}{2}(3\Lambda_\lambda+\kappa_\lambda)t}}{3\Lambda_\lambda + \kappa_\lambda}\cdot\left(3\frac{\kappa_\mathcal{A}^2 - \kappa_\mathcal{H}^2}{2\Lambda_\lambda} + \kappa_\mathcal{A} - \kappa_\mathcal{H}\right)\right]$$
$$+ \frac{\Lambda_\lambda - \kappa_\lambda}{\Lambda_\lambda}\left[\frac{t}{2}\,e^{-\frac{1}{2}(\Lambda_\lambda+\kappa_\lambda)t}\left(\frac{\kappa_\mathcal{A}^2 - \kappa_\mathcal{H}^2}{2\Lambda_\lambda} + \kappa_\mathcal{A} - \kappa_\mathcal{H}\right) - t\,e^{-\Lambda_\lambda t}\frac{\kappa_\mathcal{A}^2 - \kappa_\mathcal{H}^2}{2\Lambda_\lambda}\right]$$
$$+ \frac{e^{-\frac{1}{2}(\Lambda_\lambda+\kappa_\lambda)t} - e^{-\Lambda_\lambda t}}{\Lambda_\lambda}\left(\frac{\kappa_\mathcal{A}^2 - \kappa_\mathcal{H}^2}{2\Lambda_\lambda} - \kappa_\mathcal{A} + \kappa_\mathcal{H}\right)$$
$$+ \left[2\left(\frac{\kappa_\mathcal{A}^2 - \kappa_\mathcal{H}^2}{2\Lambda_\lambda} - \kappa_\mathcal{A} + \kappa_\mathcal{H}\right) - \frac{\Lambda_\lambda - \kappa_\lambda}{\Lambda_\lambda^2}\cdot\frac{\kappa_\mathcal{A}^2 - \kappa_\mathcal{H}^2}{2}\right]$$
$$\cdot\frac{1}{\Lambda_\lambda}\left[\frac{\Lambda_\lambda - \kappa_\lambda}{3\Lambda_\lambda + \kappa_\lambda}\left(1 - e^{-\frac{1}{2}(3\Lambda_\lambda+\kappa_\lambda)t}\right) - e^{-\frac{1}{2}(\Lambda_\lambda+\kappa_\lambda)t} + e^{-\Lambda_\lambda t}\right]. \tag{A81}$$

If $\kappa_\mathcal{A} \in\,]0,\infty[$ and $\kappa_\mathcal{H} \in [0,\infty[$ with $\kappa_\mathcal{A} \neq \kappa_\mathcal{H}$, then one gets $\lim_{\lambda\nearrow 1}\Lambda_\lambda = \lim_{\lambda\nearrow 1}\kappa_\lambda = \kappa_\mathcal{A} > 0$ which implies (A74) from (A78) to (A81). For the proof of part (b), let us correspondingly assume $\kappa_\mathcal{A} = 0$ and $\kappa_\mathcal{H} \in\,]0,\infty[$, which by (150) leads to $\kappa_\lambda = \kappa_\mathcal{H}\cdot(1 - \lambda)$, $\Lambda_\lambda = \kappa_\mathcal{H}\cdot\sqrt{1-\lambda}$ and the convergences $\lim_{\lambda\nearrow 1}\Lambda_\lambda = \lim_{\lambda\nearrow 1}\kappa_\lambda = 0$. From this, the assertions (A75), (A76), (A77) follow in a straightforward manner from (A78), (A79), (A80) – respectively – by using (parts of) the obvious relations

$$\lim_{\lambda\nearrow 1}\frac{\kappa_\lambda}{\Lambda_\lambda} = 0, \qquad \lim_{\lambda\nearrow 1}\frac{\Lambda_\lambda \pm \kappa_\lambda}{\Lambda_\lambda} = \lim_{\lambda\nearrow 1}\frac{\Lambda_\lambda - \kappa_\lambda}{\Lambda_\lambda + \kappa_\lambda} = 1\,, \tag{A82}$$

$$\lim_{\lambda\nearrow 1}\frac{1 - e^{-c_\lambda\cdot t}}{c_\lambda} = t \qquad \text{for all } c_\lambda \in \left\{\Lambda_\lambda,\,\frac{\Lambda_\lambda + \kappa_\lambda}{2},\,\frac{3\Lambda_\lambda + \kappa_\lambda}{2}\right\}. \tag{A83}$$

In order to get the last assertion in (A77), we make use of the following limits

$$\lim_{\lambda\nearrow 1}\frac{1}{\Lambda_\lambda - \kappa_\lambda} - \frac{3}{3\Lambda_\lambda + \kappa_\lambda} = \lim_{\lambda\nearrow 1}\frac{4\kappa_\mathcal{H}}{(\kappa_\mathcal{H} - \kappa_\mathcal{H}\cdot\sqrt{1-\lambda})\cdot(3\kappa_\mathcal{H} + \kappa_\mathcal{H}\cdot\sqrt{1-\lambda})} = \frac{4}{3\kappa_\mathcal{H}} \tag{A84}$$

and

$$\lim_{\lambda\nearrow 1}\frac{1}{\Lambda_\lambda}\left[\frac{1 - e^{-\frac{1}{2}(3\Lambda_\lambda+\kappa_\lambda)t}}{3\Lambda_\lambda + \kappa_\lambda} - \frac{1 - e^{-\Lambda_\lambda t}}{\Lambda_\lambda - \kappa_\lambda} + \frac{1 - e^{-\frac{1}{2}(\Lambda_\lambda+\kappa_\lambda)t}}{\Lambda_\lambda - \kappa_\lambda}\right] = 0. \tag{A85}$$

To see (A85), let us first observe that the involved limit can be rewritten as

$$\lim_{\lambda\nearrow 1}\left\{\frac{1}{\Lambda_\lambda(\Lambda_\lambda - \kappa_\lambda)}\left[\frac{1}{3} - \frac{1}{3}\,e^{-\frac{1}{2}(3\Lambda_\lambda+\kappa_\lambda)t} + e^{-\Lambda_\lambda t} - e^{-\frac{1}{2}(\Lambda_\lambda+\kappa_\lambda)t}\right]\right. \tag{A86}$$

$$\left. + \frac{1 - e^{-\frac{1}{2}(3\Lambda_\lambda+\kappa_\lambda)t}}{\Lambda_\lambda}\left[\frac{1}{3\Lambda_\lambda + \kappa_\lambda} - \frac{1}{3(\Lambda_\lambda - \kappa_\lambda)}\right]\right\}. \tag{A87}$$

Substituting $x := \sqrt{1-\lambda}$ and applying l'Hospital's rule twice, we get for the first limit (A86)

$$\lim_{x\searrow 0}\frac{\frac{1}{3} - \frac{1}{3}\,e^{-\frac{\kappa_\mathcal{H} t}{2}(3x+x^2)} + e^{-\kappa_\mathcal{H} tx} - e^{-\frac{\kappa_\mathcal{H} t}{2}(x+x^2)}}{\kappa_\mathcal{H}^2\cdot(x^2 - x^3)}$$

$$= \lim_{x\searrow 0}\frac{\frac{\kappa_\mathcal{H} t}{6}(3+2x)\,e^{-\frac{\kappa_\mathcal{H} t}{2}(3x+x^2)} - \kappa_\mathcal{H} t\,e^{-\kappa_\mathcal{H} tx} + \frac{\kappa_\mathcal{H} t}{2}(1+2x)\,e^{-\frac{\kappa_\mathcal{H} t}{2}(x+x^2)}}{\kappa_\mathcal{H}^2\cdot(2x - 3x^2)}$$

$$= \lim_{x\searrow 0}\frac{\left[-\frac{\kappa_\mathcal{H}^2 t^2}{12}(3+2x)^2 + \frac{\kappa_\mathcal{H} t}{3}\right]e^{-\frac{\kappa_\mathcal{H} t}{2}(3x+x^2)} + \kappa_\mathcal{H}^2 t^2\,e^{-\kappa_\mathcal{H} tx} - \left[\frac{\kappa_\mathcal{H}^2 t^2}{4}(1+2x)^2 - \kappa_\mathcal{H} t\right]e^{-\frac{\kappa_\mathcal{H} t}{2}(x+x^2)}}{\kappa_\mathcal{H}^2\cdot(2 - 6x)}$$

$$= \frac{1}{2\kappa_\mathcal{H}^2}\left[-\frac{3\kappa_\mathcal{H}^2 t^2}{4} + \frac{\kappa_\mathcal{H} t}{3} + \kappa_\mathcal{H}^2 t^2 - \frac{\kappa_\mathcal{H}^2 t^2}{4} + \kappa_\mathcal{H} t\right] = \frac{2t}{3\kappa_\mathcal{H}}.$$

The second limit (A87) becomes

$$\lim_{\lambda \nearrow 1} \frac{1 - e^{-\frac{1}{2}(3\Lambda_\lambda + \kappa_\lambda)t}}{3\Lambda_\lambda + \kappa_\lambda} \cdot \frac{3\Lambda_\lambda + \kappa_\lambda}{\Lambda_\lambda} \cdot \frac{-4\kappa_\mathcal{H}}{(3\kappa_\mathcal{H} + \sqrt{1 - \lambda\kappa_\mathcal{H}})(3\kappa_\mathcal{H} - 3\sqrt{1 - \lambda\kappa_\mathcal{H}})} \tag{A88}$$

and consequently (A85) follows. To proceed with the proof of (A77), we rearrange

$$\lim_{\lambda \nearrow 1} \frac{\partial U_\lambda^{(2)}(t)}{\partial \lambda} = \lim_{\lambda \nearrow 1} \left\{ \left(\frac{\Lambda_\lambda - \kappa_\lambda}{\Lambda_\lambda} \right)^2 \left[\frac{\Lambda_\lambda}{3\Lambda_\lambda + \kappa_\lambda} \left(\frac{t}{2} e^{-\frac{1}{2}(3\Lambda_\lambda + \kappa_\lambda)t} \left(-\frac{3\kappa_\mathcal{H}^2}{2\Lambda_\lambda} - \kappa_\mathcal{H} \right) \right) \right. \right.$$

$$- \frac{\Lambda_\lambda}{3\Lambda_\lambda + \kappa_\lambda} \cdot \frac{1 - e^{-\frac{1}{2}(3\Lambda_\lambda + \kappa_\lambda)t}}{3\Lambda_\lambda + \kappa_\lambda} \left(-\frac{3\kappa_\mathcal{H}^2}{2\Lambda_\lambda} - \kappa_\mathcal{H} \right) + \frac{\Lambda_\lambda}{\Lambda_\lambda - \kappa_\lambda} \frac{e^{-\frac{1}{2}(\Lambda_\lambda + \kappa_\lambda)t} - e^{-\Lambda_\lambda t}}{\Lambda_\lambda - \kappa_\lambda} \left(-\frac{\kappa_\mathcal{H}^2}{2\Lambda_\lambda} + \kappa_\mathcal{H} \right)$$

$$\left. - \frac{\Lambda_\lambda}{\Lambda_\lambda - \kappa_\lambda} \left(-\frac{t}{2} e^{-\frac{1}{2}(\Lambda_\lambda + \kappa_\lambda)t} \left(-\frac{\kappa_\mathcal{H}^2}{2\Lambda_\lambda} - \kappa_\mathcal{H} \right) - t\, e^{-\Lambda_\lambda t} \frac{\kappa_\mathcal{H}^2}{2\Lambda_\lambda} \right) \right]$$

$$\left. + \left[\frac{\Lambda_\lambda - \kappa_\lambda}{\Lambda_\lambda} \left(-\kappa_\mathcal{H}^2 + 2\Lambda_\lambda\kappa_\mathcal{H} \right) + \left(\frac{\Lambda_\lambda - \kappa_\lambda}{\Lambda_\lambda} \right)^2 \frac{\kappa_\mathcal{H}^2}{2} \right] \cdot \left[\frac{1 - e^{-\frac{1}{2}(3\Lambda_\lambda + \kappa_\lambda)t}}{\Lambda_\lambda (3\Lambda_\lambda + \kappa_\lambda)} - \frac{e^{-\frac{1}{2}(\Lambda_\lambda + \kappa_\lambda)t} - e^{-\Lambda_\lambda t}}{\Lambda_\lambda (\Lambda_\lambda - \kappa_\lambda)} \right] \right\}$$

$$= \lim_{\lambda \nearrow 1} \left\{ \left(\frac{\Lambda_\lambda - \kappa_\lambda}{\Lambda_\lambda} \right)^2 \left[\frac{\kappa_\mathcal{H}^2 t}{4} \left(-\frac{3 e^{-\frac{1}{2}(3\Lambda_\lambda + \kappa_\lambda)t}}{3\Lambda_\lambda + \kappa_\lambda} - \frac{e^{-\frac{1}{2}(\Lambda_\lambda + \kappa_\lambda)t}}{\Lambda_\lambda - \kappa_\lambda} + \frac{2 e^{-\Lambda_\lambda t}}{\Lambda_\lambda - \kappa_\lambda} \right) \right. \tag{A89}$$

$$+ \frac{\kappa_\mathcal{H}^2}{2} \left(\frac{3 \left(1 - e^{-\frac{1}{2}(3\Lambda_\lambda + \kappa_\lambda)t} \right)}{(3\Lambda_\lambda + \kappa_\lambda)^2} - \frac{1 - e^{-\Lambda_\lambda t}}{(\Lambda_\lambda - \kappa_\lambda)^2} + \frac{1 - e^{-\frac{1}{2}(\Lambda_\lambda + \kappa_\lambda)t}}{(\Lambda_\lambda - \kappa_\lambda)^2} \right) \tag{A90}$$

$$+ \kappa_\mathcal{H} \left(-\frac{\Lambda_\lambda}{3\Lambda_\lambda + \kappa_\lambda} \cdot \frac{t\, e^{-\frac{1}{2}(3\Lambda_\lambda + \kappa_\lambda)t}}{2} + \frac{\Lambda_\lambda}{3\Lambda_\lambda + \kappa_\lambda} \cdot \frac{1 - e^{-\frac{1}{2}(3\Lambda_\lambda + \kappa_\lambda)t}}{3\Lambda_\lambda + \kappa_\lambda} - \frac{\Lambda_\lambda}{\Lambda_\lambda - \kappa_\lambda} \cdot \frac{t\, e^{-\frac{1}{2}(\Lambda_\lambda + \kappa_\lambda)t}}{2} \right.$$

$$\left. \left. + \frac{\Lambda_\lambda}{\Lambda_\lambda - \kappa_\lambda} \cdot \frac{1 - e^{-\Lambda_\lambda t}}{\Lambda_\lambda - \kappa_\lambda} - \frac{\Lambda_\lambda}{\Lambda_\lambda - \kappa_\lambda} \cdot \frac{1 - e^{-\frac{1}{2}(\Lambda_\lambda + \kappa_\lambda)t}}{\Lambda_\lambda - \kappa_\lambda} \right) \right]$$

$$+ \left[\frac{\Lambda_\lambda - \kappa_\lambda}{\Lambda_\lambda} \left(-\kappa_\mathcal{H}^2 + 2\Lambda_\lambda\kappa_\mathcal{H} \right) + \left(\frac{\Lambda_\lambda - \kappa_\lambda}{\Lambda_\lambda} \right)^2 \frac{\kappa_\mathcal{H}^2}{2} \right] \cdot \left[\frac{1 - e^{-\frac{1}{2}(3\Lambda_\lambda + \kappa_\lambda)t}}{\Lambda_\lambda (3\Lambda_\lambda + \kappa_\lambda)} - \frac{e^{-\frac{1}{2}(\Lambda_\lambda + \kappa_\lambda)t} - e^{-\Lambda_\lambda t}}{\Lambda_\lambda (\Lambda_\lambda - \kappa_\lambda)} \right] \right\} .$$

$$\tag{A91}$$

By means of (A82) to (A84), the limit of the expression after the squared brackets in (A89) becomes

$$\lim_{\lambda \nearrow 1} \left\{ \frac{\kappa_\mathcal{H}^2 t}{4} \left[\frac{1 - e^{-\frac{1}{2}(\Lambda_\lambda + \kappa_\lambda)t}}{\Lambda_\lambda - \kappa_\lambda} - 2 \frac{1 - e^{-\Lambda_\lambda t}}{\Lambda_\lambda - \kappa_\lambda} + 3 \frac{1 - e^{-\frac{1}{2}(3\Lambda_\lambda + \kappa_\lambda)t}}{3\Lambda_\lambda + \kappa_\lambda} + \frac{1}{\Lambda_\lambda - \kappa_\lambda} - \frac{3}{3\Lambda_\lambda + \kappa_\lambda} \right] \right\} = \frac{\kappa_\mathcal{H} t}{3}, \tag{A92}$$

and the limit of the expression in (A90) becomes with (A85)

$$\lim_{\lambda \nearrow 1} \left\{ \frac{\Lambda_\lambda}{\Lambda_\lambda - \kappa_\lambda} \cdot \frac{\kappa_\mathcal{H}^2}{2\Lambda_\lambda} \cdot \left[\frac{1 - e^{-\frac{1}{2}(3\Lambda_\lambda + \kappa_\lambda)t}}{3\Lambda_\lambda + \kappa_\lambda} - \frac{1 - e^{-\Lambda_\lambda t}}{\Lambda_\lambda - \kappa_\lambda} + \frac{1 - e^{-\frac{1}{2}(\Lambda_\lambda + \kappa_\lambda)t}}{\Lambda_\lambda - \kappa_\lambda} \right] \right.$$

$$\left. - \frac{\kappa_\mathcal{H}^2}{2} \cdot \frac{1 - e^{-\frac{1}{2}(3\Lambda_\lambda + \kappa_\lambda)t}}{3\Lambda_\lambda + \kappa_\lambda} \cdot \left[\frac{1}{\Lambda_\lambda - \kappa_\lambda} - \frac{3}{3\Lambda_\lambda + \kappa_\lambda} \right] \right\} = -\frac{\kappa_\mathcal{H} t}{3} . \tag{A93}$$

By putting (A91)–(A93) together with (A85) we finally end up with

$$\lim_{\lambda \nearrow 1} \frac{\partial U_\lambda^{(2)}(t)}{\partial \lambda} = \left[\frac{\kappa_\mathcal{H} t}{3} - \frac{\kappa_\mathcal{H} t}{3} \right] + \kappa_\mathcal{H} \left(-\frac{t}{6} + \frac{t}{6} - \frac{t}{2} + t - \frac{t}{2} \right) + \left[-\kappa_\mathcal{H}^2 + \frac{\kappa_\mathcal{H}^2}{2} \right] \cdot 0 = 0,$$

which finishes the proof of Lemma A7. $\quad\square$

Proof of Theorem 12. Recall from (131) the approximative Poisson offspring-distribution parameter $\beta_\bullet^{(m)} := 1 - \frac{\kappa_\bullet}{\sigma^2 m}$ and Poisson immigration-distribution parameter $\alpha_\bullet^{(m)} := \beta_\bullet^{(m)} \cdot \frac{\eta}{\sigma^2}$, which is a special case of $(\beta_\mathcal{A}^{(m)}, \beta_\mathcal{H}^{(m)}, \alpha_\mathcal{A}^{(m)}, \alpha_\mathcal{H}^{(m)}) \in \mathcal{P}_\mathrm{NI} \cup \mathcal{P}_{\mathrm{SP},1}$. Let us first calculate $\lim_{m \to \infty} I\left(P_{\mathcal{A}, \lfloor \sigma^2 mt \rfloor}^{(m)} \middle\| P_{\mathcal{H}, \lfloor \sigma^2 mt \rfloor}^{(m)} \right)$ by starting from Theorem 3(a). Correspondingly, we evaluate for all $\kappa_\mathcal{A} \geq 0$, $\kappa_\mathcal{H} \geq 0$ with $\kappa_\mathcal{A} \neq \kappa_\mathcal{H}$ by a twofold application of l'Hospital's rule

$$
\lim_{m \to \infty} m^2 \cdot \left[\beta_\mathcal{A}^{(m)} \cdot \left(\log \left(\frac{\beta_\mathcal{A}^{(m)}}{\beta_\mathcal{H}^{(m)}} \right) - 1 \right) + \beta_\mathcal{H}^{(m)} \right] = \lim_{m \to \infty} \frac{-m}{2\sigma^2} \left[\kappa_\mathcal{A} \log \left(\frac{\beta_\mathcal{A}^{(m)}}{\beta_\mathcal{H}^{(m)}} \right) + \kappa_\mathcal{H} \left(1 - \frac{\beta_\mathcal{A}^{(m)}}{\beta_\mathcal{H}^{(m)}} \right) \right]
$$

$$
= \frac{1}{2\sigma^4} \cdot \lim_{m \to \infty} \frac{\beta_\mathcal{H}^{(m)} \cdot \kappa_\mathcal{A} - \beta_\mathcal{A}^{(m)} \cdot \kappa_\mathcal{H}}{\left(\beta_\mathcal{H}^{(m)} \right)^2} \cdot \left(\kappa_\mathcal{A} \cdot \frac{\beta_\mathcal{H}^{(m)}}{\beta_\mathcal{A}^{(m)}} - \kappa_\mathcal{H} \right) = \frac{(\kappa_\mathcal{A} - \kappa_\mathcal{H})^2}{2\sigma^4} . \tag{A94}
$$

Additionally there holds

$$
\lim_{m \to \infty} m \cdot (1 - \beta_\mathcal{A}^{(m)}) = \frac{\kappa_\mathcal{A}}{\sigma^2} \quad \text{and} \quad \lim_{m \to \infty} \left(\beta_\mathcal{A}^{(m)} \right)^{\lfloor \sigma^2 mt \rfloor} = \lim_{m \to \infty} \left[\left(1 - \frac{\kappa_\mathcal{A}}{\sigma^2 m} \right)^m \right]^{\lfloor \sigma^2 mt \rfloor / m} = e^{-\kappa_\mathcal{A} \cdot t} . \tag{A95}
$$

For $\kappa_\mathcal{A} > 0$, we apply the upper part of formula (69) as well as (A94) and (A95) to derive

$$
\lim_{m \to \infty} I_\lambda \left(P_{\mathcal{A}, \lfloor \sigma^2 mt \rfloor}^{(m)} \Big|\Big| P_{\mathcal{H}, \lfloor \sigma^2 mt \rfloor}^{(m)} \right) = \lim_{m \to \infty} \left[\frac{m^2 \cdot \left[\beta_\mathcal{A}^{(m)} \cdot \left(\log \left(\frac{\beta_\mathcal{A}^{(m)}}{\beta_\mathcal{H}^{(m)}} \right) - 1 \right) + \beta_\mathcal{H}^{(m)} \right]}{m \cdot (1 - \beta_\mathcal{A}^{(m)})} \right.
$$

$$
\cdot \left[\frac{X_0^{(m)}}{m} - \frac{\alpha_\mathcal{A}^{(m)}}{m \cdot (1 - \beta_\mathcal{A}^{(m)})} \right] \cdot \left(1 - \left(\beta_\mathcal{A}^{(m)} \right)^{\lfloor \sigma^2 mt \rfloor} \right)
$$

$$
+ \frac{\alpha_\mathcal{A}^{(m)}}{\beta_\mathcal{A}^{(m)} \cdot m \cdot (1 - \beta_\mathcal{A}^{(m)})} \cdot m^2 \cdot \left[\beta_\mathcal{A}^{(m)} \cdot \left(\log \left(\frac{\beta_\mathcal{A}^{(m)}}{\beta_\mathcal{H}^{(m)}} \right) - 1 \right) + \beta_\mathcal{H}^{(m)} \right] \cdot \frac{\lfloor \sigma^2 mt \rfloor}{m} \left. \right]
$$

$$
= \frac{(\kappa_\mathcal{A} - \kappa_\mathcal{H})^2}{2\sigma^2 \cdot \kappa_\mathcal{A}} \cdot \left[\left(\tilde{X}_0 - \frac{\eta}{\kappa_\mathcal{A}} \right) \cdot (1 - e^{-\kappa_\mathcal{A} \cdot t}) + \eta \cdot t \right] .
$$

For $\kappa_\mathcal{A} = 0$ (and thus $\kappa_\mathcal{H} > 0$, $\beta_\mathcal{A}^{(m)} \equiv 1$, $\alpha_\mathcal{A}^{(m)} \equiv \eta/\sigma^2$), we apply the lower part of formula (69) as well as (A94) and (A95) to obtain

$$
\lim_{m \to \infty} I_\lambda \left(P_{\mathcal{A}, \lfloor \sigma^2 mt \rfloor}^{(m)} \Big|\Big| P_{\mathcal{H}, \lfloor \sigma^2 mt \rfloor}^{(m)} \right) = \left\{ \lim_{m \to \infty} m^2 \cdot \left[\beta_\mathcal{H}^{(m)} - \log \beta_\mathcal{H}^{(m)} - 1 \right] \right.
$$

$$
\left. \cdot \left[\frac{\eta}{2\sigma^2} \cdot \frac{(\lfloor \sigma^2 mt \rfloor)^2}{m^2} + \left(\frac{X_0^{(m)}}{m} + \frac{\eta}{2\sigma^2 \cdot m} \right) \cdot \frac{\lfloor \sigma^2 mt \rfloor}{m} \right] \right\} = \frac{\kappa_\mathcal{H}^2}{2\sigma^2} \cdot \left[\frac{\eta}{2} \cdot t^2 + \tilde{X}_0 \cdot t \right] .
$$

Let us now calculate the "converse" double limit

$$
\lim_{\lambda \nearrow 1} \lim_{m \to \infty} I_\lambda \left(P_{\mathcal{A}, \lfloor \sigma^2 mt \rfloor}^{(m)} \Big|\Big| P_{\mathcal{H}, \lfloor \sigma^2 mt \rfloor}^{(m)} \right) = \lim_{\lambda \nearrow 1} \lim_{m \to \infty} \frac{1 - H_\lambda \left(P_{\mathcal{A}, \lfloor \sigma^2 mt \rfloor}^{(m)} \Big|\Big| P_{\mathcal{H}, \lfloor \sigma^2 mt \rfloor}^{(m)} \right)}{\lambda \cdot (1 - \lambda)} .
$$

This will be achieved by evaluating for each $t > 0$ the two limits

$$
\lim_{\lambda \nearrow 1} \frac{1 - D^L_{\lambda, \tilde{X}_0, t}}{\lambda \cdot (1 - \lambda)} \quad \text{and} \quad \lim_{\lambda \nearrow 1} \frac{1 - D^U_{\lambda, \tilde{X}_0, t}}{\lambda \cdot (1 - \lambda)} \tag{A96}
$$

which will turn out to coincide; the involved lower and upper bound $D^L_{\lambda, \tilde{X}_0, t}$, $D^U_{\lambda, \tilde{X}_0, t}$ defined by (153) and (154) satisfy $\lim_{\lambda \nearrow 1} D^L_{\lambda, \tilde{X}_0, t} = \lim_{\lambda \nearrow 1} D^U_{\lambda, \tilde{X}_0, t} = 1$ as an easy consequence of the limits (cf. 150)

$$
\lim_{\lambda \nearrow 1} \Lambda_\lambda = \kappa_\mathcal{A} \geq 0 \quad \text{and} \quad \lim_{\lambda \nearrow 1} \kappa_\lambda = \kappa_\mathcal{A} \geq 0 , \tag{A97}
$$

as well as the formulas (A82) and (A83) for the case $\kappa_{\mathcal{A}} = 0$. Accordingly, we compute

$$
\lim_{\lambda \nearrow 1} \frac{1 - D^L_{\lambda,\tilde{X}_0,t}}{\lambda \cdot (1-\lambda)} = \lim_{\lambda \nearrow 1} \frac{-D^L_{\lambda,\tilde{X}_0,t}}{1 - 2\lambda} \frac{\partial}{\partial \lambda} \left[-\frac{\Lambda_\lambda - \kappa_\lambda}{\sigma^2} \cdot \left[\tilde{X}_0 - \frac{\eta}{\Lambda_\lambda} \right] \cdot \left(1 - e^{-\Lambda_\lambda \cdot t} \right) - \frac{\eta}{\sigma^2} \cdot (\Lambda_\lambda - \kappa_\lambda) \cdot t \right.
$$

$$
\left. + L^{(1)}_\lambda(t) \cdot \tilde{X}_0 + \frac{\eta}{\sigma^2} \cdot L^{(2)}_\lambda(t) \right]
$$

$$
= \lim_{\lambda \nearrow 1} \left\{ -\frac{\Lambda_\lambda - \kappa_\lambda}{\sigma^2} \left[\left(\tilde{X}_0 - \frac{\eta}{\Lambda_\lambda} \right) \cdot t e^{-\Lambda_\lambda \cdot t} \cdot \frac{\partial \Lambda_\lambda}{\partial \lambda} + \left(1 - e^{-\Lambda_\lambda \cdot t} \right) \cdot \frac{\eta}{\Lambda_\lambda^2} \cdot \frac{\partial \Lambda_\lambda}{\partial \lambda} \right] \right.
$$

$$
- \frac{1}{\sigma^2} \cdot \frac{\partial}{\partial \lambda} (\Lambda_\lambda - \kappa_\lambda) \cdot \left(\tilde{X}_0 - \frac{\eta}{\Lambda_\lambda} \right) \cdot \left(1 - e^{-\Lambda_\lambda \cdot t} \right) - \frac{\eta t}{\sigma^2} \cdot \frac{\partial}{\partial \lambda} (\Lambda_\lambda - \kappa_\lambda)
$$

$$
\left. + \tilde{X}_0 \frac{\partial L^{(1)}_\lambda(t)}{\partial \lambda} + \frac{\eta}{\sigma^2} \frac{\partial L^{(2)}_\lambda(t)}{\partial \lambda} \right\}, \qquad \text{with} \tag{A98}
$$

$$
\frac{\partial \Lambda_\lambda}{\partial \lambda} = \frac{\kappa_{\mathcal{A}}^2 - \kappa_{\mathcal{H}}^2}{2 \Lambda_\lambda} \quad \text{and} \quad \frac{\partial \kappa_\lambda}{\partial \lambda} = \kappa_{\mathcal{A}} - \kappa_{\mathcal{H}} . \tag{A99}
$$

For the case $\kappa_{\mathcal{A}} > 0$, one can combine this with (A97) and (A74) to end up with

$$
\lim_{\lambda \nearrow 1} \frac{1 - D^L_{\lambda,\tilde{X}_0,t}}{\lambda \cdot (1-\lambda)} = \frac{(\kappa_{\mathcal{A}} - \kappa_{\mathcal{H}})^2}{2\sigma^2 \cdot \kappa_{\mathcal{A}}} \cdot \left[\left(\tilde{X}_0 - \frac{\eta}{\kappa_{\mathcal{A}}} \right) \cdot (1 - e^{-\kappa_{\mathcal{A}} \cdot t}) + \eta \cdot t \right]. \tag{A100}
$$

For the case $\kappa_{\mathcal{A}} = 0$, we continue the calculation (A98) by rearranging terms and by employing the Formulas (A75), (A76), (A82) and (A83) as well as the obvious relation $\frac{1}{\Lambda} - \frac{\Lambda - \kappa_\lambda}{\Lambda^2} = \frac{1}{\kappa_{\mathcal{H}}}$ and obtain

$$
\lim_{\lambda \nearrow 1} \frac{1 - D^L_{\lambda,\tilde{X}_0,t}}{\lambda \cdot (1-\lambda)} = \lim_{\lambda \nearrow 1} \left\{ \frac{\kappa_{\mathcal{H}}^2 \cdot \tilde{X}_0}{2\sigma^2} \left[\frac{\Lambda_\lambda - \kappa_\lambda}{\Lambda_\lambda} \cdot t \cdot e^{-\Lambda_\lambda t} + \frac{1 - e^{-\Lambda_\lambda t}}{\Lambda_\lambda} \right] \right.
$$

$$
+ \frac{\eta \cdot \kappa_{\mathcal{H}}^2 \cdot t}{2\sigma^2} \left[\frac{1}{\Lambda_\lambda} - \frac{\Lambda_\lambda - \kappa_\lambda}{\Lambda_\lambda^2} + \frac{\Lambda_\lambda - \kappa_\lambda}{\Lambda_\lambda} \cdot \frac{1 - e^{-\Lambda_\lambda t}}{\Lambda_\lambda} \right] - \frac{\eta \cdot \kappa_{\mathcal{H}}^2}{2\sigma^2} \cdot \frac{1 - e^{-\Lambda_\lambda t}}{\Lambda_\lambda} \left[\frac{1}{\Lambda_\lambda} - \frac{\Lambda_\lambda - \kappa_\lambda}{\Lambda_\lambda^2} \right]
$$

$$
\left. - \frac{\kappa_{\mathcal{H}} \cdot \tilde{X}_0}{\sigma^2} \left(1 - e^{-\Lambda_\lambda t} \right) + \frac{\eta \cdot \kappa_{\mathcal{H}}}{\sigma^2} \left[\frac{1 - e^{-\Lambda_\lambda t}}{\Lambda_\lambda} - t \right] + \frac{\partial L^{(1)}_\lambda(t)}{\partial \lambda} \cdot \tilde{X}_0 + \frac{\eta}{\sigma^2} \cdot \frac{\partial L^{(2)}_\lambda(t)}{\partial \lambda} \right\}
$$

$$
= \frac{\kappa_{\mathcal{H}}^2 \tilde{X}_0 t}{\sigma^2} + \frac{\eta \kappa_{\mathcal{H}}^2 t}{2\sigma^2} \left[\frac{1}{\kappa_{\mathcal{H}}} + t \right] - \frac{\eta \kappa_{\mathcal{H}} t}{2\sigma^2} - \frac{\kappa_{\mathcal{H}}^2 \tilde{X}_0 t}{2\sigma^2} - \frac{\eta \kappa_{\mathcal{H}}^2 t^2}{4\sigma^2} = \frac{\kappa_{\mathcal{H}}^2}{2\sigma^2} \cdot \left[\frac{\eta}{2} \cdot t^2 + \tilde{X}_0 \cdot t \right]. \tag{A101}
$$

Let us now turn to the second limit (A96) for which we compute analogously to (A98)

$$
\lim_{\lambda \nearrow 1} \frac{1 - D^U_{\lambda,\tilde{X}_0,t}}{\lambda \cdot (1-\lambda)} = \lim_{\lambda \nearrow 1} \frac{-D^U_{\lambda,\tilde{X}_0,t}}{1 - 2\lambda} \frac{\partial}{\partial \lambda} \left[-\frac{\Lambda_\lambda - \kappa_\lambda}{\sigma^2} \cdot \left[\tilde{X}_0 - \frac{\eta}{\frac{1}{2}(\Lambda_\lambda + \kappa_\lambda)} \right] \cdot \left(1 - e^{-\frac{1}{2}(\Lambda_\lambda + \kappa_\lambda) \cdot t} \right) \right.
$$

$$
\left. - \frac{\eta}{\sigma^2} \cdot (\Lambda_\lambda - \kappa_\lambda) \cdot t - U^{(1)}_\lambda(t) \cdot \tilde{X}_0 - \frac{\eta}{\sigma^2} \cdot U^{(2)}_\lambda(t) \right]
$$

$$
= \lim_{\lambda \nearrow 1} \left\{ -\frac{\Lambda_\lambda - \kappa_\lambda}{\sigma^2} \left[\left(\tilde{X}_0 - \frac{\eta}{\frac{1}{2}(\Lambda_\lambda + \kappa_\lambda)} \right) \cdot \frac{t}{2} \cdot e^{-\frac{1}{2}(\Lambda_\lambda + \kappa_\lambda) \cdot t} \frac{\partial}{\partial \lambda}(\Lambda_\lambda + \kappa_\lambda) \right. \right.
$$

$$
\left. + \left(1 - e^{-\frac{1}{2}(\Lambda_\lambda + \kappa_\lambda) \cdot t} \right) \cdot \frac{2 \cdot \eta}{(\Lambda_\lambda + \kappa_\lambda)^2} \cdot \frac{\partial}{\partial \lambda}(\Lambda_\lambda + \kappa_\lambda) \right]
$$

$$
- \frac{1}{\sigma^2} \cdot \frac{\partial}{\partial \lambda}(\Lambda_\lambda - \kappa_\lambda) \cdot \left(\tilde{X}_0 - \frac{\eta}{\frac{1}{2}(\Lambda_\lambda + \kappa_\lambda)} \right) \cdot \left(1 - e^{-\frac{1}{2}(\Lambda_\lambda + \kappa_\lambda) \cdot t} \right) - \frac{\eta t}{\sigma^2} \cdot \frac{\partial}{\partial \lambda}(\Lambda_\lambda - \kappa_\lambda)
$$

$$
\left. - \frac{\partial U^{(1)}_\lambda(t)}{\partial \lambda} \cdot \tilde{X}_0 - \frac{\eta}{\sigma^2} \frac{\partial U^{(2)}_\lambda(t)}{\partial \lambda} \right\}. \tag{A102}
$$

For the case $\kappa_A > 0$, one can combine this with (A97), (A99) and (A74) to end up with

$$\lim_{\lambda \nearrow 1} \frac{1 - D^U_{\lambda, \tilde{X}_0, t}}{\lambda \cdot (1 - \lambda)} = \frac{(\kappa_A - \kappa_{\mathcal{H}})^2}{2\sigma^2 \cdot \kappa_A} \cdot \left[\left(\tilde{X}_0 - \frac{\eta}{\kappa_A} \right) \cdot (1 - e^{-\kappa_A \cdot t}) + \eta \cdot t \right]. \tag{A103}$$

For the case $\kappa_A = 0$, we continue the calculation of (A102) by rearranging terms and by employing the formulas (A77), (A82) and (A83) as well as the obvious relation $\lim_{\lambda \nearrow 1} \frac{1}{\Lambda_\lambda} - \frac{\Lambda_\lambda - \kappa_\lambda}{\Lambda_\lambda (\Lambda_\lambda + \kappa_\lambda)} = \frac{2}{\kappa_{\mathcal{H}}}$ to obtain

$$
\begin{aligned}
\lim_{\lambda \nearrow 1} \frac{1 - D^U_{\lambda, \tilde{X}_0, t}}{\lambda \cdot (1 - \lambda)} = \lim_{\lambda \nearrow 1} &\left\{ \frac{t \cdot \tilde{X}_0}{4\sigma^2} \cdot \frac{\Lambda_\lambda - \kappa_\lambda}{\Lambda_\lambda} \cdot e^{-\frac{1}{2}(\Lambda_\lambda + \kappa_\lambda) \cdot t} \left(\kappa_{\mathcal{H}}^2 + 2\Lambda_\lambda \kappa_{\mathcal{H}} \right) \right. \\
&+ \frac{\tilde{X}_0}{2\sigma^2} \cdot \frac{1 - e^{-\frac{1}{2}(\Lambda_\lambda + \kappa_\lambda) \cdot t}}{\Lambda_\lambda} \left(\kappa_{\mathcal{H}}^2 - 2\Lambda_\lambda \kappa_{\mathcal{H}} \right) - \frac{\eta \cdot t}{\sigma^2} \left[\kappa_{\mathcal{H}} \left(1 + e^{-\frac{1}{2}(\Lambda_\lambda + \kappa_\lambda) \cdot t} \frac{\Lambda_\lambda - \kappa_\lambda}{\Lambda_\lambda + \kappa_\lambda} \right) \right. \\
&\left. - \frac{\kappa_{\mathcal{H}}^2}{2} \cdot \left(\frac{1}{\Lambda_\lambda} - \frac{\Lambda_\lambda - \kappa_\lambda}{\Lambda_\lambda (\Lambda_\lambda + \kappa_\lambda)} + \frac{\Lambda_\lambda - \kappa_\lambda}{\Lambda_\lambda + \kappa_\lambda} \cdot \frac{1 - e^{-\frac{1}{2}(\Lambda_\lambda + \kappa_\lambda) \cdot t}}{\Lambda_\lambda} \right) \right] \\
&+ \frac{2\eta}{\sigma^2} \cdot \frac{1 - e^{-\frac{1}{2}(\Lambda_\lambda + \kappa_\lambda) \cdot t}}{\Lambda_\lambda + \kappa_\lambda} \left[\kappa_{\mathcal{H}} \left(1 + \frac{\Lambda_\lambda - \kappa_\lambda}{\Lambda_\lambda + \kappa_\lambda} \right) - \frac{\kappa_{\mathcal{H}}^2}{2} \left(\frac{1}{\Lambda_\lambda} - \frac{\Lambda_\lambda - \kappa_\lambda}{\Lambda_\lambda (\Lambda_\lambda + \kappa_\lambda)} \right) \right] \\
&\left. - \frac{\partial U_\lambda^{(1)}(t)}{\partial \lambda} \cdot \tilde{X}_0 - \frac{\eta}{\sigma^2} \frac{\partial U_\lambda^{(2)}(t)}{\partial \lambda} \right\} \\
= &\frac{\kappa_{\mathcal{H}}^2 t \tilde{X}_0}{4\sigma^2} + \frac{\kappa_{\mathcal{H}}^2 t \tilde{X}_0}{4\sigma^2} - \frac{\eta t}{\sigma^2} \left[2\kappa_{\mathcal{H}} - \kappa_{\mathcal{H}} - \frac{\kappa_{\mathcal{H}}^2 t}{4} \right] + \frac{\eta t}{\sigma^2} [2\kappa_{\mathcal{H}} - \kappa_{\mathcal{H}}] = \frac{\kappa_{\mathcal{H}}^2}{2\sigma^2} \left[\frac{\eta}{2} \cdot t^2 + \tilde{X}_0 \cdot t \right]. \tag{A104}
\end{aligned}
$$

Since (A100) coincides with (A103) and (A101) coincides with (A104), we have finished the proof. $\square$

References

1. Liese, F.; Vajda, I. *Convex Statistical Distances*; Teubner: Leipzig, Germany, 1987.
2. Read, T.R.C.; Cressie N.A.C. *Goodness-of-Fit Statistics for Discrete Multivariate Data*; Springer: New York, NY, USA, 1988.
3. Vajda, I. *Theory of Statistical Inference and Information*; Kluwer: Dordrecht, The Netherlands, 1989.
4. Csiszár, I.; Shields, P.C. *Information Theory and Statistics: A Tutorial*; Now Publishers: Hanover, MA, USA, 2004.
5. Stummer, W. *Exponentials, Diffusions, Finance, Entropy and Information*; Shaker: Aachen, Germany, 2004.
6. Pardo, L. *Statistical Inference Based on Divergence Measures*; Chapman & Hall/CRC: Bocan Raton, FL, USA, 2006.
7. Liese, F.; Miescke, K.J. *Statistical Decision Theory: Estimation, Testing, and Selection*; Springer: New York, NY, USA, 2008.
8. Basu, A.; Shioya, H.; Park, C. *Statistical Inference: The Minimum Distance Approach*; CRC Press: Boca Raton, FL, USA, 2011.
9. Voinov V.; Nikulin, M.; Balakrishnan N. *Chi-Squared Goodness of Fit Tests with Applications*; Academic Press: Waltham, MA, USA, 2013.
10. Liese, F.; Vajda, I. On divergences and informations in statistics and information theory. *IEEE Trans. Inform. Theory* **2006**, *52*, 4394–4412. [CrossRef]
11. Vajda I.; van der Meulen, E.C. Goodness-of-fit criteria based on observations quantized by hypothetical and empirical percentiles. In *Handbook of Fitting Statistical Distributions with R*; Karian, Z.A., Dudewicz, E.J., Eds.; CRC: Heidelberg, Germany, 2010; pp. 917–994.
12. Stummer, W.; Vajda, I. On Bregman distances and divergences of probability measures. *IEEE Trans. Inform. Theory* **2012**, *58*, 1277–1288. [CrossRef]
13. Kißlinger, A.-L.; Stummer, W. Robust statistical engineering by means of scaled Bregman distances. In *Recent Advances in Robust Statistics–Theory and Applications*; Agostinelli, C., Basu, A., Filzmoser, P., Mukherjee, D., Eds.; Springer: New Delhi, India, 2016; pp. 81–113.

14. Broniatowski, M.; Stummer, W. Some universal insights on divergences for statistics, machine learning and artificial intelligence. In *Geometric Structures of Information*; Nielsen, F., Ed.; Springer: Cham, Switzerland, 2019; pp. 149–211.

15. Stummer, W.; Vajda, I. Optimal statistical decisions about some alternative financial models. *J. Econom.* **2007**, *137*, 441–471. [CrossRef]

16. Stummer, W; Lao, W. Limits of Bayesian decision related quantities of binomial asset price models. *Kybernetika* **2012**, *48*, 750–767.

17. Csiszar, I. Eine informationstheoretische Ungleichung und ihre Anwendung auf den Beweis der Ergodizität von Markoffschen Ketten. *Publ. Math. Inst. Hungar. Acad. Sci.* **1963**, *A-8*, 85–108.

18. Ali, M.S.; Silvey, D. A general class of coefficients of divergence of one distribution from another. *J. Roy. Statist. Soc. B* **1966**, *28*, 131–140. [CrossRef]

19. Morimoto, T. Markov processes and the H-theorem. *J. Phys. Soc. Jpn.* **1963**, *18*, 328–331. [CrossRef]

20. van Erven, T.; Harremoes, P. Renyi divergence and Kullback-Leibler divergence. *IEEE Trans. Inf. Theory* **2014**, *60*, 3797–3820. [CrossRef]

21. Newman, C.M. On the orthogonality of independent increment processes. In *Topics in Probability Theory*; Courant Institute of Mathematical Sciences New York University: New York, NY, USA, 1973; pp. 93–111.

22. Liese, F. Hellinger integrals of Gaussian processes with independent increments. *Stochastics* **1982**, *6*, 81–96. [CrossRef]

23. Memin, J.; Shiryayev, A.N. Distance de Hellinger-Kakutani des lois correspondant a deux processus a accroissements indépendants. *Probab. Theory Relat. Fields* **1985**, *70*, 67–89. [CrossRef]

24. Jacod, J.; Shiryaev, A.N. *Limit Theorems for Stochastic Processes*; Springer: Berlin, Germany, 1987.

25. Linkov, Y.N.; Shevlyakov, Y.A. Large deviation theorems in the hypotheses testing problems for processes with independent increments. *Theory Stoch. Process* **1998**, *4*, 198–210.

26. Liese, F. Hellinger integrals, error probabilities and contiguity of Gaussian processes with independent increments and Poisson processes. *J. Inf. Process. Cybern.* **1985**, *21*, 297–313.

27. Kabanov, Y.M.; Liptser, R.S.; Shiryaev, A.N. On the variation distance for probability measures defined on a filtered space. *Probab. Theory Relat. Fields* **1986**, *71*, 19–35. [CrossRef]

28. Liese, F. Hellinger integrals of diffusion processes. *Statistics* **1986**, *17*, 63–78. [CrossRef]

29. Vajda, I. Distances and discrimination rates for stochastic processes. *Stoch. Process. Appl.* **1990**, *35*, 47–57. [CrossRef]

30. Stummer, W. The Novikov and entropy conditions of multidimensional diffusion processes with singular drift. *Probab. Theory Relat. Fields* **1993**, *97*, 515–542. [CrossRef]

31. Stummer, W. On a statistical information measure of diffusion processes. *Stat. Decis.* **1999**, *17*, 359–376. [CrossRef]

32. Stummer, W. On a statistical information measure for a generalized Samuelson-Black-Scholes model. *Stat. Decis.* **2001**, *19*, 289–314. [CrossRef]

33. Bartoszynski, R. Branching processes and the theory of epidemics. In *Proceedings of the Fifth Berkeley Symposium on Mathematical Statistics and Probability, Vol. IV*; Le Cam, L.M., Neyman, J., Eds; University of California Press: Berkeley, CA, USA, 1967; pp. 259–269.

34. Ludwig, D. Qualitative behaviour of stochastic epidemics. *Math. Biosci.* **1975**, *23*, 47–73. [CrossRef]

35. Becker, N.G. Estimation for an epidemic model. *Biometrics* **1976**, *32*, 769–777. [CrossRef]

36. Becker, N.G. Estimation for discrete time branching processes with applications to epidemics. *Biometrics* **1977**, *33*, 515–522. [CrossRef]

37. Metz, J.A.J. The epidemic in a closed population with all susceptibles equally vulnerable; some results for large susceptible populations and small initial infections. *Acta Biotheor.* **1978**, *27*, 75–123. [CrossRef] [PubMed]

38. Heyde, C.C. On assessing the potential severity of an outbreak of a rare infectious disease. *Austral. J. Stat.* **1979**, *21*, 282–292. [CrossRef]

39. Von Bahr, B.; Martin-Löf, A. Threshold limit theorems for some epidemic processes. *Adv. Appl. Prob.* **1980**, *12*, 319–349. [CrossRef]

40. Ball, F. The threshold behaviour of epidemic models. *J. Appl. Prob.* **1983**, *20*, 227–241. [CrossRef]

41. Jacob, C. Branching processes: Their role in epidemics. *Int. J. Environ. Res. Public Health* **2010**, *7*, 1186–1204. [CrossRef]

42. Barbour, A.D.; Reinert, G. Approximating the epidemic curve. *Electron. J. Probab.* **2013**, *18*, 1–30. [CrossRef]

43. Britton, T.; Pardoux, E. Stochastic epidemics in a homogeneous community. In *Stochastic Epidemic Models*; Britton, T., Pardoux, E., Eds.; Springer: Cham, Switzerland, 2019; pp. 1–120.

44. Dion, J.P.; Gauthier, G.; Latour, A. Branching processes with immigration and integer-valued time series. *Serdica Math. J.* **1995**, *21*, 123–136.

45. Grunwald, G.K.; Hyndman, R.J.; Tedesco, L.; Tweedie, R.L. Non-Gaussian conditional linear AR(1) models. *Aust. N. Z. J. Stat.* **2000**, *42*, 479–495. [CrossRef]

46. Kedem, B.; Fokianos, K. *An Regression Models for Time Series Analysis*; Wiley: Hoboken, NJ, USA, 2002.

47. Held, L.; Höhle, M.; Hofmann, M. A statistical framework for the analysis of multivariate infectious disease surveillance counts. *Stat. Model.* **2005**, *5*, 187–199. [CrossRef]

48. Weiss, C.H. *An Introduction to Discrete-Valued Time Series*; Wiley: Hoboken, NJ, USA, 2018.

49. Feigin, P.D.; Passy, U. The geometric programming dual to the extinction probability problem in simple branching processes. *Ann. Probab.* **1981**, *9*, 498–503. [CrossRef]

50. Mordecki, E. Asymptotic mixed normality and Hellinger processes. *Stoch. Stoch. Rep.* **1994**, *48*, 129–143. [CrossRef]

51. Sriram, T.N.; Vidyashankar, A.N. Minimum Hellinger distance estimation for supercritical Galton-Watson processes. *Stat. Probab. Lett.* **2000**, *50*, 331–342. [CrossRef]

52. Guttorp, P. *Statistical Inference for Branching Processes*; Wiley: New York, NY, USA, 1991.

53. Linkov, Y.N.; Lunyova, L.A. Large deviation theorems in the hypothesis testing problems for the Galton-Watson processes with immigration. *Theory Stoch. Process* **1996**, *2*, 120–132; Erratum in *Theory Stoch. Process* **1997**, *3*, 270–285.

54. Heathcote, C.R. A branching process allowing immigration. *J. R. Stat. Soc. B* **1965**, *27*, 138–143; Erratum in: Heathcote, C.R. Corrections and comments on the paper "A branching process allowing immigration". *J. R. Stat. Soc. B* **1966**, *28*, 213–217. [CrossRef]

55. Athreya, K.B.; Ney, P.E. *Branching Processes*; Springer: New York, NY, USA, 1972.

56. Jagers, P. *Branching Processes with Biological Applications*; Wiley: London, UK, 1975.

57. Asmussen, S.; Hering, H. *Branching Processes*; Birkhäuser: Boston, MA, USA, 1983.

58. Haccou, P.; Jagers, P.; Vatutin, V.A. *Branching Processes: Variation, Growth, and Extinction of Populations*; Cambrigde University Press: Cambridge, UK, 2005.

59. Heyde, C.C.; Seneta, E. Estimation theory for growth and immigration rates in a multiplicative process. *J. Appl. Probab.* **1972**, *9*, 235–256. [CrossRef]

60. Basawa, I.V.; Rao, B.L.S. *Statistical Inference of Stochastic Processes*; Academic Press: London, UK, 1980.

61. Basawa, I.V.; Scott, D.J. *Asymptotic Optimal Inference for Non-Ergodic Models*; Springer: New York, NY, USA, 1983.

62. Sankaranarayanan, G. *Branching Processes and Its Estimation Theory*; Wiley: New Delhi, India, 1989.

63. Wei, C.Z.; Winnicki, J. Estimation of the means in the branching process with immigration. *Ann. Stat.* **1990**, *18*, 1757–1773. [CrossRef]

64. Winnicki, J. Estimation of the variances in the branching process with immigration. *Probab. Theory Relat. Fields* **1991**, *88*, 77–106. [CrossRef]

65. Yanev, N.M. Statistical inference for branching processes. In *Records and Branching Processes*; Ahsanullah, M., Yanev, G.P., Eds.; Nova Science Publishes: New York, NY, USA, 2008; pp. 147–172.

66. Harris, T.E. *The Theory of Branching Processes*; Springer: Berlin, Germany, 1963.

67. Gauthier, G.; Latour, A. Convergence forte des estimateurs des parametres d'un processus GENAR(p). *Ann. Sci. Math. Que.* **1994**, *18*, 49–71.

68. Latour, A. Existence and stochastic structure of a non-negative integer-valued autoregressive process. *J. Time Ser. Anal.* **1998**, *19*, 439–455. [CrossRef]

69. Rydberg, T.H; Shephard, N. BIN models for trade-by-trade data. Modelling the number of trades in a fixed interval of time. In *Econometric Society World Congress*; Contributed Papers No. 0740; Econometric Society: Cambridge, UK, 2000.

70. Brandt, P.T.; Williams, J.T. A linear Poisson autoregressive model: The Poisson AR(p) model. *Polit. Anal.* **2001**, *9*, 164–184. [CrossRef]

71. Heinen, A. Modelling time series count data: An autoregressive conditional Poisson model. In *Core Discussion Paper*; MPRA Paper No. 8113; University of Louvain: Louvain, Belgium, 2003; Volume 62. Available online: https://mpra.ub.uni-muenchen.de/8113 (accessed on 18 May 2020).

72. Held, L.; Hofmann, M.; Höhle, M.; Schmid, V. A two-component model for counts of infectious diseases. *Biostatistics* **2006**, *7*, 422–437. [CrossRef] [PubMed]

73. Finkenstädt, B.F.; Bjornstad, O.N; Grenfell, B.T. A stochastic model for extinction and recurrence of epidemics: Estimation and inference for measles outbreak. *Biostatistics* **2002**, *3*, 493–510. [CrossRef] [PubMed]

74. Ferland, R.; Latour, A.; Oraichi, D. Integer-valued GARCH process. *J. Time Ser. Anal.* **2006**, *27*, 923–942. [CrossRef]

75. Weiß, C.H. Modelling time series of counts with overdispersion. *Stat. Methods Appl.* **2009**, *18*, 507–519. [CrossRef]

76. Weiß, C.H. The INARCH(1) model for overdispersed time series of counts. *Comm. Stat. Sim. Comp.* **2010**, *39*, 1269–1291. [CrossRef]

77. Weiß, C.H. INARCH(1) processes: Higher-order moments and jumps. *Stat. Probab. Lett.* **2010**, *80*, 1771–1780. [CrossRef]

78. Weiß, C.H.; Testik, M.C. Detection of abrupt changes in count data time series: Cumulative sum derivations for INARCH(1) models. *J. Qual. Technol.* **2012**, *44*, 249–264. [CrossRef]

79. Kaslow, R.A.; Evans, A.S. Epidemiologic concepts and methods. In *Viral Infections of Humans*; Evans, A.S., Kaslow, R.A., Eds.; Springer: New York, NY, USA, 1997; pp. 3–58.

80. Osterholm, M.T.; Hedberg, C.W. Epidemiologic principles. In *Mandell, Douglas, and Bennett's Principles and Practice of Infectious Diseases*, 8th ed.; Bennett, J.E., Dolin, R., Blaser, M.J., Eds.; Elsevier: Philadelphia, PA, USA, 2015; pp. 146–157.

81. Grassly, N.C.; Fraser, C. Mathematical models of infectious disease transmission. *Nat. Rev.* **2008**, *6*, 477–487. [CrossRef]

82. Keeling, M.J.; Rohani, P. *Modeling Infectious Diseases in Humans and Animals*; Princeton UP: Princeton, NJ, USA, 2008.

83. Yan, P. Distribution theory stochastic processes and infectious disease modelling. In *Mathematical Epidemiology*; Brauer, F., van den Driessche, P., Wu, J., Eds.; Springer: Berlin, Germany, 2008; pp. 229–293.

84. Yan, P.; Chowell, G. *Quantitative Methods for Investigating Infectious Disease Outbreaks*; Springer: Cham, Switzerland, 2019.

85. Britton, T. Stochastic epidemic models: A survey. *Math. Biosc.* **2010**, *225*, 24–35. [CrossRef]

86. Diekmann, O.; Heesterbeek, H.; Britton, T. *Mathematical Tools for Understanding Infectious Disease Dynamics*; Princeton University Press: Princeton, NJ, USA, 2013.

87. Cummings, D.A.T.; Lessler, J. Infectious disease dynamics. In *Infectious Disease Epidemiology: Theory and Practice*; Nelson, K.E., Masters Williams, C., Eds.; Jones & Bartlett Learning: Burlington, MA, USA, 2014; pp. 131–166.

88. Just, W.; Callender, H.; Drew LaMar, M.; Toporikova, N. Transmission of infectious diseases: Data, models and simulations. In *Algebraic and Discrete Mathematical Methods of Modern Biology*; Robeva, R.S., Ed.; Elsevier: London, UK, 2015; pp. 193–215.

89. Britton, T.; Giardina, F. Introduction to statistical inference for infectious diseases. *J. Soc. Franc. Stat.* **2016**, *157*, 53–70.

90. Fine, P.E.M. The interval between successive cases of an infectious disease. *Am. J. Epidemiol.* **2003**, *158*, 1039–1047. [CrossRef] [PubMed]

91. Svensson, A. A note on generation times in epidemic models. *Math. Biosci.* **2007**, *208*, 300–311. [CrossRef] [PubMed]

92. Svensson, A. The influence of assumptions on generation time distributions in epidemic models. *Math. Biosci.* **2015**, *270*, 81–89. [CrossRef] [PubMed]

93. Wallinga, J.; Lipsitch, M. How generation intervals shape the relationship between growth rates and reproductive numbers. *Proc. R. Soc. B* **2007**, *274*, 599–604. [CrossRef]

94. Forsberg White, L.; Pagano, M. A likelihood-based method for real-time estimation of the serial interval and reproductive number of an epidemic. *Stat. Med.* **2008**, *27*, 2999–3016. [CrossRef]

95. Nishiura, H. Time variations in the generation time of an infectious disease: implications for sampling to appropriately quantify transmission potential. *Math. Biosci.* **2010**, *7*, 851–869.

96. Scalia Tomba, G.; Svensson, A.; Asikainen, T.; Giesecke, J. Some model based considerations on observing generation times for communicable diseases. *Math. Biosci.* **2010**, *223*, 24–31. [CrossRef]

97. Trichereau, J.; Verret, C.; Mayet, A.; Manet, G. Estimation of the reproductive number for A(H1N1) pdm09 influenza among the French armed forces, September 2009–March 2010. *J. Infect.* **2012**, *64*, 628–630. [CrossRef]

98. Vink, M.A.; Bootsma, M.C.J.; Wallinga, J. Serial intervals of respiratory infectious diseases: a systematic review and analysis. *Am. J. Epidemiol.* **2014**, *180*, 865–875. [CrossRef]

99. Champredon, D.; Dushoff, J. Intrinsic and realized generation intervals in infectious-disease transmission. *Proc. R. Soc. B* **2015**, *282*, 20152026. [CrossRef]

100. An der Heiden, M.; Hamouda, O. Schätzung der aktuellen Entwicklung der SARS-CoV-2-Epidemie in Deutschland— Nowcasting. *Epid. Bull.* **2020**, *17*, 10–16. (In German)

101. Ferretti, L.; Wymant, C.; Kendall, M.; Zhao, L.; Nurtay, A.; Abeler-Dörner, L.; Parker, M.; Bonsall, D.; Fraser, C. Quantifying SARS-CoV-2 transmission suggests epidemic control with digital contact tracing. *Science* **2020**, *368*, eabb6936. [CrossRef] [PubMed]

102. Ganyani, T.; Kremer, C.; Chen, D.; Torneri, A.; Faes, C.; Wallinga, J.; Hens, N. Estimating the generation interval for COVID-19 based on symptom onset data. *medRxiv Prepr.* **2020**. [CrossRef]

103. Li, M.; Liu, K.; Song, Y.; Wang, M.; Wu, J. Serial interval and generation interval for respectively the imported and local infectors estimated using reported contact-tracing data of COVID-19 in China. *medRxiv Prepr.* **2020**. [CrossRef]

104. Nishiura, H.; Linton, N.M.; Akhmetzhanov, A.R. Serial interval of novel coronavirus (COVID-19) infections. *medRxiv Prepr.* **2020**. [CrossRef]

105. Park, M.; Cook, A.R.; Lim, J.J.; Sun, X.; Dickens, B.L. A systematic review of COVID-19 epidemiology based on current evidence. *J. Clin. Med.* **2020**, *9*, 967. [CrossRef]

106. Spouge, J.L. An accurate approximation for the expected site frequency spectrum in a Galton-Watson process under an infinite sites mutation model. *Theor. Popul. Biol.* **2019**, *127*, 7–15. [CrossRef]

107. Taneyhill, D.E.; Dunn, A.M.; Hatcher, M.J. The Galton-Watson branching process as a quantitative tool in parasitology. *Parasitol. Today* **1999**, *15*, 159–165. [CrossRef]

108. Parnes, D. Analyzing the contagion effect of foreclosures as a branching process: A close look at the years that follow the Great Recession. *J. Account. Financ.* **2017**, *17*, 9–34.

109. Le Cam, L. *Asymptotic Methods in Statistical Decision Theory*; Springer: New York, NY, USA, 1986.

110. Heyde, C.C.; Johnstone, I.M. On asymptotic posterior normality for stochastic processes. *J. R. Stat. Soc. B* **1979**, *41*, 184–189. [CrossRef]

111. Johnson, R.A.; Susarla, V.; van Ryzin, J. Bayesian non-parametric estimation for age-dependent branching processes. *Stoch. Proc. Appl.* **1979**, *9*, 307–318. [CrossRef]

112. Scott, D. On posterior asymptotic normality and asymptotic normality of estimators for the Galton-Watson process. *J. R. Stat. Soc. B* **1987**, *49*, 209–214. [CrossRef]

113. Yanev, N.M.; Tsokos, C.P. Decision-theoretic estimation of the offspring mean in mortal branching processes. *Comm. Stat. Stoch. Models* **1999**, *15*, 889–902. [CrossRef]

114. Mendoza, M.; Gutierrez-Pena, E. Bayesian conjugate analysis of the Galton-Watson process. *Test* **2000**, *9*, 149–171. [CrossRef]

115. Feicht, R.; Stummer, W. An explicit nonstationary stochastic growth model. In *Economic Growth and Development (Frontiers of Economics and Globalization, Vol. 11)*; De La Grandville, O., Ed.; Emerald Group Publishing Limited: Bingley, UK, 2011; pp. 141–202.

116. Dorn, F.; Fuest, C.; Göttert, M.; Krolage, C.; Lautenbacher, S.; Link, S.; Peichl, A.; Reif, M.; Sauer, S.; Stöckli, M.; et al. Die volkswirtschaftlichen Kosten des Corona-Shutdown für Deutschland: Eine Szenarienrechnung. *ifo Schnelldienst* **2020**, *73*, 29–35. (In Germany)

117. Dorn, F.; Khailaie, S.; Stöckli, M.; Binder, S.; Lange, B.; Peichl, A.; Vanella, P.; Wollmershäuser, T.; Fuest, C.; Meyer-Hermann, M. Das gemeinsame Interesse von Gesundheit und Wirtschaft: Eine Szenarienrechnung zur Eindämmung der Corona-Pandemie. *ifo Schnelld. Dig.* **2020**, *6*, 1–9.

118. Kißlinger, A.-L.; Stummer, W. A new toolkit for robust distributional change detection. *Appl. Stoch. Models Bus. Ind.* **2018**, *34*, 682–699. [CrossRef]

119. Dehning, J.; Zierenberg, J.; Spitzner, F.P.; Wibral, M.; Neto, J.P.; Wilczek, M.; Priesemann, V. Inferring change points in the spread of COVID-19 reveals the effectiveness of interventions. *Science* **2020**, *369*, eabb9789. [CrossRef]

120. Friesen, M. Statistical surveillance. Optimality and methods. *Int. Stat. Review* **2003**, *71*, 403–434. [CrossRef]
121. Friesen, M.; Andersson, E.; Schiöler, L. Robust outbreak surveillance of epidemics in Sweden. *Stat. Med.* **2009**, *28*, 476–493. [CrossRef]
122. Brauner, J.M.; Mindermann, S.; Sharma, M.; Stephenson, A.B.; Gavenciak, T.; Johnston, D.; Salvatier, J.; Leech, G.; Besiroglu, T.; Altman, G.; et al. The effectiveness and perceived burden of nonpharmaceutical interventions against COVID-19 transmission: A modelling study with 41 countries. *medRxiv Prepr.* **2020**. [CrossRef]
123. Österreicher, F.; Vajda, I. Statistical information and discrimination. *IEEE Trans. Inform. Theory* **1993**, *39*, 1036–1039. [CrossRef]
124. De Groot, M.H. Uncertainty, information and sequential experiments. *Ann. Math. Statist.* **1962**, *33*, 404–419. [CrossRef]
125. Krafft, O.; Plachky, D. Bounds for the power of likelihood ratio tests and their asymptotic properties. *Ann. Math. Stat.* **1970**, *41*, 1646–1654. [CrossRef]
126. Basawa, I.V.; Scott, D.J. Efficient tests for branching processes. *Biometrika* **1976**, *63*, 531–536. [CrossRef]
127. Feigin, P.D. The efficiency criteria problem for stochastic processes. *Stoch. Proc. Appl.* **1978**, *6*, 115–127. [CrossRef]
128. Sweeting, T.J. On efficient tests for branching processes. *Biometrika* **1978**, *65*, 123–127. [CrossRef]
129. Linkov, Y.N. *Lectures in Mathematical Statistics, Parts 1 and 2*; American Mathematical Society: Providence, RI, USA, 2005.
130. Feller, W. Diffusion processes in genetics. In *Proceedings of the Second Berkeley Symposium on Mathematical Statistics and Probability*; Neyman, J., Ed; University of California Press: Berkeley, CA, USA, 1951; pp. 227–246.
131. Jirina, M. On Feller's branching diffusion process. *Časopis Pěst. Mat.* **1969**, *94*, 84–89.
132. Lamperti, J. Limiting distributions for branching processes. In *Proceedings of the Fifth Berkeley Symposium on Mathematical Statistics and Probability, Vol. II, Part 2*; Le Cam, L.M., Neyman, J., Eds; University of California Press: Berkeley, CA, USA, 1967; pp. 225–241.
133. Lamperti, J. The limit of a sequence of branching processes. *Z. Wahrscheinlichkeitstheorie Verw. Geb.* **1967**, *7*, 271–288. [CrossRef]
134. Lindvall, T. Convergence of critical Galton-Watson branching processes. *J. Appl. Prob.* **1972**, *9*, 445–450. [CrossRef]
135. Lindvall, T. Limit theorems for some functionals of certain Galton-Watson branching processes. *Adv. Appl. Prob.* **1974**, *6*, 309–321. [CrossRef]
136. Grimvall, A. On the convergence of sequences of branching processes. *Ann. Probab.* **1974**, *2*, 1027–1045. [CrossRef]
137. Borovkov, K.A. On the convergence of branching processes to a diffusion process. *Theor. Probab. Appl.* **1986**, *30*, 496–506. [CrossRef]
138. Ethier, S.N.; Kurtz, T.G. *Markov Processes: Characterization and Convergence*; Wiley: New York, NY, USA, 1986.
139. Durrett, R. *Stochastic Calculus*; CRC Press: Boca Raton, FL, USA, 1996.
140. Kawazu, K.; Watanabe, S. Branching processes with immigration and related limit theorems. *Theor. Probab. Appl.* **1971**, *16*, 36–54. [CrossRef]
141. Wei, C.Z.; Winnicki, J. Some asymptotic results for the branching process with immigration. *Stoch. Process. Appl.* **1989**, *31*, 261–282. [CrossRef]
142. Sriram, T.N. Invalidity of bootstrap for critical branching processes with immigration. *Ann. Stat.* **1994**, *22*, 1013–1023. [CrossRef]
143. Li, Z. Branching processes with immigration and related topics. *Front. Math. China* **2006**, *1*, 73–97. [CrossRef]
144. Dawson, D.A.; Li, Z. Skew convolution semigroups and affine Markov processes. *Ann. Probab.* **2006**, *34*, 1103–1142. [CrossRef]
145. Cox, J.C.; Ingersoll, J.E., Jr.; Ross, S.A. A theory of the term structure of interest rates. *Econometrica* **1985**, *53*, 385–407. [CrossRef]
146. Cox, J.C.; Ross, S.A. The valuation of options for alternative processes. *J. Finan. Econ.* **1976**, *3*, 145–166. [CrossRef]
147. Heston, S.L. A closed-form solution for options with stochastic volatilities with applications to bond and currency options. *Rev. Finan. Stud.* **1993**, *6*, 327–343. [CrossRef]

148. Lansky, P.; Lanska, V. Diffusion approximation of the neuronal model with synaptic reversal potentials. *Biol. Cybern.* **1987**, *56*, 19–26. [CrossRef] [PubMed]

149. Giorno, V.; Lansky, P.; Nobile, A.G.; Ricciardi, L.M. Diffusion approximation and first-passage-time problem for a model neuron. *Biol. Cybern.* **1988**, *58*, 387–404. [CrossRef] [PubMed]

150. Lanska, V.; Lansky, P.; Smith, C.E. Synaptic transmission in a diffusion model for neuron activity. *J. Theor. Biol.* **1994**, *166*, 393–406. [CrossRef] [PubMed]

151. Lansky, P.; Sacerdote, L.; Tomassetti, F. On the comparison of Feller and Ornstein-Uhlenbeck models for neural activity. *Biol. Cybern.* **1995**, *73*, 457–465. [CrossRef]

152. Ditlevsen, S.; Lansky, P. Estimation of the input parameters in the Feller neuronal model. *Phys. Rev. E* **2006**, *73*, 061910. [CrossRef]

153. Höpfner, R. On a set of data for the membrane potential in a neuron. *Math. Biosci.* **2007**, *207*, 275–301. [CrossRef]

154. Lansky, P.; Ditlevsen, S. A review of the methods for signal estimation in stochastic diffusion leaky integrate-and-fire neuronal models. *Biol. Cybern.* **2008**, *99*, 253–262. [CrossRef]

155. Pedersen, A.R. Estimating the nitrous oxide emission rate from the soil surface by means of a diffusion model. *Scand. J. Stat. Theory Appl.* **2000**, *27*, 385–403. [CrossRef]

156. Aalen, O.O.; Gjessing, H.K. Survival models based on the Ornstein-Uhlenbeck process. *Lifetime Data Anal.* **2004**, *10*, 407–423. [CrossRef]

157. Kammerer, N.B. Generalized-Relative-Entropy Type Distances Between Some Branching Processes and Their Diffusion Limits. Ph.D. Thesis, University of Erlangen-Nürnberg, Erlangen, Germany, 2011.

Article

Monitoring Parameter Change for Time Series Models of Counts Based on Minimum Density Power Divergence Estimator

Sangyeol Lee * and Dongwon Kim

Department of Statistics, Seoul National University, Seoul 08826, Korea; dongwon.k@snu.ac.kr
* Correspondence: sylee@stats.snu.ac.kr; Tel.: +82-2-880-8814

Received: 28 October 2020; Accepted: 14 November 2020; Published: 16 November 2020

Abstract: In this study, we consider an online monitoring procedure to detect a parameter change for integer-valued generalized autoregressive heteroscedastic (INGARCH) models whose conditional density of present observations over past information follows one parameter exponential family distributions. For this purpose, we use the cumulative sum (CUSUM) of score functions deduced from the objective functions, constructed for the minimum power divergence estimator (MDPDE) that includes the maximum likelihood estimator (MLE), to diminish the influence of outliers. It is well-known that compared to the MLE, the MDPDE is robust against outliers with little loss of efficiency. This robustness property is properly inherited by the proposed monitoring procedure. A simulation study and real data analysis are conducted to affirm the validity of our method.

Keywords: time series of counts; INGARCH model; SPC; CUSUM monitoring; MDPDE

1. Introduction

In this paper we consider the cumulative sum (CUSUM) monitoring procedure for detecting a parameter change in integer-valued generalized autoregressive heteroscedastic (INGARCH) models. Integer-valued time series is a core area in time series analysis that includes diverse disciplines in social, physical, engineering, and medical sciences. Both integer-valued autoregressive (INAR) time series models and the integer-valued generalized autoregressive conditional heteroscedastic (INGARCH) models have been widely studied in the literature and applied to various practical problems. Refer to McKenzie [1], Al-Osh and Alzaid [2], Ferland, Latour and Oraichi [3], Fokianos, Rahbek and Tjøstheim [4], and Weiß [5] for a general review. Poisson, negative binomial (NB), and one-parameter exponential family distributions have been widely used as underlying distributions, as seen in Davis and Wu [6], Zhu [7], Zhu [8], Jazi, Jones and Lai [9], Christou and Fokianos [10], Davis and Liu [11], Lee, Lee and Chen [12], and Chen, Khamthong and Lee [13].

Since Page [14], the CUSUM test has been a conventional tool to detect a structural change in underlying models. For a history and background, we refer to Csörgő and Horváth [15], Chen and Gupta [16], Lee, Ha, Na and Na [17], and the papers cited therein. Several authors have studied the change point test for INGARCH models, including Fokianos and Fried [18], Fokianos and Fried [19], Franke, Kirch and Kamgaing [20], Fokianos, Gombay and Hussein [21], Hudecová [22], Hudecová, Hušková and Meintanis [23], Kang and Lee [24], Lee, Lee and Chen [12], Lee, Lee and Tjøstheim [25], and Lee and Lee [26]. This CUSUM scheme has been applied not only to retrospective change point tests but also to on-line monitoring and statistical process control (SPC) problems, designed to monitoring abnormal phenomena in manufacturing processes and health care surveillance. The CUSUM control chart has been popular due to its considerable competency in early detection of anomalies. Refer to Weiß [27], Rakitzis, Maravelakis and Castagliola [28], Kim and Lee [29], and the papers cited therein. Meanwhile, Gombay and Serban [30] used the CUSUM approach based on the score vectors for independent observations, and later extended it to autoregressive processes, wherein the Type I probability error is measured for obtaining control limits instead of the conventional average run length (ARL). Their CUSUM monitoring process is based on the asymptotic property of the partial sum process generated from score vectors. Later, Huh, Kim and Lee [31] adopted their method for analyzing Poisson INGARCH models, and compared its performance with the likelihood ratio (LR)-based control chart, originally considered by Weiss and Testik [32].

In this work, taking the approach of Gombay and Serban [30] and Huh, Kim and Lee [31], we designate a robust monitoring process based on the minimum distance power divergence estimator (MDPDE) proposed by Basu, Harris, Hjort and Jones [33]. We do this because the MDPDE is well-known to be suitable for robust inference in various models, having a trade-off between efficiency and robustness controlled through the tuning parameters with little loss in asymptotic efficiency relative to the maximum likelihood estimator (MLE) (Riani, Atkinson, Corbellini and Perrotta [34]). The MDPDE method has been successfully applied to various time series models, and in particular INGARCH models (Kim and Lee [35], Kim and Lee [36]). Recently, Lee and Lee [26] and Kim and Lee [37] considered the CUSUM tests based on score vectors for the MLE and MDPDE in exponential family distribution INGARCH models. See also Kang and Song [38]. Using their results within the framework of Gombay and Serban [30] and Huh, Kim and Lee [31], we design an MDPDE-based monitoring process to detect a model parameter change in INGARCH models. Monte Carlo simulations are conducted to assess the performance of the proposed monitoring procedure. A focus is made on comparing the MDPDE-based CUSUM test with the MLE-based CUSUM test for Poisson INGARCH models to demonstrate the superiority of the former over the latter in the presence of outliers. A real data analysis of the return times of extreme events of Goldman Sachs Group (GS) stock prices is also provided to illustrate the validity of the proposed test.

The rest of the paper is organized as follows. Section 2 reviews the MDPDE for INGARCH models and Section 3 constructs the monitoring procedure for these models and investigates its asymptotic properties. Section 4 presents a simulation study and Section 5 provides a real data analysis. Section 6 concludes the paper. The proof of the main theorem is provided in Appendix A.

2. MDPDE for INGARCH Model: An Overview

In this section, we briefly review the MDPDE for INGARCH models in [36]. Let $Y_1, Y_2, \ldots$ be the observations generated from integer-valued time series models with the conditional distribution of the one-parameter exponential family:

$$Y_t | \mathcal{F}_{t-1} \sim p(y|\eta_t), \quad X_t := E(Y_t | \mathcal{F}_{t-1}) = f_\theta(X_{t-1}, Y_{t-1}), \tag{1}$$

where $\mathcal{F}_{t-1}$ is a σ-field generated by $Y_{t-1}, Y_{t-2}, \ldots$, and $f_\theta(x, y)$ is a non-negative bivariate function, depending on the parameter $\theta \in \Theta \subset \mathbb{R}^d$, and satisfies $\inf_{\theta \in \Theta} f_\theta(x, y) \geq c_*$ for some $c_* > 0$ for all x, y, and $p(\cdot | \cdot)$ is a probability mass function given by

$$p(y|\eta) = \exp\{\eta y - A(\eta)\} h(y), \quad y = 0, 1, \ldots,$$

where η is the natural parameter, $A(\eta)$ and $h(y)$ are known functions, and both A and $B = A'$ are strictly increasing. In particular, $B(\eta_t) = X_t$ and $B'(\eta_t)$ is the conditional variance of Y_t. In what follows, symbols $X_t(\theta)$ and $\eta_t(\theta) = B^{-1}(X_t(\theta))$ are also utilized to stand for X_t and η_t, respectively.

Davis and Liu [11] demonstrated that the strict stationarity and ergodicity of $\{X_t\}$, and the expression of $X_t(\theta) = f_\infty^\theta(Y_{t-1}, Y_{t-2}, \ldots)$ are allowed for some nonnegative measurable function f_∞^θ defined on $\mathbb{N}_0^\infty$ under the contraction condition: for all $x, x' \geq 0$ and $y, y' \in \mathbb{N}_0$,

$$\sup_{\theta \in \Theta} |f_\theta(x, y) - f_\theta(x', y')| \leq \lambda_1 |x - x'| + \lambda_2 |y - y'|$$

with constants $\lambda_1, \lambda_2 \geq 0$ satisfying $\lambda_1 + \lambda_2 < 1$.

Meanwhile, Basu, Harris, Hjort and Jones [33] considered the minimum distance power divergence estimator (MDPDE) for model parameters using the density power divergence d_α between two density functions g and h, defined by:

$$
d_\alpha(g,h) := \begin{cases} \int \{g^{1+\alpha}(y) - (1+\tfrac{1}{\alpha})h(y)g^\alpha(y) + \tfrac{1}{\alpha}h^{1+\alpha}(y)\}dy, & \alpha > 0, \\ \int h(y)(\log h(y) - \log g(y))dy, & \alpha = 0. \end{cases}
$$

Kim and Lee [36] studied the MDPDE for one parameter exponential family distribution INGARCH models. Given $Y_1, \ldots, Y_n$ generated from (1), the MDPDE is defined by

$$
\hat{\theta}_{\alpha,n} = \underset{\theta \in \Theta}{\operatorname{argmin}}\, \tilde{L}_{\alpha,n}(\theta) = \underset{\theta \in \Theta}{\operatorname{argmin}}\, \frac{1}{n}\sum_{t=1}^{n} \tilde{l}_{\alpha,t}(\theta), \tag{2}
$$

where

$$
\tilde{l}_{\alpha,t}(\theta) = \begin{cases} \sum_{y=0}^{\infty} p^{1+\alpha}(y|\tilde{\eta}_t(\theta)) - \left(1+\tfrac{1}{\alpha}\right)p^\alpha(Y_t|\tilde{\eta}_t(\theta)), & \alpha > 0, \\ -\log p(Y_t|\tilde{\eta}_t(\theta)), & \alpha = 0, \end{cases} \tag{3}
$$

and $\tilde{\eta}_t(\theta) = B^{-1}(\tilde{X}_t(\theta))$ is updated recursively through the equations: $\tilde{X}_t(\theta) = f_\theta(\tilde{X}_{t-1}(\theta), Y_{t-1})$, $t \geq 2$ with an initial value $\tilde{X}_1(\theta) := \tilde{X}_1$.

Below, θ_0 denotes the true value of θ and is assumed to be an interior point in the compact parameter space $\Theta \subset \mathbb{R}^d$. Moreover, it is assumed that $E\left(\sup_{\theta \in \Theta} X_1(\theta)\right)^4 < \infty$, $EY_1^4 < \infty$, $X_t(\theta) = X_t(\theta_0)$ a.s. implies $\theta = \theta_0$, and $v^T \frac{\partial X_t(\theta_0)}{\partial \theta} = 0$ a.s. implies $v = 0$. Furthermore, $\theta \mapsto X_t(\theta)$ is twice continuously differentiable with respect to θ and satisfies

$$
E\left(\sup_{\theta \in \Theta}\left\|\frac{\partial X_t(\theta)}{\partial \theta}\right\|\right)^4 < \infty \quad \text{and} \quad E\left(\sup_{\theta \in \Theta}\left\|\frac{\partial^2 X_t(\theta)}{\partial \theta \partial \theta^T}\right\|\right)^2 < \infty.
$$

Assuming

$$
\inf_{\theta \in \Theta}\inf_{0 \leq \delta \leq 1} B'((1-\delta)\eta_t(\theta) + \delta\tilde{\eta}_t(\theta)) \geq \underline{c}
$$

for some $\underline{c} > 0$, Kim and Lee [36] verified that the MDPDE is strongly consistent. Additionally, they showed that provided

$$
\sup_{\theta \in \Theta}\sup_{0 \leq \delta \leq 1} \left\{ \left|\frac{B''((1-\delta)\eta_t(\theta) + \delta\tilde{\eta}_t(\theta))}{B'((1-\delta)\eta_t(\theta) + \delta\tilde{\eta}_t(\theta))^{5/2}}\right| \leq K \text{ for some } K > 0,
$$

and

$$
\sup_{\theta \in \Theta}\left\|\frac{\partial \tilde{X}_t(\theta)}{\partial \theta} - \frac{\partial X_t(\theta)}{\partial \theta}\right\| + \left\|\frac{\partial^2 \tilde{X}_t(\theta)}{\partial \theta \partial \theta^T} - \frac{\partial^2 X_t(\theta)}{\partial \theta \partial \theta^T}\right\| \leq V\rho^t \text{ a.s.,}
$$

where V and $\rho \in (0,1)$ denote a generic integrable random variable and a constant, respectively, the symbol $\|\cdot\|$ denotes the L^2-norm for matrices and vectors, and expectation $E(\cdot)$ is taken under θ_0, the MDPDE is asymptotically normal with asymptotic variance $J_\alpha^{-1} K_\alpha J_\alpha^{-1}$ where

$$
J_\alpha = -E\left(\frac{\partial^2 l_{\alpha,t}(\theta_0)}{\partial \theta \partial \theta^T}\right), \quad K_\alpha = E\left(\frac{\partial l_{\alpha,t}(\theta_0)}{\partial \theta}\frac{\partial l_{\alpha,t}(\theta_0)}{\partial \theta^T}\right), \tag{4}
$$

and $l_{\alpha,t}(\theta)$ is the same as $\tilde{l}_{\alpha,t}(\theta)$ with $\tilde{\eta}_t(\theta)$ in (3) replaced by $\eta_t(\theta)$.

Moreover, additionally assuming

$$\sup_{\theta \in \Theta} \sup_{0 \le \delta \le 1} \left| \frac{B^{(3)}((1-\delta)\eta_t(\theta) + \delta\tilde{\eta}_t(\theta))}{B'((1-\delta)\eta_t(\theta) + \delta\tilde{\eta}_t(\theta))^4} \right| \le M \text{ for some } M > 0,$$

Kim and Lee [37] showed that the CUSUM test statistics designed for detecting a change in θ have the limiting null distribution of the sup of a Brownian bridge. In practice, $\alpha \in (0, 1]$ is often harnessed and an optimal α can be selected through the method of Warwick [39] and Warwick and Jones [40]; see Remark 1 of Kim and Lee [36].

In the literature, the following linear INGARCH model has been frequently used:

$$Y_t | \mathcal{F}_{t-1} \sim p(y|\eta_t), \quad X_t = \omega + a X_{t-1} + b Y_{t-1},$$

where $X_t = B(\eta_t) = E(Y_t | \mathcal{F}_{t-1})$ and $\theta = (\omega, a, b)^T$ satisfy $\omega > 0$ and $a + b < 1$. Here, we assume that θ_0 is an interior of a compact neighborhood $\Theta = \{\theta = (\omega, a, b)^T \in \mathbb{R}_+^3 : 0 < \omega_1 \le \omega \le \omega_2, \epsilon \le a + b \le 1 - \epsilon\}$ for some $0 < \omega_1 < \omega_2$, $\epsilon > 0$. Moreover, the Poisson INGARCH(1,1) model with $Y_t | \mathcal{F}_{t-1} \sim \text{Poisson}(X_t)$ and the NB-INGARCH(1,1) model with $Y_t | \mathcal{F}_{t-1} \sim \text{NB}(r, p_t)$, $X_t = \frac{r(1-p_t)}{p_t}$, where $\text{NB}(r, p)$ denotes a negative binomial (NB) distribution with parameters $r \in \mathbb{N}$ and $p \in (0, 1)$, satisfy the aforementioned regularity conditions. Those conditions should be checked analytically when one aims to use a specific distribution as the conditional distribution of the INGARCH model. In this case, a goodness of fit test could be conducted to check the adequacy of the assumed underlying distribution (Fokianos and Neumann [41]).

3. MDPDE-Based Monitoring Process

In this section, we consider a monitoring process detecting a significant change in the underlying models based on sequentially observed time series $Y_1, \ldots, Y_n$ following Model (1), given a training sample $Y_1', \ldots, Y_m'$ from Model (1), where $m = m(n)$ is a sequence of positive integers that diverges to ∞ as n tends to ∞. For this task, we set up the following hypotheses:

$$H_0 : \theta \text{ does not change over } t = 1, \ldots, n \text{ vs. } H_1 : \text{ not } H_0.$$

We first consider the case that θ_0 is known a priori from a past experience. Then we consider the monitoring process using the process $\hat{W}_{k,0} = \hat{K}_\alpha^{-1/2} \sum_{t=1}^k \frac{\partial \tilde{l}_{\alpha,t}(\theta_0)}{\partial \theta}$, $k = 1, \ldots, n$, constructed as

$$\hat{T}_{n,0}^{min} := \max_{1 \le k \le n} \hat{T}_{n,0}^{min}(k) = \max_{1 \le k \le n} \frac{1}{\sqrt{n}} \left\| \min_{j \le k} \hat{W}_{j,0} - \hat{W}_{k,0} \right\|_{max}, \tag{5}$$

$$\hat{T}_{n,0}^{max} := \max_{1 \le k \le n} \hat{T}_{n,0}^{max}(k) = \max_{1 \le k \le n} \frac{1}{\sqrt{n}} \left\| \max_{j \le k} \hat{W}_{j,0} - \hat{W}_{k,0} \right\|_{max},$$

$$\hat{T}_{n,0}^{cusum} := \max_{1 \le k \le n} \hat{T}_{n,0}'(k) = \max_{1 \le k \le n} \max_{1 \le i < j \le k} \frac{1}{\sqrt{n}} \left\| \left(\frac{i}{j} \right) \hat{W}_{j,0} - \hat{W}_{i,0} \right\|,$$

where $\frac{\partial \tilde{l}_{\alpha,t}}{\partial \theta}$ is the score vector as in (3) based on $Y_1, \ldots, Y_n$ and

$$\hat{K}_\alpha = \frac{1}{m} \sum_{t=1}^m \frac{\partial \tilde{l}_{\alpha,t}'(\theta_0)}{\partial \theta^T} \frac{\partial \tilde{l}_{\alpha,t}'(\theta_0)}{\partial \theta^T}, \tag{6}$$

where $\frac{\partial \tilde{l}_{\alpha,t}'}{\partial \theta}$ is the score vector based on the training sample. Here, the notation $\max_{1 \le i \le k} \mathbf{z}_i$ with $\mathbf{z}_i = (z_{i,1}, \ldots, z_{i,d})^T \in \mathbb{R}^d$ is defined to be the vector with the jth entry equal to $\max_{1 \le i \le k} z_{j,i}$ for $j = 1, \ldots, d$, and $\|\mathbf{z}\|_{max} = \max_{1 \le i \le k} |z_i|$ for $\mathbf{z} = (z_1, \ldots, z_d)^T \in \mathbb{R}^d$. Similar versions of $\hat{T}_{n,0}^{max}$ and $\hat{T}_{n,0}^{cusum}$ based on MLE have been considered by Gombay and Serban [30] and Huh, Kim and Lee

[31] for the AR and Poisson INGARCH models, while $\hat{T}_{n,0}^{min}$ is newly considered here. An anomaly is signaled at k when $\hat{T}_{n,0}^{min}(k)$, $\hat{T}_{n,0}^{max}(k)$, or $\hat{T}_{n,0}^{cusum}(k)$ get out of a control limit for some $k = 1, \ldots, n$, and the control limit can be determined using the convergence result in Theorem 1 addressed below.

Next, we consider the situation that θ_0 is unknown and must be estimated in the construction of the monitoring process in (5). We employ a monitoring process constructed based on $\hat{W}_k = \hat{K}_{\alpha,m}^{-1/2} \sum_{t=1}^{k} \frac{\partial \tilde{l}_{\alpha,t}(\hat{\theta}_{\alpha,m})}{\partial \theta}$, where $\hat{\theta}_{\alpha,m}$ is the MDPDE of θ_0 obtained from the training sample and

$$\widehat{K}_{\alpha,m} = \frac{1}{m} \sum_{t=1}^{m} \frac{\partial \tilde{l}_{\alpha,t}(\hat{\theta}_{\alpha,m})}{\partial \theta} \frac{\partial \tilde{l}_{\alpha,t}(\hat{\theta}_{\alpha,m})}{\partial \theta^T},$$

which is obtained by substituting θ_0 in K_α in (6) with $\hat{\theta}_{\alpha,m}$, namely,

$$\hat{T}_n^{min} := \max_{1 \le k \le n} \hat{T}_n^{min}(k) = \max_{1 \le k \le n} \frac{1}{\sqrt{n}} \left\| \min_{j \le k} \hat{W}_j - \hat{W}_k \right\|_{max}, \tag{7}$$

$$\hat{T}_n^{max} := \max_{1 \le k \le n} \hat{T}_n^{max}(k) = \max_{1 \le k \le n} \frac{1}{\sqrt{n}} \left\| \max_{j \le k} \hat{W}_j - \hat{W}_k \right\|_{max},$$

$$\hat{T}_n^{cusum} := \max_{1 \le k \le n} \hat{T}_n^{cusum}(k) = \max_{1 \le k \le n} \max_{1 \le i < j \le k} \frac{1}{\sqrt{n}} \left\| \left(\frac{i}{j}\right) \hat{W}_{j,0} - \hat{W}_{i,0} \right\|.$$

An anomaly is detected at k when $\hat{T}_n^{min}(k)$, $\hat{T}_n^{max}(k)$, or $\hat{T}_n^{cusum}(k)$ get out of the control limit for some $k = 1, \ldots, n$. The control limit can be determined theoretically using the asymptotic result in Theorem 1 addressed below. For this task, we investigate the asymptotic behavior of the monitoring processes $\hat{T}_n^{min}$, $\hat{T}_n^{max}$, and $\hat{T}_n^{cusum}$ defined below.

Let $W_k = K_\alpha^{-1/2} \sum_{t=1}^{k} \frac{\partial l_{\alpha,t}(\theta_0)}{\partial \theta}$, where K_α and $\frac{\partial l_{\alpha,t}}{\partial \theta}$ are the ones in (4), and

$$T_n^{min} = \max_{1 \le k \le n} \frac{1}{\sqrt{n}} \left\| \min_{j \le k} W_j - W_k \right\|_{max},$$

$$T_n^{max} = \max_{1 \le k \le n} \frac{1}{\sqrt{n}} \left\| \max_{j \le k} W_j - W_k \right\|_{max},$$

$$T_n^{cusum} = \max_{1 \le k \le n} \max_{1 \le i < j \le k} \frac{1}{\sqrt{n}} \left\| \left(\frac{i}{j}\right) W_j - W_i \right\|.$$

Using Donsker's invariance principle for martingale differences (Billingsley [42]) and the fact that $\sup_{0 \le s \le t} B(s) - B(t) = |B(t)|$ in distribution for any standard Brownian motion B, we obtain

$$T_n^{max} \xrightarrow{d} T := \sup_{0 \le s \le 1} \|\mathcal{B}_d(s)\|_{max}, \tag{8}$$

where $\mathcal{B}_d$ and denote a d-dimensional standard Brownian motion, so that

$$T_n^{min} \xrightarrow{d} T = \sup_{0 \le s \le 1} \|\mathcal{B}_d(s)\|_{max}$$

as T_n^{min} behaves asymptotically similarly to T_n^{max}. Meanwhile, we can see that

$$T_n^{cusum} \xrightarrow{d} T' = \sup_{0 < s \le s' \le 1} \left\| \frac{s}{s'} \mathcal{B}_d^\circ(s') - \mathcal{B}_d^\circ(s) \right\|, \tag{9}$$

where $\mathcal{B}_d^\circ$ is a d-dimensional Brownian bridge.

Using the above facts, we are led to attain the following theorem, whose proof is provided in the Appendix A.

Theorem 1. *Assume that* **(A.1)–(A.11)** *hold. Then, under H_0, as $n \to \infty$, $\hat{T}_{n,0}^{min}$ and $\hat{T}_{n,0}^{max}$ converge to T in distribution, and the same holds for $\hat{T}_n^{min}$ and $\hat{T}_n^{max}$ if $m/n \to \infty$. Moreover, $\hat{T}_{n,0}^{cusum}$ converges to T' in distribution as $n \to \infty$, and so does $\hat{T}_n^{cusum}$ if $m/n \to \lambda \in (0, \infty)$.*

The result in Theorem 1 can be used to determine a control limit for the monitoring process. Given significance level $0 < \alpha < 1$, we take c and c' satisfying $P(T \geq c) = P(T' \geq c') = \alpha$. In particular, $P(T \geq c) = 1 - (P(\sup_{0 \leq s \leq 1} |B(s)| \leq c))^d$, so that c can be obtained from the fact that $P(\sup_{0 \leq s \leq 1} |B(s)| \geq c) = 1 - (1 - \alpha)^{1/d}$. The performance of the proposed CUSUM monitoring methods is evaluated in our simulation study, focusing on $\hat{T}_n^{cusum}$, $\hat{T}_{n,0}^{min}$, and $\hat{T}_n^{min}$. (We do not report the result for $\hat{T}_{n,0}^{max}$ and $\hat{T}_n^{max}$, as these do not perform well compared to the others in most cases). Therein, a parametric bootstrap is adopted in obtaining control limits to reduce the parameter estimation effect, which can be more problematic when m is not so large compared to n, and the MDPDE from the training sample is used to generate the bootstrap sample.

4. Simulation Results

In this section, we compare the performance of the CUSUM monitoring processes $\hat{T}_n^{cusum}$, $\hat{T}_{n,0}^{min}$, and $\hat{T}_n^{min}$ in three different experimental environments for the Poisson INGARCH(1,1) model as follows:

$$Y_t \mid \mathcal{F}_{t-1} \sim \text{Poisson}\,(X_t), \quad X_t = \omega + a X_{t-1} + b Y_{t-1}.$$

For the comparison, we compute the empirical sizes and powers at the nominal level of 0.05 for $m = n = 500, 1000$ with 1000 implications. For the critical value of $\hat{T}_{n,0}^{min}$, we use 2.633, which is the 0.95th quantile of $\sup_{0 \leq s \leq 1} \|\mathcal{B}_3(s)\|_{max}$. However, for $\hat{T}_n^{cusum}$ and $\hat{T}_n^{min}$, we use the critical values obtained from a parametric bootstrap method, as the MDPDE $\hat{\theta}_{\alpha,m}$ might cause some size distortions. In implementation, the warp-bootstrap method is utilized to save computing times (Giacomini, Politis, and White [43]).

-Part 1. We compare the performance of MLE- and MDPDE-based monitoring processes ($\alpha = 0, 0.1, 0.2, 0.3$) by calculating the size and power for the four different cases of changing parameter from (ω_0, a_0, b_0) to (ω_1, a_1, b_1) when the parameter change is assumed to occur at $[n/2]$.

Case 1: $\omega_1 = (1 + \delta)\omega_0, a_1 = (1 + \delta)a_0, b_1 = (1 + \delta)b_0$; that is, all parameters change;

Case 2: $\omega_1 = (1 + \delta)\omega_0, a_1 = a_0, b_1 = b_0$; that is, only ω changes;

Case 3: $\omega_1 = \omega_0, a_1 = (1 + \delta)a_0, b_1 = b_0$; that is, only a changes;

Case 4: $\omega_1 = \omega_0, a_1 = a_0, b_1 = (1 + \delta)b_0$; that is, only b changes.

-Part 2. We examine the size and power for the same settings as in Part 1 when the change occurs at $[n/4]$.

-Part 3. We compare the performance of MLE- and MDPDE-based monitoring processes ($\alpha = 0, 0.1, 0.2, 0.3$) for the same settings as in Part 1 when outliers exist in the time series, wherein the parameter change is assumed to occur at $[n/2]$. In this case time series samples are generated from $(1 - p_t)Y_t + p_t Z_t$ where Y_t is the INGARCH process with the parameters as in Part 1, p_t are iid Bernoulli random variables with success probability p, and Z_t are iid Poisson variables wit intensity $\lambda > 0$. Here, $\{Y_t\}$, $\{p_t\}$ and $\{Z_t\}$ are all independent.

Figure 1 shows how the parameter change affects the pattern of the Poisson INGARCH(1,1) time series (Case 3) with $\theta_0 = (2, 0.3, 0.3)$, $\tau = 500$, and $\delta = 0$ for the left panel and $\delta = 0.5$ for the right panel. As $EY_t = \frac{\omega}{1-a-b}$, we can see that parameter change causes a mean shift. Tables 1–3 list the size and powers for Part 1 (τ therein stands for the location of the change point) and show no severe size distortions and reasonably good powers for $\delta \geq 0.5$. In particular, $\hat{T}_n^{cusum}$ and $\hat{T}_{n,0}^{min}$ largely outperform $\hat{T}_n^{min}$ in terms of power. However, as seen in Tables 4–8, the power of $\hat{T}_n^{min}$ in Part 2 appears to increase up to that of $\hat{T}_{n,0}^{min}$. In both Part 1 and Part 2, different α do not affect the size much, but a larger α tends to diminish the power. This appeals to our intuition, as the MLE is more efficient in the presence of no outliers.

Table 1. Empirical sizes and powers in Case 1 for the Poisson INGARCH(1,1) model when no outliers exist with $\theta_0 = (2, 0.1, 0.2)$.

	α	n	τ	$\delta :$	0	0.25	0.5	0.75	1
$\hat{T}_{n,0}^{min}$	0	500	250		0.035	0.541	1	1	1
$\hat{T}_{n}^{min}$	0	500	250		0.036	0.428	1	1	1
$\hat{T}_{n}^{cusum}$	0	500	250		0.048	0.997	1	1	1
$\hat{T}_{n,0}^{min}$	0	1000	500		0.042	0.791	1	1	1
$\hat{T}_{n}^{min}$	0	1000	500		0.049	0.682	1	1	1
$\hat{T}_{n}^{cusum}$	0	1000	500		0.052	1	1	1	1
$\hat{T}_{n,0}^{min}$	0.1	500	250		0.035	0.523	1	1	1
$\hat{T}_{n}^{min}$	0.1	500	250		0.036	0.398	1	1	1
$\hat{T}_{n}^{cusum}$	0.1	500	250		0.043	0.995	1	1	1
$\hat{T}_{n,0}^{min}$	0.1	1000	500		0.042	0.78	1	1	1
$\hat{T}_{n}^{min}$	0.1	1000	500		0.051	0.642	1	1	1
$\hat{T}_{n}^{cusum}$	0.1	1000	500		0.056	1	1	1	1
$\hat{T}_{n,0}^{min}$	0.2	500	250		0.035	0.493	1	1	1
$\hat{T}_{n}^{min}$	0.2	500	250		0.038	0.361	1	1	1
$\hat{T}_{n}^{cusum}$	0.2	500	250		0.041	0.994	1	1	1
$\hat{T}_{n,0}^{min}$	0.2	1000	500		0.04	0.757	1	1	1
$\hat{T}_{n}^{min}$	0.2	1000	500		0.048	0.589	1	1	1
$\hat{T}_{n}^{cusum}$	0.2	1000	500		0.066	1	1	1	1
$\hat{T}_{n,0}^{min}$	0.3	500	250		0.035	0.465	1	1	1
$\hat{T}_{n}^{min}$	0.3	500	250		0.042	0.332	1	1	1
$\hat{T}_{n}^{cusum}$	0.3	500	250		0.036	0.992	1	1	1
$\hat{T}_{n,0}^{min}$	0.3	1000	500		0.034	0.718	1	1	1
$\hat{T}_{n}^{min}$	0.3	1000	500		0.047	0.551	1	1	1
$\hat{T}_{n}^{cusum}$	0.3	1000	500		0.064	1	1	1	1

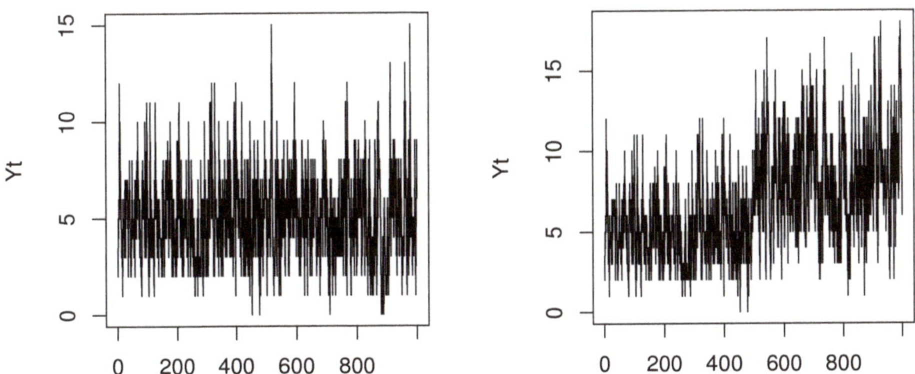

Figure 1. Plots of the Poisson INGARCH(1,1) time series (Case 3) with $\theta_0 = (2, 0.3, 0.3)$, $\tau = 500$ and $\delta = 0$ for the left panel and $\delta = 0.5$ for the right panel.

Meanwhile, Tables 9–12 show that the outliers undermine the performance of the MLE-based monitoring processes in terms of both size and power; namely, size distortions are notable and the power decreases to a certain extent. This result particularly indicates that $\hat{T}_n^{cusum}$ is improved when the MDPDE with $\alpha > 0$ is used, which demonstrates the efficacy of the MDPDE in the monitoring process. By contrast, the size of $\hat{T}_n^{min}$ significantly increases when $\alpha > 0$, indicating that $\hat{T}_n^{min}$ is unstable; see Figure 2. Although not reported here, we also examined the performance of the same monitoring processes for NB INGARCH(1,1) models. The result for this case showed a similar pattern to the Poisson INGARCH(1,1) case. All our findings strongly affirm that $\hat{T}_n^{cusum}$ is the most favorable among the monitoring methods considered in this study.

Table 2. Empirical sizes and powers in Case 2 for the Poisson INGARCH(1,1) model when no outliers exist with $\theta_0 = (2, 0.6, 0.2)$.

	α	n	τ	$\delta:$	0	$-1/5$	$-1/3$	$-3/7$	$-1/2$
$\hat{T}_{n,0}^{min}$	0	500	250		0.05	0.983	1	1	1
$\hat{T}_n^{min}$	0	500	250		0.06	0.86	0.999	1	1
$\hat{T}_n^{cusum}$	0	500	250		0.049	0.893	0.999	1	1
$\hat{T}_{n,0}^{min}$	0	1000	500		0.052	1	1	1	1
$\hat{T}_n^{min}$	0	1000	500		0.053	0.98	1	1	1
$\hat{T}_n^{cusum}$	0	1000	500		0.059	0.997	1	1	1
$\hat{T}_{n,0}^{min}$	0.1	500	250		0.047	0.984	1	1	1
$\hat{T}_n^{min}$	0.1	500	250		0.058	0.871	1	1	1
$\hat{T}_n^{cusum}$	0.1	500	250		0.046	0.9	1	1	1
$\hat{T}_{n,0}^{min}$	0.1	1000	500		0.048	1	1	1	1
$\hat{T}_n^{min}$	0.1	1000	500		0.041	0.977	1	1	1
$\hat{T}_n^{cusum}$	0.1	1000	500		0.051	0.996	1	1	1
$\hat{T}_{n,0}^{min}$	0.2	500	250		0.045	0.986	1	1	1
$\hat{T}_n^{min}$	0.2	500	250		0.05	0.852	0.999	1	1
$\hat{T}_n^{cusum}$	0.2	500	250		0.043	0.904	1	1	1
$\hat{T}_{n,0}^{min}$	0.2	1000	500		0.052	1	1	1	1
$\hat{T}_n^{min}$	0.2	1000	500		0.04	0.973	1	1	1
$\hat{T}_n^{cusum}$	0.2	1000	500		0.054	0.997	1	1	1
$\hat{T}_{n,0}^{min}$	0.3	500	250		0.04	0.985	1	1	1
$\hat{T}_n^{min}$	0.3	500	250		0.043	0.845	0.999	1	1
$\hat{T}_n^{cusum}$	0.3	500	250		0.048	0.912	1	1	1
$\hat{T}_{n,0}^{min}$	0.3	1000	500		0.05	1	1	1	1
$\hat{T}_n^{min}$	0.3	1000	500		0.052	0.978	1	1	1
$\hat{T}_n^{cusum}$	0.3	1000	500		0.053	0.996	1	1	1

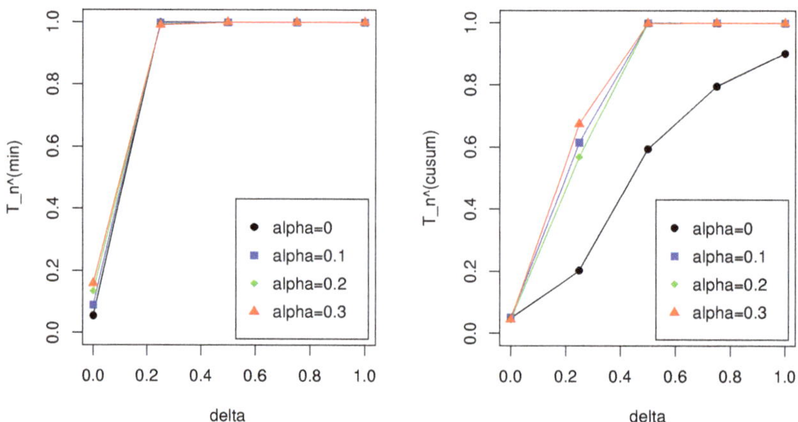

Figure 2. Plots of the sizes and powers in Table 10 (Part 3, Case 2) for $n = 1000$. The left panel is for $\hat{T}_n^{min}$ and the right panel is for $\hat{T}_n^{cusum}$.

Table 3. Empirical sizes and powers in Case 3 for the Poisson INGARCH(1,1) model when no outliers exist with $\theta_0 = (2, 0.3, 0.3)$.

	α	n	τ	$\delta:$	0	0.25	0.5	0.75	1
$\hat{T}_{n,0}^{min}$	0	500	250		0.046	0.309	0.999	1	1
$\hat{T}_{n}^{min}$	0	500	250		0.043	0.216	0.993	1	1
$\hat{T}_{n}^{cusum}$	0	500	250		0.047	0.685	1	1	1
$\hat{T}_{n,0}^{min}$	0	1000	500		0.039	0.473	1	1	1
$\hat{T}_{n}^{min}$	0	1000	500		0.041	0.337	1	1	1
$\hat{T}_{n}^{cusum}$	0	1000	500		0.057	0.969	1	1	1
$\hat{T}_{n,0}^{min}$	0.1	500	250		0.044	0.292	0.999	1	1
$\hat{T}_{n}^{min}$	0.1	500	250		0.046	0.208	0.992	1	1
$\hat{T}_{n}^{cusum}$	0.1	500	250		0.054	0.696	1	1	1
$\hat{T}_{n,0}^{min}$	0.1	1000	500		0.046	0.458	1	1	1
$\hat{T}_{n}^{min}$	0.1	1000	500		0.047	0.314	1	1	1
$\hat{T}_{n}^{cusum}$	0.1	1000	500		0.062	0.965	1	1	1
$\hat{T}_{n,0}^{min}$	0.2	500	250		0.046	0.266	0.998	1	1
$\hat{T}_{n}^{min}$	0.2	500	250		0.05	0.192	0.99	1	1
$\hat{T}_{n}^{cusum}$	0.2	500	250		0.048	0.696	1	1	1
$\hat{T}_{n,0}^{min}$	0.2	1000	500		0.044	0.44	1	1	1
$\hat{T}_{n}^{min}$	0.2	1000	500		0.042	0.287	1	1	1
$\hat{T}_{n}^{cusum}$	0.2	1000	500		0.067	0.962	1	1	1
$\hat{T}_{n,0}^{min}$	0.3	500	250		0.041	0.244	0.998	1	1
$\hat{T}_{n}^{min}$	0.3	500	250		0.051	0.179	0.986	1	1
$\hat{T}_{n}^{cusum}$	0.3	500	250		0.051	0.669	1	1	1
$\hat{T}_{n,0}^{min}$	0.3	1000	500		0.04	0.412	1	1	1
$\hat{T}_{n}^{min}$	0.3	1000	500		0.045	0.267	1	1	1
$\hat{T}_{n}^{cusum}$	0.3	1000	500		0.055	0.956	1	1	1

Table 4. Empirical sizes and powers in Case 4 for the Poisson INGARCH(1,1) model when no outliers exist with $\theta_0 = (1, 0.4, 0.4)$.

	α	n	τ	$\delta:$	0	$-1/5$	$-1/3$	$-3/7$	$-1/2$
$\hat{T}_{n,0}^{min}$	0	500	250		0.044	0.687	0.991	1	1
$\hat{T}_{n}^{min}$	0	500	250		0.049	0.345	0.75	0.941	0.986
$\hat{T}_{n}^{cusum}$	0	500	250		0.058	0.364	0.828	0.957	0.991
$\hat{T}_{n,0}^{min}$	0	1000	500		0.038	0.946	1	1	1
$\hat{T}_{n}^{min}$	0	1000	500		0.039	0.626	0.969	1	1
$\hat{T}_{n}^{cusum}$	0	1000	500		0.058	0.796	0.998	1	1
$\hat{T}_{n,0}^{min}$	0.1	500	250		0.044	0.688	0.99	1	1
$\hat{T}_{n}^{min}$	0.1	500	250		0.054	0.349	0.752	0.938	0.985
$\hat{T}_{n}^{cusum}$	0.1	500	250		0.06	0.376	0.841	0.964	0.993
$\hat{T}_{n,0}^{min}$	0.1	1000	500		0.042	0.945	1	1	1
$\hat{T}_{n}^{min}$	0.1	1000	500		0.042	0.616	0.966	0.999	1
$\hat{T}_{n}^{cusum}$	0.1	1000	500		0.053	0.782	0.997	1	1
$\hat{T}_{n,0}^{min}$	0.2	500	250		0.047	0.686	0.989	0.999	1
$\hat{T}_{n}^{min}$	0.2	500	250		0.056	0.357	0.757	0.939	0.986
$\hat{T}_{n}^{cusum}$	0.2	500	250		0.056	0.378	0.832	0.965	0.991
$\hat{T}_{n,0}^{min}$	0.2	1000	500		0.042	0.94	1	1	1
$\hat{T}_{n}^{min}$	0.2	1000	500		0.039	0.597	0.965	0.999	1
$\hat{T}_{n}^{cusum}$	0.2	1000	500		0.059	0.793	0.997	1	1
$\hat{T}_{n,0}^{min}$	0.3	500	250		0.049	0.677	0.985	0.999	1
$\hat{T}_{n}^{min}$	0.3	500	250		0.048	0.321	0.721	0.917	0.977
$\hat{T}_{n}^{cusum}$	0.3	500	250		0.054	0.381	0.831	0.963	0.991
$\hat{T}_{n,0}^{min}$	0.3	1000	500		0.043	0.931	1	1	1
$\hat{T}_{n}^{min}$	0.3	1000	500		0.047	0.606	0.962	0.999	1
$\hat{T}_{n}^{cusum}$	0.3	1000	500		0.064	0.792	0.997	1	1

Table 5. Empirical sizes and powers in Case 1 for the Poisson INGARCH(1,1) model when no outliers exist with $\theta_0 = (2, 0.1, 0.2)$.

	α	n	τ	$\delta:$	0	0.25	0.5	0.75	1
$\hat{T}_{n,0}^{min}$	0	500	125		0.035	0.759	1	1	1
$\hat{T}_n^{min}$	0	500	125		0.036	0.636	1	1	1
$\hat{T}_n^{cusum}$	0	500	125		0.048	0.983	1	1	1
$\hat{T}_{n,0}^{min}$	0	1000	250		0.042	0.936	1	1	1
$\hat{T}_n^{min}$	0	1000	250		0.049	0.874	1	1	1
$\hat{T}_n^{cusum}$	0	1000	250		0.052	1	1	1	1
$\hat{T}_{n,0}^{min}$	0.1	500	125		0.035	0.739	1	1	1
$\hat{T}_n^{min}$	0.1	500	125		0.036	0.606	1	1	1
$\hat{T}_n^{cusum}$	0.1	500	125		0.043	0.981	1	1	1
$\hat{T}_{n,0}^{min}$	0.1	1000	250		0.042	0.938	1	1	1
$\hat{T}_n^{min}$	0.1	1000	250		0.051	0.861	1	1	1
$\hat{T}_n^{cusum}$	0.1	1000	250		0.056	1	1	1	1
$\hat{T}_{n,0}^{min}$	0.2	500	125		0.035	0.716	1	1	1
$\hat{T}_n^{min}$	0.2	500	125		0.038	0.57	1	1	1
$\hat{T}_n^{cusum}$	0.2	500	125		0.041	0.981	1	1	1
$\hat{T}_{n,0}^{min}$	0.2	1000	250		0.04	0.936	1	1	1
$\hat{T}_n^{min}$	0.2	1000	250		0.048	0.842	1	1	1
$\hat{T}_n^{cusum}$	0.2	1000	250		0.066	1	1	1	1
$\hat{T}_{n,0}^{min}$	0.3	500	125		0.035	0.693	1	1	1
$\hat{T}_n^{min}$	0.3	500	125		0.042	0.542	1	1	1
$\hat{T}_n^{cusum}$	0.3	500	125		0.036	0.976	1	1	1
$\hat{T}_{n,0}^{min}$	0.3	1000	250		0.034	0.93	1	1	1
$\hat{T}_n^{min}$	0.3	1000	250		0.047	0.828	1	1	1
$\hat{T}_n^{cusum}$	0.3	1000	250		0.064	1	1	1	1

Table 6. Empirical sizes and powers Case 2 for the Poisson INGARCH(1,1) model when no outliers exist with $\theta_0 = (2, 0.6, 0.2)$.

	α	n	τ	$\delta:$	0	$-1/5$	$-1/3$	$-3/7$	$-1/2$
$\hat{T}_{n,0}^{min}$	0	500	125		0.05	0.999	1	1	1
$\hat{T}_n^{min}$	0	500	125		0.06	0.969	1	1	1
$\hat{T}_n^{cusum}$	0	500	125		0.049	0.844	1	1	1
$\hat{T}_{n,0}^{min}$	0	1000	250		0.052	1	1	1	1
$\hat{T}_n^{min}$	0	1000	250		0.053	1	1	1	1
$\hat{T}_n^{cusum}$	0	1000	250		0.059	0.988	1	1	1
$\hat{T}_{n,0}^{min}$	0.1	500	125		0.047	1	1	1	1
$\hat{T}_n^{min}$	0.1	500	125		0.058	0.971	1	1	1
$\hat{T}_n^{cusum}$	0.1	500	125		0.046	0.85	1	1	1
$\hat{T}_{n,0}^{min}$	0.1	1000	250		0.048	1	1	1	1
$\hat{T}_n^{min}$	0.1	1000	250		0.041	1	1	1	1
$\hat{T}_n^{cusum}$	0.1	1000	250		0.051	0.988	1	1	1
$\hat{T}_{n,0}^{min}$	0.2	500	125		0.045	1	1	1	1
$\hat{T}_n^{min}$	0.2	500	125		0.05	0.967	1	1	1
$\hat{T}_n^{cusum}$	0.2	500	125		0.043	0.845	1	1	1
$\hat{T}_{n,0}^{min}$	0.2	1000	250		0.052	1	1	1	1
$\hat{T}_n^{min}$	0.2	1000	250		0.04	1	1	1	1
$\hat{T}_n^{cusum}$	0.2	1000	250		0.054	0.991	1	1	1
$\hat{T}_{n,0}^{min}$	0.3	500	125		0.04	1	1	1	1
$\hat{T}_n^{min}$	0.3	500	125		0.043	0.962	1	1	1
$\hat{T}_n^{cusum}$	0.3	500	125		0.048	0.863	1	1	1
$\hat{T}_{n,0}^{min}$	0.3	1000	250		0.05	1	1	1	1
$\hat{T}_n^{min}$	0.3	1000	250		0.052	1	1	1	1
$\hat{T}_n^{cusum}$	0.3	1000	250		0.053	0.986	1	1	1

Table 7. Empirical sizes and powers Case 3 for the Poisson INGARCH(1,1) model when no outliers exist with $\theta_0 = (2, 0.3, 0.3)$.

	α	n	τ	$\delta:$	0	0.25	0.5	0.75	1
$\hat{T}^{min}_{n,0}$	0	500	125		0.046	0.488	1	1	1
$\hat{T}^{min}_{n}$	0	500	125		0.043	0.33	0.999	1	1
$\hat{T}^{cusum}_{n}$	0	500	125		0.047	0.614	1	1	1
$\hat{T}^{min}_{n,0}$	0	1000	250		0.039	0.716	1	1	1
$\hat{T}^{min}_{n}$	0	1000	250		0.041	0.554	1	1	1
$\hat{T}^{cusum}_{n}$	0	1000	250		0.057	0.916	1	1	1
$\hat{T}^{min}_{n,0}$	0.1	500	125		0.044	0.455	1	1	1
$\hat{T}^{min}_{n}$	0.1	500	125		0.046	0.314	0.999	1	1
$\hat{T}^{cusum}_{n}$	0.1	500	125		0.054	0.614	1	1	1
$\hat{T}^{min}_{n,0}$	0.1	1000	250		0.046	0.706	1	1	1
$\hat{T}^{min}_{n}$	0.1	1000	250		0.047	0.531	1	1	1
$\hat{T}^{cusum}_{n}$	0.1	1000	250		0.062	0.914	1	1	1
$\hat{T}^{min}_{n,0}$	0.2	500	125		0.046	0.434	1	1	1
$\hat{T}^{min}_{n}$	0.2	500	125		0.05	0.295	0.999	1	1
$\hat{T}^{cusum}_{n}$	0.2	500	125		0.048	0.601	0.999	1	1
$\hat{T}^{min}_{n,0}$	0.2	1000	250		0.044	0.701	1	1	1
$\hat{T}^{min}_{n}$	0.2	1000	250		0.042	0.505	1	1	1
$\hat{T}^{cusum}_{n}$	0.2	1000	250		0.067	0.901	1	1	1
$\hat{T}^{min}_{n,0}$	0.3	500	125		0.041	0.416	1	1	1
$\hat{T}^{min}_{n}$	0.3	500	125		0.051	0.283	0.999	1	1
$\hat{T}^{cusum}_{n}$	0.3	500	125		0.051	0.573	0.999	1	1
$\hat{T}^{min}_{n,0}$	0.3	1000	250		0.04	0.684	1	1	1
$\hat{T}^{min}_{n}$	0.3	1000	250		0.045	0.485	1	1	1
$\hat{T}^{cusum}_{n}$	0.3	1000	250		0.055	0.869	1	1	1

Table 8. Empirical sizes and powers in Case 4 for the Poisson INGARCH(1,1) model when no outliers exist with $\theta_0 = (1, 0.4, 0.4)$.

	α	n	τ	$\delta:$	0	$-1/5$	$-1/3$	$-3/7$	$-1/2$
$\hat{T}^{min}_{n,0}$	0	500	125		0.044	0.958	1	1	1
$\hat{T}^{min}_{n}$	0	500	125		0.049	0.559	0.937	0.995	0.999
$\hat{T}^{cusum}_{n}$	0	500	125		0.058	0.242	0.636	0.869	0.943
$\hat{T}^{min}_{n,0}$	0	1000	250		0.038	0.998	1	1	1
$\hat{T}^{min}_{n}$	0	1000	250		0.039	0.887	1	1	1
$\hat{T}^{cusum}_{n}$	0	1000	250		0.058	0.543	0.961	0.998	1
$\hat{T}^{min}_{n,0}$	0.1	500	125		0.044	0.955	1	1	1
$\hat{T}^{min}_{n}$	0.1	500	125		0.054	0.565	0.937	0.994	0.999
$\hat{T}^{cusum}_{n}$	0.1	500	125		0.06	0.283	0.667	0.881	0.953
$\hat{T}^{min}_{n,0}$	0.1	1000	250		0.042	0.999	1	1	1
$\hat{T}^{min}_{n}$	0.1	1000	250		0.042	0.883	1	1	1
$\hat{T}^{cusum}_{n}$	0.1	1000	250		0.053	0.542	0.96	0.998	1
$\hat{T}^{min}_{n,0}$	0.2	500	125		0.047	0.95	1	1	1
$\hat{T}^{min}_{n}$	0.2	500	125		0.056	0.574	0.941	0.992	0.999
$\hat{T}^{cusum}_{n}$	0.2	500	125		0.056	0.291	0.669	0.88	0.951
$\hat{T}^{min}_{n,0}$	0.2	1000	250		0.042	0.999	1	1	1
$\hat{T}^{min}_{n}$	0.2	1000	250		0.039	0.873	0.997	1	1
$\hat{T}^{cusum}_{n}$	0.2	1000	250		0.059	0.56	0.965	0.999	1
$\hat{T}^{min}_{n,0}$	0.3	500	125		0.049	0.945	1	1	1
$\hat{T}^{min}_{n}$	0.3	500	125		0.048	0.535	0.931	0.987	0.997
$\hat{T}^{cusum}_{n}$	0.3	500	125		0.054	0.294	0.662	0.873	0.95
$\hat{T}^{min}_{n,0}$	0.3	1000	250		0.043	0.999	1	1	1
$\hat{T}^{min}_{n}$	0.3	1000	250		0.047	0.885	0.996	1	1
$\hat{T}^{cusum}_{n}$	0.3	1000	250		0.064	0.569	0.967	0.998	1

Table 9. Empirical sizes and powers in Case 1 for the Poisson INGARCH(1,1) model when $\theta_0 = (2, 0.1, 0.2)$, $p = 0.1$ and $\lambda = 10$.

	α	n	τ	$\delta:$	0	0.25	0.5	0.75	1
$\hat{T}_n^{min}$	0	500	250		0.065	0.058	0.145	0.8	0.958
$\hat{T}_n^{cusum}$	0	500	250		0.048	0.047	0.066	0.512	0.997
$\hat{T}_n^{min}$	0	1000	500		0.061	0.058	0.367	0.958	0.991
$\hat{T}_n^{cusum}$	0	1000	500		0.053	0.053	0.095	0.978	1
$\hat{T}_n^{min}$	0.1	500	250		0.042	0.039	0.23	0.891	0.962
$\hat{T}_n^{cusum}$	0.1	500	250		0.035	0.037	0.122	0.897	1
$\hat{T}_n^{min}$	0.1	1000	500		0.056	0.046	0.653	0.979	0.995
$\hat{T}_n^{cusum}$	0.1	1000	500		0.053	0.054	0.963	1	1
$\hat{T}_n^{min}$	0.2	500	250		0.036	0.032	0.162	0.842	0.951
$\hat{T}_n^{cusum}$	0.2	500	250		0.035	0.036	0.111	0.804	1
$\hat{T}_n^{min}$	0.2	1000	500		0.026	0.025	0.454	0.976	0.993
$\hat{T}_n^{cusum}$	0.2	1000	500		0.023	0.023	0.514	1	1
$\hat{T}_n^{min}$	0.3	500	250		0.032	0.034	0.201	0.855	0.95
$\hat{T}_n^{cusum}$	0.3	500	250		0.032	0.032	0.114	0.771	0.979
$\hat{T}_n^{min}$	0.3	1000	500		0.024	0.02	0.485	0.973	0.991
$\hat{T}_n^{cusum}$	0.3	1000	500		0.021	0.021	0.284	0.999	1

Table 10. Empirical sizes and powers in Case 2 for the Poisson INGARCH(1,1) model when $\theta_0 = (2, 0.6, 0.2)$, $p = 0.1$ and $\lambda = 30$.

	α	n	τ	$\delta:$	0	0.25	0.5	0.75	1
$\hat{T}_n^{min}$	0	500	250		0.08	0.975	1	1	1
$\hat{T}_n^{cusum}$	0	500	250		0.065	0.11	0.194	0.329	0.456
$\hat{T}_n^{min}$	0	1000	500		0.055	1	1	1	1
$\hat{T}_n^{cusum}$	0	1000	500		0.05	0.203	0.594	0.795	0.902
$\hat{T}_n^{min}$	0.1	500	250		0.057	0.935	0.999	1	1
$\hat{T}_n^{cusum}$	0.1	500	250		0.062	0.169	0.666	0.927	0.993
$\hat{T}_n^{min}$	0.1	1000	500		0.091	0.999	1	1	1
$\hat{T}_n^{cusum}$	0.1	1000	500		0.052	0.615	1	1	1
$\hat{T}_n^{min}$	0.2	500	250		0.054	0.875	0.998	0.999	1
$\hat{T}_n^{cusum}$	0.2	500	250		0.043	0.069	0.309	0.663	0.784
$\hat{T}_n^{min}$	0.2	1000	500		0.135	0.993	1	1	1
$\hat{T}_n^{cusum}$	0.2	1000	500		0.046	0.569	0.999	1	1
$\hat{T}_n^{min}$	0.3	500	250		0.063	0.896	0.998	0.999	1
$\hat{T}_n^{cusum}$	0.3	500	250		0.046	0.086	0.455	0.763	0.853
$\hat{T}_n^{min}$	0.3	1000	500		0.159	0.992	1	1	1
$\hat{T}_n^{cusum}$	0.3	1000	500		0.047	0.675	0.999	1	1

Table 11. Empirical sizes and powers in Case 3 for the Poisson INGARCH(1,1) model when $\theta_0 = (2, 0.3, 0.3)$, $p = 0.1$ and $\lambda = 30$.

	α	n	τ	$\delta:$	0	0.25	0.5	0.75	1
$\hat{T}_n^{min}$	0	500	250		0.074	0.118	0.069	0.127	0.885
$\hat{T}_n^{cusum}$	0	500	250		0.062	0.064	0.06	0.068	0.777
$\hat{T}_n^{min}$	0	1000	500		0.062	0.213	0.058	0.257	0.935
$\hat{T}_n^{cusum}$	0	1000	500		0.049	0.05	0.049	0.057	0.992
$\hat{T}_n^{min}$	0.1	500	250		0.036	0.033	0.041	0.516	0.914
$\hat{T}_n^{cusum}$	0.1	500	250		0.037	0.037	0.04	0.268	0.961
$\hat{T}_n^{min}$	0.1	1000	500		0.029	0.026	0.03	0.824	0.963
$\hat{T}_n^{cusum}$	0.1	1000	500		0.023	0.023	0.025	0.859	1
$\hat{T}_n^{min}$	0.2	500	250		0.038	0.034	0.038	0.487	0.865
$\hat{T}_n^{cusum}$	0.2	500	250		0.04	0.042	0.046	0.321	0.612
$\hat{T}_n^{min}$	0.2	1000	500		0.019	0.017	0.018	0.725	0.922
$\hat{T}_n^{cusum}$	0.2	1000	500		0.015	0.015	0.015	0.244	0.616
$\hat{T}_n^{min}$	0.3	500	250		0.035	0.032	0.036	0.351	0.661
$\hat{T}_n^{cusum}$	0.3	500	250		0.039	0.039	0.042	0.13	0.211
$\hat{T}_n^{min}$	0.3	1000	500		0.02	0.016	0.017	0.684	0.893
$\hat{T}_n^{cusum}$	0.3	1000	500		0.012	0.012	0.012	0.085	0.161

Table 12. Empirical sizes and powers in Case 4 for the Poisson INGARCH(1,1) model when $\theta_0 = (1, 0.4, 0.4)$, $p = 0.1$ and $\lambda = 30$.

	α	n	τ	$\delta:$	0	0.25	0.5	0.75	1
$\hat{T}_n^{min}$	0	500	250		0.05	0.796	0.958	0.989	0.996
$\hat{T}_n^{cusum}$	0	500	250		0.048	0.078	0.118	0.173	0.219
$\hat{T}_n^{min}$	0	1000	500		0.032	1	1	1	1
$\hat{T}_n^{cusum}$	0	1000	500		0.043	0.613	0.874	0.931	0.957
$\hat{T}_n^{min}$	0.1	500	250		0.085	0.712	0.97	0.997	1
$\hat{T}_n^{cusum}$	0.1	500	250		0.04	0.065	0.243	0.466	0.647
$\hat{T}_n^{min}$	0.1	1000	500		0.242	0.978	0.999	1	1
$\hat{T}_n^{cusum}$	0.1	1000	500		0.069	0.916	0.999	1	1
$\hat{T}_n^{min}$	0.2	500	250		0.078	0.677	0.96	0.995	0.999
$\hat{T}_n^{cusum}$	0.2	500	250		0.032	0.069	0.284	0.535	0.735
$\hat{T}_n^{min}$	0.2	1000	500		0.229	0.965	0.999	1	1
$\hat{T}_n^{cusum}$	0.2	1000	500		0.047	0.836	0.999	1	1
$\hat{T}_n^{min}$	0.3	500	250		0.06	0.642	0.947	0.993	0.999
$\hat{T}_n^{cusum}$	0.3	500	250		0.027	0.08	0.332	0.621	0.807
$\hat{T}_n^{min}$	0.3	1000	500		0.201	0.962	0.999	1	1
$\hat{T}_n^{cusum}$	0.3	1000	500		0.027	0.749	0.999	1	1

5. Real Data Analysis

In this section, we apply $\hat{T}_n^{cusum}$ to a real dataset, using the extreme events of the daily log-returns of GS stock from 2 July 2007 to 28 February 2020. Davis and Liu [11] and Kim and Lee [37] used the GS stock datasets with different periods, but their works were focused on parameter estimation and the retrospective change point test. For the task of online monitoring, we first calculated the hitting times, $\tau_1, \tau_2, \ldots$, for which the log-returns of GS stock fall outside the 0.05 and 0.95 quantiles of the data, and generated the time series of counts $Y_t = \tau_t - \tau_{t-1} \geq 0$, $t = 1, \ldots, 319$. Figure 3 plots Y_t and exhibits the presence of a number of outliers.

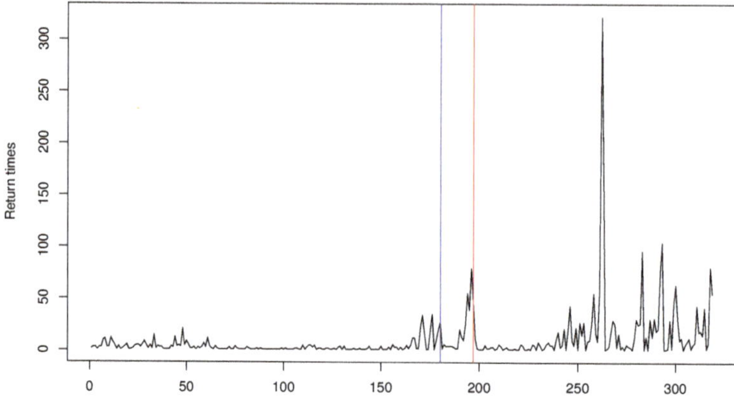

Figure 3. Plot of the return times of extreme events for Goldman Sachs Group stock.

Fitting the Poisson INGARCH(1,1) model to the whole observations, we have the MLE of $(\hat{\omega}, \hat{a}, \hat{b}) = (1.969, 0.152, 0.664)$ and the MDPDE of $(\hat{\omega}, \hat{a}, \hat{b}) = (1.213, 0.144, 0.472)$ when $\alpha = 0.1$ is used. The significant difference between the two estimates is seemingly due to the presence of outliers. Using $Y_t, t = 1, \ldots, 150$ as a training sample and viewing $Y_t, t \geq 151$ as sequentially observed testing data, we implement the monitoring process $\hat{T}_n^{cusum}$ with $\alpha = 0, 0.1$ to detect a parameter change. Subsequently, an anomaly is detected when $t = 180$ for $\alpha = 0$ (blue vertical line) and $t = 197$ for $\alpha = 0.1$ (red vertical line), which indicates that the monitoring process based on the MLE is more sensitive to relatively smaller outliers lying around $t = 180$, while that based on MDPDE is more robust to those outliers and detects a more significant change around $t = 197$, ignoring smaller ones. Obviously, we can see from Figure 3 that Y_t has a pattern with more fluctuations after $t = 180$. Our finding affirms the adequacy of the MDPDE-based monitoring process in the presence of outliers.

6. Concluding Remarks

In this work, we studied the robust on-line monitoring process based on MDPDE for detecting a parameter change in INGARCH models. For this task, we adopted the CUSUM process based on the score functions, which were originally constructed for obtaining the MDPDE. Our simulation study and real data analysis confirmed the validity of the proposed method. Here, we focused on the monitoring process within the framework of Gombay and Serban [30] and Huh, Kim and Lee [31]. However, one can also consider a different monitoring scheme, for example as in Na, Lee and Lee [44], and conduct a comparison study, which is left as our future project.

Author Contributions: Conceptualization, S.L.; project administration, S.L.; methodology and derivation, S.L.; data curation, D.K.; empirical analysis, D.K.; writing, S.L.; funding acquisition, S.L. All authors read and agreed to the published version of the manuscript.

Funding: This research was supported by the Basic Science Research Program through the National Research Foundation of Korea funded by the Ministry of Science, ICT and Future Planning (Grant 2018R1A2A2A05019433).

Acknowledgments: We would like to thank the Editor and two anonymous referees for their careful reading and valuable comments.

Abbreviations

The following abbreviations are used in this manuscript:

CUSUM	cumulative sum
INGARCH	integer-valued generalized autoregressive conditionally heteroscedastic
INAR	integer-valued autoregressive
MDPDE	minimum density power divergence etimator
MLE	maximum likelihood estimator
SPC	statistical process control

Appendix A

Proof of Theorem 1. We first verify that $\hat{T}_n^{max}$ converges to T in distribution; the cases of $\hat{T}_{n,0}^{min}$, $\hat{T}_{n,0}^{max}$, and $\hat{T}_n^{min}$ can be similarly handled and the proofs for these are omitted. As $\hat{\theta}_{\alpha,m}$ converges to θ_0 and

$$E\left(\sup_{\theta\in\Theta}\left\|\frac{\partial^2 l_{\alpha,t}(\theta)}{\partial\theta\partial\theta^T}-\frac{\partial^2 l_{\alpha,t}(\theta_0)}{\partial\theta\partial\theta^T}\right\|\right)<\infty,$$

we have that for any sequence θ_n^* converging to θ_0 a.s.,

$$\frac{1}{n}\sum_{t=1}^{n}\frac{\partial^2 l_{\alpha,t}(\theta_n^*)}{\partial\theta\partial\theta^T}\to -J_\alpha \tag{A1}$$

in probability. Then, using the mean value theorem and ergodicity, owing to (A1), we have

$$\max_{1\leq k\leq n}\max_{1\leq j\leq k}\left\|\frac{1}{\sqrt{n}}\sum_{t=1}^{j}\frac{\partial l_{\alpha,t}(\hat{\theta}_{\alpha,m})}{\partial\theta}-\frac{1}{\sqrt{n}}\sum_{t=1}^{k}\frac{\partial l_{\alpha,t}(\hat{\theta}_{\alpha,m})}{\partial\theta}\right.$$

$$\left.-\left\{\frac{1}{\sqrt{n}}\sum_{t=1}^{j}\frac{\partial l_{\alpha,t}(\theta_0)}{\partial\theta}-\frac{1}{\sqrt{n}}\sum_{t=1}^{k}\frac{\partial l_{\alpha,t}(\theta_0)}{\partial\theta}\right\}\right\|$$

$$\leq \sqrt{m}\|\hat{\theta}_{\alpha,m}-\theta_0\|\sqrt{\frac{n}{m}}\max_{1\leq j,k\leq n}\left\|\left(\frac{j}{n}\right)\frac{1}{j}\sum_{t=1}^{j}\frac{\partial^2 l_{\alpha,t}(\theta_n^*)}{\partial\theta\partial\theta^T}-\left(\frac{k}{n}\right)\frac{1}{k}\sum_{t=1}^{k}\frac{\partial^2 l_{\alpha,t}(\theta_n^{**})}{\partial\theta\partial\theta^T}\right\|$$

$$=o_P(1), \tag{A2}$$

where $\hat{\theta}_n^*$ and $\hat{\theta}_n^{**}$ are intermediate points between θ_0 and $\hat{\theta}_{\alpha,m}$. Hence, since $\widehat{K}_{\alpha,m}$ is a consistent estimator of K_α (Lemma A5 of Kim and Lee [36]) and

$$\sup_{\theta\in\Theta}\max_{1\leq k\leq n}\left\|\frac{1}{\sqrt{n}}\sum_{t=1}^{k}\frac{\partial \tilde{l}_{\alpha,t}(\theta)}{\partial\theta}-\frac{1}{\sqrt{n}}\sum_{t=1}^{k}\frac{\partial l_{\alpha,t}(\theta)}{\partial\theta}\right\|=o_P(1) \tag{A3}$$

(Lemma 6 of Kim and Lee, 2019), we get $\hat{T}_n^{max}-T_n^{max}=o_P(1)$ and $\hat{T}_n^{max}$ converges to T in distribution owing to (9).

Next, we deal with $\hat{T}_n^{cusum}$. The case of $\hat{T}_{n,0}^{cusum}$ can be similarly handled. Similarly to (A2), we can see that

$$\max_{1\leq k\leq n}\left\|\frac{1}{\sqrt{n}}\sum_{t=1}^{k}\frac{\partial l_{\alpha,t}(\hat{\theta}_{\alpha,m})}{\partial\theta}-\frac{1}{\sqrt{n}}\left(\frac{k}{n}\right)\sum_{t=1}^{n}\frac{\partial l_{\alpha,t}(\hat{\theta}_{\alpha,m})}{\partial\theta}\right.$$

$$\left.-\frac{1}{\sqrt{n}}\sum_{t=1}^{k}\frac{\partial l_{\alpha,t}(\theta_0)}{\partial\theta}-\frac{1}{\sqrt{n}}\left(\frac{k}{n}\right)\sum_{t=1}^{n}\frac{\partial l_{\alpha,t}(\theta_0)}{\partial\theta}\right\|=o_P(1). \tag{A4}$$

Then, using the arguments as in (A3) and (A4), we can see that

$$\max_{1\le k\le n} \frac{1}{\sqrt{n}} \left\| \hat{W}_k - \left(\frac{k}{n}\right)\hat{W}_k - W_k + \left(\frac{k}{n}\right)W_k \right\| = o_P(1),$$

which implies $\hat{T}_n^{cusum} - T_n^{cusum} = o_P(1)$ and $\hat{T}_n^{cusum} \xrightarrow{d} T'$ holds owing to (9). This completes the proof. $\square$

References

1. McKenzie, E. Some simple models for discrete variate time series. *J. Am. Water Resour. Assoc.* **1985**, *21*, 645–650. [CrossRef]
2. Al-Osh, M.A.; Alzaid, A.A. First order integer-valued autoregressive (INAR(1)) process. *J. Time Ser. Anal.* **1987**, *8*, 261–275. [CrossRef]
3. Ferland, R.; Latour, A.; Oraichi, D. Integer-valued GARCH processes. *J. Time Ser. Anal.* **2006**, *27*, 923–942. [CrossRef]
4. Fokianos, K.; Rahbek, A.; Tjøstheim, D. Poisson autoregression. *J. Am. Stat. Assoc.* **2009**, *104*, 1430–1439. [CrossRef]
5. Weiß, C.H. *An Introduction to Discrete-Valued Time Series*; Wiley: New York, NY, USA, 2018.
6. Davis, R. A.; Wu, R. A negative binomial model for time series of counts. *Biometrika* **2009**, *96*, 735–749. [CrossRef]
7. Zhu, F. Modeling overdispersed or underdispersed count data with generalized poisson integer-valued garch models. *J. Math. Anal. Appl.* **2012**, *389*, 58–71. [CrossRef]
8. Zhu, F. Zero-inflated Poisson and negative binomial integer-valued GARCH models. *J. Stat. Plan. Infer.* **2012**, *142*, 826–839. [CrossRef]
9. Jazi, M.A.; Jones, G.; Lai, C. First-order integer valued AR processes with zero inflated poisson innovations. *J. Time Ser. Anal.* **2012**, *33*, 954–963. [CrossRef]
10. Christou, V.; Fokianos, K. Quasi-likelihood inference for negative binomial time series models. *J. Time Ser. Anal.* **2014**, *35*, 55–78. [CrossRef]
11. Davis, R. A.; Liu, H. Theory and inference for a class of observation-driven models with application to time series of counts. *Stat. Sin.* **2016**, *26*, 1673–1707.
12. Lee, S.; Lee, Y.; Chen, C.W.S. Parameter change test for zero-inflated generalized Poisson autoregressive models. *Statistics* **2016**, *50*, 540–557. [CrossRef]
13. Chen, C.W.S.; Khamthong, K.; Lee, S. Markov switching integer-valued generalized autoregressive conditional heteroscedastic models for dengue counts. *J. Roy. Stat. Soc. C* **2019**, *68*, 963–983. [CrossRef]
14. Page, E.S. A test for a change in a parameter occurring at an unknown point. *Biometrika* **1955**, *42*, 523–527. [CrossRef]
15. Csörgő, M.; Horváth, L. *Limit Theorems in Change-Point Analysis.*; John Wiley & Sons Inc.: New York, NY, USA, 1997.
16. Chen, J.; Gupta, A.K. *Parametric Statistical Change Point Analysis with Applications to Genetics, Medicine, and Finance*; Wiley: New York, NY, USA, 2012.
17. Lee, S.; Ha, J.; Na, O.; Na, S. The CUSUM test for parameter change in time series models. *Scand. J. Stat.* **2003**, *30*, 781–796. [CrossRef]
18. Fokianos, K.; Fried, R. Interventions in INGARCH processes. *J. Time Ser. Anal.* **2010**, *31*, 210–225. [CrossRef]
19. Fokianos, K. and Fried, R. Interventions in log-linear Poisson autoregression. *Stat. Model.* **2012**, *12*, 299–322. [CrossRef]
20. Franke, J.; Kirch, C.; Kamgaing, J.T. Changepoints in times series of counts. *J. Time Ser. Anal.* **2012**, *33*, 757–770. [CrossRef]
21. Fokianos, K.; Gombay, E.; Hussein, A. Retrospective change detection for binary time series models. *J. Stat. Plan. Infer.* **2014**, *145*, 102–112. [CrossRef]
22. Hudecová, Š. Structural changes in autoregressive models for binary time series. *J. Stat. Plan. Infer.* **2013**, *143*, 1744–1752. [CrossRef]

23. Hudecová, Š.; Hušková, M.; Meintanis, S. G. Tests for structural changes in time series of counts. *Scand. J. Stat.* **2017**, *44*, 843–865. [CrossRef]
24. Kang, J.; Lee, S. Parameter change test for Poisson autoregressive models. *Scand. J. Stat.* **2014**, *41*, 1136–1152. [CrossRef]
25. Lee, Y.; Lee, S.; Tjøstheim, D. Asymptotic normality and parameter change test for bivariate Poisson INGARCH models. *Test* **2018**, *27*, 52–69. [CrossRef]
26. Lee, Y.; Lee, S. CUSUM tests for general nonlinear inter-valued GARCH models: comparison study. *Ann. Inst. Stat. Math.* **2019**, *71*, 1033–1057. [CrossRef]
27. Weiß, C.H. SPC method for time-dependent processes of counts - a literature review. *Cogent Math.* **2015**, *2*, 111–116.
28. Rakitzis, A.C.; Maravelakis, P.E.; Castagliola, P. CUSUM control charts for the monitoring of zero-inflated binomial processes. *Qual. Rel. Eng. Int.* **2016**, *32*, 465–483. [CrossRef]
29. Kim, H.; Lee, S. Improved CUSUM monitoring of Markov counting process with frequent zeros. *Qual. Rel. Eng. Int.* **2019**, *35*, 2371–2394. [CrossRef]
30. Gombay, E.; Serban, D. Monitoring parameter change in AR(p) time series models. *J. Multi. Anal.* **2009**, *100*, 715–725. [CrossRef]
31. Huh, J.; Kim, H.; Lee, S. Monitoring parameter shift with Poisson integer-valued GARCH models. *J. Stat. Comp. Sim.* **2017**, *87*, 1754–1766. [CrossRef]
32. Weiß, C.H.; Testik, M.C. CUSUM monitoring of first-order integer-valued autoregressive processes of Poisson counts. *J. Qual. Tech.* **2009**, *41*, 389–400. [CrossRef]
33. Basu, A.; Harris, I.R.; Hjort, N. L.; Jones, M.C. Robust and efficient estimation by minimizing a density power divergence. *Biometrika* **1998**, *85*, 549–559. [CrossRef]
34. Riani, M.; Atkinson, A.C.; Corbellini, A.; Perrotta, D. Robust Regression with Density Power Divergence: Theory, Comparisons, and Data Analysis. *Entroby* **2020**, *22*, 339. [CrossRef]
35. Kim, B.; Lee, S. Robust estimation for zero-inflated Poisson autoregressive models based on density power divergence. *J. Stat. Comput. Simul.* **2017**, *87*, 2981–2996. [CrossRef]
36. Kim, B.; Lee, S. Robust estimation for general integer-valued time series models. *Ann. Inst. Stat. Math.* **2020**, *72*, 1371–1396. [CrossRef]
37. Kim, B.; Lee, S. Robust change point test for general integer-valued time series models based on density power divergence. *Entropy* **2020**, *22*, 493. [CrossRef]
38. Kang, J.; Song, J. A robust approach for testing parameter change in Poisson autoregressive models. *J. Korean Stat. Soc.* **2020**, in press. [CrossRef]
39. Warwick, J. A data-based method for selecting tuning parameters in minimum distance estimators. *Comput. Stat. Data Anal.* **2005**, *48*, 571–585. [CrossRef]
40. Warwick, J.; Jones, M.C. Choosing a robustness tuning parameter. *J. Stat. Comput. Simul.* **2005**, *75*, 581–588. [CrossRef]
41. Fokianos, K.; Neumann, M.H. A goodness of fit test for Poisson count processes. *Electron. J. Stat.* **2013**, *7*, 793–819. [CrossRef]
42. Billingsley, P. *Convergence of Probability Measures*, 2nd ed.; Wiley: New York, NY, USA, 1999.
43. Giacomini, R.; Politis, D.N.; White, H. A warp-speed method for conducting Monte Carlo experiments involving bootstrap estimators. *Econ. Theory* **2013**,*29*, 567–589. [CrossRef]
44. Na, O.; Lee, Y.; Lee, S. Monitoring parameter change in time series models. *Stat. Meth. Appl.* **2011**, *20*, 171–199. [CrossRef]

Article

Robust Change Point Test for General Integer-Valued Time Series Models Based on Density Power Divergence

Byungsoo Kim [1,*] **and Sangyeol Lee** [2]

[1] Department of Statistics, Yeungnam University, Gyeongsan 38541, Korea
[2] Department of Statistics, Seoul National University, Seoul 08826, Korea; sylee@stats.snu.ac.kr
* Correspondence: bkim@yu.ac.kr

Received: 17 February 2020; Accepted: 24 April 2020; Published: 24 April 2020

Abstract: In this study, we consider the problem of testing for a parameter change in general integer-valued time series models whose conditional distribution belongs to the one-parameter exponential family when the data are contaminated by outliers. In particular, we use a robust change point test based on density power divergence (DPD) as the objective function of the minimum density power divergence estimator (MDPDE). The results show that under regularity conditions, the limiting null distribution of the DPD-based test is a function of a Brownian bridge. Monte Carlo simulations are conducted to evaluate the performance of the proposed test and show that the test inherits the robust properties of the MDPDE and DPD. Lastly, we demonstrate the proposed test using a real data analysis of the return times of extreme events related to Goldman Sachs Group stock.

Keywords: integer-valued time series; one-parameter exponential family; minimum density power divergence estimator; density power divergence; robust change point test

1. Introduction

Integer-valued time series models have received widespread attention from researchers and practitioners in diverse research areas. Since the works of McKenzie [1] as well as Al-Osh and Alzaid [2], integer-valued autoregressive (INAR) models have gained popularity in the analysis of correlated time series of counts. Later, as an alternative, Ferland et al. [3] proposed using Poisson integer-valued generalized autoregressive conditional heteroscedastic (INGARCH) models (see Engle [4] and Bollerslev [5]). Since then, INGARCH models have been studied by many authors, such as Fokianos et al. [6], who developed Poisson autoregressive (Poisson AR) models with nonlinear specifications for their intensity processes. The Poisson assumption on INGARCH models has been extended to include negative binomial INGARCH (NB-INGARCH) models (Davis and Wu [7] and Christou and Fokianos [8]), zero-inflated generalized Poisson INGARCH models (Zhu [9,10] and Lee et al. [11]), and one-parameter exponential distribution AR models (Davis and Liu [12]). The latter are also known as general integer-valued time series models and have been studied by, among others, Diop and Kengne [13] and Lee and Lee [14], who considered change point tests for these models.

The change point problem is a core issue in time series analysis because changes can occur in underlying model parameters owing to critical events or policy changes, and ignoring such changes can result in false conclusions. Numerous studies exist on change point analysis in time series models; refer to Kang and Lee [15] and Lee and Lee [14], and the articles cited therein, for the background and history of change points in integer-valued time series models. Lee and Lee [14] conducted a comparison study of the performance of various cumulative sum (CUSUM) tests using score vectors and residuals through the Monte Carlo simulations. In their work, the conditional maximum likelihood estimator (CMLE) is used for the parameter estimation and also the construction of the CUSUM tests.

However, the CMLE is often damaged by outliers, and so is the performance of the CMLE-based CUSUM test. In general, outliers easily mislead the CUSUM test since they can be mistakenly taken for abrupt changes; in the opposite, they can misidentify change points in their presence on time series. Among the robust estimation methods, we adopt the minimum density power divergence estimator (MDPDE) approach—proposed by Basu et al. [16]—as a remedy and propose to use the density power divergence (DPD)-based test as a robust change point test.

The MDPDE method is well known for consistently making robust inferences in various situations, and the trade-off between efficiency and robustness is managed via the tuning parameter. Basu et al. [16] introduced the MDPDE using the independent and identically distributed observations, and later, Ghosh and Basu [17] extended their method to the independent but not identically distributed samples. For earlier works in the context of time series, see Lee and Song [18], Kim and Lee [19], Kang and Lee [20], and Kim and Lee [21], who deal with the MDPDE for GARCH models, multivariate times series, and (zero-inflated) Poisson AR models. Kim and Lee [22] demonstrated that the MDPDE for general integer-valued time series models has strong robust properties, with little loss in asymptotic efficiency relative to the CMLE. This motivates us to use the MDPDE to construct a robust change point test for general integer-valued time series models. More precisely, we anticipate that the robust property of the MDPDE would be inherited to the proposed change point test, so that the influence of outliers should be reduced when performing a parameter change test in the presence of outliers. Although the problem of testing for a parameter change in integer-valued time series models has been investigated by many researchers, the testing procedure for observations with outliers has not been widely studied. This motivates us to develop a MDPDE-based robust change point test for general integer-valued time series models.

Kang and Song [23] proposed an estimate-based robust CUSUM test that uses the MDPDE to detect parameter changes in Poisson AR models. However, this type of test is known to suffer from severe size distortions, especially when the true parameter lies at the boundary of the parameter space. Thus, we use the test deduced based on an empirical version of the DPD, which is the objective function of the MDPDE. Song and Kang [24] and Kang and Song [25] applied DPD-based change point tests in GARCH models and Poisson AR models, respectively. However, the DPD approach basically shares the same spirit as the score-based CUSUM test of Lee and Lee [14] (see Remark 3 in Section 2.2), in that both are based on derivatives of objective functions. Thus, the idea is easily adapted to one-parameter exponential family AR models. As for a parameter change test for independent samples based on divergence measures, see Batsidis et al. [26,27], who consider the ϕ-divergence as a measure. We also refer to Martín and Pardo [28], who point out the importance of a Wald-type test based on DPD in dealing with the change point problem.

Monte Carlo simulations are conducted to evaluate the performance of the proposed test. Here, we compare the DPD-based test and the score-based CUSUM test to demonstrate the superiority of the proposed test in the presence of outliers. Then, we provide a real data analysis of the return times of extreme events related to Goldman Sachs Group (GS) stock to illustrate the proposed test. The paper proceeds as follows. Section 2 constructs the DPD-based change point test for general integer-valued time series models, and states its weak convergence theorem. Section 3 presents a simulation study and a real data analysis. Section 4 concludes the paper. All proofs are provided in the Appendix A.

2. Construction of the MDPDE and Change Point Test

2.1. MDPDE for General Integer-Valued Time Series Models

Let $Y_1, Y_2, \ldots$ be the observations generated from general integer-valued time series models with the conditional distribution of the one-parameter exponential family:

$$Y_t | \mathcal{F}_{t-1} \sim p(y|\eta_t), \quad X_t := E(Y_t | \mathcal{F}_{t-1}) = f_\theta(X_{t-1}, Y_{t-1}), \tag{1}$$

where $\mathcal{F}_{t-1}$ is a σ-field generated by $Y_{t-1}, Y_{t-2}, \ldots$ and $f_\theta(x, y)$ is a non-negative bivariate function defined on $[0, \infty) \times \mathbb{N}_0$, $\mathbb{N}_0 = \mathbb{N} \cup \{0\}$, depending on the parameter $\theta \in \Theta \subset \mathbb{R}^d$, and satisfies $\inf_{\theta \in \Theta} f_\theta(x, y) \geq x^*$ for some $x^* > 0$ for all x, y. Here, $p(\cdot|\cdot)$ is a probability mass function, given by

$$p(y|\eta) = \exp\{\eta y - A(\eta)\} h(y), \quad y \geq 0,$$

where η is the natural parameter and $A(\eta)$ and $h(y)$ are known functions. This distribution family includes several famous discrete distributions, such as the Poisson, negative binomial, and binomial distributions. If $B(\eta) = A'(\eta)$, $B(\eta_t)$ and $B'(\eta_t)$ become the conditional mean and variance of Y_t, and $X_t = B(\eta_t)$. The derivative of $A(\eta)$ exists for the exponential family; see Lehmann and Casella [29]. Since $B'(\eta_t) = Var(Y_t|\mathcal{F}_{t-1}) > 0$, $B(\eta)$ is strictly increasing, and since $B(\eta_t) = E(Y_t|\mathcal{F}_{t-1}) > 0$, $A(\eta)$ is also strictly increasing. To emphasize the role of θ, we also use $X_t(\theta)$ and $\eta_t(\theta) = B^{-1}(X_t(\theta))$ to stand for X_t and η_t, respectively.

Davis and Liu [12] showed that the assumption below ensures the strict stationarity and ergodicity of $\{(X_t, Y_t)\}$:

(A0) For all $x, x' \geq 0$ and $y, y' \in \mathbb{N}_0$,

$$\sup_{\theta \in \Theta} |f_\theta(x, y) - f_\theta(x', y')| \leq w_1|x - x'| + w_2|y - y'|,$$

where $w_1, w_2 \geq 0$ satisfy $w_1 + w_2 < 1$.

They also demonstrated that there exists a measurable function $f_\infty^\theta : \mathbb{N}_0^\infty \to [0, \infty)$, such that $X_t(\theta) = f_\infty^\theta(Y_{t-1}, Y_{t-2}, \ldots)$ almost surely (a.s.).

Meanwhile, the DPD d_α between two density functions g and h is defined as

$$d_\alpha(g, h) := \begin{cases} \int \{g^{1+\alpha}(y) - (1 + \frac{1}{\alpha})h(y)g^\alpha(y) + \frac{1}{\alpha}h^{1+\alpha}(y)\} dy, & \alpha > 0, \\ \int h(y)(\log h(y) - \log g(y)) dy, & \alpha = 0. \end{cases}$$

For a parametric family $\{G_\theta, \theta \in \Theta\}$ with densities given by $\{g_\theta\}$ and a distribution H with density h, the minimum DPD functional $T_\alpha(H)$ is defined by $d_\alpha(h, g_{T_\alpha(H)}) = \min_{\theta \in \Theta} d_\alpha(h, g_\theta)$. In particular, if $H = G_{\theta_0} \in \{G_\theta\}$, $T_\alpha(G_{\theta_0}) = \theta_0$. Then, given a random sample $Y_1, \ldots, Y_n$ with unknown density h, the MDPDE is defined by

$$\hat{\theta}_{\alpha,n} = \operatorname*{argmin}_{\theta \in \Theta} L_{\alpha,n}(\theta),$$

where $L_{\alpha,n}(\theta) = \frac{1}{n} \sum_{t=1}^n l_{\alpha,t}(\theta)$ and

$$l_{\alpha,t}(\theta) = \begin{cases} \int g_\theta^{1+\alpha}(y) dy - \left(1 + \frac{1}{\alpha}\right) g_\theta^\alpha(Y_t), & \alpha > 0, \\ -\log g_\theta(Y_t), & \alpha = 0. \end{cases}$$

When $\alpha = 0$ and 1, the MDPDE becomes the MLE and the L^2-distance estimator, respectively. Basu et al. [16] revealed that $\hat{\theta}_{\alpha,n}$ is consistent for $T_\alpha(H)$ and asymptotically normal. Furthermore, the estimator is robust against outliers, but still exhibits high efficiency when the true distribution belongs to a parametric family $\{G_\theta\}$ and α is close to zero. The tuning parameter α controls the trade-off between robustness and asymptotic efficiency. A large α escalates the robustness while a small α yields greater efficiency. The conditional version of the MDPDE is defined similarly (cf. Section 2 of Kim and Lee [22]).

For $Y_1, \ldots, Y_n$ generated from (1), the MDPDE for general integer-valued time series models is defined as

$$\hat{\theta}_{\alpha,n} = \operatorname*{argmin}_{\theta \in \Theta} \tilde{L}_{\alpha,n}(\theta) = \operatorname*{argmin}_{\theta \in \Theta} \frac{1}{n} \sum_{t=1}^n \tilde{l}_{\alpha,t}(\theta), \tag{2}$$

where

$$\tilde{l}_{\alpha,t}(\theta) = \begin{cases} \sum_{y=0}^{\infty} p^{1+\alpha}(y|\tilde{\eta}_t(\theta)) - \left(1 + \frac{1}{\alpha}\right) p^{\alpha}(Y_t|\tilde{\eta}_t(\theta)), & \alpha > 0, \\ - \log p(Y_t|\tilde{\eta}_t(\theta)), & \alpha = 0, \end{cases} \tag{3}$$

and $\tilde{\eta}_t(\theta) = B^{-1}(\tilde{X}_t(\theta))$ is updated recursively using the following equations:

$$\tilde{X}_t(\theta) = f_\theta(\tilde{X}_{t-1}(\theta), Y_{t-1}), \; t = 2, 3, \dots, \; \tilde{X}_1(\theta) = \tilde{X}_1,$$

with an arbitrarily chosen initial value $\tilde{X}_1$. The MDPDE with $\alpha = 0$ becomes the CMLE from (3).

Kim and Lee [22] showed that under the regularity conditions **(A0)**–**(A9)** stated below, the MDPDE is strongly consistent and asymptotically normal. Conditions **(A10)** and **(A11)** are imposed to derive the limiting null distribution of the DPD-based change point test in Section 2.2. Below, V and $\rho \in (0, 1)$ represent a generic integrable random variable and a constant, respectively; the symbol $\|\cdot\|$ denotes the L^2-norm for matrices and vectors; and $E(\cdot)$ is taken under θ_0, where θ_0 denotes the true value of θ.

(A1) θ_0 is an interior point in the compact parameter space $\Theta \subset \mathbb{R}^d$.
(A2) $E\left(\sup_{\theta \in \Theta} X_1(\theta)\right)^4 < \infty$.
(A3) $\inf_{\theta \in \Theta} \inf_{0 \leq \delta \leq 1} B'((1 - \delta)\eta_t(\theta) + \delta\tilde{\eta}_t(\theta)) \geq \underline{c}$ for some $\underline{c} > 0$.
(A4) $EY_1^4 < \infty$.
(A5) If there exists $t \geq 1$, such that $X_t(\theta) = X_t(\theta_0)$ a.s., then $\theta = \theta_0$.
(A6) $\sup_{\theta \in \Theta} \sup_{0 \leq \delta \leq 1} \left| \frac{B''((1-\delta)\eta_t(\theta)+\delta\tilde{\eta}_t(\theta))}{B'((1-\delta)\eta_t(\theta)+\delta\tilde{\eta}_t(\theta))^{5/2}} \right| \leq K$ for some $K > 0$.
(A7) The mapping $\theta \mapsto f_\infty^\theta$ is twice continuously differentiable with respect to θ, and satisfies

$$E\left(\sup_{\theta \in \Theta} \left\| \frac{\partial f_\infty^\theta(Y_0, Y_{-1}, \dots)}{\partial \theta} \right\|\right)^4 < \infty \quad \text{and} \quad E\left(\sup_{\theta \in \Theta} \left\| \frac{\partial^2 f_\infty^\theta(Y_0, Y_{-1}, \dots)}{\partial\theta\partial\theta^T} \right\|\right)^2 < \infty.$$

(A8) $\sup_{\theta \in \Theta} \left\| \frac{\partial \tilde{X}_t(\theta)}{\partial \theta} - \frac{\partial X_t(\theta)}{\partial \theta} \right\| \leq V\rho^t$ a.s.
(A9) $v^T \frac{\partial X_t(\theta_0)}{\partial \theta} = 0$ a.s. implies $v = 0$.
(A10) $\sup_{\theta \in \Theta} \left\| \frac{\partial^2 \tilde{X}_t(\theta)}{\partial\theta\partial\theta^T} - \frac{\partial^2 X_t(\theta)}{\partial\theta\partial\theta^T} \right\| \leq V\rho^t$ a.s.
(A11) $\sup_{\theta \in \Theta} \sup_{0 \leq \delta \leq 1} \left| \frac{B^{(3)}((1-\delta)\eta_t(\theta)+\delta\tilde{\eta}_t(\theta))}{B'((1-\delta)\eta_t(\theta)+\delta\tilde{\eta}_t(\theta))^4} \right| \leq M$ for some $M > 0$.

Proposition 1. *Under* **(A0)**–**(A5)**, *$\hat{\theta}_{\alpha,n} \longrightarrow \theta_0$ a.s. as $n \to \infty$, and further, under* **(A0)**–**(A9)**,

$$\sqrt{n}(\hat{\theta}_{\alpha,n} - \theta_0) \xrightarrow{d} N(0, J_\alpha^{-1}K_\alpha J_\alpha^{-1}) \; as \; n \to \infty,$$

where

$$J_\alpha = -E\left(\frac{\partial^2 l_{\alpha,t}(\theta_0)}{\partial\theta\partial\theta^T}\right), \; K_\alpha = E\left(\frac{\partial l_{\alpha,t}(\theta_0)}{\partial \theta} \frac{\partial l_{\alpha,t}(\theta_0)}{\partial\theta^T}\right)$$

and $l_{\alpha,t}(\theta)$ is defined by substituting $\eta_t(\theta)$ for $\tilde{\eta}_t(\theta)$ in (3).

Remark 1. *In our empirical study, discussed in Section 3.2, we select an optimal α using the method of Warwick [30] and Warwick and Jones [31]. We choose α that minimizes the trace of the estimated asymptotic mean squared error ($\widehat{AMSE}$):*

$$\widehat{AMSE} = (\hat{\theta}_{\alpha,n} - \hat{\theta}_{1,n})(\hat{\theta}_{\alpha,n} - \hat{\theta}_{1,n})^T + \widehat{As.var}(\hat{\theta}_{\alpha,n}),$$

where $\hat{\theta}_{1,n}$ is the MDPDE with $\alpha = 1$ and $\widehat{As.var}(\hat{\theta}_{\alpha,n})$ is the estimate of the asymptotic variance of $\hat{\theta}_{\alpha,n}$, computed as

$$\widehat{As.var}(\hat{\theta}_{\alpha,n}) = \left(\sum_{t=1}^{n} \frac{\partial^2 \tilde{l}_{\alpha,t}(\hat{\theta}_{\alpha,n})}{\partial\theta\partial\theta^T} \right)^{-1} \left(\sum_{t=1}^{n} \frac{\partial \tilde{l}_{\alpha,t}(\hat{\theta}_{\alpha,n})}{\partial\theta} \frac{\partial \tilde{l}_{\alpha,t}(\hat{\theta}_{\alpha,n})}{\partial\theta^T} \right) \left(\sum_{t=1}^{n} \frac{\partial^2 \tilde{l}_{\alpha,t}(\hat{\theta}_{\alpha,n})}{\partial\theta\partial\theta^T} \right)^{-1}.$$

Remark 2. *Instead of* **(A6)**, *Kim and Lee* [22] *assumed*

$$\sup_{\theta\in\Theta}\sup_{0\leq\delta\leq1} \left| \frac{B''((1-\delta)\eta_t(\theta) + \delta\tilde{\eta}_t(\theta))}{B'((1-\delta)\eta_t(\theta) + \delta\tilde{\eta}_t(\theta))^3} \right| \leq K \text{ for some } K > 0$$

to prove Proposition 1. *Note that this condition is satisfied directly if* **(A3)** *and* **(A6)** *hold. In our study, we alter the above condition to* **(A6)** *to prove Lemma* A1 *in the Appendix* A, *which is needed to obtain the limiting null distribution of the DPD-based change point test in Section* 2.2.

The following INGARCH(1,1) models are typical examples of general integer-valued time series models:

$$Y_t|\mathcal{F}_{t-1} \sim p(y|\eta_t), \quad X_t = d + aX_{t-1} + bY_{t-1},$$

where $X_t = B(\eta_t) = E(Y_t|\mathcal{F}_{t-1})$, $\theta = (d,a,b)^T \in \Theta \subset (0,\infty) \times [0,\infty)^2$ with $a+b < 1$, and Θ is compact. Condition **(A0)** trivially holds, and the process $\{(X_t, Y_t), t \geq 1\}$ has a strictly stationary and ergodic solution. Condition **(A1)** can be replaced with the following:

(A1)′ The true parameter θ_0 lies in a compact neighborhood $\Theta \in \mathbb{R}_+^3$ of θ_0, where

$$\Theta \in \{\theta = (d,a,b)^T \in \mathbb{R}_+^3 : 0 < d_L \leq d \leq d_U, \epsilon \leq a+b \leq 1-\epsilon\} \text{ for some } d_L, d_U, \epsilon > 0.$$

Moreover, we can express

$$X_t(\theta) = \frac{d}{1-a} + b\sum_{k=0}^{\infty} a^k Y_{t-k-1} \quad \text{and} \quad \tilde{X}_t(\theta) = \frac{d}{1-a} + b\sum_{k=0}^{t-2} a^k Y_{t-k-1},$$

where the initial value $\tilde{X}_1$ is taken as $d/(1-a)$ for simplicity. Based on the above and **(A4)**, the conditions **(A2)**, **(A5)**, and **(A7)–(A10)** are all satisfied for INGARCH(1,1) models, as proven by Theorem 3 of Kang and Lee [15]. Kim and Lee [22] showed recently that the following Poisson and negative binomial INGARCH(1,1) models satisfy **(A3)** and **(A4)**. Furthermore, following the arguments presented in Section 3.2 of their study, **(A6)** holds for these models as well. Below, we show that **(A11)** holds for Poisson and negative binomial INGARCH(1,1) models.

- *Poisson INGARCH(1,1) model:*

$$Y_t|\mathcal{F}_{t-1} \sim \text{Poisson}(X_t), \quad X_t = d + aX_{t-1} + bY_{t-1}.$$

In this model, $\eta_t(\theta) = \log(X_t(\theta))$ and $A(\eta_t(\theta)) = e^{\eta_t(\theta)}$. Since $B'(\eta) = B^{(3)}(\eta)$, **(A11)** holds owing to **(A3)**.

- *NB-INGARCH(1,1) model:*

$$Y_t|\mathcal{F}_{t-1} \sim \text{NB}(r, p_t), \quad X_t = \frac{r(1-p_t)}{p_t} = d + aX_{t-1} + bY_{t-1},$$

where $\text{NB}(r,p)$ denotes a negative binomial distribution with parameters $r \in \mathbb{N}$ and $p \in (0,1)$. To be more specific, it counts the number of failures before the r-th success occurs in a sequence of Bernoulli trials with success probability p. Here, r is assumed to be known. In this model, $\eta_t(\theta) = \log(X_t(\theta)/(X_t(\theta) + r))$ and $A(\eta_t(\theta)) = r\log(r/(1 - e^{\eta_t(\theta)}))$. From the fact that $B'(\eta) = re^{\eta}/(1 - e^{\eta})^2$ and $B^{(3)}(\eta) = re^{\eta}(e^{2\eta} + 4e^{\eta} + 1)/(1 - e^{\eta})^4$, we have $B^{(3)}(\eta)/B'(\eta)^4 = (1 - e^{\eta})^4(e^{2\eta} + 4e^{\eta} + 1)/r^3e^{3\eta}$, which is positive and strictly decreasing on $\eta < 0$. Moreover, since $d_L/(d_L + r) \leq e^{\eta_t(\theta)} < 1$, it holds that

$$\frac{B^{(3)}(\eta_t(\theta))}{B'(\eta_t(\theta))^4} \leq \frac{6(1 - d_L/(d_L + r))^4}{r^3(d_L/(d_L + r))^3} = \frac{6r}{d_L^3(d_L + r)}$$

and $B^{(3)}(\tilde{\eta}_t(\theta))/B'(\tilde{\eta}_t(\theta))^4$ also has the same upper bound. Hence, **(A11)** is satisfied.

In addition to the above models, general integer-valued time series models also include nonlinear models, such as the integer-valued threshold GARCH (INTGARCH) model:

$$Y_t|\mathcal{F}_{t-1} \sim \text{Poisson}(X_t), \quad X_t = d + aX_{t-1} + b_1 \max(Y_{t-1} - l, 0) + b_2 \min(Y_{t-1}, l),$$

where $\theta = (d, a, b_1, b_2)^T \in \Theta \subset (0, \infty) \times [0, \infty)^3$ with $a + \max(b_1, b_2) < 1$, Θ is compact, and l is a non-negative integer value. For more details, see Remark 3 in Kim and Lee [22].

2.2. DPD-Based Change Point Test

As a robust test for parameter changes in general integer-valued time series models, we propose a DPD-based test for the following hypotheses:

$$H_0 : \theta \text{ does not change over } Y_1, \cdots, Y_n \text{ vs. } H_1 : \text{not } H_0.$$

To construct the test, we employ the objective function of the MDPDE. That is, our test is constructed using the empirical version of the DPD. Let $\tilde{L}_{\alpha,n}$ be that in (2). To implement our test, we employ the following test statistic:

$$\hat{T}_n^\alpha := \max_{1 \leq k \leq n} \frac{k^2}{n} \frac{\partial \tilde{L}_{\alpha,k}(\hat{\theta}_{\alpha,n})}{\partial \theta^T} \hat{K}_\alpha^{-1} \frac{\partial \tilde{L}_{\alpha,k}(\hat{\theta}_{\alpha,n})}{\partial \theta},$$

where

$$\hat{K}_\alpha = \frac{1}{n} \sum_{t=1}^n \frac{\partial \tilde{l}_{\alpha,t}(\hat{\theta}_{\alpha,n})}{\partial \theta} \frac{\partial \tilde{l}_{\alpha,t}(\hat{\theta}_{\alpha,n})}{\partial \theta^T}$$

is a consistent estimator of K_α. For the consistency of $\hat{K}_\alpha$, see Lemma A5 in Appendix A.

Using the mean value theorem (MVT), we have the following, for each $s \in [0, 1]$,

$$\frac{[ns]}{\sqrt{n}} \frac{\partial \tilde{L}_{\alpha,[ns]}(\hat{\theta}_{\alpha,n})}{\partial \theta} = \frac{[ns]}{\sqrt{n}} \frac{\partial \tilde{L}_{\alpha,[ns]}(\theta_0)}{\partial \theta} + \frac{[ns]}{n} \frac{\partial^2 \tilde{L}_{\alpha,[ns]}(\theta^*_{\alpha,n,s})}{\partial \theta \partial \theta^T} \sqrt{n}(\hat{\theta}_{\alpha,n} - \theta_0), \tag{4}$$

where $\theta^*_{\alpha,n,s}$ is an intermediate point between $\hat{\theta}_{\alpha,n}$ and θ_0. From $\partial \tilde{L}_{\alpha,n}(\hat{\theta}_{\alpha,n})/\partial \theta = 0$, we obtain that, for $s = 1$,

$$0 = \sqrt{n} \frac{\partial \tilde{L}_{\alpha,n}(\theta_0)}{\partial \theta} + \frac{\partial^2 \tilde{L}_{\alpha,n}(\theta^*_{\alpha,n,1})}{\partial \theta \partial \theta^T} \sqrt{n}(\hat{\theta}_{\alpha,n} - \theta_0).$$

Furthermore, since J_α is nonsingular (cf. proof of Lemma 7 in Kim and Lee [22]), this can be expressed as

$$\begin{aligned}
\sqrt{n}(\hat{\theta}_{\alpha,n} - \theta_0) &= J_\alpha^{-1} \sqrt{n} \frac{\partial \tilde{L}_{\alpha,n}(\theta_0)}{\partial \theta} + J_\alpha^{-1} \frac{\partial^2 \tilde{L}_{\alpha,n}(\theta^*_{\alpha,n,1})}{\partial \theta \partial \theta^T} \sqrt{n}(\hat{\theta}_{\alpha,n} - \theta_0) + \sqrt{n}(\hat{\theta}_{\alpha,n} - \theta_0) \\
&= J_\alpha^{-1} \sqrt{n} \frac{\partial \tilde{L}_{\alpha,n}(\theta_0)}{\partial \theta} + J_\alpha^{-1} \left(\frac{\partial^2 \tilde{L}_{\alpha,n}(\theta^*_{\alpha,n,1})}{\partial \theta \partial \theta^T} + J_\alpha \right) \sqrt{n}(\hat{\theta}_{\alpha,n} - \theta_0).
\end{aligned}$$

Substituting the above into (4) yields

$$
\frac{[ns]}{\sqrt{n}} \frac{\partial \widetilde{L}_{\alpha,[ns]}(\hat{\theta}_{\alpha,n})}{\partial \theta} = \frac{[ns]}{\sqrt{n}} \frac{\partial \widetilde{L}_{\alpha,[ns]}(\theta_0)}{\partial \theta} + \frac{[ns]}{n} \frac{\partial^2 \widetilde{L}_{\alpha,[ns]}(\theta^*_{\alpha,n,s})}{\partial \theta \partial \theta^T} J_\alpha^{-1} \sqrt{n} \frac{\partial \widetilde{L}_{\alpha,n}(\theta_0)}{\partial \theta}
$$

$$
+ \frac{[ns]}{n} \frac{\partial^2 \widetilde{L}_{\alpha,[ns]}(\theta^*_{\alpha,n,s})}{\partial \theta \partial \theta^T} J_\alpha^{-1} \left(\frac{\partial^2 \widetilde{L}_{\alpha,n}(\theta^*_{\alpha,n,1})}{\partial \theta \partial \theta^T} + J_\alpha \right) \sqrt{n}(\hat{\theta}_{\alpha,n} - \theta_0). \tag{5}
$$

In Appendix A, we show that the first two terms on the right-hand side of (5) converge weakly to $K_\alpha^{1/2} B_d^o(s)$, where B_d^o is a d-dimensional standard Brownian bridge and the last term is asymptotically negligible. Therefore, we obtain the following theorem.

Theorem 1. *Suppose that conditions* **(A0)–(A11)** *hold. Then, under* H_0, *we have*

$$
K_\alpha^{-1/2} \frac{[ns]}{\sqrt{n}} \frac{\partial \widetilde{L}_{\alpha,[ns]}(\hat{\theta}_{\alpha,n})}{\partial \theta} \xrightarrow{w} B_d^o(s).
$$

Therefore,

$$
\widehat{T}_n^\alpha \xrightarrow{d} \sup_{0 \le s \le 1} \| B_d^o(s) \|^2.
$$

We reject H_0 if $\widehat{T}_n^\alpha$ is large; see Table 1 of Lee et al. [32] for the critical values. When a change point is detected, its location is estimated as

$$
\underset{1 \le k \le n}{\mathrm{argmax}} \, \frac{k^2}{n} \frac{\partial \widetilde{L}_{\alpha,k}(\hat{\theta}_{\alpha,n})}{\partial \theta^T} \widehat{K}_\alpha^{-1} \frac{\partial \widetilde{L}_{\alpha,k}(\hat{\theta}_{\alpha,n})}{\partial \theta}.
$$

Remark 3. *The proposed test* $\widehat{T}_n^\alpha$ *with* $\alpha = 0$ *is the same as the score-vector-based CUSUM test proposed by Lee and Lee [14], given by*

$$
\widehat{T}_n^{score} = \max_{1 \le k \le n} \frac{1}{n} \left(\sum_{t=1}^{k} \frac{\partial \widetilde{l}_{0,t}(\hat{\theta}_{0,n})}{\partial \theta^T} \right) \widehat{I}_n^{-1} \left(\sum_{t=1}^{k} \frac{\partial \widetilde{l}_{0,t}(\hat{\theta}_{0,n})}{\partial \theta} \right),
$$

where $\widetilde{l}_{0,t}(\theta)$ *is defined in* (3), $\hat{\theta}_{0,n}$ *is the CMLE, and* $\widehat{I}_n = n^{-1} \sum_{t=1}^{n} \partial^2 \widetilde{l}_{0,t}(\hat{\theta}_{0,n}) / \partial \theta \partial \theta^T$. *In the next section, we compare the performance of* $\widehat{T}_n^\alpha$ *with that of* $\widehat{T}_n^{score}$ *in the presence of outliers.*

3. Empirical Studies

3.1. Simulation

In this section, we evaluate the performance of the proposed test $\widehat{T}_n^\alpha$ (with $\alpha > 0$) through simulations, focusing on the comparison with $\widehat{T}_n^{score}$. First, we consider the Poisson INGARCH models:

$$
Y_t | \mathcal{F}_{t-1} \sim \mathrm{Poisson}(X_t), \quad X_t = d + a X_{t-1} + b Y_{t-1}, \tag{6}
$$

where X_1 is set to 0 for the data generation and $\widetilde{X}_1$ is set as the sample mean of the data. The sample sizes considered are $n = 500$ and 1000, with 1000 repetitions for each simulation. For the comparison, we examine the empirical size and power at the nominal level of 0.05, which has a corresponding critical value of 3.004. To calculate the empirical size and power for each test, we consider cases with $\theta = (d, a, b) = (1, 0.2, 0.2), (1, 0.2, 0.4), (1, 0.2, 0.7)$ and those in which $\theta = (d, a, b) = (1, 0.2, 0.2)$ changes to $\theta' = (d', a', b') = (1.5, 0.2, 0.2), (1, 0.4, 0.2), (1, 0.2, 0.4)$ at the middle time $t = [n/2]$, respectively.

Table 1 presents the results when the data are not contaminated by outliers, showing that both tests ($\widehat{T}_n^{score}$ and $\widehat{T}_n^\alpha$) exhibit reasonable size, even when $a + b$ is close to 1. When $n = 500$, $\widehat{T}_n^{score}$ outperforms $\widehat{T}_n^\alpha$ in terms of power; however, as the sample size increases to $n = 1000$, $\widehat{T}_n^\alpha$ exhibits similar power to that of $\widehat{T}_n^{score}$, particularly when α is small. The power of $\widehat{T}_n^\alpha$ tends to decrease as α increases, confirming that an MDPDE with large α results in a loss of efficiency.

Table 1. Empirical sizes and powers for Poisson integer-valued generalized autoregressive conditional heteroscedastic (INGARCH)(1,1) models when no outliers exist.

				$\widehat{T}_n^\alpha$ with α				
	$\theta = (d, a, b)$	n	$\widehat{T}_n^{score}$	$\alpha = 0.1$	$\alpha = 0.2$	$\alpha = 0.3$	$\alpha = 0.5$	$\alpha = 1$
	$(1, 0.2, 0.2)$	500	0.084	0.053	0.059	0.059	0.058	0.059
		1000	0.065	0.047	0.053	0.053	0.051	0.059
Sizes	$(1, 0.2, 0.4)$	500	0.049	0.040	0.043	0.045	0.047	0.047
		1000	0.033	0.039	0.045	0.047	0.050	0.053
	$(1, 0.2, 0.7)$	500	0.031	0.028	0.030	0.029	0.029	0.034
		1000	0.050	0.051	0.047	0.044	0.046	0.051
	$\theta' = (d', a', b')$	n	$\theta = (d, a, b) = (1, 0.2, 0.2)$ changes to $\theta' = (d', a', b')$					
	$(1.5, 0.2, 0.2)$	500	0.836	0.776	0.764	0.741	0.687	0.525
		1000	0.912	0.914	0.911	0.910	0.901	0.871
Powers	$(1, 0.4, 0.2)$	500	0.782	0.704	0.695	0.661	0.591	0.454
		1000	0.951	0.942	0.939	0.937	0.917	0.886
	$(1, 0.2, 0.4)$	500	0.819	0.804	0.800	0.795	0.736	0.634
		1000	0.996	0.996	0.996	0.993	0.991	0.978

To evaluate the robustness of the proposed test, we assume that contaminated data $Y_{c,t}$ are observed instead of Y_t in (6) (cf. Fried et al. [33]):

$$Y_{c,t} = Y_t + P_t Y_{o,t},\tag{7}$$

where P_t are independent and identically distributed (iid) Bernoulli random variables with success probability p and $Y_{o,t}$ are iid Poisson random variables with mean γ. We assume that Y_t, P_t, and $Y_{o,t}$ are all independent. In this simulation, we consider the cases $p = 0.01, 0.03$ and $\gamma = 5, 10$. The results are reported in Tables 2–5, showing that $\widehat{T}_n^{score}$ suffers from size distortions that become more severe as either p or γ increase. In contrast, $\widehat{T}_n^\alpha$ compensates for this defect remarkably well, yielding comparable power to that of $\widehat{T}_n^{score}$ when $n = 1000$. This indicates that as more data are contaminated by outliers, $\widehat{T}_n^\alpha$ increasingly outperforms $\widehat{T}_n^{score}$.

Table 2. Empirical sizes and powers for Poisson INGARCH(1,1) models when $p = 0.01$ and $\gamma = 5$.

	$\theta = (d, a, b)$	n	$\widehat{T}_n^{score}$	$\widehat{T}_n^\alpha$ with α				
				$\alpha = 0.1$	$\alpha = 0.2$	$\alpha = 0.3$	$\alpha = 0.5$	$\alpha = 1$
Sizes	(1, 0.2, 0.2)	500	0.108	0.048	0.046	0.046	0.052	0.057
		1000	0.110	0.048	0.044	0.041	0.050	0.053
	(1, 0.2, 0.4)	500	0.070	0.041	0.041	0.046	0.046	0.049
		1000	0.078	0.041	0.042	0.041	0.045	0.043
	(1, 0.2, 0.7)	500	0.057	0.035	0.039	0.038	0.045	0.045
		1000	0.061	0.041	0.042	0.045	0.044	0.049
	$\theta' = (d', a', b')$	n	$\theta = (d, a, b) = (1, 0.2, 0.2)$ changes to $\theta' = (d', a', b')$					
Powers	(1.5, 0.2, 0.2)	500	0.792	0.736	0.735	0.723	0.676	0.569
		1000	0.901	0.898	0.903	0.903	0.896	0.856
	(1, 0.4, 0.2)	500	0.766	0.684	0.686	0.667	0.626	0.525
		1000	0.944	0.934	0.935	0.931	0.915	0.864
	(1, 0.2, 0.4)	500	0.871	0.806	0.804	0.787	0.752	0.647
		1000	0.997	0.993	0.993	0.992	0.990	0.960

Table 3. Empirical sizes and powers for Poisson INGARCH(1,1) models when $p = 0.01$ and $\gamma = 10$.

	$\theta = (d, a, b)$	n	$\widehat{T}_n^{score}$	$\widehat{T}_n^\alpha$ with α				
				$\alpha = 0.1$	$\alpha = 0.2$	$\alpha = 0.3$	$\alpha = 0.5$	$\alpha = 1$
Sizes	(1, 0.2, 0.2)	500	0.246	0.069	0.075	0.070	0.069	0.079
		1000	0.317	0.071	0.062	0.070	0.070	0.062
	(1, 0.2, 0.4)	500	0.234	0.053	0.060	0.061	0.052	0.051
		1000	0.262	0.059	0.070	0.072	0.071	0.060
	(1, 0.2, 0.7)	500	0.127	0.040	0.040	0.037	0.041	0.038
		1000	0.115	0.045	0.044	0.049	0.048	0.050
	$\theta' = (d', a', b')$	n	$\theta = (d, a, b) = (1, 0.2, 0.2)$ changes to $\theta' = (d', a', b')$					
Powers	(1.5, 0.2, 0.2)	500	0.840	0.785	0.791	0.769	0.742	0.649
		1000	0.874	0.863	0.874	0.869	0.868	0.862
	(1, 0.4, 0.2)	500	0.835	0.743	0.759	0.740	0.694	0.590
		1000	0.911	0.910	0.913	0.908	0.902	0.879
	(1, 0.2, 0.4)	500	0.920	0.829	0.835	0.831	0.787	0.697
		1000	0.997	0.992	0.995	0.997	0.994	0.965

Table 4. Empirical sizes and powers for Poisson INGARCH(1,1) models when $p = 0.03$ and $\gamma = 5$.

	$\theta = (d, a, b)$	n	$\widehat{T}_n^{score}$	$\widehat{T}_n^\alpha$ with α				
				$\alpha = 0.1$	$\alpha = 0.2$	$\alpha = 0.3$	$\alpha = 0.5$	$\alpha = 1$
Sizes	(1, 0.2, 0.2)	500	0.213	0.060	0.059	0.058	0.062	0.074
		1000	0.229	0.052	0.055	0.063	0.062	0.061
	(1, 0.2, 0.4)	500	0.176	0.052	0.057	0.064	0.066	0.060
		1000	0.173	0.047	0.055	0.054	0.055	0.059
	(1, 0.2, 0.7)	500	0.073	0.030	0.039	0.037	0.037	0.045
		1000	0.086	0.039	0.035	0.040	0.042	0.039
	$\theta' = (d', a', b')$	n	$\theta = (d, a, b) = (1, 0.2, 0.2)$ changes to $\theta' = (d', a', b')$					
Powers	(1.5, 0.2, 0.2)	500	0.804	0.693	0.715	0.709	0.687	0.616
		1000	0.867	0.859	0.867	0.867	0.859	0.847
	(1, 0.4, 0.2)	500	0.786	0.662	0.693	0.681	0.634	0.561
		1000	0.908	0.896	0.903	0.899	0.893	0.868
	(1, 0.2, 0.4)	500	0.915	0.787	0.797	0.792	0.773	0.672
		1000	0.998	0.994	0.995	0.993	0.986	0.965

Table 5. Empirical sizes and powers for Poisson INGARCH(1,1) models when $p = 0.03$ and $\gamma = 10$.

| | $\theta = (d, a, b)$ | n | $\widehat{T}_n^{score}$ | $\widehat{T}_n^{\alpha}$ with α | | | | |
				$\alpha = 0.1$	$\alpha = 0.2$	$\alpha = 0.3$	$\alpha = 0.5$	$\alpha = 1$
	(1, 0.2, 0.2)	500	0.475	0.083	0.082	0.083	0.091	0.102
		1000	0.592	0.092	0.097	0.104	0.109	0.097
Sizes	(1, 0.2, 0.4)	500	0.556	0.071	0.078	0.080	0.068	0.065
		1000	0.621	0.092	0.113	0.115	0.108	0.071
	(1, 0.2, 0.7)	500	0.296	0.050	0.056	0.056	0.053	0.040
		1000	0.289	0.060	0.062	0.057	0.060	0.055
	$\theta' = (d', a', b')$	n	$\theta = (d, a, b) = (1, 0.2, 0.2)$ changes to $\theta' = (d', a', b')$					
	(1.5, 0.2, 0.2)	500	0.834	0.760	0.800	0.801	0.782	0.719
		1000	0.889	0.821	0.852	0.867	0.860	0.866
Powers	(1, 0.4, 0.2)	500	0.850	0.738	0.783	0.786	0.759	0.688
		1000	0.897	0.848	0.887	0.889	0.895	0.880
	(1, 0.2, 0.4)	500	0.951	0.817	0.847	0.842	0.815	0.728
		1000	0.997	0.991	0.992	0.992	0.983	0.969

Next, we consider the following NB-INGARCH(1,1) models:

$$Y_t | \mathcal{F}_{t-1} \sim \text{NB}(r, p_t), \quad X_t = \frac{r(1 - p_t)}{p_t} = d + a X_{t-1} + b Y_{t-1}, \tag{8}$$

where X_1 and $\widetilde{X}_1$ are 0 and the sample mean of the data, respectively. We set $r = 10$, and use the same parameter settings as in the Poisson INGARCH model case. In order to evaluate the robustness of the test, we observe contaminated data $Y_{c,t}$, as in (7), where Y_t are generated from (8), P_t are iid Bernoulli random variables with success probability p, and $Y_{o,t}$ are iid NB$(10, \kappa)$ random variables. We consider the cases $p = 0.01$, 0.03 and $\kappa = 0.6$, 0.5. The results are reported in Tables 6–10, showing similar results to those in Tables 1–5. Our findings show that the DPD-based test performs reasonably well in terms of both size and power, regardless of the existence of outliers. In addition, we confirm that the proposed test outperforms the score-based CUSUM test when the data are contaminated by outliers.

Table 6. Empirical sizes and powers for negative binomial INGARCH (NB-INGARCH)(1,1) models when no outliers exist.

| | $\theta = (d, a, b)$ | n | $\widehat{T}_n^{score}$ | $\widehat{T}_n^{\alpha}$ with α | | | | |
				$\alpha = 0.1$	$\alpha = 0.2$	$\alpha = 0.3$	$\alpha = 0.5$	$\alpha = 1$
	(1, 0.2, 0.2)	500	0.076	0.050	0.052	0.054	0.061	0.071
		1000	0.061	0.055	0.052	0.052	0.055	0.059
Sizes	(1, 0.2, 0.4)	500	0.040	0.041	0.038	0.040	0.045	0.048
		1000	0.049	0.053	0.056	0.057	0.062	0.060
	(1, 0.2, 0.7)	500	0.047	0.046	0.043	0.038	0.042	0.043
		1000	0.041	0.044	0.048	0.048	0.047	0.043
	$\theta' = (d', a', b')$	n	$\theta = (d, a, b) = (1, 0.2, 0.2)$ changes to $\theta' = (d', a', b')$					
	(1.5, 0.2, 0.2)	500	0.821	0.759	0.735	0.706	0.640	0.505
		1000	0.953	0.942	0.936	0.932	0.919	0.881
Powers	(1, 0.4, 0.2)	500	0.759	0.689	0.646	0.611	0.558	0.454
		1000	0.967	0.964	0.959	0.955	0.940	0.881
	(1, 0.2, 0.4)	500	0.733	0.719	0.718	0.702	0.650	0.544
		1000	0.984	0.984	0.981	0.975	0.961	0.908

Table 7. Empirical sizes and powers for NB-INGARCH(1,1) models when $p = 0.01$ and $\kappa = 0.6$.

	$\theta = (d, a, b)$	n	$\widehat{T}_n^{score}$	$\alpha = 0.1$	$\alpha = 0.2$	$\alpha = 0.3$	$\alpha = 0.5$	$\alpha = 1$
					$\widehat{T}_n^{\alpha}$ with α			
	$(1, 0.2, 0.2)$	500	0.158	0.062	0.066	0.066	0.071	0.071
		1000	0.173	0.069	0.066	0.067	0.068	0.061
Sizes	$(1, 0.2, 0.4)$	500	0.105	0.045	0.045	0.049	0.047	0.039
		1000	0.112	0.058	0.058	0.062	0.057	0.047
	$(1, 0.2, 0.7)$	500	0.045	0.031	0.035	0.038	0.041	0.038
		1000	0.065	0.042	0.045	0.044	0.041	0.045
	$\theta' = (d', a', b')$	n	$\theta = (d, a, b) = (1, 0.2, 0.2)$ changes to $\theta' = (d', a', b')$					
	$(1.5, 0.2, 0.2)$	500	0.803	0.705	0.714	0.695	0.647	0.516
		1000	0.945	0.931	0.931	0.930	0.921	0.909
Powers	$(1, 0.4, 0.2)$	500	0.757	0.648	0.645	0.626	0.579	0.464
		1000	0.959	0.958	0.952	0.947	0.930	0.895
	$(1, 0.2, 0.4)$	500	0.807	0.704	0.716	0.710	0.659	0.574
		1000	0.985	0.978	0.980	0.979	0.969	0.935

Table 8. Empirical sizes and powers for NB-INGARCH(1,1) models when $p = 0.01$ and $\kappa = 0.5$.

	$\theta = (d, a, b)$	n	$\widehat{T}_n^{score}$	$\alpha = 0.1$	$\alpha = 0.2$	$\alpha = 0.3$	$\alpha = 0.5$	$\alpha = 1$
					$\widehat{T}_n^{\alpha}$ with α			
	$(1, 0.2, 0.2)$	500	0.258	0.069	0.069	0.070	0.076	0.080
		1000	0.292	0.061	0.061	0.057	0.058	0.068
Sizes	$(1, 0.2, 0.4)$	500	0.177	0.048	0.048	0.052	0.057	0.058
		1000	0.236	0.072	0.079	0.081	0.073	0.074
	$(1, 0.2, 0.7)$	500	0.095	0.048	0.054	0.058	0.060	0.055
		1000	0.097	0.049	0.050	0.050	0.050	0.051
	$\theta' = (d', a', b')$	n	$\theta = (d, a, b) = (1, 0.2, 0.2)$ changes to $\theta' = (d', a', b')$					
	$(1.5, 0.2, 0.2)$	500	0.840	0.771	0.768	0.740	0.688	0.599
		1000	0.923	0.924	0.932	0.926	0.925	0.897
Powers	$(1, 0.4, 0.2)$	500	0.808	0.704	0.709	0.673	0.634	0.536
		1000	0.938	0.946	0.946	0.943	0.935	0.898
	$(1, 0.2, 0.4)$	500	0.842	0.723	0.740	0.735	0.696	0.586
		1000	0.997	0.989	0.984	0.977	0.972	0.923

Table 9. Empirical sizes and powers for NB-INGARCH(1,1) models when $p = 0.03$ and $\kappa = 0.6$.

	$\theta = (d, a, b)$	n	$\widehat{T}_n^{score}$	$\alpha = 0.1$	$\alpha = 0.2$	$\alpha = 0.3$	$\alpha = 0.5$	$\alpha = 1$
					$\widehat{T}_n^{\alpha}$ with α			
	$(1, 0.2, 0.2)$	500	0.289	0.079	0.077	0.076	0.086	0.069
		1000	0.328	0.060	0.068	0.068	0.077	0.075
Sizes	$(1, 0.2, 0.4)$	500	0.228	0.051	0.054	0.051	0.052	0.047
		1000	0.246	0.054	0.064	0.066	0.064	0.059
	$(1, 0.2, 0.7)$	500	0.090	0.035	0.040	0.040	0.044	0.036
		1000	0.108	0.058	0.053	0.052	0.050	0.040
	$\theta' = (d', a', b')$	n	$\theta = (d, a, b) = (1, 0.2, 0.2)$ changes to $\theta' = (d', a', b')$					
	$(1.5, 0.2, 0.2)$	500	0.818	0.685	0.705	0.702	0.675	0.582
		1000	0.925	0.892	0.900	0.899	0.905	0.909
Powers	$(1, 0.4, 0.2)$	500	0.806	0.637	0.666	0.664	0.627	0.522
		1000	0.938	0.927	0.926	0.922	0.913	0.896
	$(1, 0.2, 0.4)$	500	0.870	0.690	0.734	0.731	0.704	0.604
		1000	0.990	0.976	0.978	0.974	0.969	0.931

Table 10. Empirical sizes and powers for NB-INGARCH(1,1) models when $p = 0.03$ and $\kappa = 0.5$.

				$\widehat{T}_n^\alpha$ with α				
	$\theta = (d, a, b)$	n	$\widehat{T}_n^{score}$	$\alpha = 0.1$	$\alpha = 0.2$	$\alpha = 0.3$	$\alpha = 0.5$	$\alpha = 1$
	$(1, 0.2, 0.2)$	500	0.469	0.085	0.088	0.100	0.102	0.096
		1000	0.563	0.075	0.088	0.097	0.105	0.100
Sizes	$(1, 0.2, 0.4)$	500	0.506	0.068	0.071	0.076	0.081	0.072
		1000	0.532	0.089	0.096	0.101	0.089	0.078
	$(1, 0.2, 0.7)$	500	0.188	0.054	0.066	0.072	0.066	0.061
		1000	0.207	0.053	0.051	0.064	0.069	0.059
	$\theta' = (d', a', b')$	n	$\theta = (d, a, b) = (1, 0.2, 0.2)$ changes to $\theta' = (d', a', b')$					
	$(1.5, 0.2, 0.2)$	500	0.879	0.749	0.784	0.797	0.758	0.687
		1000	0.930	0.880	0.889	0.893	0.889	0.886
Powers	$(1, 0.4, 0.2)$	500	0.867	0.698	0.766	0.756	0.734	0.636
		1000	0.948	0.891	0.900	0.906	0.906	0.889
	$(1, 0.2, 0.4)$	500	0.927	0.735	0.770	0.770	0.743	0.639
		1000	0.995	0.977	0.984	0.981	0.971	0.944

3.2. Real Data Analysis

In this section, we demonstrate the validity of $\widehat{T}_n^\alpha$ using a real data analysis. To this end, we analyze the return times of extreme events related to GS stock, which are constructed based on the daily log-returns for the period of 5 May 1999 to 15 March 2012. Davis and Liu [12] and Kim and Lee [22] previously investigated this data set in their works on geometric INGARCH(1,1) models (i.e., NB-INGARCH(1,1) models with $r = 1$).

We first compute the hitting times, $\tau_1, \tau_2, \ldots$, for which the log-returns of GS stock fall outside the 0.05 and 0.95 quantiles of the data. The return times of these extreme events are calculated as $Y_t = \tau_t - \tau_{t-1}$. Figure 1 plots Y_t, $t = 1, \ldots, 323$. The figure shows that the data include large observations; for example, a sample variance of 1106 with a sample mean of 10.01 indicates the existence of aberrant observations.

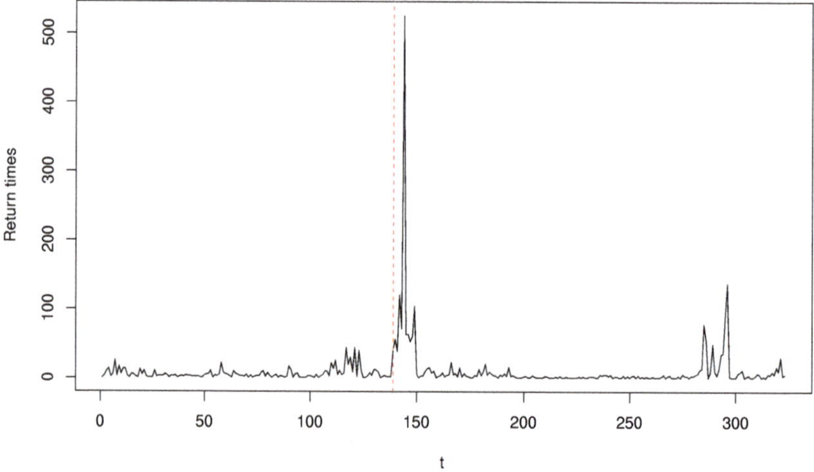

Figure 1. Plot of the return times of extreme events for Goldman Sachs Group (GS) stock.

Since $Y_t \geq 1$, we consider a geometric distribution that counts the total number of trials, rather than the number of failures, to fit the following geometric INGARCH(1,1) models to the data:

$$Y_t | \mathcal{F}_{t-1} \sim \text{Geo}(p_t), \quad X_t = \frac{1}{p_t} = d + aX_{t-1} + bY_{t-1},$$

where $\widetilde{X}_1$ is set as the sample mean of the data. Kim and Lee [22] showed that the optimal α for the MDPDE is 0.25, using the criterion provided in Remark 1. The results for the parameter estimation are summarized in Table 11 for $\alpha = 0$ (CMLE) and 0.25 (MDPDE with optimal α); figures in parentheses denote the standard errors of the corresponding estimates. We observe that, compared with the CMLE, the MDPDE with $\alpha = 0.25$ is quite different and has smaller standard errors.

Table 11. Parameter estimates for geometric INGARCH(1,1) models.

α	$\hat{d}$	$\hat{a}$	$\hat{b}$	$\widehat{\text{AMSE}}$
0(CMLE)	0.526(0.406)	0.490(0.175)	0.483(0.156)	0.623
0.25	0.432(0.242)	0.518(0.129)	0.418(0.115)	0.398

Next, we use $\widehat{T}_n^{score}$ and $\widehat{T}_n^{0.25}$ ($\widehat{T}_n^{\alpha}$ with $\alpha = 0.25$) to perform a parameter change test at the nominal level of 0.05 (the corresponding critical value is 3.004). Let $\widehat{T}_n^{score} = \max_{1 \leq k \leq n} SCORE_{k,n}$ and $\widehat{T}_n^{0.25} = \max_{1 \leq k \leq n} DPD_{k,n}$. The left and right panels of Figure 2 display $SCORE_{k,n}$ and $DPD_{k,n}$, respectively. For most k, $DPD_{k,n}$ appears to be smaller than $SCORE_{k,n}$, which is definitely attributed to the robustness of the MDPDE and DPD. We obtain $\widehat{T}_n^{score} = 5.136$, which suggests the existence of a parameter change. In Figures 1 and 2, the red, vertical, dashed line represents the location of a change when $\widehat{T}_n^{score}$ is applied. However, this result is not so reliable because $\widehat{T}_n^{score}$ can signal a change point affected by outliers as seen in the previous section, and the change point is truly detected at the occurrence time of an outlier in this case. In contrast, $\widehat{T}_n^{0.25}$ yields a value of 1.219, indicating that no change point exists. This result clearly demonstrates that outliers can severely affect parameter estimates and change point tests by mistakenly identifying a change point. Our findings confirm that the DPD-based change point test provides a functional and robust alternative to the score-based CUSUM test in the presence of outliers.

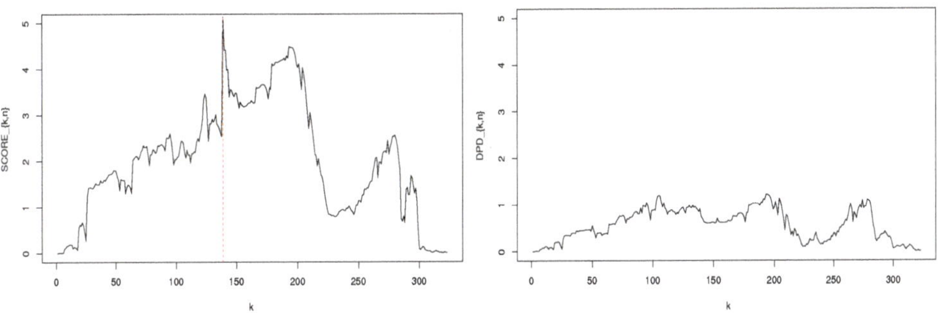

Figure 2. Plots of $SCORE_{k,n}$ and $DPD_{k,n}$.

4. Conclusions

In this study, we developed a DPD-based robust change point test for general integer-valued time series models with a conditional distribution that belongs to the one-parameter exponential family. We provided regularity conditions under which the proposed test converges weakly to the function of a Brownian bridge. The simulation study showed that the DPD-based test produces reasonable sizes and powers regardless of the existence of outliers, whereas the score-based CUSUM test suffers from severe size distortions when the data are contaminated by outliers. In the real data analysis using the return times of extreme events related to GS stock, the score-based CUSUM test supported the existence a parameter change, due to the influence of outliers, while the DPD-based test did not detect a change point because of its robust property. This result confirms the validity of the proposed test as a robust test in practice. It is noteworthy that the DPD-based test can be feasibly extended to other parametric models as far as the asymptotic properties of the MDPDE for the models are validated. We leave the issue of extension to other models as our future study.

Author Contributions: Conceptualization, B.K. and S.L.; software, B.K.; methodology, B.K. and S.L.; formal analysis, B.K. and S.L.; data curation, B.K.; writing—original draft preparation, B.K. and S.L.; funding acquisition, B.K. and S.L. All authors have read and agreed to the published version of the manuscript.

Funding: This work was supported by the National Research Foundation of Korea (NRF) grant funded by the Korea government (MSIT) (NRF-2019R1C1C1004662) (B. Kim) and (NRF-2018R1A2A2A05019433) (S. Lee).

Conflicts of Interest: The authors declare no conflict of interest.

Appendix A

In this appendix, we prove Theorem 1 for $\alpha > 0$; refer to Lee and Lee [14] for the case of $\alpha = 0$. The following properties of the probability mass function of the non-negative integer-valued exponential family are useful for proving Lemma A1. For all $y \in \mathbb{N}_0$ and $\eta \in \mathbb{R}$:

(E1) $0 < p(y|\eta) < 1$,
(E2) $\sum_{y=0}^{\infty} p(y|\eta) = 1$,
(E3) $\sum_{y=0}^{\infty} y p(y|\eta) = B(\eta)$,
(E4) $\sum_{y=0}^{\infty} y^2 p(y|\eta) = B'(\eta) + B(\eta)^2$,
(E5) $\sum_{y=0}^{\infty} y^3 p(y|\eta) = B''(\eta) + 3B'(\eta)B(\eta) + B(\eta)^3$.

Throughout this section, we denote $L_{\alpha,n}(\theta) = n^{-1} \sum_{t=1}^{n} l_{\alpha,t}(\theta)$ and employ the notation $\eta_t = \eta_t(\theta)$, $\tilde{\eta}_t = \tilde{\eta}_t(\theta)$, and $\eta_t^0 = \eta_t(\theta_0)$ for brevity. Furthermore, if we define two functions $h_\alpha(\eta)$ and $m_\alpha(\eta)$ as

$$h_\alpha(\eta) = \sum_{y=0}^{\infty} p(y|\eta)^{1+\alpha} \frac{y - B(\eta)}{B'(\eta)} - p(Y_t|\eta)^\alpha \frac{Y_t - B(\eta)}{B'(\eta)},$$

$$m_\alpha(\eta) = \sum_{y=0}^{\infty} p(y|\eta)^{1+\alpha} \left[(1+\alpha) \left(\frac{y - B(\eta)}{B'(\eta)} \right)^2 - \frac{B''(\eta)}{B'(\eta)^2} \frac{y - B(\eta)}{B'(\eta)} - \frac{1}{B'(\eta)} \right]$$
$$- p(Y_t|\eta)^\alpha \left[\alpha \left(\frac{Y_t - B(\eta)}{B'(\eta)} \right)^2 - \frac{B''(\eta)}{B'(\eta)^2} \frac{Y_t - B(\eta)}{B'(\eta)} - \frac{1}{B'(\eta)} \right],$$

we obtain

$$\frac{\partial l_{\alpha,t}(\theta)}{\partial \theta} = (1+\alpha) h_\alpha(\eta_t) \frac{\partial X_t(\theta)}{\partial \theta},$$

$$\frac{\partial^2 l_{\alpha,t}(\theta)}{\partial \theta \partial \theta^T} = (1+\alpha) \left(h_\alpha(\eta_t) \frac{\partial^2 X_t(\theta)}{\partial \theta \partial \theta^T} + m_\alpha(\eta_t) \frac{\partial X_t(\theta)}{\partial \theta} \frac{\partial X_t(\theta)}{\partial \theta^T} \right).$$

Lemma A1. *Suppose that conditions* **(A3)**, **(A6)**, *and* **(A11)** *hold. Then, we have*

$$|h_\alpha(\eta_t)| \leq \frac{1}{\underline{c}}(Y_t + 3X_t(\theta)),$$

$$|h_\alpha(\tilde{\eta}_t)| \leq \frac{1}{\underline{c}}(Y_t + 3X_t(\theta) + 3|X_t(\theta) - \tilde{X}_t(\theta)|),$$

$$|m_\alpha(\eta_t)| \leq \frac{\alpha}{\underline{c}^2}Y_t^2 + \frac{K}{\underline{c}^{1/2}}Y_t + \frac{\alpha}{\underline{c}^2}X_t(\theta)^2 + \frac{3K}{\underline{c}^{1/2}}X_t(\theta) + \frac{3+\alpha}{\underline{c}},$$

$$|h_\alpha(\eta_t) - h_\alpha(\tilde{\eta}_t)| \leq \left[\frac{\alpha}{\underline{c}^2}Y_t^2 + \frac{K}{\underline{c}^{1/2}}Y_t + \frac{2\alpha}{\underline{c}^2}\left(X_t(\theta)^2 + |X_t(\theta) - \tilde{X}_t(\theta)|^2\right)\right.$$
$$\left. + \frac{3K}{\underline{c}^{1/2}}\left(X_t(\theta) + |X_t(\theta) - \tilde{X}_t(\theta)|\right) + \frac{3+\alpha}{\underline{c}}\right]|X_t(\theta) - \tilde{X}_t(\theta)|,$$

$$|m_\alpha(\tilde{\eta}_t)| \leq \frac{\alpha}{\underline{c}^2}Y_t^2 + \frac{K}{\underline{c}^{1/2}}Y_t + \frac{2\alpha}{\underline{c}^2}\left(X_t(\theta)^2 + |X_t(\theta) - \tilde{X}_t(\theta)|^2\right)$$
$$ + \frac{3K}{\underline{c}^{1/2}}\left(X_t(\theta) + |X_t(\theta) - \tilde{X}_t(\theta)|\right) + \frac{3+\alpha}{\underline{c}},$$

$$|m_\alpha(\eta_t) - m_\alpha(\tilde{\eta}_t)| \leq \left[\frac{\alpha^2}{\underline{c}^3}Y_t^3 + \frac{3\alpha K}{\underline{c}^{3/2}}Y_t^2 + \left(\frac{3\alpha}{\underline{c}^2} + M + 3K^2\right)Y_t\right.$$
$$ + \frac{4(3\alpha^2 + 4\alpha + 2)}{\underline{c}^3}\left(X_t(\theta)^3 + |X_t(\theta) - \tilde{X}_t(\theta)|^3\right)$$
$$ + \frac{6\alpha K}{\underline{c}^{3/2}}\left(X_t(\theta)^2 + |X_t(\theta) - \tilde{X}_t(\theta)|^2\right)$$
$$ + 3\left(\frac{\alpha^2 + 5\alpha + 3}{\underline{c}^2} + M + 3K^2\right)\left(X_t(\theta) + |X_t(\theta) - \tilde{X}_t(\theta)|\right)$$
$$\left. + \frac{(\alpha^2 + 5\alpha + 8)K}{\underline{c}^{1/2}}\right]|X_t(\theta) - \tilde{X}_t(\theta)|.$$

Proof. The proofs for the first four parts of the lemma can be found in Lemma 4 of Kim and Lee [22]. The fifth part is obtained directly from the third part, together with the fact that $\widetilde{X}_t(\theta) \leq |X_t(\theta) - \widetilde{X}_t(\theta)| + X_t(\theta)$.

By the MVT, (E1)–(E5), **(A3)**, **(A6)**, and **(A11)**, we have

$$
\begin{aligned}
&|m_\alpha(\eta_t) - m_\alpha(\tilde{\eta}_t)| \\
&= \left| \frac{\partial m_\alpha(B^{-1}(X_t^*(\theta)))}{\partial X_t(\theta)} \right| |X_t(\theta) - \widetilde{X}_t(\theta)| \\
&= \left| \frac{\partial m_\alpha(\eta_t^*)}{\partial \eta_t} \frac{1}{B'(\eta_t^*)} \right| |X_t(\theta) - \widetilde{X}_t(\theta)| \\
&= \frac{1}{B'(\eta_t^*)} \left| \sum_{y=0}^{\infty} p(y|\eta_t^*)^{1+\alpha} \left[(1+\alpha)^2 \frac{1}{B'(\eta_t^*)^2}(y - B(\eta_t^*))^3 - 3(1+\alpha)\frac{B''(\eta_t^*)}{B'(\eta_t^*)^3}(y - B(\eta_t^*))^2 \right. \right. \\
&\quad \left. + \left(-3(1+\alpha)\frac{1}{B'(\eta_t^*)} - \frac{B^{(3)}(\eta_t^*)}{B'(\eta_t^*)^3} + 3\frac{B''(\eta_t^*)^2}{B'(\eta_t^*)^4} \right)(y - B(\eta_t^*)) + 2\frac{B''(\eta_t^*)}{B'(\eta_t^*)^2} \right] \\
&\quad - p(Y_t|\eta_t^*)^\alpha \left[\alpha^2 \frac{1}{B'(\eta_t^*)^2}(Y_t - B(\eta_t^*))^3 - 3\alpha\frac{B''(\eta_t^*)}{B'(\eta_t^*)^3}(Y_t - B(\eta_t^*))^2 \right. \\
&\quad \left. \left. + \left(-3\alpha\frac{1}{B'(\eta_t^*)} - \frac{B^{(3)}(\eta_t^*)}{B'(\eta_t^*)^3} + 3\frac{B''(\eta_t^*)^2}{B'(\eta_t^*)^4} \right)(Y_t - B(\eta_t^*)) + 2\frac{B''(\eta_t^*)}{B'(\eta_t^*)^2} \right] \right| |X_t(\theta) - \widetilde{X}_t(\theta)| \\
&\leq \left[(1+\alpha)^2 \frac{1}{B'(\eta_t^*)^3}\left(B''(\eta_t^*) + 3B'(\eta_t^*)B(\eta_t^*) + B(\eta_t^*)^3 + B(\eta_t^*)^3 \right) + 3(1+\alpha)\left| \frac{B''(\eta_t^*)}{B'(\eta_t^*)^3} \right| \right. \\
&\quad + \left(3(1+\alpha)\frac{1}{B'(\eta_t^*)^2} + \left| \frac{B^{(3)}(\eta_t^*)}{B'(\eta_t^*)^4} \right| + 3\frac{B''(\eta_t^*)^2}{B'(\eta_t^*)^5} \right)(B(\eta_t^*) + B(\eta_t^*)) + 2\left| \frac{B''(\eta_t^*)}{B'(\eta_t^*)^3} \right| \\
&\quad + \alpha^2 \frac{1}{B'(\eta_t^*)^3}(Y_t^3 + B(\eta_t^*)^3) + 3\alpha\left| \frac{B''(\eta_t^*)}{B'(\eta_t^*)^4} \right|(Y_t^2 + B(\eta_t^*)^2) \\
&\quad \left. + \left(3\alpha\frac{1}{B'(\eta_t^*)^2} + \left| \frac{B^{(3)}(\eta_t^*)}{B'(\eta_t^*)^4} \right| + 3\frac{B''(\eta_t^*)^2}{B'(\eta_t^*)^5} \right)(Y_t + B(\eta_t^*)) + 2\left| \frac{B''(\eta_t^*)}{B'(\eta_t^*)^3} \right| \right] |X_t(\theta) - \widetilde{X}_t(\theta)| \\
&\leq \left[\frac{\alpha^2}{\underline{c}^3}Y_t^3 + \frac{3\alpha K}{\underline{c}^{3/2}}Y_t^2 + \left(\frac{3\alpha}{\underline{c}^2} + M + 3K^2 \right)Y_t + \frac{3\alpha^2 + 4\alpha + 2}{\underline{c}^3}B(\eta_t^*)^3 + \frac{3\alpha K}{\underline{c}^{3/2}}B(\eta_t^*)^2 \right. \\
&\quad \left. + \left(\frac{3\alpha^2 + 15\alpha + 9}{\underline{c}^2} + 3M + 9K^2 \right)B(\eta_t^*) + \frac{(\alpha^2 + 5\alpha + 8)K}{\underline{c}^{1/2}} \right] |X_t(\theta) - \widetilde{X}_t(\theta)|,
\end{aligned}
$$

where $X_t^*(\theta)$ is an intermediate point between $X_t(\theta)$ and $\widetilde{X}_t(\theta)$, and $\eta_t^* = B^{-1}(X_t^*(\theta))$. Note that since B^{-1} is strictly increasing, η_t^* lies between $B^{-1}(X_t(\theta)) = \eta_t$ and $B^{-1}(\widetilde{X}_t(\theta)) = \tilde{\eta}_t$. Then, because $B(\eta_t^*) \leq B(\eta_t) + |B(\eta_t) - B(\tilde{\eta}_t)|$, the last part of the lemma is established. $\square$

Lemma A2. *Suppose that conditions* **(A0)**–**(A11)** *hold. Then, under H_0, we have as $n \to \infty$,*

$$
\frac{1}{n}\sum_{t=1}^{n} \sup_{\theta\in\Theta} \left\| \frac{\partial^2 l_{\alpha,t}(\theta)}{\partial\theta\partial\theta^T} - \frac{\partial^2 \tilde{l}_{\alpha,t}(\theta)}{\partial\theta\partial\theta^T} \right\| = o(1) \ \text{a.s.}
$$

and

$$
\frac{1}{n}\sum_{t=1}^{n} \sup_{\theta\in\Theta} \left\| \frac{\partial l_{\alpha,t}(\theta)}{\partial\theta}\frac{\partial l_{\alpha,t}(\theta)}{\partial\theta^T} - \frac{\partial \tilde{l}_{\alpha,t}(\theta)}{\partial\theta}\frac{\partial \tilde{l}_{\alpha,t}(\theta)}{\partial\theta^T} \right\| = o(1) \ \text{a.s.}
$$

Proof. It is sufficient to show that as $t \to \infty$,

$$
\sup_{\theta\in\Theta} \left\| \frac{\partial^2 l_{\alpha,t}(\theta)}{\partial\theta\partial\theta^T} - \frac{\partial^2 \tilde{l}_{\alpha,t}(\theta)}{\partial\theta\partial\theta^T} \right\| = o(1) \ \text{a.s.}
$$

and

$$\sup_{\theta\in\Theta}\left\|\frac{\partial l_{\alpha,t}(\theta)}{\partial\theta}\frac{\partial l_{\alpha,t}(\theta)}{\partial\theta^T}-\frac{\partial \tilde{l}_{\alpha,t}(\theta)}{\partial\theta}\frac{\partial \tilde{l}_{\alpha,t}(\theta)}{\partial\theta^T}\right\|=o(1)\ \text{ a.s.}$$

Note that we can write

$$\frac{1}{1+\alpha}\sup_{\theta\in\Theta}\left\|\frac{\partial^2 l_{\alpha,t}(\theta)}{\partial\theta\partial\theta^T}-\frac{\partial^2 \tilde{l}_{\alpha,t}(\theta)}{\partial\theta\partial\theta^T}\right\|$$

$$\leq \sup_{\theta\in\Theta}\left\|h_\alpha(\tilde{\eta}_t)\left(\frac{\partial^2 X_t(\theta)}{\partial\theta\partial\theta^T}-\frac{\partial^2 \tilde{X}_t(\theta)}{\partial\theta\partial\theta^T}\right)\right\|+\sup_{\theta\in\Theta}\left\|(h_\alpha(\eta_t)-h_\alpha(\tilde{\eta}_t))\frac{\partial^2 X_t(\theta)}{\partial\theta\partial\theta^T}\right\|$$

$$+\sup_{\theta\in\Theta}\left\|(m_\alpha(\eta_t)-m_\alpha(\tilde{\eta}_t))\frac{\partial X_t(\theta)}{\partial\theta}\frac{\partial X_t(\theta)}{\partial\theta^T}\right\|+\sup_{\theta\in\Theta}\left\|m_\alpha(\tilde{\eta}_t)\frac{\partial X_t(\theta)}{\partial\theta}\left(\frac{\partial X_t(\theta)}{\partial\theta^T}-\frac{\partial \tilde{X}_t(\theta)}{\partial\theta^T}\right)\right\|$$

$$+\sup_{\theta\in\Theta}\left\|m_\alpha(\tilde{\eta}_t)\left(\frac{\partial X_t(\theta)}{\partial\theta}-\frac{\partial \tilde{X}_t(\theta)}{\partial\theta}\right)\left(\frac{\partial \tilde{X}_t(\theta)}{\partial\theta^T}-\frac{\partial X_t(\theta)}{\partial\theta^T}\right)\right\|$$

$$+\sup_{\theta\in\Theta}\left\|m_\alpha(\tilde{\eta}_t)\left(\frac{\partial X_t(\theta)}{\partial\theta}-\frac{\partial \tilde{X}_t(\theta)}{\partial\theta}\right)\frac{\partial X_t(\theta)}{\partial\theta^T}\right\|$$

$$\leq \sup_{\theta\in\Theta}|h_\alpha(\tilde{\eta}_t)|\sup_{\theta\in\Theta}\left\|\frac{\partial^2 X_t(\theta)}{\partial\theta\partial\theta^T}-\frac{\partial^2 \tilde{X}_t(\theta)}{\partial\theta\partial\theta^T}\right\|+\sup_{\theta\in\Theta}|h_\alpha(\eta_t)-h_\alpha(\tilde{\eta}_t)|\sup_{\theta\in\Theta}\left\|\frac{\partial^2 X_t(\theta)}{\partial\theta\partial\theta^T}\right\|$$

$$+\sup_{\theta\in\Theta}|m_\alpha(\eta_t)-m_\alpha(\tilde{\eta}_t)|\left(\sup_{\theta\in\Theta}\left\|\frac{\partial X_t(\theta)}{\partial\theta}\right\|\right)^2+2\sup_{\theta\in\Theta}|m_\alpha(\tilde{\eta}_t)|\sup_{\theta\in\Theta}\left\|\frac{\partial X_t(\theta)}{\partial\theta}\right\|\sup_{\theta\in\Theta}\left\|\frac{\partial X_t(\theta)}{\partial\theta}-\frac{\partial \tilde{X}_t(\theta)}{\partial\theta}\right\|$$

$$+\sup_{\theta\in\Theta}|m_\alpha(\tilde{\eta}_t)|\left(\sup_{\theta\in\Theta}\left\|\frac{\partial X_t(\theta)}{\partial\theta}-\frac{\partial \tilde{X}_t(\theta)}{\partial\theta}\right\|\right)^2.$$

Using Lemma 2.1 of Straumann and Mikosch [34], together with Lemma A1, **(A2)**, **(A4)**, **(A7)**, **(A8)**, **(A10)**, and Lemma 1 of Kim and Lee [22], the right-hand side of the last inequality converges to 0 a.s. as $t\to\infty$. Hence, the first part of the lemma is verified.

Similarly, we have

$$\frac{1}{(1+\alpha)^2}\sup_{\theta\in\Theta}\left\|\frac{\partial l_{\alpha,t}(\theta)}{\partial\theta}\frac{\partial l_{\alpha,t}(\theta)}{\partial\theta^T}-\frac{\partial \tilde{l}_{\alpha,t}(\theta)}{\partial\theta}\frac{\partial \tilde{l}_{\alpha,t}(\theta)}{\partial\theta^T}\right\|$$

$$\leq \sup_{\theta\in\Theta}\left\|(h_\alpha(\eta_t)^2-h_\alpha(\tilde{\eta}_t)^2)\frac{\partial X_t(\theta)}{\partial\theta}\frac{\partial X_t(\theta)}{\partial\theta^T}\right\|+\sup_{\theta\in\Theta}\left\|h_\alpha(\tilde{\eta}_t)^2\frac{\partial X_t(\theta)}{\partial\theta}\left(\frac{\partial X_t(\theta)}{\partial\theta^T}-\frac{\partial \tilde{X}_t(\theta)}{\partial\theta^T}\right)\right\|$$

$$+\sup_{\theta\in\Theta}\left\|h_\alpha(\tilde{\eta}_t)^2\left(\frac{\partial X_t(\theta)}{\partial\theta}-\frac{\partial \tilde{X}_t(\theta)}{\partial\theta}\right)\left(\frac{\partial \tilde{X}_t(\theta)}{\partial\theta^T}-\frac{\partial X_t(\theta)}{\partial\theta^T}\right)\right\|$$

$$+\sup_{\theta\in\Theta}\left\|h_\alpha(\tilde{\eta}_t)^2\left(\frac{\partial X_t(\theta)}{\partial\theta}-\frac{\partial \tilde{X}_t(\theta)}{\partial\theta}\right)\frac{\partial X_t(\theta)}{\partial\theta^T}\right\|$$

$$\leq \sup_{\theta\in\Theta}|h_\alpha(\eta_t)-h_\alpha(\tilde{\eta}_t)|\left(\sup_{\theta\in\Theta}|h_\alpha(\eta_t)|+\sup_{\theta\in\Theta}|h_\alpha(\tilde{\eta}_t)|\right)\left(\sup_{\theta\in\Theta}\left\|\frac{\partial X_t(\theta)}{\partial\theta}\right\|\right)^2$$

$$+2\sup_{\theta\in\Theta}|h_\alpha(\tilde{\eta}_t)^2|\sup_{\theta\in\Theta}\left\|\frac{\partial X_t(\theta)}{\partial\theta}\right\|\sup_{\theta\in\Theta}\left\|\frac{\partial X_t(\theta)}{\partial\theta}-\frac{\partial \tilde{X}_t(\theta)}{\partial\theta}\right\|$$

$$+\sup_{\theta\in\Theta}|h_\alpha(\tilde{\eta}_t)^2|\left(\sup_{\theta\in\Theta}\left\|\frac{\partial X_t(\theta)}{\partial\theta}-\frac{\partial \tilde{X}_t(\theta)}{\partial\theta}\right\|\right)^2,$$

and the right-hand side of the last inequality also converges to 0 a.s. from Lemma 2.1 of Straumann and Mikosch [34]. Therefore, the lemma is asserted. $\square$

Lemma A3. *Suppose that conditions* **(A0)–(A11)** *hold. Then, under* H_0*, we have as* $n \to \infty$,

$$K_\alpha^{-1/2} \frac{[ns]}{\sqrt{n}} \frac{\partial \widetilde{L}_{\alpha,[ns]}(\theta_0)}{\partial \theta} \xrightarrow{w} B_d(s),$$

where B_d *is a d-dimensional Brownian motion.*

Proof. First, we show that K_α is nonsingular. Since $Var[h_\alpha(\eta_t^0)|\mathcal{F}_{t-1}] = Var[p(Y_t|\eta_t^0)^\alpha(Y_t - B(\eta_t^0))/B'(\eta_t^0)|\mathcal{F}_{t-1}] > 0$, we have $E(h_\alpha(\eta_t^0)^2|\mathcal{F}_{t-1}) > [E(h_\alpha(\eta_t^0)|\mathcal{F}_{t-1})]^2 = 0$. Hence, it holds that for $v \in \mathbb{R}^d / \{0\}$,

$$v^T K_\alpha v = (1+\alpha)^2 E\left[h_\alpha(\eta_t^0)^2 \left(v^T \frac{\partial X_t(\theta_0)}{\partial \theta}\right)^2\right] = (1+\alpha)^2 E\left[E(h_\alpha(\eta_t^0)^2|\mathcal{F}_{t-1}) \left(v^T \frac{\partial X_t(\theta_0)}{\partial \theta}\right)^2\right] > 0,$$

from **(A9)**, which implies that K_α is nonsingular.

Note that

$$E\left(\frac{\partial l_{\alpha,t}(\theta_0)}{\partial \theta}\bigg|\mathcal{F}_{t-1}\right) = (1+\alpha)\frac{\partial X_t(\theta_0)}{\partial \theta} E(h_\alpha(\eta_t^0)|\mathcal{F}_{t-1}) = 0,$$

and K_α is finite from Lemma 5 of Kim and Lee [22]. Since $\partial l_{\alpha,t}(\theta_0)/\partial \theta$ is stationary and ergodic, it holds from the functional central limit theorem for martingales (cf. Section 18 in Billingsley [35]) that

$$K_\alpha^{-1/2} \frac{[ns]}{\sqrt{n}} \frac{\partial L_{\alpha,[ns]}(\theta_0)}{\partial \theta} = K_\alpha^{-1/2} \frac{1}{\sqrt{n}} \sum_{t=1}^{[ns]} \frac{\partial l_{\alpha,t}(\theta_0)}{\partial \theta} \xrightarrow{w} B_d(s).$$

Furthermore, we can show that

$$\sup_{0 \leq s \leq 1} \frac{[ns]}{\sqrt{n}} \left\| \frac{\partial L_{\alpha,[ns]}(\theta_0)}{\partial \theta} - \frac{\partial \widetilde{L}_{\alpha,[ns]}(\theta_0)}{\partial \theta} \right\| \leq \frac{1}{\sqrt{n}} \sum_{t=1}^{n} \left\| \frac{\partial l_{\alpha,t}(\theta_0)}{\partial \theta} - \frac{\partial \widetilde{l}_{\alpha,t}(\theta_0)}{\partial \theta} \right\| = o(1) \text{ a.s.,}$$

from Lemma 6 of Kim and Lee [22]. Hence, the lemma is verified. $\square$

Lemma A4. *Suppose that conditions* **(A0)–(A11)** *hold. Then, under* H_0*, we have as* $n \to \infty$,

$$\max_{1 \leq k \leq n} \frac{k}{n} \left\| \frac{\partial^2 \widetilde{L}_{\alpha,k}(\bar{\theta}_{\alpha,n,k})}{\partial \theta \partial \theta^T} + J_\alpha \right\| = o(1) \text{ a.s.,}$$

where $\{\bar{\theta}_{\alpha,n,k}|1 \leq k \leq n, n \geq 1\}$ *is any double array of* Θ*-valued random vectors satisfying* $\|\bar{\theta}_{\alpha,n,k} - \theta_0\| \leq \|\hat{\theta}_{\alpha,n} - \theta_0\|$.

Proof. From Lemma 5 of Kim and Lee [22], it holds that

$$E\left(\sup_{\theta \in \Theta} \left\| \frac{\partial^2 l_{\alpha,t}(\theta)}{\partial \theta \partial \theta^T} - \frac{\partial^2 l_{\alpha,t}(\theta_0)}{\partial \theta \partial \theta^T} \right\| \right) < \infty.$$

Since $\partial^2 l_{\alpha,t}(\theta)/\partial \theta \partial \theta^T$ is continuous in θ, for any $\epsilon > 0$, we can take a neighborhood $N_\epsilon(\theta_0)$, such that

$$E\left(\sup_{\theta \in N_\epsilon(\theta_0)} \left\| \frac{\partial^2 l_{\alpha,t}(\theta)}{\partial \theta \partial \theta^T} - \frac{\partial^2 l_{\alpha,t}(\theta_0)}{\partial \theta \partial \theta^T} \right\| \right) < \epsilon \tag{A1}$$

by decreasing the neighborhood to θ_0. Since $\hat{\theta}_{\alpha,n}$ converges to θ_0 a.s. by Proposition 1, we can write that for sufficiently large n,

$$
\max_{1 \le k \le n} \frac{k}{n} \left\| \frac{\partial^2 \widetilde{L}_{\alpha,k}(\bar{\theta}_{\alpha,n,k})}{\partial\theta\partial\theta^T} + J_\alpha \right\|
$$

$$
\le \quad \max_{1 \le k \le n} \frac{k}{n} \left\| \frac{\partial^2 \widetilde{L}_{\alpha,k}(\bar{\theta}_{\alpha,n,k})}{\partial\theta\partial\theta^T} - \frac{\partial^2 L_{\alpha,k}(\bar{\theta}_{\alpha,n,k})}{\partial\theta\partial\theta^T} \right\| + \max_{1 \le k \le n} \frac{k}{n} \left\| \frac{\partial^2 L_{\alpha,k}(\bar{\theta}_{\alpha,n,k})}{\partial\theta\partial\theta^T} - \frac{\partial^2 L_{\alpha,k}(\theta_0)}{\partial\theta\partial\theta^T} \right\|
$$

$$
+ \max_{1 \le k \le n} \frac{k}{n} \left\| \frac{\partial^2 L_{\alpha,k}(\theta_0)}{\partial\theta\partial\theta^T} + J_\alpha \right\|
$$

$$
\le \quad \frac{1}{n} \sum_{t=1}^{n} \sup_{\theta \in N_\epsilon(\theta_0)} \left\| \frac{\partial^2 \widetilde{l}_{\alpha,t}(\theta)}{\partial\theta\partial\theta^T} - \frac{\partial^2 l_{\alpha,t}(\theta)}{\partial\theta\partial\theta^T} \right\| + \frac{1}{n} \sum_{t=1}^{n} \sup_{\theta \in N_\epsilon(\theta_0)} \left\| \frac{\partial^2 l_{\alpha,t}(\theta)}{\partial\theta\partial\theta^T} - \frac{\partial^2 l_{\alpha,t}(\theta_0)}{\partial\theta\partial\theta^T} \right\|
$$

$$
+ \max_{1 \le k \le n} \frac{k}{n} \left\| \frac{\partial^2 L_{\alpha,k}(\theta_0)}{\partial\theta\partial\theta^T} + J_\alpha \right\|
$$

$$
:= \quad I_n + II_n + III_n \quad \text{a.s.}
$$

By Lemma A2, $I_n = o(1)$ a.s. By using (A1) and the stationarity and ergodicity of $\partial^2 l_{\alpha,t}(\theta)/\partial\theta\partial\theta^T$, we have

$$
\lim_{n \to \infty} II_n = E\left(\sup_{\theta \in N_\epsilon(\theta_0)} \left\| \frac{\partial^2 l_{\alpha,t}(\theta)}{\partial\theta\partial\theta^T} - \frac{\partial^2 l_{\alpha,t}(\theta_0)}{\partial\theta\partial\theta^T} \right\| \right) < \epsilon \quad \text{a.s.}
$$

Finally, since $\|\partial^2 L_{\alpha,n}(\theta_0)/\partial\theta\partial\theta^T + J_\alpha\|$ converges to 0 a.s., we can show that

$$
\max_{1 \le k \le \sqrt{n}} \frac{k}{n} \left\| \frac{\partial^2 L_{\alpha,k}(\theta_0)}{\partial\theta\partial\theta^T} + J_\alpha \right\| \le \frac{1}{\sqrt{n}} \sup_{1 \le k} \left\| \frac{\partial^2 L_{\alpha,k}(\theta_0)}{\partial\theta\partial\theta^T} + J_\alpha \right\| = o(1) \quad \text{a.s.,}
$$

and

$$
\max_{\sqrt{n} \le k \le n} \frac{k}{n} \left\| \frac{\partial^2 L_{\alpha,k}(\theta_0)}{\partial\theta\partial\theta^T} + J_\alpha \right\| \le \max_{\sqrt{n} \le k \le n} \left\| \frac{\partial^2 L_{\alpha,k}(\theta_0)}{\partial\theta\partial\theta^T} + J_\alpha \right\| = o(1) \quad \text{a.s.,}
$$

which assert $III_n = o(1)$ a.s. Therefore, the lemma is established. $\square$

Proof of Theorem 1. First, we show that

$$
\frac{[ns]}{\sqrt{n}} \frac{\partial \widetilde{L}_{\alpha,[ns]}(\theta_0)}{\partial\theta} + \frac{[ns]}{n} \frac{\partial^2 \widetilde{L}_{\alpha,[ns]}(\theta^*_{\alpha,n,s})}{\partial\theta\partial\theta^T} J_\alpha^{-1} \sqrt{n} \frac{\partial \widetilde{L}_{\alpha,n}(\theta_0)}{\partial\theta} \xrightarrow{w} K_\alpha^{1/2} B_d^o(s). \tag{A2}
$$

From Lemma A3, we have

$$
\frac{[ns]}{\sqrt{n}} \frac{\partial \widetilde{L}_{\alpha,[ns]}(\theta_0)}{\partial\theta} - \frac{[ns]}{n} \sqrt{n} \frac{\partial \widetilde{L}_{\alpha,n}(\theta_0)}{\partial\theta} \xrightarrow{w} K_\alpha^{1/2} B_d^o(s).
$$

Since $\sqrt{n}\partial\widetilde{L}_{\alpha,n}(\theta_0)/\partial\theta = O_p(1)$ by Lemma A3 with $s = 1$, using Lemma A4, it holds that

$$
\sup_{0 \le s \le 1} \frac{[ns]}{n} \left\| \frac{\partial^2 \widetilde{L}_{\alpha,[ns]}(\theta^*_{\alpha,n,s})}{\partial\theta\partial\theta^T} J_\alpha^{-1} \sqrt{n} \frac{\partial \widetilde{L}_{\alpha,n}(\theta_0)}{\partial\theta} + \sqrt{n} \frac{\partial \widetilde{L}_{\alpha,n}(\theta_0)}{\partial\theta} \right\|
$$

$$
\le \quad \left\| J_\alpha^{-1} \sqrt{n} \frac{\partial \widetilde{L}_{\alpha,n}(\theta_0)}{\partial\theta} \right\| \max_{1 \le k \le n} \frac{k}{n} \left\| \frac{\partial^2 \widetilde{L}_{\alpha,k}(\theta^*_{\alpha,n,k})}{\partial\theta\partial\theta^T} + J_\alpha \right\|
$$

$$
= \quad o_p(1),
$$

where $\theta^*_{\alpha,n,k}$ denotes that corresponding to $\theta^*_{\alpha,n,s}$ when $[ns] = k$. Hence, (A2) is verified.

Next, from Lemma A4, we have

$$\sup_{0\leq s\leq 1} \frac{[ns]}{n} \left\| \frac{\partial^2 \widetilde{L}_{\alpha,[ns]}(\theta^*_{\alpha,n,s})}{\partial\theta\partial\theta^T} \right\| \leq \max_{1\leq k\leq n} \frac{k}{n} \left\| \frac{\partial^2 \widetilde{L}_{\alpha,k}(\theta^*_{\alpha,n,k})}{\partial\theta\partial\theta^T} + J_\alpha \right\| + \|J_\alpha\| = O_p(1)$$

and

$$\left\| \frac{\partial^2 \widetilde{L}_{\alpha,n}(\theta^*_{\alpha,n,1})}{\partial\theta\partial\theta^T} + J_\alpha \right\| \leq \max_{1\leq k\leq n} \frac{k}{n} \left\| \frac{\partial^2 \widetilde{L}_{\alpha,k}(\theta^*_{\alpha,n,k})}{\partial\theta\partial\theta^T} + J_\alpha \right\| = o(1) \ \text{a.s.}$$

Then, since $\sqrt{n}(\hat{\theta}_{\alpha,n} - \theta_0) = O_p(1)$ by Proposition 1, we have

$$\sup_{0\leq s\leq 1} \frac{[ns]}{n} \left\| \frac{\partial^2 \widetilde{L}_{\alpha,[ns]}(\theta^*_{\alpha,n,s})}{\partial\theta\partial\theta^T} \left(\frac{\partial^2 \widetilde{L}_{\alpha,n}(\theta^*_{\alpha,n,1})}{\partial\theta\partial\theta^T} + J_\alpha \right) \sqrt{n}(\hat{\theta}_{\alpha,n} - \theta_0) \right\| = o_p(1). \tag{A3}$$

Therefore, from (5), (A2), and (A3), the theorem is validated. $\square$

Lemma A5. *Suppose that conditions* **(A0)**–**(A11)** *hold. Then, under H_0, we have as $n \to \infty$,*

$$\frac{1}{n}\sum_{t=1}^n \frac{\partial \widetilde{l}_{\alpha,t}(\hat{\theta}_{\alpha,n})}{\partial\theta} \frac{\partial \widetilde{l}_{\alpha,t}(\hat{\theta}_{\alpha,n})}{\partial\theta^T} \xrightarrow{a.s.} K_\alpha.$$

Proof. In a similar way to Lemma A4, from Lemma 5 of Kim and Lee [22], we can also take a neighborhood $N_\epsilon(\theta_0)$, such that

$$\lim_{n\to\infty} \frac{1}{n}\sum_{t=1}^n \sup_{\theta\in N_\epsilon(\theta_0)} \left\| \frac{\partial l_{\alpha,t}(\theta)}{\partial\theta} \frac{\partial l_{\alpha,t}(\theta)}{\partial\theta^T} - \frac{\partial l_{\alpha,t}(\theta_0)}{\partial\theta} \frac{\partial l_{\alpha,t}(\theta_0)}{\partial\theta^T} \right\|$$

$$= E\left(\sup_{\theta\in N_\epsilon(\theta_0)} \left\| \frac{\partial l_{\alpha,t}(\theta)}{\partial\theta} \frac{\partial l_{\alpha,t}(\theta)}{\partial\theta^T} - \frac{\partial l_{\alpha,t}(\theta_0)}{\partial\theta} \frac{\partial l_{\alpha,t}(\theta_0)}{\partial\theta^T} \right\| \right) < \epsilon \ \text{a.s.} \tag{A4}$$

Note that we can write

$$\left\| \frac{1}{n}\sum_{t=1}^n \frac{\partial \widetilde{l}_{\alpha,t}(\hat{\theta}_{\alpha,n})}{\partial\theta} \frac{\partial \widetilde{l}_{\alpha,t}(\hat{\theta}_{\alpha,n})}{\partial\theta^T} - E\left(\frac{\partial l_{\alpha,t}(\theta_0)}{\partial\theta} \frac{\partial l_{\alpha,t}(\theta_0)}{\partial\theta^T} \right) \right\|$$

$$\leq \left\| \frac{1}{n}\sum_{t=1}^n \frac{\partial \widetilde{l}_{\alpha,t}(\hat{\theta}_{\alpha,n})}{\partial\theta} \frac{\partial \widetilde{l}_{\alpha,t}(\hat{\theta}_{\alpha,n})}{\partial\theta^T} - \frac{1}{n}\sum_{t=1}^n \frac{\partial l_{\alpha,t}(\hat{\theta}_{\alpha,n})}{\partial\theta} \frac{\partial l_{\alpha,t}(\hat{\theta}_{\alpha,n})}{\partial\theta^T} \right\|$$

$$+ \left\| \frac{1}{n}\sum_{t=1}^n \frac{\partial l_{\alpha,t}(\hat{\theta}_{\alpha,n})}{\partial\theta} \frac{\partial l_{\alpha,t}(\hat{\theta}_{\alpha,n})}{\partial\theta^T} - \frac{1}{n}\sum_{t=1}^n \frac{\partial l_{\alpha,t}(\theta_0)}{\partial\theta} \frac{\partial l_{\alpha,t}(\theta_0)}{\partial\theta^T} \right\|$$

$$+ \left\| \frac{1}{n}\sum_{t=1}^n \frac{\partial l_{\alpha,t}(\theta_0)}{\partial\theta} \frac{\partial l_{\alpha,t}(\theta_0)}{\partial\theta^T} - E\left(\frac{\partial l_{\alpha,t}(\theta_0)}{\partial\theta} \frac{\partial l_{\alpha,t}(\theta_0)}{\partial\theta^T} \right) \right\|$$

$$:= I_n + II_n + III_n.$$

By Lemma A2,

$$I_n \leq \frac{1}{n}\sum_{t=1}^n \sup_{\theta\in\Theta} \left\| \frac{\partial \widetilde{l}_{\alpha,t}(\theta)}{\partial\theta} \frac{\partial \widetilde{l}_{\alpha,t}(\theta)}{\partial\theta^T} - \frac{\partial l_{\alpha,t}(\theta)}{\partial\theta} \frac{\partial l_{\alpha,t}(\theta)}{\partial\theta^T} \right\| = o(1) \ \text{a.s.}$$

Since $\hat{\theta}_{\alpha,n}$ converges to θ_0 a.s. by Proposition 1, from (A4), we have

$$\lim_{n\to\infty} II_n \leq \lim_{n\to\infty} \frac{1}{n} \sum_{t=1}^{n} \sup_{\theta \in N_\epsilon(\theta_0)} \left\| \frac{\partial l_{\alpha,t}(\theta)}{\partial\theta} \frac{\partial l_{\alpha,t}(\theta)}{\partial\theta^T} - \frac{\partial l_{\alpha,t}(\theta_0)}{\partial\theta} \frac{\partial l_{\alpha,t}(\theta_0)}{\partial\theta^T} \right\| < \epsilon \ \text{ a.s.}$$

Finally, by the ergodic theorem, $III_n = o(1)$ a.s. Therefore, the lemma is established. $\square$

References

1. McKenzie, E. Some simple models for discrete variate time series. *J. Am. Water Resour. Assoc.* **1985**, *21*, 645–650. [CrossRef]
2. Al-Osh, M.A.; Alzaid, A.A. First order integer-valued autoregressive (INAR(1)) process. *J. Time Ser. Anal.* **1987**, *8*, 261–275. [CrossRef]
3. Ferland, R.; Latour, A.; Oraichi, D. Integer-valued GARCH processes. *J. Time Ser. Anal.* **2006**, *27*, 923–942. [CrossRef]
4. Engle, R.F. Autoregressive conditional heteroskedasticity with estimates of the variance of United Kingdom inflation. *Econometrica* **1982**, *50*, 987–1007. [CrossRef]
5. Bollerslev, T. Generalized autoregressive conditional heteroskedasticity. *J. Econom.* **1986**, *31*, 307–327. [CrossRef]
6. Fokianos, K.; Rahbek, A.; Tjøstheim, D. Poisson autoregression. *J. Am. Stat. Assoc.* **2009**, *104*, 1430–1439. [CrossRef]
7. Davis, R.A.; Wu, R. A negative binomial model for time series of counts. *Biometrika* **2009**, *96*, 735–749. [CrossRef]
8. Christou, V.; Fokianos, K. Quasi-likelihood inference for negative binomial time series models. *J. Time Ser. Anal.* **2014**, *35*, 55–78. [CrossRef]
9. Zhu, F. Modeling overdispersed or underdispersed count data with generalized poisson integer-valued garch models. *J. Math. Anal. Appl.* **2012**, *389*, 58–71. [CrossRef]
10. Zhu, F. Zero-inflated Poisson and negative binomial integer-valued GARCH models. *J. Stat. Plan. Infer.* **2012**, *142*, 826–839. [CrossRef]
11. Lee, S.; Lee, Y.; Chen, C.W.S. Parameter change test for zero-inflated generalized Poisson autoregressive models. *Statistics* **2016**, *50*, 540–557. [CrossRef]
12. Davis, R.A.; Liu, H. Theory and inference for a class of observation-driven models with application to time series of counts. *Stat. Sin.* **2016**, *26*, 1673–1707.
13. Diop, M.L.; Kengne, W. Testing parameter change in general integer-valued time series. *J. Time Ser. Anal.* **2017**, *38*, 880–894. [CrossRef]
14. Lee, Y.; Lee, S. CUSUM test for general nonlinear integer-valued GARCH models: Comparison study. *Ann. Inst. Stat. Math.* **2019**, *71*, 1033–1057. [CrossRef]
15. Kang, J.; Lee, S. Parameter change test for Poisson autoregressive models. *Scand. J. Stat.* **2014**, *41*, 1136–1152. [CrossRef]
16. Basu, A.; Harris, I.R.; Hjort, N.L.; Jones, M.C. Robust and efficient estimation by minimizing a density power divergence. *Biometrika* **1998**, *85*, 549–559. [CrossRef]
17. Ghosh, A.; Basu, A. Robust estimation for independent non-homogeneous observations using density power divergence with applications to linear regression. *Electron. J. Stat.* **2013**, *7*, 2420–2456. [CrossRef]
18. Lee, S.; Song, J. Minimum density power divergence estimator for GARCH models. *Test* **2009**, *18*, 316–341. [CrossRef]
19. Kim, B.; Lee, S. Robust estimation for the covariance matrix of multivariate time series based on normal mixtures. *Comput. Stat. Data Anal.* **2013**, *57*, 125–140. [CrossRef]
20. Kang, J.; Lee, S. Minimum density power divergence estimator for Poisson autoregressive models. *Comput. Stat. Data Anal.* **2014**, *80*, 44–56. [CrossRef]
21. Kim, B.; Lee, S. Robust estimation for zero-inflated Poisson autoregressive models based on density power divergence. *J. Stat. Comput. Simul.* **2017**, *87*, 2981–2996. [CrossRef]
22. Kim, B.; Lee, S. Robust estimation for general integer-valued time series models. *Ann. Inst. Stat. Math.* **2020**, in press.

23. Kang, J.; Song, J. Robust parameter change test for Poisson autoregressive models. *Stat. Probab. Lett.* **2015**, *104*, 14–21. [CrossRef]

24. Song, J.; Kang, J. Test for parameter change in the presence of outliers: The density power divergence based approach. *arXiv* **2019**, arXiv:1907.00004.

25. Kang, J.; Song, J. A robust approach for testing parameter change in Poisson autoregressive models. *arXiv* **2019**, arXiv:1908.11466.

26. Batsidis, A.; Horváth, L.; Martín, N.; Pardo, L.; Zografos, K. Change-point detection in multinomial data using phi-divergence test statistics. *J. Multivar. Anal.* **2013**, *118*, 53–66. [CrossRef]

27. Batsidis, A.; Martín, N.; Pardo, L.; Zografos, K. ϕ-divergence based procedure for parametric change point problems. *Methodol. Comput. Appl. Probab.* **2016**, *18*, 21–35. [CrossRef]

28. Martín, N.; Pardo, L. Comment on: Extensions of some classical methods in change point analysis. *Test* **2014**, *23*, 279–282. [CrossRef]

29. Lehmann, E.; Casella, G. *Theory of Point Estimation*, 2nd ed.; Springer: New York, NY, USA, 1998.

30. Warwick, J. A data-based method for selecting tuning parameters in minimum distance estimators. *Comput. Stat. Data Anal.* **2005**, *48*, 571–585. [CrossRef]

31. Warwick, J.; Jones, M.C. Choosing a robustness tuning parameter. *J. Stat. Comput. Simul.* **2005**, *75*, 581–588. [CrossRef]

32. Lee, S.; Ha, J.; Na, O.; Na, S. The cusum test for parameter change in time series models. *Scand. J. Stat.* **2003**, *30*, 781–796. [CrossRef]

33. Fried, R.; Agueusop, I.; Bornkamp, B.; Fokianos, K.; Fruth, J.; Ickstadt, K. Retrospective Bayesian outlier detection in INGARCH series. *Stat. Comput.* **2015**, *25*, 365–374. [CrossRef]

34. Straumann, D.; Mikosch, T. Quasi-maximum-likelihood estimation in conditionally heteroscedastic time series: A stochastic recurrence equations approach. *Ann. Stat.* **2006**, *34*, 2449–2495. [CrossRef]

35. Billingsley, P. *Convergence of Probability Measures*, 2nd ed.; Wiley: New York, NY, USA, 1999.

Article

Robust Regression with Density Power Divergence: Theory, Comparisons, and Data Analysis

Marco Riani [1], Anthony C. Atkinson [2], Aldo Corbellini [1] and Domenico Perrotta [3,*]

[1] Dipartimento di Scienze Economiche e Aziendale and Interdepartmental Centre for Robust Statistics, Università di Parma, I43125 Parma, Italy; mriani@unipr.it (M.R.); aldo.corbellini@unipr.it (A.C.)

[2] The London School of Economics, London WC2A 2AE, UK; a.c.atkinson@lse.ac.uk

[3] European Commission, Joint Research Centre, 21027 Ispra, Italy

* Correspondence: domenico.perrotta@ec.europa.eu

Received: 22 February 2020; Accepted: 26 March 2020; Published: 31 March 2020

Abstract: Minimum density power divergence estimation provides a general framework for robust statistics, depending on a parameter α, which determines the robustness properties of the method. The usual estimation method is numerical minimization of the power divergence. The paper considers the special case of linear regression. We developed an alternative estimation procedure using the methods of S-estimation. The rho function so obtained is proportional to one minus a suitably scaled normal density raised to the power α. We used the theory of S-estimation to determine the asymptotic efficiency and breakdown point for this new form of S-estimation. Two sets of comparisons were made. In one, S power divergence is compared with other S-estimators using four distinct rho functions. Plots of efficiency against breakdown point show that the properties of S power divergence are close to those of Tukey's biweight. The second set of comparisons is between S power divergence estimation and numerical minimization. Monitoring these two procedures in terms of breakdown point shows that the numerical minimization yields a procedure with larger robust residuals and a lower empirical breakdown point, thus providing an estimate of α leading to more efficient parameter estimates.

Keywords: estimation of α; monitoring; numerical minimization; S-estimation; Tukey's biweight

1. Introduction

Basu et al. [1] introduced a general form of robust estimation based on minimizing a density power divergence. The family of procedures, and so the robustness properties, depend on the value of a parameter α. In this paper, we consider normal theory regression. We use standard methods for the analysis of robust procedures, in particular S-estimation (Riani et al. [2]), to find the theoretical breakdown point and efficiency of power divergence regression as a function of α. We use these results to make comparisons with theoretical properties of other robust methods, for example, S-estimation using Tukey's biweight. We introduce a data-driven method for the estimation of α from monitoring residuals over a range of values of α and so find the empirical efficiency and breakdown point of power density estimation for several regression examples. One surprising conclusion is that, for normal theory models, the rho function for the power divergence is one minus a suitably scaled standard normal density raised to the power α.

The paper is structured as follows. The next section introduces minimum density power divergence estimation and the related estimating equations for normal theory linear regression. The important problem of estimating α is mentioned. The first part of Section 3 reviews S-estimation in the linear regression model, and the second part, Section 3.2, rewrites power divergence estimation of the regression parameter β in the form of S-estimation, derives the rho function, and so finds the asymptotic breakdown point (bdp) of the procedure. Section 3.2.2 gives the asymptotic efficiency of this

S-estimation at the Gaussian model and finds the weight function used in fitting data. Comparisons are given with some well known rho and weight functions. In Section 4, plots of asymptotic efficiency against asymptotic bdp are used to compare the properties of several S-estimators, including Tukey's biweight. Section 5 compares methods through the analysis of data. An alternative to S power divergence is the original suggestion of Basu et al. [1] to use Brute Force (BF) minimization (our acronym, not theirs). Comparisons on simulated and real data show the superiority of BF power divergence to the S-estimator. In particular, monitoring the plots of residuals as α varies may lead to a clear indication of the minimum value of α for which a robust fit is obtained. Thus, the empirical breakdown point of BF power divergence estimation can be found, leading to the most efficient robust estimation for each specific data set.

2. Minimum Density Power Divergence Estimation

Basu et al. [1] define the power divergence between two densities $f(z)$ and $g(z)$, a function of a single parameter α, as

$$d_\alpha\{g(z), f(z)\} = \int \left\{ f^{1+\alpha}(z) - \left(1 + \frac{1}{\alpha}\right) f^\alpha(z)g(z) + \frac{1}{\alpha} g^{1+\alpha}(z) \right\} dz, \quad \alpha > 0 \tag{1}$$

$$d_0\{g(z), f(z)\} = \int g(z) \log\left\{ \frac{g(z)}{f(z)} \right\} dz.$$

The parameter α controls the trade-off between efficiency and robustness for the power divergence estimator. The limit as $\alpha \to 0$ is a version of the Kullback-Leibler divergence. The value $\alpha = 1$ leads to squared L_2 estimation, an analysis of which is given by Scott [3].

Let g be the density function of the process generating the data. Given an independent and identically distributed sample $y_1, \ldots, y_n$ is available from G, Basu et al. [1] model the unknown $g(z)$ with the density $f_\theta(y)$ by minimizing $d_\alpha\{g(z), f_\theta(y)\}$. Since the third term of the divergence is independent of θ, the power divergence estimator of θ can be found by minimizing

$$\int f_\theta^{1+\alpha}(z)dz - \left(1 + \frac{1}{\alpha}\right) \frac{1}{n} \sum_{i=1}^{n} f_\theta^\alpha(y_i), \tag{2}$$

in which the empirical distribution G_n is used to approximate the unknown distribution G, thus avoiding the necessity for density estimation.

Basu et al. [1] develop their method only for random samples from the normal, exponential and Poisson distributions. For the normal distribution, Equation (2) is minimized over both the mean μ and the variance σ^2. The extension to normal theory regression models is in Ghosh and Basu [4].

As usual in a regression framework, we define y_i to be the response variable, which is related to the values of a set of $p - 1$ explanatory variables $x_{i1}, \ldots, x_{ip-1}$ by the relationship

$$y_i = \beta' x_i + \epsilon_i \qquad i = 1, \ldots, n, \tag{3}$$

where, including an intercept, $\beta' = (\beta_0, \beta_1, \ldots, \beta_{p-1})$ and $x_i = (1, x_{i1}, \ldots, x_{ip-1})'$. Let $\sigma^2 = \text{var}(\epsilon_i)$, which is assumed to be constant for all $i = 1, \ldots, n$. We also take the quantities in x_i to be fixed and assume that $x_1, \ldots, x_n$ are not collinear. The case $p = 1$ corresponds to that of a univariate response without predictors. We call σ the scale of the distribution of the error term ϵ_i, when its density takes the form

$$\sigma^{-1} f\left(\frac{\epsilon}{\sigma}\right).$$

When f is the normal distribution with mean, as in Equation (3), and variance σ^2, Durio and Isaia [5] and Ghosh and Basu [4] show that the function, as in Equation (2), to be minimized becomes

$$\frac{1}{(2\pi)^{\alpha/2}\sigma^\alpha\sqrt{1+\alpha}} - \frac{1+\alpha}{\alpha}\frac{1}{(2\pi)^{\alpha/2}\sigma^\alpha}\frac{1}{n}\sum_{i=1}^{n} e^{-\alpha(y_i-x_i'\beta)^2/2\sigma^2}. \tag{4}$$

The partial derivative of Equation (4), with respect to β_j, provides the estimating equation for β:

$$\sum_{i=1}^{n} x_{ij}(y_i - x_i'\beta)e^{-\alpha(y_i-x_i'\beta)^2/2\sigma^2}, \quad (j = 1,\ldots,p). \tag{5}$$

When $\alpha = 0$, Equation (5) becomes the equation for non-robust ordinary least squares. For $\alpha > 0$ we have weighted least squares of the kind associated in the next section with M estimation. Ghosh and Basu [4] also give the estimating equation for σ^2 which we will however not be using in our theoretical development.

An important aspect is the estimation of α. Durio and Isaia [5] test for changes in the estimates of the parameters β as a function of α, while Warwick and Jones [6] and Ghosh and Basu [7] estimate the mean squared error of the parameter estimates as α changes. In Section 5, we monitor changes in the pattern of residuals to choose the minimum value of α for which a robust fit is obtained, so leading to the most efficient parameter estimates.

3. Robust Regression

3.1. M and S Estimation

Basu et al. [1] find estimates of the parameters of the linear model by simultaneous minimization of Equation (4) as a function of β and σ^2. In this section, we recall the theory of M and S estimation, which we use in Section 3.2 to describe properties of the S power divergence estimator. In Section 5, we provide a numerical comparison of the BF minimization and S-estimation approaches.

The M-estimator of the regression parameters, which is scale equivariant (i.e., independent of the units of measurement), is defined by

$$\hat{\beta}_M = \min_{\beta \in \mathbb{R}^p} \sum_{i=1}^{n} \rho\left(\frac{r_i}{s}\right), \tag{6}$$

where $r_i = y_i - \beta'x_i$ is the i-th residual and ρ is a function with suitable properties and s is an estimate of σ. For least squares $\rho(x) = x^2$. For robust estimation $\rho(x) < x^2$ for sufficiently large absolute values of x. We also write $r_i(\beta)$ to emphasize the dependence of r_i on β.

These definitions do not depend on how σ is estimated. Clearly, if we want to keep the M-estimate robust, s should also be a robust estimate. We assume that the same ρ is used in the estimation of β and σ, which is customary in practice. In order to have a consistent scale estimate for normally distributed observations, we require

$$E_{\Phi_{0,1}}\left[\rho\left(\frac{r_i}{s}\right)\right] = K, \tag{7}$$

where $\Phi_{0,1}$ is the cdf of the standard normal distribution. To see consistency, notice that $E_{\Phi_{0,1}}(\rho) = K$ implies

$$\frac{E_{\Phi_{0,\sigma^2}}[\rho]}{K} = \frac{K\sigma^2}{K} = \sigma^2.$$

An M-estimator of scale in Equation (3), say s, is defined to be the solution to the equation

$$\frac{1}{n}\sum_{i=1}^{n}\rho\left(\frac{r_i}{s}\right) = \frac{1}{n}\sum_{i=1}^{n}\rho\left(\frac{y_i - \beta'x_i}{s}\right) = K. \tag{8}$$

Equation (8) is solved, at least in principle, among all $(\beta, \sigma) \in \Re^p \times (0, \infty)$, where $0 < K < \sup \rho$. Rousseeuw and Yohai [8] defined S-estimators by minimization of the dispersion s of the residuals

$$\hat{\beta}_S = \min_{\beta \in \Re^p} s\{r_1(\beta), \ldots, r_n(\beta)\} \tag{9}$$

with final scale estimate

$$\hat{\sigma}_S = s\{r_1(\hat{\beta}_S), \ldots, r_n(\hat{\beta}_S)\}.$$

The dispersion s is defined as the solution of Equation (8). The S-estimates, therefore, can be thought as self-scaled M-estimates whose scale is estimated simultaneously with the regression parameters. Note, in fact, that when the scale and the regression estimates are simultaneously estimated, S-estimators for regression also satisfy (for example, Maronna et al. [9], p. 131)

$$\hat{\beta}_S = \min_{\beta \in \Re^p} \sum_{i=1}^{n} \rho\left(\frac{r_i}{s}\right). \tag{10}$$

The estimator of β in Equation (9) is called an S-estimator because it is derived from a scale statistic in an implicit way.

The function ρ is the key to many important properties of M and S estimates. Rousseeuw and Leroy [10] (p. 139) show that, if the function ρ satisfies the following conditions:

1. It is symmetric and continuously differentiable, and $\rho(0) = 0$;
2. there exists a $c > 0$ such that ρ is strictly increasing on $[0, c]$ and constant on $[c, \infty)$; and
3. it is such that

$$K / \rho(c) = \text{bdp} \quad \text{with} \quad 0 < \text{bdp} \leq 0.5, \tag{11}$$

then the asymptotic breakdown point of the S-estimator tends to bdp when $n \to \infty$. Note that if $\rho(c)$ is normalized in such a way that $\sup \rho(c) = 1$, the constant K becomes exactly equal to the breakdown point of the S-estimator.

3.2. S Estimation for Power Divergence Regression

3.2.1. The Breakdown Point and the Rho Function

The function ρ is used in the estimation of β for a given estimate s. With $x = r/s$ it follows from the function to be minimized in Equation (4) that $\rho(x) \propto -\exp(-\alpha x^2/2)$. If we scale this function so that $\sup \rho_\alpha(x) = 1$ and $\rho_\alpha(0) = 0$, we obtain

$$\rho_\alpha(x) = 1 - \exp(-\alpha x^2/2). \tag{12}$$

This is a trivial reparameterization of an otherwise unreferenced rho function attributed to Welsh.

The panels of Figure 1 show plots of $\rho_\alpha(x)$ for several values of α. For $\alpha = 1$, the efficiency is 0.65, and the breakdown point is 0.29. As α decreases, the procedure becomes less robust but more efficient. Table 1 gives values of α, bdp, and *eff* for three frequently used values of each quantity; these values being given in bold. The left-hand panel of Figure 1 is for the three bold values of bdp, and the right-hand panel for the three values of *eff*. The rho functions for high efficiency are appreciably flatter than those for high bdp.

Since ρ_α is scaled, the breakdown point, bdp, is given by $E_{\Phi_{0,1}}[\rho_\alpha(x)]$. Then,

$$
\begin{aligned}
E_{\Phi_{0,1}}[\rho_\alpha(x)] &= 1 - E\left[\exp(-\alpha x^2/2)\right], \\
&= 1 - \int \exp(-\alpha x^2/2)dx, \\
&= 1 - (2\pi)^{\alpha/2} \int \phi_{0,1}^\alpha(x)\phi_{0,1}(x)dx.
\end{aligned} \tag{13}
$$

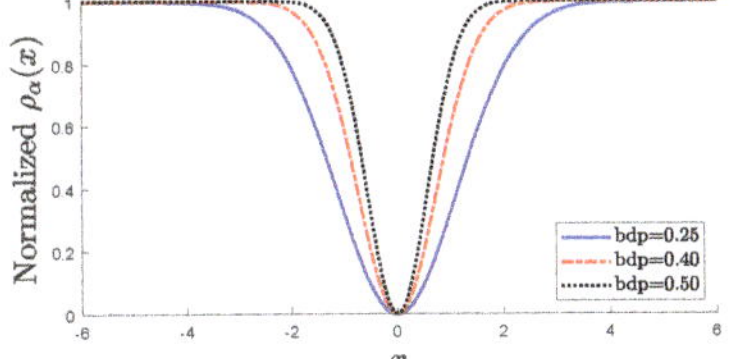
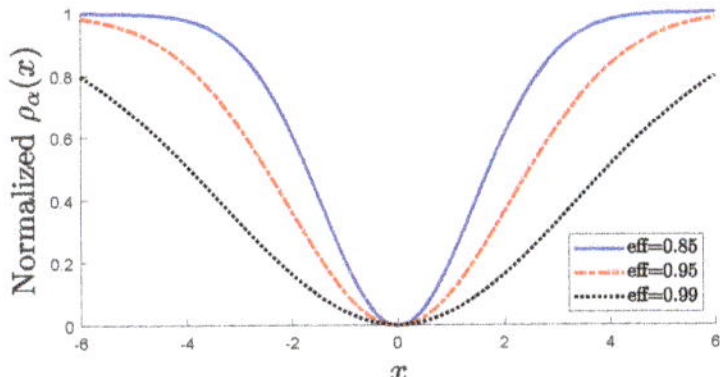

Figure 1. Dependence of $\rho_\alpha(x)$ on α, for frequently used values of robustness properties in Table 1. Left-hand panel, three values of breakdown point (bdp); right-hand panel, three values of *eff*.

From the useful general expression in Section 3.2 of Basu et al. [11] that

$$\int \phi_{m,s}^\alpha(x)\phi_{c,d}(x)dx = \frac{\exp\left[-\alpha(c-m)^2/\{2(s^2+\alpha d^2)\}\right]}{(2\pi)^{\alpha/2}s^\alpha\left(1+\frac{\alpha d^2}{s^2}\right)^{0.5}},$$

we obtain

$$E_{\Phi_{0,1}} = \mathrm{bdp} = 1 - \frac{1}{\sqrt{1+\alpha}}. \tag{14}$$

Our expression for the breakdown point comes from S-estimation, reflecting breakdown in the estimate of β under the customary assumption that σ is known. This is different from the value of

$$\frac{\alpha}{(1+\alpha)^{3/2}} \tag{15}$$

in Section 3.2 of Basu et al. [11], who consider the joint breakdown of the estimates of β and σ when "location explodes" and "scale implodes". While the expression in Equation (14) increases monotonically in the interval $\alpha = [0,3]$, Equation (15) increases monotonically in the smaller interval $\alpha = [0,2]$ and then slightly decreases.

To fit a model to data, we specify the desired asymptotic breakdown point, when the value of α from inverting the expression in Equation (14) is

$$\alpha = \frac{1}{(1-\mathrm{bdp})^2} - 1.$$

For example, for 50% breakdown, $\alpha = 3$.

Table 1. S power divergence. Values of α, bdp, and *eff* for three frequently used values of each in bold.

α	bdp	*eff*
0	**0**	**1**
0.5	0.1835	0.8381
1	0.2929	0.6495
0.7778	**0.25**	0.7271
1.7778	**0.4**	0.4536
3	**0.5**	0.2894
0.4715	0.1756	**0.85**
0.3522	0.14	**0.9**
0.2245	0.0963	**0.95**
0.089	0.0417	**0.99**

3.2.2. Efficiency, the Psi Function and the Influence Function

Other basic properties of the robust estimator follow from derivatives of $\rho_\alpha(x)$. For power density

$$\psi_\alpha(x) = \rho'_\alpha(x) = \alpha x \exp(-\alpha x^2/2)$$

and

$$\psi'_\alpha(x) = \alpha(1 - \alpha x^2)\exp(-\alpha x^2/2).$$

Figure 2 shows, for three values of α, a plot of $\psi_\alpha(x)$ (which is proportional to the Influence Function, see Maronna et al. [9] (p. 123)). As α decreases, the figure shows the curve becomes flatter.

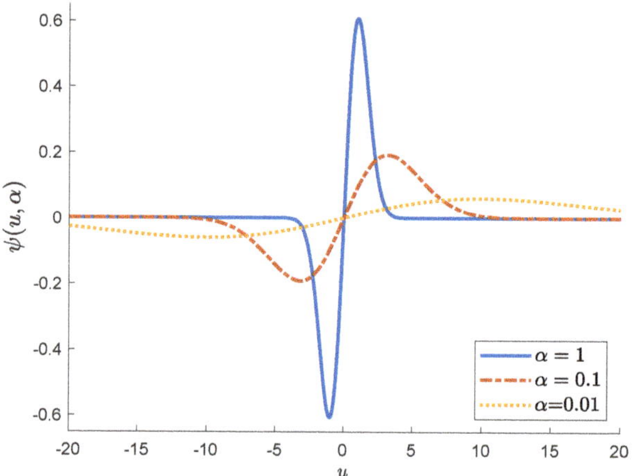

Figure 2. S power divergence; ψ function, proportional to the influence function.

From, for example, Rousseeuw and Leroy [10] (p. 142), the asymptotic efficiency *eff* of the S-estimator at the Gaussian model is

$$eff = \frac{\{\int \psi'(x)d\Phi(x)\}^2}{\int \psi^2(x)d\Phi(x)}. \tag{16}$$

For $\rho_\alpha(x)$,

$$E[\psi_\alpha^2(x)] = \alpha^2(2\pi)^{\alpha/2}\int x^2 \phi_{0,1}^{2\alpha x+1}dx. \tag{17}$$

Since

$$\int x^2 \phi_{0,1}^n dx = \frac{1}{n^3(2\pi)^{n-1}},$$

Equation (17) becomes

$$E[\psi_\alpha^2(x)] = \alpha^2 \frac{1}{(2\alpha+1)^3}.$$

To find the numerator of the efficiency

$$\begin{aligned}
E[\psi'_\alpha(x)] &= \alpha(2\pi)^{\alpha/2}\int \phi_{0,1}^{\alpha+1}dx - \alpha^2(2\pi)^{\alpha/2}\int x^2\phi_{0,1}^{\alpha+1}dx, \\
&= \frac{\alpha}{\sqrt{1+\alpha}} - \frac{\alpha^2}{\sqrt{(1+\alpha)^3}}, \\
&= \frac{\alpha}{\sqrt{(1+\alpha)^3}}.
\end{aligned} \tag{18}$$

Combining these pieces, we obtain

$$eff = \frac{\sqrt{(1+2\alpha)^3}}{(1+\alpha)^3},$$ (19)

agreeing with the expression for the asymptotic variance of the estimate of the mean μ of a univariate normal sample given in Section 4.2 of Basu et al. [1], a few values of which are tabulated in their Table 1. Inversion of Equation (19) yields

$$\alpha = (1 - F + \sqrt{1 - F})/F,$$

where $F = eff^{2/3}$.

The algorithm for S-estimation is complicated, involving weighted regression. Rousseeuw and Leroy [10] (pp. 207–208) provide a sketch. More details are in Salibian-Barrera and Yohai [12]. A central part is weighted regression, with weights

$$w(x) = \psi(x)/x.$$

Figure 3 plots the weight functions for power divergence and five other rho functions: Tukey's biweight [13], Hampel's [14] (p. 150), Huber's [15], the optimal (Yohai and Zamar [16]), and hyperbolic tangent (Hampel et al. [14] (p. 328)), all scaled to have efficiency 0.95.

Details of the functions are in the Appendix A. The similarity of the power divergence weights to those of the Tukey biweight is outstanding, although the biweight is exactly zero at $x = c$, which in this case is equal to 4.6851. For this x coordinate, the power divergence weight (when $eff = 0.95$) is 0.0851. Both have a curved shape for small values of $|x|$, unlike the Hampel and hyperbolic weights. We note that the procedure for finding the tuning constant α for the power divergence estimator, given a prefixed value of breakdown point or efficiency, is not iterative. This is distinct from all the other rho functions listed above (apart from that of Huber), for which iterative procedures are required.

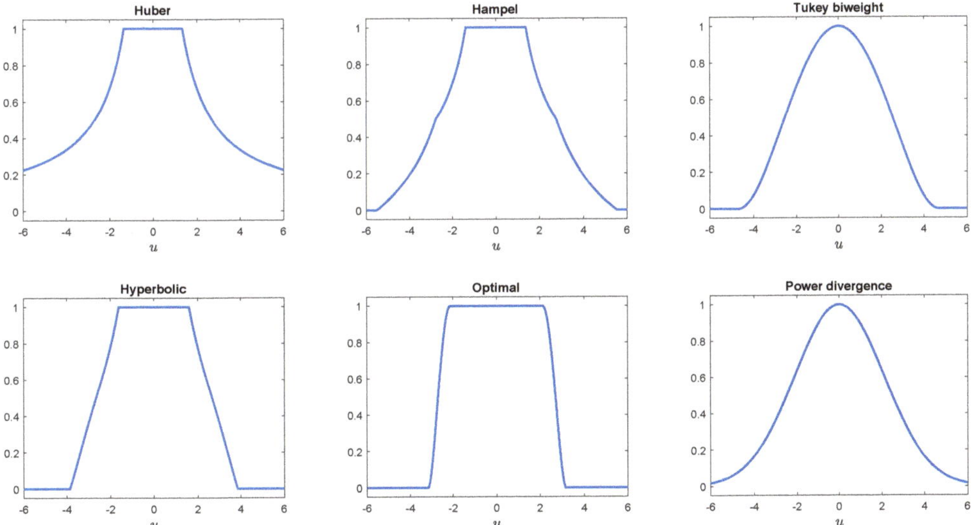

Figure 3. The weight function $\psi(x)/x$ for six S-estimators.

4. Comparisons of Asymptotic Properties

The basic properties of S power divergence are the asymptotic breakdown point, as in Equation (14), and the asymptotic efficiency, as in Equation (19). Figure 4 shows these two properties as functions of α over the range $0 \leq \alpha \leq 3$. As bdp increases from zero towards 0.5, *eff* decreases from 1 to 0.2894. These are generic shapes for robust estimators, quantifying the trade-off between robustness and efficiency. Figure 5 shows plots of efficiency against breakdown point for S power divergence and four of the other ρ functions of Figure 3 (the Huber function being excluded because it has a zero breakdown point). In order to generate these curves, we fix a particular value of breakdown point and find the associated tuning constant α for PD or c for the other estimators (the details are in the Appendix). In the case of the Hampel ρ functions, the three extra parameters c_1, c_2, and c_3 have been set equal to 2, 4, and 8. For the hyperbolic tangent estimator the extra parameter k, which reflects the log of the change of variance sensitivity of the M-estimator, has been set equal to 4.5. Given the value of the tuning constant, we found the corresponding value of the efficiency.

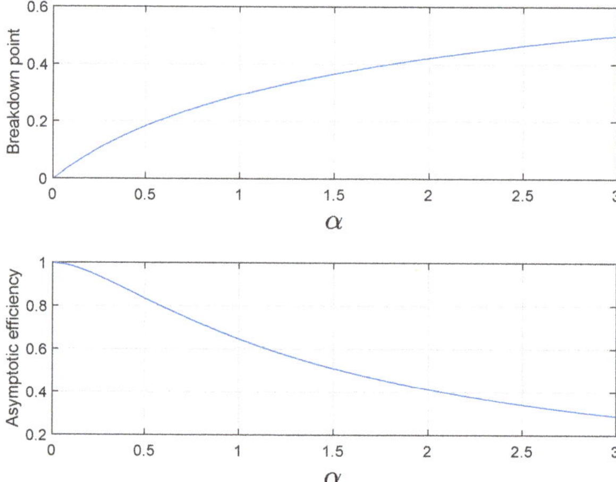

Figure 4. S power divergence: breakdown point and efficiency as functions of α.

It is clear from the figure that the general asymptotic performance of the five methods is similar. The optimal function is best for small bdp but worst for values slightly larger than 0.25. The situation for Hampel is the reverse, being worst for small bdp and best for bdp values above approximately 0.4. For small bdp, the power divergence is the second worst but behaves much like the hyperbolic and biweight functions for larger values of bdp. For 50% bdp (as the inset in the figure shows), the ordering is (we give the exact numbers in parenthesis) hyperbolic (0.3019), Hampel (0.2924), power divergence (0.2894), biweight (0.2868), and last the optimal (0.2428). Hössjer [17] proves that, for normal theory linear models, the maximum efficiency when bdp = 0.5 is 0.329.

Some further insight into the balance between breakdown point and efficiency comes from varying the parameters of the Hampel and hyperbolic functions. In Figure 5, the parameters for the Hampel were $c_1 = 2$, $c_2 = 4$, and $c_3 = 8$. The left-hand panel of Figure 6 compares the breakdown point and efficiency of Hampel's rho function with these values to those when $c_1 = 1.5$, $c_2 = 3.5$, and $c_3 = 8$. The original procedure is better for breakdown point less than around 0.3, with the modified version being slightly better for larger values. For the hyperbolic rho function in the right-hand panel the freely variable parameter, other than c, is k. The curves for three values of k are shown in the right-hand panel of Figure 6. The difference is largest for small values of bdp, when $k = 6$ has the highest efficiency. In other words, imposing a looser constraint in the change of variance parameter produces higher efficiency for small values of bdp. For breakdown points near 0.5, the order is reversed, with $k = 6$ being the least efficient, although, in this region, the differences are less than for low bdp.

The conclusion from this figure reinforces that from Figure 5; no one rho function has the highest breakdown point and efficiency over the whole range of bdp from 0 to 0.5. These results also implicitly show that the choice of the ρ function is not a crucial aspect since all (provided they are bounded) have similar behavior in terms of breakdown point and efficiency. These theoretical results are in line with the empirical findings in Salini et al. [18], where it is shown that the size of the test for outlier detection is much more affected by the choice of the requested level of efficiency or breakdown point than by the choice of the ρ function.

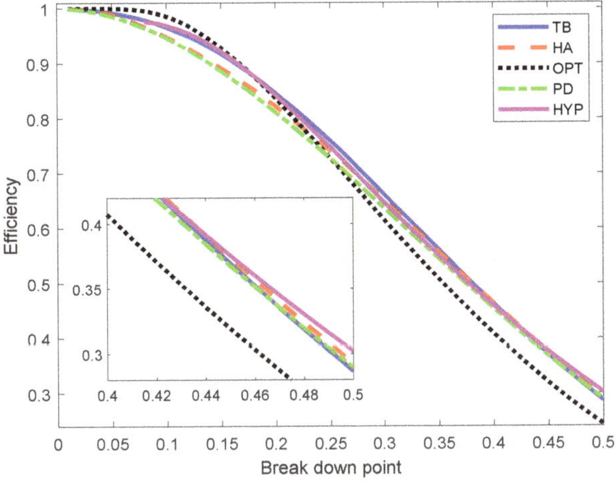

Figure 5. Breakdown point and efficiency as parameters vary for five rho functions: TB = Tukey biweight; HA = Hampel; OPT = optimal; PD = power divergence and HYP = hyperbolic. The inset is a zoom of the main figure for high breakdown point.

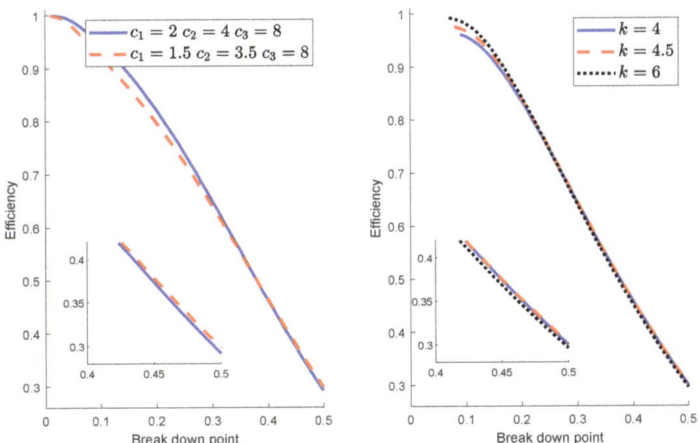

Figure 6. Breakdown point and efficiency as parameters vary for the Hampel and hyperbolic rho functions.

It is hard to reconcile the conclusions from these graphs with the statement in the opening paragraph of Jones et al. [19] that "quite small values of α were found to afford considerable robustness while retaining very high efficiency relative to maximum likelihood". Although it may be argued that S power divergence has good properties as a robust procedure, the figure shows that these fully agree with those for other S estimators. We now turn from asymptotics to data analysis to allow non-asymptotic comparisons and analysis of the 'brute force' approach to power divergence estimation.

5. Monitoring and Comparisons with Data

In order to compare the finite sample properties of robust estimators in regression, Riani et al. [20] introduced the idea of monitoring the properties of robust analyses as tuning constants are changed. For power divergence, this would be the value of α, or equivalently changes in nominal values of bdp or *eff*, which are how the range of monitored values was specified for other ρ functions. The most incisive information comes from looking at displays of residuals. Typically, for contaminated data, these display many outliers for very robust analyses, which suddenly are much reduced in magnitude at a specific value of the tuning constant. At this point, the procedure becomes close to maximum likelihood including the outliers. The sharp transition between the two regions allows estimation of the empirical breakdown point and so to the robust analysis with the highest efficiency. The monitoring process starts with bdp=0.5, which is the maximum fraction of contamination that an affine equivariant estimator can resist.

To illustrate this structure, we re-analyze regression data from Atkinson and Riani [21] (Table A2) comparing S power divergence with the BF version, using numerical minimization. We start monitoring from a bdp of 50% and use the very robust version of Least Median of Squares regression (Rousseeuw [22]) to provide initial estimates of β and σ^2. After this initial minimization for $\alpha = 3$, successive minimizations for lower values of α start from the estimates for the immediately higher value of α.

The regression data consist of 60 response observations and three explanatory variables. The scatter-plot matrix of the data does not reveal any outlying observations. The upper panel of Figure 7 is the monitoring plot of the residuals for BF power divergence as α goes from 3 to 0. There is a very clear transition from the robust analysis in the left-hand part of the plot to the non-robust analysis in the right-hand part, which occurs just before bdp = 0.21, giving an empirical breakdown point of 0.23. What is striking about this figure, apart from the clear transition point, is the distinct near constancy of the residuals in the two parts of the plot.

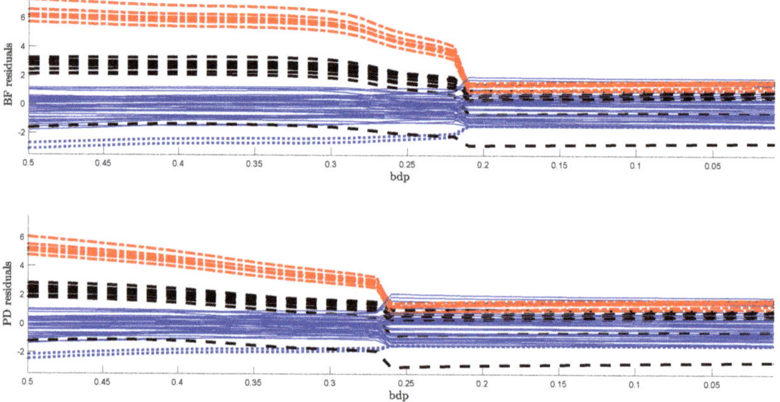

Figure 7. Regression data: residuals as bdp decreases. Upper panel, Brute Force (BF)-estimation, lower panel S-estimation.

The lower panel of Figure 7 is the same plot but for the analysis using S power divergence. The conclusion is similar, with an empirical breakdown point of 0.27, higher than that in the upper panel; BF therefore provides more efficient estimates. Although the residuals in the non-robust right-hand part are constant, those from the robust analysis decrease in magnitude as the analysis becomes less robust. This effect is caused by the gradual increase in the estimate of σ^2 as the analysis becomes less robust. A monitoring plot of the two estimates of σ is in the left-hand panel of Figure 8. The BF estimate is indeed virtually constant up to a bdp of nearly 0.3, increasing more rapidly to bdp = 0.2 with a jump corresponding to the switch from robust to non-robust analysis. At this point, it is close to that from S-estimation, which has been continually increasing. Both estimates of course coincide when bdp = 0, that is, for non-robust least squares.

These plots show the importance of the empirical breakdown point, found as α, and hence bdp, decrease. We monitor at values $\alpha_i, i = 1, \ldots, n_\alpha$, corresponding to breakdown values bdp_i. In our examples, $n_\alpha = 50$. At each i, we calculate a property of the fit, $\mathcal{P}_i$ and find the difference $\mathcal{D}_i = |\mathcal{P}_i - \mathcal{P}_{i-1}|$. Let the empirical breakdown point be bdp*. Then,

Definition 1. *The empirical breakdown point bdp* = bdp_{i*}, where*

$$i^* = \arg\max \mathcal{D}_i, i = 1, \ldots, n_\alpha - 1.$$

Some choices of the property $\mathcal{P}_i$ are

1. The residual sum of squares.
2. Changes in the parameter estimates $\hat{\beta}_i$ or $\hat{\sigma}$.
3. Measures of correlation between successive sets of residuals, rather than the sum of squares (Riani et al. [20]).

This definition is for fixed finite n. If there are m outliers with responses $y'_j = y_j + \Delta_j, j = 1, \ldots, m$, determination of bdp* is sharp as $\Delta_j \to \infty$. As $\Delta_j \to 0$, a threshold should be applied in the calculation of i^*.

We ran a number of simulations and studied the monitoring plots. For a data set of 100 observations without outliers, the trajectories of the residuals were smooth and uneventful, although a similar structure was observed to that of Figure 7: the residuals from BF were sensibly constant until around $\alpha = 1$ and then began gently to become less extreme. On the other hand, the S residuals steadily decreased in magnitude. The plot of the estimates of σ was similar to that of the left-hand panel of Figure 8. As is correct in the absence of outliers, neither plot of residuals nor σ indicated the need for robust analysis.

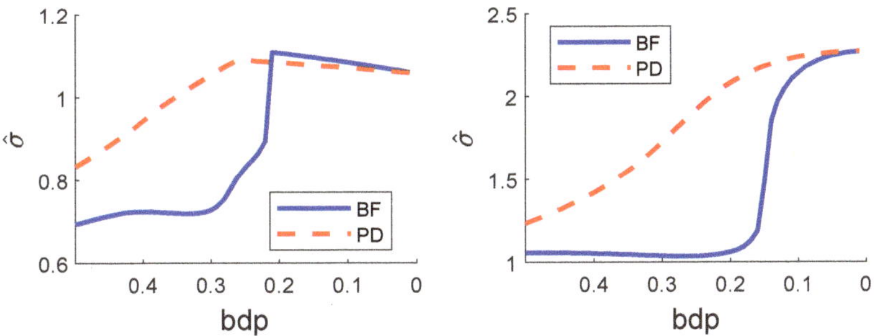

Figure 8. Comparison of estimates of σ as bdp decreases. Left-hand panel, regression data: right-hand panel, data with moderate outliers.

When the outliers in our simulations were very remote, both methods clearly indicated the outliers, although the monitoring plot for S estimation, unlike that using BF, did not show a sharp transition between two regions. The challenge for robust methods is when the outliers are less remote. As an example, we again simulated 100 observations with $\sigma^2 = 1$, but now a value of 5 was added to 20 responses. The two panels of Figure 9 show the resulting monitoring plots. Both display the same set of scaled residuals for 50% bdp, although those from BF are larger in magnitude. BF shows relatively sharp transitions at a breakdown point of 0.16, whereas S estimation shows a gradual decrease in the magnitude of the residuals as bdp (α) decreases. The right-hand panel of Figure 8 plots the two estimates of σ. As in the results for the regression data, the estimate from S-estimation increases gradually as bdp decreases, but the BF estimates are sensibly constant until a bdp around 0.16, when there is a distinct increase due to non-robust estimation.

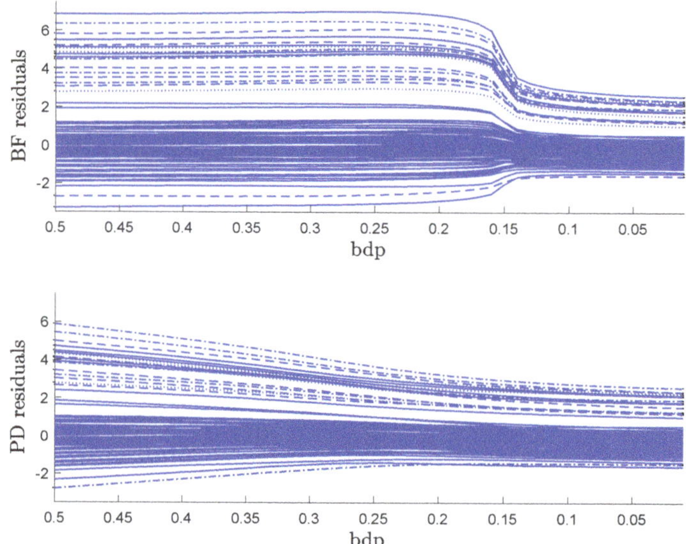

Figure 9. Data with moderate outliers: residuals as bdp decreases. Upper panel, BF-estimation; lower panel S-estimation.

Our results in Section 3.2.1 and 4 indicate the close relationship between Tukey's biweight and the power density rho functions. This is illustrated by the plot for S estimation using the biweight on these data, which we do not show here, which is indistinguishable from that using the power divergence ρ.

As a final larger data example, we analyze 509 observations on the amount spent by loyalty card holders at a supermarket chain in Northern Italy, introduced by Atkinson and Riani [23], who recommended a Box-Cox transformation for the response with $\lambda = 1/3$. Perrotta et al. [24] showed that a value of $\lambda = 0.4$ is to be preferred. We used this value in our analysis. The monitoring plot of residuals from BF power divergence is in Figure 10. It shows stable trajectories of the residuals for many values of α. A change starts around bdp = 0.17, indicating this as the empirical bdp. Again, S power divergence, which we do not show, reveals the same extreme observations, but fails to provide a sharp transition, so that the empirical breakdown point for efficient analysis is again not easily determined.

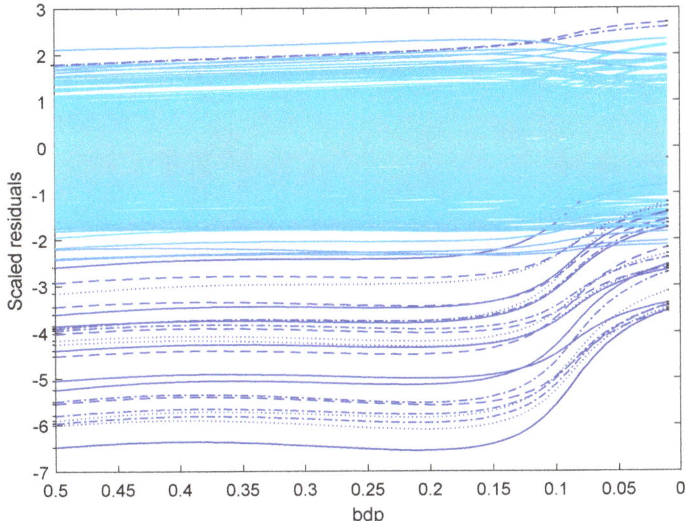

Figure 10. Loyalty card data: residuals for BF-estimation as bdp decreases.

6. Discussion

We have used the estimating equation for the linear parameters β to recast power divergence estimation in the context of S-estimation. This leads straightforwardly to calculations of asymptotic bdp and efficiency. This form of the power density estimate has asymptotic properties close to those of S estimation using Tukey's biweight.

An alternative to power divergence S-estimation is brute-force numerical minimization. The non-asymptotic comparison of the two procedures has been performed with monitoring plots of residuals as bdp varies, providing fits changing from very robust to maximum likelihood. S power divergence estimation has properties very similar to those of S-estimation with Tukey's biweight. In both, there is often a smooth decrease in the magnitude of the residuals as bdp decreases. On the other hand, BF minimization produces monitoring plots which show a clearer break between robust and non-robust fits, leading to estimation of an empirical breakdown point and so to the most efficient robust estimates.

One conclusion is that BF estimation provides more informative analyses than power density S-estimation. However, the results of monitoring regression in Riani et al. [20] show that the comparative behavior of estimators depends on the particular data set being analyzed. Figure 7 shows that S-estimation may produce monitoring plots with a sharp change, and further examples are in Riani et al. [20]. Other methods providing a sharp change, and so guidance to efficient analysis, are the Forward Search [25] and Least Trimmed Squares [22]. It remains to be seen how BF power divergence compares with these other methods, both statistically and on larger, more complicated models, such as linear mixed models, generalized linear models, or nonlinear models.

Author Contributions: All authors contributed equally to the manuscript. All authors have read and agreed to the published version of the manuscript.

Funding: This research benefits from the HPC (High Performance Computing) facility of the University of Parma. We acknowledge financial support from the University of Parma project "Statistics for fraud detection, with applications to trade data and financial statement", from the Department of Statistics of the London School of Economics and from the 2014-2020 Institutional Research Programme of the Joint Research Centre of the European Commission.

Acknowledgments: We are grateful to Leandro Pardo and Nirian Martin for the invitation to contribute to the special issue of *Entropy* on 'Robust Procedures for Estimating and Testing in the Framework of Divergence Measures'.

Conflicts of Interest: The authors declare no conflict of interest.

Appendix A. Rho Functions

In this appendix, we summarize the characteristics of the ρ functions which have been used in the paper. Since the hyperbolic tangent estimator is rarely used and, as far as we know, is not implemented in any statistical package, we describe this estimator in greater detail.

The first ρ-function was proposed in Huber (1964):

$$\rho(u) = \begin{cases} (u^2/2) & |u/c| \leq 1 \\ c|u| - c^2/2 & |u/c| > 1. \end{cases}$$

It is easily seen that this ρ function is unbounded and, therefore, the corresponding estimator has a zero breakdown point.

Perhaps the most popular ρ function for redescending M and S-estimates is **Tukey's Biweight function** [13]:

$$\rho(u) = \begin{cases} \frac{u^2}{2} - \frac{u^4}{2c^2} + \frac{u^6}{6c^4} & \text{if } |u| \leq c \\ \frac{c^2}{6} & \text{if } |u| > c, \end{cases} \tag{A1}$$

the first derivative of which vanishes outside the interval $[-c, +c]$. Therefore, for this function c is the crucial tuning constant, determining the efficiency or, equivalently, the breakdown point.

Hampel's ρ function [14] (p. 150) has a similar, but less smooth, shape.

$$\rho(u) = \begin{cases} \frac{1}{2}u^2 & \text{if } |u/c| \leq c_1 \\ c_1|u| - \frac{1}{2}c_1^2 & \text{if } c_1 < |u/c| \leq c_2 \\ c_1 \frac{c_3|u| - \frac{1}{2}u^2}{c_3 - c_2} & \text{if } c_2 < |u/c| \leq c_3 \\ c_1(c_2 + c_3 - c_1) & \text{if } |u/c| > c_3. \end{cases} \tag{A2}$$

The first derivative is piece-wise linear and vanishes outside the interval $[-c_3, +c_3]$. The crucial tuning constant is c_3. Huber and Ronchetti [26] (p. 101) suggest that the slope between c_2 and c_3 should not be too steep.

Yohai and Zamar [16] introduced a ρ function which minimizes the asymptotic variance of the regression M-estimate, subject to a bound on a robustness measure called contamination sensitivity. Therefore, this function is called the **optimal ρ function**.

$$\rho(u) = \begin{cases} 1.3846 \left(\frac{u}{c}\right)^2 & \text{if } |u| \leq \frac{2}{3}c \\ 0.5514 - 2.6917 \left(\frac{u}{c}\right)^2 + 10.7668 \left(\frac{u}{c}\right)^4 - 11.6640 \left(\frac{u}{c}\right)^6 + \\ \quad +4.0375 \left(\frac{u}{c}\right)^8 & \text{if } \frac{2}{3}c < |u| \leq c \\ 1 & \text{if } |u| > c. \end{cases} \tag{A3}$$

Now, the first derivative vanishes outside the interval $[-c, +c]$. The resulting M-estimate minimizes the maximum bias under contamination distributions (locally for a small fraction of contamination), subject to achieving a desired nominal asymptotic efficiency when the data are normally distributed.

Hampel et al. [14] (p. 328) considered another optimization problem, by minimizing the asymptotic variance of the regression M-estimate, subject to a bound on the supremum of the Change of Variance Curve (CVC) of the estimate. The CVC describes the infinitesimal increment of the logarithm of the variance of the M estimator—that is by the reciprocal of Equation (16)—in the vicinity of the null normal model, in the same way that the influence function reflects the infinitesimal asymptotic bias. This leads to the **Hyperbolic Tangent ρ function**, which, for suitable constants $c, k, A, B,$ and d, is defined as

$$\rho(u) = \begin{cases} \frac{1}{2}u^2 & \text{if } |u| \leq d \\ \frac{d^2}{2} - 2\frac{A}{B}\ln\cosh[\frac{1}{2}\sqrt{\frac{(k-1)B^2}{A}}(c-|u|)] + \\ \quad +2\frac{A}{B}\ln\cosh[\frac{1}{2}\sqrt{\frac{(k-1)B^2}{A}}(c-d)] & \text{if } d \leq |u| \leq c \\ \frac{d^2}{2} + 2\frac{A}{B}\ln\cosh[\frac{1}{2}\sqrt{\frac{(k-1)B^2}{A}}(c-d)] & \text{if } |u| > c, \end{cases} \tag{A4}$$

where $0 < d < c$ is such that

$$d = \sqrt{[A(k-1)]}\tanh[\frac{1}{2}\sqrt{\frac{(k-1)B^2}{A}}(c-d)]. \tag{A5}$$

Parameters A and B are found as:

$$A = E[\psi^2(x)] \quad \text{and} \quad B = E[\psi'(x)].$$

The value of d is found by applying the Newton-Raphson method to Equation (A5). New values of A and B are obtained (through numerical integration) and the procedure is iterated to convergence. For additional details, see Hampel et al. [27]. The parameter k is defined as

$$k = \sup_x\{CVC(\psi, x)\}.$$

In Figures 3 and 5, we used a value of 4.5 for k. The right-hand panel of Figure 6 shows that, for values of bdp close to 0.5, higher efficiencies are obtained when stronger constraints are imposed on the value of CVC by decreasing k. Conversely, smaller efficiencies result for small values of bdp. Figure A1 shows the ψ function of the hyperbolic tangent estimator for two different values of k. Note that $A, B,$ and d (and, consequently, also bdp and *eff*) are automatically determined after fixing k and c.

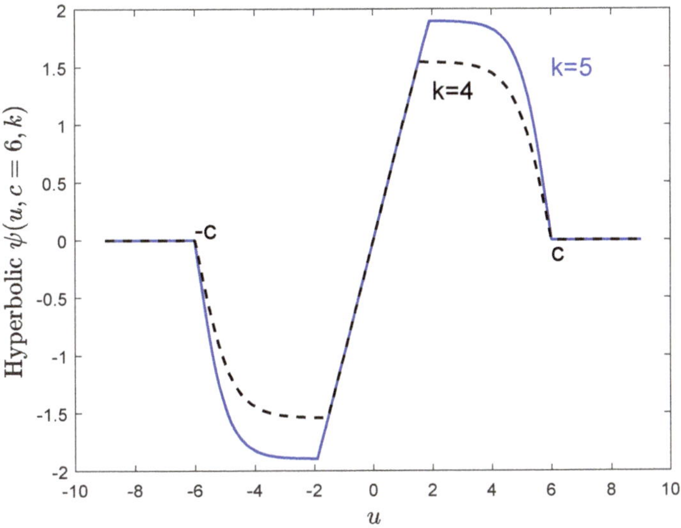

Figure A1. Hyperbolic tangent ψ function for two values of the parameter k.

We have illustrated the use of the power divergence ρ function in regression. But all these ρ functions can also be used for the estimation of robust location and covariance in the analysis of multivariate data. In this case, the scaled residuals u are replaced by scaled Mahalanobis distances.

All the functions $\rho(x)$, $\psi(x)$, $w(x) = \psi(x)/x$, $\psi'(x)$, and $\psi(x)x$ described in this appendix have been implemented in the FSDA MATLAB toolbox, which is freely downloadable from the file exchange of Mathworks. Each .m file has associated HTML documentation which is also present at web address "http://rosa.unipr.it/FSDA". The prefixes of the different links which have been used are "HU", "TB", "OPT", "HA", "HYP", and "PD". The suffixes for the different ingredients are "rho", "psi", "wei", "psider", and "psix". For example, to see the corresponding documentation for the hyperbolic ρ function, visit "http://rosa.unipr.it/FSDA/HYPrho.html". For the corresponding documentation of the derivative of the ψ function of Hampel, see "http://rosa.unipr.it/FSDA/HApsider.html". The routines for finding the constant c associated with a particular value of the breakdown point end with the suffix bdp. For example, to compute the constant c associated with the Tukey biweight for a given bdp, type "http://rosa.unipr.it/FSDA/TBbdp.html". The routines to find the constant c associated with a particular value of the efficiency end with the suffix eff. Finally, the routines which, given a particular value of c compute bdp and *eff*, end with the suffix c. For example, to compute bdp and *eff* for the power divergence estimator given c, call function PDc (the corresponding documentation is on the web at "http://rosa.unipr.it/FSDA/PDc.html").

References

1. Basu, A.; Harris, I.R.; Hjort, N.L.; Jones, M.C. Robust and efficient estimation by minimizing a density power divergence. *Biometrika* **1998**, *85*, 549–559. [CrossRef]
2. Riani, M.; Cerioli, A.; Torti, F. On consistency factors and efficiency of robust S-estimators. *TEST* **2014**, *23*, 356–387. [CrossRef]
3. Scott, D.W. Parametric Statistical Modeling by Minimum Integrated Square Error. *Technometrics* **2001**, *43*, 274–285. [CrossRef]
4. Ghosh, A.; Basu, A. Robust estimation for independent non-homogeneous observations using density power divergence with applications to linear regression. *Electron. J. Stat.* **2013**, *7*, 2420–2456. [CrossRef]
5. Durio, A.; Isaia, E.D. The minimum density power divergence approach in building robust regression models. *Informatica (Lithuania)* **2011**, *22*, 43–56.
6. Warwick, J.; Jones, M.C. Choosing a robustness tuning parameter. *J. Stat. Comput. Simul.* **2005**, *75*, 581–588. [CrossRef]
7. Ghosh, A.; Basu, A. Robust estimation for non-homogeneous data and the selection of the optimal tuning parameter: the density power divergence approach. *J. Appl. Stat.* **2015**, *42*, 2056 – 2072. [CrossRef]
8. Rousseeuw, P.J.; Yohai, V.J. Robust regression by means of S-estimators. In *Robust and Nonlinear Time Series Analysis: Lecture Notes in Statistics 26*; Franke, J., Härdle, W., Martin, R.D., Eds.; Springer: New York, NY, USA, 1984; pp. 256–272.
9. Maronna, R.A.; Martin, R.D.; Yohai, V.J. *Robust Statistics: Theory and Methods*; Wiley: Chichester, UK, 2006.
10. Rousseeuw, P.J.; Leroy, A.M. *Robust Regression and Outlier Detection*; Wiley: New York, NY, USA, 1987.
11. Basu, A.; Harris, I.R.; Hjort, N.L.; Jones, M.C. *Robust and Efficient Estimation by Minimising a Density Power Divergence*; Technical Report, 7; Department of Mathematics, University of Oslo: Oslo, Norway, 1997.
12. Salibian-Barrera, M.; Yohai, V. A fast algorithm for S-regression estimates. *J. Comput. Graph. Stat.* **2006**, *15*, 414–427. [CrossRef]
13. Beaton, A.E.; Tukey, J.W. The fitting of power series, meaning polynomials, illustrated on band-spectroscopic data. *Technometrics* **1974**, *16*, 147–185. [CrossRef]
14. Hampel, F.; Ronchetti, E.M.; Rousseeuw, P.; Stahel, W.A. *Robust Statistics*; Wiley: New York, NY, USA, 1986.
15. Huber, P.J. Robust Regression: Asymptotics, Conjectures and Monte Carlo. *Ann. Stat.* **1973**, *1*, 799–821. [CrossRef]
16. Yohai, V.J.; Zamar, R.H. Optimal locally robust M-estimates of regression. *J. Stat. Plan. Inference* **1997**, *64*, 309–323. [CrossRef]
17. Hössjer, O. On the optimality of S-estimators. *Stat. Probabil. Lett.* **1992**, *14*, 413–419. [CrossRef]
18. Salini, S.; Cerioli, A.; Laurini, F.; Riani, M. Reliable Robust Regression Diagnostics. *Int. Stat. Rev.* **2015**, *84*, 99–127. [CrossRef]

19. Jones, M.C.; Hjort, N.L.; Harris, I.R.; Basu, A. A comparison of related density-based minimum divergence estimators. *Biometrika* **2001**, *88*, 865–873. [CrossRef]
20. Riani, M.; Cerioli, A.; Atkinson, A.C.; Perrotta, D. Monitoring Robust Regression. *Electron. J. Stat.* **2014**, *8*, 642–673. [CrossRef]
21. Atkinson, A.C.; Riani, M. *Robust Diagnostic Regression Analysis*; Springer: New York, NY, USA, 2000.
22. Rousseeuw, P.J. Least median of squares regression. *J. Am. Stat. Assoc.* **1984**, *79*, 871–880. [CrossRef]
23. Atkinson, A.C.; Riani, M. Distribution theory and simulations for tests of outliers in regression. *J. Comput. Graph. Stat.* **2006**, *15*, 460–476. [CrossRef]
24. Perrotta, D.; Riani, M.; Torti, F. New robust dynamic plots for regression mixture detection. *Adv. Data Anal. Classi.* **2009**, *3*, 263–279. doi:10.1007/s11634-009-0050-y. [CrossRef]
25. Atkinson, A.C.; Riani, M.; Cerioli, A. The Forward Search: theory and data analysis (with discussion). *J. Korean Stat. Soc.* **2010**, *39*, 117–134. doi:10.1016/j.jkss.2010.02.007. [CrossRef]
26. Huber, P.J.; Ronchetti, E.M. *Robust Statistics*, 2nd ed.; Wiley: New York, NY, USA, 2009.
27. Hampel, F.; Rousseeuw, P.; Ronchetti, E. The change-of-variance curve and optimal redescending M-estimators. *J. Am. Stat. Assoc.* **1985**, *76*, 643–648. [CrossRef]

Article

Robust Model Selection Criteria Based on Pseudodistances

Aida Toma [1,2,*], Alex Karagrigoriou [3] and Paschalini Trentou [3]

[1] Department of Applied Mathematics, Bucharest University of Economic Studies, 010164 Bucharest, Romania
[2] "Gh. Mihoc - C. Iacob" Institute of Mathematical Statistics and Applied Mathematics, Romanian Academy, 010164 Bucharest, Romania
[3] Department of Statistics and Actuarial-Financial Mathematics, Lab of Statistics and Data Analysis, University of the Aegean, 83200 Karlovasi, Greece; alex.karagrigoriou@aegean.gr (A.K.); sasm17025@sas.aegean.gr (P.T.)
* Correspondence: aida.toma@csie.ase.ro

Received: 17 February 2020; Accepted: 3 March 2020; Published: 6 March 2020

Abstract: In this paper, we introduce a new class of robust model selection criteria. These criteria are defined by estimators of the expected overall discrepancy using pseudodistances and the minimum pseudodistance principle. Theoretical properties of these criteria are proved, namely asymptotic unbiasedness, robustness, consistency, as well as the limit laws. The case of the linear regression models is studied and a specific pseudodistance based criterion is proposed. Monte Carlo simulations and applications for real data are presented in order to exemplify the performance of the new methodology. These examples show that the new selection criterion for regression models is a good competitor of some well known criteria and may have superior performance, especially in the case of small and contaminated samples.

Keywords: model selection; minimum pseudodistance estimation; Robustness

1. Introduction

Model selection is fundamental to the practical applications of statistics and there is a substantial literature on this issue. Classical model selection criteria include, among others, the C_p-criterion, the Akaike Information Criterion (AIC), based on the Kullback-Leibler divergence, and the Bayesian Information Criterion (BIC) as well as a General Information Criterion (GIC) which corresponds to a general class of criteria which also estimates the Kullback-Leibler divergence. These criteria have been proposed respectively in [1–4], and represent powerful tools for choosing the best model among different candidate models that can be used to fit a given data set. On the other hand, many classical procedures for model selection are extremely sensitive to outliers and to other departures from the distributional assumptions of the model. Robust versions of classical model selection criteria, which are not strongly affected by outliers, have been proposed for example in [5–7]. Some recent proposals for robust model selection are criteria based on divergences and minimum divergence estimators. We recall here, the Divergence Information Criteria (DIC) based on the density power divergences introduced in [8], the Modified Divergence Information Criteria (MDIC) introduced in [9] and the criteria based on minimum dual divergence estimators introduced in [10].

The interest on statistical methods based on divergence measures has grown significantly in recent years. For a wide variety of models, statistical methods based on divergences have high model efficiency and are also robust, representing attractive alternatives to the classical methods. We refer to the monographs [11,12] for an excellent presentation of such methods, for their importance and applications. The pseudodistances that we use in the present paper were originally introduced in [13], where they are called "type-0" divergences, and corresponding minimum divergence estimators have been studied. They are also presented and extensively studied in [14] where they are called γ-divergences, as well as in [15] in the context of decomposable pseudodistances. Like divergences, the pseudodistances are not mathematical metrics in the strict sense of the term. They satisfy two properties, namely the nonnegativity and the fact that the pseudodistance between two probability measures equals to zero if and only if the two measures are equal. The divergences are moreover characterized by the information processing property, that is, the complete invariance with respect to statistically sufficient transformations of the observation space. In general, a pseudodistance may not satisfy this property. We have adopted the term pseudodistance for this reason, but in literature we can also encounter the other terms mentioned above.

The pseudodistances that we consider in this paper have also been used to define robustness and efficiency measures, as well as the corresponding optimal robust M-estimators following the Hampel's infinitesimal approach in [16]. The minimum pseudodistance estimators for general parametric models have been studied in [15] and consist of minimizing an empirical version of a pseudodistance between the assumed theoretical model and the true model underlying the data. These estimators have the advantage of not requiring any prior smoothing and conciliate robustness with high efficiency, providing a high degree of stability under model misspecification, often with a minimal loss in model efficiency. Such estimators are also defined and studied in the case of the multivariate normal model, as well as for linear regression models in [17,18], where applications for portfolio optimization models are also presented.

In the present paper we propose new criteria for model selection, based on pseudodistances and on minimum pseudodistance estimators. These new criteria have robustness properties, are asymptotically unbiased, consistent and compare well with some other known model selection criteria, even for small samples.

The paper is organized as follows—Section 2 is devoted to minimum pseudodistance estimators and to their asymptotic properties, which will be needed in the next sections. Section 3 presents new estimators of the expected overall discrepancy using pseudodistances, together with corresponding theoretical properties including robustness, consistency and limit laws. The new asymptotically unbiased model selection criteria are presented in Section 3.3, where the case of the univariate normal model and the case of linear regression models are investigated. Applications based on Monte Carlo simulations and on real data, illustrating the performance of the new methodology in the case of linear regression models, are included in Section 4.

2. Minimum Pseudodistance Estimators

The construction of new model selection criteria is based on using the following family of pseudodistances (see [15]). For two probability measures P and Q admitting densities p and q respectively with respect to the Lebesgue measure, the family of pseudodistances of order $\gamma > 0$ is defined by

$$R_\gamma(P, Q) = \frac{1}{\gamma + 1} \ln \left(\int p^\gamma dP \right) + \frac{1}{\gamma(\gamma + 1)} \ln \left(\int q^\gamma dQ \right) - \frac{1}{\gamma} \ln \left(\int p^\gamma dQ \right) \tag{1}$$

and satisfies the limit relation

$$\lim_{\gamma \to 0} R_\gamma(P, Q) = R_0(P, Q), \tag{2}$$

where $R_0(P, Q) := \int \ln \frac{q}{p} dQ$ is the modified Kullback-Leibler divergence.

Let (P_θ) be a parametric model indexed by $\theta \in \Theta$, where Θ is a d-dimensional parameter space, and p_θ be the corresponding densities with respect to the Lebesgue measure λ. Let $X_1, \ldots, X_n$ be a random sample on P_{θ_0}, $\theta_0 \in \Theta$. For $\gamma > 0$ fixed, a minimum pseudodistance estimator of the unknown parameter θ_0 from the law P_{θ_0} is defined by replacing the measure P_{θ_0} in the pseudodistance $R_\gamma(P_\theta, P_{\theta_0})$ by the empirical measure P_n pertaining to the sample, and then minimizing this empirical quantity with respect to θ on the parameter space. Since the middle term in $R_\gamma(P_\theta, P_{\theta_0})$ does not depend on θ, these estimators are defined by

$$\widehat{\theta}_n = \arg\min_{\theta \in \Theta} \left\{ \frac{1}{\gamma+1} \ln\left(\int p_\theta^{\gamma+1} d\lambda\right) - \frac{1}{\gamma} \ln\left(\frac{1}{n} \sum_{i=1}^{n} p_\theta^\gamma(X_i)\right), \right\} \tag{3}$$

or equivalently as

$$\widehat{\theta}_n = \arg\max_{\theta \in \Theta} \{ C_\gamma(\theta)^{-1} \cdot \frac{1}{n} \sum_{i=1}^{n} p_\theta^\gamma(X_i) \}, \tag{4}$$

where $C_\gamma(\theta) = (\int p_\theta^{\gamma+1} d\lambda)^{\gamma/(\gamma+1)}$. Denoting $h(x,\theta) := C_\gamma(\theta)^{-1} \cdot p_\theta^\gamma(x)$, these estimators can be written as

$$\widehat{\theta}_n = \arg\max_{\theta \in \Theta} \frac{1}{n} \sum_{i=1}^{n} h(X_i, \theta). \tag{5}$$

The optimum given above need not be uniquely defined.

On the other hand,

$$\arg\max_{\theta \in \Theta} \int h(x,\theta) dP_{\theta_0}(x) = \theta_0 \tag{6}$$

and here θ_0 is the unique optimizer, since $R_\gamma(P_\theta, P_{\theta_0}) = 0$ implies $\theta = \theta_0$.

Define

$$R_\gamma(\theta_0) := \max_{\theta \in \Theta} \int h(x,\theta) dP_{\theta_0}(x) = \int h(x,\theta_0) dP_{\theta_0}(x).$$

An estimator of $R_\gamma(\theta_0)$ is defined by

$$\widehat{R}_\gamma(\theta_0) := \max_{\theta \in \Theta} \int h(x,\theta) dP_n(x) = \max_{\theta \in \Theta} \frac{1}{n} \sum_{i=1}^{n} h(X_i, \theta) = \frac{1}{n} \sum_{i=1}^{n} h(X_i, \widehat{\theta}_n). \tag{7}$$

The following regularity conditions of the model will be assumed throughout the rest of the paper.

(C1) The density $p_\theta(x)$ has continuous partial derivatives with respect to θ up to the third order (for all x λ-a.e.).

(C2) There exists a neighborhood N_{θ_0} of θ_0 such that the first-, the second- and the third- order partial derivatives with respect to θ of $h(x,\theta)$ are dominated on N_{θ_0} by some P_{θ_0}-integrable functions.

(C3) The integrals $\int [\frac{\partial^2}{\partial \theta^2} h(x,\theta)]_{\theta=\theta_0} dP_{\theta_0}(x)$ and $\int [\frac{\partial}{\partial \theta} h(x,\theta)]_{\theta=\theta_0} [\frac{\partial}{\partial \theta} h(x,\theta)]_{\theta=\theta_0}^t dP_{\theta_0}(x)$ exist.

Theorem 1. *Assume that conditions (C1), (C2) and (C3) are fulfilled. Then*

(a) *Let $B := \left\{ \theta \in \Theta; \|\theta - \theta_0\| \leq n^{-1/3} \right\}$. Then, as $n \to \infty$, with probability one, the function $\theta \mapsto \frac{1}{n} \sum_{i=1}^{n} h(X_i, \theta)$ attains a local maximal value at some point $\widehat{\theta}_n$ in the interior of B, which implies that the estimator $\widehat{\theta}_n$ is $n^{1/3}$-consistent.*

(b) $\sqrt{n} \left(\widehat{\theta}_n - \theta_0 \right)$ *converges in distribution to a centered multivariate normal random variable with covariance matrix*

$$V = S^{-1} M S^{-1}, \tag{8}$$

where $S := - \int [\frac{\partial^2}{\partial \theta^2} h(x,\theta)]_{\theta=\theta_0} dP_{\theta_0}(x)$ and $M := \int [\frac{\partial}{\partial \theta} h(x,\theta)]_{\theta=\theta_0} [\frac{\partial}{\partial \theta} h(x,\theta)]_{\theta=\theta_0}^t dP_{\theta_0}(x)$.

(c) $\sqrt{n} \left(\widehat{R}_\gamma(\theta_0) - R_\gamma(\theta_0) \right)$ *converges in distribution to a centered normal variable with variance $\sigma^2(\theta_0) = \int h(x,\theta_0)^2 dP_{\theta_0}(x) - \left(\int h(x,\theta_0) dP_{\theta_0}(x) \right)^2$.*

We refer to [15] for details regarding these estimators and for the proofs of the above asymptotic properties.

3. Model Selection Criteria Based on Pseudodistances

Model selection is a method for selecting the best model among candidate models that can be used to fit a given data set. A model selection criterion can be considered as an approximately unbiased estimator of the expected overall discrepancy, a nonnegative quantity which measures the distance between the true unknown model and a fitted approximating model. If the value of the criterion is small, then the approximated candidate model can be chosen. In the following, by applying the same methodology used for AIC, we construct new criteria for model selection using pseudodistances (1) and minimum pseudodistance estimators.

Let $X_1, \ldots, X_n$ be a random sample from the distribution associated with the true model Q with density q and let p_θ be the density of a candidate model P_θ from a parametric family (P_θ), where $\theta \in \Theta \subset \mathbb{R}^d$.

3.1. The Expected Overall Discrepancy

For $\gamma > 0$ fixed, we consider the quantity

$$W_\theta = \frac{1}{\gamma + 1} \ln \left(\int p_\theta^{\gamma+1} d\lambda \right) - \frac{1}{\gamma} \ln \left(\int p_\theta^\gamma q d\lambda \right), \tag{9}$$

which is the same as the pseudodistance $R_\gamma(P_\theta, Q)$ without the middle term that remains constant irrespectively of the model (P_θ) used.

The target theoretical quantity that will be approximated by an asymptotically unbiased estimator is given by

$$E[W_{\widehat{\theta}_n}] = E[W_\theta | \theta = \widehat{\theta}_n], \tag{10}$$

where $\widehat{\theta}_n$ is a minimum pseudodistance estimator defined as in (3). The same pseudodistance is used for both W_θ and $\widehat{\theta}_n$. The quantity (10) can be seen as an average distance between Q and (P_θ) up to a constant and is called *the expected overall discrepancy* between Q and (P_θ).

The next Lemma gives the gradient vector and the Hessian matrix of W_θ and is useful for the evaluation of $E[W_{\widehat{\theta}_n}]$ through Taylor expansion.

Throughout this paper, for a scalar function $\varphi_\theta(\cdot)$, the quantity $\frac{\partial}{\partial \theta} \varphi_\theta(\cdot)$ denotes the d-dimensional gradient vector of $\varphi_\theta(\cdot)$ with respect to the vector θ and $\frac{\partial^2}{\partial \theta^2} \varphi_\theta(\cdot)$ denotes the corresponding $d \times d$ Hessian matrix. We also use the notations $\dot{\varphi}_\theta$ and $\ddot{\varphi}_\theta$ for the first and the second order derivatives of φ_θ with respect to θ.

We assume the following conditions allowing derivation under the integral sign:

(C4) There exists a neighborhood N_θ of θ such that

$$\int \sup_{t \in N_\theta} \left\| \frac{\partial}{\partial t} p_t^{\gamma+1} \right\| d\lambda < \infty, \quad \int \sup_{t \in N_\theta} \left\| \frac{\partial}{\partial t} [p_t^\gamma \dot{p}_t] \right\| d\lambda < \infty.$$

(C5) There exists a neighborhood N_θ of θ such that

$$\int \sup_{t \in N_\theta} \left\| \frac{\partial}{\partial t} p_t^\gamma \right\| q d\lambda < \infty, \quad \int \sup_{t \in N_\theta} \left\| \frac{\partial}{\partial t} [p_t^{\gamma-1} \dot{p}_t] \right\| q d\lambda < \infty.$$

Lemma 1. *Under (C4) and (C5), the gradient vector and the Hessian matrix of W_θ are*

$$\frac{\partial}{\partial \theta} W_\theta = \frac{\int p_\theta^\gamma \dot{p}_\theta d\lambda}{\int p_\theta^{\gamma+1} d\lambda} - \frac{\int p_\theta^{\gamma-1} \dot{p}_\theta q d\lambda}{\int p_\theta^\gamma q d\lambda} \tag{11}$$

$$\frac{\partial^2}{\partial\theta^2}W_\theta = \frac{[\gamma\int p_\theta^{\gamma-1}\dot p_\theta\dot p_\theta^t\,d\lambda + \int p_\theta^\gamma\ddot p_\theta\,d\lambda]\int p_\theta^{\gamma+1}\,d\lambda - (\gamma+1)\int p_\theta^\gamma\dot p_\theta\,d\lambda(\int p_\theta^\gamma\dot p_\theta\,d\lambda)^t}{(\int p_\theta^{\gamma+1}\,d\lambda)^2}$$

$$- \frac{[(\gamma-1)\int p_\theta^{\gamma-2}\dot p_\theta\dot p_\theta^t q\,d\lambda + \int p_\theta^{\gamma-1}\ddot p_\theta q\,d\lambda]\int p_\theta^\gamma q\,d\lambda - \gamma\int p_\theta^{\gamma-1}\dot p_\theta q\,d\lambda(\int p_\theta^{\gamma-1}\dot p_\theta q\,d\lambda)^t}{(\int p_\theta^\gamma q\,d\lambda)^2}.$$

When the true model Q belongs to the parametric model (P_θ), hence $Q = P_{\theta_0}$ and $q = p_{\theta_0}$, the gradient vector and the Hessian matrix of W_θ simplify to

$$\left[\frac{\partial}{\partial\theta}W_\theta\right]_{\theta=\theta_0} = 0 \tag{12}$$

$$\left[\frac{\partial^2}{\partial\theta^2}W_\theta\right]_{\theta=\theta_0} = M_\gamma(\theta_0) \tag{13}$$

where

$$M_\gamma(\theta_0) := \frac{(\int p_{\theta_0}^{\gamma-1}\dot p_{\theta_0}\dot p_{\theta_0}^t\,d\lambda)(\int p_{\theta_0}^{\gamma+1}\,d\lambda) - (\int p_{\theta_0}^\gamma\dot p_{\theta_0}\,d\lambda)(\int p_{\theta_0}^\gamma\dot p_{\theta_0}\,d\lambda)^t}{(\int p_{\theta_0}^{\gamma+1}\,d\lambda)^2}. \tag{14}$$

In the following Propositions we suppose that the true model Q belongs to the parametric model (P_θ), hence $Q = P_{\theta_0}$, $q = p_{\theta_0}$ and θ_0 is the value of the parameter corresponding to the true model $Q = P_{\theta_0}$. We also say that θ_0 is the true value of the parameter.

Proposition 1. *When the true model Q belongs to the parametric model (P_θ), assuming that (C4) and (C5) are fulfilled for $q = p_{\theta_0}$ and $\theta = \theta_0$, the expected overall discrepancy is given by*

$$E[W_{\widehat\theta_n}] = W_{\theta_0} + \frac{1}{2}E[(\widehat\theta_n - \theta_0)^t M_\gamma(\theta_0)(\widehat\theta_n - \theta_0)] + E[R_n], \tag{15}$$

where $R_n = o(\|\widehat\theta_n - \theta_0\|^2)$, $M_\gamma(\theta_0)$ is given by (14).

3.2. Estimation of the Expected Overall Discrepancy

In this section, we introduce an estimator of the expected overall discrepancy, under the hypothesis that the true model Q belongs to the parametric model (P_θ). Hence, $Q = P_{\theta_0}$ and the unknown parameter θ_0 will be estimated by a minimum pseudodistance estimator $\widehat\theta_n$.

For a given $\theta \in \Theta$, a natural estimator of W_θ is defined by

$$Q_\theta := \frac{1}{\gamma+1}\ln\left(\int p_\theta^{\gamma+1}\,d\lambda\right) - \frac{1}{\gamma}\ln\left(\frac{1}{n}\sum_{i=1}^n p_\theta^\gamma(X_i)\right). \tag{16}$$

Lemma 2. *Assuming (C4), the gradient vector and the Hessian matrix of Q_θ are given by*

$$\frac{\partial}{\partial\theta}Q_\theta = \frac{\int p_\theta^\gamma\dot p_\theta\,d\lambda}{\int p_\theta^{\gamma+1}\,d\lambda} - \frac{\sum_{i=1}^n p_\theta^{\gamma-1}(X_i)\dot p_\theta(X_i)}{\sum_{i=1}^n p_\theta^\gamma(X_i)}$$

$$\frac{\partial^2}{\partial\theta^2}Q_\theta = \frac{[\gamma\int p_\theta^{\gamma-1}\dot p_\theta\dot p_\theta^t\,d\lambda + \int p_\theta^\gamma\ddot p_\theta\,d\lambda]\int p_\theta^{\gamma+1}\,d\lambda - (\gamma+1)\int p_\theta^\gamma\dot p_\theta\,d\lambda(\int p_\theta^\gamma\dot p_\theta\,d\lambda)^t}{(\int p_\theta^{\gamma+1}\,d\lambda)^2} -$$

$$- \frac{[(\gamma-1)\sum_{i=1}^n p_\theta^{\gamma-2}(X_i)\dot p_\theta(X_i)\dot p_\theta(X_i)^t + \sum_{i=1}^n p_\theta^{\gamma-1}(X_i)\ddot p_\theta(X_i)]\sum_{i=1}^n p_\theta^\gamma(X_i)}{(\sum_{i=1}^n p_\theta^\gamma(X_i))^2}$$

$$\frac{\gamma(\sum_{i=1}^n p_\theta^{\gamma-1}(X_i)\dot p_\theta(X_i))(\sum_{i=1}^n p_\theta^{\gamma-1}(X_i)\dot p_\theta(X_i))^t}{(\sum_{i=1}^n p_\theta^\gamma(X_i))^2}.$$

Proposition 2. *When the true model Q belongs to the parametric model (P_θ), by imposing the conditions (C1)-(C5), it holds*

$$E[Q_{\theta_0}] = E[Q_{\widehat{\theta}_n}] + \frac{1}{2}E[(\theta_0 - \widehat{\theta}_n)^t M_\gamma(\theta_0)(\theta_0 - \widehat{\theta}_n)] + E[R_n], \tag{17}$$

where $R_n = o(\|\widehat{\theta}_n - \theta_0\|^2)$.

The following result allows to define an asymptotically unbiased estimator of the expected overall discrepancy.

Proposition 3. *When the true model Q belongs to the parametric model (P_θ), under (C1)-(C5), it holds*

$$\begin{aligned} E[W_{\widehat{\theta}_n}] &= E[Q_{\widehat{\theta}_n}] + E[(\theta_0 - \widehat{\theta}_n)^t M_\gamma(\theta_0)(\theta_0 - \widehat{\theta}_n)] + \\ &\quad + \frac{1}{2\gamma n}\left[1 - \frac{\int p_{\theta_0}^{2\gamma+1}d\lambda}{\left(\int p_{\theta_0}^{\gamma+1}d\lambda\right)^2}\right] + E[R_n] + \frac{1}{\gamma}E[R'_n], \end{aligned} \tag{18}$$

where $R_n = o(\|\widehat{\theta}_n - \theta_0\|^2)$ and $R'_n = o(\|\frac{1}{n}\sum_{i=1}^n p_{\theta_0}^\gamma(X_i) - \int p_{\theta_0}^{\gamma+1}d\lambda\|^2)$.

3.2.1. Limit Properties of the Estimator $Q_{\widehat{\theta}_n}$

Under the hypothesis that the true model Q belongs to the family of models (P_θ), hence $Q = P_{\theta_0}$, we prove the consistency and the asymptotic normality for the estimator $Q_{\widehat{\theta}_n}$.

Note that

$$\begin{aligned} Q_{\widehat{\theta}_n} &= \frac{1}{\gamma+1}\ln\left(\int p_{\widehat{\theta}_n}^{\gamma+1}d\lambda\right) - \frac{1}{\gamma}\ln\left(\frac{1}{n}\sum_{i=1}^n p_{\widehat{\theta}_n}^\gamma(X_i)\right) \tag{19} \\ &= -\ln\left[\frac{\frac{1}{n}\sum_{i=1}^n p_{\widehat{\theta}_n}(X_i)}{(\int p_{\widehat{\theta}_n}^{\gamma+1}d\lambda)^{\frac{\gamma}{\gamma+1}}}\right]^{\frac{1}{\gamma}} = -\ln[\widehat{R}_\gamma(\theta_0)]^{\frac{1}{\gamma}}, \tag{20} \end{aligned}$$

where $\int p_{\widehat{\theta}_n}^{\gamma+1}d\lambda = \left[\int p_\theta^{\gamma+1}d\lambda\right]_{\theta=\widehat{\theta}_n}$ and $\widehat{R}_\gamma(\theta_0)$ is given by (7).

First we prove that $\widehat{R}_\gamma(\theta_0)$ is a consistent estimator of $R_\gamma(\theta_0)$. Indeed, using Theorem 1 and the fact that $\int \frac{\partial}{\partial\theta}h(x,\theta_0)dP_{\theta_0}(x) = 0$, a Taylor expansion of $\frac{1}{n}\sum_{i=1}^n h(X_i,\theta)$ in $\widehat{\theta}_n$ around θ_0 gives

$$\widehat{R}_\gamma(\theta_0) = \frac{1}{n}\sum_{i=1}^n h(X_i,\theta_0) + o_P(n^{-1/2}). \tag{21}$$

Using the weak law of large numbers,

$$\frac{1}{n}\sum_{i=1}^n h(X_i,\theta_0) = R_\gamma(\theta_0) + o_P(1). \tag{22}$$

Combining (21) and (22), we obtain that $\widehat{R}_\gamma(\theta_0)$ converges to $R_\gamma(\theta_0)$ in probability.

Then, using the continuous mapping theorem, since $g(t) = -\ln t^{\frac{1}{\gamma}}$ is a continuous function, we get

$$Q_{\widehat{\theta}_n} = -\ln[\widehat{R}_\gamma(\theta_0)]^{\frac{1}{\gamma}} \to -\ln[R_\gamma(\theta_0)]^{\frac{1}{\gamma}} = W_{\theta_0}$$

in probability.

On the other hand, using the asymptotic normality of the estimator $\widehat{R}_\gamma(\theta_0)$ (according to Theorem 1 (c)) together with the univariate delta method, we obtain the asymptotic normality of $Q_{\widehat{\theta}_n}$. The Proposition below summarizes the above asymptotic results.

Proposition 4. *Under (C1)-(C3), when $Q = P_{\theta_0}$, it holds*

(a) $Q_{\widehat{\theta}_n}$ converges to W_{θ_0} in probability.

(b) $\sqrt{n}(Q_{\widehat{\theta}_n} - W_{\theta_0})$ converges in distribution to a centered univariate normal random variable with variance $\frac{\sigma^2(\theta_0)}{\gamma^2 R_\gamma(\theta_0)^2}$, $\sigma^2(\theta_0)$ being defined in Theorem 1.

3.2.2. Robustness Properties of the Estimator $Q_{\widehat{\theta}_n}$

The influence function is a useful tool for describing robustness of an estimator. Recall that, a map T defined on a set of probability measures and parameter space valued is a statistical functional corresponding to an estimator $\widehat{\theta}_n$ of the parameter θ, whenever $\widehat{\theta}_n = T(P_n)$, where P_n is the empirical measure associated to the sample. The influence function of T at P_θ is defined by

$$\mathrm{IF}(x; T, P_\theta) := \left. \frac{\partial T(\widetilde{P}_{\varepsilon x})}{\partial \varepsilon} \right|_{\varepsilon=0},$$

where $\widetilde{P}_{\varepsilon x} := (1 - \varepsilon)P_\theta + \varepsilon \delta_x$, $\varepsilon > 0$, δ_x being the Dirac measure putting all mass at x. The gross error sensitivity of the estimator is defined by

$$\gamma^*(T, P_\theta) = \sup_x \|\mathrm{IF}(x; T, P_\theta)\|.$$

Whenever the influence function is bounded with respect to x, the corresponding estimator is called B-robust (see [19]).

In what follows, for a given $\gamma > 0$, we derive the influence function of the estimator $Q_{\widehat{\theta}_n}$. The statistical functional associated with this estimator, which we denote by U, is defined by

$$U(P) := \frac{1}{\gamma + 1} \ln \left(\int p_{T(P)}^{\gamma+1} d\lambda \right) - \frac{1}{\gamma} \ln \left(\int p_{T(P)}^\gamma dP \right),$$

where T is the statistical functional corresponding to the used minimum pseudodistance estimator estimator $\widehat{\theta}_n$, namely

$$T(P) := \arg\sup_\theta C_\gamma(\theta)^{-1} \int p_\theta^\gamma dP$$

where $C_\gamma(\theta) = (\int p_\theta^{\gamma+1} d\lambda)^{\gamma/(\gamma+1)}$.

Due to the Fisher consistency of the functional T, according to (6), we have $T(P_{\theta_0}) = \theta_0$ which implies that $U(P_{\theta_0}) = W_{\theta_0}$.

Proposition 5. *When $Q = P_{\theta_0}$, the influence function of $Q_{\widehat{\theta}_n}$ is given by*

$$\mathrm{IF}(x; U, P_{\theta_0}) = \frac{1}{\gamma} \left[1 - \frac{p_{\theta_0}^\gamma(x)}{\int p_{\theta_0}^{\gamma+1} d\lambda} \right]. \tag{23}$$

Note that the influence function of the estimator $Q_{\widehat{\theta}_n}$ does not depend on the estimator $\widehat{\theta}_n$, but depends on the used pseudodistance. Usually, $p_{\theta_0}^\gamma(x)$ is bounded with respect to x and therefore $Q_{\widehat{\theta}_n}$ is a robust estimator with respect to W_{θ_0}.

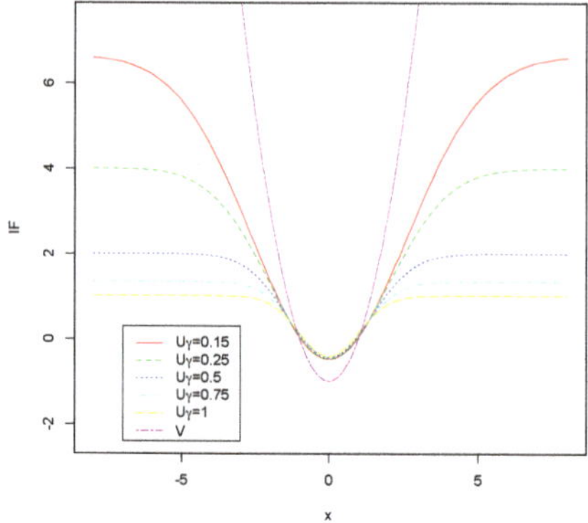

Figure 1. Influence functions in the case of the normal model.

For comparison at the level of the influence function, we consider the AIC criterion which is defined by

$$AIC = -2\ln(\mathcal{L}(\widehat{\theta}_n)) + 2d,$$

where $\mathcal{L}(\widehat{\theta}_n)$ is the maximum value of the likelihood function for the model, $\widehat{\theta}_n$ the maximum likelihood estimator and d the dimension of the parameter. The statistical functional corresponding to the statistic $-2\ln(\mathcal{L}(\widehat{\theta}_n))$ is

$$V(P) = -2\int \ln p_{T(P)}dP$$

where T here is the statistical functional corresponding to the maximum likelihood estimator. The influence function of the functional V is given by

$$IF(x; V, P_{\theta_0}) = 2\left[\int \ln p_{\theta_0}dP_{\theta_0} - \ln p_{\theta_0}(x)\right]. \tag{24}$$

This influence function is not bounded with respect to x, therefore the statistic $-2\ln(\mathcal{L}(\widehat{\theta}_n))$ is not robust.

For example, in the case of the univariate normal model, for a positive γ, the influence function (23) writes as

$$IF(x; U, P_{\theta_0}) = \frac{1}{\gamma}\left(1 - \sqrt{\gamma+1}\cdot\exp\left(-\frac{\gamma}{2}\left(\frac{x-m}{\sigma}\right)^2\right)\right) \tag{25}$$

while the influence function (24) writes as

$$IF(x; V, P_{\theta_0}) = \left(\frac{x-m}{\sigma}\right)^2 - \frac{2m^2}{\sigma^2} - 1 \tag{26}$$

(here $\theta_0 = (m, \sigma)$). For all the pseudodistances, the influence function (25) is bounded with respect to x, therefore the selection criteria based on the statistic $Q_{\widehat{\theta}_n}$ will be robust. On the other hand, the influence function (26) is not bounded with respect to x, showing the non robustness of AIC in this case. Moreover, the gross error sensitivities corresponding to these influence functions are $\gamma^*(U, P_{\theta_0}) = \frac{1}{\gamma}$ and $\gamma^*(V, P_{\theta_0}) = \infty$. These results show that, in the case of the normal model, when γ increases the gross error sensitivity decreases. Therefore, larger values of γ are associated with more robust procedures. For the particular case $m = 0$ and $\sigma = 1$, the influence functions (25) and (26) are represented in Figure 1.

3.3. Model Selection Criteria Using Pseudodistances

3.3.1. The Case of Univariate Normal Family

The criteria that we propose in this section correspond to the case where the candidate model is a univariate normal model from the family of normal models (P_θ) indexed by $\theta = (\mu, \sigma)$. We also suppose that the true model Q belongs to (P_θ).

In the case of the univariate normal model, $M_\gamma(\theta_0)$ defined in (14) expresses as

$$M_\gamma(\theta_0) = \frac{(\gamma+1)^2}{(2\gamma+1)^{3/2}} A(\gamma) V^{-1}, \tag{27}$$

where V is the asymptotic covariance matrix given by (8) and the matrix $A(\gamma)$ is given by

$$A(\gamma) = \begin{pmatrix} 1 & 0 \\ 0 & \frac{3\gamma^2 + 4\gamma + 2}{2(2\gamma+1)} \end{pmatrix}.$$

For small positive values of γ, the matrix $A(\gamma)$ can be approximated by the identity matrix I.

According to Theorem 1, $\sqrt{n}(\widehat{\theta}_n - \theta_0)$ is asymptotically multivariate normal and then the statistic $n(\theta_0 - \widehat{\theta}_n)^t V^{-1}(\theta_0 - \widehat{\theta}_n)$ has approximately a χ_d^2 distribution. For large n, it holds

$$E[(\theta_0 - \widehat{\theta}_n)^t M_\gamma(\theta_0)(\theta_0 - \widehat{\theta}_n)] \approx \frac{(\gamma+1)^2}{(2\gamma+1)^{3/2}} \cdot \frac{d}{n}. \tag{28}$$

Also, for the normal model, it holds

$$\frac{\int p_{\theta_0}^{2\gamma+1} d\lambda}{\left(\int p_{\theta_0}^{\gamma+1} d\lambda \right)^2} = \frac{\gamma+1}{\sqrt{2\gamma+1}}. \tag{29}$$

Therefore, (18) becomes

$$E[W_{\widehat{\theta}_n}] \cong E[Q_{\widehat{\theta}_n}] + \frac{(\gamma+1)^2}{(2\gamma+1)^{3/2}} \cdot \frac{d}{n} + \frac{1}{2\gamma n}\left[1 - \frac{\gamma+1}{\sqrt{2\gamma+1}}\right] + E[R_n] + \frac{1}{\gamma}E[R_n']. \tag{30}$$

Using the central limit theorem and asymptotic properties of $\widehat{\theta}_n$ given in Theorem 1, the following hold

$$n \cdot o(\|\widehat{\theta}_n - \theta_0\|^2) = o_P(1), \tag{31}$$

$$n \cdot o(\|\frac{1}{n}\sum_{i=1}^{n} p_{\theta_0}^\gamma(X_i) - \int p_{\theta_0}^{\gamma+1} d\lambda\|^2) = o_P(1). \tag{32}$$

Using (30), (31) and (32) we obtain:

Proposition 6. *For the univariate normal family, an asymptotically unbiased estimator of the expected overall discrepancy is given by*

$$Q_{\widehat{\theta}_n} + \frac{(\gamma+1)^2}{(2\gamma+1)^{3/2}} \cdot \frac{d}{n} + \frac{1}{2\gamma n}\left[1 - \frac{\gamma+1}{\sqrt{2\gamma+1}}\right], \tag{33}$$

where $\widehat{\theta}_n$ is a minimum pseudodistance estimator given by (3).

Under the hypothesis that (P_θ) is the univariate normal model, as we supposed in this subsection, the function h writes as

$$h(x,\theta) = (\sqrt{\gamma+1})^{\gamma/(\gamma+1)} \cdot (\sigma\sqrt{2\pi})^{-\gamma/(\gamma+1)} \cdot \exp\left(-\frac{\gamma}{2}\left(\frac{x-m}{\sigma}\right)^2\right) \tag{34}$$

and it can be easily checked that all the conditions (C1)–(C5) are fulfilled. Therefore we can use all results presented in the preceding subsections, such that Proposition 6 is fully justified.

Moreover, the selection criteria based on (33) are consistent on the basis of Proposition 4. It should also be noted that the bias correction term in (33) decreases slowly as the parameter γ increases staying always very close to zero ($\sim 10^{-2}$). As expected, the larger the sample size the smaller the bias correction. As we saw in Section 3.2.2, since the gross error sensitivity of $Q_{\hat{\theta}_n}$ is $\gamma^*(U, P_{\theta_0}) = \frac{1}{\gamma}$, larger values of γ are associated with more robust procedures. On the other hand, the approximation of $A(\gamma)$ with the identity matrix is realized for values of γ close to zero. Thus, positive values of γ smaller than 0.5 for example could represent choices satisfying the robustness requirement and the approximation of $A(\gamma)$ through the identity matrix, approximation which is necessary to construct the criterion in this case.

3.3.2. The Case of Linear Regression Models

In the following, we adapt the pseudodistance based model selection criterion in the case of linear regression models. Consider the linear regression model

$$Y = \alpha + \beta^t X + e \tag{35}$$

where $e \sim \mathcal{N}(0,\sigma)$ and e is independent of X. Suppose we have a sample given by the i.i.d. random vectors $Z_i = (X_i, Y_i)$, $i = 1, ..., n$, such that $Y_i = \alpha + \beta^t X_i + e_i$.

We consider the joint distribution of the entire data and write a pseudodistance between the theoretical model and the true model corresponding to the data. Let P_θ, $\theta := (\alpha, \beta, \sigma)$, be the probability measure associated to the theoretical model given by the random vector $Z = (X, Y)$ and Q the probability measure associated to the true model corresponding to the data. Denote by p_θ, respectively by q the corresponding densities. For $\gamma > 0$, the pseudodistance between P_θ and Q is defined by

$$R_\gamma(P_\theta, Q) \quad := \quad \frac{1}{\gamma+1} \ln\left(\int p_\theta^\gamma(x,y)dP_\theta(x,y)\right) + \frac{1}{\gamma(\gamma+1)} \ln\left(\int q^\gamma(x,y)dQ(x,y)\right) - $$
$$- \frac{1}{\gamma} \ln\left(\int p_\theta^\gamma(x,y)dQ(x,y)\right). \tag{36}$$

Similar to [18], since the middle term above does not depend on P_θ, a minimum pseudodistance estimator of the parameter $\theta_0 = (\alpha_0, \beta_0, \sigma_0)$ is defined by

$$\hat{\theta}_n = (\hat{\alpha}_n, \hat{\beta}_n, \hat{\sigma}_n) = \arg\min_{\alpha,\beta,\sigma}\left\{\frac{1}{\gamma+1}\ln\left(\int p_\theta^\gamma(x,y)dP_\theta(x,y)\right) - \frac{1}{\gamma}\ln\left(\int p_\theta^\gamma(x,y)dP_n(x,y)\right)\right\}, \tag{37}$$

where P_n is the empirical measure associated with the sample. This estimator can be written as

$$\hat{\theta}_n = (\hat{\alpha}_n, \hat{\beta}_n, \hat{\sigma}_n) = \arg\min_{\alpha,\beta,\sigma}\left\{\frac{1}{\gamma+1}\ln\left(\int \phi_\sigma^{\gamma+1}(e)de\right) - \frac{1}{\gamma}\ln\left(\frac{1}{n}\sum_{i=1}^n \phi_\sigma^\gamma(Y_i - \alpha - \beta^t X_i)\right)\right\}, \tag{38}$$

where ϕ_σ is the density of the random variable $e \sim \mathcal{N}(0,\sigma)$. Then, the estimator $Q_{\hat{\theta}_n}$ can be written as

$$Q_{\hat{\theta}_n} = \min_{\alpha,\beta,\sigma}\left\{\frac{1}{\gamma+1}\ln\left(\frac{1}{(\sigma\sqrt{2\pi})^\gamma\sqrt{\gamma+1}}\right) - \frac{1}{\gamma}\ln\left(\frac{1}{n}\sum_{i=1}^n \frac{1}{(\sigma\sqrt{2\pi})^\gamma}\cdot\exp\left(-\frac{\gamma}{2\sigma^2}(Y_i - \alpha - \beta^t X_i)^2\right)\right)\right\}. \tag{39}$$

In order to construct an asymptotic unbiased estimator of the expected overall discrepancy in the case of the linear regression models, we evaluated the second and the third terms from (18).

For values of γ close to 0 (γ smaller than 0.3), we found the following approximation of the matrix $M_\gamma(\theta_0)$

$$M_\gamma(\theta_0) \simeq \frac{(\gamma+1)^2}{(2\gamma+1)^{3/2}} V^{-1} \begin{pmatrix} I & 0 \\ 0 & \frac{3\gamma^2+4\gamma+2}{2\gamma+1}, \end{pmatrix} \tag{40}$$

where V is the asymptotic covariance matrix of $\widehat{\theta}_n$ and I is the identity matrix. We refer to [15] for the asymptotic properties of the minimum pseudodistance estimators in the case of linear regression models. Since $\sqrt{n}(\widehat{\theta}_n - \theta_0)$ is asymptotically multivariate normal distributed, using the χ^2 distribution, we obtain the approximation

$$E[(\widehat{\theta}_n - \theta_0)^t M_\gamma(\theta_0)(\widehat{\theta}_n - \theta_0)] \simeq \frac{1}{n} \cdot \frac{(\gamma+1)^2}{(2\gamma+1)^{3/2}} \left[(d-1) + \frac{3\gamma^2+4\gamma+2}{2(\gamma+1)(2\gamma+1)} \right]. \tag{41}$$

Also, the third term in (18) is given by

$$\frac{1}{2\gamma n} \left[1 - \left(\frac{\gamma+1}{\sqrt{2\gamma+1}} \right)^d \right]. \tag{42}$$

Then, according to Proposition 3, an asymptotically unbiased estimator of the expected overall discrepancy is given by

$$Q_{\widehat{\theta}_n} + \frac{1}{n} \cdot \frac{(\gamma+1)^2}{(2\gamma+1)^{3/2}} \left[(d-1) + \frac{3\gamma^2+4\gamma+2}{2(\gamma+1)(2\gamma+1)} \right] + \frac{1}{2\gamma n} \left[1 - \left(\frac{\gamma+1}{\sqrt{2\gamma+1}} \right)^d \right], \tag{43}$$

where $Q_{\widehat{\theta}_n}$ is given by (39). Note that, using the asymptotic properties of $\widehat{\theta}_n$ and the central limit theorem, the last two terms in (18) of Proposition 3 are $o_P(\frac{1}{n})$.

When we compare different linear regression models, as in Section 4 below, we can ignore the terms depending only on n and γ in (43). Therefore, we can use as model selection criterion the simplified expression

$$Q_{\widehat{\theta}_n} + \frac{(\gamma+1)^2}{(\sqrt{2\gamma+1})^3} \cdot \frac{d}{n} - \frac{1}{2\gamma n} \left(\frac{\gamma+1}{\sqrt{2\gamma+1}} \right)^d, \tag{44}$$

which we call Pseudodistance based Information Criterion (PIC).

4. Applications

4.1. Simulation Study

In order to illustrate the performance of the PIC criterion (44) in the case of linear regression models, we performed a simulation study using for comparison the model selection criteria AIC, BIC and MDIC. These criteria are defined respectively by

$$AIC = n \log \widehat{\sigma}_p^2 + 2(p+2)$$

$$BIC = n \log \widehat{\sigma}_p{}^2 + (p+2) \log n,$$

where n the sample size, p the number of covariates of the model and $\widehat{\sigma}_p^2$ the classical unbiased estimator of the variance of the model,

$$MDIC = n MQ_{\widehat{\theta}} + (2\pi)^{-\alpha/2}(1+\alpha)^{2+p/2} p$$

with $\alpha = 0.25$ and

$$MQ_{\hat{\theta}} = -\left[(1 + \alpha^{-1})\frac{1}{n}\sum_{n=1}^{n} f_{\hat{\theta}}^{\alpha}(X_i)\right],$$

where $\hat{\theta}$ is a consistent estimate of the vector of unknown parameters involved in the model with p covariates and $f_{\hat{\theta}}$ is the associated probability density function. Note that MDIC is based on the well known BHHJ family of divergence measures indexed by a parameter $\alpha > 0$ and on the minimum divergence estimating method for robust parameter estimation (see [20]). The value of $\alpha = 0.25$ was found in [9] to be an ideal one for a great variety of settings. The above three criteria have been chosen to be used in this comparative study with PIC not only due to their popularity, but also due to their special characteristics. Indeed, AIC is the classical representative of asymptotically efficient criteria, BIC is known to be consistent, while MDIC is associated with robust estimations (see e.g., [20–23]).

Let X_1, X_2, X_3, X_4 be four variables following respectively the normal distributions $\mathcal{N}(0,3)$, $\mathcal{N}(1,3), \mathcal{N}(2,3)$ and $\mathcal{N}(3,3)$. We consider the model

$$Y = a_0 + a_1 X_1 + a_2 X_2 + \varepsilon$$

with $a_0 = a_1 = a_2 = 1$ and $\varepsilon \sim \mathcal{N}(0,1)$. This is the uncontaminated model. In order to evaluate the robustness of the new PIC criterion, we also consider the contaminated model

$$Y = d_1(a_0 + a_1 X_1 + a_2 X_2 + \varepsilon) + d_2(a_0 + a_1 X_1 + a_2 X_2 + \varepsilon^*)$$

where $\varepsilon^* \sim \mathcal{N}(5,1)$ and $d_1, d_2 \in [0,1]$ such that $d_1 + d_2 = 1$. Note that for $d_1 = 1$ and $d_2 = 0$ the uncontaminated model is obtained.

The simulated data corresponding to the contaminated model are

$$Y_i = d_1(1 + X_{1,i} + X_{2,i} + \varepsilon_i) + d_2(1 + X_{1,i} + X_{2,i} + \varepsilon_i^*),$$

for $i = 1, \ldots, n$, where $X_{1,i}, X_{2,i}, \varepsilon_i, \varepsilon_i^*$ are values of the variables $X_1, X_2, \varepsilon, \varepsilon^*$ independently generated from the normal distributions $\mathcal{N}(1,3), \mathcal{N}(2,3), \mathcal{N}(0,1), \mathcal{N}(5,1)$ correspondingly.

With a set of four possible regressors there are $2^4 - 1 = 15$ possible model specifications that include at least one regressor. These 15 possible models constitute the set of candidate models in our study. More precisely, this set contains the full model (X_1, X_2, X_3, X_4) given by

$$Y = b_0 + b_1 X_1 + b_2 X_2 + b_3 X_3 + b_4 X_4 + \varepsilon$$

as well as all 14 possible subsets of the full model consisting of one (X_{j_1}), two (X_{j_1}, X_{j_2}) and three $(X_{j_1}, X_{j_2}, X_{j_3})$ of the four regressors X_1, X_2, X_3 and X_4, with $j_1 \neq j_2 \neq j_3, j_i \in \{1,2,3,4\}$ and $i = 1,2,3$.

In our simulation study, for several values of the parameter γ associated with the pseudodistance, we compared the new criterion PIC with the other model selection criteria. Different levels of contamination and different sample sizes have been considered. In the examples presented in this work, $d_1 \in \{0.8, 0.9, 0.95, 1\}$ and $n \in \{20, 50, 100\}$. Additional examples for $n = 30, 75, 200, 500$ have been analyzed (results not shown) with similar findings (see below). For each setting, fifty experiments were performed in order to select the best model among the available candidate models. In the framework of each of the fifty experiments, on the basis of the simulated observations, the value of each of the above model selection criteria was calculated for each of the 15 possible models. Then, for each criterion, the 15 candidate models were ranked from 1st to 15th according to the value of the criterion. The model chosen by a given criterion is the one for which the value of the criterion is the lowest among all the 15 candidate models.

Tables 1–12 present the proportions of models selected by the considered criteria. Among the 15 candidate models only 4 were chosen at least once. These four models are the same in all instances and appear in the 2nd column of all Tables.

For small sample sizes ($n = 20$, $n = 30$) the criteria PIC and MDIC yield the best results. When the level of contamination is 10% or 20%, the PIC criterion yields very good results and beats the other competitors almost all the time. When the level of contamination is small, for example 5% or when there is no contamination, the two criteria are comparable, in the sense that in many cases the proportions of selected models by the two criteria are very close, so that sometimes PIC wins and sometimes MDIC wins. Tables 1–4 present these results for $n = 20$, but similar results are obtained for $n = 30$, too.

For medium sample sizes ($n = 50$, $n = 75$), the criteria PIC and BIC yield the best results. The results for $n = 50$ are given in Tables 5–8. Note that the PIC criterion yields the best results for 0% and 10% contamination. For the other levels of contamination, there are values of γ for which PIC is the best among all the considered criteria. On the other hand, in most cases when BIC wins, the proportions of selections of the true model by BIC and PIC are close.

When the sample size is large ($n = 100$, $n = 200$, $n = 500$), BIC generally yields better results than PIC which stays relatively close behind, but sometimes BIC and PIC have the same performance. Tables 9–12 present the results obtained for $n = 100$.

Thus, the new PIC criterion works very well for small to medium sample sizes and for levels of contamination up to 20%, but falls behind BIC for large sample sizes. Note that for contaminated data, PIC with $\gamma = 0.15$ prevails in most of the considered cases. On the other hand, for uncontaminated data, it is PIC with $\gamma = 0.2$ that prevails in all the considered instances. It is also worth mentioning that PIC with $\gamma = 0.3$ appears to behave very satisfactorily in most cases irrespectively of the proportion of contamination (0%–20%) and the sample size. Observe also that in all cases, AIC has the highest overestimation rate which is somehow expected (see [24]).

Although the consistency is the main focus of the applications presented in this work, one should point out that if prediction is part of the objective of a regression analysis, then model selection carried out using criteria such as the ones used in this work, have desirable properties. In fact, the case of finite-dimensional normal regression models has been shown to be associated with satisfactory prediction errors for criteria such as AIC and BIC (see [25]). Furthermore, it should be pointed out that in many instances PIC has a behavior quite similar to the above criteria by choosing the same models. Also, according to the presented simulation results, the proportion of choosing the true model by PIC is always better than the proportion of choosing the true model by AIC (even in the case of non contaminated data) and sometimes it is better than the proportion of choosing the true model by BIC. These results imply a satisfactory prediction ability for the proposed PIC criterion.

In conclusion, the new PIC criterion is a good competitor of the well known model selection criteria AIC, BIC and MDIC and may have superior performance especially in the case of small and contaminated samples.

Table 1. Proportions of selected models by the considered criteria ($n = 20$, $d_1 = 0.8$).

Criteria	Variables	$\gamma = 0.01$	$\gamma = 0.05$	$\gamma = 0.1$	$\gamma = 0.15$	$\gamma = 0.2$	$\gamma = 0.25$	$\gamma = 0.3$
PIC	X_1, X_2	90	84	88	84	92	90	86
	X_1, X_2, X_3	(10)	(16)	(12)	(16)	(8)	(10)	(14)
	X_1, X_2, X_4							
	X_1, X_2, X_3, X_4							
AIC	X_1, X_2	62	56	52	56	66	56	60
	X_1, X_2, X_3	(38)	(44)	(48)	(44)	(34)	(44)	(40)
	X_1, X_2, X_4							
	X_1, X_2, X_3, X_4							
BIC	X_1, X_2	74	76	60	74	72	68	70
	X_1, X_2, X_3	(26)	(24)	(40)	(26)	(28)	(32)	(30)
	X_1, X_2, X_4							
	X_1, X_2, X_3, X_4							
MDIC	X_1, X_2	86	**86**	64	78	84	80	74
	X_1, X_2, X_3	(14)	(14)	(36)	(22)	(16)	(20)	(26)
	X_1, X_2, X_4							
	X_1, X_2, X_3, X_4							

Table 2. Proportions of selected models by the considered criteria ($n = 20$, $d_1 = 0.9$).

Criteria	Variables	$\gamma = 0.01$	$\gamma = 0.05$	$\gamma = 0.1$	$\gamma = 0.15$	$\gamma = 0.2$	$\gamma = 0.25$	$\gamma = 0.3$
PIC	X_1, X_2	80	**84**	**90**	82	82	**80**	80
	X_1, X_2, X_3	(20)	(16)	(10)	(18)	(18)	(20)	(20)
	X_1, X_2, X_4							
	X_1, X_2, X_3, X_4							
AIC	X_1, X_2	60	52	56	62	64	54	52
	X_1, X_2, X_3	(40)	(48)	(44)	(38)	(36)	(46)	(48)
	X_1, X_2, X_4							
	X_1, X_2, X_3, X_4							
BIC	X_1, X_2	76	70	78	72	84	76	76
	X_1, X_2, X_3	(24)	(30)	(22)	(28)	(16)	(24)	(24)
	X_1, X_2, X_4							
	X_1, X_2, X_3, X_4							
MDIC	X_1, X_2	**86**	76	88	74	**92**	78	**86**
	X_1, X_2, X_3	(14)	(24)	(12)	(26)	(8)	(22)	(14)
	X_1, X_2, X_4							
	X_1, X_2, X_3, X_4							

Table 3. Proportions of selected models by the considered criteria ($n = 20$, $d_1 = 0.95$).

Criteria	Variables	$\gamma = 0.01$	$\gamma = 0.05$	$\gamma = 0.1$	$\gamma = 0.15$	$\gamma = 0.2$	$\gamma = 0.25$	$\gamma = 0.3$
PIC	X_1, X_2	82	88	80	94	82	88	86
	X_1, X_2, X_3	(18)	(12)	(20)	(6)	(18)	(12)	(14)
	X_1, X_2, X_4							
	X_1, X_2, X_3, X_4							
AIC	X_1, X_2	78	50	66	70	66	64	66
	X_1, X_2, X_3	(22)	(50)	(34)	(30)	(34)	(36)	(34)
	X_1, X_2, X_4							
	X_1, X_2, X_3, X_4							
BIC	X_1, X_2	84	64	74	84	84	76	82
	X_1, X_2, X_3	(16)	(36)	(26)	(16)	(16)	(24)	(18)
	X_1, X_2, X_4							
	X_1, X_2, X_3, X_4							
MDIC	X_1, X_2	90	74	82	88	88	80	88
	X_1, X_2, X_3	(10)	(26)	(18)	(12)	(12)	(20)	(12)
	X_1, X_2, X_4							
	X_1, X_2, X_3, X_4							

Table 4. Proportions of selected models by the considered criteria ($n = 20$, $d_1 = 1$).

Criteria	Variables	$\gamma = 0.01$	$\gamma = 0.05$	$\gamma = 0.1$	$\gamma = 0.15$	$\gamma = 0.2$	$\gamma = 0.25$	$\gamma = 0.3$
PIC	X_1, X_2	86	86	86	86	88	82	92
	X_1, X_2, X_3	(14)	(14)	(14)	(14)	(12)	(18)	(8)
	X_1, X_2, X_4							
	X_1, X_2, X_3, X_4							
AIC	X_1, X_2	64	74	62	58	64	62	70
	X_1, X_2, X_3	(36)	(26)	(38)	(42)	(36)	(38)	(30)
	X_1, X_2, X_4							
	X_1, X_2, X_3, X_4							
BIC	X_1, X_2	78	90	78	80	82	80	74
	X_1, X_2, X_3	(22)	(10)	(22)	(20)	(18)	(20)	(26)
	X_1, X_2, X_4							
	X_1, X_2, X_3, X_4							
MDIC	X_1, X_2	84	92	88	88	88	88	80
	X_1, X_2, X_3	(16)	(8)	(12)	(12)	(12)	(12)	(20)
	X_1, X_2, X_4							
	X_1, X_2, X_3, X_4							

Table 5. Proportions of selected models by the considered criteria ($n = 50$, $d_1 = 0.8$).

Criteria	Variables	$\gamma = 0.01$	$\gamma = 0.05$	$\gamma = 0.1$	$\gamma = 0.15$	$\gamma = 0.2$	$\gamma = 0.25$	$\gamma = 0.3$
PIC	X_1, X_2	86	96	94	90	88	86	90
	X_1, X_2, X_3	(14)	(4)	(6)	(10)	(12)	(14)	(10)
	X_1, X_2, X_4							
	X_1, X_2, X_3, X_4							
AIC	X_1, X_2	74	64	82	62	64	78	72
	X_1, X_2, X_3	(26)	(36)	(18)	(38)	(36)	(22)	(28)
	X_1, X_2, X_4							
	X_1, X_2, X_3, X_4							
BIC	X_1, X_2	94	86	96	86	90	88	90
	X_1, X_2, X_3	(6)	(14)	(4)	(14)	(10)	(12)	(10)
	X_1, X_2, X_4							
	X_1, X_2, X_3, X_4							
MDIC	X_1, X_2	94	82	98	82	86	88	90
	X_1, X_2, X_3	(6)	(18)	(2)	(18)	(14)	(12)	(10)
	X_1, X_2, X_4							
	X_1, X_2, X_3, X_4							

Table 6. Proportions of selected models by the considered criteria ($n = 50$, $d_1 = 0.9$).

Criteria	Variables	$\gamma = 0.01$	$\gamma = 0.05$	$\gamma = 0.1$	$\gamma = 0.15$	$\gamma = 0.2$	$\gamma = 0.25$	$\gamma = 0.3$
PIC	X_1, X_2	92	88	92	90	82	94	86
	X_1, X_2, X_3	(8)	(12)	(8)	(10)	(18)	(6)	(14)
	X_1, X_2, X_4							
	X_1, X_2, X_3, X_4							
AIC	X_1, X_2	70	64	62	64	66	74	72
	X_1, X_2, X_3	(30)	(36)	(38)	(36)	(34)	(26)	(28)
	X_1, X_2, X_4							
	X_1, X_2, X_3, X_4							
BIC	X_1, X_2	92	88	82	92	88	88	86
	X_1, X_2, X_3	(8)	(12)	(18)	(8)	(12)	(12)	(14)
	X_1, X_2, X_4							
	X_1, X_2, X_3, X_4							
MDIC	X_1, X_2	92	86	76	88	84	88	86
	X_1, X_2, X_3	(8)	(14)	(24)	(12)	(16)	(12)	(14)
	X_1, X_2, X_4							
	X_1, X_2, X_3, X_4							

Table 7. Proportions of selected models by the considered criteria ($n = 50$, $d_1 = 0.95$).

Criteria	Variables	$\gamma = 0.01$	$\gamma = 0.05$	$\gamma = 0.1$	$\gamma = 0.15$	$\gamma = 0.2$	$\gamma = 0.25$	$\gamma = 0.3$
PIC	X_1, X_2	**94**	**92**	**92**	**88**	84	90	**88**
	X_1, X_2, X_3	(6)	(8)	(8)	(12)	(16)	(10)	(12)
	X_1, X_2, X_4							
	X_1, X_2, X_3, X_4							
AIC	X_1, X_2	70	62	66	68	70	72	58
	X_1, X_2, X_3	(30)	(38)	(34)	(32)	(30)	(28)	(42)
	X_1, X_2, X_4							
	X_1, X_2, X_3, X_4							
BIC	X_1, X_2	**96**	82	**92**	86	**92**	**92**	86
	X_1, X_2, X_3	(4)	(18)	(8)	(14)	(8)	(8)	(14)
	X_1, X_2, X_4							
	X_1, X_2, X_3, X_4							
MDIC	X_1, X_2	90	78	88	86	86	90	82
	X_1, X_2, X_3	(10)	(22)	(12)	(14)	(14)	(10)	(18)
	X_1, X_2, X_4							
	X_1, X_2, X_3, X_4							

Table 8. Proportions of selected models by the considered criteria ($n = 50$, $d_1 = 1$).

Criteria	Variables	$\gamma = 0.01$	$\gamma = 0.05$	$\gamma = 0.1$	$\gamma = 0.15$	$\gamma = 0.2$	$\gamma = 0.25$	$\gamma = 0.3$
PIC	X_1, X_2	**94**	**90**	80	84	**90**	**94**	**88**
	X_1, X_2, X_3	(6)	(10)	(20)	(16)	(10)	(6)	(12)
	X_1, X_2, X_4							
	X_1, X_2, X_3, X_4							
AIC	X_1, X_2	64	68	62	68	66	64	62
	X_1, X_2, X_3	(34)	(32)	(38)	(32)	(34)	(36)	(38)
	X_1, X_2, X_4							
	X_1, X_2, X_3, X_4							
BIC	X_1, X_2	86	86	**86**	**90**	86	**94**	82
	X_1, X_2, X_3	(14)	(14)	(14)	(10)	(14)	(6)	(18)
	X_1, X_2, X_4							
	X_1, X_2, X_3, X_4							
MDIC	X_1, X_2	84	84	82	88	84	90	82
	X_1, X_2, X_3	(16)	(16)	(18)	(12)	(16)	(10)	(18)
	X_1, X_2, X_4							
	X_1, X_2, X_3, X_4							

Table 9. Proportions of selected models by the considered criteria ($n = 100$, $d_1 = 0.8$).

Criteria	Variables	$\gamma = 0.01$	$\gamma = 0.05$	$\gamma = 0.1$	$\gamma = 0.15$	$\gamma = 0.2$	$\gamma = 0.25$	$\gamma = 0.3$
PIC	X_1, X_2	**94**	94	94	92	88	88	**94**
	X_1, X_2, X_3	(6)	(6)	(6)	(8)	(12)	(12)	(6)
	X_1, X_2, X_4							
	X_1, X_2, X_3, X_4							
AIC	X_1, X_2	70	82	78	70	68	68	72
	X_1, X_2, X_3	(30)	(18)	(22)	(30)	(32)	(32)	(28)
	X_1, X_2, X_4							
	X_1, X_2, X_3, X_4							
BIC	X_1, X_2	90	**96**	98	90	**96**	94	88
	X_1, X_2, X_3	(10)	(4)	(2)	(10)	(4)	(6)	(12)
	X_1, X_2, X_4							
	X_1, X_2, X_3, X_4							
MDIC	X_1, X_2	86	96	92	86	92	90	88
	X_1, X_2, X_3	(14)	(4)	(8)	(14)	(8)	(10)	(12)
	X_1, X_2, X_4							
	X_1, X_2, X_3, X_4							

Table 10. Proportions of selected models by the considered criteria ($n = 100$, $d_1 = 0.9$).

Criteria	Variables	$\gamma = 0.01$	$\gamma = 0.05$	$\gamma = 0.1$	$\gamma = 0.15$	$\gamma = 0.2$	$\gamma = 0.25$	$\gamma = 0.3$
PIC	X_1, X_2	88	92	**96**	88	88	88	86
	X_1, X_2, X_3	(12)	(8)	(4)	(12)	(12)	(12)	(14)
	X_1, X_2, X_4							
	X_1, X_2, X_3, X_4							
AIC	X_1, X_2	68	72	78	66	70	78	60
	X_1, X_2, X_3	(32)	(28)	(22)	(34)	(30)	(22)	(40)
	X_1, X_2, X_4							
	X_1, X_2, X_3, X_4							
BIC	X_1, X_2	**98**	**98**	96	**88**	92	**94**	92
	X_1, X_2, X_3	(2)	(2)	(4)	(12)	(8)	(6)	(8)
	X_1, X_2, X_4							
	X_1, X_2, X_3, X_4							
MDIC	X_1, X_2	90	90	**96**	84	82	90	82
	X_1, X_2, X_3	(10)	(10)	(4)	(16)	(18)	(10)	(18)
	X_1, X_2, X_4							
	X_1, X_2, X_3, X_4							

Table 11. Proportions of selected models by the considered criteria ($n = 100$, $d_1 = 0.95$).

Criteria	Variables	$\gamma = 0.01$	$\gamma = 0.05$	$\gamma = 0.1$	$\gamma = 0.15$	$\gamma = 0.2$	$\gamma = 0.25$	$\gamma = 0.3$
PIC	X_1, X_2	90	88	92	90	98	96	92
	X_1, X_2, X_3	(10)	(12)	(8)	(10)	(2)	(4)	(8)
	X_1, X_2, X_4							
	X_1, X_2, X_3, X_4							
AIC	X_1, X_2	70	78	78	66	82	68	68
	X_1, X_2, X_3	(30)	(22)	(22)	(34)	(18)	(32)	(32)
	X_1, X_2, X_4							
	X_1, X_2, X_3, X_4							
BIC	X_1, X_2	96	92	92	94	96	94	88
	X_1, X_2, X_3	(4)	(8)	(8)	(6)	(4)	(6)	(12)
	X_1, X_2, X_4							
	X_1, X_2, X_3, X_4							
MDIC	X_1, X_2	90	88	82	90	94	84	88
	X_1, X_2, X_3	(10)	(12)	(18)	(10)	(6)	(16)	(12)
	X_1, X_2, X_4							
	X_1, X_2, X_3, X_4							

Table 12. Proportions of the selected models by the considered criteria ($n = 100$, $d_1 = 1$).

Criteria	Variables	$\gamma = 0.01$	$\gamma = 0.05$	$\gamma = 0.1$	$\gamma = 0.15$	$\gamma = 0.2$	$\gamma = 0.25$	$\gamma = 0.3$
PIC	X_1, X_2	94	96	92	92	96	90	94
	X_1, X_2, X_3	(6)	(4)	(8)	(8)	(4)	(10)	(6)
	X_1, X_2, X_4							
	X_1, X_2, X_3, X_4							
AIC	X_1, X_2	78	74	72	74	70	62	74
	X_1, X_2, X_3	(22)	(26)	(28)	(26)	(30)	(38)	(26)
	X_1, X_2, X_4							
	X_1, X_2, X_3, X_4							
BIC	X_1, X_2	96	100	92	96	94	90	100
	X_1, X_2, X_3	(4)		(8)	(4)	(6)	(10)	
	X_1, X_2, X_4							
	X_1, X_2, X_3, X_4							
MDIC	X_1, X_2	94	92	86	90	86	80	94
	X_1, X_2, X_3	(6)	(8)	(14)	(10)	(14)	(20)	(6)
	X_1, X_2, X_4							
	X_1, X_2, X_3, X_4							

4.2. Real Data Example

In order to illustrate the proposed method, we used the Hald cement data (see [26]) which represent a popular example for multiple linear regression. This example concern the heat evolved in calories per gram of cement Y as a function of the amount of each of four ingredient in the mix: tricalcium aluminate (X_1), tricalcium silicate (X_2), tetracalcium alumino-ferrite (X_3) and dicalcium silicate (X_4). The data are presented in Table 13.

Table 13. Hald cement data.

X_1	X_2	X_3	X_4	Y
7	26	6	60	78.5
1	29	15	52	74.3
11	56	8	20	104.3
11	31	8	47	87.6
7	52	6	33	95.9
11	55	9	22	109.2
3	71	17	6	102.7
1	31	22	44	72.5
2	54	18	22	93.1
21	47	4	26	115.9
1	40	23	34	83.8
11	66	9	12	113.3
10	68	8	12	109.4

Since 4 variables are available, there are 15 possible candidate models (involving at least one regressor) for this data set. Note that the 4 single-variable models should be excluded from the analysis, because cement involves a mixture of at least two components that react chemically (see [27], p. 102). The model selection criteria that have been used are PIC for several values of γ, AIC, BIC and MDIC with $\alpha = 0.25$. Table 14 shows the model selected by each of the considered criteria.

Table 14. Selected models by model selection criteria.

Criteria	Variables
PIC, $\gamma = 0.05$	X_1, X_2, X_4
PIC, $\gamma = 0.15$	X_1, X_2, X_4
PIC, $\gamma = 0.2$	X_1, X_2, X_3
PIC, $\gamma = 0.25$	X_1, X_2, X_4
PIC, $\gamma = 0.3$	X_1, X_2, X_4
AIC	X_1, X_2, X_4
BIC	X_1, X_2
MDIC	X_1, X_2, X_3

Observe that, in this example, PIC behaves similarly to AIC and MDIC having a slight tendency of overestimation. Note that for this specific dataset the collinearity is quite strong with X_1 and X_3 as well as X_2 and X_4 being seriously correlated. It should be pointed out that the model (X_1, X_2, X_4) is chosen not only by AIC and PIC, but also by C_p Mallows' criterion ([1]) with (X_1, X_2, X_3) coming very close second. Note further that (X_1, X_2, X_4) has also been chosen by cross validation ([28], p. 33) and PRESS ([26], p. 325). Finally, it is worth noticing that these two models share the highest adjusted R^2 values which are almost identical (0.976 for (X_1, X_2, X_4) and 0.974 for (X_1, X_2, X_3)) making the distinction between them extremely hard. Thus, in this example, the new PIC criterion gives results as good as other recognized classical model selection criteria.

5. Conclusions

In this work, by applying the same methodology as for AIC to a family of pseudodistances, we constructed new model selection criteria using minimum pseudodistance estimators. We proved theoretical properties of these criteria including asymptotic unbiasedness, robustness, consistency, as well as the limit laws. The case of the linear regression models was studied in detail and specific selection criteria based on pseudodistance are proposed.

For linear regression models, a comparative study based on Monte Carlo simulations illustrate the performance of the new methodology. Thus, for small sample sizes, the criteria PIC and MDIC yield the best results and in many cases PIC wins, for example when the level of contamination is 10%

or 20%. For medium sample sizes, the criteria PIC and BIC yield the best results. When the sample size is large, BIC generally yields better results than PIC which stays relatively close behind, but sometimes BIC and PIC have the same performance.

Based on the results of the simulation study and on the real data example, we conclude that the new PIC criterion is a good competitor of the well known criteria AIC, BIC and MDIC with an overall performance which is very satisfactory for all possible settings according to the sample size and contamination rate. Also PIC may have superior performance, especially in the case of small and contaminated samples.

An important issue that needs further investigation is the choice of the appropriate value for the parameter γ associated to the procedure. The findings of the presented simulation study show that, for contaminated data, the value $\gamma = 0.15$ leads to very good results, irrespectively of the sample size. Also, $\gamma = 0.3$ produces overall very satisfactory results, irrespectively of the sample size and the contamination rate. We hope to explore further and provide a clear solution to this problem, in a future work. We also intend to extend this methodology to other type of models including nonlinear or time series models.

Author Contributions: A.T. conceived the methodology, obtained the theoretical results. A.T., A.K. and P.T. conceived the application part. A.K. and P.T. implemented the method in R and obtained the numerical results. All authors wrote the paper. All authors have read and approved the final manuscript.

Funding: This research received no external funding.

Acknowledgments: The work of the first author was partially supported by a grant of the Romanian National Authority for Scientific Research, CNCS-UEFISCDI, project number PN-II-RU-TE-2012-3-0007. The work of the third author was completed as part of the activities of the Laboratory of Statistics and Data Analysis of the University of the Aegean.

Conflicts of Interest: The authors declare no conflict of interst.

Abbreviations

The following abbreviations are used in this manuscript:

AIC	Akaike Information Criterion
BIC	Bayesian Information Criterion
GIC	General Information Criterion
DIC	Divergence Information Criterion
MDIC	Modified Divergence Information Criterion
PIC	Pseudodistance based Information Criterion
BHHJ family of measures	Basu, Harris, Hjort and Jones family of measures

Appendix A

Proof of Proposition 1. Using a Taylor expansion of W_θ around the true parameter θ_0 and taking $\theta = \widehat{\theta}_n$, on the basis of (12) and (13) we obtain

$$W_{\widehat{\theta}_n} = W_{\theta_0} + \frac{1}{2}(\widehat{\theta}_n - \theta_0)^t M_\gamma(\theta_0)(\widehat{\theta}_n - \theta_0) + o(\|\widehat{\theta}_n - \theta_0\|^2). \tag{A1}$$

Then (15) holds.

$\square$

Proof of Proposition 2. Using a Taylor expansion of Q_θ around to $\widehat{\theta}_n$ and taking $\theta = \theta_0$, we obtain

$$Q_{\theta_0} = Q_{\widehat{\theta}_n} + \left[\frac{\partial}{\partial \theta} Q_\theta\right]_{\theta=\widehat{\theta}_n}^t (\theta_0 - \widehat{\theta}_n) + \frac{1}{2}(\theta_0 - \widehat{\theta}_n)^t \left[\frac{\partial^2}{\partial \theta^2} Q_\theta\right]_{\theta=\widehat{\theta}_n} (\theta_0 - \widehat{\theta}_n) + o(\|\widehat{\theta}_n - \theta_0\|^2). \tag{A2}$$

Note that $\left[\frac{\partial}{\partial \theta} Q_\theta\right]_{\theta=\widehat{\theta}_n} = 0$ by the very definition of $\widehat{\theta}_n$.

By applying the weak law of large numbers and the continuous mapping theorem, we get

$$\left[\frac{\partial^2}{\partial\theta^2}Q_\theta\right]_{\theta=\theta_0} - \left[\frac{\partial^2}{\partial\theta^2}W_\theta\right]_{\theta=\theta_0} \xrightarrow{P} 0 \tag{A3}$$

and using (13)

$$\left[\frac{\partial^2}{\partial\theta^2}Q_\theta\right]_{\theta=\theta_0} - M_\gamma(\theta_0) \xrightarrow{P} 0. \tag{A4}$$

Then, using the consistency of $\widehat{\theta}_n$ and (A4), we obtain

$$\left[\frac{\partial^2}{\partial\theta^2}Q_\theta\right]_{\theta=\widehat{\theta}_n} = M_\gamma(\theta_0) + o_P(1). \tag{A5}$$

Consequently,

$$Q_{\theta_0} = Q_{\widehat{\theta}_n} + \frac{1}{2}(\theta_0 - \widehat{\theta}_n)^t M_\gamma(\theta_0)(\theta_0 - \widehat{\theta}_n) + o(\|\widehat{\theta}_n - \theta_0\|^2) \tag{A6}$$

and we deduce (17). $\square$

Proof of Proposition 3. Using Proposition 1 and Proposition 2, we obtain

$$E[W_{\widehat{\theta}_n}] = E[Q_{\widehat{\theta}_n}] + E[(\theta_0 - \widehat{\theta}_n)^t M_\gamma(\theta_0)(\theta_0 - \widehat{\theta}_n)] - E[Q_{\theta_0}] + W_{\theta_0} + E[R_n] \tag{A7}$$

where $R_n = o(\|\widehat{\theta}_n - \theta_0\|^2)$.

In order to evaluate $W_{\theta_0} - E[Q_{\theta_0}]$, note that

$$Q_{\theta_0} - W_{\theta_0} = -\frac{1}{\gamma}\left[\ln\left(\frac{1}{n}\sum_{i=1}^n p_{\theta_0}^\gamma(X_i)\right) - \ln\left(\int p_{\theta_0}^{\gamma+1}d\lambda\right)\right]. \tag{A8}$$

A Taylor expansion of the function $\ln x$ around to $\int p_{\theta_0}^{\gamma+1}d\lambda$ yields

$$\begin{aligned}
\ln\left(\frac{1}{n}\sum_{i=1}^n p_{\theta_0}^\gamma(X_i)\right) &= \ln\left(\int p_{\theta_0}^{\gamma+1}d\lambda\right) + \frac{1}{\int p_{\theta_0}^{\gamma+1}d\lambda}\left[\frac{1}{n}\sum_{i=1}^n p_{\theta_0}^\gamma(X_i) - \int p_{\theta_0}^{\gamma+1}d\lambda\right] - \\
&\quad -\frac{1}{2}\cdot\frac{1}{(\int p_{\theta_0}^{\gamma+1}d\lambda)^2}\left[\frac{1}{n}\sum_{i=1}^n p_{\theta_0}^\gamma(X_i) - \int p_{\theta_0}^{\gamma+1}d\lambda\right]^2 + \\
&\quad +o(\|\frac{1}{n}\sum_{i=1}^n p_{\theta_0}^\gamma(X_i) - \int p_{\theta_0}^{\gamma+1}d\lambda\|^2).
\end{aligned} \tag{A9}$$

Then

$$\begin{aligned}
E[Q_{\theta_0} - W_{\theta_0}] &= -\frac{1}{\gamma}E\left[\ln\left(\frac{1}{n}\sum_{i=1}^n p_{\theta_0}^\gamma(X_i)\right) - \ln\left(\int p_{\theta_0}^{\gamma+1}d\lambda\right)\right] \\
&= -\frac{1}{\gamma}\left\{\frac{1}{\int p_{\theta_0}^{\gamma+1}d\lambda}E\left[\frac{1}{n}\sum_{i=1}^n p_{\theta_0}^\gamma(X_i) - \int p_{\theta_0}^{\gamma+1}d\lambda\right] - \right. \\
&\quad \left. -\frac{1}{2}\cdot\frac{1}{(\int p_{\theta_0}^{\gamma+1}d\lambda)^2}E\left[\left(\frac{1}{n}\sum_{i=1}^n p_{\theta_0}^\gamma(X_i) - \int p_{\theta_0}^{\gamma+1}d\lambda\right)^2\right] + E[R'_n]\right\}
\end{aligned}$$

where $R'_n = o(\|\frac{1}{n}\sum_{i=1}^n p_{\theta_0}^\gamma(X_i) - \int p_{\theta_0}^{\gamma+1}d\lambda\|^2)$.

On the other hand,

$$\mathrm{E}\left[\left(\frac{1}{n}\sum_{i=1}^{n}p_{\theta_0}^{\gamma}(X_i) - \int p_{\theta_0}^{\gamma+1}\mathrm{d}\lambda\right)^2\right] = \mathrm{Var}\left[\frac{1}{n}\sum_{i=1}^{n}p_{\theta_0}^{\gamma}(X_i)\right] = \frac{1}{n}\mathrm{Var}\left[p_{\theta_0}^{\gamma}(X)\right]$$

$$= \frac{1}{n}\left\{\mathrm{E}[p_{\theta_0}^{2\gamma}(X)] - \mathrm{E}[p_{\theta_0}^{\gamma}(X)]^2\right\}$$

$$= \frac{\int p_{\theta_0}^{2\gamma+1}\mathrm{d}\lambda - (\int p_{\theta_0}^{\gamma+1}\mathrm{d}\lambda)^2}{n}. \tag{A10}$$

Consequently,

$$\mathrm{E}[Q_{\theta_0}] - W_{\theta_0} = -\frac{1}{2\gamma n}\left[1 - \frac{\int p_{\theta_0}^{2\gamma+1}\mathrm{d}\lambda}{\left(\int p_{\theta_0}^{\gamma+1}\mathrm{d}\lambda\right)^2}\right] - \frac{1}{\gamma}E\left[R_n'\right]. \tag{A11}$$

Using (A7) and (A11), we obtain (18). $\square$

Proof of Proposition 5. For the contaminated model $\widetilde{P}_{\varepsilon x} = (1-\varepsilon)P_{\theta_0} + \varepsilon\delta_x$, it holds

$$U(\widetilde{P}_{\varepsilon x}) = \frac{1}{\gamma+1}\ln\left(\int p_{T(\widetilde{P}_{\varepsilon x})}^{\gamma+1}\mathrm{d}\lambda\right) - \frac{1}{\gamma}\ln\left(\int p_{T(\widetilde{P}_{\varepsilon x})}^{\gamma}\mathrm{d}\widetilde{P}_{\varepsilon x}\right). \tag{A12}$$

Derivation with respect to ε yields

$$\frac{\partial}{\partial\varepsilon}[U(\widetilde{P}_{\varepsilon x})]_{\varepsilon=0} = \frac{1}{\int p_{\theta_0}^{\gamma+1}\mathrm{d}\lambda}\cdot\int p_{\theta_0}^{\gamma}\dot{p}_{\theta_0}\mathrm{d}\lambda\cdot\mathrm{IF}(x;T,P_{\theta_0}) -$$

$$-\frac{1}{\gamma}\cdot\frac{1}{\int p_{\theta_0}^{\gamma+1}\mathrm{d}\lambda}\left\{-\int p_{\theta_0}^{\gamma+1}\mathrm{d}\lambda + \gamma\cdot\int p_{\theta_0}^{\gamma}\dot{p}_{\theta_0}\mathrm{d}\lambda\cdot\mathrm{IF}(x;T,P_{\theta_0}) + p_{\theta_0}^{\gamma}(x)\right\}$$

$$= \frac{1}{\gamma}\cdot\left[1 - \frac{p_{\theta_0}^{\gamma}(x)}{\int p_{\theta_0}^{\gamma+1}\mathrm{d}\lambda}\right].$$

Thus we obtain

$$\mathrm{IF}(x;U,P_{\theta_0}) = \frac{1}{\gamma}\left[1 - \frac{p_{\theta_0}^{\gamma}(x)}{\int p_{\theta_0}^{\gamma+1}\mathrm{d}\lambda}\right]. \tag{A13}$$

$\square$

References

1. Mallows, C.L. Some comments on Cp. *Technometrics* **1973**, *15*, 661–675.
2. Akaike, H. Information theory and an extension of the maximum likelihood principle. In *Proceedings of the Second International Symposium on Information Theory Petrov*; Springer: Berlin/Heidelberger, Germany, 1973; pp. 267–281.
3. Schwarz, G. Estimating the dimension of a model. *Ann. Stat.* **1978**, *6*, 461–464. [CrossRef]
4. Konishi, S.; Kitagawa, G. Generalised information criteria in model selection. *Biometrika* **1996**, *83*, 875–890. [CrossRef]
5. Ronchetti, E. Robust model selection in regression. *Statist. Probab. Lett.* **1985**, *3*, 21–23. [CrossRef]
6. Ronchetti, E.; Staudte, R.G. A robust version of Mallows' Cp. *J. Am. Stat. Assoc.* **1994**, *89*, 550–559.
7. Agostinelli, C. Robust model selection in regression via weighted likelihood estimating equations. *Stat. Probab. Lett.* **2002**, *76*, 1930–1934. [CrossRef]
8. Mattheou, K.; Lee, S.; Karagrigoriou, A. A model selection criterion based on the BHHJ measure of divergence. *J. Stat. Plann. Inf.* **2009**, *139*, 228–235. [CrossRef]

9. Mantalos, P.; Mattheou, K.; Karagrigoriou, A. An improved divergence information criterion for the determination of the order of an AR process. *Commun. Stat.-Simul. Comput.* **2010**, *39*, 865–879. [CrossRef]

10. Toma, A. Model selection criteria using divergences. *Entropy* **2014**, *16*, 2686–2698. [CrossRef]

11. Pardo, L. *Statistical Inference Based on Divergence Measures*; Chapmann & Hall: London, UK, 2006.

12. Basu, A.; Shioya, H.; Park, C. *Statistical Inference: The Minimum Distance Approach*; Chapmann & Hall: London, UK, 2011.

13. Jones, M.C.; Hjort, N.L.; Harris, I.R.; Basu, A. A comparison of related density-based minimum divergence estimators. *Biometrika* **2001**, *88*, 865–873. [CrossRef]

14. Fujisawa, H.; Eguchi, S. Robust parameter estimation with a small bias against heavy contamination. *J. Multivar. Anal.* **2008**, *99*, 2053–2081. [CrossRef]

15. Broniatowski, M.; Toma, A.; Vajda, I. Decomposable pseudodistances and applications in statistical estimation. *J. Stat. Plan. Infer.* **2012**, *142*, 2574–2585. [CrossRef]

16. Toma, A.; Leoni-Aubin, S. Optimal robust M-estimators using Renyi pseudodistances. *J. Multivar. Anal.* **2013**, *115*, 359–373. [CrossRef]

17. Toma, A.; Leoni-Aubin, S. Robust portfolio optimization using pseudodistances. *PLoS ONE* **2015**, *10*, 1–26. [CrossRef]

18. Toma, A.; Fulga, C. Robust estimation for the single index model using pseudodistances. *Entropy* **2018**, *20*, 374. [CrossRef]

19. Hampel, F.R.; Ronchetti, E.; Rousseeuw, P.J.; Stahel, W. *Robust Statistics: The Approach Based on Influence Functions*; Wiley Blackwell: Hoboken, NJ, USA, 1986.

20. Basu, A.; Harris, I.R.; Hjort, N.L.; Jones, M.C. Robust and efficient estimation by minimising a density power divergence. *Biometrika* **1998**, *85*, 549–559. [CrossRef]

21. Karagrigoriou, A. Asymptotic efficiecy of the order selection of a nongaussian AR process. *Stat. Sin.* **1997**, *7*, 407–423.

22. Vonta, F.; Karagrigoriou, A. Generalized measures of divergence in survival analysis and reliability. *J. Appl. Prob.* **2010**, *47*, 216–234. [CrossRef]

23. Karagrigoriou, A.; Mattheou, K.; Vonta, F. On asymptotic properties of AIC variants with applications. *Open J. Stat.* **2011**, *1*, 105–109. [CrossRef]

24. Shibata, R. Selection of the order of an autoregressive model by Akaike's information criterion. *Biometrika* **1976**, *63*, 117–126. [CrossRef]

25. Speed, T.P.; Yu. B. Model selection and prediction: normal regression. *Ann. Inst. Stat. Math.* **1993**, *45*, 35–54. [CrossRef]

26. Draper, N.R.; Smith, H. *Applied Regression Analysis*, 2nd ed.; Wiley Blackwell: Hoboken, NJ, USA, 1981.

27. Burnham, K.P.; Anderson, D.R. *Model Selection and Multimodel Inference: A Practical Information-Theoretic Approach*; Springer: Berlin/Heidelberger, Germany, 2002.

28. Hjorth, J.S.U. *Computer Intensive Statistical Methods: Validation, Model Selection and Bootstrap*; Chapman and Hall: London, UK, 1994.

Article

Model Selection in a Composite Likelihood Framework Based on Density Power Divergence

Elena Castilla [1,*], **Nirian Martín** [2], **Leandro Pardo** [1] **and Konstantinos Zografos** [3]

[1]   Interdisciplinary Mathematics Institute and Department of Statistics and O.R. I, Complutense University of Madrid, 28040 Madrid, Spain; lpardo@mat.ucm.es

[2]   Interdisciplinary Mathematics Institute and Department of Financial and Actuarial Economics & Statistics, Complutense University of Madrid, 28003 Madrid, Spain; nirian@estad.ucm.es

[3]   Department of Mathematics, University of Ioannina, 45110 Ioannina, Greece; kzograf@uoi.gr

*   Correspondence: elecasti@ucm.es

Received: 22 January 2020; Accepted: 25 February 2020; Published: 27 February 2020

Abstract: This paper presents a model selection criterion in a composite likelihood framework based on density power divergence measures and in the composite minimum density power divergence estimators, which depends on an tuning parameter α. After introducing such a criterion, some asymptotic properties are established. We present a simulation study and two numerical examples in order to point out the robustness properties of the introduced model selection criterion.

Keywords: composite likelihood; composite minimum density power divergence estimators; model selection

---

1. Introduction

Composite likelihood inference is an important approach to deal with those real situations of large data sets or very complex models, in which classical likelihood methods are computationally difficult, or even, not possible to manage. Composite likelihood methods have been successfully used in many applications concerning, for example, genetics ([1]), generalized linear mixed models ([2]), spatial statistics ([3–5]), frailty models ([6]), multivariate survival analysis ([7,8]), etc.

Let us introduce the problem, adopting here the notation by [9]. Let $\{f(\cdot;\boldsymbol{\theta}), \boldsymbol{\theta} \in \Theta \subseteq \mathbb{R}^p, p \geq 1\}$ be a parametric identifiable family of distributions for an observation $\boldsymbol{y} = (y_1, ..., y_m)^T$, a realization of a random m-vector $\boldsymbol{Y}$. In this setting, the composite likelihood function based on K different marginal or conditional distributions has the form

$$\mathcal{CL}(\boldsymbol{\theta}, \boldsymbol{y}) = \prod_{k=1}^{K} \left(f_{A_k}(y_j, j \in A_k; \boldsymbol{\theta}) \right)^{w_k}$$

and the corresponding composite log-density

$$\log \mathcal{CL}(\boldsymbol{\theta}, \boldsymbol{y}) = \sum_{k=1}^{K} w_k \ell_{A_k}(\boldsymbol{\theta}, \boldsymbol{y}), \tag{1}$$

with $\ell_{A_k}(\boldsymbol{\theta}, \boldsymbol{y}) = \log f_{A_k}(y_j, j \in A_k; \boldsymbol{\theta})$, where $\{A_k\}_{k=1}^K$ is a family of sets of indices associated either with marginal or conditional distributions involving some $y_j, j \in \{1, ..., m\}$ and $w_k, k = 1, ..., K$ are non-negative and known weights. If the weights are all equal, then they can be ignored. In this case, all the statistical procedures give equivalent results. The composite maximum likelihood estimator (CMLE), $\widehat{\boldsymbol{\theta}}_c$, is obtained by maximizing, in respect to $\boldsymbol{\theta} \in \Theta$, the expression (1).

The CMLE is consistent and asymptotically normal and, based on it, we can establish hypothesis testing procedures in a similar way to the classical likelihood ratio test, Wald test or Rao's score test.

A development of the asymptotic theory of the CMLE including its application to obtain the composite ratio statistics, Wald-type tests and Rao score tests in the context of composite likelihood can be seen in [10]. However, in [11–13] is shown that the CMLE and the derived testing procedures present an important lack of robustness. In this sense, [11–13] derived some new distance-based estimators and tests with good robustness behaviour without an important loss of efficiency. In this paper, we are going to consider the composite minimum density power divergence estimator (CMDPDE), introduced in [12], in order to present a model selection criterion in a composite likelihood framework.

Model selection criteria, for summarizing data evidence in favor of a model, is a very well studied subject in statistical literature, overall in the context of full likelihood. The construction of such criteria requires a measure of similarity between two models, which are typically described in terms of their distributions. This can be achieved if an unbiased estimator of the expected overall discrepancy is found, which measures the statistical distance between the true, but unknown model, and the entertained model. Therefore, the model with the smallest value of the criterion is the most preferable model. The use of divergence measures, in particular Kullback–Leibler divergence ([14]), to measure this discrepancy, is the main idea of some of the most known criteria: Akaike Information Criterion (AIC, [15,16]), the criterion proposed by Takeuchi (TIC, [17]) and other modifications of AIC [18]. DIC criterion, based on the density power divergence (DPD), was presented in [19] and, recently, [20] presented a local BHHJ power divergence information criterion following [21]. In the context of the composite likelihood there are some criteria based on Kullback–Leibler divergence, see for instance [22–24] and references therein. To the best of our knowledge only Kullback–Leibler divergence was used to develop model selection criteria in a composite likelihood framework. To fill this gap, our interest is now focused on DPD.

In this paper, we present a new information criterion for model selection in the framework of composite likelihood based on DPD measure. This divergence measure, introduced and studied in the case of complete likelihood by [25], has been considered previously in [12,13] in the context of composite likelihood. In these papers, a new estimator, the CMDPDE, was introduced and its robustness in relation to the CMLE as well as the robustness of some families of test statistics were studied, but the problem of model selection was not considered. This problem is considered in this paper. The criterion introduced in this paper will be called composite likelihood DIC criterion (CLDIC). The motivation of considering a criterion based on DPD instead of Kullback–Leibler divergence is due to the robustness of the procedures based on DPD in statistical inference, not only in the context of full likelihood [25,26], but also in the context of composite likelihood [12,13]. In Section 2, the CMDPDE is presented and some properties of this estimator are discussed. The new model selection criterion, CLDIC, based on CMDPDE is introduced in Section 3 and some of its asymptotic properties are studied. A simulation study is carried out in Section 4 and some numerical examples are presented in Section 5. Finally, some concluding remarks are presented in Section 6.

2. Composite Minimum Density Power Divergence Estimator

Given two probability density functions g and f, associated with two m-dimensional random variables respectively, the DPD ([25]) measures a statistical distance between g and f by

$$d_\alpha(g,f) = \int_{\mathbb{R}^m} \left\{ f(y)^{1+\alpha} - \left(1 + \frac{1}{\alpha}\right) f(y)^\alpha g(y) + \frac{1}{\alpha} g(y)^{1+\alpha} \right\} dy, \tag{2}$$

for $\alpha > 0$, while for $\alpha = 0$ it is defined by

$$d_0(g,f) = \lim_{\alpha \to 0^+} d_\alpha(g,f) = d_{KL}(g,f),$$

where $d_{KL}(g, f)$ is the Kullback–Leibler divergence (see, for example, [26]). For $\alpha = 1$, the expression (2) leads to the L_2 distance $L_2(g, f) = \int_{\mathbb{R}^m} (f(y) - g(y))^2 \, dy$. It is also interesting to note that (2) is a special case of the so-called Bregman divergence

$$\int_{\mathbb{R}^m} \left[T(g(y)) - T(f(y)) - \{g(y) - f(y)\} T'(f(y)) \right] dy. \tag{3}$$

If we consider $T(l) = \frac{1}{\alpha} l^{1+\alpha}$ in (3), we get $d_\alpha(g, f)$. The parameter α controls the trade-off between robustness and asymptotic efficiency of the parameter estimates which are the minimizers of this family of divergences. For more details about this family of divergence measures we refer to [27].

Let now $Y_1, ..., Y_n$ be independent and identically distributed replications of Y which are characterized by the true but unknown distribution g. Taking into account that the true model g is unknown, suppose that $\Xi = \{f(\cdot; \theta), \theta \in \Theta \subseteq \mathbb{R}^p, p \geq 1\}$ is a parametric identifiable family of candidate distributions to describe the observations $y_1, ..., y_n$. Then, the DPD between the true model g and the composite likelihood function, $\mathcal{CL}(\theta, \cdot)$, associated to the parametric model $f(\cdot; \theta)$ is defined as

$$d_\alpha(g(\cdot), \mathcal{CL}(\theta, \cdot)) = \int_{\mathbb{R}^m} \left\{ \mathcal{CL}(\theta, y)^{1+\alpha} - \left(1 + \frac{1}{\alpha}\right) \mathcal{CL}(\theta, y)^\alpha g(y) + \frac{1}{\alpha} g(y)^{1+\alpha} \right\} dy, \tag{4}$$

for $\alpha > 0$, while for $\alpha = 0$ we have $d_{KL}(g(\cdot), \mathcal{CL}(\theta, \cdot))$, which is defined by

$$d_{KL}(g(\cdot), \mathcal{CL}(\theta, \cdot)) = \int_{\mathbb{R}^m} g(y) \log \frac{g(y)}{\mathcal{CL}(\theta, y)} dy. \tag{5}$$

In Section 3, we are going to introduce and study the CLDIC criterion based on (4).

Let

$$\{M_k\}_{k \in \{1,...,\ell\}} \tag{6}$$

be a family of candidate models to govern the observations $Y_1, ..., Y_n$. We shall assume that the true model is included in $\{M_k\}_{k \in \{1,...,\ell\}}$. For a specific $k = 1, ..., \ell$, the parametric model M_k is described by the composite likelihood function

$$\mathcal{CL}(\theta, \cdot), \quad \theta \in \Theta_k \subset \mathbb{R}^k.$$

In this setting, it is quite clear that the most suitable candidate model to describe the observations is the model that minimizes the DPD in (4). However, the unknown parameter θ is included in it, so it is not possible to use directly this measure for the choice of the most suitable model. A way to overcome this problem is to plug-in, in (4), the unknown parameter θ by an estimator which is desirable to obey some nice properties, like consistency and asymptotic normality. Based on this point, the CMDPDE, introduced in [12], can be used. This estimator is described in the sequel for the sake of completeness.

If we denote the kernel of (4) as

$$W_\alpha(\theta) = \int_{\mathbb{R}^m} \mathcal{CL}(\theta, y)^{1+\alpha} d\,y - \left(1 + \frac{1}{\alpha}\right) \int_{\mathbb{R}^m} \mathcal{CL}(\theta, y)^\alpha g(y) dy, \tag{7}$$

we can write

$$d_\alpha(g(\cdot), \mathcal{CL}(\theta, \cdot)) = W_\alpha(\theta) + \frac{1}{\alpha} \int_{\mathbb{R}^m} g(y)^{1+\alpha} dy$$

and the term $\frac{1}{\alpha} \int_{\mathbb{R}^m} g(y)^{1+\alpha} dy$ does not depend on θ and could be ignored in (9). A natural estimator of $W_\alpha(\theta)$, given in (7), can be obtained by observing that the last integral in (7), can be expressed in the form $\int_{\mathbb{R}^m} \mathcal{CL}(\theta, y)^\alpha dG(y)$, for G the distribution function corresponding to g. Hence, if the empirical distribution function of $Y_1, ..., Y_n$ will be exploited, this last integral is approximated by $\frac{1}{n} \sum_{i=1}^{n} C\mathcal{L}(\theta, Y_i)^\alpha$, i.e.,

$$W_{n,\alpha}(\theta) = \int_{\mathbb{R}^m} C\mathcal{L}(\theta, y)^{\alpha+1} dy - \left(1 + \frac{1}{\alpha}\right) \frac{1}{n} \sum_{i=1}^{n} C\mathcal{L}(\theta, Y_i)^\alpha. \tag{8}$$

Definition 1. *The CMDPDE of θ, $\widehat{\theta}_c^\alpha$, is defined, for $\alpha > 0$, by*

$$\widehat{\theta}_c^\alpha = \arg \min_{\theta \in \Theta} W_{n,\alpha}(\theta). \tag{9}$$

We shall denote the score of the composite likelihood by

$$u(\theta, y) = \frac{\partial \log C\mathcal{L}(\theta, y)}{\partial \theta}. \tag{10}$$

Let θ_0 be the true value of the parameter θ. In [12], it was shown that the asymptotic distribution of $\widehat{\theta}_c^\alpha$ is given by

$$\sqrt{n}(\widehat{\theta}_c^\alpha - \theta_0) \xrightarrow[n \to \infty]{\mathcal{L}} \mathcal{N}\left(0_p, H_\alpha(\theta_0)^{-1} J_\alpha(\theta_0) H_\alpha(\theta_0)^{-1}\right),$$

being

$$H_\alpha(\theta) = \int_{\mathbb{R}^m} C\mathcal{L}(\theta, y)^{\alpha+1} u(\theta, y) u(\theta, y)^T dy \tag{11}$$

and

$$J_\alpha(\theta) = \int_{\mathbb{R}^m} C\mathcal{L}(\theta, y)^{2\alpha+1} u(\theta, y) u(\theta, y)^T dy$$
$$- \int_{\mathbb{R}^m} C\mathcal{L}(\theta, y)^{\alpha+1} u(\theta, y) dy \int_{\mathbb{R}^m} u(\theta, y)^T C\mathcal{L}(\theta, y)^{1+\alpha} dy. \tag{12}$$

Remark 1. *For $\alpha = 0$ we get the CMLE of θ*

$$\widehat{\theta}_c = \arg \min_{\theta \in \Theta} \left(-\frac{1}{n} \sum_{i=1}^{n} \log C\mathcal{L}(\theta, y_i)\right). \tag{13}$$

At the same time it is well-known that

$$\sqrt{n}(\widehat{\theta}_c - \theta) \xrightarrow[n \to \infty]{\mathcal{L}} \mathcal{N}\left(0_p, G_*(\theta)^{-1}\right),$$

where $G_*(\theta)$ denotes the Godambe information matrix defined by $G_*(\theta) = H(\theta) J(\theta)^{-1} H(\theta)$, with $H(\theta)$ being the sensitivity or Hessian matrix and $J(\theta)$ being the variability matrix, defined, respectively, by

$$H(\theta) = E_\theta\left[-\frac{\partial}{\partial \theta} u(\theta, Y)^T\right], \quad J(\theta) = E_\theta\left[u(\theta, Y) u(\theta, Y)^T\right].$$

3. A New Model Selection Criterion

In order to describe the CLDIC criterion we consider the model M_k given in (6). Following standard methodology (cf. [28], pp. 240), the most suitable candidate model to describe the data $Y_1, ..., Y_n$ is the model that minimizes the expected estimated DPD

$$E_{Y_1,...,Y_n}\left[d_\alpha(g(\cdot), C\mathcal{L}(\widehat{\theta}_c^\alpha, \cdot))\right], \tag{14}$$

subject to the assumption that the unknown model g is belonging to Ξ, i.e., the true model is included in $\{M_s\}_{s \in \{1,...,\ell\}}$ and taking into account that $\widehat{\theta}_c^\alpha$, defined in (9), is a consistent and asymptotic normally distributed estimator of $\boldsymbol{\theta}$. However, this expected value is still depending on the unknown parameter $\boldsymbol{\theta}$. So, as a criterion, it should be used an asymptotically unbiased estimator of (14), for $g \in \Xi$.

The most appropriate model to select is the model which minimizes the expected value

$$E_{Y_1,\dots,Y_n}\left[W_\alpha\left(\widehat{\boldsymbol{\theta}}_c^\alpha\right)\right].$$

This expected value is still depending on the unknown parameter $\boldsymbol{\theta}$. So, an asymptotically unbiased estimator of the above expected value could be the basis of a selection criterion, for $g \in \Xi$. In order to proceed with the derivation of such an asymptotically unbiased estimator of $E_{Y_1,\dots,Y_n}\left[W_\alpha\left(\widehat{\boldsymbol{\theta}}_c^\alpha\right)\right]$. The empirical version of $W_\alpha\left(\boldsymbol{\theta}\right)$, in (7), is $W_{n,\alpha}(\boldsymbol{\theta})$, given in (8), and plays a central role in the development of the model selection criterion on the basis of the next theorem which expresses the expected value $E_{Y_1,\dots,Y_n}\left[W_\alpha\left(\widehat{\boldsymbol{\theta}}_c^\alpha\right)\right]$ by means of the respective expected value of $W_{n,\alpha}(\widehat{\boldsymbol{\theta}}_c^\alpha)$, in an asymptotically equivalent way.

Theorem 1. *If the true distribution g belongs to the parametric family Ξ and $\boldsymbol{\theta}_0$ denotes the true value of the parameter $\boldsymbol{\theta}$, then we have*

$$E_{Y_1,\dots,Y_n}\left[W_\alpha(\widehat{\boldsymbol{\theta}}_c^\alpha)\right] = E_{Y_1,\dots,Y_n}\left[W_{n,\alpha}(\widehat{\boldsymbol{\theta}}_\alpha) + \frac{\alpha+1}{n}trace\left(J_\alpha\left(\boldsymbol{\theta}_0\right)H_\alpha\left(\boldsymbol{\theta}_0\right)^{-1}\right)\right] + o_p(1)$$

with $H_\alpha\left(\boldsymbol{\theta}\right)$ and $J_\alpha\left(\boldsymbol{\theta}\right)$ given in (11) and (12), respectively.

Based on the above theorem, the proof of which is presented in a full detail in the Appendix A, an asymptotic unbiased estimator of $E_{Y_1,\dots,Y_n}\left[W_\alpha(\widehat{\boldsymbol{\theta}}_c^\alpha)\right]$ is given by

$$W_{n,\alpha}(\widehat{\boldsymbol{\theta}}_c^\alpha) + \frac{\alpha+1}{n}trace\left(J_\alpha(\widehat{\boldsymbol{\theta}}_c^\alpha)H_\alpha(\widehat{\boldsymbol{\theta}}_c^\alpha)^{-1}\right).$$

This ascertainment is the basis and a strong motivation for the next definition which introduces the model selection criterion.

Definition 2. *Let $\{M_k\}_{k\in\{1,\dots,\ell\}}$ be candidate models for the observations $Y_1,\dots,Y_n$. The selected model M^* verifies*

$$M^* = \min_{k\in\{1,\dots,\ell\}} CLDIC_\alpha\left(M_k\right),$$

where

$$CLDIC_\alpha\left(M_k\right) = W_{n,\alpha}(\widehat{\boldsymbol{\theta}}_c^\alpha) + \frac{\alpha+1}{n}trace\left(J_\alpha(\widehat{\boldsymbol{\theta}}_c^\alpha)H_\alpha(\widehat{\boldsymbol{\theta}}_c^\alpha)^{-1}\right),$$

$W_{n,\alpha}\left(\boldsymbol{\theta}\right)$ was given in (8) and $J_\alpha\left(\boldsymbol{\theta}\right)$ and $H_\alpha\left(\boldsymbol{\theta}\right)$ were defined in (11) and (12), respectively.

The next remark summarizes the model selection criterion in the case $\alpha = 0$ and it therefore extends, in a sense, the pioneer and classic AIC.

Remark 2. *For $\alpha = 0$ we have,*

$$d_{KL}(g(\cdot),C\mathcal{L}(\boldsymbol{\theta},\cdot)) = W_0(\boldsymbol{\theta}) + \int_{\mathbb{R}^n} g(y)\log g(y)dy$$

with $W_0(\boldsymbol{\theta}) = -\int_{\mathbb{R}^n}\log C\mathcal{L}(\boldsymbol{\theta},y)g(y)dy$. Therefore, the most appropriate model which should be selected, is the model which minimizes the expected value

$$E_{Y_1,\dots,Y_n}\left[W_0(\widehat{\boldsymbol{\theta}}_c)\right], \tag{15}$$

where $\widehat{\boldsymbol{\theta}}_c$ is the CMLE of $\boldsymbol{\theta}_0$ defined in (9).

The expected value (15) is still depending on the unknown parameter $\boldsymbol{\theta}$. A natural estimator of $W_0(\widehat{\boldsymbol{\theta}}_c)$ can be obtained by replacing the distribution function G, of g, by the empirical distribution function based on $Y_1, \ldots, Y_n$,

$$W_{n,0}(\boldsymbol{\theta}) = -\frac{1}{n} \sum_{i=1}^{n} \log \mathcal{CL}(\boldsymbol{\theta}, y_i).$$

Based on it, we select the model M^* that verifies

$$M^* = \min_{k \in \{1,\ldots,\ell\}} CLDIC_0(M_k),$$

with

$$CLDIC_0(M_k) = H_{n,0}(\widehat{\boldsymbol{\theta}}_c) + \frac{1}{n} trace\left(\mathbf{J}(\widehat{\boldsymbol{\theta}}_c) \mathbf{H}(\widehat{\boldsymbol{\theta}}_c)^{-1} \right),$$

where $\mathbf{J}(\widehat{\boldsymbol{\theta}}_c)$ and $\mathbf{H}(\widehat{\boldsymbol{\theta}}_c)$ are defined in Remark 1. In a manner, quite similar to that of the previous theorem, it can be established that $CLDIC_0(M_k)$ is an asymptotic unbiased estimator of $E_{Y_1,\ldots,Y_n}\left[W_0(\widehat{\boldsymbol{\theta}}_c) \right]$.

This would be the model selection criterion in a composite likelihood framework based on Kullback–Leibler divergence. We can observe that this criterion coincides with the criterion given in [22] as a generalization of the classical criterion of Akaike, which will be referred from now as Composite Akaike Information Criterion (CAIC).

4. Numerical Simulations

4.1. Scenario 1: Two-Component Mixed Model

We are starting with a simulation example, which is motivated and follows ideas from the paper [29] and the Example 4.1 in [20] which will compare the behaviour of the proposed criteria with the CAIC criterion, for $\alpha = 0$ (see Remark 2).

Consider the random vector $Y = (Y_1, Y_2, Y_3, Y_4)^T$ from an unknown density g and let now $Y_1, \ldots, Y_n$ be independent and identically distributed replications of Y which are described by the true but unknown distribution g. Taking into account that the true model g is unknown, suppose that $\{f(\cdot; \boldsymbol{\theta}), \boldsymbol{\theta} \in \Theta \subseteq \mathbb{R}^p, p \geq 1\}$ is a parametric identifiable family of candidate distributions to describe the observations $y_1, \ldots, y_n$. Let also $\mathcal{CL}(\boldsymbol{\theta}, y)$ denotes the composite likelihood function associated to the parametric model $f(\cdot; \boldsymbol{\theta})$.

We consider the problem of choosing (on the basis of n independent and identically distributed replications $y_1, \ldots, y_n$ of $Y = (Y_1, Y_2, Y_3, Y_4)^T$) between a 4-variate normal distribution, $\mathcal{N}(\boldsymbol{\mu}^N, \boldsymbol{\Sigma})$, with $\boldsymbol{\mu}^N = (\mu_1^N, \mu_2^N, \mu_3^N, \mu_4^N)^T$ and

$$\boldsymbol{\Sigma} = \begin{pmatrix} 1 & \rho & 2\rho & 2\rho \\ \rho & 1 & 2\rho & 2\rho \\ 2\rho & 2\rho & 1 & \rho \\ 2\rho & 2\rho & \rho & 1 \end{pmatrix},$$

and a 4-variate t-distribution with v degrees of freedom, $t_v(\boldsymbol{\mu}^{t_v}, \boldsymbol{\Sigma}^*)$, with different location parameters $\boldsymbol{\mu}^{t_v} = (\mu_1^{t_v}, \mu_2^{t_v}, \mu_3^{t_v}, \mu_4^{t_v})^T$ and same variance-covariance matrix $\boldsymbol{\Sigma}$, and density,

$$C_m |\boldsymbol{\Sigma}^*|^{-1/2} \left[1 + \frac{1}{v}(y - \boldsymbol{\mu}^{t_v})^T (\boldsymbol{\Sigma}^*)^{-1}(y - \boldsymbol{\mu}^{t_v}) \right]^{-(v+m)/2},$$

with $\boldsymbol{\Sigma}^* = \frac{v-2}{v}\boldsymbol{\Sigma}$, $C_m = (\pi v)^{-m/2} \frac{\Gamma[(v+m)/2]}{\Gamma(v/2)}$ and $m = 4$.

Consider the composite likelihood function,

$$\mathcal{CL}N(\rho, y) = f_{A_1}^N(y; \rho) f_{A_2}^N(y; \rho),$$

with $f_{A_1}^N(\boldsymbol{y};\rho) = f_{12}^N(y_1,y_2;\mu_1^N,\mu_2^N;\rho)$ and $f_{A_2}^N(\boldsymbol{y};\rho) = f_{34}^N(y_3,y_4;\mu_3^N,\mu_4^N;\rho)$, where f_{12}^N and f_{34}^N are the densities of the marginals of Y, i.e., bivariate normal distributions with mean vectors $(\mu_1^N,\mu_2^N)^T$ and $(\mu_3^N,\mu_4^N)^T$, respectively, and common variance-covariance matrix

$$\Sigma_0 = \begin{pmatrix} 1 & \rho \\ \rho & 1 \end{pmatrix}.$$

In a similar manner consider the composite likelihood

$$\mathcal{C}\mathcal{L}t_v(\rho,\boldsymbol{y}) = f_{A_1}^{t_v}(\boldsymbol{y};\rho)f_{A_2}^{t_v}(\boldsymbol{y};\rho),$$

with $f_{A_1}^{t_v}(\boldsymbol{y};\rho) = f_{12}^{t_v}(y_1,y_2;\mu_1^{t_v},\mu_2^{t_v};\rho)$ and $f_{A_2}^{t_v}(\boldsymbol{y};\rho) = f_{34}^{t_v}(y_3,y_4;\mu_3^{t_v},\mu_4^{t_v};\rho)$, where $f_{12}^{t_v}$ and $f_{34}^{t_v}$ are the densities of the marginals of Y, i.e., bivariate t-distributions with mean vectors $(\mu_1^{t_v},\mu_2^{t_v})^T$ and $(\mu_3^{t_v},\mu_4^{t_v})^T$, respectively, and common variance-covariance matrix

$$\Sigma_0 = \begin{pmatrix} 1 & \rho \\ \rho & 1 \end{pmatrix}.$$

Under this formulation, the simulation study follows in the next two scenarios.

4.1.1. Scenario 1a

Following Example 4.1 in [20], the steps of the simulation study are the following:

- Generate 1000 samples of size $n = 5,7,10,20,40,50,70,100$ from a two component mixture of two 4-variate distributions, namely, a 4-variate normal and a 4-variate t-distribution,

$$h_\omega(\boldsymbol{y}) = \omega N\left(\boldsymbol{\mu}^N,\boldsymbol{\Sigma}\right) + (1-\omega)t_v\left(\boldsymbol{\mu}^{t_v},\boldsymbol{\Sigma}^*\right), \quad 0 \le \omega \le 1,$$

  with $\boldsymbol{\mu}^N = (0,0,0.5,0)$ and $\boldsymbol{\mu}^{t_v} = (3.2,1.5,0.5,2)$, for $\omega = 0,0.25,0.45,0.5,0.55,0.75,1$, $v = 5,10,30$ degrees of freedom and with specific values of $\rho = -0.15,-0.10,0.10$. As pointed out in [29], taking into account that Σ should be semi-positive definite, the following condition is imposed: $-\frac{1}{5} \le \rho \le \frac{1}{3}$.
- Estimate the common parameter ρ, separately in each model, by using the CMDPDE estimator for different values of the tuning parameter $\alpha = 0,0.3$. The composite density which corresponds to the mixture $h_\omega(\boldsymbol{y})$ is defined by

$$\mathcal{C}\mathcal{L}(\rho,\boldsymbol{y}) = \omega\mathcal{C}\mathcal{L}N(\rho,\boldsymbol{y}) + (1-\omega)\mathcal{C}\mathcal{L}t_v(\rho,\boldsymbol{y}), \quad 0 \le \omega \le 1,$$

  and it is used to obtain the CMDPDE estimator, $\widehat{\rho}$, of ρ.
- Define the mixture composite likelihood function

$$\mathcal{C}\mathcal{L}(\widehat{\rho},\boldsymbol{y}) = \omega\mathcal{C}\mathcal{L}N(\widehat{\rho},\boldsymbol{y}) + (1-\omega)\mathcal{C}\mathcal{L}t_v(\widehat{\rho},\boldsymbol{y}), \quad 0 \le \omega \le 1.$$

- Calculate $CLDIC_\alpha(M_k)$, the value of the model selection criterion considered in this paper, for the two candidate models, with

$$CLDIC_\alpha(M_k) = W_{n,\alpha}(\widehat{\rho}) + \frac{\alpha+1}{n}trace\left(\boldsymbol{J}_\alpha(\widehat{\rho})\,\boldsymbol{H}_\alpha(\widehat{\rho})^{-1}\right).$$

  An explanation of how to obtain this value for the both candidate models is given in Appendix B.
- Compute the times that the 4-variate normal model was selected.

Results are summarized in Table 1. Extreme values of $\omega = 0,1$ represent the times that the 4-variate normal model was selected under the 4-variate t-distribution and 4-variate normal distribution, respectively. This means that, for $\omega = 1$, the perfect discrimination will be achieved when 1000 of the 1000 simulated samples are correctly assigned, while for $\omega = 0$, the more near to 0, the better discrimination of the criterion. $\omega = 0.5$ means that each sample was generated both from the normal and t-distribution in the same proportion.

Table 1. Main results, Scenario 1a.

ω	$\alpha = 0$ (CAIC)							$\alpha = 0.3$						
	0	**0.25**	**0.45**	**0.5**	**0.55**	**0.75**	**1**	**0**	**0.25**	**0.45**	**0.5**	**0.55**	**0.75**	**1**
						$v = 5, \rho = -0.15$								
n = 5	0	1	269	499	713	996	1000	0	0	273	498	712	1000	1000
7	0	1	246	504	758	998	1000	0	1	220	511	738	999	1000
10	0	0	202	482	775	1000	1000	0	0	185	467	771	1000	1000
20	0	0	114	486	871	1000	1000	0	0	112	473	866	1000	1000
40	0	0	41	459	947	1000	1000	0	0	54	496	954	1000	1000
50	0	0	21	475	964	1000	1000	0	0	41	556	986	1000	1000
70	0	0	9	461	985	1000	1000	0	0	48	656	995	1000	1000
100	0	0	5	472	992	1000	1000	0	0	142	885	1000	1000	1000
						$v = 10, \rho = -0.15$								
5	0	3	222	445	688	996	1000	0	3	218	433	688	997	1000
7	0	1	191	439	720	1000	1000	0	0	179	431	690	999	1000
10	0	0	163	432	747	1000	1000	0	0	152	402	725	1000	1000
20	0	0	59	399	819	1000	1000	0	0	49	361	773	1000	1000
40	0	0	19	336	912	1000	1000	0	0	12	326	899	1000	1000
50	0	0	6	362	936	1000	1000	0	0	10	334	925	1000	1000
70	0	0	1	292	960	999	1000	0	0	2	356	973	1000	1000
100	0	0	0	301	983	1000	1000	0	0	1	531	992	1000	1000
						$v = 30, \rho = -0.15$								
5	0	4	237	423	677	997	1000	0	2	235	413	656	996	1000
7	0	0	155	394	689	1000	1000	0	0	141	379	677	999	1000
10	0	0	144	413	719	1000	1000	0	0	134	393	701	1000	1000
20	0	0	57	351	801	1000	1000	0	0	40	311	764	1000	1000
40	0	0	11	296	904	1000	1000	0	0	8	263	882	1000	1000
50	0	0	6	271	918	1000	1000	0	0	3	253	903	1000	1000
70	0	0	1	225	942	1000	1000	0	0	0	229	941	1000	1000
100	0	0	0	208	978	1000	1000	0	0	0	303	989	1000	1000
						$v = 10, \rho = -0.10$								
5	0	4	242	464	680	996	1000	0	3	238	459	682	999	1000
7	0	0	187	461	733	997	1000	0	0	199	457	731	998	1000
10	0	0	162	445	738	1000	1000	0	0	165	407	713	1000	1000
20	0	0	62	378	807	1000	1000	0	0	59	354	789	1000	1000
40	0	0	19	357	902	999	1000	0	0	14	333	895	1000	1000
50	0	0	6	325	932	1000	1000	0	0	8	325	931	1000	1000
70	0	0	2	305	954	1000	1000	0	0	6	367	967	1000	1000
100	0	0	0	307	979	1000	1000	0	0	2	507	993	1000	1000
						$v = 10, \rho = 0.10$								
5	0	11	268	459	669	991	1000	1	11	268	478	680	993	1000
7	0	1	211	456	720	999	1000	0	3	207	464	716	998	1000
10	0	0	168	423	704	1000	1000	0	0	162	403	702	1000	1000
20	0	0	86	360	789	1000	999	0	0	89	357	786	1000	1000
40	0	0	35	367	893	1000	1000	0	0	38	398	896	1000	1000
50	0	0	19	331	886	1000	1000	0	0	19	360	913	1000	1000
70	0	0	11	311	933	1000	1000	0	0	16	379	963	1000	1000
100	0	0	2	276	969	1000	1000	0	0	7	490	985	1000	1000

4.1.2. Scenario 1b

Same Scenario is evaluated under the more-closed means $\mu^N = (0, 1.5, 0.5, -0.75)$ and $\mu^{t_\nu} = (0, 1.5, 0.5, 2)$ for moderate-large sample sizes and $\alpha \in \{0, 0.2, 0.4\}$. Here $\nu = 5$ and $\rho = -0.15$. Results are shown in Table 2. In this case, the models under consideration are more similar, so it would be understandable that the CLDIC criterion did not discriminate in such as good way.

Table 2. Main results, Scenario 1b.

	$\alpha = 0$ (CAIC)				$\alpha = 0.2$				$\alpha = 0.4$			
	0	0.25	0.75	1	0	0.25	0.75	1	0	0.25	0.75	1
$n = 40$	0	0	39	731	0	0	537	961	0	0	580	949
50	0	0	24	732	0	0	859	990	0	0	944	994
60	0	0	14	772	0	0	999	1000	0	1	999	1000
70	0	0	9	734	0	0	999	1000	0	27	999	1000
80	0	0	5	770	0	1	1000	1000	0	326	1000	1000
90	0	0	4	782	0	23	1000	1000	2	794	1000	1000
100	0	0	4	802	0	173	1000	1000	26	978	1000	1000

4.2. Scenario 2: Three-Component Mixed Model

Now, we consider a mixed model composed on two 4-variate normal distributions and a 4-variate t-distribution with $\nu = 10$ degrees of freedom. The three distributions have common variance-covariance matrix, as in the previous scenario, with unknown $\rho = -0.15$ and different but known means $\mu_1^N = (0, 0, 0.5, 0)$, $\mu_2^N = (0, 1.5, 0.5, 0)$ and $\mu^t = (0, 1.5, 0.5, 2)$. The model is defined by

$$\omega \mathcal{N}(\mu_1^N, \Sigma) + \lambda \mathcal{N}(\mu_2^N, \Sigma) + (1 - \omega - \lambda)t_{\nu=10}(\mu^t, \Sigma^*), \quad 0 \leq \omega, \lambda, \omega + \lambda \leq 1,$$

with Σ being again a common variance-covariance matrix with unknown parameter ρ of the form

$$\Sigma = \begin{pmatrix} 1 & \rho & 2\rho & 2\rho \\ \rho & 1 & 2\rho & 2\rho \\ 2\rho & 2\rho & 1 & \rho \\ 2\rho & 2\rho & \rho & 1 \end{pmatrix}.$$

Following the same steps that in the first scenario, we generate 1000 samples of the three-component mixture for different sample sizes $n = 5, 7, 10, 20, 40, 50, 70, 100$ and different values of ω and λ. Then, we consider the problem of choosing among one of the two 4-variate normal distributions and the 4-variate t-distribution through the CLDIC criterion, for different values of the tuning parameter $\alpha = 0, 0.3, 0.5, 0.7$. See Table 3 for results. Here, the normal models are denoted by N1 and N2, respectively, while the 4-variate t-distribution is denoted by MT. The first three cases evaluate the selected model under these multivariate distributions. In the last two scenarios, a mixed model is considered as the true distribution.

4.3. Discussion of Results

In Scenario 1a, two well-differentiated multivariate models are considered. In this case CLDIC criterion works in a very efficient way, with an almost-perfect discrimination for extreme values of ω. The good behaviour is also observed for not so extreme values of ω, such as $\omega = 0.55$ or 0.45. We can not observe a significant difference in the choice of α.

In Scenario 1b we consider closer models, which affect the discrimination power of the CLDIC. However, in this case, we do observe great differences when considering different α. While the discrimination power of CLDIC for $\alpha = 0$ (CAIC) and $\omega = 1$ is around 75%, for $\alpha = 0.2$ or $\alpha = 0.4$

the behaviour is excellent. This happens also for large but not extreme values of ω, such as $\omega = 0.75$. However, a medium value of α turns into a worse discrimination for low values of ω.

Table 3. Main results, Scenario 2.

Model *	$\alpha = 0$ (CAIC)			$\alpha = 0.3$			$\alpha = 0.5$			$\alpha = 0.7$		
	N1	**N2**	**MT**	**N1**	**N2**	**MT**	**N1**	**N2**	**MT**	**N1**	**N2**	**MT**
True model: $\mathcal{N}(\mu_1^N, \Sigma)$												
$n = 5$	957	24	19	950	16	34	939	23	38	936	28	36
7	970	19	11	966	13	24	961	13	26	950	22	28
10	993	3	4	986	4	10	979	6	15	971	6	23
20	1000	0	0	1000	0	0	998	0	2	997	0	3
40	1000	0	0	1000	0	0	1000	0	0	1000	0	0
50	1000	0	0	1000	0	0	1000	0	0	1000	0	0
70	1000	0	0	1000	0	0	1000	0	0	1000	0	0
100	1000	0	0	1000	0	0	1000	0	0	999	0	0
True model: $\mathcal{N}(\mu_2^N, \Sigma)$												
5	29	638	333	34	610	356	38	639	323	50	646	304
7	15	622	363	13	589	398	17	599	384	28	627	345
10	6	610	384	5	540	455	5	540	455	11	586	403
20	1	612	387	1	518	481	1	472	527	1	527	472
40	0	566	434	0	650	350	0	590	410	0	614	386
50	0	561	439	0	804	196	0	797	203	0	835	165
70	0	584	416	0	987	13	0	994	6	0	998	2
100	0	520	480	0	1000	0	0	1000	0	0	1000	0
True model: $t_{\nu=10}(\mu^t, \Sigma)$												
5	2	15	983	1	6	993	1	8	991	3	15	982
7	0	3	997	0	1	999	2	2	996	0	4	996
10	0	1	999	0	2	998	0	2	998	0	3	997
20	0	0	1000	0	0	1000	0	0	1000	0	0	1000
40	0	0	1000	0	0	1000	0	0	1000	0	0	1000
50	0	0	1000	0	0	1000	0	0	1000	0	0	1000
70	0	0	1000	0	0	1000	0	0	1000	0	0	1000
100	0	0	1000	0	0	1000	0	4	996	0	296	704
True model: $0.7\mathcal{N}(\mu_2^N, \Sigma) + 0.3 t_{\nu=10}(\mu^t, \Sigma)$												
5	6	384	610	6	375	619	4	401	595	11	452	537
7	1	331	668	1	294	705	1	317	682	1	373	626
10	1	261	738	1	218	781	1	253	746	1	306	693
20	0	109	891	0	101	899	0	107	893	0	141	859
40	0	26	974	0	126	874	0	122	878	0	166	834
50	0	13	987	0	311	689	0	345	655	0	445	555
70	0	6	994	0	948	52	0	982	18	0	994	6
100	0	2	998	0	1000	0	0	1000	0	0	999	1
True model: $\frac{1}{3}\mathcal{N}(\mu_1^N, \Sigma) + \frac{1}{3}\mathcal{N}(\mu_2^N, \Sigma) + \frac{1}{3}t_{\nu=10}(\mu^t, \Sigma)$												
5	127	377	496	121	363	516	107	392	501	107	424	469
7	87	357	556	70	339	591	66	356	578	63	396	541
10	69	326	605	61	314	625	56	330	614	45	381	574
20	37	259	704	25	298	677	17	337	646	15	349	636
40	7	145	848	9	452	539	4	508	488	1	469	530
50	2	122	876	5	744	251	3	814	183	3	853	144
70	0	99	901	4	996	0	4	996	0	4	996	0
100	0	36	964	355	645	0	645	355	0	856	144	0

* Here the model candidates are expressed as N1, N2, MT to denote $\mathcal{N}(\mu_1^N, \Sigma)$, $\mathcal{N}(\mu_2^N, \Sigma)$ and $t_{10}(\mu^t, \Sigma)$, respectively.

Scenario 2 deals with three different models, two multivariate normal and one multivariate *t* (N1, N2 and MT, respectively). The second normal distribution is closer to MT in terms of means. While CLDIC criterion discriminate well between N1 and N2 and between N1 and MT, it has difficulties in distinguishing N2 an MT distributions, overall for small samples sizes and $\alpha = 0$.

It seems, therefore, that when we have well-discriminated models, CLDIC criterion works very well, independently of the sample size and the tuning parameter α considered. Dealing with closer models leads, as expected, to worst results, overall for $\alpha = 0$ (CAIC).

Note that the behaviour of Wald-type and Rao tests based on CMDPDEs was studied in [12,13] through extensive simulation studies.

5. Numerical Examples

5.1. Choice of the Tuning Parameter

In the previous sections, we have seen that CLDIC criterion works generally very well, independently of α, but that some values present a better behaviour, overall when distinguishing similar models. In these situations, it appears that values close to 0.2 or 0.3 work well, while CAIC criterion presents a worse behaviour. A data-driven approach for the choice of the tuning parameter which would be helpful in practice. The approach of [30] was adapted In [13], for the choice of the optimum α in CMDPDEs. This approach consisted on minimizing the estimated mean squared error by means of a pilot estimator, θ^P. This approximation is given by

$$\widehat{MSE}_\alpha = (\widehat{\theta}_c^\alpha - \theta^P)^T (\widehat{\theta}_c^\alpha - \theta^P) + \frac{1}{n}\mathrm{Trace}\left(H_\alpha^{-1}(\widehat{\theta}_c^\alpha)J_\alpha(\widehat{\theta}_c^\alpha)H_\alpha^{-1}(\widehat{\theta}_c^\alpha)\right), \tag{16}$$

where $H_\alpha(\theta)$ and $J_\alpha(\theta)$ are given in (11) and (12). The optimum α will be the one that minimizes expression (16). The choice of the pilot estimator is probably one of the major drawbacks of this approach, as it may lead to a choice of α too close to that used for the pilot estimator. A pilot estimator with $\alpha \approx 0.4$, was proposed in [13] after some simulations, in concordance with [30], where the initial choice of a pilot is suggested to be a robust one in order to obtain the best results in terms of robustness.

5.2. Iris Data

The Iris data (Fisher, [31]) includes 3 categories of 50 sample values each, where each category refers to a type of iris plant: *setosa*, *versicolor* and *virginica*. Each plant is categorized in its class and described by other 4 variables: (1) sepal length, (2) sepal width, (3) petal length and (4) petal width. This is one of the most known data sets for discriminant analysis. [32] proposed the use of a Gaussian finite mixture for modeling Iris data, in which each known class is modeled by a single Gaussian term with the same variance-covariance matrix. The resulting model is as follows

$$f(x) = \frac{1}{3}\mathcal{N}(\mu_1, \Sigma) + \frac{1}{3}\mathcal{N}(\mu_2, \Sigma) + \frac{1}{3}\mathcal{N}(\mu_3, \Sigma), \tag{17}$$

with

$$\mu_1 = (\mu_{11}, \mu_{12}, \mu_{13}, \mu_{14})^T, \quad \mu_2 = (\mu_{21}, \mu_{22}, \mu_{23}, \mu_{24})^T, \quad \mu_3 = (\mu_{31}, \mu_{32}, \mu_{33}, \mu_{34})^T$$

and

$$\Sigma = \begin{pmatrix} \sigma_1^2 & \sigma_{12} & \sigma_{13} & \sigma_{14} \\ \sigma_{21} & \sigma_2^2 & \sigma_{23} & \sigma_{24} \\ \sigma_{31} & \sigma_{32} & \sigma_3^2 & \sigma_{34} \\ \sigma_{41} & \sigma_{42} & \sigma_{43} & \sigma_4^2 \end{pmatrix}.$$

Exact values can be obtained by *MclustDA()* function of *mclust* package in R Software ([32]).

We propose a composite likelihood approach to modeling (17) where we suppose independence between the two first and two last variables. This is

$$f_{CL}(\boldsymbol{y}) = \frac{1}{3}CLN_1 + \frac{1}{3}CLN_2 + \frac{1}{3}CLN_3, \tag{18}$$

with

$$CLN_i = f^N_{A_{i1}}(\rho_{12}, \boldsymbol{y}) f^N_{A_{i2}}(\rho_{34}, \boldsymbol{y}),$$

where $f^N_{A_{i1}}(\rho_{12}, \boldsymbol{y}) = f^N_{A_{i1}}(\rho_{12}, \mu_{i1}, \mu_{i2}, \boldsymbol{\Sigma}_{A_1}, \boldsymbol{y})$ and $f^N_{A_{i2}}(\rho_{34}, \boldsymbol{y}) = f^N_{A_{i2}}(\rho_{34}, \mu_{i3}, \mu_{i4}, \boldsymbol{\Sigma}_{A_2}, \boldsymbol{y})$, $i = 1, 2, 3$ are bivariate normals with variance-covariance matrices

$$\boldsymbol{\Sigma}_{A_1} = \begin{pmatrix} \sigma_1^2 & \rho_{12}\sigma_1\sigma_2 \\ \rho_{12}\sigma_1\sigma_2 & \sigma_2^2 \end{pmatrix}, \quad \boldsymbol{\Sigma}_{A_2} = \begin{pmatrix} \sigma_3^2 & \rho_{34}\sigma_3\sigma_4 \\ \rho_{34}\sigma_3\sigma_4 & \sigma_4^2 \end{pmatrix}.$$

We are going to evaluate the behavior of the CLDIC criterion proposed in previous sections. After estimating parameters ρ_{12} and ρ_{34} in (18), we consider 10 different subsets of the IRIS data:

- SE subset: 50 first observations, corresponding to Setosa plants ($n = 50$).
- VE subset: 50 second observations, corresponding to Versicolor plants ($n = 50$).
- VI subset: 50 last observations, corresponding to Virginica plants ($n = 50$).
- SE(VE) subset: SE subset with 2 first observations of VE subset ($n = 52$).
  Equivalently: SE(VI), VE(SE), VE(VI), VI(SE) and VI(VE).
- VI(SE+VE) subset: VI subset with 2 first observations of SE and VE subsets ($n = 54$).

In Table 4, chosen models for each one of the subsets are obtained by the proposed CLDIC criterion. When a "pure" subset is considered, all the tuning parameters lead to optimal decisions, but when a "contaminated" subset is under consideration, only $\alpha = 0.2, 0.3$ have an optimal response in all the cases.

Table 4. Selected model in each of the subsets. Iris data.

α	SE	VE	VI	SE(VE)	SE(VI)	VE(SE)	VE(VI)	VI(SE)	VI(VE)	VI(SE+VE)
0 (CAIC)	CN1	CN2	CN3	CN1	CN1	CN1*	CN2	CN1*	CN3	CN3
0.2	CN1	CN2	CN3	CN1	CN1	CN2	CN2	CN3	CN3	CN3
0.3	CN1	CN2	CN3	CN1	CN1	CN2	CN2	CN3	CN3	CN3
0.4	CN1	CN2	CN3	CN1	CN1	CN2	CN2	CN1*	CN3	CN3
0.5	CN1	CN2	CN3	CN1	CN1	CN2	CN2	CN1*	CN3	CN3
0.8	CN1	CN2	CN3	CN1	CN1	CN2	CN2	CN1*	CN3	CN3
0.22	CN1	CN2	CN3	CN1	CN1	CN2	CN2	CN3	CN3	CN3

We now apply the ad hoc approach presented in Section 5.1 for selecting the tuning parameter α in a composite likelihood framework. Applying this procedure to our data set though a grid search of length 100 and by means of a pilot estimator with $\alpha = 0.4$ leads to the optimal tuning parameter $\alpha = 0.22$, what is in concordance with the obtained results (see Table 5). We can see that the use of other pilot estimators would not affect very much to the final decission.

Table 5. Selected α for different pilot estimators, ad-hoc tuning parameter selection procedure. Iris and Wine data

	α_{pilot}	0	0.1	0.2	0.3	0.4	0.5	0.6	0.7	0.8	0.9	1
Iris	α_{opt}	0.31	0.17	0.20	0.21	**0.22**	0.23	0.24	0.24	0.25	0.25	0.25
Wine	α_{opt}	0.45	0.46	0.47	0.49	**0.51**	0.53	0.55	0.56	0.56	0.56	0.57

5.3. Wine Data

We now work with Wine data ([33]), which contain a chemical analysis of 178 Italian wines from three different cultivars (Barolo, Grignolino, Barbera) yielded 13 measurements. In order to illustrate our criterion, we will work with only first four explanatory variables: Alcohol, Malic, Ash and Alkalinity. As in the previous section, we adjust a Gaussian mixture model with weights, in this case: $59/178$, $72/178$ and $47/178$ corresponding to Barolo, Grignolino and Barbera classes, respectively. We now consider these 10 different subsets of the Wine data:

- BO subset: 20 first observations of Barolo wines ($n = 20$).
- GR subset: 20 first observations of Grignolino wines ($n = 20$).
- BA subset: 20 first observations of Barbera wines ($n = 20$).
- BO(GR) subset: BO subset with 5 first observations of GR subset ($n = 25$).
  Equivalently: BO(BA), GR(BO), GR(BA), BA(BO) and BA(GR).
- BA(BO+GR) subset: BA subset with 3 first observations of BO and GR subsets ($n = 26$).

We can observe how, for medium values of α, the discrimination is perfect (see Table 6). Applying ad-hoc tuning parameter choice procedure we obtain $\alpha_{opt} \approx 0.51$, with a perfect discrimination again (Table 5).

Table 6. Selected model in each of the subsets. Wine data.

α	BO	GR	BA	BO(GR)	BO(BA)	GR(BO)	GR(BA)	BA(BO)	BA(GR)	BA(BO+GR)
0 (CAIC)	CN1	CN2	CN3	CN1	CN1	CN2	CN2	CN3	CN3	CN2*
0.2	CN1	CN2	CN3	CN1	CN1	CN2	CN2	CN3	CN3	CN3
0.3	CN1	CN2	CN3	CN1	CN1	CN2	CN2	CN3	CN3	CN3
0.4	CN1	CN2	CN3	CN1	CN1	CN2	CN2	CN3	CN3	CN3
0.5	CN1	CN2	CN3	CN1	CN1	CN2	CN2	CN3	CN3	CN3
0.8	CN1	CN2	CN3	CN1	CN1	CN2	CN2	CN2*	CN2*	CN3
0.51	CN1	CN2	CN3	CN1	CN1	CN2	CN2	CN3	CN3	CN3

6. Conclusions and Future Research

In this paper, we have addressed the problem of model selection in the framework of composite likelihood methodology, on the basis of the DPD as a measure of the closeness of the composite density and the true model that drives the data. In this context, an information criterion is introduced and studied which is defined by means of composite minimum distance type estimators of the unknown parameters, well-known for having nice robustness properties. Thanks to a simulation study, we have shown that the proposed here model selection criterion works well in practice and mainly that the use of CMDPDE makes the criterion more robust than the criteria based on the classic CMLE and the Kullback–Leibler divergence, given in [22]. The analysis of two real data examples of the literature illustrate on how the model selection criterion, presented here, can be applied in practical cases. This paper is a part of a series of papers by the authors where composite likelihood ideas and methods are harmonically weaved with divergence theoretic methods in order to develop statistical inference (estimation and testing of hypotheses) and model selection criteria, as well. We envision future work in some directions. The development of change point methodology on the basis of composite density with CMDPDE and divergence measures would be maybe an appealing problem for a future research on the topic. However, all the information theoretic methods developed on the basis of the composite likelihood depend on the choice of the family of sets $\{A_k\}_{k=1}^K$, appeared in Formula (1). A question is raised at this point: how the information theoretic procedures developed on the basis of the composite likelihood are affected by this family of sets? It is an appealing problem which deserves also investigation in a future work.

Author Contributions: Conceptualization, E.C., N.M., L.P. and K.Z.; Methodology, E.C., N.M., L.P. and K.Z.; Software, E.C., N.M., L.P. and K.Z.; Validation, E.C., N.M., L.P. and K.Z.; Formal Analysis, E.C., N.M., L.P. and K.Z.; Investigation, E.C., N.M., L.P. and K.Z.; Resources, E.C., N.M., L.P. and K.Z.; Data Curation, E.C., N.M., L.P. and K.Z.; Writing—Original Draft Preparation, E.C., N.M., L.P. and K.Z.; Writing—Review & Editing, E.C., N.M., L.P. and K.Z.; Visualization, E.C., N.M., L.P. and K.Z.; Supervision, E.C., N.M., L.P. and K.Z. All authors have read and agreed to the published version of the manuscript.

Funding: This research is partially supported by Grant PGC2018-095194-B-I00 and Grant FPU16/03104 from Ministerio de Ciencia, Innovación y Universidades (Spain). E. Castilla, N. Martín and L. Pardo are members of the Instituto de Matemática Interdisciplinar, Complutense University of Madrid.

Acknowledgments: The authors would like to thank the Editor and Reviewers for taking their precious time to make several valuable comments on the manuscript.

Conflicts of Interest: The authors declare no conflict of interest.

Abbreviations

The following abbreviations are used in this manuscript:

MLE	Maximum likelihood estimator
CMLE	Composite maximum likelihood estimator
CLDIC	Composite likelihood DIC
DPD	Density power divergence
MDPDE	Minimum density power divergence estimator
CMDPDE	Composite minimum density power divergence estimator
AIC	Akaike Information Criterion
CAIC	Composite Akaike Information Criterion
TIC	Takeuchi Information Criterion

Appendix A. Proof of Theorem 1

Proof. A Taylor expansion of $W_\alpha(\theta)$ around the true parameter θ_0 and evaluated in $\theta = \widehat{\theta}_c^\alpha$, gives

$$
W_\alpha\left(\widehat{\theta}_c^\alpha\right) = W_\alpha(\theta_0) + \left(\frac{\partial W_\alpha(\theta)}{\partial \theta}\right)_{\theta=\theta_0}\left(\widehat{\theta}_c^\alpha - \theta_0\right)
$$
$$
+ \frac{1}{2}\left(\widehat{\theta}_c^\alpha - \theta_0\right)^T \left(\frac{\partial^2 W_\alpha(\theta)}{\partial\theta\partial\theta^T}\right)_{\theta=\theta_0}\left(\widehat{\theta}_c^\alpha - \theta_0\right) + o\left(\left\|\left(\widehat{\theta}_c^\alpha - \theta_0\right)\right\|^2\right).
$$

Now,

$$
\frac{\partial W_\alpha(\theta)}{\partial \theta} = \int_{\mathbb{R}^m} (1+\alpha)\, C\mathcal{L}(\theta,y)^\alpha \frac{\partial C\mathcal{L}(\theta,y)}{\partial \theta}dy - \left(1+\frac{1}{\alpha}\right)\alpha \int_{\mathbb{R}^m} C\mathcal{L}(\theta,y)^{\alpha-1}\frac{\partial C\mathcal{L}(\theta,y)}{\partial \theta}g(y)dy
$$
$$
= (1+\alpha)\int_{\mathbb{R}^m} C\mathcal{L}(\theta,y)^{\alpha+1}u(\theta,y)\,dy - (1+\alpha)\int_{\mathbb{R}^m} C\mathcal{L}(\theta,y)^\alpha u(\theta,y)\,g(y)dy.
$$

It is clear that if the true distribution g belongs to the parameter family $f(.;\theta), \theta \in \Theta$ and θ_0 denotes the true value of the parameter θ, we get

$$
\left(\frac{\partial W_\alpha(\theta)}{\partial \theta}\right)_{\theta=\theta_0} = 0.
$$

Now we are going to get

$$\frac{\partial^2 W_\alpha\left(\boldsymbol{\theta}\right)}{\partial\boldsymbol{\theta}\partial\boldsymbol{\theta}^T} = (1+\alpha)\left\{ \int_{\mathbb{R}^m} (1+\alpha)\,CL(\boldsymbol{\theta},\boldsymbol{y})^{\alpha+1} \boldsymbol{u}\left(\boldsymbol{\theta},\boldsymbol{y}\right)\boldsymbol{u}\left(\boldsymbol{\theta},\boldsymbol{y}\right)^T dy \right.$$

$$- \int_{\mathbb{R}^m} CL(\boldsymbol{\theta},\boldsymbol{y})^{\alpha+1}\left(-\frac{\partial^2 \log CL(\boldsymbol{\theta},\boldsymbol{y})}{\partial\boldsymbol{\theta}\partial\boldsymbol{\theta}^T}\right)dy$$

$$\left. -\alpha \int_{\mathbb{R}^m} CL(\boldsymbol{\theta},\boldsymbol{y})^{\alpha}\boldsymbol{u}\left(\boldsymbol{\theta},\boldsymbol{y}\right)\boldsymbol{u}\left(\boldsymbol{\theta},\boldsymbol{y}\right)^T g(\boldsymbol{y})dy + \int_{\mathbb{R}^m} CL(\boldsymbol{\theta},\boldsymbol{y})^{\alpha}\left(-\frac{\partial^2 \log CL(\boldsymbol{\theta},\boldsymbol{y})}{\partial\boldsymbol{\theta}\partial\boldsymbol{\theta}^T}\right)g(\boldsymbol{y})dy \right\}.$$

If the true distribution g belongs to the parameter family $f_\theta(\cdot\,;\boldsymbol{\theta})$, $\boldsymbol{\theta}\in\Theta$ and $\boldsymbol{\theta}_0$ denotes the true value of the parameter $\boldsymbol{\theta}$, verifies,

$$\left(\frac{\partial^2 W_\alpha\left(\boldsymbol{\theta}\right)}{\partial\boldsymbol{\theta}\partial\boldsymbol{\theta}^T}\right)_{\boldsymbol{\theta}=\boldsymbol{\theta}_0} = (1+\alpha)\int_{\mathbb{R}^m} CL(\boldsymbol{\theta}_0,\boldsymbol{y})^{\alpha+1}\boldsymbol{u}\left(\boldsymbol{\theta}_0,\boldsymbol{y}\right)\boldsymbol{u}\left(\boldsymbol{\theta}_0,\boldsymbol{y}\right)^T dy$$
$$= (1+\alpha)\,H_\alpha\left(\boldsymbol{\theta}_0\right).$$

Therefore,

$$nW_\alpha\left(\widehat{\boldsymbol{\theta}}_c^\alpha\right) = nW_\alpha\left(\boldsymbol{\theta}_0\right) + \frac{(1+\alpha)}{2}\sqrt{n}\left(\widehat{\boldsymbol{\theta}}_c^\alpha-\boldsymbol{\theta}_0\right)^T H_\alpha\left(\boldsymbol{\theta}_0\right)\sqrt{n}\left(\widehat{\boldsymbol{\theta}}_c^\alpha-\boldsymbol{\theta}_0\right) + no\left(\left\|\left(\widehat{\boldsymbol{\theta}}_c^\alpha-\boldsymbol{\theta}_0\right)\right\|^2\right).$$

But

$$\sqrt{n}\left(\widehat{\boldsymbol{\theta}}_c^\alpha-\boldsymbol{\theta}_0\right)\xrightarrow[n\to\infty]{L} N\left(\boldsymbol{0}, H_\alpha(\boldsymbol{\theta}_0)^{-1}J_\alpha(\boldsymbol{\theta}_0)H_\alpha(\boldsymbol{\theta}_0)^{-1}\right),$$

and $no\left(\left\|\left(\widehat{\boldsymbol{\theta}}_c^\alpha-\boldsymbol{\theta}_0\right)\right\|^2\right) = o(O_p(1)) = o_p(1)$.

The asymptotic distribution of the quadratic form $\sqrt{n}\left(\widehat{\boldsymbol{\theta}}_c^\alpha-\boldsymbol{\theta}_0\right)^T H_\alpha\left(\boldsymbol{\theta}_0\right)\sqrt{n}\left(\widehat{\boldsymbol{\theta}}_c^\alpha-\boldsymbol{\theta}_0\right)$, verifies

$$\sqrt{n}\left(\widehat{\boldsymbol{\theta}}_c^\alpha-\boldsymbol{\theta}_0\right)^T H_\alpha\left(\boldsymbol{\theta}_0\right)\sqrt{n}\left(\widehat{\boldsymbol{\theta}}_c^\alpha-\boldsymbol{\theta}_0\right)\xrightarrow[n\to\infty]{L}\sum_{r=1}^{k}\lambda_r Z_r^2$$

being $\lambda_r,\, r=1,...,k$, the eigenvalues of the matrix

$$H_\alpha\left(\boldsymbol{\theta}_0\right)H_\alpha(\boldsymbol{\theta}_0)^{-1}J_\alpha(\boldsymbol{\theta}_0)H_\alpha(\boldsymbol{\theta}_0)^{-1} = J_\alpha(\boldsymbol{\theta}_0)H_\alpha(\boldsymbol{\theta}_0)^{-1}$$

and Z_r are independent normal random variable of mean zero and variance 1. Therefore,

$$E_{Y_1,...,Y_n}\left[\sqrt{n}\left(\widehat{\boldsymbol{\theta}}_c^a-\boldsymbol{\theta}_0\right)^T H_\alpha\left(\boldsymbol{\theta}_0\right)\sqrt{n}\left(\widehat{\boldsymbol{\theta}}_c^a-\boldsymbol{\theta}_0\right)\right] = \sum_{r=1}^{k}\lambda_r + o_p(1)$$
$$= trace\left(J_\alpha(\boldsymbol{\theta}_0)H_\alpha(\boldsymbol{\theta}_0)^{-1}\right) + o_p(1)$$

and

$$E_{Y_1,...,Y_n}\left[nW_\alpha(\widehat{\boldsymbol{\theta}}_c^\alpha)\right] = nW_\alpha\left(\boldsymbol{\theta}_0\right) + \frac{(1+\alpha)}{2}trace\left(J_\alpha(\boldsymbol{\theta}_0)H_\alpha(\boldsymbol{\theta}_0)^{-1}\right) + o_p(1).$$

Now a Taylor expansion of $W_{n,\alpha}\left(\boldsymbol{\theta}\right)$, around $\widehat{\boldsymbol{\theta}}_c^\alpha$ and evaluated at $\boldsymbol{\theta}=\boldsymbol{\theta}_0$ gives

$$W_{n,\alpha}(\theta_0) = W_{n,\alpha}(\widehat{\theta}_c^\alpha) + \left(\frac{H_{n,\alpha}(\theta)}{\partial\theta}\right)_{\theta=\widehat{\theta}_c^\alpha}\left(\theta_0 - \widehat{\theta}_c^\alpha\right)$$
$$+ \frac{1}{2}\left(\theta_0 - \widehat{\theta}_c^\alpha\right)^T \left(\frac{\partial^2 W_{n,\alpha}(\theta)}{\partial\theta}\partial\theta^T\right)_{\theta=\widehat{\theta}_c^\alpha}\left(\theta_0 - \widehat{\theta}_c^\alpha\right) + o\left(\left\|\theta_0 - \widehat{\theta}_c^\alpha\right\|^2\right).$$

But

$$\frac{W_{n,\alpha}(\theta)}{\partial\theta} = (\alpha+1)\int_{\mathbb{R}^m} C\mathcal{L}(\theta,y)^{\alpha+1} u(\theta,y)\,dy - (\alpha+1)\frac{1}{n}\sum_{k=1}^{n} C\mathcal{L}(\theta,y_k)^\alpha u(\theta,y_k)$$

therefore

$$\left(\frac{W_{n,\alpha}(\theta)}{\partial\theta}\right)_{\theta=\widehat{\theta}_c^\alpha} \xrightarrow[n\to\infty]{P} 0.$$

On the other hand

$$\frac{\partial^2 W_{n,\alpha}(\theta)}{\partial\theta\,\partial\theta^T} = (1+\alpha)\left\{\int_{\mathbb{R}^m}(1+\alpha)\,C\mathcal{L}(\theta,y)^{\alpha+1} u(\theta,y)^T u(\theta,y)\,dy + \int_{\mathbb{R}^m} C\mathcal{L}(\theta,y)^{\alpha+1}\frac{\partial u(\theta,y)}{\partial\theta^T}dy\right.$$
$$\left. -\frac{1}{n}\sum_{i=1}^{n}\alpha C\mathcal{L}(\theta,y_i)^\alpha u(\theta,y_i)^T u(\theta,y_i) - \frac{1}{n}\sum_{i=1}^{n} C\mathcal{L}(\theta,y_i)^\alpha \frac{\partial u(\theta,y_i)}{\partial\theta^T}\right\}.$$

But

$$\frac{1}{n}\sum_{i=1}^{n} C\mathcal{L}(\theta,y_i)^\alpha u(\theta,y_i)^T u(\theta,y_i) \xrightarrow[n\to\infty]{P} \int_{\mathbb{R}^m} C\mathcal{L}(\theta,y)^{\alpha+1} u(\theta,y)^T u(\theta,y)\,dy$$

and

$$\frac{1}{n}\sum_{i=1}^{n} C\mathcal{L}(\theta,y_i)^\alpha \frac{\partial u(\theta,y_i)}{\partial\theta^T} \xrightarrow[n\to\infty]{P} \int_{\mathbb{R}^m} C\mathcal{L}(\theta,y)^{\alpha+1}\frac{\partial u(\theta,y)}{\partial\theta^T}dy.$$

Therefore

$$\left(\frac{\partial^2 H_{n,\alpha}(\theta)}{\partial\theta\partial\theta^T}\right)_{\theta=\widehat{\theta}_c^\alpha} \xrightarrow[n\to\infty]{P} (1+\alpha)\,H_\alpha(\theta_0).$$

We can now write

$$nW_{n,\alpha}(\theta_0) = nW_{n,\alpha}(\widehat{\theta}_c^\alpha) + \frac{(1+\alpha)}{2}\sqrt{n}\left(\theta_0 - \widehat{\theta}_c^\alpha\right)^T H_\alpha(\theta_0)\sqrt{n}\left(\theta_0 - \widehat{\theta}_c^\alpha\right) + o_p(1).$$

It is clear that

$$E_{Y_1,\dots,Y_n}\left[\sqrt{n}\left(\theta_0 - \widehat{\theta}_c^\alpha\right)^T H_\alpha(\theta_0)\sqrt{n}\left(\theta_0 - \widehat{\theta}_c^\alpha\right)\right] = \sum_{r=1}^{k}\lambda_r + o_p(1)$$
$$= trace\left(J_\alpha(\theta_0)H_\alpha(\theta_0)^{-1}\right) + o_p(1).$$

Then

$$E_{Y_1,\dots,Y_n}\left[nW_{n,\alpha}(\boldsymbol{\theta}_0)\right] = E_{Y_1,\dots,Y_n}\left[nW_{n,\alpha}(\widehat{\boldsymbol{\theta}}_c^\alpha)\right] + \frac{(1+\alpha)}{2}\,trace\left(J_\alpha(\boldsymbol{\theta}_0)H_\alpha(\boldsymbol{\theta}_0)^{-1}\right) + o_p(1)$$

and, on the other hand, it is clear that

$$E_{Y_1,\dots,Y_n}\left[W_{n,\alpha}(\boldsymbol{\theta}_0)\right] = W_\alpha(\boldsymbol{\theta}_0).$$

Therefore,

$$E_{Y_1,\dots,Y_n}\left[nW_\alpha(\widehat{\boldsymbol{\theta}}_c^\alpha)\right] = nW_\alpha(\boldsymbol{\theta}_0) + \frac{(1+\alpha)}{2}\,trace\left(J_\alpha(\boldsymbol{\theta}_0)H_\alpha(\boldsymbol{\theta}_0)^{-1}\right) + o_p(1)$$

$$= E_{Y_1,\dots,Y_n}\left[nW_{n,\alpha}(\boldsymbol{\theta}_0)\right] + \frac{(1+\alpha)}{2}\,trace\left(J_\alpha(\boldsymbol{\theta}_0)H_\alpha(\boldsymbol{\theta}_0)^{-1}\right) + o_p(1)$$

$$= E_{Y_1,\dots,Y_n}\left[nW_{n,\alpha}(\widehat{\boldsymbol{\theta}}_c^\alpha)\right] + \frac{(1+\alpha)}{2}\,trace\left(J_\alpha(\boldsymbol{\theta}_0)H_\alpha(\boldsymbol{\theta}_0)^{-1}\right)$$

$$+ \frac{(1+\alpha)}{2}\,trace\left(J_\alpha(\boldsymbol{\theta}_0)H_\alpha(\boldsymbol{\theta}_0)^{-1}\right) + o_p(1)$$

$$= E_{Y_1,\dots,Y_n}\left[nW_{n,\alpha}(\widehat{\boldsymbol{\theta}}_c^\alpha)\right] + (1+\alpha)\,trace\left(J_\alpha(\boldsymbol{\theta}_0)H_\alpha(\boldsymbol{\theta}_0)^{-1}\right) + o_p(1).$$

Hence $nW_{n,\alpha}(\widehat{\boldsymbol{\theta}}_c^\alpha) + (1+\alpha)\,trace\left(J_\alpha(\boldsymbol{\theta}_0)H_\alpha(\boldsymbol{\theta}_0)^{-1}\right)$ is an asymptotic unbiased estimator of

$$E_{Y_1,\dots,Y_n}\left[nW_\alpha(\widehat{\boldsymbol{\theta}}_c^\alpha)\right].$$

$\square$

Appendix B. Computation of the CLDIC in Section 4.1

We have to compute

$$CLDIC\,(M_k) = W_{n,\alpha}\,(\widehat{\rho}) + \frac{\alpha+1}{n}\frac{J_\alpha\,(\widehat{\rho})}{H_\alpha\,(\widehat{\rho})},$$

where

$$W_{n,\alpha}\,(\widehat{\rho}) = \int_{\mathbb{R}^4} \mathcal{CL}(\widehat{\rho},y)^{\alpha+1}dy - (1-\alpha^{-1})\frac{1}{n}\sum_{i=1}^{n}\mathcal{CL}(\widehat{\rho},y_i)^\alpha, \tag{A1}$$

$$J_\alpha(\widehat{\rho}) = \int_{\mathbb{R}^4} \mathcal{CL}(\widehat{\rho},y)^{2\alpha+1}u(\widehat{\rho},y)^2 dy - \left(\int_{\mathbb{R}^4}\mathcal{CL}(\widehat{\rho},y)^{\alpha+1}u(\widehat{\rho},y)dy\right)^2, \tag{A2}$$

$$H_\alpha(\widehat{\rho}) = -\int_{\mathbb{R}^4}\mathcal{CL}(\widehat{\rho},y)^{\alpha+1}u(\widehat{\rho},y)^2 dy, \tag{A3}$$

for our candidate models, namely, composite normal and composite 4-variate *t*-distribution. As commented in Section 4.1, we consider a composite likelihood function based on the product of two bivariate distributions with common variance-covariance matrix. It is therefore, necessary in this example, to obtain values (A1), (A2) and (A3) for both composite normal and composite *t*-distributions. However, as stated in [10], while the sensitivity and variability matrices can be sometimes be evaluated explicitly, it is more usual to use empirical estimates. Following this comment, in the current example, we compute Equations (A1), (A2) and (A3) empirically through the sample data using

$$\widehat{W}_{n,\alpha}(\widehat{\rho}) = \sum_{i=1}^{n} CL(\widehat{\rho}, y_i)^{\alpha+1} - (1 - \alpha^{-1})\frac{1}{n}\sum_{i=1}^{n} CL(\widehat{\rho}, y_i)^{\alpha},$$

$$\widehat{J}_{\alpha}(\widehat{\rho}) = \sum_{i=1}^{n} CL(\widehat{\rho}, y_i)^{2\alpha+1} u(\widehat{\rho}, y_i)^2 - \left(\sum_{i=1}^{n} CL(\widehat{\rho}, y_i)^{\alpha+1} u(\widehat{\rho}, y_i)\right)^2$$

$$\widehat{H}_{\alpha}(\widehat{\rho}) = -\sum_{i=1}^{n} CL(\widehat{\rho}, y_i)^{\alpha+1} u(\widehat{\rho}, y_i)^2.$$

Now, we obtain the score of the composite likelihood $u(\widehat{\rho}, y_i)$ explicitly for both cases. By equation (A.5) in [12],

$$u^N(\widehat{\rho}, y_i) = \frac{\widehat{\rho}}{1 - \widehat{\rho}^2}\left[2 + \frac{1}{\widehat{\rho}}(t_{1i}t_{2i} + t_{3i}t_{4i})\right.$$
$$\left. - \frac{1}{1 - \widehat{\rho}^2}\left(t_{1i}^2 - 2\widehat{\rho}t_{1i}t_{2i} + t_{2i}^2\right) - \frac{1}{1 - \widehat{\rho}^2}\left(t_{3i}^2 - 2\widehat{\rho}t_{3i}t_{4i} + t_{4i}^2\right)\right],$$

with $t_{ji} = y_{ji} - \mu_j$, $j = 1, \ldots, 4$. On the other hand, we want to compute $u^{t_\nu}(\widehat{\rho}, y_i)$.

$$u^{t_\nu}(\widehat{\rho}, y_i) = \frac{\partial CL^{t_\nu}(\widehat{\rho}, y_i)}{\partial\widehat{\rho}} = \frac{\partial \log CL^{t_\nu}(\widehat{\rho}, y_i)}{\partial\widehat{\rho}} = \frac{1}{CL^{t_\nu}(\widehat{\rho}, y_i)}\frac{\partial CL^{t_\nu}(\widehat{\rho}, y_i)}{\partial\widehat{\rho}}$$
$$= \frac{1}{f_{12}^{t_\nu}(y_i; \widehat{\rho})f_{34}^{t_\nu}(y_i; \widehat{\rho})}\left[\frac{\partial}{\partial\widehat{\rho}}f_{12}^{t_\nu}(y_i; \widehat{\rho})f_{34}^{t_\nu}(y_i; \widehat{\rho})\right]$$
$$= \frac{1}{f_{12}^{t_\nu}(y_i; \widehat{\rho})f_{34}^{t_\nu}(y_i; \widehat{\rho})}\left[\left(\frac{\partial}{\partial\widehat{\rho}}f_{12}^{t_\nu}(y_i; \widehat{\rho})\right)f_{34}^{t_\nu}(y_i; \widehat{\rho}) + f_{12}^{t_\nu}(y_i; \widehat{\rho})\left(\frac{\partial}{\partial\widehat{\rho}}f_{34}^{t_\nu}(y_i; \widehat{\rho})\right)\right]$$
$$= \frac{1}{f_{12}^{t_\nu}(y_i; \widehat{\rho})}\left(\frac{\partial}{\partial\widehat{\rho}}f_{12}^{t_\nu}(y_i; \widehat{\rho})\right) + \frac{1}{f_{34}^{t_\nu}(y_i; \widehat{\rho})}\left(\frac{\partial}{\partial\widehat{\rho}}f_{34}^{t_\nu}(y_i; \widehat{\rho})\right).$$

Now, it can be shown that

$$\frac{\partial f_{12}^{t_\nu}(y_i; \widehat{\rho})}{\partial\widehat{\rho}} = f_{12}^{t_\nu}(y_i; \widehat{\rho})\frac{\nu\left[(\nu-2)\widehat{\rho}^3 - t_{1i}t_{2i}\nu\widehat{\rho}^2 + \left((t_{1i}^2 + t_{2i}^2 - 1)\nu + t_{2i}^2 + t_{1i}^2 + 2\right)\widehat{\rho} - t_{1i}t_{2i}\nu - 2t_{1i}t_{2i}\right]}{(1 - \widehat{\rho}^2)\left[(\nu-2)\widehat{\rho}^2 + 2t_{1i}t_{2i}\widehat{\rho} - \nu - t_{1i}^2 - t_{2i}^2 + 2\right]}$$

and

$$\frac{\partial f_{34}^{t_\nu}(y_i; \widehat{\rho})}{\partial\widehat{\rho}} = f_{34}^{t_\nu}(y_i; \widehat{\rho})\frac{\nu\left[(\nu-2)\widehat{\rho}^3 - t_{3i}t_{4i}\nu\widehat{\rho}^2 + \left((t_{3i}^2 + t_{4i}^2 - 1)\nu + t_{4i}^2 + t_{3i}^2 + 2\right)\widehat{\rho} - t_{1i}t_{4i}\nu - 2t_{3i}t_{4i}\right]}{(1 - \widehat{\rho}^2)\left[(\nu-2)\widehat{\rho}^2 + 2t_{3i}t_{4i}\widehat{\rho} - \nu - t_{3i}^2 - t_{4i}^2 + 2\right]}.$$

References

1. Fearnhead, P.; Donnelly, P. Approximate likelihood methods for estimating local recombination rates. *J. R. Stat. Soc. Ser. B (Stat. Methodol.)* **2002**, *64*, 657–680. [CrossRef]
2. Renard, ; D.; Molenberghs, G.; Geys, H. A pairwise likelihood approach to estimation in multilevel probit models. *J. Comput. Stat. Data Anal.* **2004**, *44*, 649–667. [CrossRef]
3. Hjort, N.L.; Omre, H. Topics in spatial statistics. *Scand. J. Stat.* **1994**, *21*, 289–357.
4. Heagerty, P.J.; Lele, S.R. A composite likelihood approach to binary spatial data. *J. Am. Stat. Assoc.* **1998**, *93*, 1099–1111. [CrossRef]
5. Varin, C.; Host, G.; Skare, O. Pairwise likelihood inference in spatial generalized linear mixed models. *Comput. Stat. Data Anal.* **2005**, *49*, 1173–1191 [CrossRef]
6. Henderson, R.; Shimakura, S. A serially correlated gamma frailty model for longitudinal count data. *Biometrika* **2003**, *90*, 355–366. [CrossRef]
7. Parner, E.T. A composite likelihood approach to multivariate survival data. *Scand. J. Stat.* **2001**, *28*, 295–302. [CrossRef]
8. Li, Y.; Lin, X. Semiparametric Normal Transformation Models for Spatially Correlated Survival Data. *J. Am. Stat. Assoc.* **2006**, *101*, 593–603. [CrossRef]
9. Joe, H.; Reid, N.; Somg, P.X.; Firth, D.; Varin, C. Composite Likelihood Methods. Report on the Workshop on Composite Likelihood. 2012. Available online: http://www.birs.ca/events/2012/5-day-workshops/12w5046 (accessed on 23 July 2019).
10. Varin, C.; Reid, N.; Firth, D. An overview of composite likelihood methods. *Statist. Sin.* **2011**, *21*, 5–42.
11. Martín, N.; Pardo, L.; Zografos, K. On divergence tests for composite hypotheses under composite likelihood. *Stat. Pap.* **2019**, *60*, 1883–1919. [CrossRef]
12. Castilla, E.; Martin, N.; Pardo, L.; Zografos, K. Composite Likelihood Methods Based on Minimum Density Power Divergence Estimator. *Entropy* **2018**, *20*, 18. [CrossRef]
13. Castilla, E.; Martin, N.; Pardo, L.; Zografos, K. Composite likelihood methods: Rao-type tests based on composite minimum density power divergence estimator. *Stat. Pap.* **2019**. [CrossRef]
14. Kullback, S. *Information Theory and Statistics*; Wiley: New York, NY, USA, 1959.
15. Akaike, H. Information theory and an extension of the maximum likelihood principle. In *2nd International Symposium on Information Theory*; Petrov, B.N., Csaki, F., Eds.; Akademiai Kiado: Budapest, Hungary, 1973; pp. 267–281 .
16. Akaike, H. A new look at the statistical model identification. *IEEE Trans. Autom. Control* **1974**, *19*, 716–723. [CrossRef]
17. Takeuchi, K. Distribution of information statistics and criteria for adequacy of models. *Math. Sci.* **1976**, *153*, 12–18. (In Japanese)
18. Murari, A.; Peluso, E.; Cianfrani, F.; Gaudio, P.; Lungaroni, M. On the Use of Entropy to Improve Model Selection Criteria. *Entropy* **2019**, *21*, 394. [CrossRef]
19. Mattheou, K.; Lee, S.; Karagrigoriou, A. A model selection criterion based on the BHHJ measure of divergence. *J. Stat. Plan. Inference* **2009**, *139*, 228–235. [CrossRef]
20. Avlogiaris, G.; Micheas, A.; Zografos, K. A criterion for local model selection. *Shankhya* **2019**, *81*, 406–444. [CrossRef]
21. Avlogiaris, G.; Micheas, A.; Zografos, K. On local divergences between two probability measures. *Metrika* **2016**, *79*, 303–333. [CrossRef]
22. Varin, C.; Vidoni, P. A note on composite likelihood inference and model selection. *Biometrika* **2005**, *92*, 519–528. [CrossRef]
23. Gao, X.; Song, P.X.K. Composite likelihood Bayesian information criteria for model selection in high-dimensional data. *J. Am. Stat. Assoc.* **2010**, *105*, 1531–1540. [CrossRef]
24. Ng, C.T.; Joe, H. Model comparison with composite likelihood information criteria. *Bernoulli* **2014**, *20*, 1738–1764. [CrossRef]
25. Basu, A.; Harris, I.R.; Hjort, N.L.; Jones, M.C. Robust and efficient estimation by minimizing a density power divergence. *Biometrika* **1998**, *85*, 549–559. [CrossRef]
26. Pardo, L. *Statistical Inference Based on Divergence Measures*; Chapman & Hall CRC Press: Boca Raton, FL, USA, 2006.

27. Basu, A.; Shioya, H.; Park, C. *Statistical Inference. The Minimum Distance Approach*; Chapman & Hall/CRC: Boca Raton, FL, USA, 2011.

28. Burham, K.P.; Anderson, D.R. *Model Selection and Multinomial Inference: A Practical Information-Theoretic Approach*; Springer: New York, NY, USA, 2002.

29. Xu, X., Reid, N. On the robustness of maximum composite estimate. *J. Stat. Plan. Inference* **2011**, *141*, 3047–3054. [CrossRef]

30. Warwick, J.; Jones, M.C. Choosing a robustness tuning parameter. *J. Stat. Comput. Simul.* **2005**, *75*, 581–588. [CrossRef]

31. Fisher, R.A. The use of multiple measurements in taxonomic problems. *Ann. Eugenics.* **1936**, *7*, 179–188. [CrossRef]

32. Fraley, A.; Raftery, E.; Murphy, T.B.; Scrucca, L. *MCLUST Version 4 for R: Normal Mixture Modeling for Model-based Clustering, Classification, and Density Estimation*; Technical Report 597; Department of Statistics, University of Washington: Seattle, WA, USA, 2012.

33. Forina, M.; Lanteri, S.; Armanino, C.; Leardi, R. PARVUS: An *Extendable Package of Programs for Data Exploration, Classification, and Correlation*; Institute of Pharmaceutical and Food Analysis Technologies: Genoa, Italy, 1998.

Article

Convergence Rates for Empirical Estimation of Binary Classification Bounds

Salimeh Yasaei Sekeh [1,*]**, Morteza Noshad** [2]**, Kevin R. Moon** [3] **and Alfred O. Hero** [2]

[1]   School of Computing and Information Science, University of Maine, Orono, ME 04469, USA
[2]   Department of Electrical Engineering and Computer Science, University of Michigan, Ann Arbor, MI 48109, USA; noshad@umich.edu (M.N.); hero@umich.edu (A.O.H.)
[3]   Department of Mathematics and Statistics, Utah State University, Logan, UT 84322, USA; kevin.moon@usu.edu
*   Correspondence: salimeh.yasaei@maine.edu

Received: 5 November 2019; Accepted: 15 November 2019; Published: 23 November 2019

Abstract: Bounding the best achievable error probability for binary classification problems is relevant to many applications including machine learning, signal processing, and information theory. Many bounds on the Bayes binary classification error rate depend on information divergences between the pair of class distributions. Recently, the Henze–Penrose (HP) divergence has been proposed for bounding classification error probability. We consider the problem of empirically estimating the HP-divergence from random samples. We derive a bound on the convergence rate for the Friedman–Rafsky (FR) estimator of the HP-divergence, which is related to a multivariate runs statistic for testing between two distributions. The FR estimator is derived from a multicolored Euclidean minimal spanning tree (MST) that spans the merged samples. We obtain a concentration inequality for the Friedman–Rafsky estimator of the Henze–Penrose divergence. We validate our results experimentally and illustrate their application to real datasets.

Keywords: classification; Bayes error rate; Henze–Penrose divergence; Friedman–Rafsky test statistic; convergence rates; bias and variance trade-off; concentration bounds; minimal spanning trees

1. Introduction

Divergence measures between probability density functions are used in many signal processing applications including classification, segmentation, source separation, and clustering (see [1–3]). For more applications of divergence measures, we refer to [4].

In classification problems, the Bayes error rate is the expected risk for the Bayes classifier, which assigns a given feature vector $\mathbf{x}$ to the class with the highest posterior probability. The Bayes error rate is the lowest possible error rate of any classifier for a particular joint distribution. Mathematically, let $\mathbf{x}_1, \mathbf{x}_2, \ldots, \mathbf{x}_N \in \mathbb{R}^d$ be realizations of random vector $\mathbf{X}$ and class labels $S \in \{0, 1\}$, with prior probabilities $p = P(S = 0)$ and $q = P(S = 1)$, such that $p + q = 1$. Given conditional probability densities $f_0(\mathbf{x})$ and $f_1(\mathbf{x})$, the Bayes error rate is given by

$$\epsilon = \int_{\mathbb{R}^d} \min \big\{ p f_0(\mathbf{x}), q f_1(\mathbf{x}) \big\} \mathrm{d}\mathbf{x}. \tag{1}$$

The Bayes error rate provides a measure of classification difficulty. Thus, when known, the Bayes error rate can be used to guide the user in the choice of classifier and tuning parameter selection. In practice, the Bayes error is rarely known and must be estimated from data. Estimation of the Bayes error rate is difficult due to the nonsmooth min function within the integral in (1). Thus, research has focused on deriving tight bounds on the Bayes error rate based on smooth relaxations of the min function. Many of these bounds can be expressed in terms of divergence measures such as the Bhattacharyya [5] and Jensen–Shannon [6]. Tighter bounds on the Bayes error rate can be obtained using an important divergence measure known as the Henze–Penrose (HP) divergence [7,8].

Many techniques have been developed for estimating divergence measures. These methods can be broadly classified into two categories: (i) plug-in estimators in which we estimate the probability densities and then plug them in the divergence function [9–12], (ii) entropic graph approaches, in which the relationship between the divergence function and a graph functional in Euclidean space is derived [8,13]. Examples of plug-in methods include k-nearest neighbor (K-NN) and Kernel density estimator (KDE) divergence estimators. Examples of entropic graph approaches include methods based on minimal spanning trees (MST), K-nearest neighbors graphs (K-NNG), minimal matching graphs (MMG), traveling salesman problem (TSP), and their power-weighted variants.

Disadvantages of plug-in estimators are that these methods often require assumptions on the support set boundary and are more computationally complex than direct graph-based approaches. Thus, for practical and computational reasons, the asymptotic behavior of entropic graph approaches has been of great interest. Asymptotic analysis has been used to justify graph based approaches. For instance, in [14], the authors showed that a cross match statistic based on optimal weighted matching converges to the the HP-divergence. In [15], a more complex approach based on the K-NNG was proposed that also converges to the HP-divergence.

The first contribution of our paper is that we obtain a bound on the convergence rates for the Friedman and Rafsky (FR) estimator of the HP-divergence, which is based on a multivariate extension of the non-parametric run length test of equality of distributions. This estimator is constructed using a multicolored MST on the labeled training set where MST edges connecting samples with dichotomous labels are colored differently from edges connecting identically labeled samples. While previous works have investigated the FR test statistic in the context of estimating the HP-divergence (see [8,16]), to the best of our knowledge, its minimax MSE convergence rate has not been previously derived. The bound on convergence rate is established by using the umbrella theorem of [17], for which we define a dual version of the multicolor MST. The proposed dual MST in this work is different than the standard dual MST introduced by Yukich in [17]. We show that the bias rate of the FR estimator is bounded by a function of N, η and d, as $O\left((N)^{-\eta^2/(d(\eta+1))}\right)$, where N is the total sample size, d is the dimension of the data samples $d \geq 2$, and η is the Hölder smoothness parameter $0 < \eta \leq 1$. We also obtain the variance rate bound as $O\left((N)^{-1}\right)$.

The second contribution of our paper is a new concentration bound for the FR test statistic. The bound is obtained by establishing a growth bound and a smoothness condition for the multicolored MST. Since the FR test statistic is not a Euclidean functional, we cannot use the standard subadditivity and superadditivity approaches of [17–19]. Our concentration inequality is derived using a different Hamming distance approach and a dual graph to the multicolored MST.

We experimentally validate our theoretic results. We compare the MSE theory and simulation in three experiments with various dimensions $d = 2, 4, 8$. We observe that, in all three experiments, as sample size increases, the MSE rate decreases and, for higher dimensions, the rate is slower. In all sets of experiments, our theory matches the experimental results. Furthermore, we illustrate the application of our results on estimation of the Bayes error rate on three real datasets.

1.1. Related Work

Much research on minimal graphs has focused on the use of Euclidean functionals for signal processing and statistics applications such as image registration [20,21], pattern matching [22], and non-parametric divergence estimation [23]. A K-NNG-based estimator of Rényi and f-divergence measures has been proposed in [13]. Additional examples of direct estimators of divergence measures include statistic based on the nonparametric two sample problem, the Smirnov maximum deviation test [24], and the Wald–Wolfowitz [25] runs test, which have been studied in [26].

Many entropic graph estimators such as MST, K-NNG, MMG, and TSP have been considered for multivariate data from a single probability density f. In particular, the normalized weight function of graph constructions all converge almost surely to the Rényi entropy of f [17,27]. For N uniformly distributed points, the MSE is $O(N^{-1/d})$ [28,29]. Later, Hero et al. [30,31] reported bounds on L_γ-norm bias convergence rates of power-weighted Euclidean weight functionals of order γ for densities f belonging to the space of Hölder continuous functions $\Sigma_d(\eta, K)$ as $O(N^{-\alpha\eta/(\alpha\eta+1)\, 1/d})$, where $0 < \eta \le 1, d \ge 1, \gamma \in (1, d)$, and $\alpha = (d - \gamma)/d$. In this work, we derive a bound on convergence rate of FR estimator for the HP-divergence when the density functions belong to the Hölder class, $\Sigma_d(\eta, K)$, for $0 < \eta \le 1, d \ge 2$ [32]. Note that throughout the paper we assume the density functions are absolutely continuous and bounded with support on the unit cube $[0, 1]^d$.

In [28], Yukich introduced the general framework of continuous and quasi-additive Euclidean functionals. This has led to many convergence rate bounds of entropic graph divergence estimators.

The framework of [28] is as follows: Let F be finite subset of points in $[0, 1]^d, d \ge 2$, drawn from an underlying density. A real-valued function L_γ defined on F is called a Euclidean functional of order γ if it is of the form $L_\gamma(F) = \min_{E \in \mathcal{E}} \sum_{e \in E} |e(F)|^\gamma$, where $\mathcal{E}$ is a set of graphs, e is an edge in the graph E, $|e|$ is the Euclidean length of e, and γ is called the edge exponent or power-weighting constant. The MST, TSP, and MMG are some examples for which $\gamma = 1$.

Following this framework, we show that the FR test statistic satisfies the required continuity and quasi-additivity properties to obtain similar convergence rates to those predicted in [28]. What distinguishes our work from previous work is that the count of dichotomous edges in the multicolored MST is not Euclidean. Therefore, the results in [17,27,30,31] are not directly applicable.

Using the isoperimetric approach, Talagrand [33] showed that, when the Euclidean functional L_γ is based on the MST or TSP, then the functional L_γ for derived random vertices uniformly distributed in a hypercube $[0, 1]^d$ is concentrated around its mean. Namely, with high probability, the functional L_γ and its mean do not differ by more than $C(N \log N)^{(d-\gamma)/2d}$. In this paper, we establish concentration bounds for the FR statistic: with high probability $1 - \delta$, the FR statistic differs from its mean by not more than $O\left((N)^{(d-1)/d}\left(\log(C/\delta)\right)^{(d-1)/d}\right)$, where C is a function of N and d.

1.2. Organization

This paper is organized as follows. In Section 2, we first introduce the HP-divergence and the FR multivariate test statistic. We then present the bias and variance rates of the FR-based estimator of HP-divergence followed by the concentration bounds and the minimax MSE convergence rate. Section 3 provides simulations that validate the theory. All proofs and relevant lemmas are given in the Appendices A–E.

Throughout the paper, we denote expectation by $\mathbb{E}$ and variance by abbreviation Var. Bold face type indicates random variables. In this paper, when we say number of samples we mean number of observations.

2. The Henze–Penrose Divergence Measure

Consider parameters $p \in (0, 1)$ and $q = 1 - p$. We focus on estimating the HP-divergence measure between distributions f_0 and f_1 with domain $\mathbb{R}^d$ defined by

$$D_p(f_0, f_1) = \frac{1}{4pq}\left[\int \frac{(pf_0(\mathbf{x}) - qf_1(\mathbf{x}))^2}{pf_0(\mathbf{x}) + qf_1(\mathbf{x})}\, d\mathbf{x} - (p - q)^2\right]. \tag{2}$$

It can be verified that this measure is bounded between 0 and 1 and, if $f_0(\mathbf{x}) = f_1(\mathbf{x})$, then $D_p = 0$. In contrast with some other divergences such as the Kullback–Liebler [34] and Rényi divergences [35], the HP-divergence is symmetrical, i.e., $D_p(f_0, f_1) = D_q(f_1, f_0)$. By invoking relation (3) in [8],

$$\int \frac{(pf_0(\mathbf{x}) - qf_1(\mathbf{x}))^2}{pf_0(\mathbf{x}) + qf_1(\mathbf{x})}\, d\mathbf{x} = 1 - 4pq A_p(f_0, f_1),$$

where

$$A_p(f_0, f_1) \;=\; \int \frac{f_0(\mathbf{x})f_1(\mathbf{x})}{pf_0(\mathbf{x}) + qf_1(\mathbf{x})}\, \mathrm{d}\mathbf{x} = \mathbb{E}_{f_0}\left[\left(p\,\frac{f_0(\mathbf{X})}{f_1(\mathbf{X})} + q\right)^{-1}\right],$$

$$u_p(f_0, f_1) \;=\; 1 - 4pq\, A_p(f_0, f_1),$$

one can rewrite D_p in the alternative form:

$$D_p(f_0, f_1) = 1 - A_p(f_0, f_1) = \frac{u_p(f_0, f_1)}{4pq} - \frac{(p-q)^2}{4pq}.$$

Throughout the paper, we refer to $A_p(f_0, f_1)$ as the HP-integral. The HP-divergence measure belongs to the class of ϕ-divergences [36]. For the special case $p = 0.5$, the divergence (2) becomes the symmetric χ^2-divergence and is similar to the Rukhin f-divergence. See [37,38].

2.1. The Multivariate Runs Test Statistic

The MST is a graph of minimum weight among all graphs $\mathcal{E}$ that span n vertices. The MST has many applications including pattern recognition [39], clustering [40], nonparametric regression [41], and testing of randomness [42]. In this section, we focus on the FR multivariate two sample test statistic constructed from the MST.

Assume that sample realizations from f_0 and f_1, denoted by $\mathfrak{X}_m \in \mathbb{R}^{m \times d}$ and $\mathfrak{Y}_n \in \mathbb{R}^{n \times d}$, respectively, are available. Construct an MST spanning the samples from both f_0 and f_1 and color the edges in the MST that connect dichotomous samples green and color the remaining edges black. The FR test statistic $\mathfrak{R}_{m,n} := \mathfrak{R}_{m,n}(\mathfrak{X}_m, \mathfrak{Y}_n)$ is the number of green edges in the MST. Note that the test assumes a unique MST, therefore all inter point distances between data points must be distinct. We recall the following theorem from [7,8]:

Theorem 1. *As $m \to \infty$ and $n \to \infty$ such that $\dfrac{m}{n+m} \to p$ and $\dfrac{n}{n+m} \to q$,*

$$1 - \mathfrak{R}_{m,n}(\mathfrak{X}_m, \mathfrak{Y}_n)\,\frac{m+n}{2mn} \to D_p(f_0, f_1), \quad a.s. \tag{3}$$

In the next section, we obtain bounds on the MSE convergence rates of the FR approximation for HP-divergence between densities that belong to $\Sigma_d(\eta, K)$, the class of Hölder continuous functions with Lipschitz constant K and smoothness parameter $0 < \eta \le 1$ [32]:

Definition 1 (Hölder class). *Let $\mathcal{X} \subset \mathbb{R}^d$ be a compact space. The Hölder class $\Sigma_d(\eta, K)$, with η-Hölder parameter, of functions with the L_d-norm, consists of the functions g that satisfy*

$$\left\{g : \left\| g(\mathbf{z}) - p_{\mathbf{x}}^{\lfloor \eta \rfloor}(\mathbf{z}) \right\|_d \le K \left\| \mathbf{x} - \mathbf{z} \right\|_{d}^{\eta}, \; \mathbf{x}, \mathbf{z} \in \mathcal{X}\right\}, \tag{4}$$

where $p_{\mathbf{x}}^{k}(\mathbf{z})$ is the Taylor polynomial (multinomial) of g of order k expanded about the point $\mathbf{x}$ and $\lfloor \eta \rfloor$ is defined as the greatest integer strictly less than η.

In what follows, we will use both notations $\mathfrak{R}_{m,n}$ and $\mathfrak{R}_{m,n}(\mathfrak{X}_m, \mathfrak{Y}_n)$ for the FR statistic over the combined samples.

2.2. Convergence Rates

In this subsection, we obtain the mean convergence rate bounds for general non-uniform Lebesgue densities f_0 and f_1 belonging to the Hölder class $\Sigma_d(\eta, K)$. Since the expectation of $\mathfrak{R}_{m,n}$ can be closely approximated by the sum of the expectation of the FR statistic constructed on a dense partition of

$[0,1]^d$, $\mathfrak{R}_{m,n}$ is a quasi-additive functional in mean. The family of bounds (A16) in Appendix B enables us to achieve the minimax convergence rate for the mean under the Hölder class assumption with smoothness parameter $0 < \eta \leq 1, d \geq 2$:

Theorem 2 (Convergence Rate of the Mean). *Let $d \geq 2$, and $\mathfrak{R}_{m,n}$ be the FR statistic for samples drawn from Hölder continuous and bounded density functions f_0 and f_1 in $\Sigma_d(\eta, K)$. Then, for $d \geq 2$,*

$$\left| \frac{\mathbb{E}\left[\mathfrak{R}_{m,n}\right]}{m+n} - 2pq \int \frac{f_0(\mathbf{x})f_1(\mathbf{x})}{pf_0(\mathbf{x}) + qf_1(\mathbf{x})} \, d\mathbf{x} \right| \leq O\left((m+n)^{-\eta^2/(d(\eta+1))} \right). \tag{5}$$

This bound holds over the class of Lebesgue densities $f_0, f_1 \in \Sigma_d(\eta, K), 0 < \eta \leq 1$. Note that this assumption can be relaxed to $f_0 \in \Sigma_d^s(\eta, K_0)$ and $f_1 \in \Sigma_d^s(\eta, K_1)$ that is Lebesgue densities f_0 and f_1 belong to the Strong Hölder class with the same Hölder parameter η and different constants K_0 and K_1, respectively.

The following variance bound uses the Efron–Stein inequality [43]. Note that in Theorem 3 we do not impose any strict assumptions. We only assume that the density functions are absolutely continuous and bounded with support on the unit cube $[0,1]^d$. Appendix C contains the proof.

Theorem 3. *The variance of the HP-integral estimator based on the FR statistic, $\mathfrak{R}_{m,n}/(m+n)$ is bounded by*

$$Var\left(\frac{\mathfrak{R}_{m,n}(\mathfrak{X}_m, \mathfrak{Y}_n)}{m+n} \right) \leq \frac{32 \, c_d^2 \, q}{(m+n)}, \tag{6}$$

where the constant c_d depends only on d.

By combining Theorems 2 and 3, we obtain the MSE rate of the form $O\left(m+n \right)^{-\eta^2/(d(\eta+1))} + O\left((m+n)^{-1} \right)$. Figure 1 indicates a heat map showing the MSE rate as a function of d and $N = m = n$. The heat map shows that the MSE rate of the FR test statistic-based estimator given in (3) is small for large sample size N.

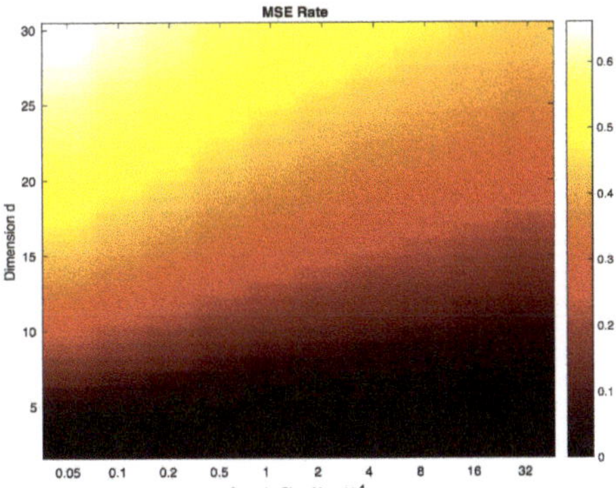

Figure 1. Heat map of the theoretical MSE rate of the FR estimator of the HP-divergence based on Theorems 2 and 3 as a function of dimension and sample size when $N = m = n$. Note the color transition (MSE) as sample size increases for high dimension. For fixed sample size N, the MSE rate degrades in higher dimensions.

2.3. Proof Sketch of Theorem 2

In this subsection, we first establish subadditivity and superadditivity properties of the FR statistic, which will be employed to derive the MSE convergence rate bound. This will establish that the mean of the FR test statistic is a quasi-additive functional:

Theorem 4. *Let $\mathfrak{R}_{m,n}(\mathfrak{X}_m, \mathfrak{Y}_n)$ be the number of edges that link nodes from differently labeled samples $\mathfrak{X}_m = \{X_1, \ldots, X_m\}$ and $\mathfrak{Y}_n = \{Y_1, \ldots, Y_n\}$ in $[0,1]^d$. Partition $[0,1]^d$ into l^d equal volume subcubes Q_i such that m_i and n_i are the number of samples from $\{X_1, \ldots, X_m\}$ and $\{Y_1, \ldots, Y_n\}$, respectively, that fall into the partition Q_i. Then, there exists a constant c_1 such that*

$$\mathbb{E}\Big[\mathfrak{R}_{m,n}(\mathfrak{X}_m, \mathfrak{Y}_n)\Big] \leq \sum_{i=1}^{l^d} \mathbb{E}\Big[\mathfrak{R}_{m_i,n_i}\big((\mathfrak{X}_m, \mathfrak{Y}_n) \cap Q_i\big)\Big] + 2\,c_1\,l^{d-1}\,(m+n)^{1/d}. \tag{7}$$

Here, $\mathfrak{R}_{m_i,n_i}$ is the number of dichotomous edges in partition Q_i. Conversely, for the same conditions as above on partitions Q_i, there exists a constant c_2 such that

$$\mathbb{E}\Big[\mathfrak{R}_{m,n}(\mathfrak{X}_m, \mathfrak{Y}_n)\Big] \geq \sum_{i=1}^{l^d} \mathbb{E}\Big[\mathfrak{R}_{m_i,n_i}\big((\mathfrak{X}_m, \mathfrak{Y}_n) \cap Q_i\big)\Big] - 2\,c_2\,l^{d-1}\,(m+n)^{1/d}. \tag{8}$$

The inequalities (7) and (8) are inspired by corresponding inequalities in [30,31]. The full proof is given in Appendix A. The key result in the proof is the inequality:

$$\mathfrak{R}_{m,n}(\mathfrak{X}_m, \mathfrak{Y}_n) \leq \sum_{i=1}^{l^d} \mathfrak{R}_{m_i,n_i}\big((\mathfrak{X}_m, \mathfrak{Y}_n) \cap Q_i\big) + 2|D|,$$

where $|D|$ indicates the number of all edges of the MST which intersect two different partitions.

Furthermore, we adapt the theory developed in [17,30] to derive the MSE convergence rate of the FR statistic-based estimator by defining a dual MST and dual FR statistic, denoted by MST* and $\mathfrak{R}^*_{m,n}$ respectively (see Figure 2):

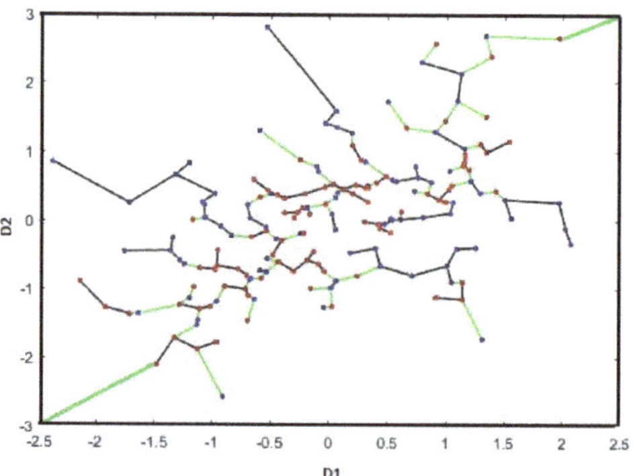

Figure 2. The dual MST spanning the merged set $\mathfrak{X}_m$ (blue points) and $\mathfrak{Y}_n$ (red points) drawn from two Gaussian distributions. The dual FR statistic ($\mathfrak{R}^*_{m,n}$) is the number of edges in the MST* (contains nodes in $\mathfrak{X}_m \cup \mathfrak{Y}_n \cup \{2$ corner points$\}$) that connect samples from different color nodes and corners (denoted in green). Black edges are the non-dichotomous edges in the MST*.

Definition 2 (Dual MST, MST* and dual FR statistic $\mathfrak{R}^*_{m,n}$). *Let $\mathbb{F}_i$ be the set of corner points of the subsection Q_i for $1 \leq i \leq l^d$. Then, we define $\mathrm{MST}^*(\mathfrak{X}_m \cup \mathfrak{Y}_n \cap Q_i)$ as the boundary MST graph of partition Q_i [17], which contains $\mathfrak{X}_m$ and $\mathfrak{Y}_n$ points falling inside the section Q_i and those corner points in $\mathbb{F}_i$ which minimize total MST length. Notice it is allowed to connect the MSTs in Q_i and Q_j through points strictly* *contained in Q_i and Q_j and corner points are taken into account under condition of minimizing total MST length. Another word, the dual MST can connect the points in $Q_i \cup Q_j$ by direct edges to pair to another point in $Q_i \cup Q_j$ or the corner the corner points (we assume that all corner points are connected) in order to minimize the total length. To clarify this, assume that there are two points in $Q_i \cup Q_j$, then the dual MST consists of the two edges connecting these points to the corner if they are closed to a corner point; otherwise, dual MST consists of an edge connecting one to another. Furthermore, we define $\mathfrak{R}^*_{m,n}(\mathfrak{X}_m, \mathfrak{Y}_n \cap Q_i)$ as the number of edges in an MST* graph connecting nodes from different samples and number of edges connecting to the corner points. Note that the edges connected to the corner nodes (regardless of the type of points) are always counted in dual FR test statistic $\mathfrak{R}^*_{m,n}$.*

In Appendix B, we show that the dual FR test statistic is a quasi-additive functional in mean and $\mathfrak{R}^*_{m,n}(\mathfrak{X}_m, \mathfrak{Y}_n) \geq \mathfrak{R}_{m,n}(\mathfrak{X}_m, \mathfrak{Y}_n)$. This property holds true since MST$(\mathfrak{X}_m, \mathfrak{Y}_n)$ and MST$^*(\mathfrak{X}_m, \mathfrak{Y}_n)$ graphs can only be different in the edges connected to the corner nodes, and in $\mathfrak{R}^*(\mathfrak{X}_m, \mathfrak{Y}_n)$ we take all of the edges between these nodes and corner nodes into account.

To prove Theorem 2, we partition $[0, 1]^d$ into l^d subcubes. Then, by applying Theorem 4 and the dual MST, we derive the bias rate in terms of partition parameter l (see (A16) in Theorem A1). See Appendix B and Appendix E for details. According to (A16), for $d \geq 2$, and $l = 1, 2, \ldots$, the slowest rates as a function of l are $l^d(m + n)^{\eta/d}$ and $l^{-\eta d}$. Therefore, we obtain an l-independent bound by letting l be a function of $m + n$ that minimizes the maximum of these rates i.e.,

$$l(m + n) = arg \min_{l} \max \left\{ l^d(m + n)^{-\eta/d}, l^{-\eta d} \right\}.$$

The full proof of the bound in (2) is given in Appendix B.

2.4. Concentration Bounds

Another main contribution of our work in this part is to provide an exponential inequality convergence bound derived for the FR estimator of the HP-divergence. The error of this estimator can be decomposed into a bias term and a variance-like term via the triangle inequality:

$$\left| \mathfrak{R}_{m,n} - \int \frac{f_0(\mathbf{x}) f_1(\mathbf{x})}{p f_0(\mathbf{x}) + q f_1(\mathbf{x})} d\mathbf{x} \right| \leq \underbrace{\left| \mathfrak{R}_{m,n} - \mathbb{E}[\mathfrak{R}_{m,n}] \right|}_{\text{variance-like term}}$$

$$+ \underbrace{\left| \mathbb{E}[\mathfrak{R}_{m,n}] - \int \frac{f_0(\mathbf{x}) f_1(\mathbf{x})}{p f_0(\mathbf{x}) + q f_1(\mathbf{x})} d\mathbf{x} \right|}_{\text{bias term}}.$$

The bias bound was given in Theorem 2. Therefore, we focus on an exponential concentration bound for the variance-like term. One application of concentration bounds is to employ these bounds to compare confidence intervals on the HP-divergence measure in terms of the FR estimator. In [44,45], the authors provided an exponential inequality convergence bound for an estimator of Rény divergence for a smooth Hölder class of densities on the d-dimensional unite cube $[0, 1]^d$. We show that if $\mathfrak{X}_m$ and $\mathfrak{Y}_n$ are the set of m and n points drawn from any two distributions f_0 and f_1, respectively, the FR criteria $\mathfrak{R}_{m,n}$ is tightly concentrated. Namely, we establish that, with high probability, $\mathfrak{R}_{m,n}$ is within

$$1 - O\left((m + n)^{-2/d} \epsilon^{*2} \right)$$

of its expected value, where ϵ^* is the solution of the following convex optimization problem:

$$\min_{\epsilon \geq 0} \quad C'_{m,n}(\epsilon) \, \exp\left(\frac{-(t/(2\epsilon))^{d/(d-1)}}{(m+n)\tilde{C}}\right) \tag{9}$$

$$\text{subject to} \quad \epsilon \geq O\big(7^{d+1}(m+n)^{1/d}\big),$$

where $\tilde{C} = 8(4)^{d/(d-1)}$ and

$$C'_{m,n}(\epsilon) = 8\left(1 - O\big((m+n)^{-2/d}\epsilon^2\big)\right)^{-2}. \tag{10}$$

Note that, under the assumption $(m+n)^{1/d} \simeq 1$, $C'_{m,n}(\epsilon)$ becomes a constant depending only on ϵ by $8\left(1 - (c\,\epsilon^2)\right)^{-2}$, where c is a constant. This is inferred from Theorems 5 and 6 below as $(m+n)^{1/d} \simeq 1$. See Appendix D, specifically Lemmas A8–A12 for more detail. Indeed, we first show the concentration around the median. A median is by definition any real number M_e that satisfies the inequalities $P(X \leq M_e) \geq 1/2$ and $P(X \geq M_e) \geq 1/2$. To derive the concentration results, the properties of growth bounds and smoothness for $\mathfrak{R}_{m,n}$, given in Appendix D, are exploited.

Theorem 5 (Concentration around the median). *Let M_e be a median of $\mathfrak{R}_{m,n}$ which implies that $P(\mathfrak{R}_{m,n} \leq M_e) \geq 1/2$. Recall ϵ^* from (9) then we have*

$$P\left(\left|\mathfrak{R}_{m,n}(\mathfrak{X}_m, \mathfrak{Y}_n) - M_e\right| \geq t\right) \leq C'_{m,n}(\epsilon^*) \, \exp\left(\frac{-(t/\epsilon^*)^{d/(d-1)}}{(m+n)\tilde{C}}\right), \tag{11}$$

where $\tilde{C} = 8(4)^{d/(d-1)}$.

Theorem 6 (Concentration of $\mathfrak{R}_{m,n}$ around the mean). *Let $\mathfrak{R}_{m,n}$ be the FR statistic. Then,*

$$P\left(\left|\mathfrak{R}_{m,n} - \mathbb{E}[\mathfrak{R}_{m,n}]\right| \geq t\right) \leq C'_{m,n}(\epsilon^*) \exp\left(\frac{-(t/(2\epsilon^*))^{d/(d-1)}}{(m+n)\,\tilde{C}}\right). \tag{12}$$

Here, $\tilde{C} = 8(4)^{d/(d-1)}$ and the explicit form for $C'_{m,n}(\epsilon^*)$ is given by (10) when $\epsilon = \epsilon^*$.

See Appendix D for full proofs of Theorems 5 and 6. Here, we sketch the proofs. The proof of the concentration inequality for $\mathfrak{R}_{m,n}$, Theorem 6, requires involving the median M_e, where $P(\mathfrak{R}_{m,n} \leq M_e) \geq 1/2$, inside the probability term by using

$$\left|\mathfrak{R}_{m,n} - \mathbb{E}[\mathfrak{R}_{m,n}]\right| \leq \left|\mathfrak{R}_{m,n} - M_e\right| + \left|\mathbb{E}[\mathfrak{R}_{m,n}] - M_e\right|.$$

To prove the expressions for the concentration around the median, Theorem 5, we first consider the h^d uniform partitions of $[0,1]^d$, with edges parallel to the coordinate axes having edge lengths h^{-1} and volumes h^{-d}. Then, by applying the Markov inequality, we show that with at least probability $1 - \left(\delta^h_{m,n}/\epsilon\right)$, where $\delta^h_{m,n} = O\big(h^{d-1}(m+n)^{1/d}\big)$, the FR statistic $\mathfrak{R}_{m,n}$ is subadditive with 2ϵ threshold. Afterward, owing to the induction method [17], the growth bound can be derived with at least probability $1 - \left(h\,\delta^h_{m,n}/\epsilon\right)$. The growth bound explains that with high probability there exists a constant depending on ϵ and h, $C_{\epsilon,h}$, such that $\mathfrak{R}_{m,n} \leq C_{\epsilon,h}(m\,n)^{1-1/d}$. Applying the law of total probability and semi-isoperimetric inequality (A108) in Lemma A11 gives us (A35). By considering the solution to convex optimization problem (9), i.e., ϵ^* and optimal $h = 7$ the claimed results (11) and (12) are derived. The only constraint here is that ϵ is lower bounded by a function of $\delta^h_{m,n} = O\big(h^{d-1}(m+n)^{1/d}\big)$.

Next, we provide a bound for the variance-like term with high probability at least $1 - \delta$. According to the previous results, we expect that this bound depends on ϵ^*, d, m and n. The proof is short and is given in Appendix D.

Theorem 7 (Variance-like bound for $\mathfrak{R}_{m,n}$). *Let $\mathfrak{R}_{m,n}$ be the FR statistic. With at least probability $1 - \delta$, we have*

$$|\mathfrak{R}_{m,n} - \mathbb{E}[\mathfrak{R}_{m,n}]| \leq O\left(\epsilon^* (m+n)^{(d-1)/d}\left(\log\left(C'_{m,n}(\epsilon^*)/\delta\right)\right)^{(d-1)/d}\right). \tag{13}$$

or, equivalently,

$$\left|\frac{\mathfrak{R}_{m,n}}{m+n} - \frac{\mathbb{E}[\mathfrak{R}_{m,n}]}{m+n}\right| \leq O\left(\epsilon^* (m+n)^{-1/d}\left(\log\left(C'_{m,n}(\epsilon^*)/\delta\right)\right)^{(d-1)/d}\right), \tag{14}$$

where $C'_{m,n}(\epsilon^*)$ depends on m, n, and d is given in (10) when $\epsilon = \epsilon^*$.

3. Numerical Experiments

3.1. Simulation Study

In this section, we apply the FR statistic estimate of the HP-divergence to both simulated and real data sets. We present results of a simulation study that evaluates the proposed bound on the MSE. We numerically validate the theory stated in Sections 2.2 and 2.4 using multiple simulations. In the first set of simulations, we consider two multivariate Normal random vectors **X**, **Y** and perform three experiments $d = 2, 4, 8$, to analyze the FR test statistic-based estimator performance as the sample sizes m, n increase. For the three dimensions $d = 2, 4, 8$, we generate samples from two normal distributions with identity covariance and shifted means: $\mu_1 = [0,0]$, $\mu_2 = [1,0]$ and $\mu_1 = [0,0,0,0]$, $\mu_2 = [1,0,0,0]$ and $\mu_1 = [0,0,\ldots,0]$, $\mu_2 = [1,0,\ldots,0]$ when $d = 2$, $d = 4$ and $d = 8$, respectively. For all of the following experiments, the sample sizes for each class are equal ($m = n$).

We vary $N = m = n$ up to 800. From Figure 3, we deduce that, when the sample size increases, the MSE decreases such that for higher dimensions the rate is slower. Furthermore, we compare the experiments with the theory in Figure 3. Our theory generally matches the experimental results. However, the MSE for the experiments tends to decrease to zero faster than the theoretical bound. Since the Gaussian distribution has a smooth density, this suggests that a tighter bound on the MSE may be possible by imposing stricter assumptions on the density smoothness as in [12].

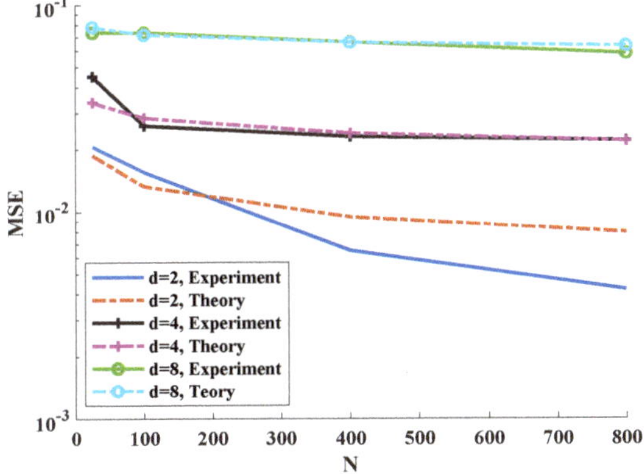

Figure 3. Comparison of the bound on the MSE theory and experiments for $d = 2, 4, 8$ standard Gaussian random vectors versus sample size from 100 trials.

In our next simulation, we compare three bivariate cases: first, we generate samples from a standard Normal distribution. Second, we consider a distinct smooth class of distributions i.e., binomial Gamma density with standard parameters and dependency coefficient $\rho = 0.5$. Third, we generate samples from Standard t-student distributions. Our goal in this experiment is to compare the MSE of the HP-divergence estimator between two identical distributions, $f_0 = f_1$, when f_0 is one of the Gamma, Normal, and t-student density function. In Figure 4, we observe that the MSE decreases as N increases for all three distributions.

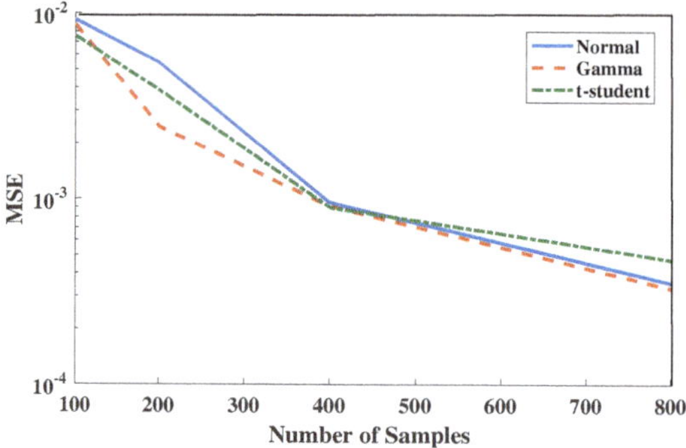

Figure 4. Comparison of experimentally predicted MSE of the FR-statistic as a function of sample size $m = n$ in various distributions Standard Normal, Gamma ($\alpha_1 = \alpha_2 = 1$, $\beta_1 = \beta_2 = 1$, $\rho = 0.5$) and Standard t-Student.

3.2. Real Datasets

We now show the results of applying the FR test statistic to estimate the HP-divergence using three different real datasets [46]:

- Human Activity Recognition (HAR), Wearable Computing, Classification of Body Postures and Movements (PUC-Rio): This dataset contains five classes (sitting-down, standing-up, standing, walking, and sitting) collected on eight hours of activities of four healthy subjects.
- Skin Segmentation dataset (SKIN): The skin dataset is collected by randomly sampling B,G,R values from face images of various age groups (young, middle, and old), race groups (white, black, and asian), and genders obtained from the FERET and PAL databases [47].
- Sensorless Drive Diagnosis (ENGIN) dataset: In this dataset, features are extracted from electric current drive signals. The drive has intact and defective components. The dataset contains 11 different classes with different conditions. Each condition has been measured several times under 12 different operating conditions, e.g., different speeds, load moments, and load forces.

We focus on two classes from each of the HAR, SKIN, and ENGIN datasets, specifically, for HAR dataset two classes "sitting" and "standing" and for SKIN dataset the classes "Skin" and "Non-skin" are considered. In the ENGIN dataset, the drive has intact and defective components, which results in 11 different classes with different conditions. We choose conditions 1 and 2.

In the first experiment, we computed the HP-divergence using KDE plug-in estimator and then the MSE for the FR test statistic estimator is derived as the sample size $N = m = n$ increases. We used 95% confidence interval as the error bars. We observe in Figure 5 that the estimated HP-divergence ranges in $[0, 1]$, which is one of the HP-divergence properties [8]. Interestingly, when N increases the HP-divergence tends to 1 for all HAR, SKIN, and ENGIN datasets. Note that in this set of experiments we have repeated the experiments on independent parts of the datasets to obtain the error bars. Figure 6 shows that the MSE expectedly decreases as the sample size grows for all three datasets. Here, we have used the KDE plug-in estimator [12], implemented on the all available samples, to determine the true HP-divergence. Furthermore, according to Figure 6, the FR test statistic-based estimator suggests that the Bayes error rate is larger for the SKIN dataset compared to the HAR and ENGIN datasets.

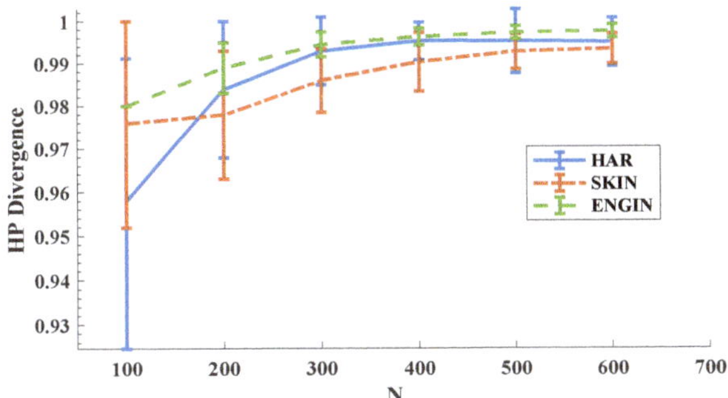

Figure 5. HP-divergence vs. sample size for three real datasets HAR, SKIN, and ENGIN.

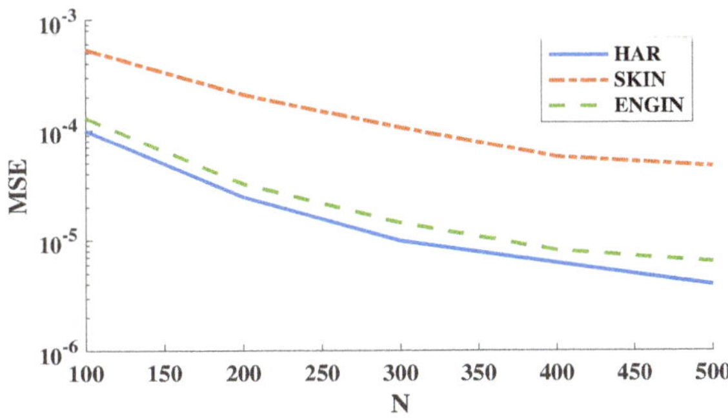

Figure 6. The empirical MSE vs. sample size. The empirical MSE of the FR estimator for all three datasets HAR, SKIN, and ENGIN decreases for larger sample size N.

In our next experiment, we add the first six features (dimensions) in order to our datasets and evaluate the FR test statistic's performance as the HP-divergence estimator. Surprisingly, the estimated HP-divergence doesn't change for the HAR sample; however, big changes are observed for the SKIN and ENGIN samples (see Figure 7).

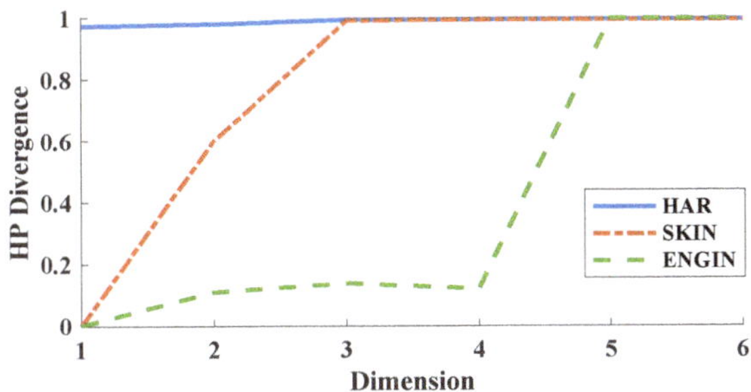

Figure 7. HP-divergence vs. dimension for three datasets HAR, SKIN, and ENGIN.

Finally, we apply the concentration bounds on the FR test statistic (i.e., Theorems 6 and 7) and compute theoretical implicit variance-like bound for the FR criteria with $\delta = 0.05$ error for the real datasets ENGIN, HAR, and SKIN. Since datasets ENGIN, HAR, and SKIN have the equal total sample size $N = m + n = 1200$ and different dimensions $d = 14, 12, 4$, respectively; here, we first intend to compare the concentration bound (13) on the FR statistic in terms of dimension d when $\delta = 0.05$. For real datasets ENGIN, HAR, and SKIN, we obtain

$$P\left(\left|\mathfrak{R}_{m,n} - \mathbb{E}[\mathfrak{R}_{m,n}]\right| \leq \xi\right) \geq 0.95,$$

where $\xi = \xi'[0.257, 0.005, 0.6 \times 10^{-11}]$, respectively, and ξ' is a constant not dependent on d. One observes that as the dimension decreases the interval becomes significantly tighter. However, this could not be generally correct and computing bound (13) precisely requires the knowledge of distributions and unknown constants. In Table 1, we compute the standard variance-like bound by applying the percentiles technique and observe that the bound threshold is not monotonic in terms of dimension d. Table 1 shows the FR test statistic, HP-divergence estimate (denoted by $\mathfrak{R}_{m,n}$, $\widehat{D}_p$, respectively), and standard variance-like interval for the FR statistic using the three real datasets HAR, SKIN, and ENGIN.

Table 1. $\mathfrak{R}_{m,n}$, $\widehat{D}_p$, m, and n are the FR test statistic, HP-divergence estimates using $\mathfrak{R}_{m,n}$, and sample sizes for two classes, respectively.

	FR Test Statistic				
Dataset	$\mathbb{E}[\mathfrak{R}_{m,n}]$	$\widehat{D}_p$	m	n	**Variance-Like Interval**
HAR	3	0.995	600	600	(2.994,3.006)
SKIN	4.2	0.993	600	600	(4.196,4.204)
ENGIN	1.8	0.997	600	600	(1.798,1.802)

4. Conclusions

We derived a bound on the MSE convergence rate for the Friedman–Rafsky estimator of the Henze–Penrose divergence assuming the densities are sufficiently smooth. We employed a partitioning strategy to derive the bias rate which depends on the number of partitions, the sample size $m + n$, the Hölder smoothness parameter η, and the dimension d. However, by using the optimal partition number, we derived the MSE convergence rate only in terms of $m + n$, η, and d. We validated our proposed MSE convergence rate using simulations and illustrated the approach for the meta-learning problem of estimating the HP-divergence for three real-world data sets. We also provided concentration bounds around the median and mean of the estimator. These bounds explicitly provide the rate that the FR statistic approaches its median/mean with high probability, not only as a function of the number of samples, m, n, but also in terms of the dimension of the space d. By using these results, we explored the asymptotic behavior of a variance-like rate in terms of m, n, and d.

Author Contributions: Conceptualization, S.Y.S., M.N. and A.O.H.; methodology, S.Y.S. and M.N.; software, S.Y.S. and M.N.; validation, S.Y.S., M.N., K.R.M. and A.O.H.; formal analysis, S.Y.S., M.N. and K.R.M.; investigation, S.Y.S. and M.N.; resources, S.Y.S. and M.N.; data curation, M.N.; writing—original draft preparation, S.Y.S.; writing—review and editing, M.N., K.R.M. and A.O.H.; supervision, A.O.H.; project administration, A.O.H.; funding acquisition, A.O.H.

Funding: The work presented in this paper was partially supported by ARO grant W911NF-15-1-0479 and DOE grants DE-NA0002534 and DE-NA0003921.

Conflicts of Interest: The authors declare no conflict of interest. The founding sponsors had no role in the design of the study; in the collection, analyses, or interpretation of data; in the writing of the manuscript, and in the decision to publish the results.

Abbreviations

HP   Henze-Penrose
BER   Bayes error rate
MST   Minimal Spanning Tree
FR   Friedman-Rafsky
MSE   Mean squared error

Appendix A. Proof of Theorem 4

In this section, we prove the subadditivity and superadditivity for the mean of FR test statistic. For this, first we need to illustrate the following lemma.

Lemma A1. *Let $\{Q_i\}_{i=1}^{l^d}$ be a uniform partition of $[0,1]^d$ into l^d subcubes Q_i with edges parallel to the coordinate axes having edge lengths l^{-1} and volumes l^{-d}. Let D_{ij} be the set of edges of MST graph between Q_i and Q_j with cardinality $|D_{ij}|$, then for $|D|$ defined as the sum of $|D_{ij}|$ for all $i, j = 1, \ldots, l^d$, $i \neq j$, we have $\mathbb{E}|D| = O(l^{d-1} \, n^{1/d})$, or more explicitly*

$$\mathbb{E}[|D|] \leq C' l^{d-1} n^{1/d} + O(l^{d-1} n^{(1/d)-s}), \tag{A1}$$

where $\eta > 0$ is the Hölder smoothness parameter and

$$s = \frac{(1 - 1/d)\eta}{d\,((1 - 1/d)\eta + 1)}.$$

Here, and in what follows, denote $\Xi_{MST}(\mathcal{X}_n)$ the length of the shortest spanning tree on $\mathcal{X}_n = \{\mathbf{X}_1, \ldots, \mathbf{X}_n\}$, namely

$$\Xi_{MST}(\mathcal{X}_n) := \min_T \sum_{e \in T} |e|,$$

where the minimum is over all spanning trees T of the vertex set $\mathcal{X}_n$. Using the subadditivity relation for Ξ_{MST} in [17], with the uniform partition of $[0,1]^d$ into l^d subcubes Q_i with edges parallel to the coordinate axes having edge lengths l^{-1} and volumes l^{-d}, we have

$$\Xi_{MST}(\mathcal{X}_n) \leq \sum_{i=1}^{l^d} \Xi_{MST}(\mathcal{X}_n \cap Q_i) + C\,l^{d-1}, \tag{A2}$$

where C is constant. Denote D the set of all edges of $MST\left(\bigcup_{i=1}^{M} Q_i\right)$ that intersect two different subcubes Q_i and Q_j with cardinality $|D|$. Let $|e_i|$ be the length of i-th edge in set D. We can write

$$\sum_{i \in |D|} |e_i| \leq C l^{d-1} \quad \text{and} \quad \mathbb{E} \sum_{i \in |D|} |e_i| \leq C l^{d-1},$$

also we know that

$$\mathbb{E} \sum_{i \in |D|} |e_i| = \mathbb{E}_D \sum_{i \in |D|} \mathbb{E}\big[|e_i|\,\big|\,D\big]. \tag{A3}$$

Note that using the result from ([31], Proposition 3), for some constants C_{i1} and C_{i2}, we have

$$\mathbb{E}|e_i| \leq C_{i1} n^{-1/d} + C_{i2} n^{-(1/d)-s}, \quad i \in |D|. \tag{A4}$$

Now, let $C_1 = \max_i \{C_{i1}\}$ and $C_2 = \max_i \{C_{i2}\}$, hence we can bound the expectation (A3) as

$$\mathbb{E}|D| \left(C_1 n^{-1/d} + C_2 (n^{-(1/d)-s}) \right) \leq C l^{d-1},$$

which implies

$$\mathbb{E}|D| \leq \left(C_1 n^{-1/d} + O(n^{-(1/d)-s}) \right)$$

$$\leq C' l^{d-1} n^{1/d} + O(l^{d-1} n^{(1/d)-s}).$$

To aim toward the goal (7), we partition $[0,1]^d$ into $M := l^d$ subcubes Q_i of side $1/l$. Recalling Lemma 2.1 in [48], we therefore have the set inclusion:

$$MST\left(\bigcup_{i=1}^{M} Q_i \right) \subset \bigcup_{i=1}^{M} MST(Q_i) \cup D, \tag{A5}$$

where D is defined as in Lemma A1. Let m_i and n_i be the number of sample $\{X_1, \ldots, X_m\}$ and $\{Y_1, \ldots, Y_n\}$ respectively falling into the partition Q_i, such that $\sum_i m_i = m$ and $\sum_i n_i = n$. Introduce sets A and B as

$$A := MST\left(\bigcup_{i=1}^{M} Q_i \right), \quad B := \bigcup_{i=1}^{M} MST(Q_i).$$

Since set B has fewer edges than set A, thus (A5) implies that the difference set of B and A contains at most $2|D|$ edges, where $|D|$ is the number of edges in D. On the other word,

$$|A \Delta B| \leq |A - B| + |B - A| = |D| + |B - A|$$

$$= |D| + (|B| - |B \cap A| \leq |D| + (|A| - |B \cap A|) = 2|D|.$$

The number of edge linked nodes from different samples in set A is bounded by the number of edge linked nodes from different samples in set B plus $2|D|$:

$$\mathfrak{R}_{m,n}(\mathfrak{X}_m, \mathfrak{Y}_n) \leq \sum_{i=1}^{M} \mathfrak{R}_{m_i,n_i}\left((\mathfrak{X}_m, \mathfrak{Y}_n) \cap Q_i \right) + 2|D|. \tag{A6}$$

Here, $\mathfrak{R}_{m_i,n_i}$ stands with the number edge linked nodes from different samples in partition Q_i, M. Next, we address the reader to Lemma A1, where it has been shown that there is a constant c such that $\mathbb{E}|D| \leq c\, l^{d-1}(m+n)^{1/d}$. This concludes the claimed assertion (7). Now, to accomplish the proof, the lower bound term in (8) is obtained with similar methodology and the set inclusion:

$$\bigcup_{i=1}^{M} MST(Q_i) \subset MST\left(\bigcup_{i=1}^{M} Q_i \right) \cup D. \tag{A7}$$

This completes the proof.

Appendix B. Proof of Theorem 2

As many of continuous subadditive functionals on $[0,1]^d$, in the case of the FR statistic, there exists a dual superadditive functional $\mathfrak{R}^*_{m,n}$ based on dual MST, MST*, proposed in Definition 2. Note that, in the MST* graph, the degrees of the corner points are bounded by c_d, where it only depends on dimension d, and is the bound for degree of every node in MST graph. The following properties hold true for dual FR test statistic, $\mathfrak{R}^*_{m,n}$:

Lemma A2. *Given samples $\mathfrak{X}_m = \{\mathbf{X}_1, \ldots, \mathbf{X}_m\}$ and $\mathfrak{Y}_n = \{\mathbf{Y}_1, \ldots, \mathbf{Y}_n\}$, the following inequalities hold true:*

(i) For constant c_d which depends on d:

$$\mathfrak{R}^*_{m,n}(\mathfrak{X}_m, \mathfrak{Y}_n) \leq \mathfrak{R}_{m,n}(\mathfrak{X}_m, \mathfrak{Y}_n) + c_d\, 2^d,$$

$$\mathfrak{R}_{m,n}(\mathfrak{X}_m, \mathfrak{Y}_n) \leq \mathfrak{R}^*_{m,n}(\mathfrak{X}_m, \mathfrak{Y}_n). \tag{A8}$$

(ii) (Subadditivity on $\mathbb{E}[\mathfrak{R}^*_{m,n}]$ and Superadditivity) Partition $[0,1]^d$ into l^d subcubes Q_i such that m_i, n_i be the number of sample $\mathfrak{X}_m = \{\mathbf{X}_1, \ldots, \mathbf{X}_m\}$ and $\mathfrak{Y}_n = \{\mathbf{Y}_1, \ldots, \mathbf{Y}_n\}$ respectively falling into the partition Q_i with dual $\mathfrak{R}^*_{m_i, n_i}$. Then, we have

$$\mathbb{E}\left[\mathfrak{R}^*_{m,n}(\mathfrak{X}_m, \mathfrak{Y}_n)\right] \leq \sum_{i=1}^{l^d} \mathbb{E}\left[\mathfrak{R}^*_{m_i,n_i}((\mathfrak{X}_m, \mathfrak{Y}_n) \cap Q_i)\right] + c\, l^{d-1}\, (m+n)^{1/d},$$

$$\mathfrak{R}^*_{m,n}(\mathfrak{X}_m, \mathfrak{Y}_n) \geq \sum_{i=1}^{l^d} \mathfrak{R}^*_{m_i,n_i}((\mathfrak{X}_m, \mathfrak{Y}_n) \cap Q_i) - 2^d c_d l^d, \tag{A9}$$

where c is a constant.

(i) Consider the nodes connected to the corner points. Since $\mathrm{MST}(\mathfrak{X}_m, \mathfrak{Y}_n)$ and $\mathrm{MST}^*(\mathfrak{X}_m, \mathfrak{Y}_n)$ can only be different in the edges connected to these nodes, and in $\mathfrak{R}^*(\mathfrak{X}_m, \mathfrak{Y}_n)$ we take all of the edges between these nodes and corner nodes into account, so we obviously have the second relation in (A8). In addition, for the first inequality in (A8), it is enough to say that the total number of edges connected to the corner nodes is upper bounded by $2^d c_d$.

(ii) Let $|D^*|$ be the set of edges of the MST^* graph which intersect two different partitions. Since MST and MST^* are only different in edges of points connected to the corners and edges crossing different partitions. Therefore, $|D^*| \leq |D|$. By eliminating one edge in set D in the worse scenario we would face two possibilities: either the corresponding node is connected to the corner which is counted anyways or any other point in MST graph which wouldn't change the FR test statistic. This implies the following subadditivity relation:

$$\mathfrak{R}^*_{m,n}(\mathfrak{X}_m, \mathfrak{Y}_n) - |D| \leq \sum_{i=1}^{l^d} \mathfrak{R}^*_{m_i,n_i}((\mathfrak{X}_m, \mathfrak{Y}_n) \cap Q_i).$$

Further from Lemma A1, we know that there is a constant c such that $\mathbb{E}|D| \leq c\, l^{d-1}\, (m+n)^{1/d}$. Hence, the first inequality in (A9) is obtained. Next, consider $|D^*_c|$ which represents the total number of edges from both samples only connected to the all corners points in MST^* graph. Therefore, one can easily claim:

$$\mathfrak{R}^*_{m,n}(\mathfrak{X}_m, \mathfrak{Y}_n) \geq \sum_{i=1}^{l^d} \mathfrak{R}^*_{m_i,n_i}((\mathfrak{X}_m, \mathfrak{Y}_n) \cap Q_i) - |D^*_c|.$$

In addition, we know that $|D^*_c| \leq 2^d l^d c_d$ where c_d stands with the largest possible degree of any vertex. One can write

$$\mathfrak{R}^*_{m,n}(\mathfrak{X}_m, \mathfrak{Y}_n) \geq \sum_{i=1}^{l^d} \mathfrak{R}^*_{m_i,n_i}((\mathfrak{X}_m, \mathfrak{Y}_n) \cap Q_i) - 2^d c_d l^d.$$

The following list of Lemmas A3, A4 and A6 are inspired from [49] and are required to prove Theorem A1. See Appendix E for their proofs.

Lemma A3. *Let $g(\mathbf{x})$ be a density function with support $[0,1]^d$ and belong to the Hölder class $\Sigma_d(\eta, L)$, $0 < \eta \leq 1$, stated in Definition 1. In addition, assume that $P(\mathbf{x})$ is a η-Hölder smooth function, such that its* *absolute value is bounded from above by a constant. Define the quantized density function with parameter l and constants ϕ_i as*

$$\widehat{g}(\mathbf{x}) = \sum_{i=1}^{M} \phi_i \mathbf{1}\{\mathbf{x} \in Q_i\}, \quad \text{where } \phi_i = l^d \int_{Q_i} g(\mathbf{x})\, d\mathbf{x}. \tag{A10}$$

Let $M = l^d$ and $Q_i = \{\mathbf{x}, \mathbf{x}_i : \|\mathbf{x} - \mathbf{x}_i\| < l^{-d}\}$. Then,

$$\int \left\| (g(\mathbf{x}) - \widehat{g}(\mathbf{x})) P(\mathbf{x}) \right\| d\mathbf{x} \leq O(l^{-d\eta}). \tag{A11}$$

Lemma A4. *Denote $\Delta(\mathbf{x}, \mathcal{S})$ the degree of vertex $\mathbf{x} \in \mathcal{S}$ in the MST over set $\mathcal{S}$ with the n number of vertices. For given function $P(\mathbf{x}, \mathbf{x})$, one obtains*

$$\int P(\mathbf{x}, \mathbf{x}) g(\mathbf{x}) \mathbb{E}[\Delta(\mathbf{x}, \mathcal{S})]\, d\mathbf{x} = 2 \int P(\mathbf{x}, \mathbf{x}) g(\mathbf{x})\, d\mathbf{x} + \varsigma_\eta(l, n), \tag{A12}$$

where, for constant $\eta > 0$,

$$\varsigma_\eta(l, n) = \left(O(l/n) - 2\, l^d/n \right) \int g(\mathbf{x}) P(\mathbf{x}, \mathbf{x})\, d\mathbf{x} + O(l^{-d\eta}). \tag{A13}$$

Lemma A5. *Assume that, for given k, $g_k(\mathbf{x})$ is a bounded function belong to $\Sigma_d(\eta, L)$. Let $P : \mathbb{R}^d \times \mathbb{R}^d \mapsto [0,1]$ be a symmetric, smooth, jointly measurable function, such that, given k, for almost every $\mathbf{x} \in \mathbb{R}^d$, $P(\mathbf{x}, .)$ is measurable with $\mathbf{x}$ a Lebesgue point of the function $g_k(.)P(\mathbf{x}, .)$. Assume that the first derivative P is bounded. For each k, let $\mathbf{Z}_1^k, \mathbf{Z}_2^k, \ldots, \mathbf{Z}_k^k$ be an independent d-dimensional variable with common density function g_k. Set $\mathcal{3}_k = \{\mathbf{Z}_1^k, \mathbf{Z}_2^k \ldots, \mathbf{Z}_k^k\}$ and $\mathcal{3}_k^{\mathbf{x}} = \{\mathbf{x}, \mathbf{Z}_2^k, \mathbf{Z}_3^k \ldots, \mathbf{Z}_k^k\}$. Then,*

$$\mathbb{E}\left[\sum_{j=2}^{k} P(\mathbf{x}, \mathbf{Z}_j^k) \mathbf{1}\{(\mathbf{x}, \mathbf{Z}_j^k) \in MST(\mathcal{3}_k^{\mathbf{x}})\} \right] = P(\mathbf{x}, \mathbf{x})\, \mathbb{E}[\Delta(\mathbf{x}, \mathcal{3}_k^{\mathbf{x}})] + \left\{ O(k^{-\eta/d}) + O(k^{-1/d}) \right\}. \tag{A14}$$

Lemma A6. *Consider the notations and assumptions in Lemma A5. Then,*

$$\left| k^{-1} \sum_{1 \leq i < j \leq k} P(\mathbf{Z}_i^k, \mathbf{Z}_j^k) \mathbf{1}\{(\mathbf{Z}_i^k, \mathbf{Z}_j^k) \in MST(\mathcal{3}_k)\} - \int_{\mathbb{R}^d} P(\mathbf{x}, \mathbf{x}) g_k(\mathbf{x})\, d\mathbf{x} \right| \tag{A15}$$
$$\leq \varsigma_\eta(l, k) + O(k^{-\eta/d}) + O(k^{-1/d}).$$

Here, $MST(\mathcal{S})$ denotes the MST graph over nice and finite set $\mathcal{S} \subset \mathbb{R}^d$ and η is the smoothness Hölder parameter. Note that $\varsigma_\eta(l, k)$ is given as before in Lemma A4 (A13).

Theorem A1. *Assume $\mathfrak{R}_{m,n} := \mathfrak{R}(\mathfrak{X}_m, \mathfrak{Y}_n)$ denotes the FR test statistic and densities f_0 and f_1 belong to the Hölder class $\Sigma_d(\eta, L)$, $0 < \eta \leq 1$. Then, the rate for the bias of the $\mathfrak{R}_{m,n}$ estimator for $d \geq 2$ is of the form:*

$$\left| \frac{\mathbb{E}[\mathfrak{R}_{m,n}]}{m+n} - 2pq \int \frac{f_0(\mathbf{x}) f_1(\mathbf{x})}{p f_0(\mathbf{x}) + q f_1(\mathbf{x})}\, d\mathbf{x} \right| \leq O\big(l^d (m+n)^{-\eta/d}\big) + O(l^{-d\eta}). \tag{A16}$$

The proof and a more explicit form for the bound (A16) are given in Appendix E.

Now, we are at the position to prove the assertion in (5). Without loss of generality, assume that $(m+n)l^{-d} > 1$. In the range $d \geq 2$ and $0 < \eta \leq 1$, we select l as a function of $m+n$ to be the sequence increasing in $m+n$ which minimizes the maximum of these rates:

$$l(m+n) = arg\ \min_l \max \left\{ l^d (m+n)^{-\eta/d},\ l^{-\eta d} \right\}.$$

The solution $l = l(m + n)$ occurs when $l^d(m + n)^{-\eta/d} = l^{-\eta d}$, or equivalently $l = \lfloor (m + n)^{\eta/(d^2(\eta+1))} \rfloor$. Substitute this into l in the bound (A16), the RHS expression in (5) for $d \geq 2$ is established.

Appendix C. Proof of Theorems 3

To bound the variance, we will apply one of the first concentration inequalities which was proved by Efron and Stein [43] and further was improved by Steele [18].

Lemma A7 (The Efron–Stein Inequality). *Let $\mathfrak{X}_m = \{\mathbf{X}_1, \ldots, \mathbf{X}_m\}$ be a random vector on the space $\mathcal{S}$. Let $\mathfrak{X}' = \{\mathbf{X}_1', \ldots, \mathbf{X}_m'\}$ be the copy of random vector $\mathfrak{X}_m$. Then, if $f : \mathcal{S} \times \cdots \times \mathcal{S} \to \mathbb{R}$, we have*

$$\mathbb{V}[f(\mathfrak{X}_m)] \leq \frac{1}{2} \sum_{i=1}^{m} \mathbb{E}\left[\left(f(\mathbf{X}_1, \ldots, \mathbf{X}_m) - f(\mathbf{X}_1, \ldots, \mathbf{X}_i', \ldots, \mathbf{X}_m) \right)^2 \right]. \tag{A17}$$

Consider two set of nodes $\mathbf{X}_i$, $1 \leq i \leq m$ and $\mathbf{Y}_j$ for $1 \leq j \leq n$. Without loss of generality, assume that $m < n$. Then, consider the $n - m$ virtual random points $\mathbf{X}_{m+1}, \ldots, \mathbf{X}_n$ with the same distribution as $\mathbf{X}_i$, and define $\mathbf{Z}_i := (\mathbf{X}_i, \mathbf{Y}_i)$. Now, for using the Efron–Stein inequality on set $\mathfrak{Z}_n = \{\mathbf{Z}_1, \ldots, \mathbf{Z}_n\}$, we involve another independent copy of $\mathfrak{Z}_n$ as $\mathfrak{Z}_n' = \{\mathbf{Z}_1', \ldots, \mathbf{Z}_n'\}$, and define $\mathfrak{Z}_n^{(i)} := (\mathbf{Z}_1, \ldots, \mathbf{Z}_{i-1}, \mathbf{Z}_i', \mathbf{Z}_{i+1}, \ldots, \mathbf{Z}_n)$, then $\mathfrak{Z}_n^{(1)}$ becomes $(\mathbf{Z}_1', \mathbf{Z}_2, \ldots, \mathbf{Z}_n) = \{(\mathbf{X}_1', \mathbf{Y}_1'), (\mathbf{X}_2, \mathbf{Y}_2), \ldots, (\mathbf{X}_m, \mathbf{Y}_n)\} =: (\mathfrak{X}_m^{(1)}, \mathfrak{Y}_n^{(1)})$ where $(\mathbf{X}_1', \mathbf{Y}_1')$ is independent copy of $(\mathbf{X}_1, \mathbf{Y}_1)$. Next, define the function $r_{m,n}(\mathfrak{Z}_n) := \mathfrak{R}_{m,n}/(m + n)$, which means that we discard the random samples $\mathbf{X}_{m+1}, \ldots, \mathbf{X}_n$, and find the previously defined $\mathfrak{R}_{m,n}$ function on the nodes $\mathbf{X}_i$, $1 \leq i \leq m$ and $\mathbf{Y}_j$ for $1 \leq j \leq n$, and multiply by some coefficient to normalize it. Then, according to the Efron–Stein inequality, we have

$$Var(r_{m,n}(\mathfrak{Z}_n)) \leq \frac{1}{2} \sum_{i=1}^{n} \mathbb{E}\left[(r_{m,n}(\mathfrak{Z}_n) - r_{m,n}(\mathfrak{Z}_n^{(i)}))^2 \right].$$

Now, we can divide the RHS as

$$\frac{1}{2} \sum_{i=1}^{n} \mathbb{E}\left[(r_{m,n}(\mathfrak{Z}_n) - r_{m,n}(\mathfrak{Z}_n^{(i)}))^2 \right] = \frac{1}{2} \sum_{i=1}^{m} \mathbb{E}\left[(r_{m,n}(\mathfrak{Z}_n) - r_{m,n}(\mathfrak{Z}_n^{(i)}))^2 \right]$$

$$+ \frac{1}{2} \sum_{i=m+1}^{n} \mathbb{E}\left[(r_{m,n}(\mathfrak{Z}_n) - r_{m,n}(\mathfrak{Z}_n^{(i)}))^2 \right]. \tag{A18}$$

The first summand becomes

$$= \frac{1}{2} \sum_{i=1}^{m} \mathbb{E}\left[(r_{m,n}(\mathfrak{Z}_n) - r_{m,n}(\mathfrak{Z}_n^{(i)}))^2 \right] = \frac{m}{2 \, (m + n)^2} \mathbb{E}\left[(\mathfrak{R}_{m,n}(\mathfrak{X}_m, \mathfrak{Y}_n) - \mathfrak{R}_{m,n}(\mathfrak{X}_m^{(1)}, \mathfrak{Y}_n^{(1)}))^2 \right],$$

which can also be upper bounded as follows:

$$\left| \mathfrak{R}_{m,n}(\mathfrak{X}_m, \mathfrak{Y}_n) - \mathfrak{R}_{m,n}(\mathfrak{X}_m^{(1)}, \mathfrak{Y}_n^{(1)}) \right| \leq \left| \mathfrak{R}_{m,n}(\mathfrak{X}_m, \mathfrak{Y}_n) - \mathfrak{R}_{m,n}(\mathfrak{X}_m^{(1)}, \mathfrak{Y}_n) \right|$$

$$+ \left| \mathfrak{R}(\mathfrak{X}_m^{(1)}, \mathfrak{Y}_n) - \mathfrak{R}_{m,n}(\mathfrak{X}_m^{(1)}, \mathfrak{Y}_n^{(1)}) \right|. \tag{A19}$$

For deriving an upper bound on the second line in (A19), we should observe how much changing a point's position modifies the amount of $\mathfrak{R}_{m,n}(\mathfrak{X}_m, \mathfrak{Y}_n)$. We consider two steps of changing $\mathbf{X}_1$'s position: we first remove it from the graph, and then add it to the new position. Removing it would change $\mathfrak{R}_{m,n}(\mathfrak{X}_m, \mathfrak{Y}_n)$ at most by $2\,c_d$ because X_1 has a degree of at most c_d, and c_d edges will be removed from the MST graph, and c_d edges will be added to it. Similarly, adding $\mathbf{X}_1$ to the new position will affect $\mathfrak{R}_{m,n}(\mathfrak{X}_{m,n}, \mathfrak{Y}_{m,n})$ at most by $2c_d$. Thus, we have

$$\left| \mathfrak{R}_{m,n}(\mathfrak{X}_m, \mathfrak{Y}_n) - \mathfrak{R}_{m,n}(\mathfrak{X}_m^{(1)}, \mathfrak{Y}_n) \right| \leq 4\, c_d,$$

and we can also similarly reason that

$$\left| \mathfrak{R}_{m,n}(\mathfrak{X}_m^{(1)}, \mathfrak{Y}_n) - \mathfrak{R}_{m,n}(\mathfrak{X}_m^{(1)}, \mathfrak{Y}_n^{(1)}) \right| \leq 4\, c_d.$$

Therefore, totally we would have

$$\left| \mathfrak{R}_{m,n}(\mathfrak{X}_m, \mathfrak{Y}_n) - \mathfrak{R}_{m,n}(\mathfrak{X}_m^{(1)}, \mathfrak{Y}_n^{(1)}) \right| \leq 8\, c_d.$$

Furthermore, the second summand in (A18) becomes

$$= \frac{1}{2} \sum_{i=m+1}^{n} \mathbb{E}\left[(r_{m,n}(\mathfrak{Z}_n) - r_{m,n}(\mathfrak{Z}_n^{(i)}))^2 \right] = K_{m,n} \mathbb{E}\left[(\mathfrak{R}_{m,n}(\mathfrak{X}_m, \mathfrak{Y}_n) - \mathfrak{R}_{m,n}(\mathfrak{X}_m^{(m+1)}, \mathfrak{Y}_n^{(m+1)}))^2 \right],$$

where $K_{m,n} = \frac{n-m}{2\,(m+n)^2}$. Since, in $(\mathfrak{X}_m^{(m+1)}, \mathfrak{Y}_n^{(m+1)})$, the point $\mathbf{X}'_{m+1}$ is a copy of virtual random point $\mathbf{X}_{m+1}$, therefore this point doesn't change the FR test statistic $\mathfrak{R}_{m,n}$. In addition, following the above arguments, we have

$$\left| \mathfrak{R}_{m,n}(\mathfrak{X}_m, \mathfrak{Y}_n) - \mathfrak{R}_{m,n}(\mathfrak{X}_m, \mathfrak{Y}_n^{(m+1)}) \right| \leq 4\, c_d.$$

Hence, we can bound the variance as below:

$$Var(r_{m,n}(\mathfrak{Z}_n)) \leq \frac{8c_d^2(n-m)}{(m+n)^2} + \frac{32\, c_d^2\, m}{(m+n)^2}. \tag{A20}$$

Combining all results with the fact that $\dfrac{n}{m+n} \to q$ concludes the proof.

Appendix D. Proof of Theorems 5–7

We will need the following prominent results for the proofs.

Lemma A8. *For $h = 1, 2, \ldots$, let $\delta_{m,n}^h$ be the function $c\, h^{d-1}(m+n)^{1/d}$, where c is a constant. Then, for $\epsilon > 0$, we have*

$$P\left(\mathfrak{R}_{m,n}(\mathfrak{X}_m, \mathfrak{Y}_n) \leq \sum_{i=1}^{h^d} \mathfrak{R}_{m_i, n_i}(\mathfrak{X}_{m_i}, \mathfrak{Y}_{n_i}) + 2\epsilon \right) \geq \frac{\epsilon - \delta_{m,n}^h}{\epsilon}. \tag{A21}$$

Note that, in the case $\epsilon \leq \delta_{m,n}^h$, the above claimed inequality becomes trivial.

The subadditivity property for FR test statistic $\mathfrak{R}_{m,n}$ in Lemma A8, as well as Euclidean functionals, leads to several non-trivial consequences. The growth bound was first explored by Rhee (1993b) [50], and as is illustrated in [17,27] has a wide range of applications. In this paper, we investigate the probabilistic growth bound for $\mathfrak{R}_{m,n}$. This observation will lead us to our main goal in this appendix that is providing the proof of Theorem 6. For what follows, we will use $\delta_{m,n}^h$ notation for the expression $O\big(h^{d-1}(m+n)^{1/d}\big)$.

Lemma A9 (Growth bounds for $\mathfrak{R}_{m,n}$). *Let $\mathfrak{R}_{m,n}$ be the FR test statistic. Then, for given non-negative ϵ, such that $\epsilon \geq h^2\, \delta^h_{m,n}$, with at least probability $g(\epsilon) := 1 - \dfrac{h\, \delta^h_{m,n}}{\epsilon}$, $h = 2, 3, \ldots$, we have*

$$\mathfrak{R}_{m,n}(\mathfrak{X}_m, \mathfrak{Y}_n) \leq c''_{\epsilon,h}\, (\#\mathfrak{X}_m\, \#\mathfrak{Y}_n)^{1-1/d}. \tag{A22}$$

Here, $c''_{\epsilon,h} = O\left(\dfrac{\epsilon}{h^{d-1}-1}\right)$ depending only on ϵ and h.

The complexity of $\mathfrak{R}_{m,n}$'s behavior and the need to pursue the proof encouraged us to explore the smoothness condition for $\mathfrak{R}_{m,n}$. In fact, this is where both subadditivity and superadditivity for the FR statistic are used together and become more important.

Lemma A10 (Smoothness for $\mathfrak{R}_{m,n}$). *Given observations of*

$$\mathfrak{X}_m := (\mathfrak{X}_{m'}, \mathfrak{X}_{m''}) = \{\mathbf{X}_1, \ldots, \mathbf{X}_{m'}, \mathbf{X}_{m'+1}, \ldots, \mathbf{X}_m\},$$

where $m' + m'' = m$ and $\mathfrak{Y}_n := (\mathfrak{Y}_{n'}, \mathfrak{Y}_{n''}) = \{\mathbf{Y}_1, \ldots, \mathbf{Y}_{n'}, \mathbf{Y}_{n'+1}, \ldots, \mathbf{Y}_n\}$, where $n' + n'' = n$, denote $\mathfrak{R}_{m,n}(\mathfrak{X}_m, \mathfrak{Y}_n)$ as before, the number of edges of $\mathrm{MST}(\mathfrak{X}_m, \mathfrak{Y}_n)$ which connect a point of $\mathfrak{X}_m$ to a point of $\mathfrak{Y}_n$. Then, for given integer $h \geq 2$, for all $(\mathfrak{X}_n, \mathfrak{Y}_m) \in [0,1]^d$, $\epsilon > h^2\, \delta^h_{m,n}$ where $\delta^h_{m,n} = O\big(h^{d-1}(m+n)^{1/d}\big)$, we have

$$P\left(\left| \mathfrak{R}_{m,n}(\mathfrak{X}_m, \mathfrak{Y}_n) - \mathfrak{R}_{m',n'}(\mathfrak{X}_{m'}, \mathfrak{Y}_{n'}) \right| \leq \tilde{c}_{\epsilon,h}\, (\#\mathfrak{X}_{m''}\, \#\mathfrak{Y}_{n''})^{1-1/d} \right)$$
$$\geq 1 - \frac{2h\, \delta^h_{m,n}}{\epsilon}, \tag{A23}$$

where $\tilde{c}_{\epsilon,h} = O\left(\dfrac{\epsilon}{h^{d-1}-1}\right)$.

Remark: Using Lemma A10, we can imply the continuty property, i.e., for all observations $(\mathfrak{X}_m, \mathfrak{Y}_n)$ and $(\mathfrak{X}_{m'}, \mathfrak{Y}_{n'})$, with at least probability $2\, g(\epsilon) - 1$, one obtains

$$\left| \mathfrak{R}_{m,n}(\mathfrak{X}_m, \mathfrak{Y}_n) - \mathfrak{R}_{m',n'}(\mathfrak{X}_{m'}, \mathfrak{Y}_{n'}) \right|$$
$$\leq c^*_{\epsilon,h}\, (\#(\mathfrak{X}_m \triangle \mathfrak{X}_{m'})\, \#(\mathfrak{Y}_n \triangle \mathfrak{Y}_{n'}))^{1-1/d}, \tag{A24}$$

for given $\epsilon > 0$, $c^*_{\epsilon,h} = O\left(\dfrac{\epsilon}{h^{d-1}-1}\right)$, $h \geq 2$. Here, $\mathfrak{X}_m \triangle \mathfrak{X}_{m'}$ denotes symmetric difference of observations $\mathfrak{X}_m$ and $\mathfrak{X}_{m'}$.

The path to approach the assertions (11) and (12) proceeds via semi-isoperimentic inequality for the $\mathfrak{R}_{m,n}$ involving the Hamming distance.

Lemma A11 (Semi-Isoperimetry). *Let μ be a measure on $[0,1]^d$; μ^n denotes the product measure on space $([0,1]^d)^n$. In addition, let M_e denotes a median of $\mathfrak{R}_{m,n}$. Set*

$$\mathbb{A} := \left\{ \mathfrak{X}_m \in ([0,1]^d)^m, \mathfrak{Y}_n \in ([0,1]^d)^n; \mathfrak{R}_{m,n}(\mathfrak{X}_m, \mathfrak{Y}_n) \leq M_e \right\}. \tag{A25}$$

Following the notations in [17], $H(\mathbf{x}, \mathbf{x}') = \#\{i, x_i \neq x'_i)$ and $\phi_{\mathbb{A}}(\mathbf{x}') + \phi_{\mathbb{A}}(\mathbf{y}') = \min\{H(\mathbf{x}, \mathbf{x}') + H(\mathbf{y}, \mathbf{y}') : \mathbf{x}, \mathbf{y} \in \mathbb{A}\}$ and $\phi_{\mathbb{A}}(\mathbf{x}') \phi_{\mathbb{A}}(\mathbf{y}') = \min\{H(\mathbf{x}, \mathbf{x}') H(\mathbf{y}, \mathbf{y}') : \mathbf{x}, \mathbf{y} \in \mathbb{A}\}$. Then,

$$
\mu^{m+n}\left(\left\{\mathbf{x}' \in ([0,1]^d)^m, \mathbf{y}' \in ([0,1]^d)^n : \phi_{\mathbb{A}}(\mathbf{x}')\,\phi_{\mathbb{A}}(\mathbf{y}') \geq t\right\}\right)
$$
$$
\leq 4\exp\left(\frac{-t}{8(m+n)}\right).
\tag{A26}
$$

Now, we continue by providing the proof of Theorem 5. Recall (A25) and denote

$$
\mathbb{F}_{\mathbf{x}} := \left\{\mathbf{x}_i, i = 1, \ldots, m, \mathbf{x}_i = \mathbf{x}'_i\right\},
$$
$$
\mathbb{F}_{\mathbf{y}} := \left\{\mathbf{y}_j, j = 1, \ldots, n, \mathbf{y}_j = \mathbf{y}'_j\right\},
$$
$$
\text{and}
$$
$$
\mathbb{G}_{\mathbf{x}} := \left\{\mathbf{x}_i, i = 1, \ldots, m, \mathbf{x}_i \neq \mathbf{x}'_i\right\},
$$
$$
\mathbb{G}_{\mathbf{y}} := \left\{\mathbf{y}_j, j = 1, \ldots, n, \mathbf{y}_j \neq \mathbf{y}'_j\right\}.
$$

In addition, for given integer h, define events $\mathbb{B}$, $\mathbb{B}'$ by

$$
\mathbb{B} := \left\{\left|\mathfrak{R}_{m,n}(\mathcal{X}'_m, \mathcal{Y}'_n) - \mathfrak{R}(\mathbb{F}_{\mathbf{x}}, \mathbb{F}_{\mathbf{y}})\right| \leq c_{\epsilon,h}\left(\#\mathbb{G}_{\mathbf{x}}\,\#\mathbb{G}_{\mathbf{y}}\right)^{1-1/d}\right\},
$$
$$
\mathbb{B}' := \left\{\left|\mathfrak{R}(\mathbb{F}_{\mathbf{x}}, \mathbb{F}_{\mathbf{y}}) - \mathfrak{R}_{m,n}(\mathcal{X}_m, \mathcal{Y}_n)\right| \leq c_{\epsilon,h}\left(\#\mathbb{G}_{\mathbf{x}}\,\#\mathbb{G}_{\mathbf{y}}\right)^{1-1/d}\right\},
$$

where $c_{\epsilon,h}$ is a constant. By virtue of smoothness property, Lemma A10, for $\epsilon \geq h^2 \delta^h_{m,n}$, we know $P(\mathbb{B}) \geq 2g(\epsilon) - 1$ and $P(\mathbb{B}') \geq 2g(\epsilon) - 1$. On the other hand, we have

$$
\mathfrak{R}_{m,n}(\mathcal{X}'_m, \mathcal{Y}'_n) \leq \left|\mathfrak{R}_{m,n}(\mathcal{X}'_m, \mathcal{Y}'_n) - \mathfrak{R}(\mathbb{F}_{\mathbf{x}}, \mathbb{F}_{\mathbf{y}})\right|
$$
$$
+ \left|\mathfrak{R}(\mathbb{F}_{\mathbf{x}}, \mathbb{F}_{\mathbf{y}}) - \mathfrak{R}_{m,n}(\mathcal{X}_m, \mathcal{Y}_n)\right| + \mathfrak{R}_{m,n}(\mathcal{X}_m, \mathcal{Y}_n).
$$
$$
= |\varpi'| + |\varpi| + \mathfrak{R}_{m,n}(\mathcal{X}_m, \mathcal{Y}_n) \quad \text{(say)}.
$$

Moreover, $P(\mathfrak{R}_{m,n}(\mathcal{X}_m, \mathcal{Y}_n) \leq M_e) \geq 1/2$. Therefore, we can write

$$
1/2 \leq P\left(\mathfrak{R}_{m,n}(\mathcal{X}'_m, \mathcal{Y}'_n) \leq M_e + |\varpi'| + |\varpi|\right)
$$
$$
\leq P\left(\mathfrak{R}_{m,n}(\mathcal{X}'_m, \mathcal{Y}'_n) \leq M_e + |\varpi'| + |\varpi|\,\big|\,\mathbb{B} \cap \mathbb{B}'\right) P(\mathbb{B} \cap \mathbb{B}')
\tag{A27}
$$
$$
+ P(\mathbb{B}^c \cup \mathbb{B}'^c).
$$

Thus, we obtain

$$
P\left(\mathfrak{R}_{m,n}(\mathcal{X}'_m, \mathcal{Y}'_n) \leq M_e + 4\epsilon\,\left(\#\mathbb{G}_{\mathbf{x}}\,\#\mathbb{G}_{\mathbf{y}}\right)^{1-1/d}\right)
$$
$$
\geq \left(1/2 - 1 + P(\mathbb{B} \cap \mathbb{B}')\right)/P(\mathbb{B} \cap \mathbb{B}')
$$
$$
= 1 - \left((2\,P(\mathbb{B} \cap \mathbb{B}'))^{-1}\right).
$$

Note that $P(\mathbb{B} \cap \mathbb{B}') = P(\mathbb{B})\,P(\mathbb{B}') \geq (2\,g(\epsilon) - 1)^2$. Now, we easily claim that

$$
1 - \left((2\,P(\mathbb{B} \cap \mathbb{B}'))^{-1}\right) \geq 1 - \left((2\,(2\,g(\epsilon) - 1)^2)^{-1}\right).
\tag{A28}
$$

Thus,

$$P\left(\mathfrak{R}_{m,n}(\mathfrak{X}'_m, \mathfrak{Y}'_n) \le M_e + 4\epsilon \left(\#\mathbb{G}_{\mathbf{x}} \, \#\mathbb{G}_{\mathbf{y}}\right)^{1-1/d}\right) \ge 1 - \left(\left(2\left(2\,g(\epsilon) - 1\right)^2\right)^{-1}\right).$$

On the other word, calling $\phi_{\mathbb{A}}(\mathbf{x}')$ and $\phi_{\mathbb{A}}(\mathbf{y}')$ in Lemma A11, we get

$$P\left(\mathfrak{R}_{m,n}(\mathfrak{X}'_m, \mathfrak{Y}'_n) \le M_e + 4\epsilon \left(\phi_{\mathbb{A}}(\mathbf{x}')\,\phi_{\mathbb{A}}(\mathbf{y}')\right)^{1-1/d}\right) \ge 1 - \left(\left(2\left(2\,g(\epsilon) - 1\right)^2\right)^{-1}\right). \tag{A29}$$

Furthermore, denote event

$$\mathbb{C} := \left\{\mathfrak{R}_{m,n}(\mathfrak{X}'_m, \mathfrak{Y}'_n) \le M_e + 4\epsilon \left(\phi_{\mathbb{A}}(\mathbf{x}')\,\phi_{\mathbb{A}}(\mathbf{y}')\right)^{1-1/d}\right\}.$$

Then, we have

$$P\left(\mathfrak{R}_{m,n}(\mathfrak{X}_m, \mathfrak{Y}_n) \ge M_e + t\right) = \mu^{m+n}\left(\mathfrak{R}_{m,n}(\mathfrak{X}'_m, \mathfrak{Y}'_n) \ge M_e + t\right)$$

$$= \mu^{m+n}\left(\left(\mathfrak{R}_{m,n}(\mathfrak{X}'_m, \mathfrak{Y}'_n) \ge M_e + t\right) \big| \mathbb{C}\right) P(\mathbb{C})$$

$$+ \mu^{m+n}\left(\left(\mathfrak{R}_{m,n}(\mathfrak{X}'_m, \mathfrak{Y}'_n) \ge M_e + t\right) \big| \mathbb{C}^c\right) P(\mathbb{C}^c)$$

$$\le \mu^{m+n}\left(\left(\phi_{\mathbb{A}}(\mathbf{x}')\,\phi_{\mathbb{A}}(\mathbf{y}')\right)^{1-1/d} \ge \frac{t}{4\epsilon}\right) P(\mathbb{C})$$

$$+ \mu^{m+n}\left(\left(\mathfrak{R}_{m,n}(\mathfrak{X}'_m, \mathfrak{Y}'_n) \ge M_e + t\right) \big| \mathbb{C}^c\right) P(\mathbb{C}^c). \tag{A30}$$

Using $P(\mathbb{C}) = 1 - P(\mathbb{C}^c)$

$$= \mu^{m+n}\left(\left(\phi_{\mathbb{A}}(\mathbf{x}')\,\phi_{\mathbb{A}}(\mathbf{y}')\right)^{1-1/d} \ge \frac{t}{4\epsilon}\right)$$

$$+ P(\mathbb{C}^c)\left\{\mu^{m+n}\left(\left(\mathfrak{R}_{m,n}(\mathfrak{X}'_m, \mathfrak{Y}'_n) \ge M_e + t\right) \big| \mathbb{C}^c\right)\right.$$

$$\left. - \mu^{m+n}\left(\left(\phi_{\mathbb{A}}(\mathbf{x}')\,\phi_{\mathbb{A}}(\mathbf{y}')\right)^{1-1/d} \ge \frac{t}{4\epsilon}\right)\right\}.$$

Define set $\mathbb{K}_t = \left\{\left(\phi_{\mathbb{A}}(\mathbf{x}')\,\phi_{\mathbb{A}}(\mathbf{y}')\right)^{1-1/d} \ge \frac{t}{4\epsilon}\right\}$, so

$$\mu^{m+n}\left(\mathfrak{R}_{m,n}(\mathfrak{X}'_m, \mathfrak{Y}'_n) \ge M_e + t \big| \mathbb{C}^c\right)$$

$$= \mu^{m+n}\left(\mathfrak{R}_{m,n}(\mathfrak{X}'_m, \mathfrak{Y}'_n) \ge M_e + t \big| \mathbb{C}^c, \mathbb{K}_t\right)\mu^{m+n}(\mathbb{K}_t) + \mu^{m+n}\left(\left(\mathfrak{R}_{m,n}(\mathfrak{X}'_m, \mathfrak{Y}'_n) \ge M_e + t\right) \big| \mathbb{C}^c, \mathbb{K}_t^c\right)\mu^{m+n}(\mathbb{K}_t^c).$$

Since

$$\mu^{m+n}\left(\mathfrak{R}_{m,n}(\mathfrak{X}'_m, \mathfrak{Y}'_n) \ge M_e + t \big| \mathbb{C}^c, \mathbb{K}_t\right) = 1,$$

and

$$\mu^{m+n}\left(\mathfrak{R}_{m,n}(\mathfrak{X}'_m, \mathfrak{Y}'_n) \ge M_e + t \big| \mathbb{C}^c, \mathbb{K}_t^c\right) = \mu^{m+n}\left(\mathfrak{R}_{m,n}(\mathfrak{X}'_m, \mathfrak{Y}'_n) \ge M_e + t\right).$$

Consequently, from (A30), one can write

$$P\big(\mathfrak{R}_{m,n}(\mathfrak{X}_m,\mathfrak{Y}_n) \geq M_e + t\big)$$

$$\leq \mu^{m+n}\Big(\big(\phi_{\mathbb{A}}(\mathbf{x}')\,\phi_{\mathbb{A}}(\mathbf{y}')\big)^{1-1/d} \geq \frac{t}{4\epsilon}\Big)$$

$$+P(\mathbb{C}^c)\Big\{\mu^{m+n}\big(\mathfrak{R}_{m,n}(\mathfrak{X}'_m,\mathfrak{Y}'_n) \geq M_e + t\big)\mu^{m+n}(\mathbb{K}^c_t)\Big\} \tag{A31}$$

$$\leq \mu^{m+n}\Big(\big(\phi_{\mathbb{A}}(\mathbf{x}')\,\phi_{\mathbb{A}}(\mathbf{y}')\big)^{1-1/d} \geq \frac{t}{4\epsilon}\Big)$$

$$+\Big(\big(2\,(2\,g(\epsilon)-1)^2\big)^{-1}\Big)P\big(\mathfrak{R}_{m,n}(\mathfrak{X}_m,\mathfrak{Y}_n) \geq M_e + t\big).$$

The last inequality implies by owing to (A29) and $\mu^{m+n}(\mathbb{K}^c_t) \leq 1$. For $g(\epsilon) \geq 1/2 + 1/(2\sqrt{2})$, we have

$$1 - \Big(\big(2\,(2\,g(\epsilon)-1)^2\big)^{-1}\Big) \geq 0,$$

or equivalently this holds true when $\epsilon \geq (2h\sqrt{2}\,\delta^h_{m,n})/(\sqrt{2}-1)$. Furthermore, for $h \geq 7$, we have

$$h^2\delta^h_{m,n} \geq (2h\sqrt{2}\,\delta^h_{m,n})/(\sqrt{2}-1), \tag{A32}$$

therefore $P\big(\mathfrak{R}_{m,n}(\mathfrak{X}_m,\mathfrak{Y}_n) \geq M_e + t\big)$ is less than and equal to

$$\Big(1 - \big((2\,(2\,g(\epsilon)-1)^2)^{-1}\big)\Big)^{-1}\mu^{m+n}\Big(\big(\phi_{\mathbb{A}}(\mathbf{x}')\,\phi_{\mathbb{A}}(\mathbf{y}')\big)^{1-1/d} \geq \frac{t}{4\epsilon}\Big). \tag{A33}$$

By virtue of Lemma A11, finally we obtain

$$P\big(\mathfrak{R}_{m,n}(\mathfrak{X}_m,\mathfrak{Y}_n) \geq M_e + t\big) \leq 4\Big(1 - \big((2\,(2\,g(\epsilon)-1)^2)^{-1}\big)\Big)^{-1}\exp\Big(\frac{-t^{d/(d-1)}}{8(4\epsilon)^{d/d-1}(m+n)}\Big). \tag{A34}$$

Similarly, we can derive the same bound on $P\big(\mathfrak{R}_{m,n}(\mathfrak{X}_m,\mathfrak{Y}_n) \leq M_e - t\big)$, so we obtain

$$P\Big(\big|\mathfrak{R}_{m,n} - M_e\big| \geq t\Big) \leq C'_{m,n}(\epsilon,h)\exp\Big(\frac{-t^{d/(d-1)}}{8(4\epsilon)^{d/(d-1)}(m+n)}\Big), \tag{A35}$$

where

$$C'_{m,n}(\epsilon,h) = 8\Big(1 - 2^{-1}\Big(1 - \frac{2h\ O\big(h^{d-1}(m+n)^{1/d}\big)}{\epsilon}\Big)^{-2}\Big)^{-1}. \tag{A36}$$

We will analyze (A35) together with Theorem 6. The next lemma will be employed in Theorem 6's proof.

Lemma A12 (Deviation of the Mean and Median). *Consider M_e as a median of $\mathfrak{R}_{m,n}$. Then, for $\epsilon \geq h^2\delta^h_{m,n}$ and given $h \geq 7$, we have*

$$\Big|\mathbb{E}\big[\mathfrak{R}_{m,n}(\mathfrak{X}_m,\mathfrak{Y}_n)\big] - M_e\Big| \leq C_{m,n}(\epsilon,h)\,(m+n)^{(d-1)/d}, \tag{A37}$$

where $C_{m,n}(\epsilon,h)$ is a constant depending on ϵ, h, m, and n by

$$C_{m,n}(\epsilon,h) = C\Big(1 - \big((2\,(2\,g(\epsilon)-1)^2)^{-1}\big)\Big)^{-1}, \tag{A38}$$

where C is a constant and

$$\delta_{m,n}^h = O\big(h^{d-1}(m+n)^{1/d}\big), \quad \text{and} \quad g(\epsilon) = 1 - \frac{h\,\delta_{m,n}^h}{\epsilon}.$$

We conclude this part by pursuing our primary intension which has been the Theorem 6's proof. Observe from Theorem 5, (11) that

$$P\Big(\big|\mathfrak{R}_{m,n} - \mathbb{E}[\mathfrak{R}_{m,n}]\big| \geq t + C_{m,n}(\epsilon, l)(m+n)^{(d-1)/d}\Big)$$

$$\leq P\Big(\big|\mathfrak{R}_{m,n} - M_e\big| + \big|\mathbb{E}[\mathfrak{R}_{m,n}] - M_e\big|$$

$$\geq t + C_{m,n}(\epsilon, l)(m+n)^{(d-1)/d}\Big)$$

$$\leq P\Big(\big|\mathfrak{R}_{m,n} - M_e\big| \geq t\Big)$$

$$\leq 8\left(1 - \big((2\,(2\,g(\epsilon)-1)^2)^{-1}\big)\right)^{-1} \exp\left(\frac{-t^{d/(d-1)}}{8(4\epsilon)^{d/d-1}(m+n)}\right).$$

Note that the last bound is derived by (11). The rest of the proof is as the following: When $t \geq 2C_{m,n}(\epsilon, h)(m+n)^{(d-1)/d}$, we use

$$\Big(t - C_{m,n}(\epsilon, h)(m+n)^{(d-1)/d}\Big)^{d/(d-1)} \geq \big(t/2\big)^{d/(d-1)}.$$

Therefore, it turns out that

$$P\Big(\big|\mathfrak{R}_{m,n} - \mathbb{E}[\mathfrak{R}_{m,n}]\big| \geq t\Big)$$
$$\leq 8\left(1 - \big((2\,(2\,g(\epsilon)-1)^2)^{-1}\big)\right)^{-1} \exp\left(\frac{-t^{d/(d-1)}}{8(8\epsilon)^{d/(d-1)}(m+n)}\right). \tag{A39}$$

In other words, there exist constants $C'_{m,n}(\epsilon, h)$ depending on m, n, ϵ, and h such that

$$P\Big(\big|\mathfrak{R}_{m,n} - \mathbb{E}[\mathfrak{R}_{m,n}]\big| \geq t\Big) \leq C'_{m,n}(\epsilon, h)\exp\left(\frac{-(t/(2\epsilon))^{d/(d-1)}}{(m+n)\,\tilde{C}}\right), \tag{A40}$$

where $\tilde{C} = 8(4)^{d/(d-1)}$.

To verify the behavior of bound (A40) in terms of ϵ, observe (A35) first; it is not hard to see that this function is decreasing in ϵ. However, the function

$$\exp\left(\frac{-(t/(2\epsilon))^{d/(d-1)}}{(m+n)\tilde{C}}\right)$$

increases in ϵ. Therefore, one can not immediately infer that the bound in (12) is monotonic with respect to ϵ. For fixed $N = n + m$, d, and h, the first and second derivatives of the bound (12) with respect to ϵ are quite complicated functions. Thus, deriving an explicit optimal solution for the minimization problem with the objective function (12) is not feasible. However, in the sequel, we discuss that under conditions when t is not much larger than $N = m + n$, this bound becomes convex with respect to ϵ. Set

$$K(\epsilon) = C'_{m,n}(\epsilon, h)\,\exp\left(\frac{-B(t)}{\epsilon^{d/(d-1)}}\right), \tag{A41}$$

where $C'_{m,n}$ is given in (10) and

$$B(t) = \frac{t^{d/(d-1)}}{8\,(8)^{d/(d-1)}(N)}.$$

By taking the derivative with respect to ϵ, we have

$$\frac{dK(\epsilon)}{d\epsilon} = K(\epsilon)\left(\frac{d}{d\epsilon}\left(\log C'_{m,n}\right) + \frac{B(t)\,d/(d-1)}{\epsilon^{(2d-1)/(d-1)}}\right), \tag{A42}$$

where

$$\frac{d}{d\epsilon}\left(\log C'_{m,n}\right) = \frac{-4\,a_h\,\epsilon}{(\epsilon - 2a_h)(8a_h^2 - 8\epsilon a_h + \epsilon^2)}, \tag{A43}$$

where $a_h = h\delta^h_{m,n}$. The second derivative $K(\epsilon)$ with respect to ϵ after simplification is given as

$$\frac{d^2}{d\epsilon^2}K(\epsilon) = \left(\frac{-4\,a_h\,\epsilon}{(\epsilon - 2a_h)(8a_h^2 - 8\epsilon a_h + \epsilon^2)} + \frac{B(t)\,\bar{d}}{\epsilon^{\bar{d}+1}}\right)^2$$

$$+ K(\epsilon)\left(\frac{8a_h\,(8a_h^3 + \epsilon^2(\epsilon - 5a_h))}{(8a_h^2 - 8a_h\epsilon + \epsilon^2)^2(\epsilon - 2a_h)^2} - \frac{B(t)\bar{d}(\bar{d}+1)}{\epsilon^{\bar{d}+2}}\right), \tag{A44}$$

where $\bar{d} = d/(d-1)$. The first term in (A44) and $K(\epsilon)$ are non-negative, so $K(\epsilon)$ is convex if the second term in the second line of (A44) is non-negative. We know that $\epsilon \geq h^2\delta^h_{m,n} = h\,a_h$, when $h = 7$, we can parameterize ϵ by setting it equal to γa_h, where $\gamma \geq 7$. After simplification, $K(\epsilon)$ is convex if

$$a_h^{\bar{d}-1}\left(\gamma^{\bar{d}-1} + 3\gamma^{\bar{d}-2}\right) + B(t)\bar{d}(\bar{d}+1)$$
$$\times\left\{a_h^{-1}\left(-32\gamma^{-6} + 64\gamma^{-5} - 48\gamma^{-4} + 8\gamma^{-3} - \frac{7}{2}\gamma^{-2} + 2\gamma^{-1} - \frac{1}{8}\right)\right.$$
$$\left. + a_h^{-2}\left(32\gamma^{-6} - 64\gamma^{-5} + 40\gamma^{-4} + 8\gamma^{-3} + \frac{1}{2}\gamma^{-2}\right)\right\} \geq 0. \tag{A45}$$

This is implied if

$$0 \leq B(t)\bar{d}(\bar{d}+1)\,a_h^{-1}$$
$$\times\left(-32\gamma^{-6} + 64\gamma^{-5} - 48\gamma^{-4} + 8\gamma^{-3} - \frac{7}{2}\gamma^{-2} + 2\gamma^{-1} - \frac{1}{8}\right), \tag{A46}$$

such that $\gamma \geq 7$. One can easily check that, as $\gamma \to \infty$, then (A46) tends to $-\frac{1}{8}B(t)\bar{d}(\bar{d}+1)\,a_h^{-1}$. This term can be negligible unless we have t that is much larger than $N = m + n$ with the threshold depending on d. Here, by setting $B(t)/a_h = 1$, a rough threshold $t = O(7^{d-1}(m+n)^{1-1/d^2})$ depending on d, $m + n$ is proposed. Therefore, minimizing (A35) and (A40) with respect to ϵ when optimal $h = 7$ is a convex optimization problem. Denote ϵ^* the solution of the convex optimization problem (9). By plugging optimal h ($h = 7$) and ϵ ($\epsilon = \epsilon^*$) in (A35) and (A40), we derive (11) and (12), respectively.

In this appendix, we also analyze the bound numerically. By simulation, we observed that lower h i.e., $h = 7$ is the optimal value experimentally. Indeed, this can be verified by Theorem 11's proof. We address the reader to Lemma A8 in Appendix D and Appendix E where, as h increases, the lower bound for the probability increases too. In other words, for fixed $N = m + n$ and d, the lowest h implies the maximum bound in (A92). For this, we set $h = 7$ in our experiments. We vary the dimension d and sample size $N = m + n$ in relatively large and small ranges. In Table A1, we solve (9) for various values of d and $N = m + n$. We also compute the lower bound for ϵ i.e., $7^{d+1}N^{1/d}$ per experiment. In Table A1, we observe that as we have higher dimension the optimal value ϵ^* equals the ϵ lower bound $h^{d+1}N^{1/d}$, but this is not true for smaller dimensions with even relatively large sample size.

Table A1. d, N, ϵ^* are dimension, total sample size $m + n$, and optimal ϵ for the bound in (12). The column $h^{d+1}N^{1/d}$ represents approximately the lower bound for ϵ which is our constraint in the minimization problem and our assumption in Theorems 5 and 6. Here, we set $h = 7$.

		Concentration Bound (11)			
d	$N = m + n$	ϵ^*	t_0	$h^{d+1}N^{1/d}$	**Optimal (11)**
2	10^3	1.1424×10^4	2×10^7	1.0847×10^4	0.3439
4	10^4	1.7746×10^5	3×10^{10}	168,070	0.0895
5	550	4.7236×10^5	10^{10}	4.1559×10^5	0.9929
6	10^4	3.8727×10^6	2×10^{12}	3.8225×10^6	0.1637
8	1200	9.7899×10^7	12×10^{12}	9.7899×10^7	0.7176
10	3500	4.4718×10^9	2×10^{15}	4.4718×10^9	0.4795
15	10^8	1.1348×10^{14}	10^{24}	1.1348×10^{14}	0.9042

To validate our proposed bound in (12), we again set $h = 7$ and for $d = 4, 5, 7$ we ran experiments with sample sizes $N = m + n = 9000, 1100, 140$, respectively. Then, we solved the minimization problem to derive optimal bound for t in the range $10^{10}[1, 3]$. Note that we chose this range to have a non-trivial bound for all three curves; otherwise, the bounds partly become one. Figure A1 shows that when t increases in the given range, the optimal curves approach zero.

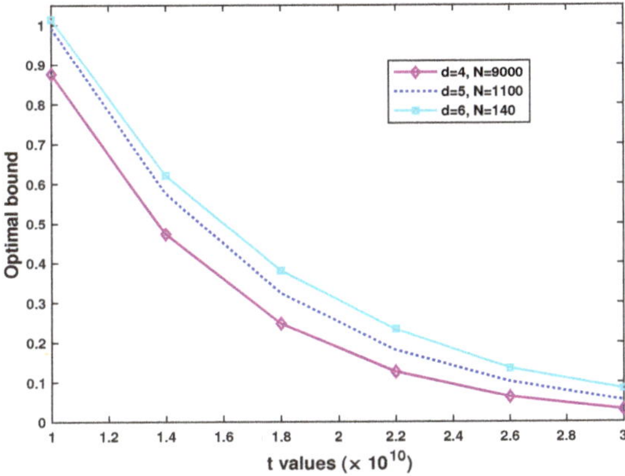

Figure A1. Optimal bound for (12), when $h = 7$ versus $t \in 10^{10}[1, 3]$. The bound decreases as t grows.

To prove the Theorem 7 in the concentration of $\Re_{m,n}$, Theorem 6, let

$$\delta = C'_{m,n}(\epsilon^*) \exp\left(\frac{-(t/(2\epsilon^*))^{d/(d-1)}}{(m+n)\,\tilde{C}} \right),$$

this implies $t = O\left(\epsilon^* (m+n)^{(d-1)/d} \left(\log\left(C'_{m,n}(\epsilon^*)/\delta \right) \right)^{(d-1)/d} \right)$ and the proofs are completed.

Appendix E. Additional Proofs

Lemma A3: Let $g(\mathbf{x})$ be a density function with support $[0, 1]^d$ and belong to the Hölder class $\Sigma_d(\eta, L)$, $0 < \eta \leq 1$, expressed in Definition 1. In addition, assume that $P(\mathbf{x})$ is a η-Hölder smooth function, such that its absolute value is bounded from above by some constants c. Define the quantized density function with parameter l and constants ϕ_i as

$$\widehat{g}(\mathbf{x}) = \sum_{i=1}^{M} \phi_i \mathbf{1}\{\mathbf{x} \in Q_i\}, \quad \text{where } \phi_i = l^d \int_{Q_i} g(\mathbf{x}) \, d\mathbf{x}, \tag{A47}$$

and $M = l^d$ and $Q_i = \{\mathbf{x}, \mathbf{x}_i : \|\mathbf{x} - \mathbf{x}_i\| < l^{-d}\}$. Then,

$$\int \left\| (g(\mathbf{x}) - \widehat{g}(\mathbf{x})) P(\mathbf{x}) \right\| \, d\mathbf{x} \leq O(l^{-d\eta}). \tag{A48}$$

Proof. By the mean value theorem, there exist points $\epsilon_i \in Q_i$ such that

$$\phi_i = l^d \int_{Q_i} g(\mathbf{x}) \, d\mathbf{x} = g(\epsilon_i).$$

Using the fact that $g \in \Sigma_d(\eta, L)$ and $P(\mathbf{x})$ is a bounded function, we have

$$\begin{aligned}
\int \left\| (g(\mathbf{x}) - \widehat{g}(\mathbf{x})) P(\mathbf{x}) \right\| \, d\mathbf{x} &= \sum_{i=1}^{M} \int_{Q_i} \left\| (g(\mathbf{x}) - \Phi_i) P(\mathbf{x}) \right\| d\mathbf{x} \\
&= \sum_{i=1}^{M} \int_{Q_i} \left\| (g(\mathbf{x}) - g(\epsilon_i)) P(\mathbf{x}) \right\| d\mathbf{x} \\
&\leq c\, L \sum_{i=1}^{M} \int_{Q_i} \left\| \mathbf{x} - \epsilon_i \right\|^{\eta} \, d\mathbf{x}.
\end{aligned}$$

Here, L is the Hölder constant. As $\mathbf{x}, \epsilon_i \in Q_i$, a sub-cube with edge length l^{-1}, then $\left\| \mathbf{x} - \epsilon_i \right\|^{\eta} = O(l^{-d\eta})$ and $\sum_{i=1}^{M} \int_{Q_i} d\mathbf{x} = 1$. This concludes the proof. $\square$

Lemma A4: Let $\Delta(\mathbf{x}, \mathcal{S})$ denote the degree of vertex $\mathbf{x} \in \mathcal{S}$ in the *MST* over set $\mathcal{S} \subset \mathbb{R}^d$ with the n number of vertices. For given function $P(\mathbf{x}, \mathbf{x})$, one yields

$$\int P(\mathbf{x}, \mathbf{x}) g(\mathbf{x}) \mathbb{E}[\Delta(\mathbf{x}, \mathcal{S})] \, d\mathbf{x} = 2 \int P(\mathbf{x}, \mathbf{x}) g(\mathbf{x}) \, d\mathbf{x} + \varsigma_\eta(l, n), \tag{A49}$$

where for constant $\eta > 0$,

$$\varsigma_\eta(l, n) = \left(O(l/n) - 2\, l^d/n \right) \int g(\mathbf{x}) P(\mathbf{x}, \mathbf{x}) \, d\mathbf{x} + O(l^{-d\eta}). \tag{A50}$$

Proof. Recall notations in Lemma A3 and

$$\left| \int g(\mathbf{x}) P(\mathbf{x}) \, d\mathbf{x} - \int \widehat{g}(\mathbf{x}) P(\mathbf{x}) \, d\mathbf{x} \right| \leq \int \left| (g(\mathbf{x}) - \widehat{g}(\mathbf{x})) P(\mathbf{x}) \right| \, d\mathbf{x}.$$

Therefore, by substituting $\widehat{g}$, defined in (A47), into g with considering its error, we have

$$\begin{aligned}
\int P(\mathbf{x}, \mathbf{x}) g(\mathbf{x}) &\mathbb{E}[\Delta(\mathbf{x}, \mathcal{S})] \, d\mathbf{x} \\
&= \int P(\mathbf{x}, \mathbf{x}) \mathbb{E}[\Delta(\mathbf{x}, \mathcal{S})] \sum_{i=1}^{M} \phi_i \mathbf{1}\{\mathbf{x} \in Q_i\} \, d\mathbf{x} + O(l^{-d\eta}) \\
&= \sum_{i=1}^{M} \phi_i \int_{Q_i} P(\mathbf{x}, \mathbf{x}) \mathbb{E}[\Delta(\mathbf{x}, \mathcal{S})] \, d\mathbf{x} + O(l^{-d\eta}).
\end{aligned} \tag{A51}$$

Here, Q_i represents as before in Lemma A3, so the RHS of (A51) becomes

$$\sum_{i=1}^{M} \phi_i \int_{Q_i} P(\mathbf{x},\mathbf{x})\mathbb{E}[\Delta(\mathbf{x},\mathcal{S}\cap Q_i)]\,d\mathbf{x} + \sum_{i=1}^{M} \phi_i \int_{Q_i} P(\mathbf{x},\mathbf{x})O(l^{1-d}/n) + O(l^{-d\eta})$$
$$= \sum_{i=1}^{M} \phi_i P(\mathbf{x}_i,\mathbf{x}_i)\frac{1}{M}\int_{Q_i} M\,\mathbb{E}[\Delta(\mathbf{x},\mathcal{S}\cap Q_i)]\,d\mathbf{x} + \sum_{i=1}^{M} \phi_i \int_{Q_i} P(\mathbf{x},\mathbf{x})O(l^{1-d}/n) + 2\,O(l^{-d\eta}). \tag{A52}$$

Now, note that $\int_{Q_i} M\,\mathbb{E}[\Delta(\mathbf{x},\mathcal{S}\cap Q_i)]\,d\mathbf{x}$ is the expectation of $\mathbb{E}[\Delta(\mathbf{x},S\cap Q_i)]$ over the nodes in Q_i, which is equal to $2 - \dfrac{2}{k_i}$, where $k_i = \dfrac{n}{M}$. Consequently, we have

$$\int P(\mathbf{x},\mathbf{x})g(\mathbf{x})\mathbb{E}[\Delta(\mathbf{x},\mathcal{S})]\,d\mathbf{x} = \left(2 - \frac{2\,M}{n}\right)\sum_{i=1}^{M} \phi_i\,P(\mathbf{x}_i,\mathbf{x}_i)\frac{1}{M} + O\left(\frac{l^{1-d}}{n}\right)\sum_{i=1}^{M} \phi_i\,P(\mathbf{x}_i,\mathbf{x}_i) + 3\,O(l^{-d\eta})$$
$$= 2\int g(\mathbf{x})P(\mathbf{x},\mathbf{x})\,d\mathbf{x} + 5\,O(l^{-d\eta}) + M\left(O\left(\frac{l^{1-d}}{n}\right) - \left(\frac{2}{n}\right)\right)\int g(\mathbf{x})P(\mathbf{x},\mathbf{x})\,d\mathbf{x}. \tag{A53}$$

This gives the assertion (A49). $\square$

Lemma A5: Assume that, for given k, $g_k(\mathbf{x})$ is a bounded function belong to $\Sigma_d(\eta, L)$. Let $P : \mathbb{R}^d \times \mathbb{R}^d \mapsto [0,1]$ be a symmetric, smooth, jointly measurable function, such that, given k, for almost every $\mathbf{x} \in \mathbb{R}^d$, $P(\mathbf{x},.)$ is measurable with $\mathbf{x}$ a Lebesgue point of the function $g_k(.)P(\mathbf{x},.)$. Assume that the first derivative P is bounded. For each k, let $\mathbf{Z}_1^k, \mathbf{Z}_2^k, \ldots, \mathbf{Z}_k^k$ be independent d-dimensional variable with common density function g_k. Set $\mathfrak{Z}_k = \{\mathbf{Z}_1^k, \mathbf{Z}_2^k \ldots, \mathbf{Z}_k^k\}$ and $\mathfrak{Z}_k^{\mathbf{x}} = \{\mathbf{x}, \mathbf{Z}_2^k, \mathbf{Z}_3^k \ldots, \mathbf{Z}_k^k\}$. Then,

$$\mathbb{E}\left[\sum_{j=2}^{k} P(\mathbf{x},\mathbf{Z}_j^k)\mathbf{1}\{(\mathbf{x},\mathbf{Z}_j^k) \in MST(\mathfrak{Z}_k^{\mathbf{x}})\}\right]$$
$$= P(\mathbf{x},\mathbf{x})\,\mathbb{E}\left[\Delta(\mathbf{x},\mathfrak{Z}_k^{\mathbf{x}})\right] + \left\{O(k^{-\eta/d}) + O(k^{-1/d})\right\}. \tag{A54}$$

Proof. Let $\mathbb{B}(\mathbf{x},r) = \{\mathbf{y} : \|\mathbf{y}-\mathbf{x}\|_d \leq r\}$. For any positive K, we can obtain:

$$\mathbb{E}\sum_{j=2}^{k}\left|P(\mathbf{x},\mathbf{Z}_j^k) - P(\mathbf{x},\mathbf{x})\right|\mathbf{1}\{\mathbf{Z}_j^k \in \mathbb{B}(\mathbf{x},Kk^{-1/d})\}$$
$$= (k-1)\int_{\mathbb{B}(\mathbf{x};Kk^{-1/d})}\left|(P(\mathbf{x},\mathbf{y})g_k(\mathbf{y}) - P(\mathbf{x},\mathbf{x})g_k(\mathbf{x})) + P(\mathbf{x},\mathbf{x})(g_k(\mathbf{x}) - g_k(\mathbf{y}))\right|\,d\mathbf{y}$$
$$\leq (k-1)\left[\int_{\mathbb{B}(\mathbf{x};Kk^{-1/d})}\left|(P(\mathbf{x},\mathbf{y})g_k(\mathbf{y}) - P(\mathbf{x},\mathbf{x})g_k(\mathbf{x}))\right|d\mathbf{y} + O(k^{-\eta/d})\mathbf{V}(\mathbb{B}(\mathbf{x},Kk^{-1/d}))\right], \tag{A55}$$

where $\mathbf{V}$ is the volume of space $\mathbb{B}$ which equals $O(k^{-1})$. Note that the above inequality appears because $g_k(\mathbf{x}) \in \Sigma_d(\eta, L)$ and $P(\mathbf{x},\mathbf{x}) \in [0,1]$. The first order Taylor series expansion of $P(\mathbf{x},\mathbf{y})$ around $\mathbf{x}$ is

$$P(\mathbf{x},\mathbf{y}) = P(\mathbf{x},\mathbf{x}) + P^{(1)}(\mathbf{x},\mathbf{x})\|\mathbf{y}-\mathbf{x}\| + o(\|\mathbf{y}-\mathbf{x}\|^2)$$
$$= P(\mathbf{x},\mathbf{x}) + O(k^{-1/d}) + o(k^{-2/d}).$$

Then, by recalling the Hölder class, we have

$$\left|P(\mathbf{x},\mathbf{y})g_k(\mathbf{y}) - P(\mathbf{x},\mathbf{x})g_k(\mathbf{x})\right| = \left|(P(\mathbf{x},\mathbf{x}) + O(k^{-1/d}))(g_k(\mathbf{x}) + O(k^{-\eta/d})) - P(\mathbf{x},\mathbf{x})g_k(\mathbf{x})\right|$$
$$= O(k^{-\eta/d}) + O(k^{-1/d}).$$

Hence, the RHS of (A55) becomes

$$(k-1)\Big[\big(O(k^{-\eta/d})+O(k^{-1/d})\big)\mathbf{V}\big(\mathbb{B}(\mathbf{x},Kk^{-1/d})\big)+O(k^{-\eta/d})\mathbf{V}\big(\mathbb{B}(\mathbf{x},Kk^{-1/d})\big)\Big]$$
$$=(k-1)\Big[O(k^{-1-\eta/d})+O(k^{-1-1/d})\Big].$$

The expression in (A54) can be obtained by choice of K. $\quad\square$

Lemma A6: Consider the notations and assumptions in Lemma A5. Then,

$$\Big|k^{-1}\sum_{1\leq i<j\leq k}P(\mathbf{Z}_i^k,\mathbf{Z}_j^k)\mathbf{1}\{(\mathbf{Z}_i^k,\mathbf{Z}_j^k)\in MST(\mathfrak{Z}_k)-\int_{\mathbb{R}^d}P(\mathbf{x},\mathbf{x})g_k(\mathbf{x})\,d\mathbf{x}\Big|$$
$$\leq \varsigma_\eta(l,k)+O(k^{-\eta/d})+O(k^{-1/d}). \tag{A56}$$

Here, $MST(\mathcal{S})$ denotes the MST graph over nice and finite set $\mathcal{S}\subset\mathbb{R}^d$ and η is the smoothness Hölder parameter. Note that $\varsigma_\eta(l,k)$ is given as before in (A50).

Proof. Following notations in [49], let $\Delta(\mathbf{x},\mathcal{S})$ denote the degree of vertex $\mathbf{x}$ in the $MST(\mathcal{S})$ graph. Moreover, let $\mathbf{x}$ be a Lebesgue point of g_k with $g_k(\mathbf{x})>0$. In addition, let $\mathfrak{Z}_k^{\mathbf{x}}$ be the point process $\{\mathbf{x},\mathbf{Z}_2^k,\mathbf{Z}_3^k,\dots,\mathbf{Z}_k^k\}$. Now, by virtue of (A55) in Lemma A5, we can write

$$\mathbb{E}\Big[\sum_{j=2}^{k}P(\mathbf{x},\mathbf{Z}_j^k)\mathbf{1}\{(\mathbf{x},\mathbf{Z}_j^k)\in MST(\mathfrak{Z}_k^{\mathbf{x}})\}\Big]=P(\mathbf{x},\mathbf{x})\,\mathbb{E}\big[\Delta(\mathbf{x},\mathfrak{Z}_k^{\mathbf{x}})\big]+\Big\{O(k^{-\eta/d})+O(k^{-1/d})\Big\}. \tag{A57}$$

On the other hand, it can be seen that

$$k^{-1}\mathbb{E}\Big[\sum_{1\leq i<j\leq k}P(\mathbf{Z}_i^k,\mathbf{Z}_j^k)\mathbf{1}\{(\mathbf{Z}_i^k,\mathbf{Z}_j^k)\in MST(\mathfrak{Z}_k)\}\Big]$$
$$=\frac{1}{2}\mathbb{E}\Big[\sum_{j=2}^{k}P(\mathbf{Z}_1^k,\mathbf{Z}_j^k)\mathbf{1}\{(\mathbf{Z}_i^k,\mathbf{Z}_j^k)\in MST(\mathfrak{Z}_k)\}\Big] \tag{A58}$$
$$=\frac{1}{2}\int g_k(\mathbf{x})\,d\mathbf{x}\,\mathbb{E}\Big[\sum_{j=2}^{k}P(\mathbf{x},\mathbf{Z}_j^k)\mathbf{1}\{(\mathbf{x},\mathbf{Z}_j^k)\in MST(\mathfrak{Z}_k)\}\Big].$$

Recalling (A57),

$$=\frac{1}{2}\int g_k(\mathbf{x})P(\mathbf{x},\mathbf{x})\mathbb{E}\big[\Delta(\mathbf{x},\mathfrak{Z}_k^{\mathbf{x}})\big]\,d\mathbf{x}+O(k^{-\eta/d})+O(k^{-1/d}). \tag{A59}$$

By virtue of Lemma A4, (A49) can be substituted into expression (A59) to obtain (A56). $\quad\square$

Theorem A1: Assume $\mathfrak{R}_{m,n}:=\mathfrak{R}(\mathfrak{X}_m,\mathfrak{Y}_n)$ denotes the FR test statistic as before. Then, the rate for the bias of the $\mathfrak{R}_{m,n}$ estimator for $0<\eta\leq 1, d\geq 2$ is of the form:

$$\Big|\frac{\mathbb{E}[\mathfrak{R}_{m,n}]}{m+n}-2pq\int\frac{f_0(\mathbf{x})f_1(\mathbf{x})}{pf_0(\mathbf{x})+qf_1(\mathbf{x})}\,d\mathbf{x}\Big|\leq O(l^d(m+n)^{-\eta/d})+O(l^{-d\eta}). \tag{A60}$$

Here, η is the Holder smoothness parameter. A more explicit form for the bound on the RHS is given in (A61) below:

$$\left| \frac{\mathbb{E}\left[\mathfrak{R}'_{m,n}(\mathfrak{X}_m, \mathfrak{Y}_n)\right]}{m+n} - \int \frac{2pq f_0(\mathbf{x}) f_1(\mathbf{x})}{p f_0(\mathbf{x}) + q f_1(\mathbf{x})} \, d\mathbf{x} \right| \leq O\left(l^d (m+n)^{-\eta/d}\right)$$

$$+ O\left(l^d (m+n)^{-1/2}\right) + 2\, c_1\, l^{d-1}(m+n)^{(1/d)-1} + c_d\, 2^d\, (m+n)^{-1}$$

$$- 2\, l^d (m+n)^{-1} \int \frac{2pq f_0(\mathbf{x}) f_1(\mathbf{x})}{p\, f_0(\mathbf{x}) + q\, f_1(\mathbf{x})} \, d\mathbf{x} + c_2\, (m+n)^{-1} l^d$$

$$+ O(l)(m+n)^{-1} \sum_{i=1}^{M} l^d(a_i)^{-1} \int \frac{2 f_0(\mathbf{x}) f_1(\mathbf{x})}{p\, f_0(\mathbf{x}) + q\, f_1(\mathbf{x})} \, d\mathbf{x} + O(l^{-d\eta}) \tag{A61}$$

$$+ O(l) \sum_{i=1}^{M} l^{d/2} \frac{\sqrt{b_i}}{a_i^2} \int \frac{2 f_0(\mathbf{x}) f_1(\mathbf{x}) \left(f_0(\mathbf{x})\sqrt{m} + f_1(\mathbf{x})\sqrt{n}\right)}{\left(m f_0(\mathbf{x}) + n f_1(\mathbf{x})\right)^2} \, d\mathbf{x}$$

$$+ \sum_{i=1}^{M} 2\, l^{-d/2} \frac{\sqrt{b_i}}{a_i^2} \int \frac{f_0(\mathbf{x}) f_1(\mathbf{x}) \left(\alpha_i \beta_i \left(m a_i f_0^2(\mathbf{x}) + n b_i f_1^2(\mathbf{x})\right)\right)^{1/2}}{\left(m f_0(\mathbf{x}) + n f_1(\mathbf{x})\right)^2 (m+n)} \, d\mathbf{x}.$$

Proof. Assume M_m and N_n be Poisson variables with mean m and n, respectively, one independent of another and of $\{X_i\}$ and $\{Y_j\}$. Let also $\mathfrak{X}'_m$ and $\mathfrak{Y}'_n$ be the Poisson processes $\{X_1, \ldots, X_{M_m}\}$ and $\{Y_1, \ldots, Y_{N_n}\}$. Set $\mathfrak{R}'_{m,n} := \mathfrak{R}_{m,n}(\mathfrak{X}'_m, \mathfrak{Y}'_n)$. Applying Lemma 1, and (12) cf. [49], we can write

$$\left| \mathfrak{R}'_{m,n} - \mathfrak{R}_{m,n} \right| \leq K_d \left(|M_m - m| + |N_n - n| \right). \tag{A62}$$

Here, K_d denotes the largest possible degree of any vertex of the MST graph in $\mathbb{R}^d$. Moreover, by the matter of Poisson variable fact and using Stirling approximation [51], we have

$$\mathbb{E}\left[|M_m - m| \right] = e^{-m} \frac{m^{m+1}}{m!} \leq e^{-m} \frac{m^{m+1}}{\sqrt{2\pi} m^{m+1/2} e^{-m}} = O\left(m^{1/2} \right). \tag{A63}$$

Similarly, $\mathbb{E}\left[|N_n - n| \right] = O(n^{1/2})$. Therefore, by (A62), one yields

$$\mathbb{E}[\mathfrak{R}_{m,n}] = \mathbb{E}[\mathfrak{R}_{m,n} - \mathfrak{R}'_{m,n}] + \mathbb{E}[\mathfrak{R}'_{m,n}] = O\left((m+n)^{1/2}\right) + \mathbb{E}[\mathfrak{R}'_{m,n}]. \tag{A64}$$

Therefore,

$$\frac{\mathbb{E}[\mathfrak{R}_{m,n}]}{m+n} = \frac{\mathbb{E}[\mathfrak{R}'_{m,n}]}{m+n} + O\left((m+n)^{-1/2}\right). \tag{A65}$$

Hence, it will suffice to obtain the rate of convergence of $\mathbb{E}\left[\mathfrak{R}'_{m,n}\right]/(m+n)$ in the RHS of (A65). For this, let m_i, n_i denote the number of Poisson process samples $\mathfrak{X}'_m$ and $\mathfrak{Y}'_n$ with the FR statistic $\mathfrak{R}'_{m,n}$, falling into partitions Q'_i with FR statistic $\mathfrak{R}'_{m_i,n_i}$. Then, by virtue of Lemma 4, we can write

$$\mathbb{E}\left[\mathfrak{R}'_{m,n}\right] \leq \sum_{i=1}^{M} \mathbb{E}\left[\mathfrak{R}'_{m_i,n_i}\right] + 2\, c_1\, l^{d-1}(m+n)^{1/d}.$$

Note that the Binomial RVs m_i, n_i are independent with marginal distributions $m_i \sim B(m, a_i l^{-d})$, $n_i \sim B(n, b_i l^{-d})$, where a_i, b_i are non-negative constants satisfying, $\forall i$, $a_i \leq b_i$ and $\sum_{i=1}^{l^d} a_i l^{-d} = \sum_{i=1}^{l^d} b_i l^{-d} = 1$. Therefore,

$$\mathbb{E}\left[\mathfrak{R}'_{m,n}\right] \leq \sum_{i=1}^{M} \mathbb{E}\left[\mathbb{E}\left[\mathfrak{R}'_{m_i,n_i} | m_i, n_i\right]\right] + 2\,c_1\,l^{d-1}(m+n)^{1/d}. \tag{A66}$$

Let us first compute the internal expectation given m_i, n_i. For this reason, given m_i, n_i, let $Z_1^{m_i,n_i}, Z_2^{m_i,n_i}, \dots$ be independent variables with common densities $g_{m_i,n_i}(\mathbf{x}) = \left(m_i f_0(\mathbf{x}) + n_i f_1(\mathbf{x})\right)/(m_i + n_i)$, $\mathbf{x} \in \mathbb{R}^d$. Moreover, let L_{m_i,n_i} be an independent Poisson variable with mean $m_i + n_i$. Denote $\mathfrak{F}'_{m_i,n_i} = \{Z_1^{m_i,n_i}, \dots, Z_{L_{m_i,n_i}}^{m_i,n_i}\}$ a non-homogeneous Poisson of rate $m_i f_0 + n_i f_1$. Let $\mathfrak{F}_{m_i,n_i}$ be the non-Poisson point process $\{Z_1^{m_i,n_i}, \dots Z_{m_i+n_i}^{m_i,n_i}\}$. Assign a mark from the set $\{1, 2\}$ to each points of $\mathfrak{F}'_{m_i,n_i}$. Let $\widetilde{\mathfrak{X}}'_{m_i}$ be the sets of points marked 1 with each probability $m_i f_0(\mathbf{x})/(m_i f_0(\mathbf{x}) + n_i f_i(\mathbf{x}))$ and let $\widetilde{\mathfrak{Y}}'_{n_i}$ be the set points with mark 2. Note that owing to the marking theorem [52], $\widetilde{\mathfrak{X}}'_{m_i}$ and $\widetilde{\mathfrak{Y}}'_{n_i}$ are independent Poisson processes with the same distribution as $\mathfrak{X}'_{m_i}$ and $\mathfrak{Y}'_{n_i}$, respectively. Considering $\widetilde{R}'_{m_i,n_i}$ as FR statistic over nodes in $\widetilde{\mathfrak{X}}'_{m_i} \cup \widetilde{\mathfrak{Y}}'_{n_i}$ we have

$$\mathbb{E}\left[\mathfrak{R}'_{m_i,n_i} | m_i, n_i\right] = \mathbb{E}\left[\widetilde{\mathfrak{R}}'_{m_i,n_i} | m_i, n_i\right].$$

Again using Lemma 1 and analogous arguments in [49] along with the fact that $\mathbb{E}\left[|M_m + N_n - m - n|\right] = O((m+n)^{1/2})$, we have

$$\mathbb{E}\left[\widetilde{\mathfrak{R}}'_{m_i,n_i} | m_i, n_i\right] = \mathbb{E}\left[\mathbb{E}\left[\widetilde{\mathfrak{R}}'_{m_i,n_i} | \mathfrak{F}'_{m_i,n_i}\right]\right]$$

$$= \mathbb{E}\left[\sum_{s<j<m_i+n_i} P_{m_i,n_i}(Z_s^{m_i,n_i}, Z_j^{m_i,n_i})\mathbf{1}\left\{(Z_s^{m_i,n_i}, Z_j^{m_i,n_i}) \in \mathfrak{F}_{m_i,n_i}\right\}\right] + O((m_i + n_i)^{1/2})).$$

Here,

$$P_{m_i,n_i}(\mathbf{x}, \mathbf{y}) := P_r\{\text{mark } x \neq \text{mark } y, (\mathbf{x}, \mathbf{y}) \in \mathfrak{F}'_{m_i,n_i}\}$$

$$= \frac{m_i f_0(\mathbf{x}) n_i f_1(\mathbf{y}) + n_i f_1(\mathbf{x}) m_i f_0(\mathbf{y})}{(m_i f_0(\mathbf{x}) + n_i f_1(\mathbf{x}))(m_i f_0(\mathbf{y}) + n_i f_1(\mathbf{y}))}.$$

By owing to Lemma A6, we obtain

$$\sum_{i=1}^{M} \mathbb{E}_{m_i,n_i} \mathbb{E}\left[\sum_{s<j<m_i+n_i} P_{m_i,n_i}(Z_s^{m_i,n_i}, Z_j^{m_i,n_i})\mathbf{1}\left\{(Z_s^{m_i,n_i}, Z_j^{m_i,n_i}) \in \mathfrak{F}_{m_i,n_i}\right\}\right] + \sum_{i=1}^{M} \mathbb{E}_{m_i,n_i}\left[O((m_i + n_i))^{1/2}\right]$$

$$= \sum_{i=1}^{M} \mathbb{E}_{m_i,n_i}\left[(m_i + n_i) \int g_{m_i,n_i}(\mathbf{x}, \mathbf{x}) P_{m_i,n_i}(\mathbf{x}, \mathbf{x})\, d\mathbf{x} + (\varsigma_\eta(l, m_i, n_i) + O((m_i + n_i)^{-\eta/d}) \right. \tag{A67}$$

$$\left. + O((m_i + n_i)^{-1/d}))(m_i + n_i)\right] + \sum_{i=1}^{M} \mathbb{E}_{m_i,n_i}\left[O((m_i + n_i)^{1/2})\right],$$

where

$$\varsigma_\eta(l, m_i, n_i) = \left(O(l/(m_i + n_i)) - 2\,l^d/(m_i + n_i)\right) \int g_{m_i,n_i}(\mathbf{x}) P_{m_i,n_i}(\mathbf{x}, \mathbf{x})\, d\mathbf{x} + O(l^{-d\eta}).$$

The expression in (A67) equals

$$\sum_{i=1}^{M} \int \mathbb{E}_{m_i,n_i}\left[\frac{2m_i n_i f_0(\mathbf{x}) f_1(\mathbf{x})}{m_i f_0(\mathbf{x}) + n_i f_1(\mathbf{x})}\right] d\mathbf{x} + \sum_{i=1}^{M} \mathbb{E}_{m_i,n_i}\left[(m_i + n_i)\, \varsigma_\eta(l, m_i, n_i)\right] \tag{A68}$$

$$+ O\left(l^d (m+n)^{1-\eta/d}\right) + O\left(l^d (m+n)^{1/2}\right).$$

Because of Jensen inequality for concave function:

$$\sum_{i=1}^{M} \mathbb{E}_{m_i,n_i}\left[O\left((m_i + n_i)^{1/2}\right)\right] = \sum_{i=1}^{M} O\left(\mathbb{E}[m_i] + \mathbb{E}[n_i]\right)^{1/2}$$

$$= \sum_{i=1}^{M} O(m a_i l^{-d} + n b_i l^{-d})^{1/2} = O\left(l^d (m+n)^{1/2}\right).$$

In addition, similarly since $\eta < d$, we have

$$\sum_{i=1}^{M} \mathbb{E}_{m_i,n_i}\left[O\left((m_i + n_i)^{1-\eta/d}\right)\right] = O\left(l^d (m+n)^{1-\eta/d}\right), \tag{A69}$$

and, for $d \geq 2$, one yields

$$\sum_{i=1}^{M} \mathbb{E}_{m_i,n_i}\left[O\left((m_i + n_i)^{1-1/d}\right)\right] = O\left(l^d (m+n)^{1-1/d}\right) = O\left(l^d (m+n)^{1/2}\right). \tag{A70}$$

Next, we state the following lemma (Lemma 1 from [30,31]), which will be used in the sequel:

Lemma A13. *Let $k(x)$ be a continuously differential function of $x \in \mathbb{R}$ which is convex and monotone decreasing over $x \geq 0$. Set $k'(x) = \dfrac{dk(x)}{dx}$. Then, for any $x_0 > 0$, we have*

$$k(x_0) + \frac{k(x_0)}{x_0}|x - x_0| \geq k(x) \geq k(x_0) - k'(x_0)|x - x_0|. \tag{A71}$$

Next, continuing the proof of (A60), we attend to find an upper bound for

$$\mathbb{E}_{m_i,n_i}\left[\frac{m_i n_i}{m_i f_0(\mathbf{x}) + n_i f_1(\mathbf{x})}\right]. \tag{A72}$$

In order to pursue this aim, in Lemma A13, consider $k(x) = \dfrac{1}{x}$ and $x_0 = \mathbb{E}_{m_i,n_i}\left[m_i f_0(\mathbf{x}) + n_i f_1(\mathbf{x})\right]$, therefore as the function $k(x)$ is decreasing and convex, one can write

$$\frac{1}{m_i f_0(\mathbf{x}) + n_i f_1(\mathbf{x})} \leq \frac{1}{\mathbb{E}_{m_i,n_i}\left[m_i f_0(\mathbf{x}) + n_i f_1(\mathbf{x})\right]} + \frac{\left|m_i f_0(\mathbf{x}) + n_i f_1(\mathbf{x}) - \mathbb{E}_{m_i,n_i}\left[m_i f_0(\mathbf{x}) + n_i f_1(\mathbf{x})\right]\right|}{\mathbb{E}^2_{m_i,n_i}\left[m_i f_0(\mathbf{x}) + n_i f_1(\mathbf{x})\right]}. \tag{A73}$$

Using the Hölder inequality implies the following inequality:

$$\mathbb{E}_{m_i,n_i}\left[\frac{m_i n_i}{m_i f_0(\mathbf{x}) + n_i f_1(\mathbf{x})}\right] \leq \frac{\mathbb{E}_{m_i,n_i}\left[m_i n_i\right]}{\mathbb{E}_{m_i,n_i}\left[m_i f_0(\mathbf{x}) + n_i f_1(\mathbf{x})\right]} \tag{A74}$$

$$+ \frac{\left(\mathbb{E}_{m_i,n_i}\left[m_i^2 n_i^2\right]\right)^{1/2}}{\mathbb{E}^2_{m_i,n_i}\left[m_i f_0(\mathbf{x}) + n_i f_1(\mathbf{x})\right]} \times \left(\mathbb{E}_{m_i,n_i}\left[m_i f_0(\mathbf{x}) + n_i f_1(\mathbf{x}) - \mathbb{E}_{m_i,n_i}\left[m_i f_0(\mathbf{x}) + n_i f_1(\mathbf{x})\right]\right]^2\right)^{1/2}.$$

As random variables m_i, n_i are independent, and because of $\mathbb{V}[m_i] \leq ma_i l^{-d}$, $\mathbb{V}[n_i] \leq nb_i l^{-d}$, we can claim that the RHS of (A74) becomes less than and equal to

$$\frac{mna_i b_i l^{-2d}}{ma_i l^{-d} f_0(\mathbf{x}) + nb_i l^{-d} f_1(\mathbf{x})} + \frac{\left(\alpha_i \beta_i \left(ma_i l^{-d} f_0^2(\mathbf{x}) + nb_i l^{-d} f_1^2(\mathbf{x})\right)\right)^{1/2}}{\left(ma_i f_0(\mathbf{x}) + nb_i f_1(\mathbf{x})\right)^2}, \tag{A75}$$

where

$$\alpha_i = ma_i l^d \left(1 - a_i l^{-d}\right) + m^2 a_i^2,$$

$$\beta_i = nb_i l^d \left(1 - b_i l^{-d}\right) + n^2 b_i^2.$$

Going back to (A66), we have

$$\mathbb{E}\left[\mathfrak{R}'_{m,n}(\mathfrak{X}_m, \mathfrak{Y}_n)\right] \leq \sum_{i=1}^{M} a_i b_i l^{-d} \int \frac{2\, mn f_0(\mathbf{x}) f_1(\mathbf{x})}{ma_i f_0(\mathbf{x}) + nb_i f_1(\mathbf{x})}\, d\mathbf{x}$$

$$+ \sum_{i=1}^{M} 2 \int \frac{f_0(\mathbf{x}) f_1(\mathbf{x}) \left(\alpha_i \beta_i \left(ma_i l^{-d} f_0^2(\mathbf{x}) + nb_i l^{-d} f_1^2(\mathbf{x})\right)\right)^{1/2}}{\left(ma_i f_0(\mathbf{x}) + nb_i f_1(\mathbf{x})\right)^2}\, d\mathbf{x} \tag{A76}$$

$$+ \sum_{i=1}^{M} \mathbb{E}_{m_i,n_i}\left[(m_i + n_i)\, \varsigma_\eta(l, m_i, n_i)\right] + O\left(l^d (m+n)^{1-\eta/d}\right)$$

$$+ O\left(l^d (m+n)^{1/2}\right) + 2c_1\, l^{d-1} (m+n)^{1/d}.$$

Finally, owing to $a_i \leq b_i$ and $\sum_{i=1}^{M} b_i l^{-d} = 1$, when $\dfrac{m}{m+n} \to p$, we have

$$\frac{\mathbb{E}\left[\mathfrak{R}'_{m,n}(\mathfrak{X}_m, \mathfrak{Y}_n)\right]}{m+n} \leq \int \frac{2\, pq f_0(\mathbf{x}) f_1(\mathbf{x})}{p f_0(\mathbf{x}) + q f_1(\mathbf{x})}\, d\mathbf{x}$$

$$+ \sum_{i=1}^{M} 2 \int \frac{f_0(\mathbf{x}) f_1(\mathbf{x}) \left(\alpha_i \beta_i \left(ma_i l^{-d} f_0^2(\mathbf{x}) + nb_i l^{-d} f_1^2(\mathbf{x})\right)\right)^{1/2}}{\left(ma_i f_0(\mathbf{x}) + nb_i f_1(\mathbf{x})\right)^2 (m+n)}\, d\mathbf{x} \tag{A77}$$

$$+ \frac{1}{m+n} \sum_{i=1}^{M} \mathbb{E}_{m_i,n_i}\left[(m_i + n_i)\, \varsigma_\eta(l, m_i, n_i)\right] + O\left(l^d (m+n)^{-\eta/d}\right)$$

$$+ O\left(l^d (m+n)^{-1/2}\right) + 2c_1\, l^{d-1} (m+n)^{(1/d)-1}.$$

Passing to Definition 2, MST*, and Lemma A2, a similar discussion as above, consider the Poisson processes samples and the FR statistic under the union of samples, denoted by $\mathfrak{R}'^*_{m,n}$, and superadditivity of dual $\mathfrak{R}^*_{m,n}$, we have

$$\mathbb{E}\left[\mathfrak{R}'^*_{m,n}(\mathfrak{X}_m, \mathfrak{Y}_n)\right] \geq \sum_{i=1}^{M} \mathbb{E}\left[\mathfrak{R}'^*_{m_i,n_i}\left((\mathfrak{X}_m, \mathfrak{Y}_n) \cap Q_i\right)\right] - c_2\, l^d$$

$$= \sum_{i=1}^{M} \mathbb{E}_{m_i,n_i}\left[\mathbb{E}\left[\mathfrak{R}'^*_{m_i,n_i}\left((\mathfrak{X}_m, \mathfrak{Y}_n) \cap Q_i\right) | m_i, n_i\right]\right] - c_2\, l^d \tag{A78}$$

$$\geq \sum_{i=1}^{M} \mathbb{E}_{m_i,n_i}\left[\mathbb{E}\left[\mathfrak{R}'_{m_i,n_i}\left((\mathfrak{X}_m, \mathfrak{Y}_n) \cap Q_i\right) | m_i, n_i\right]\right] - c_2\, l^d,$$

the last line is derived from Lemma A2, (ii), inequality (A8). Owing to the Lemma A6, (A69), and (A70), one obtains

$$
\mathbb{E}\left[\mathfrak{R}'^{*}_{m,n}(\mathfrak{X}_m,\mathfrak{Y}_n)\right] \geq \sum_{i=1}^{M} \int \mathbb{E}_{m_i,n_i}\left[\frac{2m_i n_i f_0(\mathbf{x}) f_1(\mathbf{x})}{m_i f_0(\mathbf{x}) + n_i f_1(\mathbf{x})}\right] d\mathbf{x}
$$
$$
- \sum_{i=1}^{M} \mathbb{E}_{m_i,n_i}\left[(m_i + n_i)\,\varsigma_\eta(l, m_i, n_i)\right] - O\!\left(l^d(m+n)^{1-\eta/d}\right) - O\!\left(l^d(m+n)^{1/2}\right) - c_2\, l^d.
\tag{A79}
$$

Furthermore, by using the Jenson's inequality, we get

$$
\mathbb{E}_{m_i,n_i}\left[\frac{m_i n_i}{m_i f_0(\mathbf{x}) + n_i f_1(\mathbf{x})}\right] \geq \frac{\mathbb{E}[m_i]\mathbb{E}[n_i]}{\mathbb{E}[m_i] f_0(\mathbf{x}) + \mathbb{E}[n_i] f_1(\mathbf{x})} = \frac{l^{-d}\left(m a_i n b_i\right)}{m a_i f_0(\mathbf{x}) + n b_i f_1(\mathbf{x})}.
$$

Therefore, since $a_i \leq b_i$, we can write

$$
\mathbb{E}_{m_i,n_i}\left[\frac{m_i n_i}{m_i f_0(\mathbf{x}) + n_i f_1(\mathbf{x})}\right] \geq \frac{l^{-d} mn\, a_i b_i}{b_i\left(m f_0(\mathbf{x}) + n f_1(\mathbf{x})\right)} = \frac{l^{-d} mn\, a_i}{\left(m f_0(\mathbf{x}) + n f_1(\mathbf{x})\right)}.
\tag{A80}
$$

Consequently, the RHS of (A79) becomes greater than or equal to

$$
\sum_{i=1}^{M} a_i\, l^{-d} \int \frac{2mn f_0(\mathbf{x}) f_1(\mathbf{x})}{m f_0(\mathbf{x}) + n f_1(\mathbf{x})}\, d\mathbf{x}
$$
$$
- \sum_{i=1}^{M} \mathbb{E}_{m_i,n_i}\left[(m_i + n_i)\,\varsigma_\eta(l, m_i, n_i)\right] - O\!\left(l^d(m+n)^{1-\eta/d}\right) - O\!\left(l^d(m+n)^{1/2}\right) - c_2\, l^d.
\tag{A81}
$$

Finally, since $\sum_{i=1}^{M} a_i l^{-d} = 1$ and $\dfrac{m}{m+n} \to p$, we have

$$
\frac{\mathbb{E}\left[\mathfrak{R}'^{*}_{m,n}(\mathfrak{X}_m,\mathfrak{Y}_n)\right]}{m+n} \geq \int \frac{2pq f_0(\mathbf{x}) f_1(\mathbf{x})}{p f_0(\mathbf{x}) + q f_1(\mathbf{x})}\, d\mathbf{x} - (m+n)^{-1} \sum_{i=1}^{M} \mathbb{E}_{m_i,n_i}\left[(m_i + n_i)\,\varsigma(l, m_i, n_i)\right]
$$
$$
- O\!\left(l^d(m+n)^{-\eta/d}\right) - O\!\left(l^d(m+n)^{-1/2}\right) - c_2\, l^d(m+n)^{-1}.
\tag{A82}
$$

By definition of the dual $\mathfrak{R}^{*}_{m,n}$ and (i) in Lemma A2,

$$
\frac{\mathbb{E}\left[\mathfrak{R}'_{m,n}(\mathfrak{X}_m,\mathfrak{Y}_n)\right]}{m+n} + \frac{c_d\, 2^d}{m+n} \geq \frac{\mathbb{E}\left[\mathfrak{R}'^{*}_{m,n}(\mathfrak{X}_m,\mathfrak{Y}_n)\right]}{m+n},
\tag{A83}
$$

we can imply

$$
\frac{\mathbb{E}\left[\mathfrak{R}'_{m,n}(\mathfrak{X}_m,\mathfrak{Y}_n)\right]}{m+n} \geq \int \frac{2pq f_0(\mathbf{x}) f_1(\mathbf{x})}{p f_0(\mathbf{x}) + q f_1(\mathbf{x})}\, d\mathbf{x} - (m+n)^{-1} \sum_{i=1}^{M} \mathbb{E}_{m_i,n_i}\left[(m_i + n_i)\,\varsigma_\eta(l, m_i, n_i)\right]
$$
$$
- O\!\left(l^d(m+n)^{-\eta/d}\right) - O\!\left(l^d(m+n)^{-1/2}\right) - c_2\, l^d(m+n)^{-1} - c_d\, 2^d\,(m+n)^{-1}.
\tag{A84}
$$

The combination of two lower and upper bounds (A84) and (A77) yields the following result

$$\left| \frac{\mathbb{E}\left[\mathfrak{R}'_{m,n}(\mathfrak{X}_m, \mathfrak{Y}_n)\right]}{m+n} - \int \frac{2pq f_0(\mathbf{x}) f_1(\mathbf{x})}{p f_0(\mathbf{x}) + q f_1(\mathbf{x})} \, d\mathbf{x} \right|$$

$$\leq O\big(l^d (m+n)^{-\eta/d}\big) + O\big(l^d (m+n)^{-1/2}\big) + 2\, c_1 \, l^{d-1} \, (m+n)^{(1/d)-1}$$

$$+ c_d \, 2^d \, (m+n)^{-1} + c_2 \, (m+n)^{-1} \, l^d + \frac{1}{m+n} \sum_{i=1}^{M} \mathbb{E}_{m_i, n_i}\left[(m_i + n_i)\, \varsigma_\eta(l, m_i, n_i)\right] \tag{A85}$$

$$+ \sum_{i=1}^{M} 2 \int \frac{f_0(\mathbf{x}) f_1(\mathbf{x}) \left(\alpha_i \beta_i \big(m a_i l^{-d} f_0^2(\mathbf{x}) + n b_i l^{-d} f_1^2(\mathbf{x})\big)\right)^{1/2}}{\big(m a_i f_0(\mathbf{x}) + n b_i f_1(\mathbf{x})\big)^2 (m+n)} \, d\mathbf{x}.$$

Recall $\varsigma_\eta(l, m_i, n_i)$, then we obtain

$$\sum_{i=1}^{M} \mathbb{E}_{m_i, n_i}\left[(m_i + n_i)\, \varsigma_\eta(l, m_i, n_i)\right] = \sum_{i=1}^{M} O(l) \int \mathbb{E}\left[\frac{2 m_i n_i f_0(\mathbf{x}) f_1(\mathbf{x})}{(m_i + n_i)(m_i f_0(\mathbf{x}) + n_i f_1(\mathbf{x}))}\right] d\mathbf{x}$$

$$-2\, l^d \sum_{i=1}^{M} \int \mathbb{E}\left[\frac{2 m_i n_i f_0(\mathbf{x}) f_1(\mathbf{x})}{(m_i + n_i)(m_i f_0(\mathbf{x}) + n_i f_1(\mathbf{x}))}\right] d\mathbf{x} + O(l^{-\eta}) \sum_{i=1}^{M} \mathbb{E}_{m_i, n_i}[m_i + n_i]. \tag{A86}$$

In addition, we have

$$\mathbb{E}_{m_i, n_i}\left[\frac{2 m_i n_i f_0(\mathbf{x}) f_1(\mathbf{x})}{(m_i + n_i)(m_i f_0(\mathbf{x}) + n_i f_1(\mathbf{x}))}\right] \geq \frac{1}{m+n} \mathbb{E}_{m_i, n_i}\left[\frac{2 m_i n_i f_0(\mathbf{x}) f_1(\mathbf{x})}{(m_i f_0(\mathbf{x}) + n_i f_1(\mathbf{x}))}\right]. \tag{A87}$$

This implies

$$\sum_{i=1}^{M} \int \mathbb{E}\left[\frac{2 m_i n_i f_0(\mathbf{x}) f_1(\mathbf{x})}{(m_i + n_i)(m_i f_0(\mathbf{x}) + n_i f_1(\mathbf{x}))}\right] d\mathbf{x} \geq \int \frac{2pq f_0(\mathbf{x}) f_1(\mathbf{x})}{p f_0(\mathbf{x}) + q f_1(\mathbf{x})} \, d\mathbf{x}. \tag{A88}$$

Note that the above inequality is derived from (A80) and $\dfrac{m}{m+n} \to p$. Furthermore,

$$\frac{1}{m+n} \sum_{i=1}^{M} O(l) \int \mathbb{E}_{m_i, n_i}\left[\frac{2 m_i n_i f_0(\mathbf{x}) f_1(\mathbf{x})}{(m_i + n_i)(m_i f_0(\mathbf{x}) + n_i f_1(\mathbf{x}))}\right] d\mathbf{x}$$

$$\leq \sum_{i=1}^{M} O(l) \int \mathbb{E}_{m_i, n_i}\left[\frac{2 m_i n_i f_0(\mathbf{x}) f_1(\mathbf{x})}{(m_i + n_i)^2 (m_i f_0(\mathbf{x}) + n_i f_1(\mathbf{x}))}\right] d\mathbf{x} \tag{A89}$$

$$\leq \sum_{i=1}^{M} O(l) \int \mathbb{E}_{m_i, n_i}\left[\frac{2 f_0(\mathbf{x}) f_1(\mathbf{x})}{(m_i f_0(\mathbf{x}) + n_i f_1(\mathbf{x}))}\right] d\mathbf{x}.$$

The last line holds because of $m_i n_i \leq (m_i + n_i)^2$. Going back to (A73), we can give an upper bound for the RHS of above inequality as

$$\mathbb{E}_{m_i, n_i}\left[(m_i f_0(\mathbf{x}) + n_i f_1(\mathbf{x}))^{-1}\right] \leq \big(m a_i l^{-d} f_0(\mathbf{x}) + n b_i l^{-d} f_1(\mathbf{x})\big)^{-1}$$

$$+ \left(\mathbb{E}_{m_i, n_i}\big| m_i f_0(\mathbf{x}) + n_i f_1(\mathbf{x}) - \big(\mathbb{E}[m_i] f_0(\mathbf{x}) + \mathbb{E}[n_i] f_1(\mathbf{x})\big|\right) \Big/ \big(m a_i l^{-d} f_0(\mathbf{x}) + n b_i l^{-d} f_1(\mathbf{x})\big)^2.$$

Note that we have assumed $a_i \leq b_i$ and by using Hölder inequality we write

$$\mathbb{E}_{m_i,n_i}\left[(m_i f_0(\mathbf{x}) + n_i f_1(\mathbf{x}))^{-1}\right] \leq l^d(a_i)^{-1}(m f_0(\mathbf{x}) + n f_1(\mathbf{x}))^{-1}$$

$$+\left(f_0(\mathbf{x})\sqrt{\mathbb{V}(m_i)} + f_1(\mathbf{x})\sqrt{\mathbb{V}(n_i)}\right)\Big/\left(a_i^2 l^{-d}(m f_0(\mathbf{x}) + n f_1(\mathbf{x}))^2\right) \leq l^d(a_i)^{-1}(m f_0(\mathbf{x}) + n f_1(\mathbf{x}))^{-1} \quad \text{(A90)}$$

$$+ l^{-d/2}\sqrt{b_i}\left(f_0(\mathbf{x})\sqrt{m} + f_1(\mathbf{x})\sqrt{n}\right)\Big/\left(a_i^2 l^{-d}(m f_0(\mathbf{x}) + n f_1(\mathbf{x}))^2\right).$$

As result, we have

$$\sum_{i=1}^{M} O(l) \int \mathbb{E}_{m_i,n_i}\left[\frac{2 f_0(\mathbf{x}) f_1(\mathbf{x})}{(m_i f_0(\mathbf{x}) + n_i f_1(\mathbf{x}))}\right] d\mathbf{x}$$

$$\leq \sum_{i=1}^{M} O(l) \int l^d(a_i)^{-1} \frac{2 f_0(\mathbf{x}) f_1(\mathbf{x})}{m f_0(\mathbf{x}) + n f_1(\mathbf{x})} d\mathbf{x} \quad \text{(A91)}$$

$$+ \sum_{i=1}^{M} O(l) \int l^{-d/2}\sqrt{b_i} \, \frac{2 f_0(\mathbf{x}) f_1(\mathbf{x})\left(f_0(\mathbf{x})\sqrt{m} + f_1(\mathbf{x})\sqrt{n}\right)}{a_i^2 l^{-d}\left(m f_0(\mathbf{x}) + n f_1(\mathbf{x})\right)^2} d\mathbf{x}.$$

As a consequence, owing to (A85), for $0 < \eta \leq 1$, $d > 2$, which implies $\eta \leq d - 1$, we can derive (A61). Thus, the proof can be concluded by giving the summarized bound in (A60). $\square$

Lemma A8: For $h = 1, 2, \ldots$, let $\delta_{m,n}^h$ be the function $c\, h^{d-1}(m+n)^{1/d}$. Then, for $\epsilon > 0$, we have

$$P\left(\mathfrak{R}_{m,n}(\mathfrak{X}_m, \mathfrak{Y}_n) \leq \sum_{i=1}^{h^d} \mathfrak{R}_{m_i,n_i}(\mathfrak{X}_{m_i}, \mathfrak{Y}_{n_i}) + 2\epsilon\right) \geq \frac{\epsilon - \delta_{m,n}^h}{\epsilon}. \quad \text{(A92)}$$

Note that in case $\epsilon \leq \delta_{m,n}^h$ the above claimed inequality is trivial.

Proof. Consider the cardinality of the set of all edges of $\mathrm{MST}\left(\bigcup_{i=1}^{h^d} Q_i\right)$ which intersect two different subcubes Q_i and Q_j, $|D|$. Using the Markov inequality, we can write

$$P\left(|D| \geq \epsilon\right) \leq \frac{\mathbb{E}(|D|)}{\epsilon},$$

where $\epsilon > 0$. Since $\mathbb{E}|D| \leq c\, h^{d-1}(m+n)^{1/d} := \delta_{m,n}^h$, therefore for $\epsilon > \delta_{m,n}^h$ and $h = 1, 2, \ldots$:

$$P\left(|D| \geq \epsilon\right) \leq \frac{\delta_{m,n}^h}{\epsilon}.$$

In addition, if Q_i, $i = 1, \ldots h^d$ is a partition of $[0,1]^d$ into congruent subcubes of edge length $1/h$, then

$$P\left(\sum_{i=1}^{h^d} \mathfrak{R}_{m_i,n_i}(\mathfrak{X}_m, \mathfrak{Y}_n \cap Q_i) + 2|D| \geq \sum_{i=1}^{h^d} \mathfrak{R}_{m_i,n_i}(\mathfrak{X}_m, \mathfrak{Y}_n \cap Q_i) + 2\epsilon\right) \leq \frac{\delta_{m,n}^h}{\epsilon}. \quad \text{(A93)}$$

This implies

$$P\left(\sum_{i=1}^{h^d} \mathfrak{R}_{m_i,n_i}(\mathfrak{X}_m, \mathfrak{Y}_n \cap Q_i) + 2|D| \leq \sum_{i=1}^{h^d} \mathfrak{R}_{m_i,n_i}(\mathfrak{X}_m, \mathfrak{Y}_n \cap Q_i) + 2\epsilon\right) \geq 1 - \frac{\delta_{m,n}^h}{\epsilon}. \quad \text{(A94)}$$

By subadditivity (A6), we can write

$$\mathfrak{R}_{m,n}(\mathfrak{X}_m,\mathfrak{Y}_n) \le \sum_{i=1}^{h^d} \mathfrak{R}_{m_i,n_i}(\mathfrak{X}_m,\mathfrak{Y}_n \cap Q_i) + 2|D|,$$

and this along with (A94) establishes (A92). □

Lemma A9: (Growth bounds for $\mathfrak{R}_{m,n}$) Let $\mathfrak{R}_{m,n}$ be the FR statistic. Then, for given non-negative ϵ, such that $\epsilon \ge h^2 \, \delta^h_{m,n}$, with at least probability $g(\epsilon) := 1 - \dfrac{h \, \delta^h_{m,n}}{\epsilon}$, $h = 2, 3, \ldots$, we have

$$\mathfrak{R}_{m,n}(\mathfrak{X}_m,\mathfrak{Y}_n) \le c''_{\epsilon,h} \left(\#\mathfrak{X}_m \, \#\mathfrak{Y}_n\right)^{1-1/d}. \tag{A95}$$

Here, $c''_{\epsilon,h} = O\left(\dfrac{\epsilon}{h^{d-1}-1}\right)$ depending only on ϵ, h. Note that, for $\epsilon < h^2 \, \delta^h_{m,n}$, the claim is trivial.

Proof. Without loss of generality, consider the unit cube $[0,1]^d$. For given h, if Q_i, $i = 1, \ldots h^d$ is a partition of $[0,1]^d$ into congruent subcubes of edge length $1/h$, then, by Lemma A8, we have

$$P\left(\mathfrak{R}_{m,n}(\mathfrak{X}_m,\mathfrak{Y}_n) \le \sum_{i=1}^{h^d} \mathfrak{R}_{m_i,n_i}(\mathfrak{X}_{m_i},\mathfrak{Y}_{n_i}) + 2\epsilon\right) \ge \frac{\epsilon - \delta^h_{m,n}}{\epsilon}. \tag{A96}$$

We apply the induction methodology on $\#\mathfrak{X}_m$ and $\#\mathfrak{Y}_n$. Set $c := \sup\limits_{\mathbf{x},\mathbf{y}\in[0,1]^d} \mathfrak{R}_{m,n}(\{\mathbf{x},\mathbf{y}\})$ which is finite according to assumption. Moreover, set $c_2 := \dfrac{2\epsilon}{h^{d-1}-1}$ and $c_1 := c + d \, h^{d-1} c_2$. Therefore, it is sufficient to show that for all $(\mathfrak{X}_m,\mathfrak{Y}_n) \in [0,1]^d$ with at least probability $g(\epsilon)$

$$\mathfrak{R}_{m,n}(\mathfrak{X}_m,\mathfrak{Y}_n) \le c_1\left(\#\mathfrak{X}_m \, \#\mathfrak{Y}_n\right)^{(d-1)/d}. \tag{A97}$$

Alternatively, as for the induction hypothesis, we assume the stronger bound

$$\mathfrak{R}_{m,n}(\mathfrak{X}_m,\mathfrak{Y}_n) \le c_1\left(\#\mathfrak{X}_m \, \#\mathfrak{Y}_n\right)^{(d-1)/d} - c_2 \tag{A98}$$

holds whenever $\#\mathfrak{X}_m < m$ and $\#\mathfrak{Y}_n < n$ with at least probability $g(\epsilon)$. Note that $d \ge 2$, $\epsilon > 0$ and c_1, c_2 both depend on ϵ, h. Hence,

$$c_1 - c_2 = c + c_2\left(d \, h^{d-1} - 1\right) \ge c + c_2\left(h^{d-1} - 1\right) = c + 2\epsilon \ge c,$$

which implies $P(\mathfrak{R}_{m,n} \le c_1 - c_2) \ge P(\mathfrak{R}_{m,n} \le c)$. In addition, we know that $P(\mathfrak{R}_{m,n} \le c) = 1 \ge g(\epsilon)$; therefore, the induction hypothesis holds particularly $\#\mathfrak{X}_m = 1$ and $\#\mathfrak{Y}_n = 1$. Now, consider the partition Q_i of $[0,1]^d$; therefore, for all $1 \le i \le h^d$, we have $m_i := \#(\mathfrak{X}_m \cap Q_i) < m$ and $n_i := \#(\mathfrak{Y}_n \cap Q_i) < n$ and thus, by induction hypothesis, one yields with at least probability $g(\epsilon)$

$$\mathfrak{R}_{m_i,n_i}(\mathfrak{X}_m,\mathfrak{Y}_n \cap Q_i) \le c_1 \, (m_i \, n_i)^{1-1/d} - c_2. \tag{A99}$$

Set $\mathbb{B}$ the event $\{\text{all } i : \mathfrak{R}_{m_i,n_i} \le c_1 \, (m_i \, n_i)^{1-1/d} - c_2\}$ and $\mathbb{B}_i$ stands with the event $\{\mathfrak{R}_{m_i,n_i} \le c_1 \, (m_i \, n_i)^{1-1/d} - c_2\}$. From (A96) and since Q_i's are partitions, which implies

$$P(\mathbb{B}) = \left(P(\mathbb{B}_i)\right)^{h^d} \le P(\mathbb{B}_i), \quad P(\mathbb{B}^c) = P\left(\bigcup_{i=1}^{l^d} \mathbb{B}_i^c\right) \le \sum_{i=1}^{h^d} P(\mathbb{B}_i^c) \le h^d\left(1 - g(\epsilon)\right),$$

$$\text{and} \quad P(\mathbb{B}) = \prod_{i=1}^{h^d} P(\mathbb{B}_i) \ge \left(g(\epsilon)\right)^{h^d},$$

we thus obtain

$$\frac{\epsilon - \delta^h_{m,n}}{\epsilon} \leq P\left(\mathfrak{R}_{m,n} \leq \sum_{i=1}^{h^d} \mathfrak{R}_{m_i,n_i}(\mathfrak{X}_{m_i},\mathfrak{Y}_{n_i}) + 2\epsilon \Big| \mathbb{B}\right)P(\mathbb{B}) + P\left(\mathfrak{R}_{m,n} \leq \sum_{i=1}^{h^d} \mathfrak{R}_{m_i,n_i}(\mathfrak{X}_{m_i},\mathfrak{Y}_{n_i}) + 2\epsilon \Big| \mathbb{B}^c\right)P(\mathbb{B}^c)$$

$$\leq P\left(\mathfrak{R}_{m,n} \leq \sum_{i=1}^{l^d} \mathfrak{R}_{m_i,n_i}(\mathfrak{X}_{m_i},\mathfrak{Y}_{n_i}) + 2\epsilon \Big| \mathbb{B}\right)P(\mathbb{B}) + P(\mathbb{B}^c).$$

Equivalently,

$$P\left(\mathfrak{R}_{m,n} \leq \sum_{i=1}^{h^d} \mathfrak{R}_{m_i,n_i}(\mathfrak{X}_{m_i},\mathfrak{Y}_{n_i}) + 2\epsilon \Big| \mathbb{B}\right) \geq \left(1 - \frac{\delta^h_{m,n}}{\epsilon} - 1 + P(\mathbb{B})\right)/P(\mathbb{B}) = 1 - \frac{\delta^h_{m,n}}{\epsilon\, P(\mathbb{B})}.$$

In fact, in this stage, we want to show that

$$1 - \frac{\delta^h_{m,n}}{\epsilon\, P(\mathbb{B})} \geq g(\epsilon) \quad \text{or} \quad P(\mathbb{B}) \geq \frac{\delta^h_{m,n}}{\epsilon\,(1 - g(\epsilon))}.$$

Since $P(\mathbb{B}) \geq (g(\epsilon))^{h^d}$, therefore it is sufficient to derive that $(g(\epsilon))^{h^d} \geq \dfrac{\delta^h_{m,n}}{\epsilon\,(1 - g(\epsilon))}$. Indeed, for

given $g(\epsilon) = \left(\dfrac{\epsilon - h\,\delta^h_{m,n}}{\epsilon}\right)$, we have $g(\epsilon) \leq \dfrac{\epsilon - \delta^h_{m,n}}{\epsilon}$ hence $\dfrac{\delta^h_{m,n}}{\epsilon\,(1 - g(\epsilon))} = \dfrac{1}{h} \leq 1$. Furthermore, we

know $\dfrac{1}{h} \leq 1 - \dfrac{1}{h^{(1/h^d)}}$ and since $\epsilon \geq h^2\,\delta^h_{m,n}$ this implies $\dfrac{h\,\delta^h_{m,n}}{\epsilon} \leq \dfrac{1}{h}$ and consequently

$$\frac{h\,\delta^h_{m,n}}{\epsilon} \leq 1 - \frac{1}{h^{h-d}}$$

or

$$g(\epsilon)^{h^d} = \left(\frac{\epsilon - h\,\delta^h_{m,n}}{\epsilon}\right)^{h^d} \geq \frac{1}{h} = \frac{\delta^h_{m,n}}{\epsilon\,(1 - g(\epsilon))}.$$

This implies the fact that for $\epsilon \geq h^2\delta^h_{m,n}$

$$P\left(\mathfrak{R}_{m,n} \leq \sum_{i=1}^{h^d} \left(c_1(m_i n_i)^{1-1/d} - c_2\right) + 2\epsilon\right) \geq g(\epsilon), \quad \text{where} \quad g(\epsilon) = \frac{\epsilon - h\,\delta^h_{m,n}}{\epsilon}.$$

Now, let $\gamma := \#\{i : m_i, n_i > 0\}$ and using Hölder inequality gives

$$P\left(\mathfrak{R}_{m,n}(\mathfrak{X}_m,\mathfrak{Y}_n) \leq c_1\gamma^{1/d}(m\,n)^{1-1/d} - \gamma c_2 + c_2\left(h^{d-1} - 1\right)\right) \geq g(\epsilon). \tag{A100}$$

Next, we just need to show that $c_1\gamma^{1/d}(m\,n)^{1-1/d} - \gamma c_2 + c_2\left(h^{d-1} - 1\right)$ in (A100) is less than or equal to $c_1(m\,n)^{1-1/d} - c_2$, which is equivalent to show

$$c_2\left(h^{d-1} - \gamma\right) \leq c_1(m\,n)^{1-1/d}(1 - \gamma^{1/d}).$$

We know that $m, n \geq 1$ and $c_1 \geq d\,h^{d-1}c_2$, so it is sufficient to get

$$c_2\left(h^{d-1} - \gamma\right) \leq d\,h^{d-1}c_2(1 - \gamma^{1/d}), \tag{A101}$$

choose t as $\gamma = t\,h^d$, then $0 < t \leq 1$, so (A101) becomes

$$\left(h^{-1} - t\right) \geq d\,h^{-1}\left(1 - h\,t^{1/d}\right). \tag{A102}$$

Note that the function $d\, h^{-1}(1 - h\, t^{1/d}) + t - h^{-1}$ has a minimum at $t = 1$ which implies (A101) and subsequently (A95). Hence, the proof is completed. $\square$

Lemma A10: (Smoothness for $\mathfrak{R}_{m,n}$) Given observations of

$$\mathfrak{X}_m := (\mathfrak{X}_{m'}, \mathfrak{X}_{m''}) = \{\mathbf{X}_1, \ldots, \mathbf{X}_{m'}, \mathbf{X}_{m'+1}, \ldots, \mathbf{X}_m\},$$

such that $m' + m'' = m$ and $\mathfrak{Y}_n := (\mathfrak{Y}_{n'}, \mathfrak{Y}_{n''}) = \{\mathbf{Y}_1, \ldots, \mathbf{Y}_{n'}, \mathbf{Y}_{n'+1}, \ldots, \mathbf{Y}_n\}$, where $n' + n'' = n$, denote $\mathfrak{R}_{m,n}(\mathfrak{X}_m, \mathfrak{Y}_n)$ as before, the number of edges of $\mathrm{MST}(\mathfrak{X}_m, \mathfrak{Y}_n)$ which connect a point of $\mathfrak{X}_m$ to a point of $\mathfrak{Y}_n$. Then, for integer $h \geq 2$, for all $(\mathfrak{X}_n, \mathfrak{Y}_m) \in [0,1]^d$, $\epsilon \geq h^2\, \delta_{m,n}^h$, where $\delta_{m,n}^h = O(h^{d-1}(m+n)^{1/d})$, we have

$$P\left(\left|\mathfrak{R}_{m,n}(\mathfrak{X}_m, \mathfrak{Y}_n) - \mathfrak{R}_{m',n'}(\mathfrak{X}_{m'}, \mathfrak{Y}_{n'})\right| \leq \tilde{c}_{\epsilon,h}\, (\#\mathfrak{X}_{m''}\, \#\mathfrak{Y}_{n''})^{1-1/d}\right) \geq 1 - \frac{2h\, \delta_{m,n}^h}{\epsilon}, \tag{A103}$$

where $\tilde{c}_{\epsilon,h} = O\left(\dfrac{\epsilon}{h^{d-1} - 1}\right)$. For the case $\epsilon < h^2\, \delta_{m,n}^h$, this holds trivially.

Proof. We begin with removing the edges which contain a vertex in $\mathfrak{X}_{m''}$ and $\mathfrak{Y}_{n''}$ in minimal spanning tree on $(\mathfrak{X}_m, \mathfrak{Y}_n)$. Now, since each vertex has bounded degree, say c_d, we can generate a subgraph in which has at most $c_d(\#\mathfrak{X}_{m''} + \#\mathfrak{Y}_{n''})$ components. Next, choose one vertex from each component and form the minimal spanning tree on these vertices, assuming all of them can be considered in FR test statistic, we can write

$$\mathfrak{R}_{m,n}(\mathfrak{X}_m, \mathfrak{Y}_n) \leq \mathfrak{R}_{m',n'}(\mathfrak{X}_{m'}, \mathfrak{Y}_{n'}) + c''_{\epsilon,h}\left(c_d^2\, \#\mathfrak{X}_{m''}\, \#\mathfrak{Y}_{n''}\right)^{1-1/d},$$

or equivalently

$$\leq \mathfrak{R}_{m',n'}(\mathfrak{X}_{m'}, \mathfrak{Y}_{n'}) + c_{\epsilon 1}^h\left(\#\mathfrak{X}_{m''}\, \#\mathfrak{Y}_{n''}\right)^{1-1/d}, \tag{A104}$$

with probability at least $g(\epsilon)$, where $g(\epsilon)$ is as in Lemma A9. Note that this expression is obtained from Lemma A9. In this stage, it remains to show that with at least probability $g(\epsilon)$

$$\mathfrak{R}_{m,n}(\mathfrak{X}_m, \mathfrak{Y}_n) \geq \mathfrak{R}_{m',n'}(\mathfrak{X}_{m'}, \mathfrak{Y}_{n'}) - \tilde{c}_{\epsilon,h}\left(\#\mathfrak{X}_{m''}\, \#\mathfrak{Y}_{n''}\right)^{1-1/d}, \tag{A105}$$

which, again by using the method before, with at least probability $g(\epsilon)$, one derives

$$\mathfrak{R}_{m',n'}(\mathfrak{X}_{m'}, \mathfrak{Y}_{n'}) \leq \mathfrak{R}_{m,n}(\mathfrak{X}_m, \mathfrak{Y}_n) + \hat{c}_{\epsilon,h}\left(c_d^2\,(\#\mathfrak{X}_{m''}\, \#\mathfrak{Y}_{n''})\right)^{1-1/d},$$
$$orequivalently$$
$$\leq \mathfrak{R}_{m,n}(\mathfrak{X}_m, \mathfrak{Y}_n) + c_{\epsilon 2}^h\left(\#\mathfrak{X}_{m''}\, \#\mathfrak{Y}_{n''}\right)^{1-1/d}.$$

Letting $\tilde{c}_{\epsilon,h} = \max\{c_{\epsilon 1}^h, c_{\epsilon 2}^h\}$ implies (A105). Thus,

$$P\left(\left|\mathfrak{R}_{m,n}(\mathfrak{X}_m, \mathfrak{Y}_n) - \mathfrak{R}_{m',n'}(\mathfrak{X}_{m'}, \mathfrak{Y}_{n'})\right| \geq \tilde{c}_{\epsilon,h}\left(\#\mathfrak{X}_{m''}\, \#\mathfrak{Y}_{n''}\right)^{1-1/d}\right) \leq 2 - 2\, g(\epsilon), \tag{A106}$$

Hence, the smoothness is given with at least probability $2\, g(\epsilon) - 1$ as in the statement of Lemma A10. $\square$

Lemma A11: (Semi-Isoperimetry) Let μ be a measure on $[0,1]^d$; μ^n denotes the product measure on space $([0,1]^d)^n$. In addition, let M_e denotes a median of $\mathfrak{R}_{m,n}$. Set

$$\mathbb{A} := \left\{\mathfrak{X}_m \in \left([0,1]^d\right)^m, \mathfrak{Y}_n \in \left([0,1]^d\right)^n; \mathfrak{R}_{m,n}(\mathfrak{X}_m, \mathfrak{Y}_n) \leq M_e\right\}. \tag{A107}$$

Then,

$$\mu^{m+n}\left(\left\{\mathbf{x}'\in([0,1]^d)^m,\mathbf{y}'\in([0,1]^n):\phi_{\mathbb{A}}(\mathbf{x}')\,\phi_{\mathbb{A}}(\mathbf{y}')\geq t\right\}\right)\leq 4\exp\left(\frac{-t}{8(m+n)}\right). \tag{A108}$$

Proof. Let $\phi_{\mathbb{A}}(\mathbf{z}')=\min\{H(\mathbf{z},\mathbf{z}'),\mathbf{z}\in\mathbb{A}\}$. Using Proposition 6.5 in [17], isoperimetric inequality, we have

$$\mu^{m+n}\left(\left\{\mathbf{z}'\in([0,1]^d)^{m+n}:\phi_{\mathbb{A}}(\mathbf{z}')\geq t\right\}\right)\leq 4\exp\left(\frac{-t^2}{8(m+n)}\right). \tag{A109}$$

Furthermore, we know that

$$\left(\phi_{\mathbb{A}}(\mathbf{x}')+\phi_{\mathbb{A}}(\mathbf{y}')\right)^2\geq\phi_{\mathbb{A}}(\mathbf{x}')\,\phi_{\mathbb{A}}(\mathbf{y}'),$$

hence

$$\mu^{m+n}\left(\left\{(\mathbf{x}'\in([0,1]^d)^m,\mathbf{y}'\in([0,1]^n):\phi_{\mathbb{A}}(\mathbf{x}')\phi_{\mathbb{A}}(\mathbf{y}')\geq t\right\}\right)$$

$$\leq\mu^{m+n}\left(\left\{(\mathbf{x}'\in([0,1]^d)^m,\mathbf{y}'\in([0,1]^n):(\phi_{\mathbb{A}}(\mathbf{x}')+\phi_{\mathbb{A}}(\mathbf{y}'))^2\geq t\right\}\right) \tag{A110}$$

$$=\mu^{m+n}\left(\left\{(\mathbf{x}'\in([0,1]^d)^m,\mathbf{y}'\in([0,1]^n):\phi_{\mathbb{A}}(\mathbf{x}')+\phi_{\mathbb{A}}(\mathbf{y}')\geq\sqrt{t}\right\}\right).$$

The last equality in (A110) achieves because of $\phi_{\mathbb{A}}(\mathbf{x}'),\phi_{\mathbb{A}}(\mathbf{y}')\geq 0$ and note that $\phi_{\mathbb{A}}(\mathbf{z}')\geq\phi_{\mathbb{A}}(\mathbf{x}')+\phi_{\mathbb{A}}(\mathbf{y}')$. Therefore,

$$\mu^{m+n}\left(\left\{(\mathbf{x}'\in([0,1]^d)^m,\mathbf{y}'\in([0,1]^n):\phi_{\mathbb{A}}(\mathbf{x}')+\phi_{\mathbb{A}}(\mathbf{y}')\geq\sqrt{t}\right\}\right)$$

$$\leq\mu^{m+n}\left(\left\{(\mathbf{z}'\in([0,1]^d)^{m+n}:\phi_{\mathbb{A}}(\mathbf{z}')\geq\sqrt{t}\right\}\right).$$

By recalling (A109), we derive the bound (A108). $\square$

Lemma A12: (Deviation of the Mean and Median) Consider M_e as a median of $\mathfrak{R}_{m,n}$. Then, for given $g(\epsilon)=1-\dfrac{h\,\delta_{m,n}^h}{\epsilon}$, and $\delta_{m,n}^h=O\big(h^{d-1}(m+n)^{1/d}\big)$ such that for $h\geq 7, \epsilon\geq h^2\delta_{m,n}^h$, we have

$$\left|\mathbb{E}\left[\mathfrak{R}_{m,n}(\mathfrak{X}_m,\mathfrak{Y}_n)\right]-M_e\right|\leq C_{m,n}(\epsilon,h)\,(m+n)^{(d-1)/d}, \tag{A111}$$

where $C_{m,n}(\epsilon,h)$ stands with a form depends on ϵ,h,m,n as

$$C_{m,n}(\epsilon,h)=C\left(1-\left((2\,(2\,g(\epsilon)-1)^2)^{-1}\right)\right)^{-1}, \tag{A112}$$

where C is a constant.

Proof. Following the analogous arguments in [17,53], we have

$$\left| \mathbb{E}\left[\mathfrak{R}_{m,n}(\mathfrak{X}_m, \mathfrak{Y}_n) \right] - M_e \right| \leq \mathbb{E}\left| \mathfrak{R}_{m,n}(\mathfrak{X}_m, \mathfrak{Y}_n) - M_e \right| = \int_0^\infty P\left(\left| \mathfrak{R}_{m,n}(\mathfrak{X}_m, \mathfrak{Y}_n) - M_e \right| \geq t \right) \, \mathrm{d}t$$

$$\leq 8 \left(1 - \left(1 \big/ \left(2 \left(2\, g(\epsilon) - 1 \right)^2 \right) \right) \right)^{-1} \int_0^\infty \exp\left(\frac{-t^{d/(d-1)}}{8(4\epsilon)^{d/d-1}(m+n)} \right) \, \mathrm{d}t$$

$$= C \left(1 - \left(\left(2 \left(2\, g(\epsilon) - 1 \right)^2 \right)^{-1} \right) \right)^{-1} (m+n)^{(d-1)/d},$$

$$(A113)$$

where $g(\epsilon) = 1 - \left(h\, O\!\left(h^{d-1}(m+n)^{1/d} \right) \right)\big/\epsilon$. The inequality in (A113) is implied from Theorem 5. Hence, the proof is completed. $\square$

References

1. Xuan, G.; Chia, P.; Wu, M. Bhattacharyya distance feature selection. In Proceedings of the 13th International Conference on Pattern Recognition, Vienna, Austria, 25–29 August 1996; Volume 2, pp. 195–199.
2. Hamza, A.; Krim, H. Image registration and segmentation by maximizing the Jensen-Renyi divergence. In *Energy Minimization Methods in Computer Vision and Pattern Recognition. EMMCVPR 2003*; Springer: Berlin/Heidelberg, Germany, 2003; pp. 147–163.
3. Hild, K.E.; Erdogmus, D.; Principe, J. Blind source separation using Renyi's mutual information. *IEEE Signal Process. Lett.* **2001**, *8*, 174–176. [CrossRef]
4. Basseville, M. Divergence measures for statistical data processing–An annotated bibliography. *Signal Process.* **2013**, *93*, 621–633. [CrossRef]
5. Battacharyya, A. On a measure of divergence between two multinomial populations. *Sankhy ā Indian J. Stat.* **1946**, *7*, 401–406.
6. Lin, J. Divergence Measures Based on the Shannon Entropy. *IEEE Trans. Inf. Theory* **1991**, *37*, 145–151. [CrossRef]
7. Berisha, V.; Hero, A. Empirical non-parametric estimation of the Fisher information. *IEEE Signal Process. Lett.* **2015**, *22*, 988–992. [CrossRef]
8. Berisha, V.; Wisler, A.; Hero, A.; Spanias, A. Empirically estimable classification bounds based on a nonparametric divergence measure. *IEEE Trans. Signal Process.* **2016**, *64*, 580–591. [CrossRef]
9. Moon, K.; Hero, A. Multivariate f-divergence estimation with confidence. In Proceedings of the Advances in Neural Information Processing Systems 27 (NIPS 2014), Montreal, QC, Canada, 8–13 December 2014; pp. 2420–2428.
10. Moon, K.; Hero, A. Ensemble estimation of multivariate f-divergence. In Proceedings of the IEEE International Symposium on Information Theory (ISIT), Honolulu, HI, USA, 29 June–4 July 2014; pp. 356–360.
11. Moon, K.; Sricharan, K.; Greenewald, K.; Hero, A. Improving convergence of divergence functional ensemble estimators. In Proceedings of the IEEE International Symposium on Information Theory (ISIT), Barcelona, Spain, 10–15 July 2016; pp. 1133–1137.
12. Moon, K.; Sricharan, K.; Greenewald, K.; Hero, A. Nonparametric ensemble estimation of distributional functionals. *arXiv* **2016**, arXiv:1601.06884v2.
13. Noshad, M.; Moon, K.; Yasaei Sekeh, S.; Hero, A. Direct Estimation of Information Divergence Using Nearest Neighbor Ratios. In Proceedings of the IEEE International Symposium on Information Theory (ISIT), Aachen, Germany, 25–30 June 2017.
14. Yasaei Sekeh, S.; Oselio, B.; Hero, A. A Dimension-Independent discriminant between distributions. In Proceedings of the IEEE International Conference on Acoustics, Speech and Signal Processing (ICASSP), Calgary, AB, Canada, 15–20 April 2018.
15. Noshad, M.; Hero, A. Rate-optimal Meta Learning of Classification Error. In Proceedings of the IEEE International Conference on Acoustics, Speech and Signal Processing (ICASSP), Calgary, AB, Canada, 15–20 April 2018.

16. Wisler, A.; Berisha, V.; Wei, D.; Ramamurthy, K.; Spanias, A. Empirically-estimable multi-class classification bounds. In Proceedings of the IEEE International Conference on Acoustics, Speech and Signal Processing (ICASSP), Shanghai, China, 20–25 March 2016.

17. Yukich, J. *Probability Theory of Classical Euclidean Optimization*; Lecture Notes in Mathematics; Springer: Berlin, Germany, 1998; Volume 1675.

18. Steele, J. An Efron–Stein inequality for nonsymmetric statistics. *Ann. Stat.* **1986**, *14*, 753–758. [CrossRef]

19. Aldous, D.; Steele, J.M. Asymptotic for Euclidean minimal spanning trees on random points. *Probab. Theory Relat. Fields* **1992**, *92*, 247–258. [CrossRef]

20. Ma, B.; Hero, A.; Gorman, J.; Michel, O. Image registration with minimal spanning tree algorithm. In Proceedings of the IEEE International Conference on Image Processing, Vancouver, BC, Canada, 10–13 September 2000; pp. 481–484.

21. Neemuchwala, H.; Hero, A.; Carson, P. Image registration using entropy measures and entropic graphs. *Eur. J. Signal Process.* **2005**, *85*, 277–296. [CrossRef]

22. Hero, A.; Ma, B., M.O.; Gorman, J. Applications of entropic spanning graphs. *IEEE Signal Process. Mag.* **2002**, *19*, 85–95. [CrossRef]

23. Hero, A.; Michel, O. Estimation of Rényi information divergence via pruned minimal spanning trees. In Proceedings of the IEEE Workshop on Higher Order Statistics, Caesarea, Isreal, 16 June 1999.

24. Smirnov, N. On the estimation of the discrepancy between empirical curves of distribution for two independent samples. *Bull. Mosc. Univ.* **1939**, *2*, 3–6.

25. Wald, A.; Wolfowitz, J. On a test whether two samples are from the same population. *Ann. Math. Stat.* **1940**, *11*, 147–162. [CrossRef]

26. Gibbons, J. *Nonparametric Statistical Inference*; McGraw-Hill: New York, NY, USA, 1971.

27. Singh, S.; Póczos, B. *Probability Theory and Combinatorial Optimization*; CBMF-NSF Regional Conference in Applied Mathematics; Society for Industrial and Applied Mathematics (SIAM): Philadelphia, PA, USA, 1997; Volume 69.

28. Redmond, C.; Yukich, J. Limit theorems and rates of convergence for Euclidean functionals. *Ann. Appl. Probab.* **1994**, *4*, 1057–1073. [CrossRef]

29. Redmond, C.; Yukich, J. Asymptotics for Euclidean functionals with power weighted edges. *Stoch. Process. Their Appl.* **1996**, *6*, 289–304. [CrossRef]

30. Hero, A.; Costa, J.; Ma, B. Convergence Rates of Minimal Graphs with Random Vertices. Available online: https://pdfs.semanticscholar.org/7817/308a5065aa0dd44098319eb66f81d4fa7a14.pdf (accessed on 18 November 2019).

31. Hero, A.; Costa, J.; Ma, B. *Asymptotic Relations between Minimal Graphs and Alpha-Entropy*; Tech. Rep.; Communication and Signal Processing Laboratory (CSPL), Department EECS, University of Michigan: Ann Arbor, MI, USA, 2003.

32. Lorentz, G. *Approximation of Functions*; Holt, Rinehart and Winston: New York, NY, USA, 1996.

33. Talagrand, M. Concentration of measure and isoperimetric inequalities in product spaces. *Publications Mathématiques de i'I. H. E. S.* **1995**, *81*, 73–205. [CrossRef]

34. Kullback, S.; Leibler, R. On information and sufficiency. *Ann. Math. Stat.* **1951**, *22*, 79–86. [CrossRef]

35. Rényi, A. On measures of entropy and information. In Proceedings of the Fourth Berkeley Symposium on Mathematical Statistics and Probability, Berkeley, USA, 20 June–30 July 1961; pp. 547–561.

36. Ali, S.; Silvey, S.D. A general class of coefficients of divergence of one distribution from another. *J. R. Stat. Soc. Ser. B (Methodol.)* **1996**, *28*, 131–142. [CrossRef]

37. Cha, S. Comprehensive survey on distance/similarity measures between probability density functions. *Int. J. Math. Models Methods Appl. Sci.* **2007**, *1*, 300–307.

38. Rukhin, A. Optimal estimator for the mixture parameter by the method of moments and information affinity. In Proceedings of the 12th Prague Conference on Information Theory, Prague, Czech Republic, 29 August–2 September 1994; pp. 214–219.

39. Toussaint, G. The relative neighborhood graph of a finite planar set. *Pattern Recognit.* **1980**, *12*, 261–268. [CrossRef]

40. Zahn, C. Graph-theoretical methods for detecting and describing Gestalt clusters. *IEEE Trans. Comput.* **1971**, *100*, 68–86. [CrossRef]

41. Banks, D.; Lavine, M.; Newton, H. The minimal spanning tree for nonparametric regression and structure discovery. In *Computing Science and Statistics, Proceedings of the 24th Symposium on the Interface*; Joseph Newton, H., Ed.; Interface Foundation of North America: Fairfax Station, FA, USA, 1992; pp. 370–374.
42. Hoffman, R.; Jain, A. A test of randomness based on the minimal spanning tree. *Pattern Recognit. Lett.* **1983**, *1*, 175–180. [CrossRef]
43. Efron, B.; Stein, C. The jackknife estimate of variance. *Ann. Stat.* **1981**, *9*, 586–596. [CrossRef]
44. Singh, S.; Póczos, B. Generalized exponential concentration inequality for Rényi divergence estimation. In Proceedings of the 31st International Conference on Machine Learning (ICML-14), Bejing, China, 22–24 June 2014; pp. 333–341.
45. Singh, S.; Póczos, B. Exponential concentration of a density functional estimator. In Proceedings of the 27th International Conference on Neural Information Processing Systems (NIPS 2014), Montreal, QC, Canada, 8–13 December 2014; pp. 3032–3040.
46. Lichman, M. UCI Machine Learning Repository. 2013. Available online: https://www.re3data.org/repository/r3d100010960 (accessed on 18 November 2019).
47. Bhatt, R.B.; Sharma, G.; Dhall, A.; Chaudhury, S. Efficient skin region segmentation using low complexity fuzzy decision tree model. In Proceedings of the IEEE-INDICON, Ahmedabad, India, 16–18 December 2009; pp. 1–4.
48. Steele, J.; Shepp, L.; Eddy, W. On the number of leaves of a euclidean minimal spanning tree. *J. Appl. Prob.* **1987**, *24*, 809–826. [CrossRef]
49. Henze, N.; Penrose, M. On the multivarite runs test. *Ann. Stat.* **1999**, *27*, 290–298.
50. Rhee, W. A matching problem and subadditive Euclidean funetionals. *Ann. Appl. Prob.* **1993**, *3*, 794–801. [CrossRef]
51. Whittaker, E.;Watson, G. *A Course in Modern Analysis*, 4th ed.; Cambridge University Press: New York, NY, USA, 1996.
52. Kingman, J. *Poisson Processes*; Oxford Univ. Press: Oxford, UK, 1993.
53. Pál, D.; Póczos, B.; Szapesvári, C. Estimation of Renyi entropy andmutual information based on generalized nearest-neighbor graphs. In Proceedings of the 23th International Conference on Neural Information Processing Systems (NIPS 2010), Vancouver, BC, Canada, 6–9 December 2010.

Article

Distance-Based Estimation Methods for Models for Discrete and Mixed-Scale Data

Elisavet M. Sofikitou [1], Ray Liu [2], Huipei Wang [1] and Marianthi Markatou [1,*]

[1] Department of Biostatistics, University at Buffalo, Buffalo, NY 14214, USA; esofikit@buffalo.edu (E.M.S.); huipeiwa@buffalo.edu (H.W.)

[2] Head of Oncology Data Science, AstraZeneca PLC, Gaithersburg, MD 20878, USA; ray.liu1@astrazeneca.com

* Correspondence: markatou@buffalo.edu

Abstract: Pearson residuals aid the task of identifying model misspecification because they compare the estimated, using data, model with the model assumed under the null hypothesis. We present different formulations of the Pearson residual system that account for the measurement scale of the data and study their properties. We further concentrate on the case of mixed-scale data, that is, data measured in both categorical and interval scale. We study the asymptotic properties and the robustness of minimum disparity estimators obtained in the case of mixed-scale data and exemplify the performance of the methods via simulation.

Keywords: contingency tables; disparity; mixed-scale data; pearson residuals; residual adjustment function; robustness; statistical distances

Citation: Sofikitou, E.M.; Liu, R.; Wang, H.; Markatou, M. Distance-Based Estimation Methods for Models for Discrete and Mixed-Scale Data. *Entropy* **2021**, *23*, 107. https://doi.org/10.3390/e23010107

Received: 5 December 2020
Accepted: 12 January 2021
Published: 14 January 2021

Publisher's Note: MDPI stays neutral with regard to jurisdictional claims in published maps and institutional affiliations.

1. Introduction

Minimum disparity estimation has been studied extensively in models where the scale of the data is either interval or ratio (Beran [1], Basu and Lindsay [2]). It has also been studied in the discrete outcomes case. Specifically, when the response variable is discrete and the explanatory variables are continuous, Pardo et al. [3] introduced a general class of distance estimators based on ϕ-divergence measures, the minimum ϕ-divergence estimators, and they studied their asymptotic properties. The estimators can be viewed as an extension/generalization of the Maximum Likelihood Estimator (MLE). Pardo et al. [4] used the minimum ϕ-divergence estimator in a ϕ-divergence statistic to perform goodness-of-fit tests in logistic regression models, while Pardo and Pardo [5] extended the previous works to address solving problems for testing in generalized linear models with binary scale data.

The case where data are measured on discrete scale (either on ordinal or generally categorical scale) has also attracted the interest of other researchers. For instance, Simpson [6] demonstrated that minimum Hellinger distance estimators fulfill desirable robustness properties and for this reason can be effective in the analysis of count data prone to outliers. Simpson [7] also suggested tests based on the minimum Hellinger distance for parametric inference which are robust as the density of the (parametric) model can be nonparametrically estimated. In contrast, Markatou et al. [8] used weighted likelihood equations to obtain efficient and robust estimators in discrete probability models and applied their methods to logistic regression, whereas Basu and Basu [9] considered robust penalized minimum disparity estimators for multinomial models with good small sample efficiency.

Moreover, Gupta et al. [10], Martín and Pardo [11] and Castilla et al. [12] used the minimum ϕ-divergence estimator to provide solution to testing problems in polytomous regression models. Working in a similar fashion, Martín and Pardo [13] studied the properties of the family of ϕ-divergence estimators for log-linear models with linear constraints under multinomial sampling in order to identify potential associations between various

277

variables in multi-way contingency tables. Pardo and Martín [14] presented an overview of works associated with contigency tables of symmetric structure on the basis of minimum ϕ-divergence estimators and minimum ϕ-divergence test statistics. Additional works include Pardo and Pardo [15] and Pardo et al. [16]. Alternative power divergence measures have been introduced by Basu et al. [17].

The class of f or $\phi-$divergences was originally introduced by Csiszár [18]. The structural characteristics of this class and their relationship to the concepts of efficiency and robustness were studied, for the case of discrete probability models, by Lindsay [19]. Basu and Lindsay [2] studied the properties of estimators derived by minimizing $f-$divergences between continuous models and presented examples showing the robustness results of these estimates. We also note that Tamura and Boos [20] studied the minimum Hellinger distance estimation for multivariate location and covariance. Additionally, formal robustness results were presented in Markatou et al. [8,21] in connection with the introduction of weighted likelihood estimation.

If G is a real valued, convex function, defined on $[0, \infty)$ and such that $G(u)$ converges to 0 as $u \to \infty$, $0G(0/0) = 0$, $0G(u/0) = uG_\infty$, $G_\infty = \lim_{u \to \infty}(G(u)/u)$, the class of $\phi-$divergences is defined as

$$\rho(\tau, m_{\beta_0}) = \sum G\left(\frac{\tau(t)}{m_{\beta_0}(t)}\right) m_{\beta_0}(t),$$

where $\tau(\cdot)$, $m_{\beta_0}(\cdot)$ are two probability models. Notice that we define $\rho(\tau, m_{\beta_0})$ on discrete probability models first, where $\mathscr{T} = \{0, 1, 2, \ldots, T\}$ is a discrete sample space, T possibly infinite, and $m_{\beta_0}(t) \in \mathscr{M} = \{m_\beta(t) : \beta \in \mathscr{B}\}$, $\mathscr{B}$ is the parameter space $\mathscr{B} \subseteq \mathbb{R}^d$. Furthermore, different forms of the function $G(u)$ provide different statistical distances or divergences.

We can change the argument of the function G from $\frac{\tau(t)}{m_{\beta_0}(t)}$ to $\frac{\tau(t)}{m_{\beta_0}(t)} - 1$. Then, G is a function of the Pearson residual which is defined as $\delta(t) = \frac{\tau(t)}{m_{\beta_0}(t)} - 1$, and takes values in $[-1, \infty)$. If the measurement scale is interval/ratio, then the Pearson residuals are modified to reflect and adjust for the discrepancy of scale between data, that are always discrete, and the assumed continuous probability model (see Basu and Lindsay [2]).

The Pearson residual is used by Lindsay [19], Basu and Lindsay [2] and Markatou et al. [8,21] in investigating the robustness of the minimum disparity and weighted likelihood estimators, respectively. This residual system allows one to identify distributional errors. If, in the equation of Pearson residual, we replace $\tau(t)$ with its best nonparametric representative $d(t)$, the proportion of observations in a sample with value t, then $\delta(t) = \frac{d(t)}{m_{\beta_0}(t)} - 1$. We note that the Pearson residuals are called so because $n \sum \delta^2(t) m(t)$ is Pearson's chi-squared distance. Furthermore, these residuals are not symmetric since they take values in $[-1, \infty]$ and are not standardized to have identical variances.

How does robustness fit into this picture? In the robustness literature, there is a denial of the model's truth. Following this logic, the framework based on disparities starts with goodness-of-fit by identifying a measure that assesses whether the model fits the data adequately. Then, we examine whether this measure of adequacy is robust and in what sense. A fundamental tool that assists in measuring the degree of robustness is the Pearson residual, because it measures model misspecification. That is, Pearson residuals provide information about the degree to which the specified model m_β fits the data. In this context, outliers are defined as those data points that have a low probability of occurrence under the hypothesized model. Such probabilistic outliers are called *surprising observations* (Lindsay [19]). Furthermore, the robustness of estimators obtained via minimization of the divergence measures we discuss here is indicated by the shape of the associated Residual Adjustment Function (RAF), a concept that is reviewed in Section 2. Of note is that in contingency table analysis, the generalized residual system is used for examination of sources of error in models for contingency tables, see, for example, Haberman [22], Haberman and Sinharay [23]. The concept of generalized residuals in the case of generalized linear models is discussed, for example, in Pierce and Schafer [24].

Data sets are comprised of data measured on both categorical (ordinal or nominal) scale and interval/ratio scale. We can think of these data as realizations of discrete and continuous random variables respectively. Examples of data sets that include mixed-scale data are electronic health records containing diagnostic codes (discrete) and laboratory measurements (e.g., blood pressure, alanine amino transferase (ALT) measurements on interval/ratio scale) and marketing data (customer records include income and gender information). Additional examples include data from developmental toxicology (Aerts et al. [25]), where fetal data from laboratory animals include binary, categorical and continuous outcomes. In this context, the joint density of the discrete and continuous random variables is given as $m_\beta(x,y) = f_{\beta_1}(y|x)g_{\beta_2}(x)$, where $\beta^T = (\beta_1^T, \beta_2^T)$ are parameter vectors indexing the joint, conditional on x and probability density function of x.

Work on the analysis of mixed-scale data is complicated by the fact that is difficult to identify suitable joint probability distributions to describe both measurement scales of the data, although a number of ad hoc methods to the analysis of mixed-scale data have been used in applications. Olkin and Tate [26] proposed multivariate correlation models for mixed-scale data. Copulas also provide an attractive approach to modeling the joint distribution of mixed-scale data, though copulas are less straightforward to implement, and there are subtle identifiability issues that complicate the specification of a model (Genest and Nešlehová [27]).

To formulate the joint distribution in the mixed-scale variables case one can either specify the marginal distribution of the discrete variables and the conditional distribution of the continuous variables. Alternatively, one can specify the marginal distribution of the continuous variables and the conditional distribution of the discrete variables given the continuous variables. Of note here is that the direction of factorization generally yields distinct model interpretations and results. The first approach has received much attention in the literature, in the context of the analysis of data with mixtures of categorical and continuous variables. Here, the continuous variables follow different multivariate normal distributions for each possible setting of the categorical variable values; the categorical variables then follow an arbitrary marginal multinomial distribution. This model is known in the literature as the conditional Gaussian distribution model and is central in the discussion of graphical association models with mixed-scale variables (Lauritzen and Wermuth [28]). A very special case of this model is used in our simulations.

In this paper, we develop robust methods for mixed-scale data. Specifically, Section 2 reviews basic concepts in minimum disparity estimation, Section 3 defines Pearson residuals for data measured in discrete, interval/ratio and mixed-scale, and studies their properties. Section 4 establishes the optimization problem for obtaining estimators of the model parameters, while Sections 5 and 6 establish the robustness and asymptotic properties of these estimators. Finally, Section 7 presents simulations showing the performance of these methods and Section 8 offers discussions. The Appendix A includes proofs of the theoretical results.

2. Concepts in Minimum Disparity Estimation

Beran [1] introduced a robust method to estimate the parameters of a statistical model, called minimum Hellinger distance estimation. The parameter estimator is obtained by minimizing the Hellinger distance between a parametric model density and a nonparametric density estimator. Lindsay [19] extended the aforementioned method to incorporate many other distances, and introduced the concept of the residual adjustment function in the context of minimum disparity estimation. The Minimum Distance Estimators (MDE) of a parameter vector β are obtained by minimizing over β, the distance (or disparity)

$$\rho(d, m_\beta) = \sum_x G(\delta(x))m_\beta(x), \tag{1}$$

where the assumed model m_β is a probability mass function. When the model m_β is continuous, the MDE of the parameter vector β is obtained by minimizing over β the quantity

$$\rho(f^*, m_\beta^*) = \int G(\delta(x)) m_\beta^*(x)\, dx, \tag{2}$$

where $f^*(x) = \int k(x; t, h)\, d\hat{F}(t)$, $m_\beta^*(x) = \int k(x; t, h) m_\beta(t)\, dt$, $\hat{F}$ is the empirical distribution function obtained from the data and k is a smooth family of kernel functions. One example is the normal density with mean t and standard deviation h. Furthermore, $\delta(x)$ is the Pearson residual defined as $\delta(x) = f^*(x)/m^*(x) - 1$. Lindsay [19] and Basu and Lindsay [2] discuss the efficiency and robustness properties of these estimators.

If $G(\delta) = \frac{1}{\lambda(1+\lambda)}\left\{(1+\delta)^{(\lambda+1)} - 1\right\}$ we obtain the class of power divergence measures. Notice that we have $G(0) = 0$. Different values of λ offer different measures; for example, when $\lambda = -2$ we obtain Neyman's chi-squared divided by 2 measure, while $\lambda = -1, -1/2$ return the Kullback-Leibler and Hellinger distances, respectively.

Under appropriate conditions, (1) and (2) can be written as

$$\sum A(\delta(x)) m_\beta(x) = 0,$$

or

$$\int A(\delta(x)) \nabla m_\beta^*(x)\, dx = 0,$$

where $A(\delta) = (\delta + 1)G'(\delta) - G(\delta)$ and the prime denotes differentiation with respect to δ.

Lindsay [19] has shown that the structural characteristics of the function $A(\delta)$ play an important role in the robustness and efficiency properties of these methods. Furthermore, without loss of generality, we can center and rescale $A(\delta)$, and define the RAF as follows.

Definition 1 (Lindsay [19]). *Let $A(\delta)$ be an increasing and twice differentiable function on $[-1, \infty)$ defined as*

$$A(\delta) = (\delta + 1)G'(\delta) - G(\delta),$$
$$A(0) = 0,$$
$$A'(0) = 1,$$

where G is strictly convex and twice differentiable with respect to δ on $[-1, \infty)$ with $G(0) = 0$. Then, $A(\delta)$ is called residual adjustment function.

Remark 1. *Since $A'(\delta) = (1 + \delta)G''(\delta)$, the second order differentiability of G, in addition to its strict convexity, implies that $A(\delta)$ is strictly increasing function of δ on $[-1, \infty)$. Thus, we can define $A(\delta)$ as above without changing the solutions of the aforementioned estimating equations in the discrete case (see Lindsay [19], p. 1089). In the continuous case, such standardization does not change the estimating properties of the associated disparities (see Basu and Lindsay [2], p. 687).*

Two fundamental and at the same time conflicting goals in robust statistics are the goals of robustness and efficiency. In the traditional literature on robustness, first order efficiency is sacrificed and, instead, safety of the estimation or testing method against outliers is guaranteed. Here, one adheres to the notion that information about robustness of a method is carried by the influence function. In our setting, using the influence function to characterize the robustness properties of the associated estimation procedures is misleading. Instead, the shape of the RAF, $A(\cdot)$, provides information to the extent of which our procedures can be characterized as robust. The interested reader is directed to Lindsay [19] for further discussion on this topic.

3. Pearson Residual Systems

In this section, we define various Pearson residuals, appropriate for the measurement scale of the data. We introduce our notation first.

Let (y_i, x_i), $i = 1, 2, \ldots, n$ be realizations from n independent and identically distributed random variables that follow a distribution with density $m_\beta(x, y)$. Recall that we use the word density to denote a general probability function, independently of whether the random variables X, Y are discrete, continuous or mixed. In what follows, we define different Pearson residual systems that account for the measurement scale of the data and study their properties.

Case 1: *Both X and Y are discrete.*
In this case, the pairs (y_i, x_i) follow a discrete probability mass function $m_\beta(x_i, y_i)$. Define the Pearson residual as

$$\delta(x, y) = \frac{\frac{n_{x,y}}{n}}{m_\beta(y|x)\pi_x} - 1,$$

where $\pi_x = P(X = x) = g(x)$, and $n_{x,y}$ is the number of observations in the cell with $Y = y$ and $X = x$.

Note that this definition of the Pearson residual is nonparametric on the discrete support of X. In the case of regression, one can carry out a semiparametric argument to obtain the estimators of the vector β and π_x.

We now establish that, under correct model specification, the residual $\delta(x, y)$ converges, almost surely, to zero.

Proposition 1. *When the model is correctly specified and as $n \to \infty$,*

$$\delta(x, y) \xrightarrow{a.s.} 0.$$

Proof. Write

$$\delta(x, y) = \frac{\frac{n_{x,y}}{n}}{m_\beta(y|x)\pi_x} - 1$$

$$= \frac{\frac{n_{x,y}}{n_x} \cdot \frac{n_x}{n}}{m_\beta(y|x)\pi_x} - 1.$$

Then

$$\frac{n_x}{n} = \frac{(\# \text{ of observations in the sample equal to x})}{n}$$

$$= \frac{1}{n}\sum_{i=1}^{n} I(x_i = x),$$

where $I(\cdot)$ is the indicator function. Furthermore,

$$E\left[\frac{1}{n}I(X_i = x)\right] = P(X = x) < \infty,$$

and by the strong law of large numbers

$$\frac{n_x}{n} \xrightarrow[n \to \infty]{a.s.} E[I(X = x)] = P(X = x) = \pi_x.$$

Similarly,

$$\frac{n_{x,y}}{n_x} \xrightarrow{a.s.} m_\beta(y|x),$$

therefore

$$\delta(x, y) \xrightarrow[n \to \infty]{a.s.} 0$$

under correct model specification. □

Case 2: *Y is continuous and X is discrete.*
This is the case in some *ANOVA* models. We can still define the Pearson residual in this setting as

$$\delta(x,y) = \frac{f_n(y,x)}{m_\beta(y,x)} - 1,$$

where

$$f_n(y,x) = f_n^*(y|x)g(x)$$
$$= \left\{ \int k(y,t,h)\, d\hat{F}_n(t|x) \right\} \frac{n_x}{n}$$

and

$$m_\beta(y,x) = m_\beta^*(y|x)g(x)$$
$$= \left\{ \int k(y,t,h)\, dM_\beta(t|x) \right\} \pi_x.$$

Then,

$$\delta(x,y) = \frac{f_n^*(y|X=x)\frac{n_x}{n}}{m_\beta^*(y|X=x)\pi_x} - 1.$$

Proposition 2. *Assume the model is correctly specified and $k(y,t,h)$ is a continuous function. Then,*

$$\delta(x,y) \xrightarrow[n\to\infty]{a.s.} 0.$$

Proof. Under the strong law of large numbers

$$\frac{n_x}{n} \xrightarrow[n\to\infty]{a.s.} \pi_x.$$

Under the correct model specification, continuity of the kernel function and the fact that $\hat{F}_n$ converges completely to F (implication of Glivenko-Cantelli theorem),

$$\lim_{n\to\infty} \int k(y;t,h)\, d\hat{F}_n(t|x) \to \int k(y;t,h)\, dF(t|x) = \int k(y;t,h)\, dM_\beta(t|x) = m_\beta^*(y|x)$$

(extension of Helly-Bray lemma). Therefore,

$$\frac{\frac{n_x}{n} f_n^*(y|x)}{\pi_x m_\beta^*(y|x)} \xrightarrow{a.s.} \frac{\pi_x}{\pi_x} \cdot \frac{m_\beta^*(y|x)}{m_\beta^*(y|x)} = 1$$

and hence

$$\delta(x,y) = \frac{\frac{n_x}{n} f_n^*(y|x)}{\pi_x m_\beta^*(y|x)} - 1 \xrightarrow{a.s.} 1 - 1 = 0.$$

□

Case 3: *Y is continuous and X is continuous.*
In this case, the pairs (y_i, x_i) follow a continuous probability distribution. The Pearson residual is then defined as

$$\delta(x,y) = \frac{f_n^*(y,x)}{m_\beta^*(y,x)} - 1,$$

where

$$f_n^*(x,y) = \int k(x,y;t_1,t_2)\, d\hat{F}_n(t_1,t_2),$$

$$m_\beta^*(x,y) = \int k(x,y;t_1,t_2) m_\beta(t_1,t_2)\, dt_1 dt_2.$$

As an example, we take the linear regression model with random carriers X, and $\epsilon_i \sim N(0,1)$. Furthermore, assume that the random carriers follow a normal distribution with mean vector μ and covariance matrix Σ. In this case, $y_i = x_i^T \beta + \epsilon_i$ and the quantities $z_i = (y_i - x_i^T \beta)/\sigma$ are independent, identically distributed random variables when β represents the vector of true parameters. Hence, the z_i's represent realizations of a random variable Z that has a completely known density $f(z)$. Thus,

$$m_\beta(x,y) = m_\beta(z|x) \cdot g(x), \qquad z = (y - x^T \beta)/\sigma$$

and hence

$$m_\beta^*(x,y) = m_\beta^*(y - x^T \beta | X = x) g^*(x),$$

$$m_\beta^*(y - x^T \beta | X = x) = m_\beta^*(z|x) = \int k(z,t,h)\, dM_\beta(t|x),$$

$$g^*(x) = \int k'(x,t',h') g(t')\, dt'.$$

The kernel $k(z,t,h)$ is selected so that it facilitates easy computation. Kernels that do not entail loss of information when they are used to smooth the assumed parametric model are called transparent kernels (Basu and Lindsay [2]). Basu and Lindsay [2] provide a formal definition of transparent kernels and an insightful discussion on the point of why transparent kernels do not exhibit information loss when convoluted with the hypothesized model (see Section 3.1 of Basu and Lindsay [2]).

4. Estimating Equations

In this section, we concentrate on cases 1, 2 presented in the previous section. We carefully outline the optimization problems and discuss the associated estimating equations for these two cases. The case where both X and Y are continuous has been discussed in the literature, see, for example, Markatou et al. [21].

Case 1: *Both X and Y are discrete.*
In this case, the minimum distance estimators of the parameter vector β and π_x are obtained by solving the following optimization problem

$$\min_{\beta,\pi_x} \rho(d,m_\beta) \tag{3}$$

subject to

$$\sum_x \pi_x = 1.$$

Optimization problem (3) is equivalent to the problem

$$\min \sum_{x,y} G(\delta(x,y)) m_\beta(x,y)$$

subject to

$$\sum_x \pi_x = 1.$$

The class of G functions that we use creates distances that belong in the family of ϕ-divergences.

Proposition 3. *The estimating equations for β and π_x are given as:*

$$\sum_{x,y} w(\delta(x,y))\, n_{x,y}\, u(y|x;\beta) = 0,$$

$$\sum_{x,y} w(\delta(x,y))\, n_{x,y} \left\{ \frac{I(X=x)}{\pi_x} - 1 \right\} = 0. \tag{4}$$

The function $w(\delta(x,y))$ is a weight function, such that $0 \leq w(\delta(x,y)) \leq 1$, and it is defined as

$$w(\delta(x,y)) = \min\left\{ \frac{[A(\delta(x,y))+1]^+}{\delta(x,y)+1}, 1 \right\}$$

with $[\cdot]^+$ indicating the positive part of the function $A(\delta(x,y))+1$.

Proof. The main steps of the proof are provided in the Appendix A.1. $\square$

Remark 2.

1. *The above two estimating equations can be solved with respect to β and π_x. In an iterative algorithm, we can solve the second equation* (4) *explicitly for π_x to obtain*

$$\pi_x = \frac{\sum_y w(\delta(x,y)) n_{x,y}}{\sum_{x,y} w(\delta(x,y)) n_{x,y}}.$$

   This means that if the model does not fit any of the y, observed at a particular x well, the weight for this x will drop as well.
2. *When $A(\delta(x,y)) = \delta(x,y)$ the corresponding estimating equation for β becomes $\sum_{x,y} n_{x,y} u(y|x;\beta) = 0$ and the MLE is obtained. This is because the corresponding weight function $w(\delta(x,y)) = 1$. In this case, the estimating equations for the π_xs become $\sum n_{x,y}\left[\frac{I(X=x)}{\pi_x} - 1 \right] = 0$, the estimating equations for the MLEs of π_x.*
3. *The Fisher consistency property of the function that introduces the estimates guarantees that the expectation of the corresponding estimating function is 0, under the correct model specification.*

Case 2: *Y is continuous and X is discrete.*
In this case, the estimates of the parameters β and π_x are obtained by solving the following optimization problem

$$\min_{\beta,\pi_x} \sum_x \int G(\delta(x,y)) m_\beta^*(y,x)\,dy$$

subject to

$$\sum_x \pi_x = 1.$$

In general $m_\beta^*(y,x) = m_\beta^*(y|x)\pi_x$; in the case where y, x are independent $m_\beta^*(y,x) = m_\beta^*(y)\pi_x$, and the optimization problem stated above is equivalent to

$$\min_{\beta,\pi_x} \sum_x \pi_x \int G(\delta(x,y)) m_\beta^*(y)\,dy \tag{5}$$

subject to

$$\sum_x \pi_x = 1.$$

Proposition 4. *The estimating equations for β and π_x in the case of independence of y, x are given as follows:*

$$\sum_x \pi_x \int A(\delta(x,y))\nabla_\beta m_\beta^*(y)dy = 0,$$

$$\sum_x \pi_x \int A(\delta(x,y))\left[\frac{I(X=x)}{\pi_x} - 1\right] m_\beta^*(y)dy = 0,$$

(6)

where $A(\delta)$ is the residual adjustment function (RAF) that corresponds to the function G, and $G'(\delta)$ is the derivative of G with respect to δ.

Proof. Straightforward, after differentiating the Lagrangian with respect to β and π_x. $\square$

Case 3: *Y is continuous and X is continuous.*
In this case, we refer the reader to Basu and Lindsay [2].

5. Robustness Properties

Hampel et al. [29] and Hampel [30,31] define robust statistics as the "statistics of approximate parametric models", and introduce one of the fundamental tools of robust statistics, the concept of the influence function, in order to investigate the behavior of a statistic T_n expressed as a functional $T(G)$. The influence function is a heuristic tool with the intuitive interpretation of measuring the bias caused by an infinitesimal contamination at a point x on the estimate standardized by the mass of contamination. Its formal definition is as follows:

Definition 2. *The influence function of a functional T at the distribution F is given as*

$$IF(x;T,F) = \lim_{t\to 0} \frac{T((1-t)F + t\Delta_x) - T(F)}{t},$$

in those $x \in \mathcal{X}$ where the limit exists, $0 \le t \le 1$ and Δ_x is the Dirac measure defined as

$$\Delta_x(u) = \begin{cases} 1, & u = x, \\ 0, & u \neq x. \end{cases}$$

(7)

If an estimator has a bounded influence function, the estimator is considered to be robust to outliers, that is data which is away from the pattern set by the majority of the data. The effect of bounding the influence function is the sacrifice of efficiency; estimators with bounded influence function, while are not affected by outlying points, are not fully efficient under the correct model specification.

Our goal in calculating the influence function is to show the full efficiency of the proposed estimators. That is, the influence function of the proposed estimators, under correct model specification, equals the influence function of the corresponding maximum likelihood estimators. In our context, robustness of the estimators is quantified by the associated RAFs (see Lindsay [19] and Basu and Lindsay [2]).

In what follows, we will derive the influence function of the estimators for the parameter vector β in the case where both y, x are discrete. Similar calculations provide the influence functions of estimators obtained under the remaining scenarios. To do so, we need to resort to the estimators' functional form, denoted by β_ϵ, with corresponding estimating equations

$$\sum_{s,t} w(\delta_\epsilon(s,t))u(t|s;\beta_\epsilon)d_\epsilon(s,t) = 0,$$

where $d_\epsilon(s,t) = (1-\epsilon)d(s,t) + \epsilon\Delta_{x,y}(s,t)$. The influence function is then obtained by differentiating the aforementioned estimating equations with respect to ϵ and then evaluating the derivative at $\epsilon = 0$.

Proposition 5. *The influence function of the β estimator is given by*

$$\beta_0' = [A(d)]^{-1} B(x, y; d),$$

where

$$A(d) = \sum_{s,t} [\delta_0(t) + 1] w'(\delta_0(s,t)) u(t|s; \beta_0) u^T(t|s; \beta_0) d(s,t)$$
$$- \sum_{s,t} w(\delta_0(s,t)) \nabla u(t|s; \beta_0) d(s,t),$$

$$B(x, y; d) = \sum_{s,t} \left[\frac{I(s = x, t = y)}{m_{\beta_0}(t|s) \pi_s} - \frac{d(s,t)}{m_{\beta_0}(t|s) \pi_s} w'(\delta_0(s,t)) \right] u(t|s; \beta_0) d(s,t)$$
$$- \sum_{s,t} w(\delta_0(s,t)) u(t|s; \beta_0) d(s,t) + w(\delta_0(x,y)) u(t|s; \beta_0),$$

with $u(t|s; \beta) = \nabla \ln m_\beta(t|s)$, and the subscript 0 indicates evaluation at a parametric model.

Proof. The proof is obtained via straightforward differentiation and its main steps are provided in the Appendix A.2. $\square$

Proposition 6. *Under the assumption that the model is correct, the influence function derived, reduces to the influence function of the MLE of β.*

Proof. Under the assumption that the adopted model is the correct model, the density $d(s,t)$ is $m_{\beta_0}(s,t)$, so that $\delta(s,t) = 0$. Now recall that $w(0) = 1$ and $w'(0) = 0$, so the expression $A(d)$ reduces to

$$A(d) = - \sum_{s,t} \nabla u(t|s; \beta_0) m_{\beta_0}(s,t)$$
$$= i(\beta, x, y). \tag{8}$$

Furthermore, the expression $B(x, y; d)$ reduces to $u(y|x; \beta_0)$, where we assume exchangeability of differentiation and integration and use the fact that $u(t|s; \beta_0) = u(s, t; \beta_0)$. Hence, the influence function is given as

$$i^{-1}(\beta; x, y) u(y|x; \beta_0),$$

which is exactly the influence function of the MLE. Therefore, full efficiency is preserved under the model. $\square$

6. Asymptotic Properties

In what follows, we establish asymptotic normality of the estimators in the case of discrete variables. The techniques for obtaining asymptotic normality in the mixed-scale case are similar and not presented here.

Case 1: *Both X and Y are discrete.*
Recall that the $k-$th estimating equation is given as $\sum_{x,y} w(\delta_\beta(x,y)) n_{x,y} u_k(y|x; \beta) = 0$, which can be expanded in Taylor series in the neighborhood of the true parameter β_0 to obtain:

$$\frac{1}{n} \sum_{x,y} w(\delta_\beta(x,y)) n_{x,y} u_k(y|x; \beta) \cong A_n + (\beta - \beta_0)^T B_n + \frac{1}{2} (\beta - \beta_0)^T C_n (\beta - \beta_0), \tag{9}$$

where

$$A_n = \frac{1}{n} \sum_{x,y} w(\delta_\beta(x,y)) n_{x,y} u_k(y|x; \beta_0),$$

$$B_n = \nabla_\beta \left\{ \frac{1}{n} \sum_{x,y} w(\delta_\beta(x,y)) n_{x,y} u_k(y|x; \beta) \right\} \Big|_{\beta_0},$$

(10)

C_n is a $p \times p$ Hessian matrix whose (t, e)−th element is given as

$$\frac{\partial^2}{\partial \beta_t \partial \beta_e} \left\{ \frac{1}{n} \sum_{x,y} w(\delta_\beta(x,y)) n_{x,y} u_k(y|x; \beta) \right\} \Big|_{\beta_0}.$$

Under assumptions 1–8, listed in the Appendix A.3, we have the following theorem.

Theorem 1. *The minimum disparity estimators of the parameter vector β are asymptotically normal with asymptotic variance $I^{-1}(\beta_0)$, where $I(\cdot)$ indicates the Fisher information matrix.*

7. Simulations

The simulation study presented below has two aims. The first one, is to indicate the versatility of the disparity methods for different data measurement scales. The second aim is to exemplify and study the robustness of these methods under different contamination scenarios.

Case 1: *Both X and Y are discrete.*
The Cressie-Read family of power divergence is given by

$$PWD(d, m_\beta) = \sum m_\beta(x,y) \cdot \frac{[1 + \delta(x,y)]^{\lambda+1} - 1}{\lambda(\lambda+1)} = \sum d(x,y) \cdot \frac{[d(x,y)/m_\beta(x,y)]^\lambda - 1}{\lambda(\lambda+1)},$$

where $d(x,y) = n_{x,y}/n$ is the proportion of observations with value x, y and $m_\beta(x,y) = m_\beta(y|x)\pi_x$ is the density function of the model of interest.

To evaluate the performance of our algorithmic procedure, we use the following disparity measures, that is,

Likelihood disparity $(\lambda = 0)$:
$$LD(d, m_\beta) = \sum d(x,y) \cdot \left\{ \log[d(x,y)/m_\beta(x,y)] \right\},$$

Twice-squared Hellinger's $(\lambda = -1/2)$:
$$HD(d, m_\beta) = 2 \cdot \sum \left[\sqrt{d(x,y)} - \sqrt{m_\beta(x,y)} \right]^2,$$

Pearson's chi-squared divided by 2 $(\lambda = 1)$:
$$PCS(d, m_\beta) = \sum \frac{[d(x,y) - m_\beta(x,y)]^2}{2 \cdot m_\beta(x,y)},$$

Symmetric chi-squared $\left(G(\delta(x,y)) = \frac{2[\delta(x,y)]^2}{\delta(x,y)+2} \right)$:
$$SCS(d, m_\beta) = 2 \cdot \sum \frac{[m_\beta(x,y) - d(x,y)]^2}{[m_\beta(x,y) + d(x,y)]}.$$

The data are generated in four different ways using three different sample sizes N, say $N = 100$; $N = 1000$ and $N = 10{,}000$. The data format used can be represented in a 5×5 contingency table, with $n_{i,j}, i = 1, 2, \ldots, 5; j = 1, 2, \ldots, 5$ denoting the counts in the ij-th cell, $n_{i\bullet}$ and $n_{\bullet j}$ representing the row and column totals, respectively. Furthermore, the variable x indicates columns, while y indicates the rows. In each of the aforementioned cases/scenarios, 10,000 tables were generated and that corresponds to the number of Monte Carlo (MC) replications. Our purpose is to get the mean values of the estimates of the parameters $m_\beta(y|x)$'s and π_x's along with their corresponding standard deviations (SDs). Notice that, in this setting, the estimation of π_x and $m_\beta(y|x)$ is completely nonparametric, that is, no model is assumed for estimating the marginal probabilities of X and Y.

The table was generated by using either a fixed total sample size N or fixed marginal probabilities. These two data generating schemes imply two different sampling schemes that could have generated the data with consequences for the probability model one would use. For example, with fixed total sample size the distribution of the counts is multinomial, or if the row margin is fixed in advance the distribution of the counts is a product binomial distribution. In the former case of fixed N, we explored two different scenarios: a balanced and an imbalanced one. The imbalanced scenario allows for the presence of one zero cell in the contingency table, whereas the balanced scenario does not. In the latter case of fixed marginal probabilities, the row marginal probabilities ($m_\beta(y|x)$'s) were fixed, while the column marginals (π_x's) were randomly chosen and these values were used to obtain the contingency table. In this case, we also explored a balanced and an imbalanced scenario based on whether the row marginal probabilities were chosen so that to be equal to each other or not, respectively.

Specifically, under Scenario Ia, where the total sample size N was fixed and the balanced design was exploited, none of the n_{ij}'s ($n_{ij} \neq 0$, $\forall\, i,j = 1,2,3,4,5$) was set equal to zero, with equal row and column marginal probabilities. Table 1 presents the mean of 10,000 estimates and the corresponding SDs for all four distances (PCS, HD, SCS, LD) when N is fixed under the balanced scenario. Table 1 clearly shows that all distances provide estimates approximately equal to 0.200 regardless of the sample size used. Furthermore, as the sample size increases, the SDs decrease noticeably.

In Scenario IIa, where the total sample size N was fixed and the contingency table was structured using the imbalanced design, the presence of a zero cell ($n_{11} = 0$) was allowed. The results of this scenario are presented in Table 2, where the estimates were calculated exploiting all disparity measures. For the LD, n_{11} was set equal to 10^{-8}. The presence of zero cells in contingency tables has a large history in the relevant literature on contingency tables analysis, where several options are provided for the analysis of these tables (Fienberg [32], Agresti [33], Johnson and May [34], Poon et al. [35]). From Table 2, one could infer that the different distances handle differently the zero cell. This difference is reflected in the estimate of $\hat{m}_{\beta(y_1|x)} = \hat{m}_{\beta_1}$, because it is affected by the zero value of n_{11}. The strongest control is provided by the Hellinger and symmetric chi-squared distances. All distances estimate the parameters π_{x_i} similarly, with the bias in their estimation been between 2.7% and 5.2%. The SDs are almost the same for all distances per estimate and their values are ameliorated for $N = 10,000$.

A referee suggested that in certain cases interest may be centered on smaller samples. We generated 2×3 tables with fixed total sample size of 50 and 70 observations. Tables 3 and 4 describe the results when the contingency tables were generated under a balanced and an imbalanced design with associated respective Scenarios Ib and IIb. More precisely, Table 3 presents the estimators of the marginal row and column probabilities obtained when PC, HD, SCS and LD distances are used. We notice that the increase in the sample size provides for a decrease in the overall absolute bias in estimation, defined as $\sum_{\ell=1}^{L} |\hat{\theta}_\ell - \theta_{0,\ell}|$, where $\hat{\theta}_\ell$ is the estimate of the ℓ-th component of an $L \times 1$ vector $\boldsymbol{\theta}$ and $\theta_{0,\ell}$ is the corresponding true value. In our case, $\boldsymbol{\theta}^T = (m_{\beta_1}, m_{\beta_2}, \pi_{x_1}, \pi_{x_2}, \pi_{x_3})$. This observation applies to all distances used in our calculations. Table 4 presents results associated with the imbalanced case. The generated 2×3 tables contain two empty cells ($n_{12} = n_{21} = 0$). Once again, for calculating the LD, cells $n_{12} = n_{21} = 10^{-8}$. We notice that the bias associated with the estimates is rather large for all the distances, and an increased sample size does not alleviate the observed bias. Basu and Basu [9] have proposed an empty cell penalty for the minimum power-divergence estimators. This penalty leads to estimators with improved small sample properties. See also Alin and Kurt [36] for a discussion of the need of penalization in small samples.

Table 5 provides the results obtained under Scenario III. In this case, the parameter estimates were calculated using the *PCS*, *HD*, *SCS* and *LD* distances when the 5×5 contingency table was constructed by fixing the row marginal probabilities so that they were all set at 0.20, that is, $(0.20, 0.20, 0.20, 0.20, 0.20)$. The column marginals were randomly chosen in the interval $[0, 1]$ and summed to 1. In this case, the produced column marginal probabilities were $(0.1472, 0.2365, 0.3196, 0.2370, 0.0597)$. The simulation study reveals that the estimates of the parameters $m_\beta(y|x)$'s and π_x's do not differ substantially from the respective row and column marginal probabilities for any of the four distances utilized. The SDs are approximately the same and they get lower values for larger N.

Finally, in Table 6 the data generation was done by exploiting Scenario IV, that is, by having fixed the row marginal probabilities, which were not equal to each other; while, the column marginals were randomly chosen in the interval $[0, 1]$ so that they sum to 1. In particular, the row marginal probabilities were fixed at values $(0.04, 0.20, 0.20, 0.20, 0.36)$, while the column marginals used were $(0.2171, 0.1676, 0.2347, 0.1178, 0.2628)$. When $N = 100$, the value of $\hat{m}_\beta(y_1|x) = \hat{m}_{\beta_1}$ is not approximately 0.07 and not equal to 0.04 for all distances. However, when $N = 1000$ or $N = 10,000$, we get better estimates irrespectively of the disparity measure choice. The SDs are approximately the same and they become smaller as the sample size increases.

We also notice from Tables 1, 5 and 6 that in all cases the standard deviation associated with the estimates obtained when we use other than likelihood distances, is approximately the same with the standard deviation that corresponds to the likelihood estimates, thereby showing the asymptotic efficiency of the disparity estimators.

All calculations were performed using the *R* language. Given that the problem described in this section can be viewed as a general non-linear optimization problem, the solnp function of the Rsolnp package (Ye [37]) was used to obtain the aforementioned estimates. For our calculations, we tried using a variety of different initial values ($\hat{\pi}_x^{(0)}$'s and $\hat{m}_\beta^{(0)}(y|x)$'s); we notice that no matter how the initial values were chosen, the estimates were always pretty similar and very close to the observed values ($n_{i\bullet}/N$ and $n_{\bullet j}/N$ for $i, j = 1, 2, 3, 4, 5$). Only the number of iterations needed for convergence is slightly affected. Consequently, random numbers from a Uniform distribution in the interval $[0, 1]$ were set as initial values (which were not necessarily summing to 1). The solnp function has a built-in stopping rule and there was no need to set our own stopping rule. We only set the boundary constraints to be in the interval $[0, 1]$ for all estimates which were also subject to $\sum \pi_x = \sum m_\beta(y|x) = 1$.

Other functions may also be used to obtain the estimates. For example, we used the auglag function of the nloptr package with local solvers "lbfgs" or "SLSQP" (Conn et al. [38], Birgin and Martínez [39]) which emulates Augmented Lagrangian multipliers. However, the convergence using the solnp function (the number of iterations was on average 2) was extremely faster than using the auglag function (the average number of iterations was approximately 100). For this reason, the results presented in Tables 1–6 were based only on the function solnp.

Table 1. Scenario Ia: Means and standard deviations (SDs) of 4 distances (PCS, HD, SCS, LD). A 5×5 contingency table was generated having fixed the total sample size N under a balanced design with $n_{ij} \neq 0, \ \forall \, i, j = 1, 2, 3, 4, 5$. The number of Monte Carlo (MC) replications used is 10,000.

N	Statistical Distance	Summary	Estimates Means and SDs over 10,000 Replications									
			$\hat{m}_{\beta_1}$	$\hat{m}_{\beta_2}$	$\hat{m}_{\beta_3}$	$\hat{m}_{\beta_4}$	$\hat{m}_{\beta_5}$	$\hat{\pi}_{x_1}$	$\hat{\pi}_{x_2}$	$\hat{\pi}_{x_3}$	$\hat{\pi}_{x_4}$	$\hat{\pi}_{x_5}$
100	PCS	Mean	0.199	0.199	0.201	0.201	0.200	0.201	0.200	0.199	0.200	0.201
		SD	0.038	0.041	0.039	0.039	0.039	0.038	0.038	0.037	0.038	0.038
	HD	Mean	0.199	0.200	0.200	0.200	0.201	0.200	0.200	0.200	0.200	0.200
		SD	0.037	0.041	0.037	0.037	0.037	0.037	0.037	0.035	0.036	0.037
	SCS	Mean	0.199	0.201	0.200	0.200	0.200	0.200	0.200	0.199	0.200	0.201
		SD	0.037	0.041	0.038	0.038	0.038	0.032	0.033	0.030	0.031	0.032
	LD	Mean	0.199	0.200	0.200	0.200	0.200	0.200	0.002	0.200	0.200	0.200
		SD	0.035	0.039	0.036	0.036	0.036	0.035	0.036	0.036	0.034	0.035
1000	PCS	Mean	0.200	0.200	0.200	0.200	0.200	0.200	0.200	0.200	0.200	0.200
		SD	0.014	0.015	0.016	0.016	0.014	0.017	0.015	0.015	0.013	0.016
	HD	Mean	0.200	0.200	0.200	0.200	0.200	0.200	0.200	0.200	0.200	0.200
		SD	0.013	0.015	0.013	0.013	0.013	0.013	0.012	0.012	0.012	0.013
	SCS	Mean	0.200	0.200	0.200	0.200	0.200	0.200	0.200	0.200	0.200	0.200
		SD	0.014	0.015	0.013	0.013	0.013	0.008	0.009	0.011	0.012	0.008
	LD	Mean	0.200	0.200	0.200	0.200	0.200	0.200	0.200	0.200	0.200	0.200
		SD	0.013	0.015	0.013	0.013	0.013	0.013	0.013	0.012	0.012	0.013
10,000	PCS	Mean	0.200	0.200	0.200	0.200	0.200	0.200	0.200	0.200	0.200	0.200
		SD	0.008	0.007	0.006	0.006	0.009	0.010	0.010	0.007	0.008	0.006
	HD	Mean	0.200	0.200	0.200	0.200	0.200	0.200	0.200	0.200	0.200	0.200
		SD	0.004	0.005	0.004	0.004	0.004	0.004	0.004	0.004	0.004	0.004
	SCS	Mean	0.200	0.200	0.200	0.200	0.200	0.200	0.200	0.200	0.200	0.200
		SD	0.004	0.005	0.004	0.004	0.004	0.007	0.005	0.008	0.008	0.004
	LD	Mean	0.200	0.200	0.200	0.200	0.200	0.200	0.200	0.200	0.200	0.200
		SD	0.004	0.005	0.004	0.004	0.004	0.004	0.004	0.004	0.004	0.004

Table 2. Scenario IIa Means and SDs of 4 distances (PCS, HD, SCS, LD). A 5×5 contingency table was generated having fixed the total sample size N under an imbalanced design with $n_{11} = 0$. The number of MC replications used is 10,000.

N	Statistical Distance	Summary	Estimates Means and SDs over 10,000 Replications									
			$\hat{m}_{\beta_1}$	$\hat{m}_{\beta_2}$	$\hat{m}_{\beta_3}$	$\hat{m}_{\beta_4}$	$\hat{m}_{\beta_5}$	$\hat{\pi}_{x_1}$	$\hat{\pi}_{x_2}$	$\hat{\pi}_{x_3}$	$\hat{\pi}_{x_4}$	$\hat{\pi}_{x_5}$
100	PCS	Mean	0.052	0.197	0.198	0.198	0.355	0.165	0.173	0.172	0.245	0.245
		SD	0.028	0.045	0.044	0.044	0.053	0.041	0.039	0.044	0.044	0.047
	HD	Mean	0.026	0.202	0.202	0.202	0.368	0.156	0.168	0.168	0.254	0.254
		SD	0.019	0.049	0.045	0.045	0.054	0.041	0.042	0.041	0.046	0.049
	SCS	Mean	0.033	0.209	0.209	0.209	0.340	0.166	0.172	0.171	0.245	0.246
		SD	0.022	0.047	0.045	0.045	0.051	0.036	0.036	0.033	0.038	0.040
	LD	Mean	0.040	0.200	0.200	0.200	0.360	0.160	0.170	0.170	0.250	0.250
		SD	0.020	0.043	0.040	0.040	0.048	0.037	0.038	0.036	0.042	0.044
1000	PCS	Mean	0.044	0.197	0.197	0.197	0.365	0.164	0.170	0.170	0.248	0.248
		SD	0.011	0.017	0.014	0.014	0.018	0.013	0.014	0.013	0.015	0.015
	HD	Mean	0.034	0.203	0.202	0.202	0.359	0.156	0.170	0.170	0.252	0.252
		SD	0.005	0.015	0.013	0.013	0.016	0.011	0.012	0.012	0.013	0.014
	SCS	Mean	0.038	0.210	0.210	0.210	0.332	0.166	0.169	0.169	0.248	0.248
		SD	0.006	0.015	0.014	0.014	0.016	0.014	0.013	0.011	0.013	0.014
	LD	Mean	0.040	0.200	0.200	0.200	0.360	0.160	0.170	0.170	0.250	0.250
		SD	0.006	0.015	0.013	0.013	0.016	0.012	0.012	0.011	0.013	0.014
10,000	PCS	Mean	0.044	0.197	0.196	0.196	0.367	0.164	0.170	0.170	0.248	0.248
		SD	0.002	0.006	0.007	0.007	0.010	0.007	0.006	0.005	0.007	0.008
	HD	Mean	0.034	0.203	0.202	0.202	0.359	0.156	0.171	0.171	0.252	0.252
		SD	0.002	0.005	0.004	0.004	0.005	0.004	0.004	0.004	0.004	0.005
	SCS	Mean	0.038	0.210	0.210	0.210	0.332	0.166	0.169	0.169	0.248	0.248
		SD	0.002	0.005	0.004	0.004	0.005	0.007	0.006	0.004	0.006	0.006
	LD	Mean	0.040	0.200	0.200	0.200	0.360	0.160	0.170	0.170	0.250	0.250
		SD	0.002	0.005	0.004	0.004	0.005	0.004	0.004	0.004	0.004	0.004

Table 3. Scenario Ib: Means and Biases of 4 distances (PCS, HD, SCS, LD). A 2×3 contingency table was generated having fixed the total sample size N under a balanced design with $n_{ij} \neq 0$, $\forall\, i = 1, 2$, $j = 1, 2, 3$. The number of MC replications used is 10,000.

N	Statistical Distance	Summary	Estimates Means and Biases over 10,000 Replications				
			$\hat{m}_{\beta_1}$	$\hat{m}_{\beta_2}$	$\hat{\pi}_{x_1}$	$\hat{\pi}_{x_2}$	$\hat{\pi}_{x_3}$
50	PCS	Mean	0.5008	0.4992	0.3339	0.3336	0.3325
		Abs.Biases	0.0008	0.0008	0.0006	0.0003	0.0009
		Overall Bias			0.0034		
	HD	Mean	0.5008	0.4992	0.3339	0.3335	0.3326
		Abs.Biases	0.0008	0.0008	0.0006	0.0002	0.0007
		Overall Bias			0.0031		
	SCS	Mean	0.5007	0.4993	0.3338	0.3335	0.3326
		Abs.Biases	0.0007	0.0007	0.0005	0.0002	0.0007
		Overall Bias			0.0028		
	LD	Mean	0.5008	0.4992	0.3339	0.3335	0.3326
		Abs.Biases	0.0008	0.0008	0.0006	0.0002	0.0008
		Overall Bias			0.0032		
70	PCS	Mean	0.4998	0.5002	0.3333	0.3331	0.3337
		Abs.Biases	0.0002	0.0002	0.0001	0.0003	0.0003
		Overall Bias			0.0011		
	HD	Mean	0.4998	0.5002	0.3333	0.3330	0.3336
		Abs.Biases	0.0002	0.0002	0.0000	0.0003	0.0003
		Overall Bias			0.0009		
	SCS	Mean	0.4998	0.5002	0.3334	0.3331	0.3335
		Abs.Biases	0.0002	0.0002	0.0000	0.0002	0.0002
		Overall Bias			0.0008		
	LD	Mean	0.4999	0.5001	0.3333	0.3330	0.3336
		Abs.Biases	0.0001	0.0001	0.0000	0.0003	0.0003
		Overall Bias			0.0009		

Table 4. Scenario IIb: Means and Biases of 4 distances (PCS, HD, SCS, LD). A 2×3 contingency table was generated having fixed the total sample size N under an imbalanced design with $n_{12} = n_{21} = 0$. The number of MC replications used is 10,000.

N	Statistical Distance	Summary	Estimates Means and Biases over 10,000 Replications				
			$\hat{m}_{\beta_1}$	$\hat{m}_{\beta_2}$	$\hat{\pi}_{x_1}$	$\hat{\pi}_{x_2}$	$\hat{\pi}_{x_3}$
50	PCS	Mean	0.6391	0.3609	0.3489	0.2278	0.4234
		Abs.Biases	0.0276	0.0276	0.0155	0.0611	0.0766
		Overall Bias			0.2084		
	HD	Mean	0.7815	0.2185	0.3346	0.0497	0.6157
		Abs.Biases	0.1149	0.1149	0.0013	0.1170	0.1157
		Overall Bias			0.4638		
	SCS	Mean	0.6420	0.3580	0.3510	0.2726	0.3765
		Abs.Biases	0.0247	0.0247	0.0176	0.1059	0.1235
		Overall Bias			0.2964		
	LD	Mean	0.6677	0.3323	0.3342	0.1660	0.4998
		Abs.Biases	0.0010	0.0010	0.0009	0.0007	0.0002
		Overall Bias			0.0038		
70	PCS	Mean	0.6377	0.3623	0.3483	0.2297	0.4220
		Abs.Biases	0.0290	0.0290	0.0150	0.0631	0.0780
		Overall Bias			0.2141		
	HD	Mean	0.7812	0.2188	0.3328	0.0491	0.6180
		Abs.Biases	0.1145	0.1145	0.0005	0.1175	0.1180
		Overall Bias			0.4650		
	SCS	Mean	0.6395	0.3605	0.3505	0.2739	0.3756
		Abs.Biases	0.0271	0.0271	0.0172	0.1072	0.1244
		Overall Bias			0.3030		
	LD	Mean	0.6657	0.3343	0.3331	0.1671	0.4998
		Abs.Biases	0.0010	0.0010	0.0002	0.0004	0.0002
		Overall Bias			0.0028		

Table 5. Scenario III: Means and SDs of 4 distances (PCS, HD, SCS, LD). A 5×5 contingency table was generated having fixed the row marginal probabilities at (0.20, 0.20, 0.20, 0.20, 0.20). The number of MC replications used is 10,000.

N	Statistical Distance	Summary	Estimates Means and SDs over 10,000 Replications									
			$\hat{m}_{\beta_1}$	$\hat{m}_{\beta_2}$	$\hat{m}_{\beta_3}$	$\hat{m}_{\beta_4}$	$\hat{m}_{\beta_5}$	$\hat{\pi}_{x_1}$	$\hat{\pi}_{x_2}$	$\hat{\pi}_{x_3}$	$\hat{\pi}_{x_4}$	$\hat{\pi}_{x_5}$
100	PCS	Mean	0.199	0.200	0.200	0.200	0.201	0.153	0.230	0.302	0.229	0.086
		SD	0.037	0.037	0.037	0.037	0.037	0.034	0.039	0.043	0.039	0.023
	HD	Mean	0.200	0.200	0.200	0.200	0.200	0.147	0.230	0.311	0.230	0.082
		SD	0.039	0.040	0.039	0.039	0.040	0.033	0.043	0.037	0.042	0.019
	SCS	Mean	0.200	0.200	0.200	0.200	0.200	0.153	0.230	0.302	0.230	0.085
		SD	0.039	0.085	0.038	0.038	0.038	0.033	0.039	0.043	0.039	0.022
	LD	Mean	0.200	0.200	0.200	0.200	0.200	0.150	0.230	0.307	0.230	0.083
		SD	0.038	0.038	0.038	0.038	0.038	0.033	0.041	0.045	0.040	0.019
1000	PCS	Mean	0.200	0.200	0.200	0.200	0.200	0.148	0.236	0.319	0.236	0.061
		SD	0.013	0.013	0.013	0.013	0.014	0.012	0.014	0.017	0.015	0.011
	HD	Mean	0.200	0.200	0.200	0.200	0.200	0.147	0.237	0.320	0.237	0.059
		SD	0.013	0.013	0.013	0.013	0.013	0.011	0.014	0.015	0.014	0.008
	SCS	Mean	0.200	0.200	0.200	0.200	0.200	0.148	0.236	0.319	0.237	0.060
		SD	0.015	0.015	0.015	0.015	0.015	0.011	0.014	0.016	0.014	0.013
	LD	Mean	0.200	0.200	0.200	0.200	0.200	0.147	0.237	0.320	0.237	0.059
		SD	0.013	0.013	0.013	0.013	0.013	0.011	0.014	0.015	0.013	0.008
10,000	PCS	Mean	0.200	0.200	0.200	0.200	0.200	0.147	0.236	0.320	0.237	0.060
		SD	0.006	0.006	0.006	0.006	0.006	0.008	0.006	0.011	0.006	0.008
	HD	Mean	0.200	0.200	0.200	0.200	0.200	0.147	0.236	0.320	0.237	0.060
		SD	0.004	0.004	0.004	0.004	0.004	0.004	0.004	0.005	0.004	0.002
	SCS	Mean	0.200	0.200	0.200	0.200	0.200	0.147	0.236	0.320	0.237	0.060
		SD	0.005	0.005	0.005	0.005	0.005	0.004	0.006	0.008	0.006	0.008
	LD	Mean	0.200	0.200	0.200	0.200	0.200	0.147	0.236	0.320	0.237	0.060
		SD	0.004	0.004	0.004	0.004	0.004	0.004	0.005	0.005	0.005	0.002

Table 6. Scenario IV: Means and SDs of 4 distances (PCS, HD, SCS, LD). A 5×5 contingency table was generated having fixed the row marginal probabilities at (0.04, 0.20, 0.20, 0.20, 0.36). The number of MC replications used is 10,000.

N	Statistical Distance	Summary	Estimates Means and SDs over 10,000 Replications									
			$\hat{m}_{\beta_1}$	$\hat{m}_{\beta_2}$	$\hat{m}_{\beta_3}$	$\hat{m}_{\beta_4}$	$\hat{m}_{\beta_5}$	$\hat{\pi}_{x_1}$	$\hat{\pi}_{x_2}$	$\hat{\pi}_{x_3}$	$\hat{\pi}_{x_4}$	$\hat{\pi}_{x_5}$
100	PCS	Mean	0.074	0.197	0.197	0.197	0.335	0.214	0.173	0.228	0.132	0.253
		SD	0.022	0.037	0.038	0.038	0.045	0.038	0.035	0.039	0.031	0.041
	HD	Mean	0.070	0.194	0.195	0.195	0.346	0.215	0.170	0.231	0.126	0.258
		SD	0.015	0.039	0.039	0.039	0.048	0.041	0.037	0.042	0.030	0.044
	SCS	Mean	0.074	0.194	0.195	0.195	0.342	0.214	0.173	0.229	0.131	0.253
		SD	0.015	0.039	0.039	0.039	0.048	0.038	0.035	0.040	0.030	0.041
	LD	Mean	0.071	0.195	0.196	0.196	0.342	0.214	0.172	0.230	0.128	0.256
		SD	0.015	0.037	0.038	0.038	0.046	0.040	0.036	0.041	0.030	0.042
1000	PCS	Mean	0.042	0.200	0.200	0.200	0.358	0.217	0.168	0.234	0.119	0.262
		SD	0.011	0.014	0.013	0.013	0.017	0.014	0.013	0.014	0.014	0.015
	HD	Mean	0.039	0.200	0.200	0.200	0.361	0.217	0.167	0.235	0.118	0.263
		SD	0.006	0.013	0.013	0.013	0.015	0.013	0.012	0.013	0.010	0.014
	SCS	Mean	0.039	0.200	0.200	0.200	0.361	0.217	0.168	0.234	0.118	0.263
		SD	0.007	0.013	0.013	0.013	0.016	0.016	0.013	0.014	0.010	0.015
	LD	Mean	0.040	0.200	0.200	0.200	0.360	0.217	0.167	0.235	0.118	0.263
		SD	0.006	0.013	0.013	0.013	0.015	0.013	0.012	0.013	0.010	0.014
10,000	PCS	Mean	0.040	0.200	0.200	0.200	0.360	0.217	0.167	0.235	0.118	0.263
		SD	0.008	0.005	0.007	0.007	0.009	0.006	0.005	0.005	0.007	0.006
	HD	Mean	0.040	0.200	0.200	0.200	0.360	0.217	0.167	0.235	0.118	0.263
		SD	0.002	0.004	0.004	0.004	0.005	0.004	0.004	0.004	0.003	0.004
	SCS	Mean	0.040	0.200	0.200	0.200	0.360	0.217	0.167	0.235	0.118	0.263
		SD	0.002	0.004	0.004	0.004	0.005	0.006	0.005	0.007	0.003	0.008
	LD	Mean	0.040	0.200	0.200	0.200	0.360	0.217	0.167	0.235	0.118	0.263
		SD	0.002	0.004	0.004	0.004	0.005	0.004	0.004	0.005	0.003	0.005

Case 2: *X is discrete and Y is continuous*
In this section, we are interested in solving the optimization problem (5) when X is discrete, Y is continuous and X, Y are independent of each other. To evaluate the performance of our procedure, we used Hellinger's distance, which in this case takes on the following form:

$$HD(f^*, m_\beta^*) = \int \sum_x \left[\sqrt{f_N^*(x, y)} - \sqrt{m_\beta^*(x, y)} \right]^2 dy = \int \sum_x \left[\sqrt{f_Y^*(y) \cdot \frac{n_X}{N}} - \sqrt{m_X(x) \cdot m_Y^*(y)} \right]^2 dy.$$

The aim of this simulation is to obtain the minimum Hellinger distance estimators of π_x and μ assuming (without loss of generality) that σ^2 is known to be equal to 1. All calculations were performed in R language.

For this purpose, we generated mixed-type data of size N using the package OrdNor (Amatya and Demirtas [40]). More precisely, the data are comprised of one categorical variable X with three levels and probability vector $(1/3, 1/3, 1/3)$, while the continuous part is coming from a trivariate normal distribution; symbolic $Y = (Y_1, Y_2, Y_3) \sim MVN_3(\mu, \mathbf{I}_3)$, where $\mu^T = (\mu_1, \mu_2, \mu_3)$. We used two different mean vectors: $\mu^T = (0, 0, 0)$ and $\mu^T = (0, 3, 6)$. The set of ordinal and normal variables were generated concurrently using an overall correlation matrix Σ, which consists of three components/sub-matrices: Σ_{OO}, Σ_{ON} and Σ_{NN}, with O and N corresponding to "Ordinal" and "Normal" variables, respectively. More precisely, the overall correlation matrix Σ used is the following

$$\Sigma = \begin{pmatrix} 1 & \rho_{ON} & \rho_{ON} & \rho_{ON} \\ \rho_{ON} & 1 & 0 & 0 \\ \rho_{ON} & 0 & 1 & 0 \\ \rho_{ON} & 0 & 0 & 1 \end{pmatrix},$$

where $\Sigma_{OO} = 1, \Sigma_{NN} = \mathbf{I}_3, \Sigma_{ON} = \begin{pmatrix} \rho_{ON} & \rho_{ON} & \rho_{ON} \end{pmatrix}$ and ρ_{ON} represents the polyserial correlations for the ON combinations (for more information on polyserial correlations refer to Olsson et al. [41]). Since X, Y were assumed to be independent, we set $\rho_{ON} = 0.0$. However, we also used weak correlations, say $\rho_{ON} = 0.1$ and 0.2, to investigate whether the estimates we receive in these cases remain reasonable.

The kernel function was the multivariate normal density $MVN_3(\mathbf{0}, \mathrm{H})$ with H being estimated by the data using the kde function of the ks package (Duong [42]), $m_Y^*(y)$ represented the multivariate normal density $MVN_3(\mu, \Sigma + \mathrm{H})$ and $m_X(x)$ was the multinomial mass function. This choice of smoothing parameter, stemmed from the fact that we were interested in evaluating the performance, in terms of robustness, of standard bandwidth selection.

To solve the optimization problem, the solnp function of the Rsolnp package (Ye [37]) was used. Specifically, the initial values set for the probabilities $\pi_{x_1}, \pi_{x_2}, \pi_{x_3}$ associated with the X variable were random uniform numbers in the interval $[0, 1]$, while the initial values for the means $\mu_{y_1}, \mu_{y_2}, \mu_{y_3}$ were random numbers in the interval $[Q1(Y_i), Q3(Y_i)]$ for $i = 1, 2, 3$, where $Q1$ and $Q3$ stand for the respective 25th and the 75th quantile per component of the continuous part. Following the same procedure with the one of Basu and Lindsay [2] in the univariate continuous case, here (in the mixed-case) the numerical evaluation of the integrals was also done on the basis of the Simpson's 1/3rd rule using the sintegral function of the Bolstad2 package (Bolstad [43]). Moreover, we calculated the mean values, the SDs, as well as the percentages of bias of the mean and the probability vectors for three different sample sizes: $N = 100$; $N = 1000$ and $N = 1500$ over 1000 MC replications. The bias is defined as the difference of the estimates from their "true" values, that is, $bias(\mu_{y_i}) = \hat{\mu}_{y_i} - \mu_i$ and $bias(\pi_{x_i}) = \hat{\pi}_{x_i} - 1/3$ for $i = 1, 2, 3$. The results are shown in Tables 7 and 8.

In particular, Table 7 illustrates the mean values, the SDs and the bias percentages of the corresponding minimum Hellinger distance estimators, over 1000 MC replications, for the three different sample sizes and polyserial correlations, when $\mu = (0, 0, 0)^T$. The estimates for the π_{x_i} are approximately equal to $1/3 = 0.333$, while the μ_{y_i} estimates are almost zero, even in the cases of weak correlations. When $\rho_{ON} = 0.0$, the sample size choice does not seem to affect the values of the estimates either overall or per component of X, Y variables. Specifically, we observe that the total absolute bias, computed as the sum of the individual component-wise absolute biases of the vectors $\pi^T = (\pi_1, \pi_2, \pi_3)$ and $\mu^T = (\mu_1, \mu_2, \mu_3)$ are approximately the same, with larger samples providing slightly less biases at the expense of a higher computational cost.

Table 7. Means, Absolute Biases and Overall Absolute Bias of the Hellinger's distance (HD). The data were concurrently generated with a given correlation structure (an overall correlation matrix Σ) and consist of a discrete variable X with marginal probability vector $(1/3, 1/3, 1/3)$ and a continuous vector $Y = (Y_1, Y_2, Y_3) \sim MVN_3(\mu, \mathbf{I}_3)$, where $\mu^T = (0, 0, 0)$ and $\mathbf{I}_3$ is a (3×3) identity matrix. The number of MC replications used is 1000.

ρ_{ON}	N	Summary	Estimates Means, Biases over 1000 Replications					
			$\hat{\pi}_{x_1}$	$\hat{\pi}_{x_2}$	$\hat{\pi}_{x_3}$	$\hat{\mu}_{y_1}$	$\hat{\mu}_{y_2}$	$\hat{\mu}_{y_3}$
0.0	50	Mean	0.332	0.340	0.329	0.016	0.011	−0.011
		Abs. Biases	0.001	0.007	0.004	0.016	0.011	0.011
		Overall Bias			0.050			
	100	Mean	0.330	0.350	0.320	0.017	−0.018	−0.010
		Abs. Biases	0.003	0.017	0.013	0.017	0.018	0.010
		Overall Bias			0.078			
	1000	Mean	0.324	0.337	0.339	0.001	−0.008	0.007
		Abs. Biases	0.009	0.004	0.006	0.001	0.008	0.007
		Overall Bias			0.035			
0.1	50	Mean	0.351	0.320	0.329	−0.006	0.003	0.005
		Abs. Biases	0.018	0.013	0.004	0.006	0.003	0.005
		Overall Bias			0.049			
	100	Mean	0.330	0.323	0.347	0.001	0.005	−0.004
		Abs. Biases	0.003	0.010	0.014	0.001	0.005	0.004
		Overall Bias			0.037			
	1000	Mean	0.327	0.343	0.330	−0.021	0.008	0.003
		Abs. Biases	0.006	0.010	0.003	0.021	0.008	0.003
		Overall Bias			0.051			

In Table 8, analogous results are presented with the difference that the mean vector used was $\mu = (0, 3, 6)^T$. The π_{x_i} estimates are very close to $1/3$ ($= 0.333$) for all X components, no matter which sample size or correlation is used. On the contrary, the interpretation of the μ_i estimates slightly differs in this case. We also calculated the overall absolute bias as well as the individual, per parameter, absolute biases. In this case, larger samples clearly provide estimates with smaller bias for both parameter vectors π, μ and for both cases, the case of independence as well as the case of weak correlations. However, the computational time increases.

In what follows, we also present -for illustration purposes- a small simulation example using a mixed-type, contaminated data set of size $N = 1000$, which was generated using OrdNor package setting $\rho_{ON} = 0.0$. Once again, the data were comprised of one categorical variable X with three levels and probability vector $(1/3, 1/3, 1/3)$, and a trivariate continuous vector $Y = (Y_1, Y_2, Y_3)$. The contamination is happening only in the continuous part on the basis of $\alpha \in \{1.00, 0.95, 0.90, 0.85, 0.80\}$, as follows: $Y \sim \alpha \times MVN_3(\mathbf{0}, \mathbf{I}_3) + (1 - \alpha) \times MVN_3(\mu, \mathbf{I}_3)$, where $\mu^T = (3, 3, 3)$. This means that, $N_1 = \alpha \times N$ data were generated with Y coming from multivaraiate standard normal and the remaining $N_2 = N - N_1$ subset of the data followed a multivaraiate normal distribution with mean vector $\mu^T = (3, 3, 3)$. It goes without saying that when $\alpha = 1.00$, there is no contamination. Here, we are still considering the same optimization problem with the one described above and, consequently, we are interested in evaluating the minimum Hellinger distance estimators over 1000 MC replications by examining/studying to what extend the contamination level affects these estimates.

Table 8. Means, Absolute Biases and Overall Absolute Bias of the Hellinger's distance (HD). The data were concurrently generated with a given correlation structure (an overall correlation matrix Σ) and consist of a discrete variable X with marginal probability vector $(1/3, 1/3, 1/3)$ and a continuous vector $Y = (Y_1, Y_2, Y_3) \sim MVN_3(\mu, \mathbf{I}_3)$, where $\mu^T = (0, 3, 6)$ and $\mathbf{I}_3$ is a (3×3) identity matrix. The number of MC replications used is 1000.

ρ_{ON}	N	Summary	Estimates Means, Biases over 1000 Replications					
			$\hat{\pi}_{x_1}$	$\hat{\pi}_{x_2}$	$\hat{\pi}_{x_3}$	$\hat{\mu}_{y_1}$	$\hat{\mu}_{y_2}$	$\hat{\mu}_{y_3}$
0.0	50	Mean	0.340	0.328	0.332	−0.004	2.606	5.227
		Abs. Biases	0.007	0.005	0.001	0.004	0.394	0.773
		Overall Bias			1.184			
	100	Mean	0.313	0.350	0.337	−0.004	2.777	5.593
		Abs. Biases	0.020	0.017	0.004	0.004	0.223	0.407
		Overall Bias			0.675			
	1000	Mean	0.338	0.334	0.328	0.012	2.972	5.958
		Abs. Biases	0.005	0.001	0.005	0.012	0.028	0.042
		Overall Bias			0.093			
0.1	50	Mean	0.347	0.323	0.330	−0.021	2.628	5.249
		Abs. Biases	0.014	0.010	0.003	0.021	0.372	0.751
		Overall Bias			1.171			
	100	Mean	0.317	0.343	0.340	0.017	2.817	5.615
		Abs. Biases	0.016	0.010	0.007	0.017	0.183	0.385
		Overall Bias			0.618			
	1000	Mean	0.334	0.320	0.346	−0.013	2.988	5.956
		Abs. Biases	0.001	0.013	0.013	0.013	0.012	0.044
		Overall Bias			0.096			
0.2	50	Mean	0.324	0.333	0.343	−0.004	2.589	5.240
		Abs. Biases	0.009	0.000	0.010	0.004	0.411	0.760
		Overall Bias			1.194			
	100	Mean	0.329	0.350	0.321	0.024	2.763	5.549
		Abs. Biases	0.004	0.017	0.012	0.024	0.237	0.451
		Overall Bias			0.745			
	1000	Mean	0.337	0.344	0.319	−0.011	2.971	5.951
		Abs. Biases	0.004	0.011	0.014	0.019	0.029	0.049
		Overall Bias			0.118			

As indicated from Table 9, when there is no contamination in the data ($\alpha = 1.00$), the estimates for the π_{x_i}s are almost equal to $1/3$, while the μ_y's estimates are almost equal to zero. As the data become more contaminated (i.e., the value of α decreases), the minimum disparity estimators corresponding to X variable remain pretty consistent with their true values. However, this is not the case with the estimates for the μ_{y_i}s, which deteriorate as the value of the contamination level α shifts from the target/null value, that is 1.00.

The mean parameters are estimated with reasonable bias (maximum bias is 9% for the second component of the mean) when $\alpha = 0.95$, that is the contamination is 5%. When the contamination is 10%, the bias of the mean components is relatively high but still below 19%. With higher contamination, the percentage of bias in the mean components is in the interval $[28.3\%, 47\%]$. This is the result of using standard density estimation to obtain the smoothing parameters for the different mean components. Smaller values of these component smoothing parameters result in substantial bias reduction.

Table 9. Means and SDs of the Hellinger's distance (HD). The data were concurrently generated with a given correlation structure (an overall correlation matrix Σ) and consist of a discrete variable X with marginal probability vector $(1/3, 1/3, 1/3)$ and a continuous trivariate vector $Y = (Y_1, Y_2, Y_3) \sim \alpha \times MVN_3(\mathbf{0}, \mathbf{I}_3) + (1 - \alpha) \times MVN_3(\boldsymbol{\mu}, \mathbf{I}_3)$, where $\boldsymbol{\mu}^T = (3, 3, 3)$, $\mathbf{I}_3$ is a (3×3) identity matrix and $\alpha = 1.00(0.05)0.80$ indicates the contamination level. The number of MC replications used is 1000.

ρ_{ON}	N	α	Summary	Estimates Means and SDs over 1000 Replications					
				$\hat{\pi}_{x_1}$	$\hat{\pi}_{x_2}$	$\hat{\pi}_{x_3}$	$\hat{\mu}_{y_1}$	$\hat{\mu}_{y_2}$	$\hat{\mu}_{y_3}$
0.0	1000	1.00	Mean	0.324	0.337	0.339	0.001	−0.008	0.007
			SD	0.293	0.293	0.298	0.378	0.378	0.386
		0.95	Mean	0.327	0.326	0.347	0.068	0.090	0.079
			SD	0.304	0.299	0.309	0.413	0.413	0.413
		0.90	Mean	0.318	0.331	0.351	0.188	0.170	0.189
			SD	0.300	0.305	0.306	0.443	0.450	0.436
		0.85	Mean	0.324	0.337	0.339	0.292	0.283	0.312
			SD	0.293	0.293	0.297	0.484	0.487	0.491
		0.80	Mean	0.324	0.337	0.338	0.447	0.436	0.470
			SD	0.293	0.293	0.297	0.552	0.547	0.559

We also looked at the case where the continuous model was contaminated by a trivariate normal with mean $\boldsymbol{\mu}^T = (1.5, 1.5, 1.5)$ and covariance matrix $\mathbf{I}$. In this case (results not shown), when the contamination is 5% the maximum bias of the mean components is 6.6%, while when the contamination is 10% the maximum bias of the mean components is 13.5%. Again, in this case the bandwidth parameters were obtained by fitting a unimodal density to the data.

The above results are not surprising. A judicious selection of the smoothing parameter decreases the bias of the component estimates of the mean. Agostinelli and Markatou [44] provide suggestions of how to select the smoothing parameter that can be extended and applied in this context.

8. Discussion and Conclusions

In this paper, we discuss Pearson residual systems that conform to the measurement scale of the data. We place emphasis on the mixed-scale measurements scenario, which is equivalent to having both discrete (categorical or nominal) and continuous type random variables, and obtain robust estimators of the parameters of the joint probability distribution that describes those variables. We show that, disparity methods can be used to actually control against model misspecification and the presence of outliers, and these methods provide reasonable results.

The scale and nature of measurement of the data imposes additional challenges, both computationally and statistically. Detecting outliers in this multidimensional space is an open research question (Eiras-Franco et al. [45]). The concept of outliers has a long history in the field of statistics and outlier detection methods have broad applications in many scientific fields such as security (Diehl and Hampshire [46], Portnoy et al. [47]), health care (Tran et al. [48]) and insurance (Konijn and Kowalczyk [49]) to mention just a few.

Classical outlier detection methods are largely designed for single measurement scale data. Handling mixed measurement scale is a challenge with few works coming from both, the field of statistics (Fraley and Wilkinson [50], Wilkinson [51]) and the fields of engineering and computer science (Do et al. [52], Koufakou et al. [53]). All these works use some version of a probabilistic outlier, either looking for regions in the space of data that have low density (Do et al. [52], Koufakou et al. [53]) or by attaching a probability, under a model, to the suspicious data point (Fraley and Wilkinson [50], Wilkinson [51]).

Our concept of a probabilistic outlier discussed here and expressed via the construction of appropriate Pearson residuals can unify the different measurement scales, and the class of disparity functions discussed above can provide estimators for the model parameters that are not influenced unduly by potential outliers.

One of the important parameters that controls the robustness of these methods is the smoothing parameter(s) used to compute the density estimator of the continuous part of the model. In our computations, we use standard smoothing parameters obtained from utilizing appropriate R functions for density estimation. The results show that, depending on the level of contamination and the type of contaminating probability model, the performance of the methods is satisfactory. Specifically, a small simulation study using the model reported in the caption of Table 9 shows that the overall bias associated with the mean components of the standard multivariate normal model is low when contamination with a multivariate normal model with mean components equal to 3 is less than or equal to 10%. But even in this case, when the percentage of contamination is greater than 10%, the bias increases when the smoothing parameter used is the one obtained from the R density function. Here, smaller values of the smoothing parameter guarantee reduction of the bias.

Devising rules for selecting the smoothing parameter(s) in the context of mixed-scale measurements that can guarantee robustness for larger than 5% levels of contamination may be possible. However, it is the opinion of the authors that greater levels of data inhomogeneity may indicate model failure, a case where assessing model goodness of fit is of importance.

Author Contributions: The authors of this paper have contributed as follows. *Conceptualization*: M.M.; *Methodology*: M.M., E.M.S., R.L.; *Software*: E.M.S., H.W.; *Writing-original draft presentation*: M.M., E.M.S., R.L., H.W.; *Supervision, funding acquisition and project administration*: M.M. All authors have read and agreed to the published version of the manuscript.

Funding: This research was funded by the Troup Fund, KALEIDA Health Foundation, under award number 82114, to Markatou who supported the work of the first and the third author of the paper.

Conflicts of Interest: The authors declare no conflict of interest.

Abbreviations

The following abbreviations are used in this manuscript:

ALT	Alanine Aminotransferase
HD	Twice-Squared Hellinger's Disparity
LD	Likelihood Disparity
MC	Monte Carlo Replications
MDE	Minimum Distance Estimators
MLE	Maximum Likelihood Estimator
PCS	Pearson's Chi-Squared Disparity Divided by 2
PWD	Power Divergence Disparity
RAF	Residual Adjustment Function
SCS	Symmetric Chi-Squared Disparity
SD	Standard Deviation

Appendix A

Appendix A.1. Proof of Proposition 3

Proof. The equations (4) are obtained from solving optimization problem (3). To solve this problem we need to form the corresponding Langrangian, which is

$$\sum_{x,y} G(\delta(x,y)) m_\beta(y|x) \pi_x - \lambda \left(\sum \pi_x - 1 \right).$$

(i) Let ∇_β denote gradient with respect to β. The estimators of β are obtained as solutions of the set of equations:

$$\nabla_\beta \left\{ \sum_{x,y} G(\delta(x,y)) m_\beta(y|x) \pi_x - \lambda \left(\sum \pi_x - 1 \right) \right\} = 0,$$

which can be equivalently expressed as follows,

$$\sum_{x,y} \pi_x [\nabla_\beta G(\delta(x,y))] m_\beta(y|x) + \sum_{x,y} \pi_x G(\delta(x,y)) \nabla_\beta(y|x) = 0.$$

Notice that the ∇_β of $G(\delta(x,y))$ is given by

$$\nabla_\beta G(\delta(x,y)) = -G'(\delta(x,y))(\delta(x,y)+1)\, u(y|x;\beta),$$

where the superscript "'" denote derivative with respect to δ, $\delta(x,y)$ is the Pearson residual and

$$u(y|x;\beta) = \frac{\nabla_\beta m_\beta(y|x)}{m_\beta(y|x)} = \nabla_\beta \ln[m_\beta(y|x)]$$

is the score for β in the conditional distribution of y given x. Therefore,

$$\sum_{x,y} A(\delta(x,y)) \pi_x u(y|x;\beta) m_\beta(y|x) = 0,$$

where

$$A(\delta(x,y)) = G'(\delta(x,y))[\delta(x,y)+1] - G(\delta(x,y)).$$

By making use of the fact that $\sum_x \pi_x \nabla_\beta m_\beta(y|x) = 0$, the resulting equations can represented as

$$\sum_{x,y} \frac{A(\delta(x,y))+1}{\delta(x,y)+1} n_{x,y} u(y|x;\beta) = 0,$$

or equivalently,

$$\sum_{x,y} w(\delta(x,y)) n_{x,y} u(y|x;\beta) = 0.$$

Without loss of generality, we can take,

$$w(\delta(x,y)) = \min \left\{ \frac{[A(\delta(x,y))+1]^+}{\delta(x,y)+1}, 1 \right\}, \quad w(\delta(x,y)) \le 1.$$

(ii) We now need to obtain $\hat{\pi}_x$, which can be obtained by setting the gradient of formula with respect to π_z equal to zero, that is, by the following equations:

$$\sum_y G'(\delta(z,y))[\nabla_{\pi_z}\delta(z,y)] m_\beta(y|z) \pi_z + \sum_y G(\delta(z,y)) m_\beta(y|z) - \lambda = 0.$$

Recording $A(\delta(z,y)) = G'(\delta(z,y))[\delta(z,y)+1] - G(\delta(z,y))$ and $\delta(z,y)+1 = \frac{n_{z,y}/n}{m_\beta(y|z)\pi_z}$, the above equations are reduced to,

$$\sum_y A(\delta(z,y)) m_\beta(z,y) \frac{1}{\pi_z} + \lambda = 0$$

and we readily conclude that,

$$\pi_z = -\frac{1}{\lambda} \sum_y A(\delta(z,y)) m(z,y), \forall z.$$

Furthermore, to satisfy the constraint $\sum_x \pi_x = 1$, we obtain

$$\lambda = -\sum_{x,y} A(\delta(x,y))m_\beta(x,y).$$

Therefore, we get

$$\sum_{x,y} A(\delta(x,y))m_\beta(y,x)\left[\frac{I(X=z)}{\pi_x} - 1\right] = 0$$

and by making use of the fact that $\sum_{x,y} m_\beta(x,y)\left[\frac{I(X=z)}{\pi_x} - 1\right] = 0$, the above equation can be represented as

$$\sum_{x,y} w(\delta(x,y))n_{x,y}\left[\frac{I(X=x)}{\pi_x} - 1\right] = 0$$

for any x where $I(X = x)$ is the indicator function of the event $\{X = x\}$. $\quad\square$

Appendix A.2. Proof of Proposition 5

Recall that β_ϵ is a solution of the set of estimating equation

$$\sum_{s,t} w(\delta_\epsilon(s,t))u(t|s;\beta_\epsilon)d_\epsilon(s,t) = 0, \tag{A1}$$

where $d_\epsilon(s,t) = (1-\epsilon)d(s,t) + \epsilon\nabla_{x,y}(s,t)$ and $u(t|s;\beta) = \frac{\nabla_\beta m_\beta(s,t)}{m_\beta(s,t)} = \nabla_\beta \ln[m_\beta(s,t)]$ is a p-dimensional vector.

The influence function of β is calculated by differentiating, with respect to ϵ, the quantity (A1), and evaluating the derivative at $\epsilon = 0$. Thus, we need

$$\frac{d}{d\epsilon}\left\{\sum_{s,t} w(\delta_\epsilon(s,t))u(t|s;\beta_\epsilon)d(s,t)\right.$$
$$-\epsilon\sum_{s,t} w(\delta_\epsilon(s,t))u(t|s;\beta_\epsilon)d(s,t) \tag{A2}$$
$$\left.+\epsilon\sum_{s,t} w(\delta_\epsilon(s,t))u(t|s;\beta_\epsilon)\nabla_{(x,y)}(s,t)\right\}\bigg|_{\epsilon=0} = 0.$$

Taking into account that $\delta_\epsilon(s,t) = \frac{d_\epsilon(s,t)}{m_\beta(s,t)} - 1 = \frac{d_\epsilon(s,t)}{m_\beta(t|s)\pi_s} - 1$, the aforementioned evaluation implies

$$\left\{\sum_{s,t}(\delta_0(t) + 1)w_0'(\delta_0(s,t))u(t|s;\beta_0)u^T(t|s;\beta_0)d(s,t)\right.$$
$$-\sum_{s,t} w(\delta_0(s,t))\nabla u(t|s;\beta_0)d(s,t)\bigg\}\beta_0'$$
$$=\sum_{s,t}\left\{\frac{I(s=x,y=t)}{m_{\beta_0}(t|s)\pi_s} - \frac{d(s,t)}{m_{\beta_0}(t|s)\pi_s}w'(\delta_0(s,t))\right\}u(t|s;\beta_0)d(s,t) \tag{A3}$$
$$-\sum_{s,t} w(\delta_0(s,t))u(t|s;\beta_0)d(s,t) + w(\delta_0(x,y))u(y|x;\beta_0),$$

which implies that

$$\beta_0' = IF(\beta;F) = [A(d)]^{-1}B(x,y;d).$$

Appendix A.3. Assumptions of Theorem 1

The following assumptions are needed to be able to establish asymptotic normality of the estimators.

1. The weight functions are nonnegative, bounded and differentiable with respect to δ.

2. The weight function is regular, that is, $w'(\delta)(\delta + 1)$ is bounded, where $w'(\delta)$ is the derivative of w with respect to δ.

3. $\sum_{x,y} m^{\frac{1}{2}}(x,y) E[u_k^2(y|x; \beta_0)] < \infty.$

4. The elements of the Fisher information matrix are finite and the Fisher information matrix is nonsingular.

5. $\sum_{x,y} m^{\frac{1}{2}}(x,y) E[u_i^2(y|x; \beta_0) u_j^2(y|x; \beta_0)] < \infty \ \ \forall i,j = 1, 2, \cdots, p.$

6. If β_0 denotes the true value of β, there exist functions $M_{ijk}(x)$ such that $|u_{ijk}(y|x; \beta_0)| \leq M_{ijk}(x), \forall \beta$ with $\| \beta - \beta_0 \|^2 < r(\beta_0), r(\beta_0) < 0$ and $E_{\beta_0}|M_{ijk}(y|x)| < \infty, \ \ \forall i,j,k.$

7. If β_0 denotes the true value of β, there is a neighborhood $N(\beta_0)$ such that for $\beta \in N(\beta_0)$ the quantity $|u_t(y|x; \beta_0) u_i(y|x; \beta_0) u_e(y|x; \beta_0)|$ are bounded by $M_1(y|x)$ and $M_2(y|x)$ respectively, such that their corresponding expectations are finite.

8. $A''(\delta + 1)(\delta + 1)$ is bounded, where A'' denotes the second derivative of A with respect to δ.

References

1. Beran, R. Minimum Hellinger Distance Estimates for Parametric Models. *Ann. Stat.* **1977**, *5*, 445–463. [CrossRef]
2. Basu, A.; Lindsay, B.G. Minimum Disparity Estimation for Continuous Models: Efficiency, Distributions and Robustness. *Ann. Inst. Stat. Math.* **1994**, *46*, 683–705. [CrossRef]
3. Pardo, J.A.; Pardo, L.; Pardo, M.C. Minimum ϕ-Divergence Estimator in Logistic Regression Models. *Stat. Pap.* **2005**, *47*, 91–108. [CrossRef]
4. Pardo, J.A.; Pardo, L.; Pardo, M.C. Testing In Logistic Regression Models on ϕ-Divergences Measures. *J. Stat. Plan. Inference* **2006**, *136*, 982–1006. [CrossRef]
5. Pardo, J.A.; Pardo, M.C. Minimum ϕ-Divergence Estimator and ϕ-Divergence Statistics in Generalized Linear Models with Binary Data. *Methodol. Comput. Appl. Probab.* **2008**, *10*, 357–379. [CrossRef]
6. Simpson, D.G. Minimum Hellinger Distance Estimation for the Analysis of Count Data. *J. Am. Stat. Assoc.* **1987**, *82*, 802–807. [CrossRef]
7. Simpson, D.G. Hellinger Deviance Tests: Efficiency, Breakdown Points, and Examples. *J. Am. Stat. Assoc.* **1989**, *84*, 104–113. [CrossRef]
8. Markatou, M.; Basu, A.; Lindsay, B.G. Weighted Likelihood Estimating Equations: The Discrete Case with Applications to Logistic Regression. *J. Stat. Plan. Inference* **1997**, *57*, 215–232. [CrossRef]
9. Basu, A.; Basu, S. Penalized Minimum Disparity Methods for Multinomial Models. *Stat. Sin.* **1998**, *8*, 841–860.
10. Gupta, A.K.; Nguyen, T.; Pardo, L. Inference Procedures for Polytomous Logistic Regression Models Based on ϕ-Divergence Measures. *Math. Methods Stat.* **2006**, *15*, 269–288.
11. Martín, N.; Pardo, L. New Influence Measures in Polytomous Logistic Regression Models Based on Phi-Divergence Measures. *Commun. Stat. Theory Methods* **2014**, *43*, 2311–2321. [CrossRef]
12. Castilla, E.; Ghosh, A.; Martín, N.; Pardo, L. New Robust Statistical Procedures for Polytomous Logistic Regression Models. *Biometrics* **2018**, *74*, 1282–1291. [CrossRef] [PubMed]
13. Martín, N.; Pardo, L. Minimum Phi-Divergence Estimators for Loglinear Models with Linear Constraints and Multinomial Sampling. *Stat. Pap.* **2008**, *49*, 2311–2321. [CrossRef]
14. Pardo, L.; Martín, N. Minimum Phi-Divergence Estimators and Phi-Divergence Test for Statistics in Contingency Tables with Symmetric Structure: An Overview. *Symmetry* **2010**, *2*, 1108–1120. [CrossRef]
15. Pardo, L.; Pardo, M.C. Minimum Power-Divergence Estimator in Three-Way Contingency Tables. *J. Stat. Comput. Simul.* **2003**, *73*, 819–831. [CrossRef]
16. Pardo, L.; Pardo, M.C.; Zografos, K. Minimum ϕ-Divergence Estimator for Homogeneity in Multinomial Populations. *Sankhyā: Indian J. Stat. Ser. A (1961–2002)* **2001**, *63*, 72–92.
17. Basu, A.; Harris, I.A.; Hjort, N.L.; Jones, M.C. Robust and Efficient Estimation by Minimising a Density Power Divergence. *Biometrika* **1998**, *85*, 549–559. [CrossRef]
18. Csiszár, I. Information-Type Measures of Difference of Probability Distributions and Indirect Observations. *Stud. Sci. Math. Hung.* **1967**, *25*, 299–318.
19. Lindsay, B.G. Efficiency Versus Robustness: The Case for Minimum Hellinger Distance and Related Methods. *Ann. Stat.* **1994**, *22*, 1081–1114. [CrossRef]

20. Tamura, R.N.; Boos, D.D. Minimum Hellinger Distance Estimation for Multivariate Location and Covariance. *J. Am. Stat. Assoc.* **1986**, *81*, 223–229. [CrossRef]
21. Markatou, M.; Basu, A.; Lindsay, B.G. Weighted Likelihood Equations with Bootstrap Root Search. *J. Am. Stat. Assoc.* **1998**, *93*, 740–750. [CrossRef]
22. Haberman, S.J. Generalized Residuals for Log-Linear Models. In Proceedings of the 9th International Biometrics Conference, Boston, MA, USA, 22–27 August 1976; pp. 104–122.
23. Haberman, S.J.; Sinharay, S. Generalized Residuals for General Models for Contingency Tables with Application to Item Response Theory. *J. Am. Stat. Assoc.* **2013**, *108*, 1435–1444. [CrossRef]
24. Pierce, D.A.; Schafer, D.W. Residuals in Generalized Linear Models. *J. Am. Stat. Assoc.* **1986**, *81*, 977–986. [CrossRef]
25. Aerts, M.; Molenberghs, G.; Geys, H.; Ryan, L. *Topics in Modelling of Clustered Data*; Monographs on Statistics and Applied Probability; Chapman & Hall/CRC Press: New York, NY, USA, 1986; Volume 96.
26. Olkin, I.; Tate, R.F. Multivariate Correlation Models with Mixed Discrete and Continuous Variables. *Ann. Math. Stat.* **1961**, *32*, 448–465; With correction in **1961**, *36*, 343–344. [CrossRef]
27. Genest, C.; Nešlehová, J. A Primer on Copulas for Count Data. *ASTIN Bull.* **2007**, *37*, 475–515. [CrossRef]
28. Lauritzen, S.; Wermuth, N. Graphical Models for Associations between Variables, some of which are Qualitative and some Quantitative. *Ann. Stat.* **1989**, *17*, 31–57. [CrossRef]
29. Hampel, F.R.; Ronchetti, E.M.; Rousseeuw, P.J.; Stahel, W.A. *Robust Statistics: The Approach Based on Influence Functions*; Wiley Series in Probability and Mathematical Statistics. Probability and Mathematical Statistics; Wiley: New York, NY, USA, 1986.
30. Hampel, F.R. Contributions to the Theory of Robust Estimation. Ph.D. Thesis, Department of Statistics, University of California, Berkeley, Berkeley, CA, USA, 1968. Unpublished.
31. Hampel, F.R. The Influence Curve and its Role in Robust Estimation. *J. Am. Stat. Assoc.* **1974**, *69*, 383–393. [CrossRef]
32. Fienberg, S.E. The Analysis of Incomplete Multi-Way Contingency Tables. *Biometrics* **1972**, *28*, 177–202. [CrossRef]
33. Agresti, A. *Categorical Data Analysis*, 3rd ed.; John Wiley & Sons: Hoboken, NJ, USA, 2013.
34. Johnson, W.D.; May, W.L. Combining 2 × 2 Tables That Contain Structural Zeros. *Biometrics* **1972**, *14*, 1901–1911. [CrossRef]
35. Poon, W.Y.; Tang, M.L.; Wang, S.J. Influence Measures in Contingency Tables with Application in Sampling Zeros. *Sociol. Methods Res.* **2003**, *31*, 439–452. [CrossRef]
36. Alin, A.; Kurt, S. Ordinary and Penalized Minimum Power-Divergence Estimators in Two-Way Contingency Tables. *Comput. Stat.* **2008**, *23*, 455–468. [CrossRef]
37. Ye, Y. Interior Algorithms for Linear, Quadratic, and Linearly Constrained Convex Programming. Ph.D. Thesis, Department of Engineering-Economic Systems, Stanford University, Stanford, CA, USA, 1987. Unpublished.
38. Conn, A.R.; Gould, N.I.M.; Toint, P. A Globally Convergent Augmented Lagrangian Algorithm for Optimization with General Constraints and Simple Bounds. *SIAM J. Numer. Anal.* **1991**, *28*, 545–572. [CrossRef]
39. Birgin, E.G.; Martínez, J.M. Improving Ultimate Convergence of an Augmented Lagrangian Method. *Optim. Methods Softw.* **2008**, *23*, 177–195. [CrossRef]
40. Amatya, A.; Demirtas, H. OrdNor: An R Package for Concurrent Generation of Correlated Ordinal and Normal Data. *J. Stat. Softw.* **2015**, *68*, 1–14. [CrossRef]
41. Olsson, U.; Drasgow, F.; Dorans, N.J. The Polyserial Correlation Coefficient. *Psychmetrika* **1982**, *47*, 337–347. [CrossRef]
42. Duong, T. ks: Kernel Density Estimation and Kernel Discriminant Analysis for Multivariate Data in R. *J. Stat. Softw.* **2007**, *21*, 1–16. [CrossRef]
43. Bolstad, W.M. *Understanding Computational Bayesian Statistics*; John Wiley & Sons: Hoboken, NJ, USA, 2010.
44. Agostinelli, C.; Markatou, M. Test of Hypotheses Based on the Weighted Likelihood Methodology. *Stat. Sin.* **2001**, *11*, 499–514.
45. Eiras-Franco, C.; Martínez-Rego, D.; Guijarro-Berdiñas, B.; Alonso-Betanzos, A.; Bahamonde, A. Large Scale Anomaly Detection in Mixed Numerical and Categorical Input Spaces. *Inf. Sci.* **2019**, *487*, 115–127. [CrossRef]
46. Diehl, C.; Hampshire, J. Real-Time Object Classification and Novelty Detection for Collaborative Video Surveillance. In Proceedings of the 2002 International Joint Conference on Neural Networks. IJCNN'02 (Cat. No.02CH37290), Honolulu, HI, USA, 12–17 May 2002; Volume 3, pp. 2620–2625.
47. Portnoy, L.; Eskin, E.; Stolfo, S. Intrusion Detection with Unlabeled Data Using Clustering. In Proceedings of ACM CSS Workshop on Data Mining Applied to Security (DMSA-2001), Philadelphia, PA, USA, 5–8 November 2001; pp. 5–8.
48. Tran, T.; Phung, D.; Luo, W.; Harvey, R.; Berk, M.; Venkatesh, S. An Integrated Framework for Suicide Risk Prediction. In Proceedings of the 19th ACM SIGKDD International Conference on Knowledge Discovery and Data Mining, Chicago, IL, USA, 11–14 August 2013; ACM: New York, NY, USA, 2013; pp. 1410–1418.
49. Konijn, R.M.; Kowalczyk, W. Finding Fraud in Health Insurance Data with Two-Layer Outlier Detection Approach. In Data Warehousing and Knowledge Discovery, DaWak 2011; Cuzzocrea, A., Dayal, U., Eds.; Springer: Berlin/Heidelberg, Germany, 2011; pp. 394–405.
50. Fraley, C.; Wilkinson, L. Package 'HDoutliers'. R Package, 2020. Available online: https://cran.r-project.org/web/packages/HDoutliers/index.html (accessed on 31 December 2020).
51. Wilkinson, L. Visualizing Outliers. 2016. Available online: https://www.cs.uic.edu/~wilkinson/Publications/outliers.pdf (accessed on 31 December 2020).

52. Do, K.; Tran, T.; Phung, D.; Venkatesh, S. Outlier Detection on Mixed-Type Data: An Energy-Based Approach. In *Advanced Data Mining and Applications*; Li, J., Li, X., Wang, S., Li, J., Sheng, Q.Z., Eds.; Springer: Cham, Switzerland, 2016; pp. 111–125.
53. Koufakou, A.; Georgiopoulos, M.; Anagnostopoulos, G.C. Detecting Outliers in High-Dimensional Datasets with Mixed Attributes. In Proceedings of the 2008 International Conference on Data Mining, DMIN, Las Vegas, NV, USA, 14–17 July 2008; pp. 427–433.

Article

Rare Event Analysis for Minimum Hellinger Distance Estimators via Large Deviation Theory

Anand N. Vidyashankar [1,*] and Jeffrey F. Collamore [2]

[1] Department of Statistics, George Mason University, Fairfax, VA 22030, USA
[2] Department of Mathematical Sciences, University of Copenhagen, Universitetsparken 5, DK-2100 Copenhagen Ø, Denmark; collamore@math.ku.dk
* Correspondence: avidyash@gmu.edu

Abstract: Hellinger distance has been widely used to derive objective functions that are alternatives to maximum likelihood methods. While the asymptotic distributions of these estimators have been well investigated, the probabilities of rare events induced by them are largely unknown. In this article, we analyze these rare event probabilities using large deviation theory under a potential model misspecification, in both one and higher dimensions. We show that these probabilities decay exponentially, characterizing their decay via a "rate function" which is expressed as a convex conjugate of a limiting cumulant generating function. In the analysis of the lower bound, in particular, certain geometric considerations arise that facilitate an explicit representation, also in the case when the limiting generating function is nondifferentiable. Our analysis involves the modulus of continuity properties of the affinity, which may be of independent interest.

Keywords: Hellinger distance; large deviations; divergence measures; rare event probabilities

Citation: Vidyashankar, A.N.; Collamore, J.F. Rare Event Analysis for Minimum Hellinger Distance Estimators via Large Deviation Theory. *Entropy* **2021**, *23*, 386. https://doi.org/10.3390/e23040386

Academic Editor: Leandro Pardo

Received: 22 February 2021
Accepted: 15 March 2021
Published: 24 March 2021

Publisher's Note: MDPI stays neutral with regard to jurisdictional claims in published maps and institutional affiliations.

1. Introduction

In a variety of applications, the use of divergence-based inferential methods is gaining momentum, as these methods provide robust alternatives to traditional maximum likelihood-based procedures. Since the work of [1,2], divergence-based methods have been developed for various classes of statistical models. A comprehensive treatment of these ideas is available, for instance, in [3,4]. The objective of this paper is to study the large deviation tail behavior of the minimum divergence estimators and, more specifically, the minimum Hellinger distance estimators (MHDE).

To describe the general problem, suppose $\Theta \subset \mathbb{R}^d$, and let $\mathfrak{F} = \{f_\theta(\cdot) : \theta \in \Theta\}$ denote a family of densities indexed by θ. Let $\{X_n : n \geq 1\}$ denote a class of i.i.d. random variables, postulated to have a continuous density with respect to Lebesgue measure and belonging to the family $\mathfrak{F}$, and let X be a generic element of this class. We denote by $g(\cdot)$ the true density of X.

Before providing an informal description of our results, we begin by recalling that the square of the Hellinger distance (SHD) between two densities $h_1(\cdot)$ and $h_2(\cdot)$ on $\mathbb{R}$ is given by

$$\mathrm{HD}^2(h_1, h_2) = \left\| h_1^{\frac{1}{2}} - h_2^{\frac{1}{2}} \right\|_2^2 = 2 - 2 \int_{\mathbb{R}} (h_1(x)h_2(x))^{\frac{1}{2}} dx.$$

The quantity $\int_{\mathbb{R}} (h_1(x)h_2(x))^{\frac{1}{2}} dx$ is referred to as the *affinity* between $h_1(\cdot)$ and $h_2(\cdot)$ and denoted by $\mathscr{A}(h_1, h_2)$. Hence, the SHD between the postulated density and the true density is given by $\mathrm{SHD}(\theta) = \mathrm{HD}^2(f_\theta, g)$. When Θ is compact, it is known that there exists a unique $\theta_g \in \Theta$ minimizing the $\mathrm{SHD}(\theta)$. Furthermore, when $g(\cdot) = f_{\theta_0}(\cdot)$ and $\mathfrak{F}$ satisfies an identifiability condition, it is well known that θ_g coincides with θ_0; cf. [1]. Turning to the

303

sample version, we replace $g(\cdot)$ by $g_n(\cdot)$ in the definition of SHD, obtaining the objective function $\text{SHD}_n(\boldsymbol{\theta}) = \text{HD}^2(f_{\boldsymbol{\theta}}, g_n)$ and

$$g_n(x) = \frac{1}{nb_n} \sum_{i=1}^{n} K\left(\frac{x - X_i}{b_n}\right),\tag{1}$$

where the kernel $K(\cdot)$ is a probability density function and $b_n \searrow 0$ and $nb_n \nearrow \infty$ as $n \to \infty$.

It is known that when the parameter space Θ is compact, there exists a unique $\hat{\boldsymbol{\theta}}_n \in \Theta$ minimizing $\text{SHD}_n(\boldsymbol{\theta})$, and that $\hat{\boldsymbol{\theta}}_n$ converges almost surely to $\boldsymbol{\theta}_g$ as $n \to \infty$; cf. [1]. Furthermore, under some natural assumptions,

$$n^{\frac{1}{2}}(\hat{\boldsymbol{\theta}}_n - \boldsymbol{\theta}_g) \xrightarrow{d} G,\tag{2}$$

where, under the probability measure associated with $g(\cdot)$, G is a Gaussian random vector with mean vector $\mathbf{0}$ and covariance matrix Σ_g. If $g(\cdot) = f_{\boldsymbol{\theta}_0}(\cdot)$, then the variance of G coincides with the inverse of the Fisher information matrix $\mathfrak{I}(\boldsymbol{\theta}_0)$, yielding statistical efficiency. When the true distribution $g(\cdot)$ does not belong to $\mathfrak{F}$, we will call this the "model misspecifed case," while when $g \in \mathfrak{F}$, we will say that the "postulated model" holds.

In this paper, we focus on the large deviation behavior of $\{\hat{\boldsymbol{\theta}}_n : n \geq 1\}$; namely, the asymptotic probability that the estimate $\hat{\boldsymbol{\theta}}_n$ will achieve values within a set *away* from the central tendency described in (2). We establish results of the form

$$\log P_g(\hat{\boldsymbol{\theta}}_n \in B) \approx -n \inf_{\boldsymbol{\theta} \in B} I(\boldsymbol{\theta}),\tag{3}$$

for some "rate function" I and given Borel subset $B \subset \Theta$. Similar large deviation estimates for maximum likelihood estimators (MLE) have been investigated in [5–7], and for general M-estimators in [8,9]. These results allow for a precise description of the probabilities of Type I and Type II error in both the Neymann–Pearson and likelihood ratio test frameworks. Furthermore, large deviation bounds allow one to identify the best exponential rate of decrease of Type II error amongst all tests that satisfy a bound on the Type I error, as in Stein's lemma (cf. [10]). Additional evidence of the importance of large deviation results for statistical inference has been described in [11] and in the book [12].

One of our initial goals was to derive sharp probability bounds for Type I and Type II error in the context of robust hypothesis testing using Hellinger deviance tests. This article is a first step towards this endeavor. A key issue that distinguishes our work from earlier works is that, in our case, the objective function is a nonlinear function of the smoothed empirical measure, and the analysis of this case requires more involved methods compared with those currently existing in the statistical literature on large deviations. Consistent with large deviation analysis more generally, we identify the rate function I as the convex conjugate of a certain limiting cumulant generating function, although in our problem, we uncover a subtle asymmetry between the upper and lower bounds when our limiting generating function is nondifferentiable. In the classical large deviation literature, similar asymmetries have been studied in other one-dimensional contexts (e.g. [13]), although the statistical problem is still quite different, as the dependence on the parameter $\boldsymbol{\theta}$ arises explicitly—inhibiting the use of convexity methods typically exploited in the large deviation literature—and hence requiring novel techniques.

1.1. Large Deviations

In this subsection we provide relevant definitions and properties from large deviation theory required in the sequel. In the following, $\mathbb{R}_+$ will denote the set of non-negative real numbers.

Definition 1. *A collection of probability distributions* $\{P_n : n \geq 1\}$ *on a topological space* $(\mathscr{X}, \mathscr{B})$ *is said to satisfy the weak large deviation principle if*

$$\limsup_{n \to \infty} \frac{1}{n} \log P_n(F) \leq -\inf_{x \in F} I(x), \quad \text{for all closed } F \in \mathscr{B},$$

and

$$\liminf_{n \to \infty} \frac{1}{n} \log P_n(G) \geq -\inf_{x \in G} I(x) \quad \text{for all open sets } F \in \mathscr{B}$$

for some lower semicontinuous function $I : \mathscr{X} \to [0, \infty]$*. The function* I *is called the* rate function*. If the level sets of* I *are compact, we call* I *a* good rate function *and we say that* $\{P_n\}$ *satisfies the large deviation principle (LDP).*

We begin with a brief review of large deviation results for i.i.d. random variables and empirical measures. Let $\{X_n\} \subset \mathbb{R}$ be an i.i.d. sequence of real-valued random variables, and let P_n denote the distribution of the sample mean $\bar{X}_n$. If the moment generating function of X_1 is finite in a neighborhood of the origin, then Cramér's theorem states that $\{P_n\}$ satisfies the LDP with good rate function Λ^*, where Λ^* is the convex conjugate (or Legendre–Fenchel transform) of Λ, and where $\Lambda(\alpha) = \log E[e^{\alpha X_1}]$ is the cumulant generating function of X_1 (cf. [10], Section 2.2).

Next, consider the empirical measures $\{\mu_n\}$, defined by

$$\mu_n(B) = \frac{1}{n} \sum_{i=1}^{n} I_{\{X_i \in B\}}, \quad B \in \mathscr{B}, \tag{4}$$

where $\mathscr{B}$ denotes the collection of Borel subsets of $\mathbb{R}$. It is well known (cf. [14]) that $\{\mu_n\}$ converges weakly to P, namely to the distribution of X_1. Then Sanov's theorem asserts that $\{\mu_n\}$ satisfies a large deviation principle with rate function I_P given by

$$I_P(\nu) = \begin{cases} \mathrm{KL}(\nu, P) & \text{if } \nu \ll P, \\ \infty & \text{otherwise,} \end{cases} \tag{5}$$

where $\mathrm{KL}(\nu, P)$ is the *Kullback–Leibler information* between the probability measures ν and P. When ν and P each possesses a density with respect to Lebesgue measure (say p and g, respectively), the above expression becomes

$$\mathrm{KL}(p, g) := \begin{cases} \int_S p(x) \log\left(\frac{p(x)}{g(x)}\right) d\mu(x) & \text{if } p \ll g, \\ \infty & \text{otherwise.} \end{cases} \tag{6}$$

In Sanov's theorem, the rate function I_P is defined on the space of probability measures, which is a metric space with the open sets induced by weak convergence. Extensions of Sanov's theorem to strong topologies have been investigated in the literature; cf., e.g., [15].

We now turn to a general result, which will play a central role in this paper, namely Varadhan's integral lemma (cf. [10], Theorem 4.3.1). This result will allow us to infer the scaled limit of a sequence of generating functions from the existence of the large deviation principle.

Lemma 1 (Varadhan). *Let* $\{Y_n\}$ *be a sequence of random variables taking values in a regular topological space* $(\mathscr{X}, \mathscr{B})$*, and assume that the probability law of* $\{Y_n\}$ *satisfies the LDP with good rate function* I*. Then for any bounded continuous function* $F : \mathscr{X} \to \mathbb{R}$*,*

$$\lim_{n \to \infty} \frac{1}{n} \log E[\exp(nF(Y_n))] = \sup_{x \in \mathscr{X}} \{F(x) - I(x)\}. \tag{7}$$

1.2. Minimum Hellinger Distance Estimator and Large Deviations

We first observe that the MHDE is obtained by maximizing

$$\mathscr{A}_n(\boldsymbol{\theta}) \equiv \mathscr{A}_n(\boldsymbol{\theta}, g_n) := \int_{\mathbb{R}} f_{\boldsymbol{\theta}}^{\frac{1}{2}}(x) g_n^{\frac{1}{2}}(x) d\mu(x), \tag{8}$$

which involves solving the equation $\nabla \mathscr{A}_n(\boldsymbol{\theta}) = 0$. The idea behind the large deviation analysis is to observe that the large deviation behavior of the maximizer can be extracted from that of the objective function $\nabla \mathscr{A}_n(\boldsymbol{\theta})$ near **0**. By the Gärtner–Ellis theorem (cf. [10], Section 2.3), this amounts to investigating the asymptotic behavior as $n \to \infty$ of

$$\frac{1}{a_n} \log E_g[\exp\{a_n \langle \boldsymbol{\alpha}, \nabla \mathscr{A}_n(\boldsymbol{\theta}) \rangle\}], \quad \boldsymbol{\alpha} \in \mathbb{R}^d, \tag{9}$$

where $a_n \nearrow \infty$ as $n \to \infty$. In the case of maximum likelihood estimation (MLE) or minimum contrast estimation (MCE), the objective function can be expressed as

$$\sum_{i=1}^{n} h_{\boldsymbol{\theta}}(X_i) = n \int_{\mathbb{R}} h_{\boldsymbol{\theta}}(x) d\mu_n(x), \tag{10}$$

where $\{\mu_n : n \geq 1\}$ is the empirical measure associated with $\{X_k : 1 \leq k \leq n\}$. Thus, while the objective functions associated with the MLE and MCE are linear functions of the empirical measure, the affinity is a nonlinear function of the empirical measure. This creates certain complications in identifying the rate function $I(\cdot)$ alluded to in (3). Of course, in the case of likelihood and minimum contrast estimator analysis, an explicit formula for $I(\cdot)$ ensues as the Legendre–Fenchel transform of the cumulant generating function of $h_{\boldsymbol{\theta}}(X_1)$, viz. $\log E_{\boldsymbol{\theta}_0}[\exp(\alpha h_{\boldsymbol{\theta}}(X_1))]$. One approach to evaluating the limiting generating function is to apply Varadhan's lemma as given above in (7). In the context of our problem, that requires an investigation into the large deviation principle for the density estimators $g_n(\cdot)$ viewed as elements of $L_1(S)$, viz. the space of integrable functions on S. Equivalently, we require a version of Sanov's theorem in L_1-space, which leads to certain topological considerations. The main issue here is that, when L_1 is equipped with a norm topology, the sequence of kernel density estimates $\{g_n(\cdot)\}$ possesses large deviation bounds, but the associated rate function may not have compact level sets, as is required for a typical application of Varadhan's lemma. Nonetheless, one obtains a full LDP when $L_1(S)$ is equipped with the weak topology.

The asymptotic properties of MHDE, such as consistency and asymptotic normality, are established using the norm convergence of $g_n(\cdot)$ to $g(\cdot)$. For this reason, we focus on a subclass of densities $\mathscr{G}$ (see Proposition 1 below) possessing certain equicontinuity properties where norm convergence prevails. These issues are handled in Section 2, where the precise statements of our main results can also be found. Section 3 is devoted to the proofs of the main results. Section 4 contains some concluding remarks.

2. Notation, Assumptions, and Main Results

Let $f_{\boldsymbol{\theta}}(\cdot)$ denote the postulated density of $\{X_n\}$, defined on a measure space $(\Omega, \mathscr{F})$. Let $S \subset \mathbb{R}$ denote the support of X and $s_{\boldsymbol{\theta}}(\cdot) = f_{\boldsymbol{\theta}}^{\frac{1}{2}}(\cdot)$. Let the true density of $\{X_n\}$ be given by $g(\cdot)$. Throughout the paper, we assume that the following regularity conditions hold.

Hypothesis 1. Θ *is a compact and convex subset of* $\mathbb{R}^d$.

Hypothesis 2. *The family* $\mathfrak{F}$ *is identifiable; namely, if* $\boldsymbol{\theta}_1 \neq \boldsymbol{\theta}_2$, $f_{\boldsymbol{\theta}_1}(\cdot) \neq f_{\boldsymbol{\theta}_2}(\cdot)$ *on a set of positive Lebesgue measure.*

Hypothesis 3. *For every $\boldsymbol{\theta} \in \Theta$, $s_{\boldsymbol{\theta}}$ is three times continuously differentiable with respect to all components of $\boldsymbol{\theta}$. Denote by $\nabla s_{\boldsymbol{\theta}}$ the gradient of $s_{\boldsymbol{\theta}}$ and its components by $\dot{s}_{\boldsymbol{\theta}}^{i}(\cdot)$. Let $\mathscr{H}_{\boldsymbol{\theta}}$ denote the matrix of second partial derivatives of $s_{\boldsymbol{\theta}}(\cdot)$ with respect to $\boldsymbol{\theta}$ and $\ddot{s}_{\boldsymbol{\theta}}^{ij}$ the $(i,j)^{th}$ element of $\mathscr{H}_{\boldsymbol{\theta}}$.*

Hypothesis 4. *Let the matrix of second partial derivatives of $\mathscr{A}_{n}(\boldsymbol{\theta})$ and $\mathscr{A}(\boldsymbol{\theta})$ be denoted by $H_{\mathscr{A}_{n}}(\boldsymbol{\theta})$ and $H_{\mathscr{A}}(\boldsymbol{\theta})$, respectively. Assume that $H_{\mathscr{A}_{n}}(\boldsymbol{\theta})$ and $H_{\mathscr{A}}(\boldsymbol{\theta})$ are continuous in $\boldsymbol{\theta}$ and that $H_{\mathscr{A}}(\boldsymbol{\theta})$ is positive definite for every $\boldsymbol{\theta} \in \Theta$. For $p \in \mathscr{G}$ and $\boldsymbol{\theta} \in \Theta$, let $\lambda_{\boldsymbol{\theta}}(p)$ denote the smallest eigenvalue of the matrix $\int_{S} \mathscr{H}_{\boldsymbol{\theta}}(x) p^{\frac{1}{2}}(x) dx$. Assume that $\inf\{\lambda_{\boldsymbol{\theta}}(p) : p \in \mathscr{G}\} \geq c > 0$, where c is independent of $\boldsymbol{\theta}$.*

These hypotheses on the family $\mathfrak{F}$ are generally standard and are used to establish the asymptotic properties of the MHDE. Sufficient conditions on $\mathfrak{F}$ for the validity of these hypotheses are described in [3,16], and [17]. A remark on Hypothesis 4 is warranted here. When $p = g$, this assumption is related to the positive definiteness of the Fisher information matrix. If one assumes $\mathscr{G} = \mathfrak{F}$, then this hypothesis reduces to the condition that $\inf\{\lambda_{\boldsymbol{\theta}} : \boldsymbol{\theta} \in \Theta\} \geq c > 0$, which is standard. Finally, we remark that we have not attempted to provide the weakest regularity conditions, and we do believe some of these conditions can possibly be relaxed.

Recall that the MHDE of $\boldsymbol{\theta}$ can be obtained by solving the equation

$$\nabla \mathscr{A}_{n}(\boldsymbol{\theta}) := \nabla_{\boldsymbol{\theta}} \mathscr{A}(f_{\boldsymbol{\theta}}, g_{n}) = \frac{1}{2} \int_{\mathbb{R}} u_{\boldsymbol{\theta}}(x) s_{\boldsymbol{\theta}}(x) g_{n}^{\frac{1}{2}}(x) dx = 0, \tag{11}$$

where $u_{\boldsymbol{\theta}}(x) = \nabla_{\boldsymbol{\theta}} f_{\boldsymbol{\theta}}(x)(f_{\boldsymbol{\theta}}(x))^{-1}$ is the *score function*, which is obtained using $\nabla_{\boldsymbol{\theta}} s(x; \boldsymbol{\theta}) = \frac{1}{2} u(x; \boldsymbol{\theta}) s(x; \boldsymbol{\theta})$.

We begin by providing some heuristics for the case $d = 1$. Let $\dot{\mathscr{A}}_{n}(\theta)$ denote the derivative of $\mathscr{A}_{n}(\theta)$ when $d = 1$. Let $\hat{\theta}_{n}$ denote the argzero of the function $\dot{\mathscr{A}}_{n}(\theta)$ obtained from (11) above. Let $\hat{\theta}_{n,l} = \inf\{\theta \in \Theta : \dot{\mathscr{A}}_{n}(\theta) \leq 0\}$ and $\hat{\theta}_{n,u} = \sup\{\theta \in \Theta : \dot{\mathscr{A}}_{n}(\theta) \geq 0\}$. Since $\hat{\theta}_{n,l} \leq \hat{\theta}_{n} \leq \hat{\theta}_{n,u}$, we obtain using Markov's inequality that for any $\epsilon > 0$,

$$P_{g}(\hat{\theta}_{n,l} \geq \theta_{g} + \epsilon) \leq P_{g}(\dot{\mathscr{A}}_{n}(\theta_{g} + \epsilon) \geq 0) \leq E_{g}[\exp(n\alpha\dot{\mathscr{A}}_{n}(\theta_{g} + \epsilon))], \tag{12}$$

where $\alpha > 0$. Similarly, for $\alpha < 0$, it can be seen that

$$P_{g}(\hat{\theta}_{n,u} \leq \theta_{g} - \epsilon) \leq P_{g}(\dot{\mathscr{A}}_{n}(\theta_{g} - \epsilon) \leq 0) \leq E_{g}[\exp(n\alpha\dot{\mathscr{A}}_{n}(\theta_{g} - \epsilon))]. \tag{13}$$

Thus, an evaluation of (9) will allow us to obtain the logarithmic upper bound for $\hat{\theta}_{n,l}$ and $\hat{\theta}_{n,u}$. Next, using the inequalities

$$P_{g}(\hat{\theta}_{n,l} \geq \theta_{g} + \epsilon) \;\; \leq \;\; P_{g}(\dot{\mathscr{A}}_{n}(\theta_{g} + \epsilon) \geq 0) \leq P_{g}(\hat{\theta}_{n,u} \geq \theta_{g} + \epsilon), \tag{14}$$

$$P_{g}(\hat{\theta}_{n,u} \leq \theta_{g} - \epsilon) \;\; \leq \;\; P_{g}(\dot{\mathscr{A}}_{n}(\theta_{g} - \epsilon) \leq 0) \leq P_{g}(\hat{\theta}_{n,l} \leq \theta_{g} - \epsilon), \tag{15}$$

under additional hypotheses, one can derive large deviation lower bounds for $\hat{\theta}_{n}$. Deriving these bounds for MLE and MCE is rather standard, since the objective functions and their derivatives are *linear* functionals of the empirical distribution, as stated in (10), but this is not the case for the Hellinger distance.

Observe that the probabilities in (12) and (13) represent rare-event probabilities since, under the hypotheses described previously, $\hat{\theta}_{n}$ converges to θ_{g} almost surely as $n \to \infty$. The distributional results concerning $\hat{\theta}_{n}$ rely on the continuity and differentiability properties of $\nabla \mathscr{A}_{n}(\theta)$, which depend nonlinearly on g_{n}, and the norm convergence of g_{n} to g.

Let $\mathscr{G}$ denote the collection of all probability densities with support S. By Scheffe's theorem, the pointwise convergence of g_{n} to g implies $g_{n} \overset{L_{1}}{\to} g$ as $n \to \infty$. Additionally, when $g_{n}(\cdot)$ is the kernel density estimator, then Glick's Theorem guarantees that $g_{n} \overset{L_{1}}{\to} g$ almost surely as $n \to \infty$ when $b_{n} \searrow 0$ and $n \nearrow \infty$; cf. [18]. Since the MHDE are functionals of density estimators, it is natural to expect that the large deviations of density estimators will play a significant role in our analysis. For this reason, one is forced to consider the topological issues that arise in the large deviation analysis of density estimators. Interestingly, it turns out that the weak topology on $L_1(S)$ plays a prominent role. This, in turn, leads to the question of whether certain continuity properties, which were part of the traditional theory of MHD analysis, continue to hold if $\mathscr{G}$ were viewed as a subset of $L_1(S)$ equipped with weak topology. Expectedly, while the answer in general is no (cf. [19]), Proposition 1 provides sufficient conditions on the family $\mathscr{G}$ under which one additionally obtains norm convergence.

Before proceeding, we now introduce some further regularity conditions, as follows.

Hypothesis 5. *$u_\theta s_\theta \in L_2(S)$ and is an $L_2(S)$-continuous function of θ.*

Hypothesis 6. *The family $\mathfrak{F}$ consists of bounded equicontinuous densities.*

Hypothesis 7. *The family $\mathscr{G}$ consists of bounded and equicontinuous densities.*

Hypothesis 8. *$u_\theta g \in L_2(S)$ and is an $L_2(S)$-continuous function of θ.*

Here, we note that Hypotheses 6 and 7 are related. Furthermore, if one is willing to assume that $\mathscr{G} = \mathfrak{F}$, then one does not need Hypothesis 7. On the other hand, if one believes that parametric distributions are approximations to $\mathscr{G}$, then one needs to work with Hypothesis 7. For this reason, we have maintained both of these hypotheses in our main results. Hypotheses 5 and 8 are related to finiteness of the Fisher information and are standard in the statistical literature.

Before we state the first proposition, we recall the definition of weak topology on L_1 (cf. [19]). A sequence $\{h_n : n \geq 1\}$ is said to converge weakly in L_1 if $\int_S h_n(x)b(x)dx \to \int_S h(x)b(x)dx$ as $n \to \infty$ for every $b \in L_\infty(S)$, where $L_\infty(S)$ is a class of essentially bounded functions. We assume throughout the paper that the topology on Θ is the standard topology generated by the Euclidean metric.

Proposition 1. *Let $\mathscr{G}$ denote the class of densities, equipped with the weak topology. Further assume that Hypotheses 1–7 hold. Let $\Theta \otimes \mathscr{G}$ be equipped with the product topology. Then the mapping $\nabla\mathscr{A} : \Theta \otimes \mathscr{G} \to \mathbb{R}^d$ defined by*

$$\nabla\mathscr{A}(\theta, g) := \int_\mathbb{R} u_\theta(x)s_\theta(x)g^{\frac{1}{2}}(x)dx \tag{16}$$

is jointly continuous in (θ, g). Furthermore, if $g_n \xrightarrow{w} g$, then

$$\lim_{n \to \infty} \sup_{\theta \in \Theta} \|\nabla\mathscr{A}(\theta, g_n) - \nabla\mathscr{A}(\theta, g)\| = 0. \tag{17}$$

Finally, under Hypothesis 7, the family $\mathscr{G}$ is a weakly sequentially closed subset of $L_1(S)$.

Our next result is concerned with the limit behavior of the generating function of $\nabla\mathscr{A}_n(\theta)$. In the following we use the notation $p \ll g$ to mean the probability measures associated with $p(\cdot)$ and $g(\cdot)$ are absolutely continuous.

Theorem 1. *Assume that Hypotheses 1–7 hold, and set*

$$\Lambda_{n,\theta}(\alpha) := \frac{1}{n}\log E_g[\exp(n\langle \alpha, \nabla\mathscr{A}_n(\theta)\rangle)], \quad \alpha \in \mathbb{R}^d. \tag{18}$$

Then $\Lambda_\theta(\alpha) := \lim_{n\to\infty} \Lambda_{n,\theta}(\alpha)$ exists and is a convex function given by

$$\Lambda_\theta(\alpha) = \sup_{p\in\mathcal{G}}\left\{ \int_S \langle \alpha, u_\theta(x)\rangle s_\theta(x) p^{\frac{1}{2}}(x)dx - \mathrm{KL}(p,g) \right\}, \tag{19}$$

where

$$\mathrm{KL}(p,g) = \begin{cases} \int_S p(x)\log\!\left(\frac{p(x)}{g(x)}\right)dx & \text{if } p \ll g, \\ \infty & \text{otherwise.} \end{cases} \tag{20}$$

Remark 1. *Since Λ_θ is defined via a limiting operation, it is hard to extract its qualitative properties. However, we can obtain a simple lower bound by observing that $\mathrm{KL}(p,g) = 0$ if and only if $p = g$, and an upper bound using that the Kullback–Leibler information is nonnegative. This results in the following bounds:*

$$\int_S \langle \alpha, u_\theta(x)\rangle s_\theta(x) g^{\frac{1}{2}}(x)dx \leq \Lambda_\theta(\alpha) \leq \sup_{p\in\mathcal{G}}\left[\int_S \langle \alpha, u_\theta(x)\rangle s_\theta(x) p^{\frac{1}{2}}(x)dx \right]. \tag{21}$$

Furthermore, if all densities in $\mathcal{G}$ are bounded by one, then $p^{\frac{1}{2}}(\cdot) \geq p(\cdot)$ implies

$$\Lambda_\theta(\alpha) \geq \sup_{p\in\mathcal{G}}\left\{ \int_S \langle \alpha, u_\theta(x)\rangle s_\theta(x) p(x)dx - \mathrm{KL}(p,g) \right\}. \tag{22}$$

Using a variational argument, it can be shown that the supremum on the right-hand side is attained at p^* given by

$$p^*(x) := \frac{\exp(\langle \alpha, u_\theta(x)\rangle)s_\theta(x)}{\int_S \langle \alpha, u_\theta(x)\rangle s_\theta(x)g(x)dx}; \tag{23}$$

cf. [20]. Furthermore, the maximum that results from this choice of $p^*(\cdot)$ is

$$\log \int_S \exp(\langle \alpha, u_\theta(x)\rangle)s_\theta(x)g(x)dx,$$

yielding yet another lower bound for $\Lambda_\theta(\alpha)$, although the comparison of these two lower bounds is not immediate.

Returning to our main discussion, recall from [21] that the convex conjugate of the function Λ_θ is defined by

$$\Lambda_\theta^*(x) = \sup_{\alpha\in\mathbb{R}^d}\left\{ \langle \alpha, x\rangle - \Lambda(\alpha) \right\}, \quad x \in \mathbb{R}^d. \tag{24}$$

Let $\mathfrak{D}_\theta$ denote the domain of Λ_θ; namely,

$$\mathfrak{D}_\theta = \{\alpha \in \mathbb{R}^d : \Lambda_\theta(\alpha) < \infty\}; \tag{25}$$

and let $\mathfrak{R}_\theta$ denote the range of the gradient map $\nabla\Lambda_\theta$; that is,

$$\mathfrak{R}_\theta = \left\{ x \in \mathbb{R}^d : \nabla\Lambda_\theta(\alpha) = x, \quad \text{some } \alpha \in \mathbb{R}^d \right\}.$$

We begin with the discussion of the case $d = 1$. In this case, the generating function Λ_θ reduces to

$$\Lambda_\theta(\alpha) = \sup_{p\in\mathcal{G}}\left\{ \alpha \int_S \exp(n\alpha \mathscr{A}_n(\theta))s(x;\theta) p^{\frac{1}{2}}(x)dx - \mathrm{KL}(p,g) \right\}. \tag{26}$$

By the convexity of $\Lambda_\theta(\cdot)$, this function is differentiable almost everywhere (cf. [21]), and in the proof, we would like to exploit the differentiability of this function at the point α_θ^* where it attains its minimum value. If Λ_θ is not differentiable at this point, it is helpful to consider the directional derivatives of Λ_θ. Specifically, let $\Lambda_{\theta,+}'(\cdot)$ and $\Lambda_{\theta,-}'(\cdot)$ denote the right and left derivatives of $\Lambda_\theta(\cdot)$, respectively. When $x \in \left(\Lambda_{\theta,-}'(\alpha), \Lambda_{\theta,+}'(\alpha) \right)$, then it is well known that $\Lambda_\theta^*(x) = \alpha x - \Lambda_\theta(\alpha)$, but this observation will not be sufficient to obtain a proper lower bound. For that to hold, we need a stronger condition, namely that $0 \in \mathfrak{R}_\theta$, which will only be true if Λ_θ is differentiable at its point of minimum, α_θ^*. Otherwise, the expected lower bound turns out to be $\Lambda_\theta^*(x)$, where $x = \Lambda_{\theta,+}'(\alpha_\theta^*)$; cf. [13].

We now turn to our large deviation theorem in $\mathbb{R}^1$, where we study the rare-event probabilities $P_g(\hat\theta_n \in C)$ for sets C that are away from the true value θ_g. Specifically, we establish an analogue of the LDP, but where a subtle difference arises in the lower bound in the absence of differentiability of Λ_θ.

We recall that $\hat\theta_n$ is defined using the kernel density estimator $g_n(\cdot)$ defined in (1), whose behavior is dictated by the bandwidth sequence $\{b_n\}$.

Theorem 2. *Assume $d = 1$, Hypotheses 1–8 are satisfied, and $\hat\theta_n$ is the unique zero of $\dot{\mathscr{A}}_n(\theta) = 0$. Further assume that $b_n \searrow 0$ and $nb_n \nearrow \infty$ as $n \to \infty$. Then for any closed set F not containing θ_g,*

$$\limsup_{n\to\infty} \frac{1}{n} \log P_g(\hat\theta_n \in F) \leq - \inf_{\theta \in F} \Lambda_\theta^*(0). \tag{27}$$

Moreover, for any open set G not including θ_g,

$$\liminf_{n\to\infty} \frac{1}{n} \log P_g(\hat\theta_n \in G) \geq - \inf_{\theta \in G} I(\theta), \tag{28}$$

where

$$I(\theta) = \inf\{\Lambda_\theta^*(x) : x \in \mathfrak{R}_\theta \cap [0, \infty)\}, \tag{29}$$

and the infimum is taken to be infinity if the set $\mathfrak{R}_\theta \cap [0, \infty)$ is empty.

Remark 2. *If $F = [\theta, \infty)$ where $\theta > \theta_g$, then in both the upper and lower bounds, it is sufficient to evaluate the infimum at the boundary point θ. That is,*

$$\limsup_{n\to\infty} \frac{1}{n} \log P_g(\hat\theta_n \in [\theta, \infty)) \leq -\Lambda_\theta^*(0).$$

Similarly, if $G = (\theta, \infty)$ where $\theta > \theta_g$, then

$$\liminf_{n\to\infty} \frac{1}{n} \log P_g(\hat\theta_n \in (\theta, \infty)) \geq -I(\theta).$$

Furthermore, if $\inf_\alpha \Lambda_\theta(\alpha)$ is achieved at a unique point α_θ^* and Λ_θ is differentiable at α_θ^*, then the right-hand side of (28) reduces to $\Lambda_\theta^*(0)$, i.e., the upper and lower bounds coincide and the limits exist. Since the rate function appearing in the upper and lower bounds coincide in this case, we obtain a proper LDP if the resulting rate function has the required regularity properties, in particular, $I(\theta) = \Lambda_\theta^*(0)$ is lower semicontinuous and has compact level sets.

The proof of the above theorem relies on (14) and (15) combined with Theorem 1, together with a change of measure argument characteristic of large deviation analysis. The comparison inequalities in (14) and (15) are critical to obtaining the characterizations in the above theorem, but these are essentially one-dimensional results and their analogues in higher dimensions ($d \geq 2$) are not immediate. Consequently, when Λ_θ is not differentiable, new complications arise, which lead to a slightly different, and less explicit, representation of the lower bound.

Next we establish a large deviation theorem for $\mathbb{R}^d$, generalizing the previous theorem to higher dimensions. In the following, let $\mathrm{dist}(x, G) = \inf_{y \in G} ||x - y||$ denote the distance between a point $x \in \mathbb{R}^d$ and a set $G \subset \mathbb{R}^d$.

Theorem 3. *Assume Hypotheses 1–8 are satisfied, and assume that $b_n \searrow 0$ and $nb_n \nearrow \infty$ as $n \to \infty$. Then for any closed set F not containing $\boldsymbol{\theta}_g$,*

$$\limsup_{n \to \infty} \frac{1}{n} \log P_g(\hat{\boldsymbol{\theta}}_n \in F) \leq - \inf_{\theta \in F} \Lambda_\theta^\star(0). \tag{30}$$

Moreover, for any open set G not including $\boldsymbol{\theta}_g$,

$$\liminf_{n \to \infty} \frac{1}{n} \log P_g(\hat{\boldsymbol{\theta}}_n \in G) \geq - \inf_{\theta \in G} I(\boldsymbol{\theta}), \tag{31}$$

where $I(\boldsymbol{\theta}) = \inf\{\Lambda_\theta^\star(x) : x \in \mathfrak{R}_\theta \cap B(0; c_\theta)\}$ and $c_\theta = b \, \mathrm{dist}(\theta, \Theta - G)$ for some universal constant $b \in (0, \infty)$, and the infimum is taken to be infinity if the set $\mathfrak{R}_\theta \cap B(0; c_\theta)$ is empty.

Remark 3. *As we noted for the one-dimensional case in Remark 2, under a differentiability assumption on Λ_θ, the function $I(\boldsymbol{\theta})$ can be identified as $\Lambda_\theta^\star(0)$, but in full generality, it is not immediately known that $I(\boldsymbol{\theta})$ is even nontrivial. Moreover, without differentiability, the infimum in the definition of $I(\boldsymbol{\theta})$ is more restrictive than what we encountered in the one-dimensional problem. However, if one assumes additional geometry on G, such as a translated cone structure, then one obtains improved estimates in the sense that one can take unbounded regions in the definition of $I(\boldsymbol{\theta})$, just as we saw in Theorem 2.2. For further remarks in this direction, see the discussion given after the proof of the theorem.*

3. Proofs

We turn first to Proposition 1.

Proof of Proposition 1. Since $\Theta \otimes \mathcal{G}$ is equipped with product topology, it is sufficient to show that if $\theta_n \to \theta$ and $g_n \xrightarrow{w} g$, then $\nabla \mathscr{A}_n(\boldsymbol{\theta})$ converges to $\nabla \mathscr{A}(\boldsymbol{\theta})$, where

$$\nabla \mathscr{A}(\boldsymbol{\theta}) = \int_S u_\theta(x) s_\theta(x) g^{\frac{1}{2}}(x) dx. \tag{32}$$

Let $r_\theta(x) = u_\theta(x) s_\theta(x)$, and observe that

$$\begin{aligned}
|\nabla \mathscr{A}(\theta_n, g_n) - \nabla \mathscr{A}(\theta, g)| &\leq \int_S |r_{\theta_n}(x)| |g_n^{\frac{1}{2}}(x) - g^{\frac{1}{2}}(x)| dx + \int_S |r_{\theta_n}(x) - r_\theta(x)| g^{\frac{1}{2}}(x) dx \\
&\leq ||r_\theta||_2 \mathrm{HD}(g_n, g) + \int_S |r_{\theta_n}(x) - r_\theta(x)| g^{\frac{1}{2}}(x) dx \\
&= T_{n,1} + T_{n,2}, \tag{33}
\end{aligned}$$

where the penultimate equation follows by applying the Cauchy–Schwarz inequality. Then by the Cauchy–Schwarz inequality and Hypothesis 5, $T_{n,2} \to 0$. Since Hellinger distance is dominated by the L_1-distance, in order to complete the proof, it is sufficient to show that $||g_n - g||_1 \to 0$. Now since $g_n \xrightarrow{w} g$, it follows that as $n \to \infty$,

$$G_n(x) := \int_S g_n(y) I_{\{y \leq x\}} dy \to \int_S g(y) I_{\{y \leq x\}} dy := G(x). \tag{34}$$

Evidently, $G_n(\cdot)$ and $G(\cdot)$ are nondecreasing and right continuous. Furthermore, if $x_* = \inf\{x : x \in S\}$ and $x^* = \sup\{x : x \in S\}$, then $G_n(x_*) \to G(x_*)$ and $G_n(x^*) \to G(x^*)$, where $G_n(x_*) = \lim_{x \to x_*} G_n(x)$, $G_n(x^*) = \lim_{x \to x^*} G_n(x)$, $G(x_*) = \lim_{x \to x_*} G(x)$,

$G(x^*) = \lim_{x \to x*} G(x)$. Thus G_n converges to G, which is a proper distribution function. Then by Lemma 1 of Boos [22], $g_n(\cdot)$ converges to $g(\cdot)$ uniformly on compact sets. This, in turn, implies the L_1 convergence of $g_n(\cdot)$ to $g(\cdot)$ (by Scheffe's lemma), which establishes the convergence of $T_{n,1}$ to 0, thus completing the proof of the joint continuity of $\nabla \mathscr{A}(\theta, g)$.

Next, the uniform convergence (17) follows by Hypothesis 5, since

$$
\sup_{\theta \in \Theta} |\nabla \mathscr{A}(\theta, g_n) - \nabla \mathscr{A}(\theta, g)| \leq \int_S |r_\theta(x)||g_n^{\frac{1}{2}}(x) - g^{\frac{1}{2}}(x)|dx
$$

$$
\leq \sup_{\theta \in \Theta} ||r_\theta||_2 \mathrm{HD}(g_n, g) \to 0.
$$

Finally, to prove that $\mathscr{G}$ is weakly sequentially closed, note that convergence in weak topology implies pointwise convergence, yielding $g(\cdot) \geq 0$. Noting that

$$
\int_S g(x)d\mu(x) = 1 + \int_S (g(x) - g_n(x))dx, \tag{35}
$$

it follows that $g(\cdot)$ integrates to one, using L_1 convergence, thus completing the proof of the proposition. $\square$

We now turn to the proof of Theorem 1. The proof relies on the large deviation theorem for the kernel density estimator $g_n(\cdot)$ in the weak topology of $\mathscr{G}$. The next proposition is concerned with the LDP for $\{g_n\}$ in $\mathscr{G}$, equipped with the inherited weak topology from $L_1(S)$. This issue has received considerable attention recently (cf. [23,24]), where it is established that the full LDP may *not* hold for $\{g_n\}$ in norm topology, but does hold under the weak topology.

Proposition 2. *Assume Hypotheses 1–8 and that $b_n \searrow 0$ and $nb_n \nearrow \infty$ as $n \to \infty$. Then $\{g_n\}$ satisfies the LDP in the weak topology of $L_1(S)$ with good rate function I given by*

$$
I(p) = \begin{cases} \int_S p(x) \log\left(\frac{p(x)}{g(x)}\right)dx & \text{if } g \ll p, \\ \infty & \text{otherwise.} \end{cases} \tag{36}
$$

Proof of Theorem 1. As before, let $\mathscr{G}$ be equipped with the weak topology. Set $r_\theta(x) = u_\theta(x)s_\theta(x)$, and define $F : \mathscr{G} \to \mathbb{R}$ as follows:

$$
F(h) = \int_S \langle \alpha, r_\theta(x) \rangle h^{\frac{1}{2}}(x)dx. \tag{37}
$$

By Hypothesis 5, $r_\theta \in L_2(S)$. To show that $F(\cdot)$ is continuous, let $h_n \overset{w}{\to} h$ as $n \to \infty$. Then

$$
|F(h_n) - F(h)| \leq \int_S r_\theta(x)|h_n^{\frac{1}{2}}(x) - h^{\frac{1}{2}}(x)|d\mu(x)
$$

$$
\leq ||r_\theta||_2 \mathrm{HD}(h_n, h) \leq ||r_\theta||_2 ||h_n - h||_1 \to 0 \quad \text{as} \quad n \to \infty, \tag{38}
$$

where we have used the Cauchy–Schwarz inequality that the L_1 distance dominates the Hellinger distance in (38). Now by Hypothesis 7, as in the proof of Proposition 1, we have that $||h_n - h||_1 \to 0$ as $n \to \infty$, establishing the continuity of $F(\cdot)$. Next, to show that $F(\cdot)$ is bounded, note that $\sup\{F(p) : p \in \mathscr{G}\} \leq ||r_\theta||_2$ by the Cauchy–Schwarz inequality. Then by Proposition 2, it follows by Varadhan's integral lemma (see [10], Theorem 4.3.1) that

$$\lim_{n\to\infty} \frac{1}{n}\log E[\exp(nF(g_n(x)))] = \lim_{n\to\infty}\frac{1}{n}\log E\left[\exp\left(n\int_S \langle\alpha, u_\theta(x)s_\theta(x)\rangle g_n^{\frac{1}{2}}(x)dx\right)\right]$$

$$= \sup_{p\in\mathcal{G}}\left\{\int_S \langle\alpha, u_\theta(x)\rangle s_\theta(x)p^{\frac{1}{2}}(x)dx - \mathrm{KL}(p,g)\right\}$$

$$:= \Lambda_\theta(\alpha). \tag{39}$$

This completes the proof of the theorem. $\square$

The proofs of our main results will involve probability bounds on the modulus of continuity of $\mathscr{A}_n(\theta)$ and $\nabla\mathscr{A}_n(\theta)$, respectively. Recall that the modulus of continuity $\omega(h;r)$ of a function $h : \mathbb{R}^d \to \mathbb{R}$ is given by

$$\omega(h;r) := \sup_{||x_1-x_2||\le r} |h(x_1) - h(x_2)|, \quad r > 0. \tag{40}$$

Observe that when $h(\cdot)$ is replaced by $\mathscr{A}_n(\theta)$ or $\nabla\mathscr{A}_n(\theta)$, the modulus of continuity becomes a random quantity. Our next proposition summarizes the continuity properties of $\mathscr{A}_n(\theta)$ and $\nabla\mathscr{A}_n(\theta)$ via their modulus of continuity as real-valued functionals from $\mathcal{G}$ equipped with the weak topology.

Proposition 3. *Assume that Hypotheses 1–8 hold and that $b_n \searrow 0$ and $nb_n \nearrow \infty$ as $n \to \infty$. Then, with respect to $\{\mathscr{A}_n\}$ and $\mathscr{A}$, the modulus of continuity satisfies the following relations, each with probability one:*

(i) $\lim_{n\to\infty} \omega(\mathscr{A}_n;r) = \omega(\mathscr{A},r)$; (ii) $\lim_{r\to 0}\omega(\mathscr{A}_n;r) = 0$; and (iii) $\lim_{r\to 0}\omega(\mathscr{A};r) = 0$.

Similarly, the sequence $\{\nabla\mathscr{A}_n\}$ and $\nabla\mathscr{A}$ satisfy the analogous relations with probability one; namely,

(iv) $\lim_{n\to\infty} \omega(\nabla\mathscr{A}_n;r) = \omega(\nabla\mathscr{A};r)$; (v) $\lim_{r\to 0}\omega(\nabla\mathscr{A}_n;r) = 0$; and (vi) $\lim_{r\to 0}\omega(\nabla\mathscr{A};r) = 0$.

Proof. First observe that $\mathscr{A}_n(\theta)$ converges uniformly to $\mathscr{A}(\theta)$. To see this, note that if $g_n \overset{w}{\to} g$, then by Proposition 1, it converges in L_1. Hence

$$\sup_{\theta\in\Theta} |\mathscr{A}_n(\theta) - \mathscr{A}(\theta)| \le \sup_{\theta\in\Theta}\int_{\mathbb{R}} s_\theta(x)|g_n^{\frac{1}{2}}(x) - g^{\frac{1}{2}}(x)|dx$$

$$\le ||g_n^{\frac{1}{2}}(x) - g^{\frac{1}{2}}(x)||_2 \le ||g_n - g||_1 \to 0, \tag{41}$$

where the last inequality follows using that the Hellinger distance is dominated by the L_1-distance. We now prove (i). For this we invoke the properties of the modulus of continuity. Observe that

$$\omega(\mathscr{A}_n;r) = \omega(\mathscr{A}_n - \mathscr{A} + \mathscr{A};r) \le \omega(\mathscr{A}_n - \mathscr{A};r) + \omega(\mathscr{A};r), \tag{42}$$

which yields

$$|\omega(\mathscr{A}_n;r) - \omega(\mathscr{A};r)| \le \omega(\mathscr{A}_n - \mathscr{A};r). \tag{43}$$

Next observe that

$$\omega(\mathscr{A}_n - \mathscr{A}; r) = \sup_{||\theta_1 - \theta_2|| \leq r} |(\mathscr{A}_n - \mathscr{A})(\theta_1) - (\mathscr{A}_n - \mathscr{A})(\theta_2)|$$

$$\leq 2 \sup_{\theta \in \Theta} |\mathscr{A}_n(\theta) - \mathscr{A}(\theta)| \to 0, \tag{44}$$

where the last convergence follows from the uniform convergence of $(\mathscr{A}_n - \mathscr{A})(\theta)$ to 0 as shown in (42). The proof of (iv) is similar, and specifically is obtained by using that

$$\omega(\nabla(\mathscr{A}_n - \mathscr{A}); r) \leq 2 \sup_{\theta \in \Theta} ||\nabla \mathscr{A}_n(\theta) - \nabla \mathscr{A}(\theta)|| \to 0, \tag{45}$$

where the above convergence follows from (17).

We now turn to the proof of (ii). Using the Cauchy–Schwarz inequality and the definition of Hellinger distance,

$$\omega(\mathscr{A}_n; r) = \sup_{||\theta_1 - \theta_2|| \leq r} |\mathscr{A}_n(\theta_1) - \mathscr{A}_n(\theta_2)|$$

$$= \sup_{||\theta_1 - \theta_2|| \leq r} \left| \int_{\mathbb{R}} (s_{\theta_1}(x) - s_{\theta_2}(x)) g_n^{\frac{1}{2}}(x) dx \right| \leq \mathrm{HD}(f_{\theta_1}, f_{\theta_2})$$

$$\leq \sup_{||\theta_1 - \theta_2|| \leq r} ||f_{\theta_1} - f_{\theta_2}||_1 := \omega(H; r), \tag{46}$$

where $H : (\theta_1, \theta_2) \to ||f_{\theta_1} - f_{\theta_2}||_1$ is continuous since $\mathfrak{F}$ is continuous in θ. Also, since $\Theta \times \Theta$ is compact, $H(\cdot, \cdot)$ is uniformly continuous. Since the modulus of continuity converges to 0 if and only if $H(\cdot, \cdot)$ is uniformly continuous, (ii) follows. Turning to (v), notice that, as before,

$$\omega(\nabla \mathscr{A}_n; r) \leq \sup_{||\theta_1 - \theta_2|| \leq r} ||u_{\theta_1} s_{\theta_1} - u_{\theta_2} s_{\theta_2}||_2. \tag{47}$$

Now, since $u_\theta s_\theta$ is L_2 continuous, by Hypothesis 5, the proof follows as in (ii) due to to the compactness of Θ. The proofs of (iii) and (vi) are similar to (ii) and (v), respectively, and are therefore omitted. $\square$

Proposition 4. *For any $0 < M < \infty$ and $\delta > 0$, there exists a positive number $r(M, \delta)$ such that*

$$P_g(\omega(\mathscr{A}_n; r) \geq \delta) \leq e^{-Mn} \quad and \quad P_g(\omega(\nabla \mathscr{A}_n; r) \geq \delta) \leq e^{-Mn}. \tag{48}$$

Proof. By Markov's inequality and (46), it follows that for any $\beta > 0$,

$$P_g(\omega(\mathscr{A}_n; r) \geq \delta) \leq E_g[e^{n\beta\omega(\mathscr{A}_n; r)}] e^{-n\beta\delta} \leq e^{-n\beta(\delta - \omega(H; r))}. \tag{49}$$

Since $\omega(H; r) \to 0$ as $r \searrow 0$, there exists an r_0 such that for all $r \leq r_0$, $(\delta - \omega(H; r)) > 0$. Since $\beta > 0$ is arbitrary, the proposition follows by taking $\beta = M(\delta - \omega(H; r))^{-1}$, for some $r \leq r_0$. The proof of the second inequality is similar, using (47). $\square$

Proof of Theorem 2. We begin with the proof of the upper bound. Since we assume that the equation $\mathscr{A}_n(\theta) = 0$ has a unique solution, it follows from the inequality in (12) that for any $\alpha > 0$ and $\theta > \theta_g$,

$$\limsup_{n \to \infty} \frac{1}{n} \log P_g(\hat{\theta}_n \geq \theta) \leq \limsup_{n \to \infty} \frac{1}{n} \log E_g[\exp(n\alpha \mathscr{A}_n(\theta))] = \Lambda_\theta(\alpha), \tag{50}$$

where the last equality follows by applying Theorem 1 with $d = 1$. Since the inequality holds for every $\alpha > 0$,

$$\limsup_{n \to \infty} \frac{1}{n} \log P_g(\hat{\theta}_n \geq \theta) \; \leq \; \sup_{\alpha > 0} \Lambda_\theta(\alpha) \leq \sup_{\alpha \in \mathbb{R}} \Lambda_\theta(\alpha). \tag{51}$$

Now, noticing that $\sup_{\alpha \in \mathbb{R}} \Lambda_\theta(\alpha) = -\inf_{\alpha \in \mathbb{R}} -\Lambda_\theta(\alpha) = -\Lambda_\theta^*(0)$, we then obtain

$$\limsup_{n \to \infty} \frac{1}{n} \log P_g(\hat{\theta}_n \geq \theta) \leq -\Lambda_\theta^*(0). \tag{52}$$

Similarly, for $\theta < \theta_g$, using (13), one can show by an analogous calculation that

$$\limsup_{n \to \infty} \frac{1}{n} \log P_g(\hat{\theta}_n \leq \theta) \; \leq \; -\Lambda_\theta^*(0). \tag{53}$$

Now let $\theta_1 = \inf\{\theta > \theta_g : \theta \in F\}$ and $\theta_2 = \sup\{\theta < \theta_g : \theta \in F\}$. Then

$$P_g(\hat{\theta}_n \in F) \leq P_g(\hat{\theta}_n \geq \theta_1) + P_g(\hat{\theta}_n \leq \theta_2), \tag{54}$$

and so by (52) and (53), it follows that

$$\limsup_{n \to \infty} \frac{1}{n} \log P_g(\hat{\theta}_n \in F) \leq - \min_{\theta \in \{\theta_1, \theta_2\}} \Lambda_\theta^*(0) \leq - \inf_{\theta \in F} \Lambda_\theta^*(0), \tag{55}$$

where the last step follows since F closed implies $\{\theta_1, \theta_2\} \subset F$.

Next we turn now to the proof of the lower bound. Let G be an open set, and let $\theta \in G$. Then there exists an $\epsilon > 0$ (to be chosen) such that $I_\epsilon := (\theta - \epsilon, \theta + \epsilon) \subsetneq G$. Note that

$$\{\hat{\theta}_n \in I_\epsilon\} \; = \; \{\mathscr{A}_n(\hat{\theta}_n) = 0, \, \hat{\theta}_n \in I_\epsilon\}$$
$$\supset \; \{\mathscr{A}_n(\theta) - \mathscr{A}_n(\hat{\theta}_n) \geq \delta\} \cup \{\hat{\theta}_n \in I_\epsilon, \sup_{\theta_1, \theta_2 \in I_\epsilon} |\mathscr{A}_n(\theta_1) - \mathscr{A}_n(\theta_2)| \leq \delta\}.$$

Thus,

$$\begin{aligned}
P_g(\hat{\theta}_n \in I_\epsilon) \; &\geq \; P_g(\mathscr{A}_n(\theta) - \mathscr{A}_n(\hat{\theta}_n) \geq \delta) - P_g(\hat{\theta}_n \notin I_\epsilon, \sup_{\theta_1, \theta_2 \in I_\epsilon} |\mathscr{A}_n(\theta_1) - \mathscr{A}_n(\theta_2| > \delta)) \\
&\geq \; P_g(\mathscr{A}_n(\theta) - \mathscr{A}_n(\hat{\theta}_n) \geq \delta) - P_g(\sup_{\theta_1, \theta_2 \in I_\epsilon} |\mathscr{A}_n(\theta_1) - \mathscr{A}_n(\theta_2)| > \delta) \\
&= \; P_g(\mathscr{A}_n(\theta) \geq \delta) - P_g(\sup_{\theta_1, \theta_2 \in I_\epsilon} |\mathscr{A}_n(\theta_1) - \mathscr{A}_n(\theta_2)| > \delta) \\
&= \; P_g(\mathscr{A}_n(\theta) \geq \delta) - P_g(\omega(\mathscr{A}_n; \epsilon) > \delta).
\end{aligned} \tag{56}$$

We now investigate $P_g(\mathscr{A}_n(\theta) \geq \delta)$. Let Q_n denote the distribution of $\mathscr{A}_n(\theta)$, and define $Q_{n,\alpha}$ as follows:

$$Q_{n,\alpha}(B) = \frac{1}{\Lambda_{n,\theta}(\alpha)} \int_B e^{-n\alpha y} dQ_n(y), \quad B \in \mathscr{B}. \tag{57}$$

Let $B = (x - \eta, x + \eta)$, for some $\eta > 0$, where $B \subset (\delta, \infty)$ and $x \in \mathfrak{R}_\theta$. Then

$$Q_n(B) \geq \exp\{-n\alpha x - n\eta|\alpha| + n\Lambda_{n,\theta}(\alpha)\} Q_{n,\alpha}(B). \tag{58}$$

Taking the logarithm, dividing by n, and then taking the limit as $n \to \infty$, we obtain

$$\liminf_{n \to \infty} \frac{1}{n} \log Q_n(B) \geq -\alpha x - \eta|\alpha| - \Lambda_\theta(\alpha) + \liminf_{n \to \infty} \frac{1}{n} \log Q_{n,\alpha}(B). \tag{59}$$

Now since $x \in \mathfrak{R}_\theta$, we can apply Theorem IV.1 of [25] to obtain that the last term on the right-hand side of the previous equation converges to zero. Upon letting $\eta \to 0$, it follows that

$$\liminf_{n\to\infty} \frac{1}{n} \log Q_n(B) \geq -\Lambda^*_\theta(x). \tag{60}$$

Since the above inequality holds for all $x \in \mathfrak{R}_\theta \cap (\delta, \infty)$, we conclude that

$$\lim_{n\to\infty} \frac{1}{n} \log P_g(\dot{\mathscr{A}}_n(\theta) \geq \delta) \geq -I_\delta(\theta), \tag{61}$$

where $I_\delta(\theta) = \inf_{x \in \mathfrak{R}_\theta \cap (\delta, \infty)} \Lambda^*_\theta(x)$.

By Proposition 4, choosing $M > I_\delta(\theta)$, one can find $\epsilon > 0$ such that

$$P_g(\omega(\dot{\mathscr{A}}_n; \epsilon) > \delta) \leq e^{-Mn}. \tag{62}$$

Since

$$P_g(\hat{\theta}_n \in G) \geq P_g(\dot{\mathscr{A}}_n(\theta) \geq \delta)\left(1 - \frac{P_g(\omega(\dot{\mathscr{A}}_n; \epsilon))}{P_g(\dot{\mathscr{A}}_n(\theta) \geq \delta)}\right), \tag{63}$$

by the choice of M, it follows from (61) that

$$\liminf_{n\to\infty} \frac{1}{n} \log P_g(\hat{\theta}_n \in G) \geq -I_\delta(\theta). \tag{64}$$

Taking the supremum on left- and right-hand side over all $\delta > 0$ yields the required lower bound. $\square$

Turning to the higher dimensional case, we first need the following result, which provides a uniform bound on the Hessian of the objective function $\mathscr{A}_n(\theta)$.

Lemma 2. *Under Hypotheses 1–8, there exists a finite constant $0 < C < \infty$ such that with probability one,*

$$\sup_{n\geq 1} \sup_{\theta \in \Theta} ||H_{\mathscr{A}_n}(\boldsymbol{\theta})||_2 \leq C. \tag{65}$$

Proof. This is standard. Specifically, note that the $(i,j)^{\text{th}}$ element of the matrix $H_{\mathscr{A}_n}(\boldsymbol{\theta})$ is given by

$$h_{n,ij} = \int_S \ddot{s}^{ij}_\theta(x) g^{\frac{1}{2}}_n(x)\,dx. \tag{66}$$

Next, writing down the expression for $\ddot{s}^{ij}_\theta$ in terms of the derivatives of the score function u_θ, using the Cauchy–Schwarz inequality along with Hypotheses 3, 4, 6, and 8, and the definition of the matrix norm, the lemma follows. $\square$

In the proof of the lower bound, we will take a somewhat different approach, involving the analysis of k constraints, and our strategy will be to reduce this to a problem involving a single constraint. Specifically, in (67) below, we establish that, instead of studying k constraints on a quantity $\mathscr{D}_n$ (which we are about to define), we can cast the problem in terms of a d-dimensional vector Y_n (defined in (70) below) belonging to a ball centered at 0 and of appropriate radius.

To be more precise, let $G \subset \mathbb{R}^d$ be open, and consider the probability that we obtain an estimated value $\theta \in G$. Let $\{\theta_1, \ldots, \theta_k\} \subset \Theta - G$, and for any $\delta > 0$, set

$$d_n(j) = \mathscr{A}_n(\boldsymbol{\theta}) - \mathscr{A}_n(\boldsymbol{\theta}_j) - \delta, \quad j = 1, \ldots, k$$

and $\mathscr{D}_n(\theta) = (d_n(1), \cdots d_n(k))$. If θ is chosen as the estimate, then we must have $\mathscr{A}_n(\theta) - \mathscr{A}_n(\theta_j) \geq 0$ for all j, so, in particular,

$$P_g(\hat{\theta}_n \in G) \geq P_g(\mathscr{D}_n(\theta) \geq 0) \tag{67}$$

(by which we mean that $d_n(j) \geq 0$ for all j in this last probability).

To evaluate the latter probability, observe that by a second-order Taylor expansion,

$$s_\theta(x) - s_{\theta_j}(x) = \langle \theta - \theta_j, \nabla s_\theta(x) \rangle + \frac{1}{2}(\theta - \theta_j)\mathscr{H}(x; \theta_j^*)(\theta - \theta_j)'. \tag{68}$$

Using the positive definiteness and uniform boundedness of the matrix $\int_{\mathbb{R}} \mathscr{H}(x; \theta)p^{\frac{1}{2}}(x)dx$, by Hypothesis 4, we have that for any unit vector $v \in \mathbb{R}^d$,

$$\sup_{p \in \mathscr{G}} \inf_{\eta \in \Theta} \left\{ v \left(\int_{\mathbb{R}} \mathscr{H}(x; \eta)p^{\frac{1}{2}}(x)dx \right) v' \right\} \geq c,$$

where c is a positive constant independent of v. Thus, for each j,

$$\sup_{p \in \mathscr{G}} \inf_{\eta \in \Theta} \left\{ (\theta - \theta_j) \left(\int_{\mathbb{R}} \mathscr{H}(x; \eta)p^{\frac{1}{2}}(x)dx \right) (\theta - \theta_j)' \right\} \geq c \, \|\theta - \theta_j\|^2. \tag{69}$$

Integrating with respect to $g_n^{\frac{1}{2}}(\cdot)$ and using the definition of $\mathscr{A}_n(\cdot)$, we then obtain that

$$d_n(j) = \int_{\mathbb{R}} [\langle \theta - \theta_j, \nabla s(x, \theta) \rangle] g_n^{\frac{1}{2}}(x)dx + \mathscr{R}(\theta, \theta_j), \tag{70}$$

where

$$\mathscr{R}(\theta, \theta_j) \geq c \, \|\theta - \theta_j\|^2 - \delta.$$

Let $Y_n(\theta) = (Y_{n,1}, \ldots, Y_{n,d})$, where for $s(x; \theta) := s_\theta(x)$:

$$Y_{n,j} = \int_S \frac{\partial}{\partial \theta_j} s(x; \theta) g_n^{\frac{1}{2}}(x)dx, \quad 1 \leq j \leq k. \tag{71}$$

(We have suppressed θ in the notation for $Y_{n,j}$.) Then the inequality $d_n(j) \geq 0$ corresponds to an event $\mathscr{E}_{n,j}$ described by the occurrence of the inequality

$$\left\langle \frac{\theta - \theta_j}{\|\theta - \theta_j\|}, Y_n \right\rangle \geq -c\|\theta - \theta_j\| + \delta(\|\theta - \theta_j\|)^{-1}, \tag{72}$$

where the right-hand side is always negative for small δ (since $\text{dist}(\theta, \Theta - G) > 0$) and behaves like a constant multiple of $\text{dist}(\theta, \Theta - G)$ as this distance tends to infinity. Thus, we can choose a positive constant a_δ such that

$$a_\delta \, \text{dist}(\theta, \Theta - G) \leq c\|\theta - \theta_j\| - \delta(\|\theta - \theta_j\|)^{-1}, \quad j = 1, \ldots, k,$$

and set $c_\theta(\delta) := a_\delta \, \text{dist}(\theta, \Theta - G)$. Finally, let $\tilde{\mathscr{E}}_n$ denote the event that

$$\left\langle \frac{\theta - \theta_j}{\|\theta - \theta_j\|}, Y_n \right\rangle \geq -c_\theta(\delta). \tag{73}$$

Then for all j, $\mathscr{E}_{n,j} \supset \mathring{\mathscr{E}}_n$, where we recall that $\mathscr{E}_{n,j}$ was defined via (72). Now, since the definition of the event $\mathring{\mathscr{E}}_n$ does not depend on any specific vector θ_j, one can replace the vector $(\theta - \theta_j)(||\theta - \theta_j||)^{-1}$ by any unit vector v in $\mathbb{R}^d$. Hence

$$P_g(\mathscr{D}_n \geq 0) \geq P_g(\langle v, Y_n \rangle \geq -c_\theta(\delta), \text{ for all unit vectors } v) = P_g(Y_n \in \overline{B}(0; c_\theta(\delta))), \quad (74)$$

and we now derive a large deviation lower bound for the probability on the right-hand side.

Proposition 5. *Assume that Hypotheses 1–8 hold, and suppose that G is an open subset of $\mathbb{R}^d$. Assume that $b_n \searrow 0$ and $nb_n \nearrow \infty$ as $n \to \infty$. Then for any $\theta \in G$ and $r > 0$,*

$$\lim_{n \to \infty} \frac{1}{n} \log P_g(Y_n \in B(0; r)) \geq -I_r(\theta), \quad (75)$$

where $I_r(\theta) = \inf\{\Lambda_\theta^*(x) : x \in \mathfrak{R}_\theta \cap B(0; r)\}$ and the infimum is taken to be infinity if the set $\mathfrak{R}_\theta \cap B(0; r)$ is empty.

Proof. We begin by studying the limiting generating function of Y_n. By Varadhan's integral lemma, it follows that

$$\lim_{n \to \infty} \Lambda_{n,\theta}(\alpha) := \lim_{n \to \infty} \frac{1}{n} \log E_g[\exp(n\langle \alpha, Y_n \rangle)] = \Lambda_\theta(\alpha), \quad (76)$$

where

$$\Lambda_\theta(\alpha) = \sup_{p \in \mathscr{G}} \left[\int_S \langle \alpha, \nabla s_\theta(x) \rangle p^{\frac{1}{2}}(x)dx - \mathrm{KL}(p, g) \right]. \quad (77)$$

Define the α-shifted distribution by

$$Q_{n,\alpha}(B) = \frac{1}{\Lambda_{n,\theta}(\alpha)} \int_B e^{n\langle \alpha, y \rangle} dQ_n(y), \quad (78)$$

where Q_n denotes the distribution of Y_n. Note by the convexity of $\Lambda_\theta(\alpha)$ that it is almost everywhere differentiable. Fix $x \in \mathfrak{R}_\theta \cap B(0; r)$ and choose α such that $\nabla \Lambda_\theta(\alpha) = x$. Let $\delta > 0$ be such that $B(x; \delta) \subsetneq B(0; r)$. Then

$$\begin{aligned} Q_n(B(x; \delta)) &= \exp(n\Lambda_{n,\theta}(\alpha)) \int_{B(x;\delta)} \exp(-n\langle \alpha, y \rangle) dQ_{n,\alpha}(y) \\ &\geq \exp(n(-\langle \alpha, x \rangle + \Lambda_{n,\theta}(\alpha) + ||\alpha||\delta)) Q_{n,\alpha}(B(x; \delta)), \end{aligned} \quad (79)$$

implying

$$\liminf_{n \to \infty} \frac{1}{n} \log Q_n(B(x; \delta)) \geq -\langle \alpha, x \rangle + \Lambda_\theta(\alpha) - ||\alpha||\delta + \liminf_{n \to \infty} \frac{1}{n} \log Q_{n,\alpha}(B(x; \delta)). \quad (80)$$

Now, notice that the limiting cumulant generating function of Y_n under the measure $Q_{n,\alpha}$ is given by

$$\tilde{\Lambda}_\theta(\beta) = \Lambda_\theta(\alpha + \beta) - \Lambda_\theta(\beta). \quad (81)$$

Since $\tilde{\Lambda}_\theta$ is a proper convex function, it is continuous since $\Lambda_\theta(\alpha)$ is finite in the $\mathbb{R}^d$, and moreover, by the choice of x, it is differentiable at 0. Hence Condition II.1 of [25] is satisfied. Now, using Theorem IV.1 of [25], it follows that

$$\liminf_{n \to \infty} \frac{1}{n} \log Q_{n,\alpha}(B(x; \delta)) = 0. \quad (82)$$

Substituting the above into (80), we obtain

$$\liminf_{n\to\infty} \frac{1}{n} \log P_g(Y_n \in B(0;r)) \geq -\Lambda_\theta^*(x). \tag{83}$$

Taking the supremum in $x \in \mathfrak{R}_\theta \cap B(0;r)$, the proposition follows. $\square$

Proof of Theorem 3: Upper Bound. Let F be a closed subset of Θ. Note Θ compact implies that F is compact. Let $\{B(\theta;r) : \theta \in \Theta\}$ denote an open cover of F, and let $\{B(\theta_1;r),\dots,B(\theta_k;r)\}$ denote the finite subcover. Using that $\nabla\mathscr{A}_n(\hat{\theta}_n) = 0$, we then obtain that for any $\alpha \in \mathbb{R}^d$,

$$
\begin{aligned}
P_g(\hat{\theta}_n \in F) &\leq \sum_{j=1}^{k} P_g(\hat{\theta}_n \in B(\theta_k;r)) \\
&= \sum_{j=1}^{k} E_g[\exp(n\langle \alpha, \mathscr{A}_n(\hat{\theta}_n)\rangle) I_{\{\hat{\theta}_n \in B(\theta_j;r)\}}] := \sum_{j=1}^{k} T_n(j).
\end{aligned}
\tag{84}
$$

Adding and subtracting $\nabla\mathscr{A}_n(\theta_j)$ to $\nabla\mathscr{A}_n(\theta)$ and then applying Hölder's inequality yields $T_n(j) \leq T_n(1,j,p) T_n(2,j,q)$, where

$$\log T_n(1,j,p) = \frac{1}{p} \log E_g[\exp(np\langle \alpha, \nabla\mathscr{A}_n(\theta_j)\rangle) I_{\{\theta \in B(\theta_j;r)\}}],$$

$$\log T_n(2,j,q) = \frac{1}{q} \log E_g[\exp(nq\langle \alpha, \nabla(\mathscr{A}_n(\hat{\theta}_n) - \mathscr{A}_n(\theta_j))\rangle) I_{\{\hat{\theta}_n \in B(\theta_j;r)\}}].$$

First we study $T_n(2,j,q)$. For $\hat{\theta}_n \in B(\theta_j,r_j)$ and $\theta_1,\theta_2 \in \Theta$, the Cauchy–Schwarz inequality gives

$$
\begin{aligned}
|\langle \alpha, \nabla\mathscr{A}_n(\hat{\theta}_n) - \nabla\mathscr{A}_n(\theta_j)\rangle| &\leq \|\alpha\|_2 \sup_{\theta_1,\theta_2 \in B(\theta_j,r)} \|\nabla\mathscr{A}_n(\theta_1) - \nabla\mathscr{A}_n(\theta_2)\|_2 \\
&\leq \|\alpha\|_2 |r| \sup_{\theta \in B(\theta_j,r)} \|H_{\mathscr{A}_n}(\theta)\|_2 \\
&\leq \|\alpha\|_2 |r| \max_{1\leq j\leq k} \left[\sup_{\theta \in B(\theta_j,r_j)} \|H_{\mathscr{A}_n}(\theta)\|_2 \right],
\end{aligned}
$$

where $H_{\mathscr{A}_n}(\theta)$ is the Hessian matrix consisting of the second partial derivatives of $\mathscr{A}_n(\theta)$. Hence we obtain for any $1 \leq j \leq k$ that

$$
\begin{aligned}
\frac{1}{n} \log T_n(2,j,q) &\leq r\frac{1}{nq}(nq\|\alpha\|_2) \max_{1\leq j\leq k} \left\{ \sup_{\theta \in B(\theta_j,r)} \|H_{\mathscr{A}_n}(\theta)\|_2 \right\} \\
&= r\|\alpha\|_2 \max_{1\leq j\leq k} \left\{ \sup_{\theta \in B(\theta_j,r)} \|H_{\mathscr{A}_n}(\theta)\|_2 \right\}.
\end{aligned}
\tag{85}
$$

Now by Lemma 2,

$$\limsup_{n\to\infty} \frac{1}{n} \log T_n(2,j,q) \leq Cr. \tag{86}$$

Also, for each $1 \leq j \leq k$, Theorem 1 provides that

$$\limsup_{n\to\infty} \frac{1}{n} \log T_n(1,j,p) \leq \frac{1}{p}\Lambda_{\theta_j}(p\alpha). \tag{87}$$

Thus

$$\limsup_{n\to\infty} \frac{1}{n} P_g(\hat{\theta}_n \in F) \;\le\; \max_{1\le j\le k} \limsup_{n\to\infty} \frac{1}{n} \log T_n(1,j,p) + \max_{1\le j\le k} \limsup_{n\to\infty} \frac{1}{n} \log T_n(2,j,p)$$

$$\le\; \max_{1\le j\le k} \frac{1}{p} \Lambda_{\theta_j}(p\alpha) + Cr. \tag{88}$$

Since the last inequality holds for all $p > 1$,

$$\limsup_{n\to\infty} \frac{1}{n} P_g(\hat{\theta}_n \in F) \;\le\; \max_{1\le j\le k} \frac{1}{p} \Lambda_{\theta_j}(p\alpha) + Cr$$

$$\to\; \max_{1\le j\le k} \Lambda_{\theta_j}(\alpha) + Cr \quad \text{as } p \searrow 0. \tag{89}$$

Moreover, for each j,

$$\Lambda_{\theta_j}(\alpha) \le \sup_{\alpha\in\mathbb{R}^d} \Lambda_{\theta_j}(\alpha) := -\Lambda_{\theta_j}(0).$$

Hence

$$\limsup_{n\to\infty} \frac{1}{n} P_g(\hat{\theta}_n \in F) \;\le\; \max_{1\le j\le k} -\Lambda_{\theta_j}^*(0) + Cr$$

$$\le\; -\inf_{\theta\in F} \Lambda_\theta^*(0) + Cr. \tag{90}$$

The upper bound follows by letting $r \searrow 0$. $\quad\square$

Proof of Theorem 3: Lower Bound. Let G be an open subset of Θ, and let $\theta \in G$. Then $G^c = \Theta - G$ is compact, and there exists a collection $\mathbb{T} = \{\theta_1,\ldots,\theta_k\} \subset G^c$ such that $B(\theta_1;\epsilon),\ldots,B(\theta_k;\epsilon)$ forms a finite subcover of $\Theta - G$, where $\epsilon > 0$. Since

$$\left\{ \mathscr{A}_n(\theta) \ge \sup_{t\in\mathbb{T}} \mathscr{A}_n(t) \right\} \;\supset\; \left\{ \mathscr{A}_n(\theta) \ge \max_{1\le j\le k} \mathscr{A}_n(\theta_j) + \max_{1\le j\le k} \sup_{t\in B(\theta_j;\epsilon)} [\mathscr{A}_n(t) - \mathscr{A}_n(\theta_j)] \right\}$$

$$\supset\; \left\{ \mathscr{A}_n(\theta) \ge \max_{1\le j\le k} \mathscr{A}_n(\theta_j) + \sup_{\|\theta_1-\theta_2\|<\epsilon} |\mathscr{A}_n(\theta_1) - \mathscr{A}_n(\theta_2)| \right\}, \tag{91}$$

it follows that

$$P_g \hat{\theta}_n \in G \;\ge\; P_g\left(\mathscr{A}_n(\theta) > \max_{1\le j\le k} \mathscr{A}_n(\theta_j) + \delta, \; \sup_{\|\theta_1-\theta_2\|<\epsilon} [\mathscr{A}_n(\theta_1) - \mathscr{A}_n(\theta_2)] \le \delta \right)$$

$$\ge\; J_{n,1} - J_{n,2}, \tag{92}$$

where

$$J_{n,1} := P_g\left(\mathscr{A}_n(\theta) > \max_{1\le j\le k} \mathscr{A}_n(\theta_j) + \delta \right),$$

$$J_{n,2} := P_g\left(\sup_{\|\theta_1-\theta_2\|\le r} [\mathscr{A}_n(\theta_1) - \mathscr{A}_n(\theta_2)] \ge \delta \right) := P_g(\omega(\mathscr{A}_n;\epsilon) \ge \delta).$$

We now investigate the behavior of $J_{n,1}$ and $J_{n,2}$. Starting with $J_{n,1}$, note that

$$J_{n,1} \geq P_g\left(\min_{1\leq j\leq k}(\mathscr{A}_n(\boldsymbol{\theta}) - \mathscr{A}_n(\boldsymbol{\theta}_j) - \delta) \geq 0\right) = P_g(\mathscr{D}_n \geq 0). \tag{93}$$

Now by (74), it follows that

$$J_{n,1} \geq P_g(Y_n \in B(\boldsymbol{0}; r)), \tag{94}$$

where Y_n is as in (71) and $r = c_{\boldsymbol{\theta}}(\delta)$. Applying Proposition 3.4, we obtain

$$\lim_{n\to\infty} \frac{1}{n} \log J_{n,1} \geq -I_r(\boldsymbol{\theta}), \tag{95}$$

where $I_r(\boldsymbol{\theta}) = \inf\{\Lambda_{\boldsymbol{\theta}}^*(x) : x \in \mathfrak{R}_{\boldsymbol{\theta}} \cap B(\boldsymbol{0}; r)\}$, and we now observe that r may be chosen to be $c_{\boldsymbol{\theta}} := \lim_{\delta\downarrow 0} c_{\boldsymbol{\theta}}(\delta) > 0$, where $c_{\boldsymbol{\theta}}(\delta)$ is given as in (73). Hence we may replace $I_r(\cdot)$ with $I(\cdot)$ on the right-hand side of the previous equation. Next, using Proposition 4 yields that

$$\liminf_{n\to\infty} \frac{1}{n} \log P_g(\hat{\boldsymbol{\theta}}_n \in G) \geq \liminf_{n\to\infty} \frac{1}{n} \log J_{n,1} + \lim_{n\to\infty} \log\left(1 - \frac{J_{n,2}}{J_{n,1}}\right) \geq -I(\boldsymbol{\theta}). \tag{96}$$

Finally, the required lower bound is obtained by maximizing the right-hand side over all $\boldsymbol{\theta} \in G$. $\square$

In the proof of the lower bound, it is clear that the choice of $\{\boldsymbol{\theta}_1, \ldots, \boldsymbol{\theta}_k\}$ plays a central role, and the rate function $I(\boldsymbol{\theta})$ will be minimized when k is small. As a simple example, suppose that our goal is to obtain a lower bound for $P_g(\hat{\boldsymbol{\theta}}_n \in G)$, where

$$G = \{(\theta_1, \theta_2) : \theta_1 > a_1 \text{ or } \theta_2 > a_2\} \subset \mathbb{R}^2, \quad \boldsymbol{\theta}_g \notin G,$$

which is a union of two halfspaces, This can be expressed as $a + \mathscr{C}$, where $a = (a_1, a_2)$ and $\mathscr{C} = \{(\theta_1, \theta_2) : \theta_1 > 0 \text{ or } \theta_2 > 0\}$, which is an example of a *translated* cone. Now if $\boldsymbol{\theta} \in G$, then we can find two elements which generate the entire set $\Theta - G$, in the sense that all other normalized differences lie between these two unit vectors. These two representative points are the unit vectors $\mathbf{e}_1 = (-1, 0)$ and $\mathbf{e}_2 = (0, -1)$, and all other normalized differences $(\boldsymbol{\theta} - \tilde{\boldsymbol{\theta}}/\|\boldsymbol{\theta} - \tilde{\boldsymbol{\theta}}\|$ lie between these vectors for all $\tilde{\boldsymbol{\theta}} \in \Theta - G$. Now going back to (73), we see that this equation again holds. Furthermore, (74) holds with $B(\boldsymbol{0}; c_\delta(\boldsymbol{\theta}))$ now replaced by an intersection of *two* halfspaces rather than of all halfspaces, yielding an unbounded region in the definition of $I(\boldsymbol{\theta})$. This potentially improves the quality of the lower bound compared with what is presented in the statement of Theorem 3. This idea can be potentially generalized to other sets, such as other unions of halfspaces, and so from a practical perspective, could apply somewhat generally.

4. Concluding Remarks

In this article, we have derived large deviation results for the minimum Hellinger distance estimators of a family of continuous distributions satisfying an equicontinuity condition. These results extend large deviation asymptotics for M-estimators given, e.g., in [6,9]. In contrast to the case for M-estimators, our setting is complicated due to its inherent nonlinearity, leading to complications in the proofs of both the upper and lower bounds, and an unexpected subtlety in the form of the rate function for the lower bound. Our results suggest that one can, under additional hypotheses, establish saddlepoint approximations to the density of MHDE, which would enable one to sharpen inference for small samples.

Similar results are expected to hold for discrete distributions. However, the equicontinuity condition is not required in that case, since ℓ_1, unlike $L_1(S)$, possesses the *Schur property*. Hence the LDP in the weak topology of ℓ_1 can be derived (more easily) using a standard Gärtner–Ellis argument, and utilizing this, one can, in principle, repeat all of the arguments above to derive results analogous to Theorems 2 and 3. Large deviations for other divergences under weak family regularity (such as noncompactness of the parameter space Θ)—and their connections to estimation and test efficiency—are interesting open problems requiring new techniques beyond those described in this article.

Author Contributions: Conceptualization, A.N.V. and J.F.C.; Methodology, A.N.V. and J.F.C.; Validation, A.N.V. and J.F.C.; Writing—original draft, A.N.V. and J.F.C.; Writing—review & editing, A.N.V. and J.F.C. All authors have read and agreed to the published version of the manuscript.

Funding: This research received no external funding.

Data Availability Statement: Not applicable.

Conflicts of Interest: The authors declare no conflict of interest.

References

1. Beran, R. Minimum Hellinger distance estimates for parametric models. *Ann. Stat.* **1977**, *5*, 445–463. [CrossRef]
2. Lindsay, B.G. Efficiency versus robustness: The case for minimum Hellinger distance and related methods. *Ann. Stat.* **1994**, *22*, 1081–1114. [CrossRef]
3. Basu, A.; Shioya, H.; Park, C. *Statistical Inference: The Minimum Distance Approach*; CRC Press: Boca Raton, FL, USA, 2011.
4. Pardo, L. *Statistical Inference Based on Divergence Measures*; CRC Press: Boca Raton, FL, USA, 2006.
5. Bahadur, R.R. Rates of convergence of estimates and test statistics. *Ann. Math. Stat.* **1967**, *38*, 303–324. [CrossRef]
6. Borovkov, A.A.; Mogulskii, A.A. Large Deviations and Testing Statistical Hypotheses. *Sib. Adv. Math.* **1992**, *2*, 43–72.
7. Fu, J.C. On a theorem of Bahadur on the rate of convergence of point estimators. *Ann. Stat.* **1973**, *1*, 745–749. [CrossRef]
8. Arcones, M.A. Large deviations for M-estimators. *Ann. Inst. Stat. Math.* **2006**, *58*, 21–52. [CrossRef]
9. Joutard, C. Large deviations for M-estimators. *Math. Methods Stat.* 2004, *13*, 179–200.
10. Dembo, A.; Zeitouni, O. *Large Deviations Techniques and Applications*; Springer: Berlin, Germany, 1998.
11. Puhalskii, A.; Spokoiny, V. On large-deviation efficiency in statistical inference. *Bernoulli* **1998**, *4*, 203–272. [CrossRef]
12. Nikitin, Y. *Asymptotic Efficiency of Nonparametric Tests*; Cambridge University Press: Cambridge, UK, 1995.
13. Biggins, J.; Bingham, N. Large deviations in the supercritical branching process. *Adv. Appl. Probab.* **1993**, *25*, 757–772. [CrossRef]
14. Billingsley, P. *Convergence of Probability Measures*, 2nd ed.; John Wiley & Sons, Inc.: New York, NY, USA, 1999.
15. de Acosta, A. On large deviations of empirical measures in the τ-topology. *J. Appl. Probab.* **1993**, *31*, 41–47. [CrossRef]
16. Basu, A.; Sarkar, S.; Vidyashankar, A.N. Minimum negative exponential disparity estimation in parametric models. *J. Statist. Plann. Inference* **1997**, *58*, 349–370. [CrossRef]
17. Cheng, A.-L.; Vidyashankar, A.N. Minimum Hellinger distance estimation for randomized play the winner design. *J. Statist. Plann. Inference* **2006**, *136*, 1875–1910. [CrossRef]
18. Devroye, L.; Györfi, L. *Nonparametric Density Estimation: The L_1 View*; Wiley Series in Probability and Mathematical Statistics: Tracts on Probability and Statistics; John Wiley & Sons, Inc.: New York, NY, USA, 1985.
19. Conway, J.B. *A Course in Functional Analysis*; Springer: New York, NY, USA, 1990.
20. Dupuis, P.; Ellis, R.S. *A Weak Convergence Approach to the Theory of Large Deviations*; John Wiley & Sons: New York, NY, USA, 1997.
21. Rockafellar, R.T. *Convex Analysis*; Princeton University Press: Princeton, NJ, USA, 1970.
22. Boos, D.D. A converse to Scheffé's theorem. *Ann. Stat.* **1985**, *13*, 423–427. [CrossRef]
23. Lei, L. Large Deviations for Kernel Density Estimators and Study for Random Decrement Estimator. Ph. D. Thesis, Université Blaise Pascal-Clermont-Ferrand II, Clermont-Ferrand, France, 2005.
24. Louani, D.; Maouloud, S.M.O. Some functional large deviations principles in nonparametric function estimation. *J. Theor. Probab.* **2012**, *25*, 280–309. [CrossRef]
25. Ellis, R.S. Large deviations for a general class of random vectors. *Ann. Probab.* **1984**, *12*, 1–12. [CrossRef]

MDPI

St. Alban-Anlage 66

4052 Basel

Switzerland

Tel. +41 61 683 77 34

Fax +41 61 302 89 18

www.mdpi.com

Entropy Editorial Office

E-mail: entropy@mdpi.com

www.mdpi.com/journal/entropy